# Interdisciplinary Applied Mathematics

Volume 26

*Editors*
**S.S. Antman J.E. Marsden**
**L. Sirovich S. Wiggins**

*Geophysics and Planetary Sciences*

*Imaging, Vision, and Graphics*
**D. Geman**

*Mathematical Biology*
**L. Glass,** J.D. Murray

*Mechanics and Materials*
**R.V. Kohn**

*Systems and Control*
**S.S. Sastry,** P.S. Krishnaprasad

Problems in engineering, computational science, and the physical and biological sciences are using increasingly sophisticated mathematical techniques. Thus, the bridge between the mathematical sciences and other disciplines is heavily traveled. The correspondingly increased dialog between the disciplines has led to the establishment of the series: *Interdisciplinary Applied Mathematics*.

The purpose of this series is to meet the current and future needs for the interaction between various science and technology areas on the one hand and mathematics on the other. This is done, firstly, by encouraging the ways that mathematics may be applied in traditional areas, as well as point towards new and innovative areas of applications; and, secondly, by encouraging other scientific disciplines to engage in a dialog with mathematicians outlining their problems to both access new methods and suggest innovative developments within mathematics itself.

The series will consist of monographs and high-level texts from researchers working on the interplay between mathematics and other fields of science and technology.

# Interdisciplinary Applied Mathematics

Volumes published are listed at the end of this book.

**Springer**
*New York*
*Berlin*
*Heidelberg*
*Hong Kong*
*London*
*Milan*
*Paris*
*Tokyo*

# An Invitation to 3-D Vision

## From Images to Geometric Models

Yi Ma
Stefano Soatto
Jana Košecká
S. Shankar Sastry

With 170 Illustrations

Springer

Yi Ma
Department of Electrical and
Computer Engineering
University of Illinois at
Urbana-Champaign
Urbana, IL 61801
USA
yima@uiuc.edu

Stefano Soatto
Department of Computer Science
University of California, Los Angeles
Los Angeles, CA 90095
USA
soatto@ucla.edu

Jana Košecká
Department of Computer Science
George Mason University
Fairfax, VA 22030
USA
kosecka@cs.gmu.edu

S. Shankar Sastry
Department of Electrical Engineering
and Computer Science
University of California, Berkeley
Berkeley, CA 94720
USA
sastry@eecs.berkeley.edu

*Editors*

S.S. Antman
Department of Mathematics
*and*
Institute for Physical Science
and Technology
University of Maryland
College Park, MD 20742
USA
ssa@math.umd.edu

J.E. Marsden
Control and Dynamical Systems
Mail Code 107-81
California Institute of Technology
Pasadena, CA 91125
USA
marsden@cds.caltech.edu

L. Sirovich
Division of Applied Mathematics
Brown University
Providence, RI 02912
USA
chico@camelot.mssm.edu

S. Wiggins
School of Mathematics
University of Bristol
Bristol BS8 1TW
UK
s.wiggins@bris.ac.uk

*Cover Illustration*: "Axo-GJ" 1968 (180 × 180 cm) by Victor Vasarely. Copyright Michèle Vasarely.

Mathematics Subject Classification (2000): 51U10, 68U10, 65D18

ISBN 978-1-4419-1846-8 Printed on acid-free paper.
e-ISBN 978-0-387-21779-6

© 2004 Springer-Verlag New York, Inc.
Softcover reprint of the hardcover 1st edition 2004
All rights reserved. This work may not be translated or copied in whole or in part without the written permission of the publisher (Springer-Verlag New York, Inc., 233 Spring Street, New York, NY 10013, USA), except for brief excerpts in connection with reviews or scholarly analysis. Use in connection with any form of information storage and retrieval, electronic adaptation, computer software, or by similar or dissimilar methodology now known or hereafter developed is forbidden.
The use in this publication of trade names, trademarks, service marks, and similar terms, even if they are not identified as such, is not to be taken as an expression of opinion as to whether or not they are subject to proprietary rights.

9 8 7 6 5 4 3 2 (EB)

springeronline.com

*To my mother and my father (Y.M.)*

*To Giuseppe Torresin, Engineer (S.S.)*

*To my parents (J.K.)*

*To my mother (S.S.S.)*

# Preface

This book is intended to give students at the advanced undergraduate or introductory graduate level, and researchers in computer vision, robotics and computer graphics, a self-contained introduction to the geometry of three-dimensional (3-D) vision. This is the study of the reconstruction of 3-D models of objects from a collection of 2-D images. An essential prerequisite for this book is a course in linear algebra at the advanced undergraduate level. Background knowledge in rigid-body motion, estimation and optimization will certainly improve the reader's appreciation of the material but is not critical since the first few chapters and the appendices provide a review and summary of basic notions and results on these topics.

*Our motivation*

Research monographs and books on geometric approaches to computer vision have been published recently in two batches: The first was in the mid 1990s with books on the geometry of two views, see e.g. [Faugeras, 1993, Kanatani, 1993b, Maybank, 1993, Weng et al., 1993b]. The second was more recent with books focusing on the geometry of multiple views, see e.g. [Hartley and Zisserman, 2000] and [Faugeras and Luong, 2001] as well as a more comprehensive book on computer vision [Forsyth and Ponce, 2002]. We felt that the time was ripe for synthesizing the material in a unified framework so as to provide a self-contained exposition of this subject, which can be used both for pedagogical purposes and by practitioners interested in this field. Although the approach we take in this book deviates from several other classical approaches, the techniques we use are mainly linear algebra and our book gives a comprehensive view of what is known to date on the geometry of 3-D vision. It also develops homogeneous terminology on a solid analytical foundation to enable what should be a great deal of future research in this young field.

Apart from a self-contained treatment of geometry and algebra associated with computer vision, the book covers relevant aspects of the image formation process, basic image processing, and feature extraction techniques – essentially all that one needs to know in order to build a system that can automatically generate a 3-D model from a set of 2-D images.

*Organization of the book*

This book is organized as follows: Following a brief introduction, Part I provides background material for the rest of the book. Two fundamental transformations in multiple-view geometry, namely, rigid-body motion and perspective projection, are introduced in Chapters 2 and 3, respectively. Feature extraction and correspondence are discussed in Chapter 4.

Chapters 5, 6, and 7, in Part II, cover the classic theory of two-view geometry based on the so-called epipolar constraint. Theory and algorithms are developed for both discrete and continuous motions, both general and planar scenes, both calibrated and uncalibrated camera models, and both single and multiple moving objects.

Although the epipolar constraint has been very successful in the two-view case, Part III shows that a more proper tool for studying the geometry of multiple views is the so-called *rank condition on the multiple-view matrix* (Chapter 8), which unifies all the constraints among multiple images that are known to date. The theory culminates in Chapter 9 with a unified theorem on a rank condition for arbitrarily mixed point, line, and plane features. It captures *all* possible constraints among multiple images of these geometric primitives, and serves as a key to both geometric analysis and algorithmic development. Chapter 10 uses the rank condition to reexamine and unify the study of single-view and multiple-view geometry given scene knowledge such as symmetry.

Based on the theory and conceptual algorithms developed in the early part of the book, Chapters 11 and 12, in Part IV, demonstrate practical reconstruction algorithms step-by-step, as well as discuss possible extensions of the theory covered in this book. An outline of the logical dependency among chapters is given in Figure 1.

*Curriculum options*

Drafts of this book and the exercises in it have been used to teach a one-semester course at the University of California at Berkeley, the University of Illinois at Urbana-Champaign, Washington University in St. Louis, the George Mason University and the University of Pennsylvania, and a one-quarter course at the University of California at Los Angeles. There is apparently adequate material for two semesters or three quarters of lectures. Advanced topics suggested in Part IV or chosen by the instructor can be added to the second half of the sec-

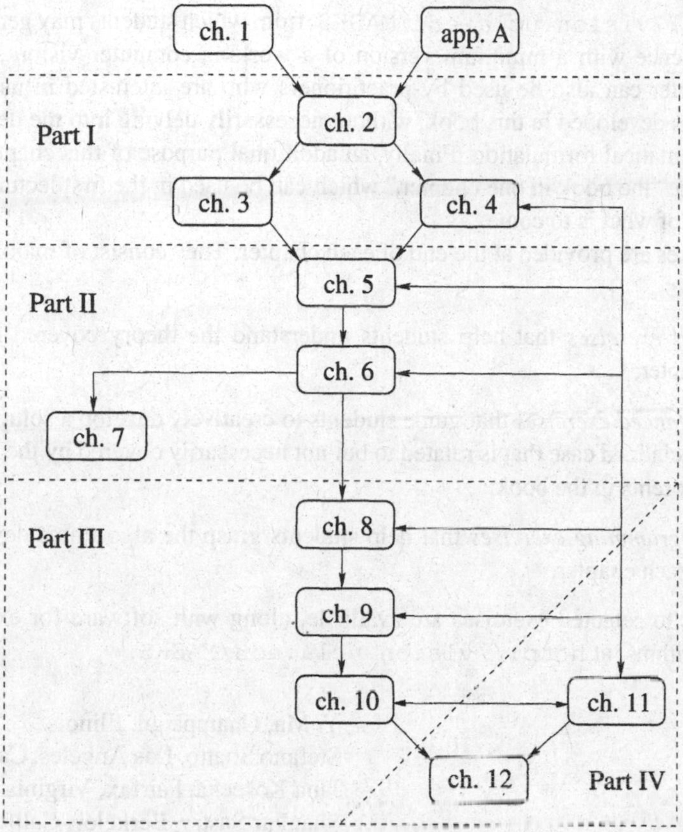


Figure 1. Organization of the book: logical dependency among parts and chapters.

ond semester if a two-semester course is offered. Below are some suggestions for course development based on this book:

1. *A one-semester course:* Appendix A, Chapters 1–6, and part of Chapters 8–10.

2. *A two-quarter course:* Chapters 1–6 for the first quarter, and Chapters 8–10, 12 for the second quarter.

3. *A two-semester course:* Appendix A and Chapters 1–6 for the first semester; Chapters 7–10 and the instructor's choice of some advanced topics from Chapter 12 for the second semester.

4. *A three-quarter sequence:* Chapters 1–6 for the first quarter, Chapters 7–10 for the second quarter, and the instructor's choice of advanced topics and projects from Chapters 11 and 12 for the third quarter.

Chapter 11 plays a special role in this book: Its purpose is to make it easy for the instructor to develop and assign experimental exercises or course projects along with other chapters being taught throughout the course. Relevant code is available at http://vision.ucla.edu/MASKS, from which students may get hands-on experience with a minimum version of a working computer vision system. This chapter can also be used by practitioners who are interested in using the algorithms developed in this book, without necessarily delving into the details of the mathematical formulation. Finally, an additional purpose of this chapter is to summarize "the book in one chapter," which can be used in the first lecture as an overview of what is to come.

Exercises are provided at the end of each chapter. They consist of mainly three types:

1. *drill exercises* that help students understand the theory covered in each chapter;
2. *advanced exercises* that guide students to creatively develop a solution to a specialized case that is related to but not necessarily covered by the general theorems in the book;
3. *programming exercises* that help students grasp the algorithms developed in each chapter.

Solutions to selected exercises are available, along with software for examples and algorithms, at http://vision.ucla.edu/MASKS.

Yi Ma, Champaign, Illinois
Stefano Soatto, Los Angeles, California
Jana Košecká, Fairfax, Virginia
Shankar Sastry, Berkeley, California
Spring, 2003

# Acknowledgments

The idea for writing this book grew during the completion of Yi Ma's doctoral dissertation at Berkeley. The book was written during the course of three years after Yi Ma graduated from Berkeley, when all the authors started teaching the material at their respective institutions. Feedback and input from students was instrumental in improving the quality of the initial manuscript. We are deeply grateful to the student input. Two students whose doctoral research especially helped us are René Vidal at Berkeley and Kun Huang at UIUC. In addition, the research projects of many other students led to the development of new material that became an integral part of this book. We thank especially Wei Hong, Yang Yang at UIUC, Omid Shakernia at Berkeley, Hailin Jin, Paolo Favaro at Washington University in St. Louis, Alessandro Chiuso now at the University of Padova, Wei Zhang at George Mason University, and Marco Zucchelli at the Royal Institute of Technology in Stockholm.

Many colleagues contributed with comments and suggestions on early drafts of the manuscript; in particular, Daniel Cremers of UCLA, Alessandro Duci of the Scuola Normale of Pisa, and attendees of the short course at the UCLA Extension in the Fall of 2002. We owe a special debt of gratitude to Camillo Taylor now at the University of Pennsylvania, Philip McLauchlan now at Imagineer Systems Inc., and Jean-Yves Bouguet now at Intel for their collaboration in developing the real-time vision algorithms presented in Chapter 12. We are grateful to Hany Farid of Dartmouth College for his feedback on the material in Chapters 3 and 4, and Serge Belongie of the University of California at San Diego for his correction to Chapter 5. Kostas Daniilidis offered a one-semester course based on a draft of the book at the University of Pennsylvania in the spring of 2003, and we are grateful to him and his students for valuable comments. We also thank Robert Fossum of the

Mathematics Department of UIUC for proofreading the final manuscript, Thomas Huang of the ECE Department of UIUC for his advice and encouragement all along, Alan Yuille of UCLA, Rama Chellappa and Yiannis Aloimonos of the University of Maryland, and Long Quan of Hong Kong University of Science & Technology for valuable references. We would also like to thank Harry Shum and Zhengyou Zhang for stimulating discussions during our visit to Microsoft Research.

Some of the seeds of this work go back to the early nineties, when Ruggero Frezza, Pietro Perona, and Giorgio Picci started looking at the problem of structure from motion within the context of systems and controls. They taught a precursor of this material in a course at the University of Padova in 1991. Appendix B was inspired by the beautiful lectures of Giorgio Picci. We owe them our sincere gratitude for their vision and for their initial efforts that sparked our interest in the field.

In the vision community, we derived a great deal from the friendly notes of Thomas Huang of UIUC, Yiannis Aloimonos of the University of Maryland, Carlo Tomasi of Duke University, Takeo Kanade of Carnegie Mellon University, and Ruzena Bajcsy now at the University of California at Berkeley.

At the same time, we have also been influenced by the work of Ernst Dickmanns of Bundeswehr University in Munich, Olivier Faugeras of INRIA, and David Mumford of Brown University, who brought the discipline of engineering and the rigor of mathematics into a field that needed both. Their work inspired many to do outstanding work, and we owe them our deepest gratitude.

We are grateful to Jitendra Malik and Alan Weinstein at Berkeley, Pietro Perona at Caltech, Berthold Horn at MIT, P. R. Kumar at UIUC, and Roger Brockett at Harvard for their openness to new points of view and willingness to listen to our initial fledgling attempts. We are also grateful to the group of researchers in the Robotics and Intelligent Machines Laboratory at Berkeley for providing a stimulating atmosphere for the exchange of ideas during our study and visits there, especially Cenk Çavuşoğlu now at Case Western University, Lara Crawford now at Xerox Palo Alto Research Center, João Hespanha now at the University of California at Santa Barbara, John Koo now at Berkeley, John Lygeros now at Cambridge University, George Pappas now at the University of Pennsylvania, Maria Prandini now at Politecnico di Milano, Claire Tomlin now at Stanford University, and Joe Yan now at the University of British Columbia.

Much of the research leading to the material of this book has been generously supported by a number of institutions, funding agencies and their program managers. A great deal of the research was supported by the Army Research Office under the grant DAAH04-96-1-0341 for a center "An Integrated Approach to Intelligent Systems" under the friendly program management of Linda Bushnell and later Hua Wang. We would also like to acknowledge the support of the Office of Naval Research and the program manager Allen Moshfegh for supporting the work on vision-based landing under contract N00014-00-1-062. We also thank the ECE Department, the Coordinated Science Laboratory, the Research Board of UIUC, and the Computer Science Departments at George Mason University and

UCLA for their generous startup funds. We also thank Behzad Kamgar-Parsi of the Office of Naval Research, Belinda King of the Air Force Office of Scientific Research, Jean-Yves Bouguet of Intel, and the Robotics and Computer Vision program of the National Science Foundation.

Finally on a personal note, Yi would like to thank his parents for their remarkable tolerance of his many broken promises for longer and more frequent visits home, and this book is dedicated to their understanding and caring. Stefano would like to thank Anthony Yezzi and Andrea Mennucci for many stimulating discussions. Jana would like to thank her parents for their never ceasing inspiration and encouragement and Frederic Raynal for his understanding, continuing support and love. Shankar would like to thank his mother and Claire Tomlin for their loving support during the preparation of this book.

# Contents

# Chapter 1
# Introduction

*All human beings by nature desire to know. A sign of this is our liking for the senses; for even apart from their usefulness we like them for themselves – especially the sense of sight, since we choose seeing above practically all the others, not only as an aid to action, but also when we have no intention of acting. The reason is that sight, more than any of the other senses, gives us knowledge of things and clarifies many differences among them.*

– Aristotle

## 1.1 Visual perception from 2-D images to 3-D models

The sense of vision plays an important role in the life of primates: it allows them to infer spatial properties of the environment that are necessary to perform crucial tasks for survival. Primates use vision to explore unfamiliar surroundings, negotiate physical space with one another, detect and recognize prey at a distance, and fetch it, all with seemingly little effort. So, why should it be so difficult to make a computer "see"? First of all, we need to agree on what it means for a computer to see. It is certainly not as simple as connecting a camera to it. Nowadays, a digital camera can deliver several "frames" per second to a computer, analogously to what the retina does with the brain. Each frame, however, is just a collection of positive numbers that measure the amount of light incident on a particular location (or "pixel") on a photosensitive surface (see, for instance, Table 3.1 and

Figures 3.2 and 3.3 in Chapter 3). How can we "interpret" these pixel values and tell whether we are looking at an apple, a tree, or our grandmother's face? To make matters worse, we can take the same exact scene (say the apple), and change the viewpoint. All the pixel values change, but we have no difficulty in interpreting the scene as being our apple. The same goes if we change the lighting around the apple (say, if we view it in candle light or under a neon lamp), or if we coat it with shiny wax.

A visual system, in broad terms, is a collection of devices that transform measurements of light into information about spatial and material properties of a scene. Among these devices, we need photosensitive sensors (say, a camera or a retina) as well as computational mechanisms (say, a computer or a brain) that allow us to extract information from the raw sensory readings. To appreciate the fact that vision is no easy computational task, it is enlightening to notice that about half of the entire cerebral cortex in primates is devoted to processing visual information [Felleman and van Essen, 1991]. So, even when we are absorbed in profound thoughts, the majority of our brain is actually busy trying to make sense of the information coming from our eyes.

## *Why is vision hard?*

Once we accept the fact that vision is not just transferring data from a camera to a computer, we can try to identify the factors that affect our visual measurements. Certainly the pixel values recorded with a camera (or the firing activity of retinal neurons) depend upon the shape of objects in the scene: if we change the shape, we notice a change in the image. Therefore, images depend on the *geometry* of the scene. However, they also depend on its *photometry*, the illumination and the material properties of objects: a bronze apple looks different than a real one; a skyline looks different on a cloudy day than in sunshine. Finally, as objects move in the scene, and their pose changes, so does our image. Therefore, visual measurements depend on the *dynamics* of the environment. In general, we do not know the shape of the scene, we do not know its material properties, and we do not know its motion. Our goal is to infer *some* representation of the world from collection of images.

Now, the complexity of the physical world is infinitely superior to the complexity of the measurements of its images.[1] Therefore, in a sense, vision is more than difficult; it is impossible. We cannot simply "invert" the image formation process, and reconstruct the "true" scene from a number of images. What we can reconstruct is at best a *model* of the world, or an "internal representation." This requires

[1] One entire dimension is lost in the projection from the 3-D world to the 2-D image. Moreover, the geometry of a scene can be described by a collection of surfaces; its photometry, by a collection of functions defined on these surfaces that describe how light interacts with the underlying material; its dynamics, by differential equations. All these entities live in infinite-dimensional spaces, and inferring them from finite-dimensional images is impossible without imposing additional constraints on the problem.

introducing assumptions, or hypotheses, on some of the unknown properties of the environment, in order to infer the others. There is no right or wrong way to do so, and modeling is a form of engineering art, which depends upon the *task* at hand. For instance, what model of the scene to infer depends on whether we want to use it to move within the environment, to visualize it from novel viewpoints, or to recognize objects or materials. In each case, some of the unknown properties are of interest, whereas the others are "nuisance factors": if we want to navigate in an unknown environment, we care for the shape and motion of obstacles, not so much for their material or for the ambient light. Nevertheless, the latter influence the measurements, and have to be dealt with somehow.

In this book, we are mostly interested in vision as a sensor for artificial systems to interact with their surroundings. Therefore, we envision machines capable of inferring the *shape* and *motion* of objects in a scene, and our book reflects this emphasis.

## *Why now?*

The idea of endowing machines with a sense of vision to have them interact with humans and the environment in a dynamic fashion is not new. In fact, this goal permeates a good portion of modern engineering, and many of the ideas discussed above can be found in the work of [Wiener, 1949] over half a century ago. So, if the problem is so hard, and if so much effort has gone toward it unsuccessfully,[2] why should we insist?

First, until just over a decade ago, there was no commercial hardware available to transfer a full-resolution image into the memory of a computer at frame rate (30 Hz), let alone to process it and do anything useful with it. This step has now been overcome; digital cameras are ubiquitous, and so are powerful computers that can process their output in real time. Second, although many of the necessary analytical tools were present in various disciplines of mathematics for quite some time, only in the past decade has the geometry of vision been understood thoroughly and explained systematically. Finally, the study of vision is ultimately driven by the demand for its use in applications that have the potential to positively impact our quality of life.

## *What for?*

Think of spending a day with your eyes closed, and all the things that you would not be able to do. Artificial vision offers the potential of relieving humans of tasks that are dangerous, monotonous, or boring, such as driving a car, surveying an underwater platform, detecting intruders in a building. In addition, they can be used to augment human capabilities or replace lost skills. Rather than discussing

[2]After all, we do not have household personal robot assistants, . . . yet.

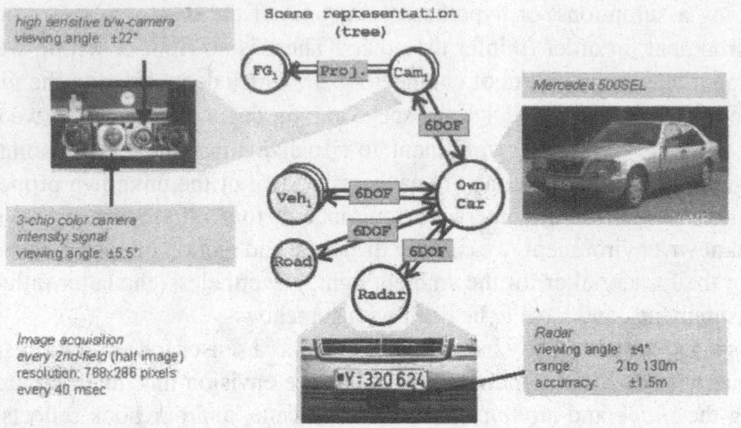


Figure 1.1. The VaMP system developed by E. D. Dickmanns and his coworkers (courtesy of E. D. Dickmanns).

this in the abstract, we would like to cite a few recent success stories that hint at the potential of vision as a sensor.

Starting from the late 1970s, E. D. Dickmanns and his coworkers have been developing vision-based systems to drive cars autonomously on public freeways. In 1984, they demonstrated a system capable of driving a small truck up to speeds of 90 km/h on an empty road. In 1994 they took this to the next level, demonstrating a system capable of driving a passenger car on a public freeway with normal traffic, reading speed signs, passing slower cars, etc., up to a speed of 180 km/h (Figure 1.1). Although for legal reasons this system has not appeared on the public market,[3] several components of it are trickling into consumer products. For instance, similar systems are used to monitor the driving behavior of long-haul trucks and wake up the driver if he or she falls asleep, and "smart cruise control" that can maintain a minimum distance from the preceding vehicle is being introduced into the market [Hofman et al., 2000]. Along the same lines, the California PATH project has been developing automated freeway systems for quite some time now (Figure 1.2). Nowadays, vision-guided helicopters or aircrafts can automatically take off, fly, or land. Figure 1.3 shows one of such systems.

Another application of vision techniques for real-time interaction is in the broadcasting of sports events. American football fans watching games on television have recently noticed the addition of a yellow line on the field that moves during the game (the "first down line"). This line, along with advertising banners on the side of the field and on the line itself, is not present in the real stadium, but is instead overlaid to live video for broadcast. In order to make the line appear to

[3]Even if an entirely automated freeway would probably decrease the number of fatal accidents dramatically, machine errors causing fatalities would not be acceptable, whereas human errors are.

Figure 1.2. Vision-based autonomous cars (courtesy of California PATH).

Figure 1.3. Berkeley unmanned aerial vehicle (UAV) with on-board vision system: Yamaha R 50 helicopter with pan/tilt camera (Sony EVI-D30) and computer box (Littleboards) hovering above a landing platform (courtesy of Intelligent Machines & Robotics Laboratory, University of California at Berkeley).

be attached to the floor, camera motion must be inferred in real-time, along with a model of the scene (e.g., the field plane, Figure 1.4).

Vision techniques are also becoming ubiquitous in entertainment. Special effects mixing real scenes with computer-generated ones require accurate registration of the camera motion in the scene during live action. While, traditionally, the entertainment industry has resisted automation, the quantity and quality of results enabled by automatic analysis of image sequences is quickly changing the course of events. Nowadays, with a hand-held camcorder, a person can record a home-made video and insert virtual objects into it (Figure 1.5). Combined with computer graphics technology, vision systems have also been used extensively to acquire 3-D models for large-scale urban areas. For example, Figure 1.6 shows a virtual 3-D model of the campus of the University of California at Berkeley acquired by one such system.

Mastering the material described in this book will give readers the fundamental knowledge necessary to implement a complete working computer vision system. In addition, a solid grasp of the underlying theory will allow them to push the

Figure 1.4. "First down line" and virtual advertising in sports events (courtesy of Princeton Video Image, Inc.).

Figure 1.5. Software is now publicly available to estimate camera motion and structure in real time from live video (courtesy of UCLA Vision Lab).

envelope of future applications. This book aims to provide the needed geometric principles for 3-D vision, as part of the endeavor to establish a solid foundation for exploring new avenues in both theoretical developments and practical applications for machine vision.

## *What this book does not do*

The considerations above should convince the reader that the study of vision is multifaceted. In this book, we do not aim at covering all of its aspects, and concentrate instead on the geometry of multiple views.

We do not address issues related to the perception of individual images, using so-called *pictorial cues* such as texture, shading, blur, and contour (Figure 1.7). Without detracting from the importance of the analysis of the pictorial cues, we concentrate on the motion and stereo cue among multiple images, where

Figure 1.6. Virtual models and views of the campus of UC Berkeley (courtesy of Paul Debevec).

Figure 1.7. Some "pictorial" cues for 3-D structure: texture (left), shading (middle), and contours (right).

the geometry is well understood and can be presented in a coherent and unified framework.

### *What this book does*

This book concentrates on the analysis of scenes that contain a number of rigidly moving objects that have "benign" photometric properties. What "benign" means will be clear in Chapters 3 and 4, although we can already say that this excludes shiny and translucent materials.

Given a number of two-dimensional (2-D) images of scenes that satisfy these assumptions, this book seeks to answer the following questions: (1) to what extent (and how) can we estimate the three-dimensional (3-D) shape of each object? (2)

to what extent can we recover the motion of each object relative to the camera? (3) to what extent can we recover a model of the geometry of the camera itself? Traditionally, these questions are referred to as the *structure from motion* problem. This book describes *algorithms* designed to address these questions, namely, to *estimate 3-D structure, motion, and camera calibration from a collection of images*. In this sense, this book teaches how to go from 2-D images to 3-D models of the geometry of the scene.

While the goal is easy to state, carrying out this program is by no means trivial, and analytical tools must be developed if we want to appreciate and exploit the subtleties of the problem. That is why a good portion of this book is devoted to developing an appropriate mathematical approach.

## 1.2 A mathematical approach

The problem of inferring 3-D information of a scene from a set of 2-D images has a long history in computer vision. By its very nature, this problem falls into the category of so-called inverse problems, which are prone to be ill-conditioned and difficult to solve in their full generality unless additional assumptions are imposed. While, in general, the task of selecting a correct mathematical model can be elusive, simple choices of representation can be made by exploiting geometric primitives such as points, lines, curves, surfaces, and volumes.

Since these geometric primitives live in three-dimensional Euclidean space, the study of Euclidean geometry and groups of transformations that preserve its properties takes a natural place in this book (Chapter 2). Perspective projection, with its roots tracing back to ancient Greek philosophers and Renaissance artists, has been widely studied in projective geometry (a branch of algebra in mathematics) and modern computer graphics. In this book, we adopt it as an ideal model for the image formation process (Chapter 3). The blend of perspective projection and matrix groups is then at the heart of *multiple-view geometry*, the study of fundamental geometric laws that govern the structure from motion problem. In this book, we will formally establish that a simple, complete and yet unifying description of the laws governing multiple images is given by certain *rank conditions* on the so-called *multiple-view matrix* (Part III). Briefly stated, a multiple-view matrix associated with a geometric feature (point, line, or plane) is exactly the 3-D information that is missing in a single 2-D image of the feature but encoded in multiple ones. The rank conditions hence impose the incidence relation that all images must correspond to the same 3-D feature. If there are multiple features in 3-D with incidence relations (e.g., intersection) among themselves, the rank conditions can also uniformly take them into account.

This simple theory essentially enables us to carry out global geometric analysis for multiple images and systematically characterize degenerate configurations, without breaking the image sequence into pairs or triplets of views. Such a uniform and global treatment allows us to utilize all geometric constraints that govern all features and all incidence relations in all images simultaneously for a consistent recovery of motion and structure from multiple views (Chapter 9). In Chapter 10, we show how this theory naturally relates the perspective projection with properties of 3-D space that are invariant under symmetry groups, which allows us to exploit the symmetric nature of many man-made and natural objects.

## 1.3 A historical perspective

Rudimentary understanding of image formation (such as the phenomenon of pinhole imaging) was present in ancient civilizations throughout the world. However, the first mathematical formulation of the notion of projection (as well as rigid-body motion) is attributed to Euclid of Alexandria in the fourth century B.C. Brunelleschi and Leon Battista Alberti studied perspective in the context of painting and architecture, and Alberti wrote the first general treatise on the laws of perspective, *Della Pictura*, in 1435.[4] Perspective projection had very much contributed to the invention of a "new," non-Greek, way of doing geometry, called projective geometry by the French mathematician Girard Desargues, for his famous "perspective theorem" published in 1648.[5] Projective geometry was later reinvented and made popular by pupils of another French mathematician, Gaspard Monge, in the eighteenth and early nineteenth century. However, this doctrine was later challenged by Felix Klein's famous 1872 Erlangen Program, which essentially established a more democratic but unified platform for modern geometry in terms of group theory.[6]

> *The rise of projective geometry made such an overwhelming impression on the geometers of the first half of the nineteenth century that they tried to fit all geometric considerations into the projective scheme.... The dictatorial regime of the projective idea in geometry was first successfully broken by the German astronomer and geometer Möbius, but the classical document of the democratic platform in geometry establishing the group of transformations as the ruling principle in any kind of geometry and yielding equal rights to independent consideration to each and any such group, is F. Klein's Erlangen program.*
>
> – Herman Weyl, *Classical Groups*

[4] Some historical records suggest that the Greek mathematician Anaxagoras of Clazomenae might have written a treatise on perspective in the fourth century B.C, but it did not survive.

[5] The theorem states that if two triangles are in perspective, the intersections of corresponding sides are collinear.

[6] Klein's synthesis of geometry is the study of the properties of a space that are invariant under a given group of transformations.

Figure 1.8. Photograph of Erwin Kruppa (courtesy of A. Kruppa).

The first work that is directly related to multiple-view geometry is believed to be a 1913 paper by the German mathematician Kruppa (see Figure 1.8).[7] He proved that two views of five points are sufficient to determine both the relative transformation between the views and the 3-D location of the points up to finitely many solutions.[8] Kruppa's proof was done in the traditional projective geometry setting [Kruppa, 1913].

These earlier theoretical developments were embraced by two disciplines that strove for the development of techniques for 3-D reconstruction from image data: photogrammetry and computer vision. While the first techniques for reconstruction in computer vision date back to the mid 1970s, the origin of a more modern treatment is traditionally attributed to Longuet-Higgins, who in 1981 first proposed a linear algorithm for structure and motion recovery from two images of a set of points, based on the so-called *epipolar constraint* [Longuet-Higgins, 1981].[9] This linear algorithm was later modified and polished by [Huang and Faugeras, 1989], pioneers of the modern geometric approach to computer vision. These early works on two-view geometry

[7]But some work on two-view geometry can be traced back to the mid nineteenth century; see [Maybank and Faugeras, 1992] and references therein.

[8]Kruppa's original claim was that there would be no more than 11 solutions. The proof and result were refined later by [Demazure, 1988, Maybank, 1990, Heyden and Sparr, 1999]. It is shown that the number of solutions is 10 (including complex solutions), instead of 11. An infinitesimal version of this theorem is proven by [Maybank, 1993].

[9]The epipolar constraint, which states that two image rays corresponding to the same 3-D point are coplanar, had been known in photogrammetry since the mid nineteenth century. To the best of our knowledge, the epipolar constraint was first formulated analytically in [Thompson, 1959]. At around the same time of Longuet-Higgins, [Tsai and Huang, 1981, Tsai and Huang, 1984] studied the motion estimation problem from a rigid planar patch.

were summarized in several monographs [Faugeras, 1993, Kanatani, 1993b, Maybank, 1993, Weng et al., 1993b]. In this book, these classic results, together with their extension to the continuous motion and multiple-body cases, are presented in Chapters 5, 6, and 7.

Extensions of the reconstruction techniques from point features to lines initiated the study of the relationships among three views, since in the line case there is no effective constraint for two views that can be used for motion and structure recovery, as we will explain in Chapter 8. The so-called trilinear constraint among three images of a line was first studied by [Spetsakis and Aloimonos, 1987, Liu and Huang, 1986, Liu et al., 1990]. This result was later unified in the two papers of [Spetsakis and Aloimonos, 1990a, Spetsakis and Aloimonos, 1990b] for both the point and line cases, together with unified algorithms. An alternative derivation of the trilinear constraint for three uncalibrated images of points was suggested in [Shashua, 1994] and its equivalence to the line case was soon pointed out by [Hartley, 1995].

While early studies in multiple-view geometry often concentrated on finding out what the minimum amount of data needed for a reconstruction is, the advance of modern computer technologies has certainly changed the focus of the investigation. Given that computers today can easily handle tens of images with hundreds and even thousands of features at a time, the main problem becomes how to discover all the information that is available from all the images and how to efficiently extract it through computation. The attempts to develop algorithms that efficiently utilize a large number of images are marked by the introduction of factorization techniques for the simplified case of orthographic projection [Tomasi and Kanade, 1992]. In spite of the restrictive assumption about the projection model, the practical implications of such factorization techniques were striking.

Unfortunately, such factorization techniques cannot be directly generalized to the case of pure perspective projection. The study of perspective images hence took a different route, almost opposite to the global approach adopted in the orthographic case. Throughout the 1990s, researchers in computer vision and image processing focused on the study of geometry of two, three or four images at a time. Numerous but equivalent forms of the constraints among two or three images of points and lines had been reformulated for various purposes of analysis and application. This line of work culminated in the publication of two recent manuscripts: [Hartley and Zisserman, 2000, Faugeras and Luong, 2001].

This book will deviate from the projective geometric doctrine. We will deal with multiple-view geometry directly in its original setting and make it a self-contained subject. Consequently, most existing results and algorithms will be reformulated and simplified in our framework. Even the classical epipolar constraint between two views will eventually be expressed in a different form, which conforms more to other constraints that are present in multiple images (Chapter 8). This approach will not only complete with little overhead the task of searching for all intrinsic constraints among multiple views, but also present them ultimately in a unified form that is much more accessible for geometric insight and algorithmic development (see Chapter 9). We believe that such an adjustment is necessary and appropriate for the development of a theoretic and algorithmic framework suitable for studying *multiple* images of *multiple* features (of different types), *multiple* incidence relations, and *multiple* kinds of assumptions on the scene.

# Part I

# Introductory Material

# Chapter 2
# Representation of a Three-Dimensional Moving Scene

*I will not define time, space, place and motion, as being well known to all.*

– Isaac Newton, *Principia Mathematica, 1687*

The study of the geometric relationship between a three-dimensional (3-D) scene and its two-dimensional (2-D) images taken from a moving camera is at heart the interplay between two fundamental sets of transformations: *Euclidean motion*, also called *rigid-body motion*, which models how the camera moves, and *perspective projection*, which describes the image formation process. Long before these two transformations were brought together in computer vision, their theory had been developed independently. The study of the principles of motion of a material body has a long history belonging to the foundations of mechanics. For our purpose, more recent noteworthy insights to the understanding of the motion of rigid objects came from Chasles and Poinsot in the early 1800s. Their findings led to the current treatment of this subject, which has since been widely adopted.

In this chapter, we will start with an introduction to three-dimensional Euclidean space as well as to rigid-body motions. The next chapter will then focus on the perspective projection model of the camera. Both chapters require familiarity with some basic notions from linear algebra, many of which are reviewed in Appendix A at the end of this book.

## 2.1 Three-dimensional Euclidean space

We will use $\mathbb{E}^3$ to denote the familiar three-dimensional Euclidean space. In general, a Euclidean space is a set whose elements satisfy the five axioms of Euclid. Analytically, three-dimensional Euclidean space can be represented globally by a Cartesian coordinate frame: every point $p \in \mathbb{E}^3$ can be identified with a point in $\mathbb{R}^3$ with three coordinates

$$\boldsymbol{X} \doteq [X_1, X_2, X_3]^T = \begin{bmatrix} X_1 \\ X_2 \\ X_3 \end{bmatrix} \quad \in \mathbb{R}^3.$$

Sometimes, we may also use $[X, Y, Z]^T$ to indicate individual coordinates instead of $[X_1, X_2, X_3]^T$. Through such an assignment of a Cartesian frame, one establishes a one-to-one correspondence between $\mathbb{E}^3$ and $\mathbb{R}^3$, which allows us to safely talk about points and their coordinates as if they were the same thing.

Cartesian coordinates are the first step toward making it possible measure distances and angles. In order to do so, $\mathbb{E}^3$ must be endowed with a *metric*. A precise definition of metric relies on the notion of *vector*.

**Definition 2.1 (Vector).** *In Euclidean space, a* vector *$v$ is determined by a pair of points $p, q \in \mathbb{E}^3$ and is defined as a directed arrow connecting $p$ to $q$, denoted $v = \overrightarrow{pq}$.*

The point $p$ is usually called the base point of the vector $v$. In coordinates, the vector $v$ is represented by the triplet $[v_1, v_2, v_3]^T \in \mathbb{R}^3$, where each coordinate is the difference between the corresponding coordinates of the two points: if $p$ has coordinates $\boldsymbol{X}$ and $q$ has coordinates $\boldsymbol{Y}$, then $v$ has coordinates[1]

$$v \doteq \boldsymbol{Y} - \boldsymbol{X} \quad \in \mathbb{R}^3.$$

The preceding definition of a vector is referred to as a *bound vector*. One can also introduce the concept of a *free vector*, a vector whose definition does not depend on its base point. If we have two pairs of points $(p, q)$ and $(p', q')$ with coordinates satisfying $\boldsymbol{Y} - \boldsymbol{X} = \boldsymbol{Y}' - \boldsymbol{X}'$, we say that they define the same free vector. Intuitively, this allows a vector $v$ to be transported in parallel anywhere in $\mathbb{E}^3$. In particular, without loss of generality, one can assume that the base point is the origin of the Cartesian frame, so that $\boldsymbol{X} = 0$ and $\boldsymbol{Y} = v$. Note, however, that this notation is confusing: $\boldsymbol{Y}$ here denotes the coordinates of a vector that happen to be the same as the coordinates of the point $q$ just because we have chosen the point $p$ to be the origin. The reader should keep in mind that points and vectors are different geometric objects. This will be important, as we will see shortly, since a rigid-body motion acts differently on points and vectors.

[1]Note that we use the same symbol $v$ for a vector and its coordinates.

The set of all free vectors forms a *linear vector space*[2] (Appendix A), with the linear combination of two vectors $v, u \in \mathbb{R}^3$ defined by

$$\alpha v + \beta u = [\alpha v_1 + \beta u_1, \alpha v_2 + \beta u_2, \alpha v_3 + \beta u_3]^T \in \mathbb{R}^3, \quad \forall \alpha, \beta \in \mathbb{R}.$$

The *Euclidean metric* for $\mathbb{E}^3$ is then defined simply by an inner product[3] (Appendix A) on the vector space $\mathbb{R}^3$. It can be shown that by a proper choice of Cartesian frame, any inner product in $\mathbb{E}^3$ can be converted to the following canonical form

$$\langle u, v \rangle \doteq u^T v = u_1 v_1 + u_2 v_2 + u_3 v_3, \quad \forall u, v \in \mathbb{R}^3. \tag{2.1}$$

This inner product is also referred to as the standard Euclidean metric. In most parts of this book (but not everywhere!) we will use the canonical inner product $\langle u, v \rangle = u^T v$. Consequently, the norm (or length) of a vector $v$ is $\|v\| \doteq \sqrt{\langle v, v \rangle} = \sqrt{v_1^2 + v_2^2 + v_3^2}$. When the inner product between two vectors is zero, i.e. $\langle u, v \rangle = 0$, they are said to be *orthogonal*.

Finally, Euclidean space $\mathbb{E}^3$ can be formally described as a space that, with respect to a Cartesian frame, can be identified with $\mathbb{R}^3$ and has a metric (on its vector space) given by the above inner product. With such a metric, one can measure not only distances between points or angles between vectors, but also calculate the length of a curve[4] or the volume of a region.

While the inner product of two vectors is a real scalar, the so-called *cross product* of two vectors is a vector as defined below.

**Definition 2.2 (Cross product).** *Given two vectors $u, v \in \mathbb{R}^3$, their* cross product *is a third vector with coordinates given by*

$$u \times v \doteq \begin{bmatrix} u_2 v_3 - u_3 v_2 \\ u_3 v_1 - u_1 v_3 \\ u_1 v_2 - u_2 v_1 \end{bmatrix} \in \mathbb{R}^3.$$

It is immediate from this definition that the cross product of two vectors is linear in each of its arguments: $u \times (\alpha v + \beta w) = \alpha u \times v + \beta u \times w$, $\forall\, \alpha, \beta \in \mathbb{R}$. Furthermore, it is immediate to verify that

$$\langle u \times v, u \rangle = \langle u \times v, v \rangle = 0, \quad u \times v = -v \times u.$$

Therefore, the cross product of two vectors is orthogonal to each of its factors, and the order of the factors defines an *orientation* (if we change the order of the factors, the cross product changes sign).

---

[2]Note that the set of points does not.

[3]In some literature, the inner product is also referred to as the "dot product."

[4]If the trajectory of a moving particle $p$ in $\mathbb{E}^3$ is described by a curve $\gamma(\cdot) : t \mapsto \boldsymbol{X}(t) \in \mathbb{R}^3, t \in [0, 1]$, then the total length of the curve is given by

$$l(\gamma(\cdot)) = \int_0^1 \|\dot{\boldsymbol{X}}(t)\| \, dt,$$

where $\dot{\boldsymbol{X}}(t) = \frac{d}{dt}(\boldsymbol{X}(t)) \in \mathbb{R}^3$ is the so-called tangent vector to the curve.

If we fix $u$, the cross product can be represented by a map from $\mathbb{R}^3$ to $\mathbb{R}^3$: $v \mapsto u \times v$. This map is linear in $v$ and therefore can be represented by a matrix (Appendix A). We denote this matrix by $\widehat{u} \in \mathbb{R}^{3\times 3}$, pronounced "$u$ hat." It is immediate to verify by substitution that this matrix is given by[5]

$$\widehat{u} \doteq \begin{bmatrix} 0 & -u_3 & u_2 \\ u_3 & 0 & -u_1 \\ -u_2 & u_1 & 0 \end{bmatrix} \in \mathbb{R}^{3\times 3}. \tag{2.2}$$

Hence, we can write $u \times v = \widehat{u}v$. Note that $\widehat{u}$ is a $3 \times 3$ skew-symmetric matrix, i.e. $\widehat{u}^T = -\widehat{u}$ (see Appendix A).

**Example 2.3 (Right-hand rule).** It is immediate to verify that for $e_1 \doteq [1,0,0]^T, e_2 \doteq [0,1,0]^T \in \mathbb{R}^3$, we have $e_1 \times e_2 = [0,0,1]^T \doteq e_3$. That is, for a standard Cartesian frame, the cross product of the principal axes $X$ and $Y$ gives the principal axis $Z$. The cross product therefore conforms to the *right-hand rule*. See Figure 2.1. ■

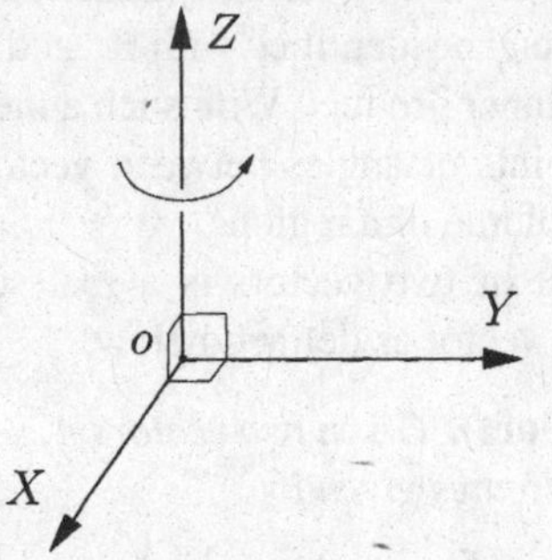


Figure 2.1. A right-handed $(X, Y, Z)$ coordinate frame.

The cross product, therefore, naturally defines a map between a vector $u$ and a $3 \times 3$ skew-symmetric matrix $\widehat{u}$. By inspection, the converse of this statement is clearly true, since we can easily identify a three-dimensional vector associated with every $3 \times 3$ skew-symmetric matrix (just extract $u_1, u_2, u_3$ from (2.2)).

**Lemma 2.4 (Skew-symmetric matrix).** *A matrix $M \in \mathbb{R}^{3\times 3}$ is skew-symmetric if and only if $M = \widehat{u}$ for some $u \in \mathbb{R}^3$.*

Therefore, the vector space $\mathbb{R}^3$ and the space of all skew-symmetric $3\times 3$ matrices, called $so(3)$,[6] are isomorphic (i.e. there exists a one-to-one map that preserves the vector space structure). The isomorphism is the so-called *hat operator*

$$\wedge : \mathbb{R}^3 \to so(3); \quad u \mapsto \widehat{u},$$

[5]In some literature, the matrix $\widehat{u}$ is denoted by $u_\times$ or $[u]_\times$.

[6]We will explain the reason for this name later in this chapter.

and its inverse map, called the *vee operator*, which extracts the components of the vector $u$ from a skew-symmetric matrix $\widehat{u}$, is given by

$$\vee : so(3) \to \mathbb{R}^3; \quad \widehat{u} \mapsto \widehat{u}^\vee = u.$$

## 2.2 Rigid-body motion

Consider an object moving in front of a camera. In order to describe its motion one should, in principle, specify the trajectory of every single point on the object, for instance, by specifying coordinates of a point as a function of time $\boldsymbol{X}(t)$. Fortunately, for rigid objects we do not need to specify the motion of every point. As we will see shortly, it is sufficient to specify the motion of one (instead of every) point, and the motion of three coordinate axes attached to that point. The reason is that for every rigid object, the distance between any two points on it does not change over time as the object moves. See Figure 2.2.

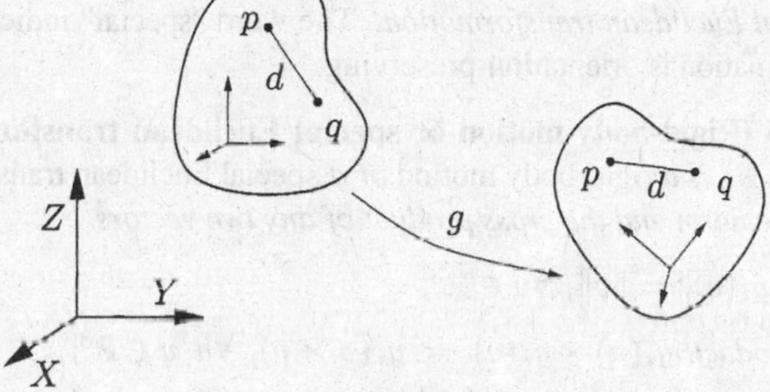


Figure 2.2. A motion of a rigid body preserves the distance $d$ between any pair of points $(p, q)$ on it.

Thus, if $\boldsymbol{X}(t)$ and $\boldsymbol{Y}(t)$ are the coordinates of any two points $p$ and $q$ on the object, respectively, the distance between them is constant:

$$\|\boldsymbol{X}(t) - \boldsymbol{Y}(t)\| \equiv \text{constant}, \quad \forall t \in \mathbb{R}. \tag{2.3}$$

A *rigid-body motion* (or rigid-body transformation) is then a family of maps that describe how the coordinates of every point on a rigid object change in time while satisfying (2.3). We denote such a map by

$$g(t) : \mathbb{R}^3 \to \mathbb{R}^3; \quad \boldsymbol{X} \mapsto g(t)(\boldsymbol{X}).$$

If instead of looking at the entire continuous path of the moving object, we concentrate on the map between its initial and final configuration, we have a *rigid-body displacement*, denoted by

$$g : \mathbb{R}^3 \to \mathbb{R}^3; \quad \boldsymbol{X} \mapsto g(\boldsymbol{X}).$$

Besides transforming the coordinates of points, $g$ also induces a transformation on vectors. Suppose that $v$ is a vector defined by two points $p$ and $q$ with coordinates

$v = \boldsymbol{Y} - \boldsymbol{X}$; then, after the transformation $g$, we obtain a new vector[7]

$$u = g_*(v) \doteq g(\boldsymbol{Y}) - g(\boldsymbol{X}).$$

Since $g$ preserves the distance between points, we have that $\|g_*(v)\| = \|v\|$ for all free vectors $v \in \mathbb{R}^3$.

A map that preserves the distance is called a *Euclidean transformation*. In the 3-D space, the set all Euclidean transformations is denoted by $E(3)$. Note that preserving distances between points is not sufficient to characterize a rigid object moving in space. In fact, there are transformations that preserve distances, and yet they are not physically realizable. For instance, the map

$$f : [X_1, X_2, X_3]^T \mapsto [X_1, X_2, -X_3]^T$$

preserves distances but not orientations. It corresponds to a reflection of points in the $XY$-plane as a double-sided mirror. To rule out this kind of maps,[8] we require that any rigid-body motion, besides preserving distances, preserves orientations as well. That is, in addition to preserving the norm of vectors, it must also preserve their cross product. The map or transformation induced by a rigid-body motion is called a *special Euclidean transformation*. The word "special" indicates the fact that a transformation is orientation-preserving.

**Definition 2.5 (Rigid-body motion or special Euclidean transformation).** *A map $g : \mathbb{R}^3 \rightarrow \mathbb{R}^3$ is a* rigid-body motion *or a* special Euclidean transformation *if it preserves the norm and the cross product of any two vectors,*

1. *norm:* $\|g_*(v)\| = \|v\|, \ \forall v \in \mathbb{R}^3$,
2. *cross product:* $g_*(u) \times g_*(v) \ = \ g_*(u \times v), \ \forall u, v \in \mathbb{R}^3$.

*The collection of all such motions or transformations is denoted by $SE(3)$.*

In the above definition of rigid-body motions, it is not immediately obvious that the angles between vectors are preserved. However, the inner product $\langle \cdot, \cdot \rangle$ can be expressed in terms of the norm $\| \cdot \|$ by the *polarization identity*

$$\boxed{\langle u, v \rangle = \frac{1}{4}\left(\|u + v\|^2 - \|u - v\|^2\right)} \tag{2.4}$$

and, since $\|u + v\| = \|g_*(u) + g_*(v)\|$, one can conclude that, for any rigid-body motion $g$,

$$\langle u, v \rangle = \langle g_*(u), g_*(v) \rangle, \quad \forall u, v \in \mathbb{R}^3. \tag{2.5}$$

In other words, a rigid-body motion can also be defined as one that preserves both the inner product and the cross product.

---

[7]The use of $g_*$ here is consistent with the so-called *push-forward map* or *differential operator* of $g$ in differential geometry, which denotes the action of a differentiable map on the tangent spaces of its domains.

[8]In Chapter 10, however, we will study the important role of reflections in multiple-view geometry.

**Example 2.6 (Triple product and volume).** From the definition of a rigid-body motion, one can show that it also preserves the so-called *triple product* among three vectors:

$$\langle g_*(u), g_*(v) \times g_*(w) \rangle = \langle u, v \times w \rangle.$$

Since the triple product corresponds to the volume of the parallelepiped spanned by the three vectors, rigid-body motion also preserves volumes. ■

How do these properties help us describe a rigid-body motion concisely? The fact that distances and orientations are preserved by a rigid-body motion means that individual points cannot move relative to each other. As a consequence, a rigid-body motion can be described by the motion of a chosen point on the body and the rotation of a coordinate frame attached to that point. In order to see this, we represent the *configuration* of a rigid body by attaching a Cartesian coordinate frame to some point on the rigid body, and we will keep track of the motion of this coordinate frame relative to a fixed *world (reference) frame*.

To this end, consider a coordinate frame, with its principal axes given by three *orthonormal* vectors $e_1, e_2, e_3 \in \mathbb{R}^3$; that is, they satisfy

$$e_i^T e_j = \delta_{ij} \doteq \begin{cases} 1 & \text{for} \quad i = j, \\ 0 & \text{for} \quad i \neq j. \end{cases} \tag{2.6}$$

The vectors are ordered so as to form a right-handed frame: $e_1 \times e_2 = e_3$. Then, after a rigid-body motion $g$, we have

$$g_*(e_i)^T g_*(e_j) = \delta_{ij}, \quad g_*(e_1) \times g_*(e_2) = g_*(e_3). \tag{2.7}$$

That is, the resulting three vectors $g_*(e_1), g_*(e_2), g_*(e_3)$ still form a right-handed orthonormal frame. Therefore, a rigid object can always be associated with a right-handed orthonormal frame, which we call the *object coordinate frame* or the *body coordinate frame*, and its rigid-body motion can be entirely specified by the motion of such a frame.

In Figure 2.3 we show an object, in this case a camera, moving relative to a world reference frame $W : (X, Y, Z)$ selected in advance. In order to specify the configuration of the camera relative to the world frame $W$, one may pick a fixed point $o$ on the camera and attach to it an object frame, in this case called a camera frame,[9] $C : (x, y, z)$. When the camera moves, the camera frame also moves along with the camera. The configuration of the camera is then determined by two components:

1. the vector between the origin $o$ of the world frame and that of the camera frame, $g(o)$, called the "translational" part and denoted by $T$;

2. the relative orientation of the camera frame $C$, with coordinate axes $(x, y, z)$, relative to the fixed world frame $W$ with coordinate axes $(X, Y, Z)$, called the "rotational" part and denoted by $R$.

---

[9]Here, to distinguish the two coordinate frames, we use lower-case $x, y, z$ for coordinates in the camera frame.

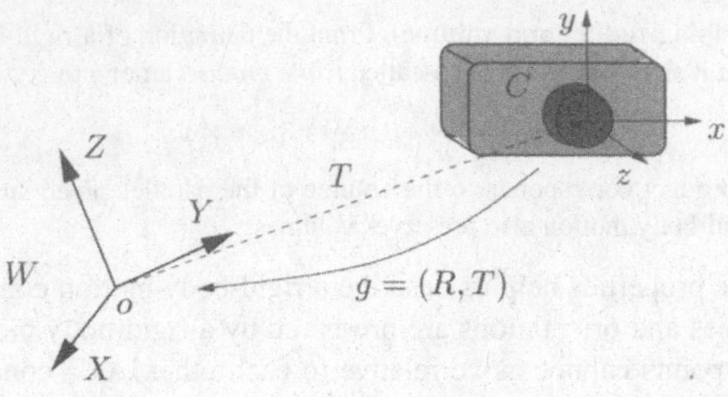


Figure 2.3. A rigid-body motion between a camera frame $C$: $(x, y, z)$ and a world coordinate frame $W$: $(X, Y, Z)$.

In the problems we consider in this book, there is no obvious choice of the world reference frame and its origin $o$. Therefore, we can choose the world frame to be attached to the camera and specify the translation and rotation of the scene relative to that frame (as long as it is rigid), or we could attach the world frame to the scene and specify the motion of the camera relative to that frame. All that matters is the *relative* motion between the scene and the camera; the choice of the world reference frame is, from the point of view of geometry, arbitrary.[10]

If we can move a rigid object (e.g., a camera) from one place to another, we can certainly reverse the action and put it back to its original position. Similarly, we can combine several motions to generate a new one. Roughly speaking, this property of invertibility and composition can be mathematically characterized by the notion of "group" (Appendix A). As we will soon see, the set of rigid-body motions is indeed a group, the so-called *special Euclidean group*. However, the abstract notion of group is not useful until we can give it an explicit representation and use it for computation. In the next few sections, we will focus on studying in detail how to represent rigid-body motions in terms of matrices.[11] More specifically, we will show that any rigid-body motion can be represented as a $4 \times 4$ matrix. For simplicity, we start with the rotational component of a rigid-body motion.

## 2.3 Rotational motion and its representations

### *2.3.1 Orthogonal matrix representation of rotations*

Suppose we have a rigid object rotating about a fixed point $o \in \mathbb{E}^3$. How do we describe its orientation relative to a chosen coordinate frame, say $W$? Without loss of generality, we may always assume that the origin of the world frame is

[10]The human vision literature, on the other hand, debates whether the primate brain maintains a view-centered or an object-centered representation of the world.

[11]The notion of matrix representation for a group is introduced in Appendix A.

the center of rotation $o$. If this is not the case, simply translate the origin to the point $o$. We now attach another coordinate frame, say $C$, to the rotating object, say a camera, with its origin also at $o$. The relation between these two coordinate frames is illustrated in Figure 2.4.

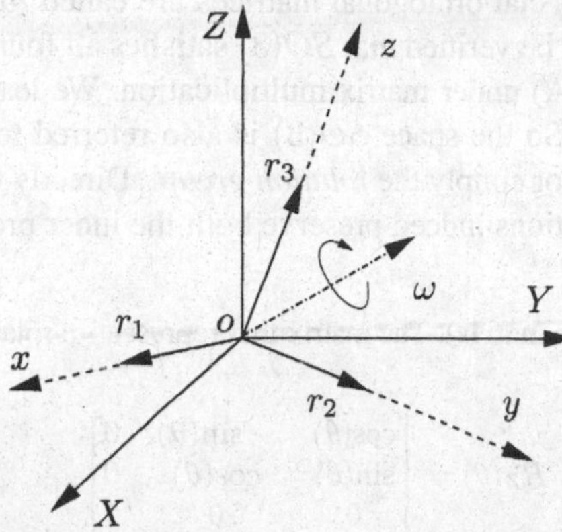


Figure 2.4. Rotation of a rigid body about a fixed point $o$ and along the axis $\omega$. The coordinate frame $W$ (solid line) is fixed, and the coordinate frame $C$ (dashed line) is attached to the rotating rigid body.

The configuration (or "orientation") of the frame $C$ relative to the frame $W$ is determined by the coordinates of the three orthonormal vectors $r_1 = g_*(e_1), r_2 = g_*(e_2), r_3 = g_*(e_3) \in \mathbb{R}^3$ relative to the world frame $W$, as shown in Figure 2.4. The three vectors $r_1, r_2, r_3$ are simply the unit vectors along the three principal axes $x, y, z$ of the frame $C$, respectively. The configuration of the rotating object is then completely determined by the $3 \times 3$ matrix

$$R_{wc} \doteq [r_1, r_2, r_3] \quad \in \mathbb{R}^{3\times 3},$$

with $r_1, r_2, r_3$ stacked in order as its three columns. Since $r_1, r_2, r_3$ form an orthonormal frame, it follows that

$$r_i^T r_j = \delta_{ij} \doteq \begin{cases} 1 & \text{for} \quad i = j, \\ 0 & \text{for} \quad i \neq j, \end{cases} \qquad \forall i, j \in \{1, 2, 3\}.$$

This can be written in matrix form as

$$R_{wc}^T R_{wc} = R_{wc} R_{wc}^T = I.$$

Any matrix that satisfies the above identity is called an *orthogonal matrix*. It follows from the above definition that the inverse of an orthogonal matrix is simply its transpose: $R_{wc}^{-1} = R_{wc}^T$. Since $r_1, r_2, r_3$ form a right-handed frame, we further have the condition that the determinant of $R_{wc}$ must be $+1$.[12] Hence $R_{wc}$ is a *special orthogonal matrix*, where as before, the word "special" indicates that it is

[12]This can easily be seen by computing the determinant of the rotation matrix $\det(R) = r_1^T(r_2 \times r_3)$, which is equal to $+1$.

orientation-preserving. The space of all such special orthogonal matrices in $\mathbb{R}^{3\times3}$ is usually denoted by

$$\boxed{SO(3) \doteq \{R \in \mathbb{R}^{3\times3} \mid R^T R = I, \det(R) = +1\}.}$$

Traditionally, $3 \times 3$ special orthogonal matrices are called *rotation matrices* for obvious reasons. It can be verified that $SO(3)$ satisfies all four axioms of a group (defined in Appendix A) under matrix multiplication. We leave the proof to the reader as an exercise. So the space $SO(3)$ is also referred to as the *special orthogonal group* of $\mathbb{R}^3$, or simply the *rotation group*. Directly from the definition, one can show that rotations indeed preserve both the inner product and the cross product of vectors.

**Example 2.7 (A rotation matrix).** The matrix that represents a rotation about the $Z$-axis by an angle $\theta$ is

$$R_Z(\theta) = \begin{bmatrix} \cos(\theta) & -\sin(\theta) & 0 \\ \sin(\theta) & \cos(\theta) & 0 \\ 0 & 0 & 1 \end{bmatrix}.$$

The reader can similarly derive matrices for rotation about the $X$-axis or the $Y$-axis. In the next section we will study how to represent a rotation about any axis. ■

Going back to Figure 2.4, every rotation matrix $R_{wc} \in SO(3)$ represents a possible configuration of the object rotated about the point $o$. Besides this, $R_{wc}$ takes another role as the matrix that represents the coordinate transformation from the frame $C$ to the frame $W$. To see this, suppose that for a given a point $p \in \mathbb{E}^3$, its coordinates with respect to the frame $W$ are $\boldsymbol{X}_w = [X_{1w}, X_{2w}, X_{3w}]^T \in \mathbb{R}^3$. Since $r_1, r_2, r_3$ also form a basis for $\mathbb{R}^3$, $\boldsymbol{X}_w$ can be expressed as a linear combination of these three vectors, say $\boldsymbol{X}_w = X_{1c}r_1 + X_{2c}r_2 + X_{3c}r_3$ with $[X_{1c}, X_{2c}, X_{3c}]^T \in \mathbb{R}^3$. Obviously, $\boldsymbol{X}_c = [X_{1c}, X_{2c}, X_{3c}]^T$ are the coordinates of the same point $p$ with respect to the frame $C$. Therefore, we have

$$\boldsymbol{X}_w = X_{1c}r_1 + X_{2c}r_2 + X_{3c}r_3 = R_{wc}\boldsymbol{X}_c.$$

In this equation, the matrix $R_{wc}$ transforms the coordinates $\boldsymbol{X}_c$ of a point $p$ relative to the frame $C$ to its coordinates $\boldsymbol{X}_w$ relative to the frame $W$. Since $R_{wc}$ is a rotation matrix, its inverse is simply its transpose,

$$\boldsymbol{X}_c = R_{wc}^{-1}\boldsymbol{X}_w = R_{wc}^T\boldsymbol{X}_w.$$

That is, the inverse transformation of a rotation is also a rotation; we call it $R_{cw}$, following an established convention, so that

$$R_{cw} = R_{wc}^{-1} = R_{wc}^T.$$

The configuration of a continuously rotating object can then be described as a trajectory $R(t) : t \mapsto SO(3)$ in the space $SO(3)$. When the starting time is not $t = 0$, the relative motion between time $t_2$ and time $t_1$ will be denoted as $R(t_2, t_1)$. The composition law of the rotation group (see Appendix A) implies

$$R(t_2, t_0) = R(t_2, t_1)R(t_1, t_0), \quad \forall t_0 < t_1 < t_2 \in \mathbb{R}.$$

For a rotating camera, the world coordinates $\boldsymbol{X}_w$ of a fixed 3-D point $p$ are transformed to its coordinates relative to the camera frame $C$ by

$$\boldsymbol{X}_c(t) = R_{cw}(t)\boldsymbol{X}_w.$$

Alternatively, if a point $p$ is fixed with respect to the camera frame has coordinates $\boldsymbol{X}_c$, its world coordinates $\boldsymbol{X}_w(t)$ as a function of $t$ are then given by

$$\boldsymbol{X}_w(t) = R_{wc}(t)\boldsymbol{X}_c.$$

### 2.3.2 *Canonical exponential coordinates for rotations*

So far, we have shown that a rotational rigid-body motion in $\mathbb{E}^3$ can be represented by a $3 \times 3$ rotation matrix $R \in SO(3)$. In the matrix representation that we have so far, each rotation matrix $R$ is described by its $3 \times 3 = 9$ entries. However, these nine entries are not free parameters because they must satisfy the constraint $R^T R = I$. This actually imposes six independent constraints on the nine entries. Hence, the dimension of the space of rotation matrices $SO(3)$ should be only three, and six parameters out of the nine are in fact redundant. In this subsection and Appendix 2.A, we will introduce a few explicit parameterizations for the space of rotation matrices.

Given a trajectory $R(t) : \mathbb{R} \to SO(3)$ that describes a continuous rotational motion, the rotation must satisfy the following constraint

$$R(t)R^T(t) = I.$$

Computing the derivative of the above equation with respect to time $t$ and noticing that the right-hand side is a constant matrix, we obtain

$$\dot{R}(t)R^T(t) + R(t)\dot{R}^T(t) = 0 \quad \Rightarrow \quad \dot{R}(t)R^T(t) = -(\dot{R}(t)R^T(t))^T.$$

The resulting equation reflects the fact that the matrix $\dot{R}(t)R^T(t) \in \mathbb{R}^{3\times 3}$ is a skew-symmetric matrix. Then, as we have seen in Lemma 2.4, there must exist a vector, say $\omega(t) \in \mathbb{R}^3$, such that

$$\dot{R}(t)R^T(t) = \widehat{\omega}(t).$$

Multiplying both sides by $R(t)$ on the right yields

$$\dot{R}(t) = \widehat{\omega}(t)R(t). \tag{2.8}$$

Notice that from the above equation, if $R(t_0) = I$ for $t = t_0$, we have $\dot{R}(t_0) = \widehat{\omega}(t_0)$. Hence, around the identity matrix $I$, a skew-symmetric matrix gives a first-order approximation to a rotation matrix:

$$R(t_0 + dt) \approx I + \widehat{\omega}(t_0)\, dt.$$

As we have anticipated, the space of all skew-symmetric matrices is denoted by

$$\boxed{so(3) \doteq \left\{\widehat{\omega} \in \mathbb{R}^{3\times 3} \mid \omega \in \mathbb{R}^3\right\},} \tag{2.9}$$

and following the above observation, it is also called the *tangent space* at the identity of the rotation group $SO(3)$.[13] If $R(t)$ is not at the identity, the tangent space at $R(t)$ is simply $so(3)$ transported to $R(t)$ by a multiplication by $R(t)$ on the right: $\dot{R}(t) = \widehat{\omega}(t)R(t)$. This also shows that, locally, elements of $SO(3)$ depend on only three parameters, $(\omega_1, \omega_2, \omega_3)$.

Having understood its local approximation, we will now use this knowledge to obtain a useful representation for the rotation matrix. Let us start by assuming that the matrix $\widehat{\omega}$ in (2.8) is constant,

$$\dot{R}(t) = \widehat{\omega}R(t). \tag{2.10}$$

In the above equation, $R(t)$ can be interpreted as the *state transition matrix* for the following linear ordinary differential equation (ODE):

$$\dot{x}(t) = \widehat{\omega}x(t), \quad x(t) \in \mathbb{R}^3. \tag{2.11}$$

It is then immediate to verify that the solution to the above ODE is given by

$$x(t) = e^{\widehat{\omega}t}x(0), \tag{2.12}$$

where $e^{\widehat{\omega}t}$ is the matrix exponential

$$e^{\widehat{\omega}t} = I + \widehat{\omega}t + \frac{(\widehat{\omega}t)^2}{2!} + \cdots + \frac{(\widehat{\omega}t)^n}{n!} + \cdots. \tag{2.13}$$

The exponential $e^{\widehat{\omega}t}$ is also often denoted by $\exp(\widehat{\omega}t)$. Due to the uniqueness of the solution to the ODE (2.11), and assuming $R(0) = I$ is the initial condition for (2.10), we must have

$$\boxed{R(t) = e^{\widehat{\omega}t}.} \tag{2.14}$$

To verify that the matrix $e^{\widehat{\omega}t}$ is indeed a rotation matrix, one can directly show from the definition of the matrix exponential that

$$(e^{\widehat{\omega}t})^{-1} = e^{-\widehat{\omega}t} = e^{\widehat{\omega}^T t} = (e^{\widehat{\omega}t})^T.$$

Hence $(e^{\widehat{\omega}t})^T e^{\widehat{\omega}t} = I$. It remains to show that $\det(e^{\widehat{\omega}t}) = +1$, and we leave this fact to the reader as an exercise (see Exercise 2.12). *A physical interpretation of equation (2.14) is that if $\|\omega\| = 1$, then $R(t) = e^{\widehat{\omega}t}$ is simply a rotation around the axis $\omega \in \mathbb{R}^3$ by an angle of $t$ radians.*[14] In general, $t$ can be absorbed into $\omega$, so we have $R = e^{\widehat{\omega}}$ for $\omega$ with arbitrary norm. So, the matrix exponential (2.13) indeed defines a map from the space $so(3)$ to $SO(3)$, the so-called *exponential map*

$$\exp: so(3) \to SO(3); \quad \widehat{\omega} \mapsto e^{\widehat{\omega}}.$$

Note that we obtained the expression (2.14) by assuming that the $\omega(t)$ in (2.8) is constant. This is, however, not always the case. A question naturally arises:

---

[13] Since $SO(3)$ is a Lie group, $so(3)$ is called its Lie algebra.

[14] We can use either $e^{\widehat{\omega}\theta}$, where $\theta$ encodes explicitly the rotation angle and $\|\omega\| = 1$, or more simply $e^{\widehat{\omega}}$ where $\|\omega\|$ encodes the rotation angle.

can every rotation matrix $R \in SO(3)$ be expressed in an exponential form as in (2.14)? The answer is yes, and the fact is stated as the following theorem.

**Theorem 2.8 (Logarithm of $SO(3)$).** *For any $R \in SO(3)$, there exists a (not necessarily unique) $\omega \in \mathbb{R}^3$ such that $R = \exp(\widehat{\omega})$. We denote the inverse of the exponential map by $\widehat{\omega} = \log(R)$.*

*Proof.* The proof of this theorem is by construction: if the rotation matrix $R \neq I$ is given as

$$R = \begin{bmatrix} r_{11} & r_{12} & r_{13} \\ r_{21} & r_{22} & r_{23} \\ r_{31} & r_{32} & r_{33} \end{bmatrix},$$

the corresponding $\omega$ is given by

$$\boxed{\|\omega\| = \cos^{-1}\left(\frac{\operatorname{trace}(R) - 1}{2}\right), \quad \frac{\omega}{\|\omega\|} = \frac{1}{2\sin(\|\omega\|)}\begin{bmatrix} r_{32} - r_{23} \\ r_{13} - r_{31} \\ r_{21} - r_{12} \end{bmatrix}.} \tag{2.15}$$

If $R = I$, then $\|\omega\| = 0$, and $\frac{\omega}{\|\omega\|}$ is not determined (and therefore can be chosen arbitrarily). □

The significance of this theorem is that any rotation matrix can be realized by rotating around some fixed axis $\omega$ by a certain angle $\|\omega\|$. However, the exponential map from $so(3)$ to $SO(3)$ is not one-to-one, since any vector of the form $2k\pi\omega$ with $k$ integer would give rise to the same $R$. This will become clear after we have introduced the so-called *Rodrigues' formula* for computing $R = e^{\widehat{\omega}}$.

From the constructive proof of Theorem 2.8, we know how to compute the exponential coordinates $\omega$ for a given rotation matrix $R \in SO(3)$. On the other hand, given $\omega$, how do we effectively compute the corresponding rotation matrix $R = e^{\widehat{\omega}}$? One can certainly use the series (2.13) from the definition. The following theorem, however, provides a very useful formula that simplifies the computation significantly.

**Theorem 2.9 (Rodrigues' formula for a rotation matrix).** *Given $\omega \in \mathbb{R}^3$, the matrix exponential $R = e^{\widehat{\omega}}$ is given by*

$$\boxed{e^{\widehat{\omega}} = I + \frac{\widehat{\omega}}{\|\omega\|}\sin(\|\omega\|) + \frac{\widehat{\omega}^2}{\|\omega\|^2}(1 - \cos(\|\omega\|)).} \tag{2.16}$$

*Proof.* Let $t = \|\omega\|$ and redefine $\omega$ to be of unit length. Then, it is immediate to verify that powers of $\widehat{\omega}$ can be reduced by the following two formulae

$$\widehat{\omega}^2 = \omega\omega^T - I, \quad \widehat{\omega}^3 = -\widehat{\omega}.$$

Hence the exponential series (2.13) can be simplified as

$$e^{\widehat{\omega}t} = I + \left(t - \frac{t^3}{3!} + \frac{t^5}{5!} - \cdots\right)\widehat{\omega} + \left(\frac{t^2}{2!} - \frac{t^4}{4!} + \frac{t^6}{6!} - \cdots\right)\widehat{\omega}^2.$$

The two sets of parentheses contain the Taylor series for $\sin(t)$ and $(1 - \cos(t))$, respectively. Thus, we have $e^{\widehat{\omega}t} = I + \widehat{\omega}\sin(t) + \widehat{\omega}^2(1 - \cos(t))$. □

Using Rodrigues' formula, it is immediate to see that if $\|\omega\| = 1, t = 2k\pi$, we have

$$e^{\widehat{\omega}2k\pi} = I$$

for all $k \in \mathbb{Z}$. Hence, for a given rotation matrix $R \in SO(3)$, there are infinitely many exponential coordinates $\omega \in \mathbb{R}^3$ such that $e^{\widehat{\omega}} = R$. The exponential map $\exp : so(3) \to SO(3)$ is therefore *not* one-to-one. It is also useful to know that the exponential map is not *commutative*, i.e. for two $\widehat{\omega}_1, \widehat{\omega}_2 \in so(3)$,

$$e^{\widehat{\omega}_1}e^{\widehat{\omega}_2} \neq e^{\widehat{\omega}_2}e^{\widehat{\omega}_1} \neq e^{\widehat{\omega}_1+\widehat{\omega}_2},$$

unless $\widehat{\omega}_1\widehat{\omega}_2 = \widehat{\omega}_2\widehat{\omega}_1$.

**Remark 2.10.** *In general, the difference between $\widehat{\omega}_1\widehat{\omega}_2$ and $\widehat{\omega}_2\widehat{\omega}_1$ is called the* Lie bracket *on $so(3)$, denoted by*

$$[\widehat{\omega}_1, \widehat{\omega}_2] = \widehat{\omega}_1\widehat{\omega}_2 - \widehat{\omega}_2\widehat{\omega}_1, \quad \forall\, \widehat{\omega}_1, \widehat{\omega}_2 \in so(3).$$

*From the definition above it can be verified that $[\widehat{\omega}_1, \widehat{\omega}_2]$ is also a skew-symmetric matrix in $so(3)$. The linear structure of $so(3)$ together with the Lie bracket form the* Lie algebra *of the (Lie) group $SO(3)$. For more details on the Lie group structure of $SO(3)$, the reader may refer to [Murray et al., 1993]. Given $\widehat{\omega}$, the set of all rotation matrices $e^{\widehat{\omega}t}, t \in \mathbb{R}$, is then a* one-parameter *subgroup of $SO(3)$, i.e. the planar rotation group $SO(2)$. The multiplication in such a subgroup is always commutative, since for the same $\omega \in \mathbb{R}^3$, we have*

$$e^{\widehat{\omega}t_1}e^{\widehat{\omega}t_2} = e^{\widehat{\omega}t_2}e^{\widehat{\omega}t_1} = e^{\widehat{\omega}(t_1+t_2)}, \quad \forall t_1, t_2 \in \mathbb{R}.$$

The exponential coordinates introduced above provide a local parameterization for rotation matrices. There are also other ways to parameterize rotation matrices, either globally or locally, among which *quaternions* and *Euler angles* (or more formally, *Lie-Cartan coordinates*) are two popular choices. We leave more detailed discussions to Appendix 2.A at the end of this chapter. We use exponential coordinates because they are simpler and more intuitive.

## 2.4 Rigid-body motion and its representations

In the previous section, we studied purely rotational rigid-body motions and how to represent and compute a rotation matrix. In this section, we will study how to represent a rigid-body motion in general, a motion with both rotation and translation.

Figure 2.5 illustrates a moving rigid object with a coordinate frame $C$ attached to it. To describe the coordinates of a point $p$ on the object with respect to the world frame $W$, it is clear from the figure that the vector $\boldsymbol{X}_w$ is simply the sum of

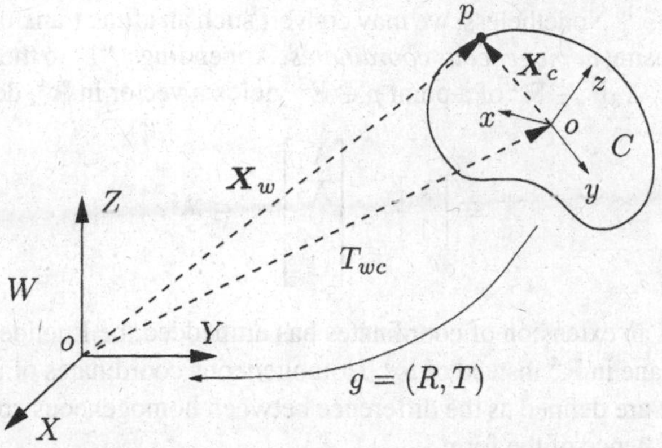


Figure 2.5. A rigid-body motion between a moving frame $C$ and a world frame $W$.

the translation $T_{wc} \in \mathbb{R}^3$ of the origin of the frame $C$ relative to that of the frame $W$ and the vector $\boldsymbol{X}_c$ but expressed relative to the frame $W$. Since $\boldsymbol{X}_c$ are the coordinates of the point $p$ relative to the frame $C$, with respect to the world frame $W$, it becomes $R_{wc}\boldsymbol{X}_c$, where $R_{wc} \in SO(3)$ is the relative rotation between the two frames. Hence, the coordinates $\boldsymbol{X}_w$ are given by

$$\boldsymbol{X}_w = R_{wc}\boldsymbol{X}_c + T_{wc}. \tag{2.17}$$

Usually, we denote the full rigid-body motion by $g_{wc} = (R_{wc}, T_{wc})$, or simply $g = (R, T)$ if the frames involved are clear from the context. Then $g$ represents not only a description of the configuration of the rigid-body object but also a transformation of coordinates between the two frames. In compact form, we write

$$\boldsymbol{X}_w = g_{wc}(\boldsymbol{X}_c).$$

The set of all possible configurations of a rigid body can then be described by the space of rigid-body motions or special Euclidean transformations

$$\boxed{SE(3) \doteq \{g = (R, T) \mid R \in SO(3), T \in \mathbb{R}^3\}}.$$

Note that $g = (R, T)$ is not yet a matrix representation for $SE(3)$.[15] To obtain such a representation, we need to introduce the so-called homogeneous coordinates. We will introduce only what is needed to carry our study of rigid-body motions.

### 2.4.1 *Homogeneous representation*

One may have already noticed from equation (2.17) that in contrast to the pure rotation case, the coordinate transformation for a full rigid-body motion is not

[15] For this to be the case, the composition of two rigid-body motions needs to be the multiplication of two matrices. See Appendix A.

linear but *affine*.[16] Nonetheless, we may convert such an affine transformation to a linear one by using *homogeneous coordinates*. Appending a "1" to the coordinates $\boldsymbol{X} = [X_1, X_2, X_3]^T \in \mathbb{R}^3$ of a point $p \in \mathbb{E}^3$ yields a vector in $\mathbb{R}^4$, denoted by

$$\bar{\boldsymbol{X}} \doteq \begin{bmatrix} \boldsymbol{X} \\ 1 \end{bmatrix} = \begin{bmatrix} X_1 \\ X_2 \\ X_3 \\ 1 \end{bmatrix} \in \mathbb{R}^4.$$

In effect, such an extension of coordinates has embedded the Euclidean space $\mathbb{E}^3$ into a hyperplane in $\mathbb{R}^4$ instead of $\mathbb{R}^3$. Homogeneous coordinates of a vector $v = \boldsymbol{X}(q) - \boldsymbol{X}(p)$ are defined as the difference between homogeneous coordinates of the two points hence of the form

$$\bar{v} \doteq \begin{bmatrix} v \\ 0 \end{bmatrix} = \begin{bmatrix} \boldsymbol{X}(q) \\ 1 \end{bmatrix} - \begin{bmatrix} \boldsymbol{X}(p) \\ 1 \end{bmatrix} = \begin{bmatrix} v_1 \\ v_2 \\ v_3 \\ 0 \end{bmatrix} \in \mathbb{R}^4.$$

Notice that in $\mathbb{R}^4$, vectors of the above form give rise to a subspace, and all linear structures of the original vectors $v \in \mathbb{R}^3$ are perfectly preserved by the new representation. Using the new notation, the (affine) transformation (2.17) can then be rewritten in a "linear" form

$$\bar{\boldsymbol{X}}_w = \begin{bmatrix} \boldsymbol{X}_w \\ 1 \end{bmatrix} = \begin{bmatrix} R_{wc} & T_{wc} \\ 0 & 1 \end{bmatrix} \begin{bmatrix} \boldsymbol{X}_c \\ 1 \end{bmatrix} \doteq \bar{g}_{wc} \bar{\boldsymbol{X}}_c,$$

where the $4 \times 4$ matrix $\bar{g}_{wc} \in \mathbb{R}^{4 \times 4}$ is called the *homogeneous representation* of the rigid-body motion $g_{wc} = (R_{wc}, T_{wc}) \in SE(3)$. In general, if $g = (R, T)$, then its homogeneous representation is

$$\bar{g} = \begin{bmatrix} R & T \\ 0 & 1 \end{bmatrix} \in \mathbb{R}^{4 \times 4}. \tag{2.18}$$

Notice that by introducing a little redundancy into the notation, we can represent a rigid-body transformation of coordinates by a linear matrix multiplication. The homogeneous representation of $g$ in (2.18) gives rise to a natural matrix representation of the special Euclidean transformations

$$\boxed{SE(3) \doteq \left\{ \bar{g} = \begin{bmatrix} R & T \\ 0 & 1 \end{bmatrix} \middle| R \in SO(3), T \in \mathbb{R}^3 \right\} \subset \mathbb{R}^{4 \times 4}.}$$

Using this representation, it is then straightforward to verify that the set $SE(3)$ indeed satisfies all the requirements of a group (Appendix A). In particular, $\forall g_1, g_2$

[16] We say that two vectors $u, v$ are related by a *linear* transformation if $u = Av$ for some matrix $A$, and by an *affine* transformation if $u = Av + b$ for some matrix $A$ and vector $b$. See Appendix A.

and $g \in SE(3)$, we have

$$\bar{g}_1\bar{g}_2 = \begin{bmatrix} R_1 & T_1 \\ 0 & 1 \end{bmatrix} \begin{bmatrix} R_2 & T_2 \\ 0 & 1 \end{bmatrix} = \begin{bmatrix} R_1R_2 & R_1T_2 + T_1 \\ 0 & 1 \end{bmatrix} \quad \in SE(3)$$

and

$$\bar{g}^{-1} = \begin{bmatrix} R & T \\ 0 & 1 \end{bmatrix}^{-1} = \begin{bmatrix} R^T & -R^TT \\ 0 & 1 \end{bmatrix} \quad \in SE(3).$$

Thus, $\bar{g}$ is indeed a matrix representation for the group of rigid-body motions according to the definition we mentioned in Section 2.2 (but given formally in Appendix A). In the homogeneous representation, the action of a rigid-body motion $g \in SE(3)$ on a vector $v = \boldsymbol{X}(q) - \boldsymbol{X}(p) \in \mathbb{R}^3$ becomes

$$\bar{g}_*(\bar{v}) = \bar{g}\bar{\boldsymbol{X}}(q) - \bar{g}\bar{\boldsymbol{X}}(p) = \bar{g}\bar{v}.$$

That is, the action is also simply represented by a matrix multiplication. In the 3-D coordinates, we have $g_*(v) = Rv$, since only rotational part affects vectors. The reader can verify that such an action preserves both the inner product and the cross product. As can be seen, *rigid motions act differently on points (rotation and translation) than they do on vectors (rotation only).*

### 2.4.2 Canonical exponential coordinates for rigid-body motions

In Section 2.3.2, we studied exponential coordinates for a rotation matrix $R \in SO(3)$. Similar coordinatization also exists for the homogeneous representation of a full rigid-body motion $g \in SE(3)$. For the rest of this section, we demonstrate how to extend the results we have developed for the rotational motion to a full rigid-body motion. The results developed here will be extensively used throughout the book. The derivation parallels the case of a pure rotation in Section 2.3.2.

Consider the motion of a continuously moving rigid body described by a trajectory on $SE(3)$: $g(t) = (R(t), T(t))$, or in the homogeneous representation

$$g(t) = \begin{bmatrix} R(t) & T(t) \\ 0 & 1 \end{bmatrix} \quad \in \mathbb{R}^{4\times4}.$$

From now on, for simplicity, whenever there is no ambiguity, we will remove the bar "$-$" to indicate a homogeneous representation and simply use $g$. We will use the same convention for points, $\boldsymbol{X}$ for $\bar{\boldsymbol{X}}$, and for vectors, $v$ for $\bar{v}$, whenever their correct dimension is clear from the context.

In analogy with the case of a pure rotation, let us first look at the structure of the matrix

$$\dot{g}(t)g^{-1}(t) = \begin{bmatrix} \dot{R}(t)R^T(t) & \dot{T}(t) - \dot{R}(t)R^T(t)T(t) \\ 0 & 0 \end{bmatrix} \quad \in \mathbb{R}^{4\times4}. \qquad (2.19)$$

From our study of the rotation matrix, we know that $\dot{R}(t)R^T(t)$ is a skew-symmetric matrix; i.e. there exists $\widehat{\omega}(t) \in so(3)$ such that $\widehat{\omega}(t) = \dot{R}(t)R^T(t)$.

Define a vector $v(t) \in \mathbb{R}^3$ such that $v(t) = \dot{T}(t) - \widehat{\omega}(t)T(t)$. Then the above equation becomes

$$\dot{g}(t)g^{-1}(t) = \begin{bmatrix} \widehat{\omega}(t) & v(t) \\ 0 & 0 \end{bmatrix} \in \mathbb{R}^{4\times 4}.$$

If we further define a matrix $\widehat{\xi} \in \mathbb{R}^{4\times 4}$ to be

$$\widehat{\xi}(t) = \begin{bmatrix} \widehat{\omega}(t) & v(t) \\ 0 & 0 \end{bmatrix},$$

then we have

$$\dot{g}(t) = \left(\dot{g}(t)g^{-1}(t)\right)g(t) = \widehat{\xi}(t)g(t), \tag{2.20}$$

where $\widehat{\xi}$ can be viewed as the "tangent vector" along the curve of $g(t)$ and can be used to approximate $g(t)$ locally:

$$g(t+dt) \approx g(t) + \widehat{\xi}(t)g(t)dt = \left(I + \widehat{\xi}(t)dt\right) g(t).$$

A $4 \times 4$ matrix of the form of $\widehat{\xi}$ is called a *twist*. The set of all twists is denoted by

$$\boxed{se(3) \doteq \left\{ \widehat{\xi} = \begin{bmatrix} \widehat{\omega} & v \\ 0 & 0 \end{bmatrix} \middle| \, \widehat{\omega} \in so(3), v \in \mathbb{R}^3 \right\} \quad \subset \mathbb{R}^{4\times 4}.}$$

The set $se(3)$ is called the tangent space (or Lie algebra) of the matrix group $SE(3)$. We also define two operators "$\vee$" and "$\wedge$" to convert between a twist $\widehat{\xi} \in se(3)$ and its *twist coordinates* $\xi \in \mathbb{R}^6$ as follows:

$$\begin{bmatrix} \widehat{\omega} & v \\ 0 & 0 \end{bmatrix}^{\vee} \doteq \begin{bmatrix} v \\ \omega \end{bmatrix} \in \mathbb{R}^6, \qquad \begin{bmatrix} v \\ \omega \end{bmatrix}^{\wedge} \doteq \begin{bmatrix} \widehat{\omega} & v \\ 0 & 0 \end{bmatrix} \in \mathbb{R}^{4\times 4}.$$

In the twist coordinates $\xi$, we will refer to $v$ as the *linear velocity* and $\omega$ as the *angular velocity*, which indicates that they are related to either the translational or the rotational part of the full motion. Let us now consider a special case of equation (2.20) when the twist $\widehat{\xi}$ is a constant matrix

$$\dot{g}(t) = \widehat{\xi}g(t).$$

We have again a time-invariant linear ordinary differential equation, which can be integrated to give

$$g(t) = e^{\widehat{\xi}t}g(0).$$

Assuming the initial condition $g(0) = I$, we may conclude that

$$\boxed{g(t) = e^{\widehat{\xi}t},}$$

where the twist exponential is

$$e^{\widehat{\xi}t} = I + \widehat{\xi}t + \frac{(\widehat{\xi}t)^2}{2!} + \cdots + \frac{(\widehat{\xi}t)^n}{n!} + \cdots. \tag{2.21}$$

By Rodrigues' formula (2.16) introduced in the previous section and additional properties of the matrix exponential, the following relationship can be established:

$$\boxed{e^{\widehat{\xi}} = \begin{bmatrix} e^{\widehat{\omega}} & \frac{(I-e^{\widehat{\omega}})\widehat{\omega}v+\omega\omega^T v}{\|\omega\|} \\ 0 & 1 \end{bmatrix}, \quad \text{if} \quad \omega \neq 0.} \tag{2.22}$$

If $\omega = 0$, the exponential is simply $e^{\widehat{\xi}} = \begin{bmatrix} I & v \\ 0 & 1 \end{bmatrix}$. It is clear from the above expression that the exponential of $\widehat{\xi}$ is indeed a rigid-body transformation matrix in $SE(3)$. Therefore, the exponential map defines a transformation from the space $se(3)$ to $SE(3)$,

$$\exp : se(3) \rightarrow SE(3); \quad \widehat{\xi} \mapsto e^{\widehat{\xi}},$$

and the twist $\widehat{\xi} \in se(3)$ is also called the *exponential coordinates* for $SE(3)$, as is $\widehat{\omega} \in so(3)$ for $SO(3)$.

Can every rigid-body motion $g \in SE(3)$ be represented in such an exponential form? The answer is yes and is formulated in the following theorem.

**Theorem 2.11 (Logarithm of $SE(3)$).** *For any $g \in SE(3)$, there exist (not necessarily unique) twist coordinates $\xi = (v, \omega)$ such that $g = \exp(\widehat{\xi})$. We denote the inverse to the exponential map by $\widehat{\xi} = \log(g)$.*

*Proof.* The proof is constructive. Suppose $g = (R, T)$. From Theorem 2.8, for the rotation matrix $R \in SO(3)$ we can always find $\omega$ such that $e^{\widehat{\omega}} = R$. If $R \neq I$, i.e. $\|\omega\| \neq 0$, from equation (2.22) we can solve for $v \in \mathbb{R}^3$ from the linear equation

$$\boxed{\frac{(I - e^{\widehat{\omega}})\widehat{\omega}v + \omega\omega^T v}{\|\omega\|} = T.} \tag{2.23}$$

If $R = I$, then $\|\omega\| = 0$. In this case, we may simply choose $\omega = 0, v = T$. □

As with the exponential coordinates for rotation matrices, the exponential map from $se(3)$ to $SE(3)$ is not one-to-one. There are usually infinitely many exponential coordinates (or twists) that correspond to every $g \in SE(3)$.

**Remark 2.12.** *As in the rotation case, the linear structure of $se(3)$, together with the closure under the Lie bracket operation*

$$[\widehat{\xi}_1, \widehat{\xi}_2] = \widehat{\xi}_1\widehat{\xi}_2 - \widehat{\xi}_2\widehat{\xi}_1 = \begin{bmatrix} \widehat{\omega_1 \times \omega_2} & \omega_1 \times v_2 - \omega_2 \times v_1 \\ 0 & 0 \end{bmatrix} \in se(3),$$

*makes $se(3)$ the Lie algebra for $SE(3)$. The two rigid-body motions $g_1 = e^{\widehat{\xi}_1}$ and $g_2 = e^{\widehat{\xi}_2}$ commute with each other, $g_1g_2 = g_2g_1$, if and only if $[\widehat{\xi}_1, \widehat{\xi}_2] = 0$.*

**Example 2.13 (Screw motions).** Screw motions are a specific class of rigid-body motions. A screw motion consists of rotation about an axis in space through an angle of $\theta$ radians, followed by translation along the same axis by an amount $d$. Define the *pitch* of the screw motion to be the ratio of translation to rotation, $h = d/\theta$ (assuming $\theta \neq 0$). If we choose a point $\boldsymbol{X}$ on the axis and $\omega \in \mathbb{R}^3$ to be a unit vector specifying the direction, the axis is the set of points $L = \{\boldsymbol{X} + \mu\omega\}$. Then the rigid-body motion given by the screw is

$$g = \begin{bmatrix} e^{\widehat{\omega}\theta} & (I - e^{\widehat{\omega}\theta})\boldsymbol{X} + h\theta\omega \\ 0 & 1 \end{bmatrix} \in SE(3). \tag{2.24}$$

The set of all screw motions along the same axis forms a subgroup $SO(2) \times \mathbb{R}$ of $SE(3)$, which we will encounter occasionally in later chapters. A statement, also known as *Chasles' theorem*, reveals a rather remarkable fact that any rigid-body motion can be realized as a rotation around a particular axis in space and translation along that axis. ■

## 2.5 Coordinate and velocity transformations

In this book, we often need to know how the coordinates of a point and its velocity change as the camera moves. This is because it is usually more convenient to choose the camera frame as the reference frame and describe both camera motion and 3-D points relative to it. Since the camera may be moving, we need to know how to transform quantities such as coordinates and velocities from one camera frame to another. In particular, we want to know how to correctly express the location and velocity of a point with respect to a moving camera. Here we introduce a convention that we will be using for the rest of this book.

*Rules of coordinate transformations*

The time $t \in \mathbb{R}$ will typically be used to index camera motion. Even in the discrete case in which a few snapshots are given, we will take $t$ to be the index of the camera position and the corresponding image. Therefore, we will use $g(t) = (R(t), T(t)) \in SE(3)$ or

$$g(t) = \begin{bmatrix} R(t) & T(t) \\ 0 & 1 \end{bmatrix} \in SE(3)$$

to denote the relative displacement between some fixed world frame $W$ and the camera frame $C$ at time $t \in \mathbb{R}$. Here we will ignore the subscript "$cw$" from the notation $g_{cw}(t)$ as long as it is clear from the context. By default, we assume $g(0) = I$, i.e. at time $t = 0$ the camera frame coincides with the world frame. So if the coordinates of a point $p \in \mathbb{E}^3$ relative to the world frame are $\boldsymbol{X}_0 = \boldsymbol{X}(0)$, its coordinates relative to the camera at time $t$ are given by

$$\boxed{\boldsymbol{X}(t) = R(t)\boldsymbol{X}_0 + T(t),} \tag{2.25}$$

or in the homogeneous representation,

$$\boldsymbol{X}(t) = g(t)\boldsymbol{X}_0. \tag{2.26}$$

If the camera is at locations $g(t_1), g(t_2), \ldots, g(t_m)$ at times $t_1, t_2, \ldots, t_m$, respectively, then the coordinates of the point $p$ are given as $\boldsymbol{X}(t_i) = g(t_i)\boldsymbol{X}_0, i =$

$1, 2, \ldots, m$, correspondingly. If it is only the position, not the time, that matters, we will often use $g_i$ as a shorthand for $g(t_i)$, and similarly $R_i$ for $R(t_i)$, $T_i$ for $T(t_i)$, and $\boldsymbol{X}_i$ for $\boldsymbol{X}(t_i)$. We hence have

$$\boxed{\boldsymbol{X}_i = R_i \boldsymbol{X}_0 + T_i.} \tag{2.27}$$

When the starting time is not $t = 0$, the relative motion between the camera at time $t_2$ and time $t_1$ will be denoted by $g(t_2, t_1) \in SE(3)$. Then we have the following relationship between coordinates of the same point $p$ at different times:

$$\boldsymbol{X}(t_2) = g(t_2, t_1)\boldsymbol{X}(t_1), \quad \forall t_1, t_2 \in \mathbb{R}.$$

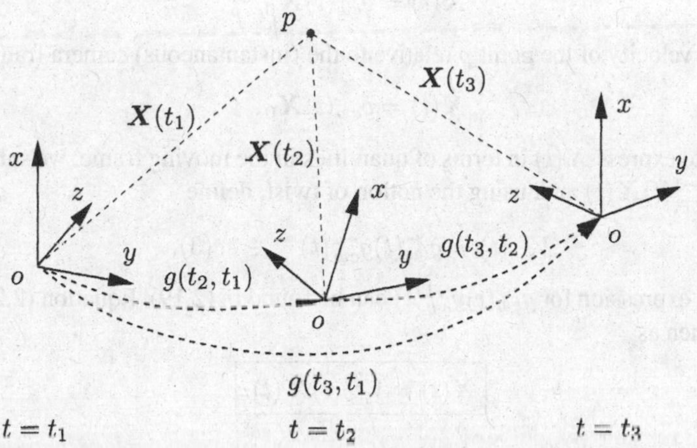


Figure 2.6. Composition of rigid-body motions. $\boldsymbol{X}(t_1), \boldsymbol{X}(t_2), \boldsymbol{X}(t_3)$ are the coordinates of the point $p$ with respect to the three camera frames at time $t = t_1, t_2, t_3$, respectively.

Now consider a third position of the camera at $t = t_3 \in \mathbb{R}$, as shown in Figure 2.6. The relative motion between the camera at $t_3$ and $t_2$ is $g(t_3, t_2)$, and that between $t_3$ and $t_1$ is $g(t_3, t_1)$. We then have the following relationship among the coordinates:

$$\boldsymbol{X}(t_3) = g(t_3, t_2)\boldsymbol{X}(t_2) = g(t_3, t_2)g(t_2, t_1)\boldsymbol{X}(t_1).$$

Comparing this with the direct relationship between the coordinates at $t_3$ and $t_1$,

$$\boldsymbol{X}(t_3) = g(t_3, t_1)\boldsymbol{X}(t_1),$$

we see that the following *composition rule* for consecutive motions must hold:

$$g(t_3, t_1) = g(t_3, t_2)g(t_2, t_1).$$

The composition rule describes the coordinates $\boldsymbol{X}$ of the point $p$ relative to any camera position if they are known with respect to a particular one. The same composition rule implies the *rule of inverse*

$$g^{-1}(t_2, t_1) = g(t_1, t_2),$$

since $g(t_2, t_1)g(t_1, t_2) = g(t_2, t_2) = I$. In cases in which time is of no physical meaning, we often use $g_{ij}$ as a shorthand for $g(t_i, t_j)$. The above composition rules then become (in the homogeneous representation)

$$\boxed{\boldsymbol{X}_i = g_{ij}\boldsymbol{X}_j, \quad g_{ik} = g_{ij}g_{jk}, \quad g_{ij}^{-1} = g_{ji}.} \tag{2.28}$$

*Rules of velocity transformation*

Having understood the transformation of coordinates, we now study how it affects velocity. We know that the coordinates $\boldsymbol{X}(t)$ of a point $p \in \mathbb{E}^3$ relative to a moving camera are a function of time $t$:

$$\boldsymbol{X}(t) = g_{cw}(t)\boldsymbol{X}_0.$$

Then the velocity of the point $p$ relative to the (instantaneous) camera frame is

$$\dot{\boldsymbol{X}}(t) = \dot{g}_{cw}(t)\boldsymbol{X}_0. \tag{2.29}$$

In order to express $\dot{\boldsymbol{X}}(t)$ in terms of quantities in the moving frame, we substitute $\boldsymbol{X}_0$ by $g_{cw}^{-1}(t)\boldsymbol{X}(t)$ and, using the notion of twist, define

$$\widehat{V}_{cw}^{c}(t) = \dot{g}_{cw}(t)g_{cw}^{-1}(t) \quad \in se(3), \tag{2.30}$$

where an expression for $\dot{g}_{cw}(t)g_{cw}^{-1}(t)$ can be found in (2.19). Equation (2.29) can be rewritten as

$$\boxed{\dot{\boldsymbol{X}}(t) = \widehat{V}_{cw}^{c}(t)\boldsymbol{X}(t).} \tag{2.31}$$

Since $\widehat{V}_{cw}^{c}(t)$ is of the form

$$\widehat{V}_{cw}^{c}(t) = \begin{bmatrix} \widehat{\omega}(t) & v(t) \\ 0 & 0 \end{bmatrix},$$

we can also write the velocity of the point in 3-D coordinates (instead of homogeneous coordinates) as

$$\boxed{\dot{\boldsymbol{X}}(t) = \widehat{\omega}(t)\boldsymbol{X}(t) + v(t).} \tag{2.32}$$

The physical interpretation of the symbol $\widehat{V}_{cw}^{c}$ is the velocity of the world frame moving relative to the camera frame, as viewed in the camera frame, as indicated by the subscript and superscript of $\widehat{V}_{cw}^{c}$. Usually, to clearly specify the physical meaning of a velocity, we need to specify the velocity of which frame is moving relative to which frame, and which frame it is viewed from. If we change the location from which we view the velocity, the expression will change accordingly. For example, suppose that a viewer is in another coordinate frame displaced relative to the camera frame by a rigid-body transformation $g \in SE(3)$. Then the coordinates of the same point $p$ relative to this frame are $\boldsymbol{Y}(t) = g\boldsymbol{X}(t)$. We compute the velocity in the new frame, and obtain

$$\dot{\boldsymbol{Y}}(t) = g\dot{g}_{cw}(t)g_{cw}^{-1}(t)g^{-1}\boldsymbol{Y}(t) = g\widehat{V}_{cw}^{c}g^{-1}\boldsymbol{Y}(t).$$

So the new velocity (or twist) is

$$\widehat{V} = g\widehat{V}^c_{cw}g^{-1}.$$

This is the same physical quantity but viewed from a different vantage point. We see that the two velocities are related through a mapping defined by the relative motion $g$; in particular,

$$ad_g : se(3) \to se(3); \quad \widehat{\xi} \mapsto g\widehat{\xi}g^{-1}.$$

This is the so-called *adjoint map* on the space $se(3)$. Using this notation in the previous example we have $\widehat{V} = ad_g(\widehat{V}^c_{cw})$. Note that the adjoint map transforms velocity from one frame to another. Using the fact that $g_{cw}(t)g_{wc}(t) = I$, it is straightforward to verify that

$$\widehat{V}^c_{cw} = \dot{g}_{cw}g^{-1}_{cw} = -g^{-1}_{wc}\dot{g}_{wc} = -g_{cw}(\dot{g}_{wc}g^{-1}_{wc})g^{-1}_{cw} = ad_{g_{cw}}(-\widehat{V}^w_{wc}).$$

Hence $\widehat{V}^c_{cw}$ can also be interpreted as the *negated* velocity of the camera moving relative to the world frame, viewed in the (instantaneous) camera frame.

## 2.6 Summary

We summarize the properties of 3-D rotations and rigid-body motions introduced in this chapter in Table 2.1.

| | Rotation $SO(3)$ | Rigid-body motion $SE(3)$ |
|---|---|---|
| Matrix representation | $R : \begin{cases} R^TR = I \\ \det(R) = 1 \end{cases}$ | $g = \begin{bmatrix} R & T \\ 0 & 1 \end{bmatrix}$ |
| Coordinates (3-D) | $\boldsymbol{X} = R\boldsymbol{X}_0$ | $\boldsymbol{X} = R\boldsymbol{X}_0 + T$ |
| Inverse | $R^{-1} = R^T$ | $g^{-1} = \begin{bmatrix} R^T & -R^TT \\ 0 & 1 \end{bmatrix}$ |
| Composition | $R_{ik} = R_{ij}R_{jk}$ | $g_{ik} = g_{ij}g_{jk}$ |
| Exp. representation | $R = \exp(\widehat{\omega})$ | $g = \exp(\widehat{\xi})$ |
| Velocity | $\dot{\boldsymbol{X}} = \widehat{\omega}\boldsymbol{X}$ | $\dot{\boldsymbol{X}} = \widehat{\omega}\boldsymbol{X} + v$ |
| Adjoint map | $\widehat{\omega} \mapsto R\widehat{\omega}R^T$ | $\widehat{\xi} \mapsto g\widehat{\xi}g^{-1}$ |

Table 2.1. Rotation and rigid-body motion in 3-D space.

## 2.7 Exercises

**Exercise 2.1 (Linear vs. nonlinear maps).** Suppose $A, B, C, X \in \mathbb{R}^{n\times n}$. Consider the following maps from $\mathbb{R}^{n\times n} \to \mathbb{R}^{n\times n}$ and determine whether they are linear or not. Give a brief proof if true and a counterexample if false:

$$\begin{array}{lll} \text{(a)} & X & \mapsto & AX + XB, \\ \text{(b)} & X & \mapsto & AX + BXC, \\ \text{(c)} & X & \mapsto & AXA - B, \\ \text{(d)} & X & \mapsto & AX + XBX. \end{array}$$

Note: A map $f : \mathbb{R}^n \to \mathbb{R}^m$, $x \mapsto f(x)$, is called linear if $f(\alpha x + \beta y) = \alpha f(x) + \beta f(y)$ for all $\alpha, \beta \in \mathbb{R}$ and $x, y \in \mathbb{R}^n$.

**Exercise 2.2 (Inner product).** Show that for any positive definite symmetric matrix $S \in \mathbb{R}^{3\times3}$, the map $\langle\cdot,\cdot\rangle_S : \mathbb{R}^3 \times \mathbb{R}^3 \to \mathbb{R}$ defined as

$$\langle u, v\rangle_S = u^T S v, \quad \forall u, v \in \mathbb{R}^3,$$

is a valid inner product on $\mathbb{R}^3$, according to the definition given in Appendix A.

**Exercise 2.3 (Group structure of $SO(3)$).** Prove that the space $SO(3)$ satisfies all four axioms in the definition of group (in Appendix A).

**Exercise 2.4 (Skew-symmetric matrices).** Given any vector $\omega = [\omega_1, \omega_2, \omega_3]^T \in \mathbb{R}^3$, we know that the matrix $\widehat{\omega}$ is skew-symmetric; i.e. $\widehat{\omega}^T = -\widehat{\omega}$. Now for any matrix $A \in \mathbb{R}^{3\times3}$ with determinant $\det(A) = 1$, show that the following equation holds:

$$A^T\widehat{\omega}A = \widehat{A^{-1}\omega}. \tag{2.33}$$

Then, in particular, if $A$ is a rotation matrix, the above equation holds.
Hint: Both $A^T\widehat{(\cdot)}A$ and $\widehat{A^{-1}(\cdot)}$ are linear maps with $\omega$ as the variable. What do you need in order to prove that two linear maps are the same?

**Exercise 2.5** Show that a matrix $M \in \mathbb{R}^{3\times3}$ is skew-symmetric if and only if $u^T M u = 0$ for every $u \in \mathbb{R}^3$.

**Exercise 2.6** Consider a $2 \times 2$ matrix

$$R_1 = \begin{bmatrix} \cos\theta & -\sin\theta \\ \sin\theta & \cos\theta \end{bmatrix}.$$

What is the determinant of the matrix? Consider another transformation matrix

$$R_2 = \begin{bmatrix} \sin\theta & \cos\theta \\ \cos\theta & -\sin\theta \end{bmatrix}.$$

Is the matrix orthogonal? What is the determinant of the matrix? Is $R_2$ a 2-D rigid-body transformation? What is the difference between $R_1$ and $R_2$ ?

**Exercise 2.7 (Rotation as a rigid-body motion).** Given a rotation matrix $R \in SO(3)$, its action on a vector $v$ is defined as $Rv$. Prove that any rotation matrix must preserve both the inner product and cross product of vectors. Hence, a rotation is indeed a rigid-body motion.

**Exercise 2.8** Show that for any nonzero vector $u \in \mathbb{R}^3$, the rank of the matrix $\widehat{u}$ is always two. That is, the three row (or column) vectors span a two-dimensional subspace of $\mathbb{R}^3$.

**Exercise 2.9 (Range and null space).** Recall that given a matrix $A \in \mathbb{R}^{m \times n}$, its *null space* is defined as a subspace of $\mathbb{R}^n$ consisting of all vectors $x \in \mathbb{R}^n$ such that $Ax = 0$. It is usually denoted by $\text{null}(A)$. The *range* of the matrix $A$ is defined as a subspace of $\mathbb{R}^m$ consisting of all vectors $y \in \mathbb{R}^m$ such that there exists some $x \in \mathbb{R}^n$ such that $y = Ax$. It is denoted by $\text{range}(A)$. In mathematical terms,

$$\begin{aligned} \text{null}(A) &\doteq \{x \in \mathbb{R}^n \mid Ax = 0\}, \\ \text{range}(A) &\doteq \{y \in \mathbb{R}^m \mid \exists x \in \mathbb{R}^n, y = Ax\}. \end{aligned}$$

1. Recall that a set of vectors $V$ is a subspace if for all vectors $x, y \in V$ and scalars $\alpha, \beta \in \mathbb{R}$, $\alpha x + \beta y$ is also a vector in $V$. Show that both $\text{null}(A)$ and $\text{range}(A)$ are indeed subspaces.

2. What are $\text{null}(\widehat{\omega})$ and $\text{range}(\widehat{\omega})$ for a nonzero vector $\omega \in \mathbb{R}^3$? Can you describe intuitively the geometric relationship between these two subspaces in $\mathbb{R}^3$? (Drawing a picture might help.)

**Exercise 2.10 (Noncommutativity of rotation matrices).** What is the matrix that represents a rotation about the $X$-axis or the $Y$-axis by an angle $\theta$? In addition to that

1. Compute the matrix $R_1$ that is the combination of a rotation about the $X$-axis by $\pi/3$ followed by a rotation about the $Z$-axis by $\pi/6$. Verify that the resulting matrix is also a rotation matrix.

2. Compute the matrix $R_2$ that is the combination of a rotation about the $Z$-axis by $\pi/6$ followed by a rotation about the $X$-axis by $\pi/3$. Are $R_1$ and $R_2$ the same? Explain why.

**Exercise 2.11** Let $R \in SO(3)$ be a rotation matrix generated by rotating about a unit vector $\omega$ by $\theta$ radians that satisfies $R = \exp(\widehat{\omega}\theta)$. Suppose $R$ is given as

$$R = \begin{bmatrix} 0.1729 & -0.1468 & 0.9739 \\ 0.9739 & 0.1729 & -0.1468 \\ -0.1468 & 0.9739 & 0.1729 \end{bmatrix}.$$

- Use the formulae given in this chapter to compute the rotation axis and the associated angle.
- Use Matlab's function `eig` to compute the eigenvalues and eigenvectors of the above rotation matrix $R$. What is the eigenvector associated with the unit eigenvalue? Give its form and explain its meaning.

**Exercise 2.12 (Properties of rotation matrices).** Let $R \in SO(3)$ be a rotation matrix generated by rotating about a unit vector $\omega \in \mathbb{R}^3$ by $\theta$ radians. That is, $R = e^{\widehat{\omega}\theta}$.

1. What are the eigenvalues and eigenvectors of $\widehat{\omega}$? You may use a computer software (e.g., Matlab) and try some examples first. If you cannot find a brute-force way to do it, can you use results from Exercise 2.4 to simplify the problem first (hint: use the relationship between trace, determinant and eigenvalues).

2. Show that the eigenvalues of $R$ are $1, e^{i\theta}, e^{-i\theta}$, where $i = \sqrt{-1}$ is the imaginary unit. What is the eigenvector that corresponds to the eigenvalue 1? This actually gives another proof for $\det(e^{\widehat{\omega}\theta}) = 1 \cdot e^{i\theta} \cdot e^{-i\theta} = +1$, not $-1$.

**Exercise 2.13 (Adjoint transformation on twist).** Given a rigid-body motion $g$ and a twist $\widehat{\xi}$,

$$g = \begin{bmatrix} R & T \\ 0 & 1 \end{bmatrix} \in SE(3), \quad \widehat{\xi} = \begin{bmatrix} \widehat{\omega} & v \\ 0 & 0 \end{bmatrix} \in se(3),$$

show that $g\widehat{\xi}g^{-1}$ is still a twist. Describe what the corresponding $\omega$ and $v$ terms have become in the new twist. The adjoint map is sort of a generalization to $R\widehat{\omega}R^T = \widehat{R\omega}$.

**Exercise 2.14** Suppose that there are three camera frames $C_0, C_1, C_2$ and the coordinate transformation from frame $C_0$ to frame $C_1$ is $(R_1, T_1)$ and from $C_0$ to $C_2$ is $(R_2, T_2)$. What is the relative coordinate transformation from $C_1$ to $C_2$ then? What about from $C_2$ to $C_1$? (Express these transformations in terms of $R_1, T_1$ and $R_2, T_2$ only.)

## 2.A Quaternions and Euler angles for rotations

For the sake of completeness, we introduce a few conventional schemes to parameterize rotation matrices, either globally or locally, that are often used in numerical computations for rotation matrices. However, we encourage the reader to use the exponential parameterizations described in this chapter.

*Quaternions*

We know that the set of complex numbers $\mathbb{C}$ can be simply defined as $\mathbb{C} = \mathbb{R} + \mathbb{R}i$ with $i^2 = -1$. Quaternions generalize complex numbers in a similar fashion. The set of quaternions, denoted by $\mathbb{H}$, is defined as

$$\mathbb{H} = \mathbb{C} + \mathbb{C}j, \quad \text{with } j^2 = -1 \text{ and } i \cdot j = -j \cdot i. \tag{2.34}$$

So, an element of $\mathbb{H}$ is of the form

$$q = q_0 + q_1 i + (q_2 + iq_3)j = q_0 + q_1 i + q_2 j + q_3 ij, \quad q_0, q_1, q_2, q_3 \in \mathbb{R}. \tag{2.35}$$

For simplicity of notation, in the literature $ij$ is sometimes denoted by $k$. In general, the *multiplication* of any two quaternions is similar to the multiplication of two complex numbers, except that the multiplication of $i$ and $j$ is *anticommutative*: $ij = -ji$. We can also similarly define the concept of *conjugation* for a quaternion:

$$q = q_0 + q_1 i + q_2 j + q_3 ij \quad \Rightarrow \quad \bar{q} = q_0 - q_1 i - q_2 j - q_3 ij. \tag{2.36}$$

It is immediate to check that

$$q\bar{q} = q_0^2 + q_1^2 + q_2^2 + q_3^2. \tag{2.37}$$

Thus, $q\bar{q}$ is simply the square of the norm $\|q\|$ of $q$ as a four-dimensional vector in $\mathbb{R}^4$. For a nonzero $q \in \mathbb{H}$, i.e. $\|q\| \neq 0$, we can further define its *inverse* to be

$$q^{-1} = \frac{\bar{q}}{\|q\|^2}. \tag{2.38}$$

The multiplication and inverse rules defined above in fact endow the space $\mathbb{R}^4$ with an algebraic structure of a *skew field*. In fact $\mathbb{H}$ is called a *Hamiltonian field*, or *quaternion field*.

One important usage of the quaternion field $\mathbb{H}$ is that we can in fact embed the rotation group $SO(3)$ into it. To see this, let us focus on a special subgroup of $\mathbb{H}$, the *unit quaternions*

$$\mathbb{S}^3 = \left\{q \in \mathbb{H} \mid \|q\|^2 = q_0^2 + q_1^2 + q_2^2 + q_3^2 = 1\right\}. \tag{2.39}$$

The set of all unit quaternions is simply the unit sphere in $\mathbb{R}^4$. To show that $\mathbb{S}^3$ is indeed a group, we simply need to prove that it is closed under the multiplication and inverse of quaternions; i.e. the multiplication of two unit quaternions is still a unit quaternion, and so is the inverse of a unit quaternion. We leave this simple fact as an exercise to the reader.

Given a rotation matrix $R = e^{\widehat{\omega}t}$ with $\|\omega\| = 1$ and $t \in \mathbb{R}$, we can associate with it a unit quaternion as follows:

$$q(R) = \cos(t/2) + \sin(t/2)(\omega_1 i + \omega_2 j + \omega_3 ij) \in \mathbb{S}^3. \tag{2.40}$$

One may verify that this association preserves the group structure between $SO(3)$ and $\mathbb{S}^3$:

$$q(R^{-1}) = q^{-1}(R),\ q(R_1R_2) = q(R_1)q(R_2),\ \forall R, R_1, R_2 \in SO(3). \tag{2.41}$$

Further study can show that this association is also *genuine*; i.e. for different rotation matrices, the associated unit quaternions are also different. In the opposite direction, given a unit quaternion $q = q_0 + q_1 i + q_2 j + q_3 ij \in \mathbb{S}^3$, we can use the following formulae to find the corresponding rotation matrix $R(q) = e^{\widehat{\omega}t}$:

$$t = 2\arccos(q_0), \quad \omega_m = \begin{cases} q_m/\sin(t/2), & t \neq 0, \\ 0, & t = 0, \end{cases} \quad m = 1, 2, 3. \tag{2.42}$$

However, one must notice that according to the above formula, there are two unit quaternions that correspond to the same rotation matrix: $R(q) = R(-q)$, as shown in Figure 2.7. Therefore, topologically, $\mathbb{S}^3$ is a double covering of $SO(3)$. So $SO(3)$ is topologically the same as a three-dimensional projective plane $\mathbb{RP}^3$.

Compared to the exponential coordinates for rotation matrices that we studied in this chapter, in using unit quaternions $\mathbb{S}^3$ to represent rotation matrices $SO(3)$, we have less redundancy: there are only two unit quaternions that corresponding to the same rotation matrix, while there are infinitely many for exponential coordinates (all related by periodicity). Furthermore, such a representation for rotation matrices is smooth, and there is no singularity, as opposed to the representation by *Euler angles*, which we will now introduce.

### *Euler angles*

Unit quaternions can be viewed as a way to *globally* parameterize rotation matrices: the parameterization works for every rotation matrix practically the same way. On the other hand, the Euler angles to be introduced below fall into the cat-

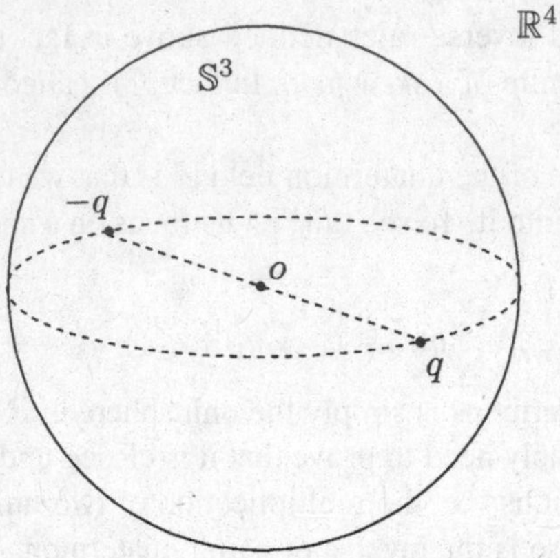


Figure 2.7. Antipodal unit quaternions $q$ and $-q$ on the unit sphere $\mathbb{S}^3 \subset \mathbb{R}^4$ correspond to the same rotation matrix.

egory of *local* parameterizations. This kind of parameterization is good for only a portion of $SO(3)$, but not for the entire space.

In the space of skew-symmetric matrices $so(3)$, pick a *basis* $(\widehat{\omega}_1, \widehat{\omega}_2, \widehat{\omega}_3)$, i.e. the three vectors $\omega_1, \omega_2, \omega_3$ are linearly independent. Define a mapping (a parameterization) from $\mathbb{R}^3$ to $SO(3)$ as

$$\alpha : (\alpha_1, \alpha_2, \alpha_3) \quad \mapsto \quad \exp(\alpha\widehat{\omega}_1 + \alpha_2\widehat{\omega}_2 + \alpha_3\widehat{\omega}_3).$$

The coordinates $(\alpha_1, \alpha_2, \alpha_3)$ are called the *Lie-Cartan coordinates of the first kind* relative to the basis $(\widehat{\omega}_1, \widehat{\omega}_2, \widehat{\omega}_3)$. Another way to parameterize the group $SO(3)$ using the same basis is to define another mapping from $\mathbb{R}^3$ to $SO(3)$ by

$$\beta : (\beta_1, \beta_2, \beta_3) \quad \mapsto \quad \exp(\beta_1\widehat{\omega}_1)\exp(\beta_2\widehat{\omega}_2)\exp(\beta_3\widehat{\omega}_3).$$

The coordinates $(\beta_1, \beta_2, \beta_3)$ are called the *Lie-Cartan coordinates of the second kind*.

In the special case in which we choose $\omega_1, \omega_2, \omega_3$ to be the principal axes $Z, Y, X$, respectively, i.e.

$$\omega_1 = [0,0,1]^T \doteq \boldsymbol{z}, \quad \omega_2 = [0,1,0]^T \doteq \boldsymbol{y}, \quad \omega_3 = [1,0,0]^T \doteq \boldsymbol{x},$$

the Lie-Cartan coordinates of the second kind then coincide with the well-known *ZYX Euler angles* parameterization, and $(\beta_1, \beta_2, \beta_3)$ are the corresponding Euler angles, called "yaw," "pitch," and "roll." The rotation matrix is defined by

$$R(\beta_1, \beta_2, \beta_3) = \exp(\beta_1\widehat{\boldsymbol{z}})\exp(\beta_2\widehat{\boldsymbol{y}})\exp(\beta_3\widehat{\boldsymbol{x}}). \tag{2.43}$$

More precisely, $R(\beta_1, \beta_2, \beta_3)$ is the multiplication of the three rotation matrices

$$\begin{bmatrix} \cos(\beta_1) & -\sin(\beta_1) & 0 \\ \sin(\beta_1) & \cos(\beta_1) & 0 \\ 0 & 0 & 1 \end{bmatrix}, \begin{bmatrix} \cos(\beta_2) & 0 & \sin(\beta_2) \\ 0 & 1 & 0 \\ -\sin(\beta_2) & 0 & \cos(\beta_2) \end{bmatrix}, \begin{bmatrix} 1 & 0 & 0 \\ 0 & \cos(\beta_3) & -\sin(\beta_3) \\ 0 & \sin(\beta_3) & \cos(\beta_3) \end{bmatrix}.$$

Similarly, we can define the $YZX$ Euler angles and the $ZYZ$ Euler angles. There are instances for which this representation becomes singular, and for certain rotation matrices, their corresponding Euler angles cannot be uniquely determined. For example, when $\beta_2 = -\pi/2$ the $ZYX$ Euler angles become singular. The presence of such singularities is expected because of the topology of the space $SO(3)$. Globally, $SO(3)$ is like a sphere in $\mathbb{R}^4$, as we know from the quaternions, and therefore any attempt to find a global (three-dimensional) coordinate chart is doomed to failure.

## Historical notes

The study of rigid-body motion mostly relies on the tools of linear algebra. Elements of screw theory can be tracked back to the early 1800s in the work of Chasles and Poinsot. The use of the exponential coordinates for rigid-body motions was introduced by [Brockett, 1984], and related formulations can be found in the classical work of [Ball, 1900] and others. The use of quaternions in robot vision was introduced by [Broida and Chellappa, 1986b, Horn, 1987]. The presentation of the material in this chapter follows the development in [Murray et al., 1993]. More details on the study of rigid-body motions as well as further references can also be found there.

# Chapter 3
# Image Formation

*And since geometry is the right foundation of all painting, I have decided to teach its rudiments and principles to all youngsters eager for art...*

– Albrecht Dürer, *The Art of Measurement, 1525*

This chapter introduces simple mathematical models of the image formation process. In a broad figurative sense, vision is the inverse problem of image formation: the latter studies how objects give rise to images, while the former attempts to use images to recover a description of objects in space. Therefore, designing vision algorithms requires first developing a suitable model of image formation. Suitable, in this context, does not necessarily mean physically accurate: the level of abstraction and complexity in modeling image formation must trade off physical constraints and mathematical simplicity in order to result in a manageable model (i.e. one that can be inverted with reasonable effort). Physical models of image formation easily exceed the level of complexity necessary and appropriate for this book, and determining the right model for the problem at hand is a form of engineering art.

It comes as no surprise, then, that the study of image formation has for centuries been in the domain of artistic reproduction and composition, more so than of mathematics and engineering. Rudimentary understanding of the geometry of image formation, which includes various models for projecting the three-dimensional world onto a plane (e.g., a canvas), is implicit in various forms of visual arts. The roots of formulating the geometry of image formation can be traced back to the work of Euclid in the fourth century B.C. Examples of partially

Figure 3.1. Frescoes from the first century B.C. in Pompeii. Partially correct perspective projection is visible in the paintings, although not all parallel lines converge to the vanishing point. The skill was lost during the middle ages, and it did not reappear in paintings until the Renaissance (image courtesy of C. Taylor).

correct perspective projection are visible in the frescoes and mosaics of Pompeii (Figure 3.1) from the first century B.C. Unfortunately, these skills seem to have been lost with the fall of the Roman empire, and it took over a thousand years for correct perspective projection to emerge in paintings again in the late fourteenth century. It was the early Renaissance painters who developed systematic methods for determining the perspective projection of three-dimensional landscapes. The first treatise on perspective, *Della Pictura*, was published by Leon Battista Alberti, who emphasized the "eye's view" of the world capturing correctly the geometry of the projection process. It is no coincidence that early attempts to formalize the rules of perspective came from artists proficient in architecture and engineering, such as Alberti and Brunelleschi. Geometry, however, is only a part of the image formation process: in order to obtain an image, we need to decide not only where to draw a point, but also what brightness value to assign to it. The interaction of light with matter is at the core of the studies of Leonardo Da Vinci in the 1500s, and his insights on perspective, shading, color, and even stereopsis are vibrantly expressed in his notes. Renaissance painters such as Caravaggio and Raphael exhibited rather sophisticated skills in rendering light and color that remain compelling to this day.[1]

In this book, we restrict our attention to the geometry of the scene, and therefore, we need a simple geometric model of image formation, which we derive

[1]There is some evidence that suggests that some Renaissance artists secretly used camera-like devices (*camera obscura*) [Hockney, 2001].

in this chapter. More complex photometric models are beyond the scope of this book; in the next two sections as well as in Appendix 3.A at the end of this chapter, we will review some of the basic notions of radiometry so that the reader can better evaluate the assumptions by which we are able to reduce image formation to a purely geometric process.

## 3.1 Representation of images

An *image*, as far as this book is concerned, is a two-dimensional brightness array.[2] In other words, it is a map $I$, defined on a compact region $\Omega$ of a two-dimensional surface, taking values in the positive real numbers. For instance, in the case of a camera, $\Omega$ is a planar, rectangular region occupied by the photographic medium or by the CCD sensor. So $I$ is a function

$$I : \Omega \subset \mathbb{R}^2 \to \mathbb{R}_+; \quad (x, y) \mapsto I(x, y). \tag{3.1}$$

Such an image (function) can be represented, for instance, using the graph of $I$ as in the example in Figure 3.2. In the case of a digital image, both the domain $\Omega$ and the range $\mathbb{R}_+$ are discretized. For instance, $\Omega = [1, 640] \times [1, 480] \subset \mathbb{Z}^2$, and $\mathbb{R}_+$ is approximated by an interval of integers $[0, 255] \subset \mathbb{Z}_+$. Such an image can be represented by an array of numbers as in Table 3.1.

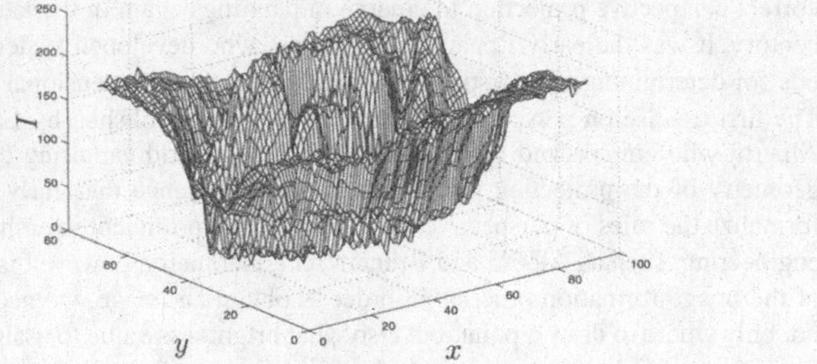


Figure 3.2. An image $I$ represented as a two-dimensional surface, the graph of $I$.

The values of the image $I$ depend upon physical properties of the scene being viewed, such as its shape, its material reflectance properties, and the distribution of the light sources. Despite the fact that Figure 3.2 and Table 3.1 do not seem very indicative of the properties of the scene they portray, this is how they are represented in a computer. A different representation of the same image that is

[2]If it is a color image, its RGB (red, green, blue) values represent three such arrays.

| | | | | | | | | | | | | | | | | | | | |
|---|---|---|---|---|---|---|---|---|---|---|---|---|---|---|---|---|---|---|---|
| 188 | 186 | 188 | 187 | 168 | 130 | 101 | 99 | 110 | 113 | 112 | 107 | 117 | 140 | 153 | 153 | 156 | 158 | 156 | 153 |
| 189 | 189 | 188 | 181 | 163 | 135 | 109 | 104 | 113 | 113 | 110 | 109 | 117 | 134 | 147 | 152 | 156 | 163 | 160 | 156 |
| 190 | 190 | 188 | 176 | 159 | 139 | 115 | 106 | 114 | 123 | 114 | 111 | 119 | 130 | 141 | 154 | 165 | 160 | 156 | 151 |
| 190 | 188 | 188 | 175 | 158 | 139 | 114 | 103 | 113 | 126 | 112 | 113 | 127 | 133 | 137 | 151 | 165 | 156 | 152 | 145 |
| 191 | 185 | 189 | 177 | 158 | 138 | 110 | 99 | 112 | 119 | 107 | 115 | 137 | 140 | 135 | 144 | 157 | 163 | 158 | 150 |
| 193 | 183 | 178 | 164 | 148 | 134 | 118 | 112 | 119 | 117 | 118 | 106 | 122 | 139 | 140 | 152 | 154 | 160 | 155 | 147 |
| 185 | 181 | 178 | 165 | 149 | 135 | 121 | 116 | 124 | 120 | 122 | 109 | 123 | 139 | 141 | 154 | 156 | 159 | 154 | 147 |
| 175 | 176 | 176 | 163 | 145 | 131 | 120 | 118 | 125 | 123 | 125 | 112 | 124 | 139 | 142 | 155 | 158 | 158 | 155 | 148 |
| 170 | 170 | 172 | 159 | 137 | 123 | 116 | 114 | 119 | 122 | 126 | 113 | 123 | 137 | 141 | 156 | 158 | 159 | 157 | 150 |
| 171 | 171 | 173 | 157 | 131 | 119 | 116 | 113 | 114 | 118 | 125 | 113 | 122 | 135 | 140 | 155 | 156 | 160 | 160 | 152 |
| 174 | 175 | 176 | 156 | 128 | 120 | 121 | 118 | 113 | 112 | 123 | 114 | 122 | 135 | 141 | 155 | 155 | 158 | 159 | 152 |
| 176 | 174 | 174 | 151 | 123 | 119 | 126 | 121 | 112 | 108 | 122 | 115 | 123 | 137 | 143 | 156 | 155 | 152 | 155 | 150 |
| 175 | 169 | 168 | 144 | 117 | 117 | 127 | 122 | 109 | 106 | 122 | 116 | 125 | 139 | 145 | 158 | 156 | 147 | 152 | 148 |
| 179 | 179 | 180 | 155 | 127 | 121 | 118 | 109 | 107 | 113 | 125 | 133 | 130 | 129 | 139 | 153 | 161 | 148 | 155 | 157 |
| 176 | 183 | 181 | 153 | 122 | 115 | 113 | 106 | 105 | 109 | 123 | 132 | 131 | 131 | 140 | 151 | 157 | 149 | 156 | 159 |
| 180 | 181 | 177 | 147 | 115 | 110 | 111 | 107 | 107 | 105 | 120 | 132 | 133 | 133 | 141 | 150 | 154 | 148 | 155 | 157 |
| 181 | 174 | 170 | 141 | 113 | 111 | 115 | 112 | 113 | 105 | 119 | 130 | 132 | 134 | 144 | 153 | 156 | 148 | 152 | 151 |
| 180 | 172 | 168 | 140 | 114 | 114 | 118 | 113 | 112 | 107 | 119 | 128 | 130 | 134 | 146 | 157 | 162 | 153 | 153 | 148 |
| 186 | 176 | 171 | 142 | 114 | 114 | 116 | 110 | 108 | 104 | 116 | 125 | 128 | 134 | 148 | 161 | 165 | 159 | 157 | 149 |
| 185 | 178 | 171 | 138 | 109 | 110 | 114 | 110 | 109 | 97 | 110 | 121 | 127 | 136 | 150 | 160 | 163 | 158 | 156 | 150 |

Table 3.1. The image $I$ represented as a two-dimensional matrix of integers (subsampled).

better suited for interpretation by the human visual system is obtained by generating a *picture*. A picture can be thought of as a scene different from the true one that produces on the imaging sensor (the eye in this case) the same image as the true one. In this sense *pictures* are "controlled illusions": they are scenes different from the true ones (they are *flat*) that produce in the eye the same image as the original scenes. A picture of the same image $I$ described in Figure 3.2 and Table 3.1 is shown in Figure 3.3. Although the latter seems more *informative* as to the content of the scene, it is merely a different representation and contains exactly the same information.

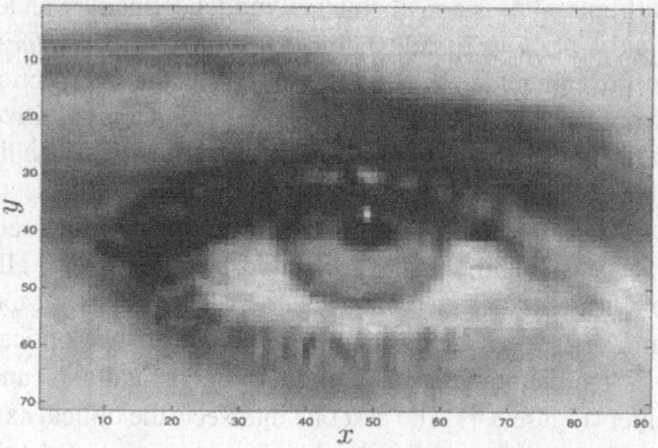


Figure 3.3. A "picture" of the image $I$ (compare with Figure 3.2 and Table 3.1).

## 3.2 Lenses, light, and basic photometry

In order to describe the image formation process, we must specify the value of $I(x, y)$ at each point $(x, y)$ in $\Omega$. Such a value $I(x, y)$ is typically called *image intensity* or *brightness*, or more formally *irradiance*. It has the units of power per unit area ($W/m^2$) and describes the energy falling onto a small patch of the imaging sensor. The irradiance at a point of coordinates $(x, y)$ is obtained by integrating energy both in time (e.g., the shutter interval in a camera, or the integration time in a CCD array) and in a region of space. The region of space that contributes to the irradiance at $(x, y)$ depends upon the shape of the object (surface) of interest, the optics of the imaging device, and it is by no means trivial to determine. In Appendix 3.A at the end of this chapter, we discuss some common simplifying assumptions to approximate it.

### 3.2.1 *Imaging through lenses*

A camera (or in general an optical system) is composed of a set of lenses used to "direct" light. By directing light we mean a controlled change in the direction of propagation, which can be performed by means of diffraction, refraction, and reflection. For the sake of simplicity, we neglect the effects of diffraction and reflection in a lens system, and we consider only refraction. Even so, a complete description of the functioning of a (purely refractive) lens is well beyond the scope of this book. Therefore, we will consider only the simplest possible model, that of a *thin lens*. For a more germane model of light propagation, the interested reader is referred to the classic textbook [Born and Wolf, 1999].

A *thin lens* (Figure 3.4) is a mathematical model defined by an axis, called the *optical axis*, and a plane perpendicular to the axis, called the *focal plane*, with a circular aperture centered at the *optical center*, i.e. the intersection of the focal plane with the optical axis. The thin lens has two parameters: its *focal length* $f$ and its *diameter* $d$. Its function is characterized by two properties. The first property is that all rays entering the aperture parallel to the optical axis intersect on the optical axis at a distance $f$ from the optical center. The point of intersection is called the *focus* of the lens (Figure 3.4). The second property is that all rays through the optical center are undeflected. Consider a point $p \in \mathbb{E}^3$ not too far from the optical axis at a distance $Z$ along the optical axis from the optical center. Now draw two rays from the point $p$: one parallel to the optical axis, and one through the optical center (Figure 3.4). The first one intersects the optical axis at the focus; the second remains undeflected (by the defining properties of the thin lens). Call $\boldsymbol{x}$ the point where the two rays intersect, and let $z$ be its distance from the optical center. By decomposing any other ray from $p$ into a component ray parallel to the optical axis and one through the optical center, we can argue that all rays from $p$ intersect at $\boldsymbol{x}$ on the opposite side of the lens. In particular, a ray from $\boldsymbol{x}$ parallel to the optical axis, must go through $p$. Using similar triangles, from Figure 3.4, we obtain the following *fundamental equation of the thin lens*:

$$\boxed{\frac{1}{Z} + \frac{1}{z} = \frac{1}{f}}.$$

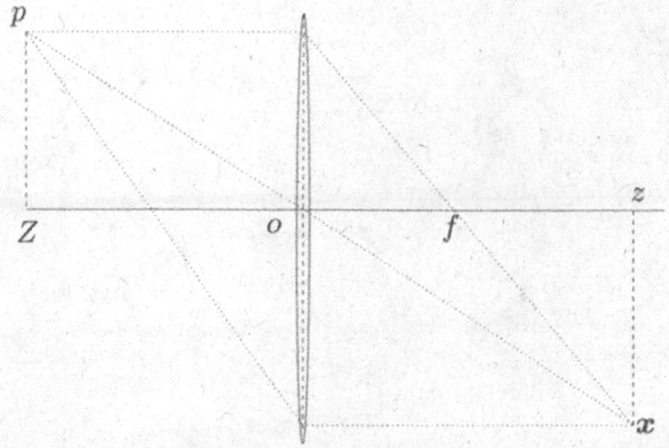


Figure 3.4. The image of the point $p$ is the point $\boldsymbol{x}$ at the intersection of rays going parallel to the optical axis and the ray through the optical center.

The point $\boldsymbol{x}$ will be called the *image*[3] of the point $p$. Therefore, under the assumption of a thin lens, the irradiance $I(\boldsymbol{x})$ at the point $\boldsymbol{x}$ with coordinates $(x, y)$ on the image plane is obtained by integrating all the energy emitted from the region of space contained in the cone determined by the geometry of the lens, as we describe in Appendix 3.A.

### 3.2.2 *Imaging through a pinhole*

If we let the aperture of a thin lens decrease to zero, all rays are forced to go through the optical center $o$, and therefore they remain undeflected. Consequently, the aperture of the cone decreases to zero, and the only points that contribute to the irradiance at the image point $\boldsymbol{x} = [x, y]^T$ are on a line through the center $o$ of the lens. If a point $p$ has coordinates $\boldsymbol{X} = [X, Y, Z]^T$ relative to a reference frame centered at the optical center $o$, with its $z$-axis being the optical axis (of the lens), then it is immediate to see from similar triangles in Figure 3.5 that the coordinates of $p$ and its image $\boldsymbol{x}$ are related by the so-called ideal *perspective projection*

$$x = -f\frac{X}{Z}, \quad y = -f\frac{Y}{Z}, \tag{3.2}$$

where $f$ is referred to as the *focal length*. Sometimes, we simply write the projection as a map $\pi$:

$$\pi : \mathbb{R}^3 \to \mathbb{R}^2; \quad \boldsymbol{X} \mapsto \boldsymbol{x}. \tag{3.3}$$

We also often write $\boldsymbol{x} = \pi(\boldsymbol{X})$. Note that any other point on the line through $o$ and $p$ projects onto the same coordinates $\boldsymbol{x} = [x, y]^T$. This imaging model is called an *ideal pinhole camera* model. It is an idealization of the thin lens model, since

[3]Here the word "image" is to be distinguished from the irradiance image $I(\boldsymbol{x})$ introduced before. Whether "image" indicates $\boldsymbol{x}$ or $I(\boldsymbol{x})$ will be made clear by the context.

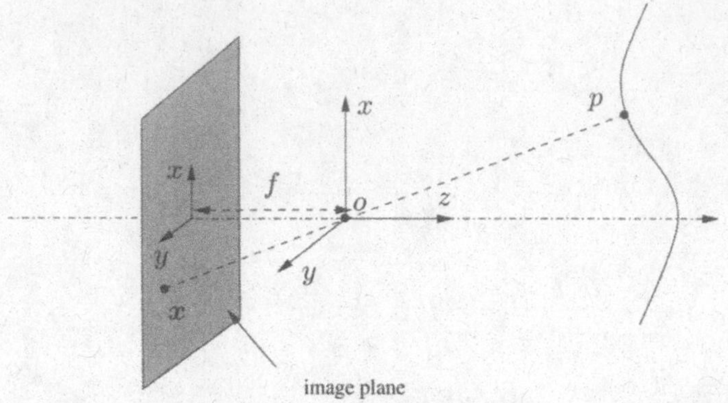


Figure 3.5. Pinhole imaging model: The image of the point $p$ is the point $\boldsymbol{x}$ at the intersection of the ray going through the optical center $o$ and an image plane at a distance $f$ away from the optical center.

when the aperture decreases, diffraction effects become dominant, and therefore the (purely refractive) thin lens model does not hold [Born and Wolf, 1999]. Furthermore, as the aperture decreases to zero, the energy going through the lens also becomes zero. Although it is possible to actually build devices that approximate pinhole cameras, from our perspective the pinhole model will be just a good geometric approximation of a well-focused imaging system.

Notice that there is a negative sign in each of the formulae (3.2). This makes the image of an object appear to be upside down on the image plane (or the retina). To eliminate this effect, we can simply flip the image: $(x, y) \mapsto (-x, -y)$. This corresponds to placing the image plane $\{z = -f\}$ in front of the optical center instead $\{z = +f\}$. In this book we will adopt this more convenient "frontal" pinhole camera model, illustrated in Figure 3.6. In this case, the image $\boldsymbol{x} = [x, y]^T$ of the point $p$ is given by

$$\boxed{x = f\frac{X}{Z}, \quad y = f\frac{Y}{Z}.} \tag{3.4}$$

We often use the same symbol, $\boldsymbol{x}$, to denote the homogeneous representation $[fX/Z, fY/Z, 1]^T \in \mathbb{R}^3$, as long as the dimension is clear from the context.[4]

In practice, the size of the image plane is usually limited, hence not every point $p$ in space will generate an image $\boldsymbol{x}$ inside the image plane. We define the *field of view* (FOV) to be the angle subtended by the spatial extent of the sensor as seen from the optical center. If $2r$ is the largest spatial extension of the sensor (e.g., the

[4]In the homogeneous representation, it is only the direction of the vector $\boldsymbol{x}$ that is important. It is not crucial to normalize the last entry to 1 (see Appendix 3.B). In fact, $\boldsymbol{x}$ can be represented by $\lambda \boldsymbol{X}$ for any nonzero $\lambda \in \mathbb{R}$ as long as we remember that any such vector uniquely determines the intersection of the image ray and the actual image plane, in this case $\{Z = f\}$.

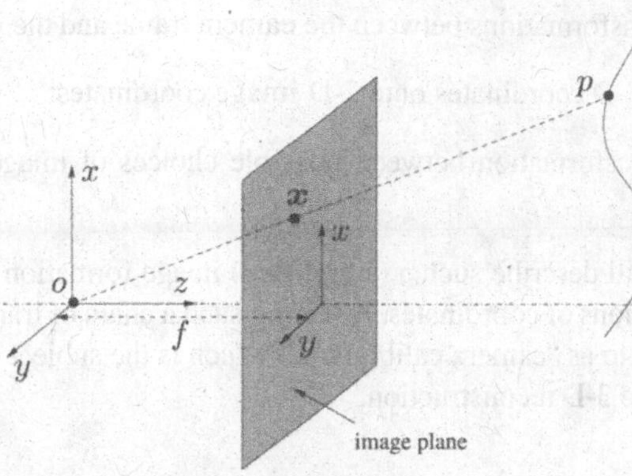


Figure 3.6. Frontal pinhole imaging model: the image of a 3-D point $p$ is the point $x$ at the intersection of the ray going through the optical center $o$ and the image plane at a distance $f$ in front of the optical center.

side of the CCD), then the field of view is $\theta = 2\arctan(r/f)$. Notice that if a flat plane is used as the image plane, the angle $\theta$ is always less than 180°.[5]

In Appendix 3.A we give a concise description of a simplified model to determine the intensity value of the image at the position $x$, $I(x)$. This depends upon the ambient light distribution, the material properties of the visible surfaces and, their geometry. There we also show under what conditions this model can be reduced to a purely geometric one, where the intensity measured at a pixel is identical to the amount of energy radiated at the corresponding point in space, independent of the vantage point, e.g., a Lambertian surface. Under these conditions, the image formation process can be reduced to tracing rays from surfaces in space to points on the image plane. How to do so is explained in the next section.

## 3.3 A geometric model of image formation

As we have mentioned in the previous section and we elaborate further in Appendix 3.A, under the assumptions of a pinhole camera model and Lambertian surfaces, one can essentially reduce the process of image formation to tracing rays from points on objects to pixels. That is, knowing which point in space projects onto which point on the image plane allows one to directly associate the radiance at the point to the irradiance of its image; see equation (3.36) in Appendix 3.A. In order to establish a precise correspondence between points in 3-D space (with respect to a fixed global reference frame) and their projected images in a 2-D image plane (with respect to a local coordinate frame), a mathematical model for this process must account for three types of transformations:

[5]In case of a spherical or ellipsoidal imaging surface, common in omnidirectional cameras, the field of view can often exceed 180°.

1. coordinate transformations between the camera frame and the world frame;
2. projection of 3-D coordinates onto 2-D image coordinates;
3. coordinate transformation between possible choices of image coordinate frame.

In this section we will describe such a (simplified) image formation process as a series of transformations of coordinates. Inverting such a chain of transformations is generally referred to as "camera calibration," which is the subject of Chapter 6 and also a key step to 3-D reconstruction.

### 3.3.1 *An ideal perspective camera*

Let us consider a generic point $p$, with coordinates $\boldsymbol{X}_0 = [X_0, Y_0, Z_0]^T \in \mathbb{R}^3$ relative to the world reference frame.[6] As we know from Chapter 2, the coordinates $\boldsymbol{X} = [X, Y, Z]^T$ of the same point $p$ relative to the camera frame are given by a rigid-body transformation $g = (R, T)$ of $\boldsymbol{X}_0$:

$$\boldsymbol{X} = R\boldsymbol{X}_0 + T \quad \in \mathbb{R}^3.$$

Adopting the frontal pinhole camera model introduced in the previous section (Figure 3.6), we see that the point $\boldsymbol{X}$ is projected onto the image plane at the point

$$\boldsymbol{x} = \begin{bmatrix} x \\ y \end{bmatrix} = \frac{f}{Z}\begin{bmatrix} X \\ Y \end{bmatrix}.$$

In homogeneous coordinates, this relationship can be written as

$$Z\begin{bmatrix} x \\ y \\ 1 \end{bmatrix} = \begin{bmatrix} f & 0 & 0 & 0 \\ 0 & f & 0 & 0 \\ 0 & 0 & 1 & 0 \end{bmatrix}\begin{bmatrix} X \\ Y \\ Z \\ 1 \end{bmatrix}. \tag{3.5}$$

We can rewrite the above equation equivalently as

$$Z\boldsymbol{x} = \begin{bmatrix} f & 0 & 0 & 0 \\ 0 & f & 0 & 0 \\ 0 & 0 & 1 & 0 \end{bmatrix}\boldsymbol{X}, \tag{3.6}$$

where $\boldsymbol{X} \doteq [X, Y, Z, 1]^T$ and $\boldsymbol{x} \doteq [x, y, 1]^T$ are now in homogeneous representation. Since the coordinate $Z$ (or the depth of the point $p$) is usually unknown, we may simply write it as an arbitrary positive scalar $\lambda \in \mathbb{R}_+$. Notice that in the

[6] We often indicate with $\boldsymbol{X}_0$ the coordinates of the point relative to the initial position of a moving camera frame.

above equation we can decompose the matrix into

$$\begin{bmatrix} f & 0 & 0 & 0 \\ 0 & f & 0 & 0 \\ 0 & 0 & 1 & 0 \end{bmatrix} = \begin{bmatrix} f & 0 & 0 \\ 0 & f & 0 \\ 0 & 0 & 1 \end{bmatrix} \begin{bmatrix} 1 & 0 & 0 & 0 \\ 0 & 1 & 0 & 0 \\ 0 & 0 & 1 & 0 \end{bmatrix}.$$

Define two matrices

$$K_f \doteq \begin{bmatrix} f & 0 & 0 \\ 0 & f & 0 \\ 0 & 0 & 1 \end{bmatrix} \in \mathbb{R}^{3\times 3}, \quad \Pi_0 \doteq \begin{bmatrix} 1 & 0 & 0 & 0 \\ 0 & 1 & 0 & 0 \\ 0 & 0 & 1 & 0 \end{bmatrix} \in \mathbb{R}^{3\times 4}. \tag{3.7}$$

The matrix $\Pi_0$ is often referred to as the *standard (or "canonical") projection matrix*. From the coordinate transformation we have for $\boldsymbol{X} = [X, Y, Z, 1]^T$,

$$\begin{bmatrix} X \\ Y \\ Z \\ 1 \end{bmatrix} = \begin{bmatrix} R & T \\ 0 & 1 \end{bmatrix} \begin{bmatrix} X_0 \\ Y_0 \\ Z_0 \\ 1 \end{bmatrix}. \tag{3.8}$$

To summarize, using the above notation, the overall geometric model for *an ideal camera* can be described as

$$\lambda \begin{bmatrix} x \\ y \\ 1 \end{bmatrix} = \begin{bmatrix} f & 0 & 0 \\ 0 & f & 0 \\ 0 & 0 & 1 \end{bmatrix} \begin{bmatrix} 1 & 0 & 0 & 0 \\ 0 & 1 & 0 & 0 \\ 0 & 0 & 1 & 0 \end{bmatrix} \begin{bmatrix} R & T \\ 0 & 1 \end{bmatrix} \begin{bmatrix} X_0 \\ Y_0 \\ Z_0 \\ 1 \end{bmatrix},$$

or in matrix form,

$$\lambda \boldsymbol{x} = K_f \Pi_0 \boldsymbol{X} = K_f \Pi_0 g \boldsymbol{X}_0. \tag{3.9}$$

If the focal length $f$ is known and hence can be normalized to 1, this model reduces to a Euclidean transformation $g$ followed by a standard projection $\Pi_0$, i.e.

$$\boxed{\lambda \boldsymbol{x} = \Pi_0 \boldsymbol{X} = \Pi_0 g \boldsymbol{X}_0.} \tag{3.10}$$

### 3.3.2 *Camera with intrinsic parameters*

The ideal model of equation (3.9) is specified relative to a very particular choice of reference frame, the "canonical retinal frame," centered at the optical center with one axis aligned with the optical axis. In practice, when one captures images with a digital camera the measurements are obtained in terms of pixels $(i, j)$, with the origin of the image coordinate frame typically in the upper-left corner of the image. In order to render the model (3.9) usable, we need to specify the relationship between the retinal plane coordinate frame and the pixel array.

The first step consists in specifying the units along the $x$- and $y$-axes: if $(x, y)$ are specified in terms of metric units (e.g., millimeters), and $(x_s, y_s)$ are scaled

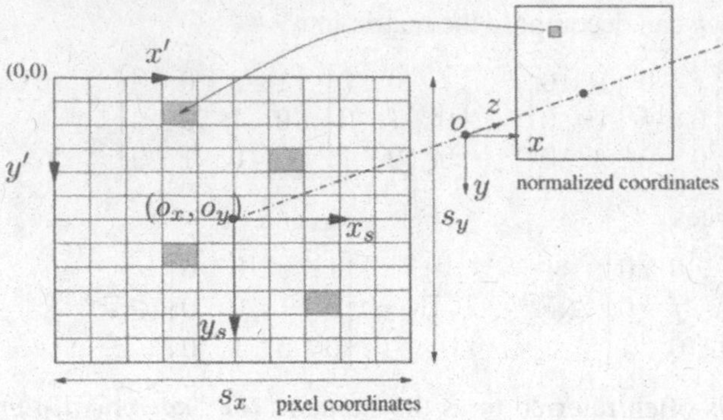


Figure 3.7. Transformation from normalized coordinates to coordinates in pixels.

versions that correspond to coordinates of the pixel, then the transformation can be described by a scaling matrix

$$\begin{bmatrix} x_s \\ y_s \end{bmatrix} = \begin{bmatrix} s_x & 0 \\ 0 & s_y \end{bmatrix} \begin{bmatrix} x \\ y \end{bmatrix} \tag{3.11}$$

that depends on the size of the pixel (in metric units) along the $x$ and $y$ directions (Figure 3.7). When $s_x = s_y$, each pixel is square. In general, they can be different, and then the pixel is rectangular. However, here $x_s$ and $y_s$ are still specified relative to the *principal point* (where the $z$-axis intersects the image plane), whereas the pixel index $(i, j)$ is conventionally specified relative to the upper-left corner, and is indicated by positive numbers. Therefore, we need to translate the origin of the reference frame to this corner (as shown in Figure 3.7),

$$x' = x_s + o_x,$$
$$y' = y_s + o_y,$$

where $(o_x, o_y)$ are the coordinates (in pixels) of the principal point relative to the image reference frame. So the actual image coordinates are given by the vector $\boldsymbol{x}' = [x', y', 1]^T$ instead of the ideal image coordinates $\boldsymbol{x} = [x, y, 1]^T$. The above steps of coordinate transformation can be written in the homogeneous representation as

$$\boldsymbol{x}' \doteq \begin{bmatrix} x' \\ y' \\ 1 \end{bmatrix} = \begin{bmatrix} s_x & 0 & o_x \\ 0 & s_y & o_y \\ 0 & 0 & 1 \end{bmatrix} \begin{bmatrix} x \\ y \\ 1 \end{bmatrix}, \tag{3.12}$$

where $x'$ and $y'$ are actual image coordinates in pixels. This is illustrated in Figure 3.7. In case the pixels are not rectangular, a more general form of the scaling matrix can be considered,

$$\begin{bmatrix} s_x & s_\theta \\ 0 & s_y \end{bmatrix} \in \mathbb{R}^{2\times 2},$$

where $s_\theta$ is called a *skew factor* and is proportional to $\cot(\theta)$, where $\theta$ is the angle between the image axes $x_s$ and $y_s$.[7] The transformation matrix in (3.12) then takes the general form

$$K_s \doteq \begin{bmatrix} s_x & s_\theta & o_x \\ 0 & s_y & o_y \\ 0 & 0 & 1 \end{bmatrix} \in \mathbb{R}^{3\times 3}. \tag{3.13}$$

In many practical applications it is common to assume that $s_\theta = 0$.

Now, combining the projection model from the previous section with the scaling and translation yields a more realistic model of a transformation between homogeneous coordinates of a 3-D point relative to the camera frame and homogeneous coordinates of its image expressed in terms of pixels,

$$\lambda \begin{bmatrix} x' \\ y' \\ 1 \end{bmatrix} = \begin{bmatrix} s_x & s_\theta & o_x \\ 0 & s_y & o_y \\ 0 & 0 & 1 \end{bmatrix} \begin{bmatrix} f & 0 & 0 \\ 0 & f & 0 \\ 0 & 0 & 1 \end{bmatrix} \begin{bmatrix} 1 & 0 & 0 & 0 \\ 0 & 1 & 0 & 0 \\ 0 & 0 & 1 & 0 \end{bmatrix} \begin{bmatrix} X \\ Y \\ Z \\ 1 \end{bmatrix}.$$

Notice that in the above equation, the effect of a real camera is in fact carried through two stages:

- The first stage is a standard perspective projection with respect to a *normalized coordinate system* (as if the focal length were $f = 1$). This is characterized by the standard projection matrix $\Pi_0 = [I, 0]$.
- The second stage is an additional transformation (on the obtained image $\boldsymbol{x}$) that depends on parameters of the camera such as the focal length $f$, the scaling factors $s_x$, $s_y$, and $s_\theta$, and the center offsets $o_x, o_y$.

The second transformation is obviously characterized by the combination of the two matrices $K_s$ and $K_f$:

$$K \doteq K_s K_f \doteq \begin{bmatrix} s_x & s_\theta & o_x \\ 0 & s_y & o_y \\ 0 & 0 & 1 \end{bmatrix} \begin{bmatrix} f & 0 & 0 \\ 0 & f & 0 \\ 0 & 0 & 1 \end{bmatrix} = \begin{bmatrix} fs_x & fs_\theta & o_x \\ 0 & fs_y & o_y \\ 0 & 0 & 1 \end{bmatrix}. \tag{3.14}$$

The coupling of $K_s$ and $K_f$ allows us to write the projection equation in the following way:

$$\lambda \boldsymbol{x}' = K\Pi_0 \boldsymbol{X} = \begin{bmatrix} fs_x & fs_\theta & o_x \\ 0 & fs_y & o_y \\ 0 & 0 & 1 \end{bmatrix} \begin{bmatrix} 1 & 0 & 0 & 0 \\ 0 & 1 & 0 & 0 \\ 0 & 0 & 1 & 0 \end{bmatrix} \begin{bmatrix} X \\ Y \\ Z \\ 1 \end{bmatrix}. \tag{3.15}$$

The constant $3 \times 4$ matrix $\Pi_0$ represents the perspective projection. The upper triangular $3 \times 3$ matrix $K$ collects all parameters that are "intrinsic" to a particular camera, and is therefore called the *intrinsic parameter matrix*, or the *calibration*

[7] Typically, the angle $\theta$ is very close to $90°$, and hence $s_\theta$ is very close to zero.

*matrix* of the camera. The entries of the matrix $K$ have the following geometric interpretation:

- $o_x$: $x$-coordinate of the principal point in pixels,
- $o_y$: $y$-coordinate of the principal point in pixels,
- $fs_x = \alpha_x$: size of unit length in horizontal pixels,
- $fs_y = \alpha_y$: size of unit length in vertical pixels,
- $\alpha_x/\alpha_y$: aspect ratio $\sigma$,
- $fs_\theta$: skew of the pixel, often close to zero.

Note that the height of the pixel is not necessarily identical to its width unless the aspect ratio $\sigma$ is equal to 1.

When the calibration matrix $K$ is known, the *calibrated* coordinates $\boldsymbol{x}$ can be obtained from the pixel coordinates $\boldsymbol{x}'$ by a simple inversion of $K$:

$$\lambda \boldsymbol{x} = \lambda K^{-1}\boldsymbol{x}' = \Pi_0 \boldsymbol{X} = \begin{bmatrix} 1 & 0 & 0 & 0 \\ 0 & 1 & 0 & 0 \\ 0 & 0 & 1 & 0 \end{bmatrix} \begin{bmatrix} X \\ Y \\ Z \\ 1 \end{bmatrix}. \tag{3.16}$$

The information about the matrix $K$ can be obtained through the process of camera calibration to be described in Chapter 6. With the effect of $K$ compensated for, equation (3.16), expressed in the normalized coordinate system, corresponds to the ideal pinhole camera model with the image plane located in front of the center of projection and the focal length $f$ equal to 1.

To summarize, the geometric relationship between a point of coordinates $\boldsymbol{X}_0 = [X_0, Y_0, Z_0, 1]^T$ relative to the world frame and its corresponding image coordinates $\boldsymbol{x}' = [x', y', 1]^T$ (in pixels) depends on the rigid-body motion $(R, T)$ between the world frame and the camera frame (sometimes referred to as the *extrinsic calibration parameters*), an ideal projection $\Pi_0$, and the camera intrinsic parameters $K$. The overall model for image formation is therefore captured by the following equation:

$$\lambda \begin{bmatrix} x' \\ y' \\ 1 \end{bmatrix} = \begin{bmatrix} fs_x & fs_\theta & o_x \\ 0 & fs_y & o_y \\ 0 & 0 & 1 \end{bmatrix} \begin{bmatrix} 1 & 0 & 0 & 0 \\ 0 & 1 & 0 & 0 \\ 0 & 0 & 1 & 0 \end{bmatrix} \begin{bmatrix} R & T \\ 0 & 1 \end{bmatrix} \begin{bmatrix} X_0 \\ Y_0 \\ Z_0 \\ 1 \end{bmatrix}.$$

In matrix form, we write

$$\lambda \boldsymbol{x}' = K\Pi_0 \boldsymbol{X} = K\Pi_0 g \boldsymbol{X}_0, \tag{3.17}$$

or equivalently,

$$\lambda \boldsymbol{x}' = K\Pi_0 \boldsymbol{X} = [KR, KT]\boldsymbol{X}_0. \tag{3.18}$$

Often, for convenience, we call the $3 \times 4$ matrix $K\Pi_0 g = [KR, KT]$ a (general) *projection matrix* $\Pi$, to be distinguished from the standard projection matrix $\Pi_0$.

Hence, the above equation can be simply written as

$$\boxed{\lambda \boldsymbol{x}' = \Pi \boldsymbol{X}_0 = K\Pi_0 g \boldsymbol{X}_0.} \tag{3.19}$$

Compared to the ideal camera model (3.10), the only change here is the standard projection matrix $\Pi_0$ being replaced by a general one $\Pi$.

At this stage, in order to explicitly see the nonlinear nature of the perspective projection equation, we can divide equation (3.19) by the scale $\lambda$ and obtain the following expressions for the image coordinates $(x', y', z')$,

$$x' = \frac{\pi_1^T \boldsymbol{X}_0}{\pi_3^T \boldsymbol{X}_0}, \quad y' = \frac{\pi_2^T \boldsymbol{X}_0}{\pi_3^T \boldsymbol{X}_0}, \quad z' = 1, \tag{3.20}$$

where $\pi_1^T, \pi_2^T, \pi_3^T \in \mathbb{R}^4$ are the three rows of the projection matrix $\Pi$.

**Example 3.1 (Spherical perspective projection).** The perspective pinhole camera model outlined above considers planar imaging surfaces. An alternative imaging surface that is also commonly used is that of a sphere, shown in Figure 3.8.

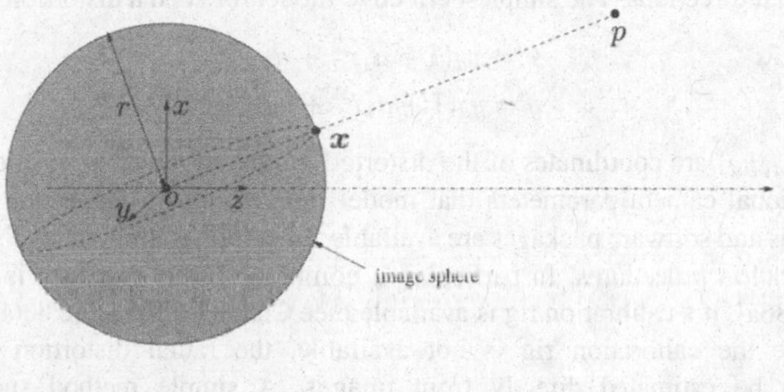


Figure 3.8. Spherical perspective projection model: the image of a 3-D point $p$ is the point $\boldsymbol{x}$ at the intersection of the ray going through the optical center $o$ and a sphere of radius $r$ around the optical center. Typically $r$ is chosen to be 1.

This choice is partly motivated by retina shapes often encountered in biological systems. For spherical projection, we simply choose the imaging surface to be the unit sphere $\mathbb{S}^2 = \{p \in \mathbb{R}^3 \mid \|\boldsymbol{X}(p)\| = 1\}$. Then, the spherical projection is defined by the map $\pi_s$ from $\mathbb{R}^3$ to $\mathbb{S}^2$:

$$\pi_s : \mathbb{R}^3 \to \mathbb{S}^2; \quad \boldsymbol{X} \mapsto \boldsymbol{x} = \frac{\boldsymbol{X}}{\|\boldsymbol{X}\|}.$$

As in the case of planar perspective projection, the relationship between pixel coordinates of a point and their 3-D metric counterpart can be expressed as

$$\lambda \boldsymbol{x}' = K\Pi_0 \boldsymbol{X} = K\Pi_0 g \boldsymbol{X}_0, \tag{3.21}$$

where the scale is given by $\lambda = \sqrt{X^2 + Y^2 + Z^2}$ in the case of spherical projection while $\lambda = Z$ in the case of planar projection. Therefore, mathematically, spherical projection and planar projection can be described by the same set of equations. The only difference is that the unknown (depth) scale $\lambda$ takes different values. ■

For convenience, we often write $\boldsymbol{x} \sim \boldsymbol{y}$ for two (homogeneous) vectors $\boldsymbol{x}$ and $\boldsymbol{y}$ equal up to a scalar factor (see Appendix 3.B for more detail). From the above example, we see that for any perspective projection we have

$$\boldsymbol{x}' \sim \Pi \boldsymbol{X}_0 = K \Pi_0 g \boldsymbol{X}_0, \tag{3.22}$$

and the shape of the imaging surface chosen does not matter. The imaging surface can be any (regular) surface as long as any ray $\overrightarrow{op}$ intersects with the surface at one point at most. For example, an entire class of ellipsoidal surfaces can be used, which leads to the so-called *catadioptric model* popular in many omnidirectional cameras. In principle, all images thus obtained contain exactly the same information.

### 3.3.3 Radial distortion

In addition to linear distortions described by the parameters in $K$, if a camera with a wide field of view is used, one can often observe significant distortion along radial directions. The simplest effective model for such a distortion is:

$$x = x_d(1 + a_1 r^2 + a_2 r^4),$$
$$y = y_d(1 + a_1 r^2 + a_2 r^4),$$

where $(x_d, y_d)$ are coordinates of the distorted points, $r^2 = x_d^2 + y_d^2$ and $a_1, a_2$ are additional camera parameters that model the amount of distortion. Several algorithms and software packages are available for compensating radial distortion via calibration procedures. In particular, a commonly used approach is that of [Tsai, 1986a], if a calibration rig is available (see Chapter 6 for more details).

In case the calibration rig is not available, the radial distortion parameters can be estimated directly from images. A simple method suggested by [Devernay and Faugeras, 1995] assumes a more general model of radial distortion:

$$\begin{aligned} \boldsymbol{x} &= c + f(r)(\boldsymbol{x}_d - c), \\ f(r) &= 1 + a_1 r + a_2 r^2 + a_3 r^3 + a_4 r^4, \end{aligned}$$

where $\boldsymbol{x}_d = [x_d, y_d]^T$ are the distorted image coordinates, $r^2 = \|\boldsymbol{x}_d - c\|^2$, $c = [c_x, c_y]^T$ is the center of the distortion, not necessarily coincident with the center of the image, and $f(r)$ is the distortion correction factor. The method assumes a set of straight lines in the world and computes the best parameters of the radial distortion model which would transform the curved images of the lines into straight segments. One can use this model to transform Figure 3.9 (left) into 3.9 (right) via preprocessing algorithms described in [Devernay and Faugeras, 1995]. Therefore, in the rest of this book we assume that radial distortion has been compensated for, and a camera is described simply by the parameter matrix $K$. The interested reader may consult classical references such as [Tsai, 1986a, Tsai, 1987, Tsai, 1989, Zhang, 1998b], which are available as software packages. Some authors have shown that radial distortion

Figure 3.9. Left: image taken by a camera with a short focal length; note that the straight lines in the scene become curved on the image. Right: image with radial distortion compensated for.

can be recovered from multiple corresponding images: a simultaneous estimation of 3-D geometry and radial distortion can be found in the more recent work of [Zhang, 1996, Stein, 1997, Fitzgibbon, 2001]. For more sophisticated lens aberration models, the reader can refer to classical references in geometric optics given at the end of this chapter.

### 3.3.4 Image, preimage, and coimage of points and lines

The preceding sections have formally established the notion of a perspective image of a point. In principle, this allows us to define an image of any other geometric entity in 3-D that can be defined as a set of points (e.g., a line or a plane). Nevertheless, as we have seen from the example of spherical projection, even for a point, there exist seemingly different representations for its image: two vectors $\boldsymbol{x} \in \mathbb{R}^3$ and $\boldsymbol{y} \in \mathbb{R}^3$ may represent the same image point as long as they are related by a nonzero scalar factor; i.e. $\boldsymbol{x} \sim \boldsymbol{y}$ (as a result of different choices in the imaging surface). To avoid possible confusion that can be caused by such different representations for the same geometric entity, we introduce a few abstract notions related to the image of a point or a line.

Consider the perspective projection of a straight line $L$ in 3-D onto the 2-D image plane (Figure 3.10). To specify a line in 3-D, we can typically specify a point $p_o$, called the base point, on the line and specify a vector $v$ that indicates the direction of the line. Suppose that $\boldsymbol{X}_o = [X_o, Y_o, Z_o, 1]^T$ are the homogeneous coordinates of the base point $p_o$ and $\boldsymbol{V} = [V_1, V_2, V_3, 0]^T \in \mathbb{R}^4$ is the homogeneous representation of $v$, relative to the camera coordinate frame. Then the (homogeneous) coordinates of any point on the line $L$ can be expressed as

$$\boldsymbol{X} = \boldsymbol{X}_o + \mu \boldsymbol{V}, \quad \mu \in \mathbb{R}.$$

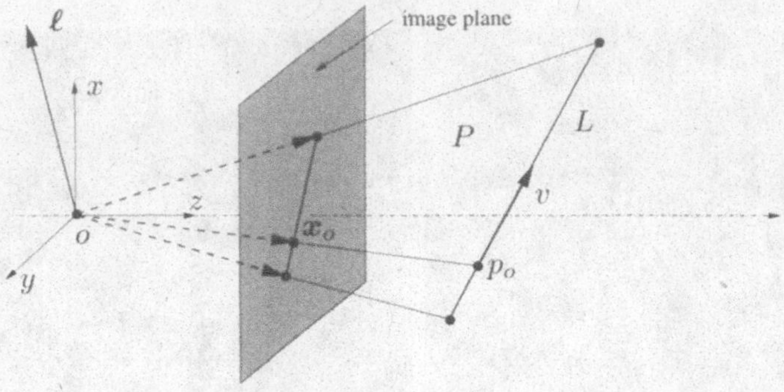


Figure 3.10. Perspective image of a line $L$ in 3-D. The collection of images of points on the line forms a plane $P$. Intersection of this plane and the image plane gives a straight line $\ell$ which is the image of the line.

Then, the image of the line $L$ is given by the collection of image points with homogeneous coordinates given by

$$\boldsymbol{x} \sim \Pi_0 \boldsymbol{X} = \Pi_0(\boldsymbol{X}_o + \mu \boldsymbol{V}) = \Pi_0 \boldsymbol{X}_o + \mu \Pi_0 \boldsymbol{V}.$$

It is easy to see that this collection of points $\{\boldsymbol{x}\}$, treated as vectors with origin at $o$, span a 2-D subspace $P$, shown in Figure 3.10. The intersection of this subspace with the image plane gives rise to a straight line in the 2-D image plane, also shown in Figure 3.10. This line is then the (physical) image of the line $L$.

Now the question is how to efficiently represent the image of the line. For this purpose, we first introduce the notion of preimage:

**Definition 3.2 (Preimage).** *A preimage of a point or a line in the image plane is the set of 3-D points that give rise to an image equal to the given point or line.*

Note that the given image is constrained to lie in the image plane, whereas the preimage lies in 3-D space. In the case of a point $\boldsymbol{x}$ on the image plane, its preimage is a one-dimensional subspace, spanned by the vector joining the point $\boldsymbol{x}$ to the camera center $o$. In the case of a line, the preimage is a plane $P$ through $o$ (hence a subspace) as shown in Figure 3.10, whose intersection with the image plane is exactly the given image line. Such a plane can be represented as the span of any two linearly independent vectors in the same subspace. Thus the preimage is really the largest set of 3-D points or lines that gives rise to the same image. The definition of a preimage can be given not only for points or lines in the image plane but also for curves or other more complicated geometric entities in the image plane as well. However, when the image is a point or a line, the preimage is a subspace, and we may also represent this subspace by its (unique) orthogonal complement in $\mathbb{R}^3$. For instance, a plane can be represented by its normal vector. This leads to the following notion of coimage:

**Definition 3.3 (Coimage).** *The coimage of a point or a line is defined to be the subspace in $\mathbb{R}^3$ that is the (unique) orthogonal complement of its preimage.*

The reader must be aware that the image, preimage, and coimage are *equivalent* representations, since they uniquely determine one another:

$$\text{image} = \text{preimage} \cap \text{image plane}, \quad \text{preimage} = \text{span(image)},$$
$$\text{preimage} = \text{coimage}^{\perp}, \quad \text{coimage} = \text{preimage}^{\perp}.$$

Since the preimage of a line $L$ is a two-dimensional subspace, its coimage is represented as the span of the normal vector to the subspace. The notation we use for this is $\boldsymbol{\ell} = [a, b, c]^T \in \mathbb{R}^3$ (Figure 3.10). If $\boldsymbol{x}$ is the image of a point $p$ on this line, then it satisfies the orthogonality equation

$$\boldsymbol{\ell}^T \boldsymbol{x} = 0. \tag{3.23}$$

Recall that we use $\widehat{u} \in \mathbb{R}^{3\times 3}$ to denote the skew-symmetric matrix associated with a vector $u \in \mathbb{R}^3$. Its column vectors span the subspace orthogonal to the vector $u$. Thus the column vectors of the matrix $\widehat{\boldsymbol{\ell}}$ span the plane that is *orthogonal* to $\boldsymbol{\ell}$; i.e. they span the preimage of the line $L$. In Figure 3.10, this means that $P = \text{span}(\widehat{\boldsymbol{\ell}})$. Similarly, if $\boldsymbol{x}$ is the image of a point $p$, its coimage is the plane orthogonal to $\boldsymbol{x}$ given by the span of the column vectors of the matrix $\widehat{\boldsymbol{x}}$. Thus, in principle, we should use the notation in Table 3.2 to represent the image, preimage, or coimage of a point and a line.

| Notation | Image | Preimage | Coimage |
|---|---|---|---|
| Point | $\text{span}(\boldsymbol{x}) \cap$ image plane | $\text{span}(\boldsymbol{x}) \subset \mathbb{R}^3$ | $\text{span}(\widehat{\boldsymbol{x}}) \subset \mathbb{R}^3$ |
| Line | $\text{span}(\widehat{\boldsymbol{\ell}}) \cap$ image plane | $\text{span}(\widehat{\boldsymbol{\ell}}) \subset \mathbb{R}^3$ | $\text{span}(\boldsymbol{\ell}) \subset \mathbb{R}^3$ |

Table 3.2. The image, preimage, and coimage of a point and a line.

Although the (physical) image of a point or a line, strictly speaking, is a notion that depends on a particular choice of imaging surface, mathematically it is more convenient to use its preimage or coimage to represent it. For instance, we will use the vector $\boldsymbol{x}$, defined up to a scalar factor, to represent the preimage (hence the image) of a point; and the vector $\boldsymbol{\ell}$, defined up to a scalar factor, to represent the coimage (hence the image) of a line. The relationships between preimage and coimage of points and lines can be expressed in terms of the vectors $\boldsymbol{x}, \boldsymbol{\ell} \in \mathbb{R}^3$ as

$$\widehat{\boldsymbol{x}}\boldsymbol{x} = 0, \quad \widehat{\boldsymbol{\ell}}\boldsymbol{\ell} = 0.$$

Often, for a simpler language, we may refer to either the preimage or coimage of points and lines as the "image" if its actual meaning is clear from the context. For instance, in Figure 3.10, we will, in future chapters, often mark in the image plane the image of the line $L$ by the same symbol $\boldsymbol{\ell}$ as the vector typically used to denote its coimage.

## 3.4 Summary

In this chapter, perspective projection is introduced as a model of the image formation for a pinhole camera. In the ideal case (e.g., when the calibration matrix $K$ is the identity), homogeneous coordinates of an image point are related to their 3-D counterparts by an unknown (depth) scale $\lambda$,

$$\lambda \boldsymbol{x} = \Pi_0 \boldsymbol{X} = \Pi_0 g \boldsymbol{X}_0.$$

If $K$ is not the identity, the standard perspective projection is augmented by an additional linear transformation $K$ on the image plane

$$\boldsymbol{x}' = K\boldsymbol{x}.$$

This yields the following relationship between coordinates of an (uncalibrated) image and their 3-D counterparts:

$$\lambda \boldsymbol{x}' = K\Pi_0 \boldsymbol{X} = K\Pi_0 g \boldsymbol{X}_0.$$

As equivalent representations for an image of a point or a line, we introduced the notions of image, preimage, and coimage, whose relationships were summarized in Table 3.2.

## 3.5 Exercises

**Exercise 3.1** Show that any point on the line through $o$ and $p$ projects onto the same image coordinates as $p$.

**Exercise 3.2** Consider a thin lens imaging a plane parallel to the lens at a distance $z$ from the focal plane. Determine the region of this plane that contributes to the image $I$ at the point $\boldsymbol{x}$. (Hint: consider first a one-dimensional imaging model, then extend to a two-dimensional image.)

**Exercise 3.3 (Field of view).** An important parameter of the imaging system is the *field of view* (FOV). The field of view is twice the angle between the optical axis ($z$-axis) and the end of the retinal plane (CCD array). Imagine having a camera system with focal length 24 mm, and retinal plane (CCD array) (16 mm $\times$ 12 mm) and that your digitizer samples your imaging surface at 500 $\times$ 500 pixels in the horizontal and vertical directions.

1. Compute the FOV.
2. Write down the relationship between the image coordinate and a point in 3-D space expressed in the camera coordinate system.
3. Describe how the size of the FOV is related to the focal length and how it affects the resolution in the image.
4. Write a software program (in Matlab) that simulates the geometry of the projection process; given the coordinates of an object with respect to the calibrated camera frame, create an image of that object. Experiment with changing the parameters of the imaging system.

**Exercise 3.4** Under the standard perspective projection (i.e. $K = I$):

1. What is the image of a sphere?
2. Characterize the objects for which the image of the centroid is the centroid of the image.

**Exercise 3.5 (Calibration matrix).** Compute the calibration matrix $K$ that represents the transformation from image $I$ to $I'$ as shown in Figure 3.11. Note that from the definition of the calibration matrix, you need to use homogeneous coordinates to represent image points. Suppose that the resulting image $I'$ is further digitized into an array of $640 \times 480$ pixels and the intensity value of each pixel is quantized to an integer in $[0, 255]$. Then how many *different* digitized images can one possibly get from such a process?

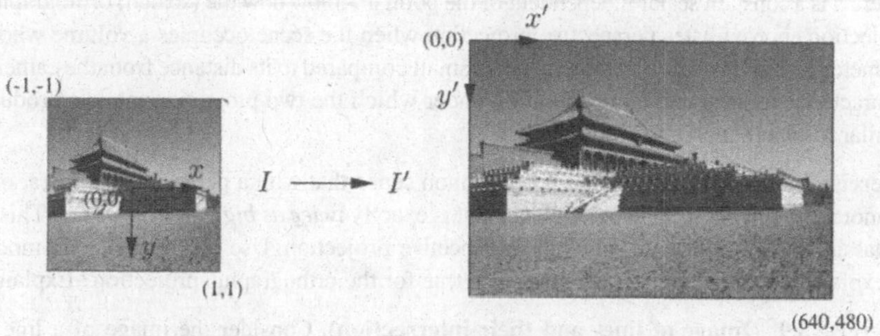


Figure 3.11. Transformation of a normalized image into pixel coordinates.

**Exercise 3.6 (Image cropping).** In this exercise, we examine the effect of cropping an image from a change of coordinate viewpoint. Compute the coordinate transformation between pixels (of same points) between the two images in Figure 3.12. Represent this transformation in homogeneous coordinates.

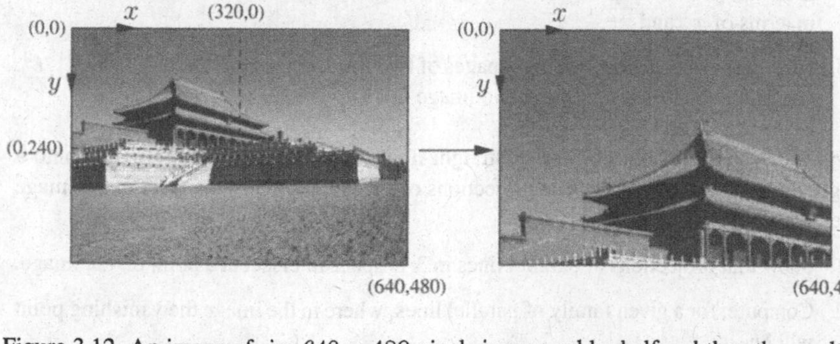


Figure 3.12. An image of size $640 \times 480$ pixels is cropped by half and then the resulting image is up-sampled and restored as a $640 \times 480$-pixel image.

**Exercise 3.7 (Approximate camera models).** The most commonly used approximation to the perspective projection model is *orthographic projection.* The light rays in the orthographic model travel along lines parallel to the optical axis. The relationship between image points and 3-D points in this case is particularly simple: $x = X; y = Y$. So, the geometric model for orthographic projection can be expressed as

$$\begin{bmatrix} x \\ y \end{bmatrix} = \begin{bmatrix} 1 & 0 & 0 \\ 0 & 1 & 0 \end{bmatrix} \begin{bmatrix} X \\ Y \\ Z \end{bmatrix}, \tag{3.24}$$

or simply in matrix form

$$\boldsymbol{x} = \Pi_o \boldsymbol{X}, \tag{3.25}$$

where $\Pi_o \doteq [I_{2\times 2}, 0] \in \mathbb{R}^{2\times 3}$. A scaled version of the orthographic model leads to the so-called *weak-perspective* model

$$\boldsymbol{x} = s\Pi_o \boldsymbol{X}, \tag{3.26}$$

where $s$ is a constant scalar independent of the point $\boldsymbol{x}$. Show how the (scaled) orthographic projection approximates perspective projection when the scene occupies a volume whose diameter (or depth variation of the scene) is small compared to its distance from the camera. Characterize at least one more condition under which the two projection models produce similar results (equal in the limit).

**Exercise 3.8 (Scale ambiguity).** It is common sense that with a perspective camera, one cannot tell an object from another object that is exactly *twice as big but twice as far*. This is a classic ambiguity introduced by the perspective projection. Use the ideal camera model to explain why this is true. Is the same also true for the orthographic projection? Explain.

**Exercise 3.9 (Image of lines and their intersection).** Consider the image of a line $L$ (Figure 3.10).

1. Show that there exists a vector in $\mathbb{R}^3$, call it $\boldsymbol{\ell}$, such that
$$\boldsymbol{\ell}^T \boldsymbol{x} = 0$$
for the image $\boldsymbol{x}$ of every point on the line $L$. What is the geometric meaning of the vector $\boldsymbol{\ell}$? (Note that the vector $\boldsymbol{\ell}$ is defined only up to an arbitrary scalar factor.)

2. If the images of two points on the line $L$ are given, say $\boldsymbol{x}^1, \boldsymbol{x}^2$, express the vector $\boldsymbol{\ell}$ in terms of $\boldsymbol{x}^1$ and $\boldsymbol{x}^2$.

3. Now suppose you are given two images of two lines, in the above vector form $\boldsymbol{\ell}^1, \boldsymbol{\ell}^2$. If $\boldsymbol{x}$ is the intersection of these two image lines, express $\boldsymbol{x}$ in terms of $\boldsymbol{\ell}^1, \boldsymbol{\ell}^2$.

**Exercise 3.10 (Vanishing points).** A straight line on the 3-D world is projected onto a straight line in the image plane. The projections of two parallel lines intersect in the image plane at the *vanishing point*.

1. Show that projections of parallel lines in 3-D space intersect at a point on the image.

2. Compute, for a given family of parallel lines, where in the image the vanishing point will be.

3. When does the vanishing point of the lines in the image plane lie at infinity (i.e. they do not intersect)?

The reader may refer to Appendix 3.B for a more formal treatment of vanishing points as well as their mathematical interpretation.

## 3.A Basic photometry with light sources and surfaces

In this section we give a concise description of a basic radiometric image formation model, and show that some simplifications are necessary in order to reduce the model to a purely geometric one, as described in this chapter. The idea is to describe how the intensity at a pixel on the image is generated. Under suitable assumptions, we show that such intensity depends only on the amount of energy radiated from visible surfaces in space and not on the vantage point.

Let $S$ be a smooth visible surface in space; we denote the tangent plane to the surface at a point $p$ by $T_pS$ and its outward unit normal vector by $\nu_p$. At each point $p \in S$ we can construct a local coordinate frame with its origin at $p$, its $z$-axis parallel to the normal vector $\nu_p$, and its $xy$-plane parallel to $T_pS$ (see Figure 3.13). Let $L$ be a smooth surface that is irradiating light, which we call the *light source*. For simplicity, we may assume that $L$ is the only source of light in space. At a point $q \in L$, we denote with $T_qS$ and $\nu_q$ the tangent plane and the outward unit normal of $L$, respectively, as shown in Figure 3.13.

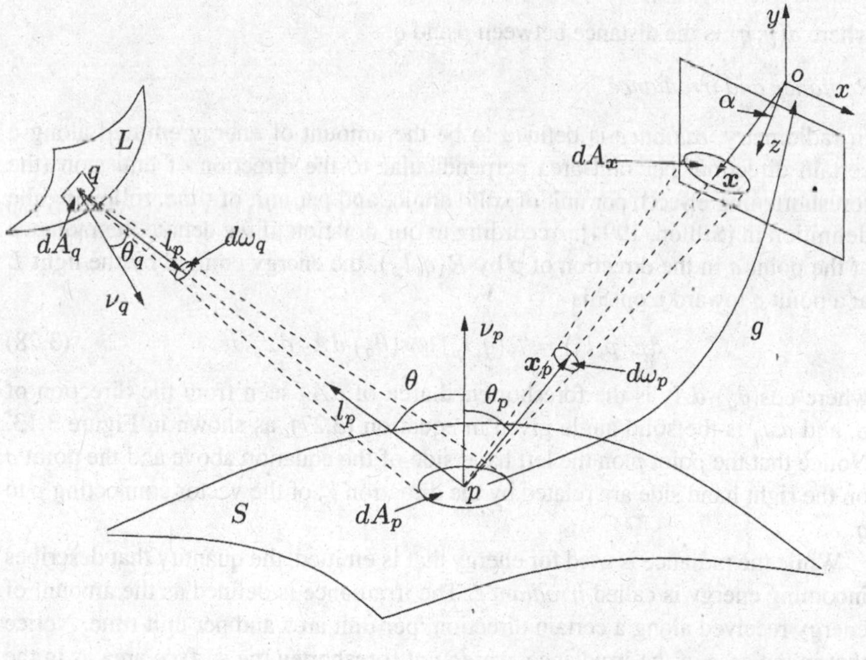


Figure 3.13. Generative model.

The change of coordinates between the local coordinate frame at $p$ and the camera frame, which we assume coincides with the world frame, is indicated by a rigid-body transformation $g$; then $g$ maps coordinates in the local coordinate frame at $p$ into those in the camera frame, and any vector $u$ in the local coordinate frame to a vector $v = g_*(u)$ in the camera frame.[8]

*Foreshortening and solid angle*

When considering interactions between a light source and a surface, we need to introduce the notion of *foreshortening* and that of *solid angle*. Foreshortening encodes how the light distribution on a surface changes as we change the surface orientation with respect to the source of illumination. In formulas, if $dA_p$ is the area element in $T_pS$, and $l_p$ is the unit vector that indicates the direction from $p$ to $q$ (see Figure 3.13), then the corresponding foreshortened area as seen from $q$ is

$$\cos(\theta)dA_p,$$

where $\theta$ is the angle between the direction $l_p$ and the normal vector $\nu_p$; i.e. $\cos(\theta) = \langle \nu_p, l_p \rangle$. A solid angle is defined to be the area of a cone cut out on a unit sphere. Then, the infinitesimal solid angle $d\omega_q$ seen from a point $q$ of the infinitesimal area $dA_p$ is

$$d\omega_q \doteq \frac{\cos(\theta)dA_p}{d(p,q)^2}, \tag{3.27}$$

where $d(p,q)$ is the distance between $p$ and $q$.

*Radiance and irradiance*

In radiometry, *radiance* is defined to be the amount of energy emitted along a certain direction, per unit area perpendicular to the direction of emission (the foreshortening effect), per unit of solid angle, and per unit of time, following the definition in [Sillion, 1994]. According to our notation, if we denote the radiance at the point $q$ in the direction of $p$ by $\mathcal{R}(q, l_p)$, the energy emitted by the light $L$ at a point $q$ toward $p$ on $S$ is

$$dE(p, l_p) \doteq \mathcal{R}(q, l_p)\cos(\theta_q)\, dA_q\, d\omega_q\, dt, \tag{3.28}$$

where $\cos(\theta_q)\, dA_q$ is the foreshortened area of $dA_q$ seen from the direction of $p$, and $d\omega_q$ is the solid angle given in equation (3.27), as shown in Figure 3.13. Notice that the point $p$ on the left hand side of the equation above and the point $q$ on the right hand side are related by the direction $l_p$ of the vector connecting $p$ to $q$.

While the radiance is used for energy that is emitted, the quantity that describes incoming energy is called *irradiance*. The irradiance is defined as the amount of energy received along a certain direction, per unit area and per unit time. Notice that in the case of the irradiance, we *do not* foreshorten the surface area as in the case of the radiance. Denote the irradiance at $p$ received in the direction $l_p$ by

[8] We recall from the previous chapter that if we represent the change of coordinates $g$ with a rotation matrix $R \in SO(3)$ and a translation vector $T$, then the action of $g$ on a point $p$ of coordinates $\mathbf{X} \in \mathbb{R}^3$ is given by $g(\mathbf{X}) \doteq R\mathbf{X} + T$, while the action of $g$ on a *vector* of coordinates $u$ is given by $g_*(u) \doteq Ru$.

$dI(p, l_p)$. By energy preservation, we have $dI(p, l_p)\, dA_p\, dt = dE(p, l_p)$. Then the radiance $\mathcal{R}$ at a point $q$ that illuminates the surface $dA_p$ along the direction $l_p$ with a solid angle $d\omega$ and the irradiance $dI$ measured at the same surface $dA_p$ received from this direction are related by

$$dI(p, l_p) = \mathcal{R}(q, l_p) \cos(\theta)\, d\omega, \tag{3.29}$$

where $d\omega = \frac{\cos(\theta_q)}{d(p,q)^2} dA_q$ is the solid angle of $dA_q$ seen from $p$.

*Bidirectional reflectance distribution function*

For many common materials, the portion of energy coming from a direction $l_p$ that is reflected onto a direction $\boldsymbol{x}_p$ (i.e. the direction of the vantage point) by the surface $S$, is described by $\beta(\boldsymbol{x}_p, l_p)$, the ***bidirectional reflectance distribution function*** (BRDF). Here both $\boldsymbol{x}_p$ and $l_p$ are vectors expressed in local coordinates at $p$. More precisely, if $d\mathcal{R}(p, \boldsymbol{x}_p, l_p)$ is the radiance emitted in the direction $\boldsymbol{x}_p$ from the irradiance $dI(p, l_p)$, the BRDF is given by the ratio

$$\beta(\boldsymbol{x}_p, l_p) \doteq \frac{d\mathcal{R}(p, \boldsymbol{x}_p, l_p)}{dI(p, l_p)} = \frac{d\mathcal{R}(p, \boldsymbol{x}_p, l_p)}{\mathcal{R}(q, l_p) \cos(\theta)\, d\omega}. \tag{3.30}$$

To obtain the total radiance at a point $p$ in the outgoing direction $\boldsymbol{x}_p$, we need to integrate the BRDF against all the incoming irradiance directions $l_p$ in the hemisphere $\Omega$ at $p$:

$$\mathcal{R}(p, \boldsymbol{x}_p) = \int_\Omega d\mathcal{R}(p, \boldsymbol{x}_p, l_p) = \int_\Omega \beta(\boldsymbol{x}_p, l_p)\, \mathcal{R}(q, l_p) \cos(\theta)\, d\omega. \tag{3.31}$$

*Lambertian surfaces*

The above model can be considerably simplified if we restrict our attention to a class of materials, called *Lambertian*, that do not change appearance depending on the viewing direction. For example, matte surfaces are to a large extent well approximated by the Lambertian model, since they diffuse light almost uniformly in all directions. Metal, mirrors, and other shiny surfaces, however, do not. Figure 3.14 illustrates a few common surface properties.

For a perfect Lambertian surface, its radiance $\mathcal{R}(p, \boldsymbol{x}_p)$ only depends on how the surface faces the light source, but not on the direction $\boldsymbol{x}_p$ from which it is viewed. Therefore, $\beta(\boldsymbol{x}_p, l_p)$ is actually independent of $\boldsymbol{x}_p$, and we can think of the radiance function as being "glued," or "painted" on the surface $S$, so that at each point $p$ the radiance $\mathcal{R}$ depends only on the surface. Hence, the perceived irradiance will depend only on which point on the surface is seen, not on in which direction it is seen. More precisely, for Lambertian surfaces, we have

$$\beta(\boldsymbol{x}_p, l_p) = \rho(p),$$

where $\rho(p) : \mathbb{R}^3 \mapsto \mathbb{R}_+$ is a scalar function. In this case, we can easily compute the *surface albedo* $\rho_a$, which is the percentage of incident irradiance reflected in

Figure 3.14. This figure demonstrates different surface properties widely used in computer graphics to model surfaces of natural objects: Lambertian, diffuse, reflective, specular (highlight), transparent with refraction, and textured. Only the (wood textured) pyramid exhibits Lambertian reflection. The ball on the right is partly ambient, diffuse, reflective and specular. The checkerboard floor is partly ambient, diffuse and reflective. The glass ball on the left is both reflective and refractive.

any direction, as

$$\begin{aligned}\rho_a(p) &= \int_\Omega \beta(\boldsymbol{x}_p, l_p)\cos(\theta_p)\, d\omega_p = \rho(p)\int_0^{2\pi}\int_0^{\frac{\pi}{2}} \cos(\theta_p)\sin(\theta_p)\, d\theta_p\, d\phi_p \\ &= \pi\rho(p),\end{aligned}$$

where $d\omega_p$, as shown in Figure 3.13, is the infinitesimal solid angle in the outgoing direction, which can be parameterized by the space angles $(\theta_p, \phi_p)$ as $d\omega_p = \sin(\theta_p)d\theta_p d\phi_p$. Hence the radiance from the point $p$ on a Lambertian surface $S$ is

$$\mathcal{R}(p) = \int_\Omega \frac{1}{\pi}\rho_a(p)\, \mathcal{R}(q, l_p)\cos(\theta)\, d\omega. \tag{3.32}$$

This equation is known as *Lambertian cosine law*. Therefore, for a Lambertian surface, the radiance $\mathcal{R}$ depends only on the surface $S$, described by its generic point $p$, and on the light source $L$, described by its radiance $\mathcal{R}(q, l_p)$.

*Image intensity for a Lambertian surface*

In order to express the direction $\boldsymbol{x}_p$ in the camera frame, we consider the change of coordinates from the local coordinate frame at the point $p$ to the camera frame: $\boldsymbol{X}(p) \doteq g(0)$ and $\boldsymbol{x} \sim g_*(\boldsymbol{x}_p)$, where we note that $g_*$ is a rotation.[9] The reader should be aware that the transformation $g$ itself depends on local shape of the

[9]The symbol $\sim$ indicates equivalence up to a scalar factor. Strictly speaking, $\boldsymbol{x}$ and $g_*(\boldsymbol{x}_p)$ do not represent the same vector, but only the same direction (they have opposite sign and different lengths). To obtain a rigorous expression, we would have to write $\boldsymbol{x} = \pi(-g_*(\boldsymbol{x}_p))$. However, these

surface at $p$, in particular its tangent plane $T_pS$ and its normal $\nu_p$ at the point $p$. We now can rewrite the expression (3.31) for the radiance in terms of the camera coordinates and obtain

$$\mathcal{R}(\boldsymbol{X}) \doteq \mathcal{R}(p, g_*^{-1}(\boldsymbol{x})), \quad \text{where} \quad \boldsymbol{x} = \pi(\boldsymbol{X}). \tag{3.33}$$

If the surface is Lambertian, the above expression simplifies to

$$\mathcal{R}(\boldsymbol{X}) = \mathcal{R}(p). \tag{3.34}$$

Suppose that our imaging sensor is well modeled by a thin lens. Then, by measuring the amount of energy received along the direction $\boldsymbol{x}$, the irradiance (or image intensity) $I$ at $\boldsymbol{x}$ can be expressed as a function of the radiance from the point $p$:

$$I(\boldsymbol{x}) = \mathcal{R}(\boldsymbol{X}) \frac{\pi}{4} \left(\frac{d}{f}\right)^2 \cos^4(\alpha), \tag{3.35}$$

where $d$ is the lens diameter, $f$ is the focal length, and $\alpha$ is the angle between the optical axis (i.e. the $z$-axis) and the image point $\boldsymbol{x}$, as shown in Figure 3.13. The quantity $\frac{d}{f}$ is called the *F-number* of the lens. A detailed derivation of the above formula can be found in [Horn, 1986] (page 208). For a Lambertian surface, we have

$$\begin{aligned} I(\boldsymbol{x}) &= \mathcal{R}(\boldsymbol{X}) \frac{\pi}{4} \left(\frac{d}{f}\right)^2 \cos^4(\alpha) = \mathcal{R}(p) \frac{\pi}{4} \left(\frac{d}{f}\right)^2 \cos^4(\alpha) \\ &= \frac{1}{4} \left(\frac{d}{f}\right)^2 \cos^4(\alpha) \int_\Omega \rho_a(p)\, \mathcal{R}(q, l_p) \cos(\theta)\, d\omega, \end{aligned}$$

where $\boldsymbol{x}$ is the image of the point $p$ taken at the vantage point $g$. Notice that in the above expression, only the angle $\alpha$ depends on the vantage point. In general, for a thin lens with a small field of view, $\alpha$ is approximately constant. Therefore, in our ideal pin-hole model, we may assume that the image intensity (i.e. irradiance) is related to the surface radiance by the *irradiance equation*:

$$\boxed{I(\boldsymbol{x}) = \gamma \mathcal{R}(p),} \tag{3.36}$$

where $\gamma \doteq \frac{\pi}{4}\left(\frac{d}{f}\right)^2 \cos^4(\alpha)$ is a constant factor that is independent of the vantage point.

In all subsequent chapters we will adopt this simple model. The fact that the irradiance $I$ does not change with the vantage point for Lambertian surfaces constitutes a fundamental condition that allows to establish correspondence across multiple images of the same object. This condition and its implications will be studied in more detail in the next chapter.

---

two vectors do represent the same ray through the camera center, and therefore we will regard them as the same.

## 3.B Image formation in the language of projective geometry

The perspective pinhole camera model described by (3.18) or (3.19) has retained the physical meaning of all parameters involved. In particular, the last entry of both $\boldsymbol{x}'$ and $\boldsymbol{X}$ is normalized to 1 so that the other entries may correspond to actual 2-D and 3-D coordinates (with respect to the metric unit chosen for respective coordinate frames). However, such a normalization is not always necessary as long as we know that it is the direction of those homogeneous vectors that matters. For instance, the two vectors

$$[X, Y, Z, 1]^T, \quad [XW, YW, ZW, W]^T \quad \in \mathbb{R}^4 \tag{3.37}$$

can be used to represent the same point in $\mathbb{R}^3$. Similarly, we can use $[x', y', z']^T$ to represent a point $[x, y, 1]^T$ on the 2-D image plane as long as $x'/z' = x$ and $y'/z' = y$. However, we may run into trouble if the last entry $W$ or $z'$ happens to be 0. To resolve this problem, we need to generalize the interpretation of homogeneous coordinates introduced in the previous chapter.

**Definition 3.4 (Projective space and its homogeneous coordinates).** *An n-dimensional* projective space $\mathbb{P}^n$ *is the set of one-dimensional subspaces (i.e. lines through the origin) of the vector space* $\mathbb{R}^{n+1}$. *A point p in* $\mathbb{P}^n$ *can then be assigned* homogeneous coordinates $\boldsymbol{X} = [x_1, x_2, \ldots, x_{n+1}]^T$ *among which at least one* $x_i$ *is nonzero. For any nonzero* $\lambda \in \mathbb{R}$ *the coordinates* $\boldsymbol{Y} = [\lambda x_1, \lambda x_2, \ldots, \lambda x_{n+1}]^T$ *represent the same point p in* $\mathbb{P}^n$. *We say that* $\boldsymbol{X}$ *and* $\boldsymbol{Y}$ *are equivalent, denoted by* $\boldsymbol{X} \sim \boldsymbol{Y}$.

**Example 3.5 (Topological models for the projective space $\mathbb{P}^2$).** Figure 3.15 demonstrates two equivalent geometric interpretations of the 2-D projective space $\mathbb{P}^2$. According

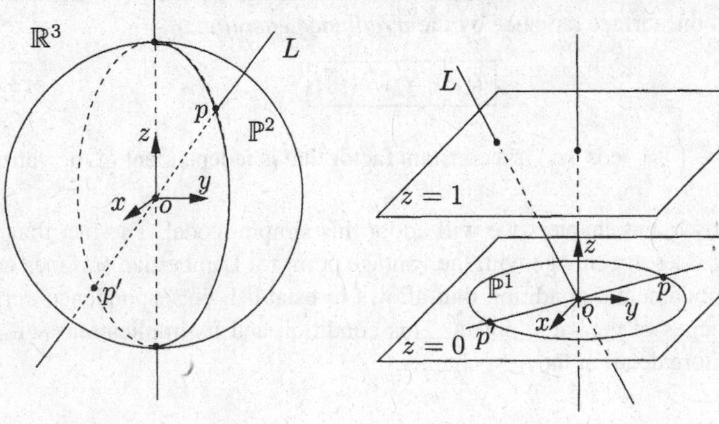


Figure 3.15. Topological models for $\mathbb{P}^2$.

to the definition, it is simply a family of 1-D lines $\{L\}$ in $\mathbb{R}^3$ through a point $o$ (typically chosen to be the origin of the coordinate frame). Hence, $\mathbb{P}^2$ can be viewed as a 2-D sphere $\mathbb{S}^2$ with any pair of antipodal points (e.g., $p$ and $p'$ in the figure) identified as one point in $\mathbb{P}^2$. On the right-hand side of Figure 3.15, lines through the center $o$ in general intersect with the plane $\{z = 1\}$ at a unique point except when they lie on the plane $\{z = 0\}$. Lines in the plane $\{z = 0\}$ simply form the 1-D projective space $\mathbb{P}^1$ (which is in fact a circle). Hence, $\mathbb{P}^2$ can be viewed as a 2-D plane $\mathbb{R}^2$ (i.e. $\{z = 1\}$) with a circle $\mathbb{P}^1$ attached. If we adopt the view that lines in the plane $\{z = 0\}$ intersect the plane $\{z = 1\}$ infinitely far, this circle $\mathbb{P}^1$ then represents a *line at infinity*. Homogeneous coordinates for a point on this circle then take the form $[x, y, 0]^T$; on the other hand, all regular points in $\mathbb{R}^2$ have coordinates $[x, y, 1]^T$. In general, any projective space $\mathbb{P}^n$ can be visualized in a similar way: $\mathbb{P}^3$ is then $\mathbb{R}^3$ with a plane $\mathbb{P}^2$ attached at infinity; and $\mathbb{P}^n$ is $\mathbb{R}^n$ with $\mathbb{P}^{n-1}$ attached at infinity, which is, however, harder to illustrate on a piece of paper. ■

Using this definition, $\mathbb{R}^n$ with its homogeneous representation can then be identified as a subset of $\mathbb{P}^n$ that includes exactly those points with coordinates $\boldsymbol{X} = [x_1, x_2, \ldots, x_{n+1}]^T$ where $x_{n+1} \neq 0$. Therefore, we can always normalize the last entry to 1 by dividing $\boldsymbol{X}$ by $x_{n+1}$ if we so wish. Then, in the pinhole camera model described by (3.18) or (3.19), $\lambda \boldsymbol{x}'$ and $\boldsymbol{x}'$ now represent the same projective point in $\mathbb{P}^2$ and therefore the same 2-D point in the image plane. Suppose that the projection matrix is

$$\Pi = K\Pi_0 g = [KR, KT] \quad \in \mathbb{R}^{3\times 4}. \tag{3.38}$$

Then the camera model simply reduces to a projection from a three-dimensional projective space $\mathbb{P}^3$ to a two-dimensional projective space $\mathbb{P}^2$,

$$\pi : \mathbb{P}^3 \to \mathbb{P}^2; \quad \boldsymbol{X}_0 \mapsto \boldsymbol{x}' \sim \Pi \boldsymbol{X}_0, \tag{3.39}$$

where $\lambda$ is omitted here, since the equivalence "$\sim$" is defined in the homogeneous sense, i.e. up to a nonzero scalar factor.

Intuitively, the remaining points in $\mathbb{P}^3$ with the fourth coordinate $x_4 = 0$ can be interpreted as points that are "infinitely far away from the origin." This is because for a very small value $\epsilon$, if we normalize the last entry of $\boldsymbol{X} = [X, Y, Z, \epsilon]^T$ to 1, it gives rise to a point in $\mathbb{R}^3$ with 3-D coordinates $\boldsymbol{X} = [X/\epsilon, Y/\epsilon, Z/\epsilon]^T$. The smaller $|\epsilon|$ is, the farther away is the point from the origin. In fact, all points with coordinates $[X, Y, Z, 0]^T$ form a two-dimensional plane described by the equation $[0, 0, 0, 1]^T \boldsymbol{X} = 0$.[10] This plane is called *plane at infinity*. We usually denote this plane by $P_\infty$. That is,

$$P_\infty \doteq \mathbb{P}^3 \setminus \mathbb{R}^3 \ (= \mathbb{P}^2).$$

Then the above imaging model (3.39) is well-defined on the entire projective space $\mathbb{P}^3$ including points in this plane at infinity. This slight generalization allows us to talk about images of points that are infinitely far away from the camera.

---

[10]It is two-dimensional because $X, Y, Z$ are not totally free: the coordinates are determined only up to a scalar factor.

**Example 3.6 (Image of points at infinity and "vanishing points").** Two parallel lines in $\mathbb{R}^3$ do not intersect. However, we can view them as intersecting at infinity. Let $V = [V_1, V_2, V_3, 0]^T \in \mathbb{R}^4$ be a (homogeneous) vector indicating the direction of two parallel lines $L^1$, $L^2$. Let $X_o^1 = [X_o^1, Y_o^1, Z_o^1, 1]^T$ and $X_o^2 = [X_o^2, Y_o^2, Z_o^2, 1]^T$ be two base points on the two lines, respectively. Then (homogeneous) coordinates of points on $L^1$ can be expressed as

$$X^1 = X_o^1 + \mu V, \quad \mu \in \mathbb{R},$$

and similarly for points on $L^2$. Then the two lines can be viewed as intersecting at a point at infinity with coordinates $V$. The "image" of this intersection, traditionally called a *vanishing point*, is simply given by

$$x' \sim \Pi V.$$

This can be shown by considering images of points on the lines and letting $\mu \to \infty$ asymptotically. If the images of these two lines are given, the image of this intersection can be easily computed or measured. Figure 3.16 shows the intersection of images of parallel lines at the vanishing point, a concept well known to Renaissance artists. ■

Figure 3.16. "The School of Athens" by Raphael (1518), a fine example of architectural perspective with a central vanishing point, marking the end of the classical Renaissance (courtesy of C. Taylor).

**Example 3.7 (Image "outside" the image plane).** Consider the standard perspective projection of a pair of parallel lines as in the previous example. We further assume that they are also parallel to the image plane, i.e. the $xy$-plane. In this case, we have

$$\Pi = \Pi_0 = [I, 0] \quad \text{and} \quad V = [V_1, V_2, 0, 0]^T.$$

Hence, the "image" of the intersection is given in homogeneous coordinates as

$$x' = [V_1, V_2, 0]^T.$$

This does not correspond to any physical point on the 2-D image plane (whose points supposedly have homogeneous coordinates of the form $[x, y, 1]^T$). It is, in fact, a vanishing point at infinity. Nevertheless, we can still treat it as a valid image point. One way is to view it as the image of a point with zero depth (i.e. with the $z$-coordinate zero). Such a problem will automatically go away if we choose the imaging surface to be an entire sphere rather than a flat plane. This is illustrated in Figure 3.17. ■

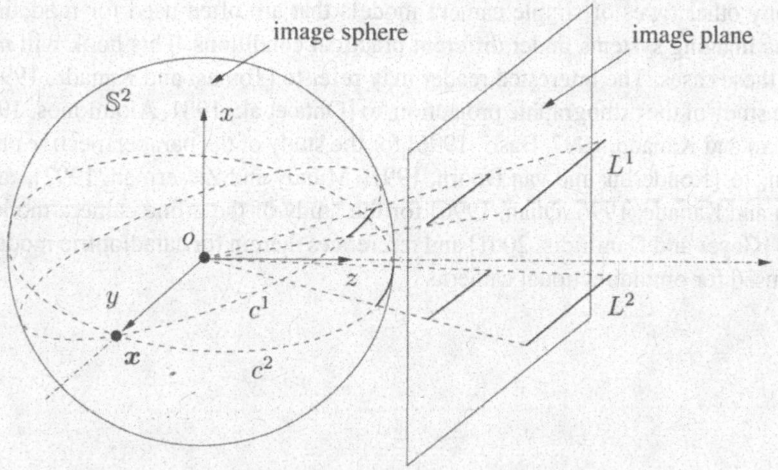


Figure 3.17. Perspective images of two parallel lines that are also parallel to the 2-D image plane. In this case they are parallel to the $y$-axis. The two image lines on the image plane are also parallel, and hence they do not intersect. On an image sphere, however, the two image circles $c^1$ and $c^2$ do intersect at the point $\boldsymbol{x}$. Clearly, $\boldsymbol{x}$ is the direction of the two image lines.

# Further readings

*Deviations from the pinhole model*

As we mentioned earlier in this chapter, the analytical study of pinhole perspective imaging dates back to the Renaissance. Nevertheless, the pinhole perspective model is a rather ideal approximation to actual CCD photosensors or film-based cameras. Before the pinhole model can be applied to such cameras, a correction is typically needed to convert them to an exact perspective device; see [Brank et al., 1993] and references therein.

In general, the pinhole perspective model is not adequate for modeling complex optical systems that involve a zoom lens or multiple lenses. For a systematic introduction to photographic optics and lens systems, we recommend the classic books [Stroebel, 1999, Born and Wolf, 1999]. For a more detailed account of models for a zoom lens, the reader may refer to [Horn, 1986, Lavest et al., 1993] and references therein. Other approaches such as using a two-plane model [Wei and Ma, 1991] have also been proposed to overcome the limitations of the pinhole model.

*Other simple camera models*

In the computer vision literature, besides the pinhole perspective model, there exist many other types of simple camera models that are often used for modeling various imaging systems under different practical conditions. This book will *not* cover these cases. The interested reader may refer to [Tomasi and Kanade, 1992] for the study of the orthographic projection, to [Ohta et al., 1981, Aloimonos, 1990, Poelman and Kanade, 1997, Basri, 1996] for the study of the paraperspective projection, to [Konderink and van Doorn, 1991, Mundy and Zisserman, 1992], and [Quan and Kanade, 1996, Quan, 1996] for the study of the affine camera model, and to [Geyer and Daniilidis, 2001] and references therein for catadioptric models often used for omnidirectional cameras.

# Chapter 4
# Image Primitives and Correspondence

*Everything should be made as simple as possible, but not simpler.*
– Albert Einstein

In previous chapters we have seen how geometric primitives, such as points and lines in space, can be transformed so that one can compute the coordinates of their "image," i.e. their projection onto the image plane. In practice, however, images are arrays of positive numbers that measure the amount of light incident on a sensor at a particular location (see Sections 3.1 and 3.2, and Appendix 3.A). So, how do we reconcile a geometric image formation model (Section 3.3) with the fact that what we measure with a camera is not points and lines, but light intensity? In other words, how do we go from measurements of light (photometry) to geometry? This is the subject of this chapter: we will show how geometric primitives can be extracted from photometric measurements and matched across different views, so that the rest of the book can concentrate on geometry.

The reader should be aware that although extracting geometric primitives at the outset is widely accepted and practiced, this approach has limitations. For one, in the process we throw away almost all the information (geometric primitives are a set of measure zero in the image). Moreover, as we will see, geometric primitives are extracted and matched by *local* analysis of the image, and are therefore prone to ambiguities and false matches. Nevertheless, *global* analysis of images to infer scene photometry as well as geometry would be computationally challenging, and it is not even clear that it would be meaningful. In fact, if we consider an object with arbitrary geometry and arbitrary photometry, one can always construct (infinitely many) objects with different geometry and different photometry that give rise to the same images. One example is the image itself: It is an object different from the true scene (it is flat) that gives rise to the same image (itself).

Therefore, in what follows we will rely on *assumptions* on the photometry of the scene in order to be able to establish *correspondence* between geometric primitives in different views. Such assumptions will allow us to use measurements of light in order to discern how points and lines are "moving" in the image. Under such assumptions, the "image motion" is related to the three-dimensional structure of the scene and its motion relative to the camera in ways that we will exploit in later chapters in order to reconstruct the geometry of the scene.

## 4.1 Correspondence of geometric features

Suppose we have available two images of a scene taken from different vantage points, for instance those in Figure 4.1. Consider the coordinates of a specific point in the left image, for instance the one indicated by a white square. It is immediate for a human observer to establish what the "corresponding" point on the right image is. The two points correspond in the sense that, presumably, they are the projection of the same point in space. Naturally, we cannot expect the pixel coordinates of the point on the left to be identical to those of the point on the right. Therefore, *the "correspondence problem" consists in establishing which point in one image corresponds to which point in another, in the sense of being the image of the same point in space.*

Figure 4.1. "Corresponding points" in two views are projections of the same point in space.

The fact that humans solve the correspondence problem so effortlessly should not lead us to think that this problem is trivial. On the contrary, humans exploit a remarkable amount of information in order to arrive at successfully declaring correspondence, including analyzing context and neighboring structures in the image and prior information on the content of the scene. If we were asked to establish correspondence by just looking at the small regions of the image enclosed in the circle and square on the left, things would get much harder: which of the regions in Figure 4.2 is the right match? Hard to tell. The task is no easier for a computer.

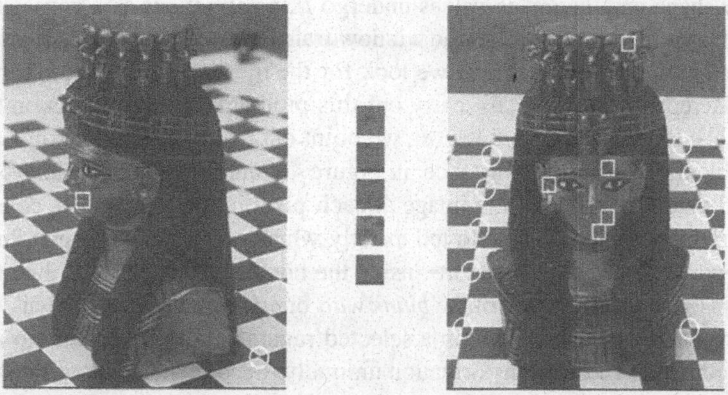

Figure 4.2. Which of these circular or square regions on the right match the ones on the left? Correspondence based on local photometric information is prone to ambiguity. The image on the right shows the corresponding positions on the image. Note that some points do not have a correspondent at all, for instance due to occlusions.

### 4.1.1 *From photometric features to geometric primitives*

Let us begin with a naive experiment. Suppose we want to establish correspondence for a pixel in position $\boldsymbol{x}_1$ in the left image in Figure 4.1. The value of the image at $\boldsymbol{x}_1$ is $I_1(\boldsymbol{x}_1)$, so we may be tempted to look for a position $\boldsymbol{x}_2$ in the right image that has the same brightness, $I_1(\boldsymbol{x}_1) = I_2(\boldsymbol{x}_2)$, which can be thought of as a "label" or "signature." Based on the discussion above, it should be obvious to the reader that this approach is doomed to failure. First, there are 307,200 pixel locations in the right image (640 × 480), each taking a value between 0 and 255 (three for red, green and blue if in color). Therefore, we can expect to find many pixel locations in $I_2$ matching the value $I_1(\boldsymbol{x}_1)$. Moreover, the actual corresponding point may not even be one of them, since measuring light intensity consists in counting photons, a process intrinsically subject to uncertainty. One way to fix this is to compare not the brightness of individual pixels, but the brightness of each pixel in a small window around the point of interest (see Figure 4.2). We can think of this as attaching to each pixel, instead of a scalar label $I(\boldsymbol{x})$ denoting the brightness of that pixel, an augmented vector label that contains the brightness of each pixel in the window: $l(\boldsymbol{x}) = \{I(\tilde{\boldsymbol{x}}) \mid \tilde{\boldsymbol{x}} \in W(\boldsymbol{x})\}$, where $W(\boldsymbol{x})$ is a window around $\boldsymbol{x}$. Now matching points is carried out by matching windows, under the assumption that each point in the window moves with the same motion (Figure 4.3). Again, due to noise we cannot expect an exact matching of labels, so we can look for the windows that minimize some discrepancy measure between their labels.

This discussion can be generalized: each point has associated with itself a support window and the value of the image at each point in the window. Both the window shape and the image values undergo *transformations* as a consequence of the change in viewpoint (e.g., the window translates, and the image intensity is corrupted by additive noise), and we look for the transformation that minimizes some discrepancy measure. We carry out this program in the next section (Section 4.1.2). Before doing so, however, we point out that this does not solve all of our problems. Consider, for instance, in Figure 4.2 the rectangular regions on the checkerboard. The value of the image at each pixel in these regions is constant, and therefore it is not possible to tell exactly which one is the corresponding region; it could be any region that fits inside the homogeneous patch of the image. This is just one manifestation of the *blank wall* or *aperture problem*, which occurs when the brightness profile within a selected region is not rich enough to allow us to recover the chosen transformation uniquely (Section 4.3.1). It will be wise, then, to restrict our attention only to those regions for which the correspondence problem can be solved. Those will be called "features," and they establish the link between photometric measurements and geometric primitives.

## 4.1.2 *Local vs. global image deformations*

In the discussion above, one can interpret matching windows, rather than points, as the local integration of intensity information, which is known to have beneficial (averaging) effects in counteracting the effects of noise. Why not, then, take this to the extreme, integrate intensity information over the entire image? After all, Chapters 2 and 3 tell us precisely how to compute the coordinates of corresponding points. Of course, the deformation undergone by the entire image cannot be captured by a simple displacement, as we will soon see. Therefore, one can envision two opposite strategies: one is to choose a complex transformation that captures the changes undergone by the entire image, or one can pick a simple transformation, and then restrict the attention to only those regions in the image whose motion can be captured, within reasonable bounds, by the chosen transformation.

As we have seen in Chapter 3, an image, for instance $I_1$, can be represented as a function defined on a compact two-dimensional region $\Omega$ taking irradiance values in the positive reals,

$$I_1 : \Omega \subset \mathbb{R}^2 \rightarrow \mathbb{R}_+; \quad \boldsymbol{x} \mapsto I_1(\boldsymbol{x}).$$

Under the simplifying assumptions in Appendix 3.A, the irradiance $I_1(\boldsymbol{x})$ is obtained by integrating the radiant energy in space along the ray $\{\lambda \boldsymbol{x},\ \lambda \in \mathbb{R}_+\}$.[1]

[1]We remind the reader that we do not differentiate in our notation an image point $\boldsymbol{x}$ from its homogeneous representation (with a "1" appended).

If the scene contains only opaque objects, then only one point, say $p$, along the projection ray contributes to the irradiance. With respect to the camera reference frame, let this point have coordinates $\boldsymbol{X} \in \mathbb{R}^3$, corresponding to a particular value of $\lambda$ determined by the first intersection of the ray with a visible surface: $\lambda \boldsymbol{x} = \boldsymbol{X}$. Let $\mathcal{R} : \mathbb{R}^3 \to \mathbb{R}_+$ be the radiance distribution of the visible surface and its value $\mathcal{R}(p)$ at $p$, i.e. the "color" of the scene at the point $p$.[2] According to Appendix 3.A, we have the *irradiance equation*

$$I_1(\boldsymbol{x}) \sim \mathcal{R}(p). \tag{4.1}$$

Now suppose that a different image of the same scene becomes available, $I_2$, for instance one taken from a different vantage point. Naturally, we get another function

$$I_2 : \Omega \subset \mathbb{R}^2 \to \mathbb{R}_+; \quad \boldsymbol{x} \mapsto I_2(\boldsymbol{x}).$$

However, $I_2(\boldsymbol{x})$ will in general be different from $I_1(\boldsymbol{x})$ at the same image location $\boldsymbol{x}$. The first step in establishing correspondence is to understand how such a difference occurs.

Let us now use the background developed in Chapters 2 and 3. Assume for now that we are imaging empty space except for a point $p$ with coordinates $\boldsymbol{X} \in \mathbb{R}^3$ that emits light with the same energy in all directions (i.e. the visible "surface," a point, is Lambertian, see Appendix 3.A). This is a simplifying assumption that we will relax in the next section. If $I_1$ and $I_2$ are images of the same scene, they must satisfy the same irradiance equation (4.1). Therefore, if $\boldsymbol{x}_1$ and $\boldsymbol{x}_2$ are the two images of the same point $p$ in the two views, respectively, we must have

$$I_2(\boldsymbol{x}_2) = I_1(\boldsymbol{x}_1) \sim \mathcal{R}(p). \tag{4.2}$$

Under these admittedly restrictive assumptions, the correspondence (or matching) problem consists in establishing the relationship between $\boldsymbol{x}_1$ and $\boldsymbol{x}_2$, i.e. verifying that the two points $\boldsymbol{x}_1$ and $\boldsymbol{x}_2$ are indeed images of the same 3-D point. Suppose that the displacement between the two camera viewpoints is a rigid-body motion $(R, T)$. From the projection equations introduced in Chapter 3, the point $\boldsymbol{x}_1$ in image $I_1$ corresponds to the point $\boldsymbol{x}_2$ in image $I_2$ if

$$\boldsymbol{x}_2 = h(\boldsymbol{x}_1) = \frac{1}{\lambda_2(\boldsymbol{X})}(R\lambda_1(\boldsymbol{X})\boldsymbol{x}_1 + T), \tag{4.3}$$

where we have emphasized the fact that the scales $\lambda_i$, $i = 1, 2$, depend on the 3-D coordinates $\boldsymbol{X}$ of the point with respect to the camera frame at their respective viewpoints. Therefore, a model for the deformation between two images of the same scene is given by an image matching constraint

$$I_1(\boldsymbol{x}_1) = I_2(h(\boldsymbol{x}_1)), \quad \forall\, \boldsymbol{x}_1 \in \Omega \cap h^{-1}(\Omega) \subset \mathbb{R}^{2\times 2}. \tag{4.4}$$

[2] In the case of gray-scale images, "color" is often used inappropriately to denote intensity.

This equation is sometime called the *brightness constancy constraint*, since it expresses the fact that given a point on an image, there exists a different (transformed) point in another image that has the same brightness.

The function $h$ describes the transformation of the domain, or "image motion," that we have described informally in the beginning of this chapter. In order to make it more suggestive of the motion of individual pixels, we could write $h$ as

$$h(\boldsymbol{x}) = \boldsymbol{x} + \Delta\boldsymbol{x}(\boldsymbol{X}), \tag{4.5}$$

where the fact that $h$ depends on the shape of the scene is made explicitly in the term $\Delta\boldsymbol{x}(\boldsymbol{X})$. Intuitively, $\Delta\boldsymbol{x}(\boldsymbol{X})$ is the displacement of the image of the same point from one view to another: $\Delta\boldsymbol{x} = \boldsymbol{x}_2 - \boldsymbol{x}_1$.[3] Note that the dependency of $h(\boldsymbol{x})$ on the position of the point $\boldsymbol{X}$ comes through the scales $\lambda_1, \lambda_2$, i.e. the depth of visible surfaces. In general, therefore, $h$ is a function in an infinite-dimensional space (the space of all surfaces in 3-D), and solving for image correspondence is as difficult as estimating the shape of visible objects.

If the scene is not Lambertian,[4] we cannot count on equation (4.4) being satisfied at all. Therefore, as we suggested in the beginning of this subsection, modeling the transformation undergone by the entire image is an extremely hard proposition.

## 4.2 Local deformation models

The problem with a global model, as described in the previous section, is that the transformation undergone by the entire image is, in general, infinite-dimensional, and finding it amounts to inferring the entire 3-D structure of the scene. Therefore, in what follows we concentrate on choosing a class of simple transformations, and then restrict our attention to "regions" of the image that can be modeled as undergoing such transformations. Such transformations occur in the domain of the image, say a window $W(\boldsymbol{x})$ around $\boldsymbol{x}$, and in the intensity values $I(\tilde{\boldsymbol{x}})$, $\tilde{\boldsymbol{x}} \in W(\boldsymbol{x})$. We examine these two instances in the next two subsections.

### 4.2.1 *Transformations of the image domain*

Here we consider three cases of increasingly rich local deformation models, starting from the simplest.

*Translational motion model*

The simplest transformation one can conceive of is one in which each point in the window undergoes the same exact motion, i.e. $\Delta\boldsymbol{x} = \text{constant}$, no longer

[3] A precise analytical expression for $\Delta\boldsymbol{x}$ will be given in the next chapter.

[4] As we explain in Appendix 3.A, a Lambertian surface is one whose appearance does not depend on the viewing direction. In other words, the radiance of the surface at any given point is the same in all directions.

Figure 4.3. Two basic types of local domain $W(\boldsymbol{x})$ deformation. Left: translational; right: affine.

depending on $\boldsymbol{X}$, or $h(\tilde{\boldsymbol{x}}) = \tilde{\boldsymbol{x}} + \Delta\boldsymbol{x}$, $\forall\, \tilde{\boldsymbol{x}} \in W(\boldsymbol{x})$, where $\Delta\boldsymbol{x} \in \mathbb{R}^2$. This model is valid only for portions of the scene that are flat and parallel to the image plane, and moving parallel to it. While one could in principle approximate smooth portions of the scene as a collection of such planar patches, their motion will in general not satisfy the model. The model is therefore only a crude approximation, valid locally in space (small windows) and in time (adjacent time instants, or small camera motion).

Although coarse, this model is at the core of most feature matching or tracking algorithms due to its simplicity and the efficiency of the resulting implementation, which we present in Section 4.3.1.

*Affine motion model*

In the affine motion model, points in the window $W(\boldsymbol{x})$ do not undergo the same motion but, instead, the motion of each point depends *linearly* on its location plus a constant offset. More precisely, we have $h(\tilde{\boldsymbol{x}}) = A\tilde{\boldsymbol{x}} + \boldsymbol{d}$, $\forall\, \tilde{\boldsymbol{x}} \in W(\boldsymbol{x})$, where $A \in \mathbb{R}^{2\times 2}$ and $\boldsymbol{d} \in \mathbb{R}^2$. This model is a good approximation for small planar patches parallel to the image plane undergoing an arbitrary translation and rotation about the optical axis, and modest rotation about an axis orthogonal to it. This model represents a convenient tradeoff between simplicity and flexibility, as we will see in Section 4.3.2. The affine and translational models are illustrated in Figure 4.3.

*Projective motion model*

An additional generalization of the affine model occurs when we consider transformations that are linear in the homogeneous coordinates, so that $h(\tilde{\boldsymbol{x}}) \sim H\tilde{\boldsymbol{x}}$, $\forall\, \tilde{\boldsymbol{x}} \in W(\boldsymbol{x})$, where $H \in \mathbb{R}^{3\times 3}$ is defined up to a scalar factor. This model, as we will see in Section 5.3, captures an arbitrary rigid-body motion of a planar patch in the scene. Since any smooth surface can be approximated arbitrarily well by a collection of planes, this model is appropriate everywhere in the image, except at discontinuities and occluding boundaries.

Whatever the transformation $h$ one chooses, in order to establish correspondence, it seems that one has to find $h$ that solves equation (4.4). It turns out that the equality (4.4) is way too much to ask, as we describe in the next section. Therefore, in Section 4.3 we will describe ways to rephrase matching as an optimization problem, which lends itself to the derivation of effective algorithms.

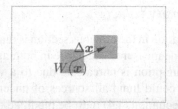

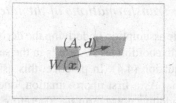

### 4.2.2 *Transformations of the intensity value*

The basic assumption underlying the derivation in the previous section is that each point with coordinates $\boldsymbol{X}$ results in the same measured irradiance in both images, as in equation (4.4). In practice, this assumption is unrealistic due to a variety of factors. As a first approximation, one could lump all sources of uncertainty into an additive noise term $n$.[5] Therefore, equation (4.4) must be modified to take into account changes in the intensity value, in addition to the deformation of the domain

$$I_1(\boldsymbol{x}_1) = I_2(h(\boldsymbol{x}_1)) + n(h(\boldsymbol{x}_1)). \tag{4.6}$$

More fundamental departures from the model (4.4) occur when one considers that points that are visible from one vantage point may become occluded from another. *Occlusions* could be represented by a factor multiplying $I_2$ that depends upon the shape of the surface ($\boldsymbol{X}$) being imaged and the viewpoint ($g$): $I_1(\boldsymbol{x}_1) = f_o(\boldsymbol{X}, g)I_2(h(\boldsymbol{x}_1)) + n(h(\boldsymbol{x}_1))$. For instance, for the case where only one point on the surface is emitting light, $f_o(\boldsymbol{X}, g) = 1$ when the point is visible, and $f_o(\boldsymbol{X}, g) = 0$ when not. This equation should make very clear the fact that associating the label $I_1$ with the point $\boldsymbol{x}_1$ is not a good idea, since the value of $I_1$ depends upon the noise $n$ and the shape of the surfaces in space $\boldsymbol{X}$, which we cannot control.

There is more: in most natural scenes, objects do not emit light of their own; rather, they reflect ambient light in a way that depends upon the properties of the material. Even in the absence of occlusions, different materials may scatter or reflect light by different amounts in different directions, violating the Lambertian assumption discussed in Appendix 3.A. In general, few materials exhibit perfect Lambertian reflection, and far more complex reflection models, such as translucent or anisotropic materials, are commonplace in natural and man-made scenes (see Figure 3.14).

## 4.3 Matching point features

From the discussion above one can conclude that point correspondence cannot be established for scenes with arbitrary reflectance properties. Even for relatively simple scenes, whose appearance does not depend on the viewpoint, one cannot establish correspondence among points due to the aperture problem. So how do we proceed? As we have hinted several times already, we proceed by *integrating* local photometric information. Instead of considering equation (4.2) in terms of points on an image, we can consider it defining correspondence in terms of *regions*. This can be done by integrating each side on a window $W(\boldsymbol{x})$ around

[5] This noise is often described statistically as a Poisson random variable (in emphasizing the nature of the counting process and enforcing the nonnegativity constraints on irradiance), or as a Gaussian random variable (in emphasizing the concurrence of multiple independent sources of uncertainty).

each point $\boldsymbol{x}$, and using the equation to characterize the correspondence at $\boldsymbol{x}$. Due to the presence of uncertainty, noise, deviations from Lambertian reflection, occlusions, etc. we can expect that equation (4.4) will be satisfied only up to uncertainty, as in (4.6). Therefore, we formulate correspondence as the solution of an optimization problem. We choose a class of transformations, and we look for the particular transformation $\hat{h}$ that minimizes the effects of noise (measured according to some criterion),[6] subject to equation (4.6), integrated over a window. For instance, we could have $\hat{h} = \arg\min_h \sum_{\tilde{\boldsymbol{x}} \in W(\boldsymbol{x})} \|n(\tilde{\boldsymbol{x}})\|^2$ subject to (4.6) or, writing $n$ explicitly,

$$\hat{h} = \arg\min_h \sum_{\tilde{\boldsymbol{x}} \in W(\boldsymbol{x})} \|I_1(\tilde{\boldsymbol{x}}) - I_2(h(\tilde{\boldsymbol{x}}))\|^2 \tag{4.7}$$

if we choose as a discrepancy measure the norm of the additive error. In the next subsections we explore a few choices of discrepancy criteria, which include the sum of squared differences and normalized cross-correlation. Before doing so, however, let us pause for a moment to consider whether the optimization problem just defined is well posed.

Consider equation (4.7), where the point $\boldsymbol{x}$ happens to fall within a region of constant intensity. Then $I_1(\tilde{\boldsymbol{x}}) =$ constant for all $\tilde{\boldsymbol{x}} \in W(\boldsymbol{x})$. The same is true for $I_2$ and therefore, the norm being minimized does not depend on $h$, and any choice of $\hat{h}$ would solve the equation. This is the "blank wall" effect, a manifestation of the so-called aperture problem. Therefore, it appears that in order for the problem (4.7) to be well posed, the intensity values inside a window have to be "rich enough."

Having this important fact in mind, we choose a class of transformations, $h$, that depends on a set of parameters $\alpha$. For instance, $\alpha = \Delta\boldsymbol{x}$ for the translational model, and $\alpha = \{A, \boldsymbol{d}\}$ for the affine motion model. With an abuse of notation we indicate the dependency of $h$ on the parameters as $h(\alpha)$. We can then define a pixel $\boldsymbol{x}$ to be a *point feature* if there exists a neighborhood $W(\boldsymbol{x})$ such that the equations

$$I_1(\tilde{\boldsymbol{x}}) = I_2(h(\tilde{\boldsymbol{x}}, \alpha)), \quad \forall \tilde{\boldsymbol{x}} \in W(\boldsymbol{x}), \tag{4.8}$$

uniquely determine the parameters $\alpha$. From the example of the blank wall, it is intuitive that such conditions would require that $I_1$ and $I_2$ at least have nonzero gradient. In the sections to follow we will derive precisely what the conditions are for the translational model. Similarly, one may define a *line feature* as a line segment with a support region and a collection of labels such that the orientation and normal displacement of the transformed line can be uniquely determined from the equation above.

In the next sections we will see how to efficiently solve the problem above for the case in which the $\alpha$ are translational or affine parameters. We first describe

---

[6] Here the "hat" symbol, $\hat{(\cdot)}$, indicates an estimated quantity (see Appendix B), not to be confused with the "wide hat," $\widehat{(\cdot)}$, used to indicate a skew-symmetric matrix.

how to compute the velocity either of a moving point (feature tracking) or at a fixed location on the pixel grid (optical flow), and then give an effective algorithm to detect point features that can be easily tracked.

The definition of feature points and lines allows us to move our discussion from pixels and images to geometric entities such as points and lines. However, as we will discuss in later chapters, this separation is more conceptual than factual. Indeed, all the constraints among geometric entities that we will derive in chapters of Part II and Part III can be rephrased in terms of constraints on the irradiance values on collections of images, under the assumption of rigidity of Chapter 2 and Lambertian reflection of Chapter 3.

### 4.3.1 *Small baseline: feature tracking and optical flow*

Consider the translational model described in the previous sections, where

$$I_1(\boldsymbol{x}) = I_2(h(\boldsymbol{x})) = I_2(\boldsymbol{x} + \Delta\boldsymbol{x}). \tag{4.9}$$

If we consider the two images as being taken from infinitesimally close vantage points, we can write a continuous version of the above constraint. In order to make the notation more suggestive, we call $t$ the time at which the first image is taken, i.e. $I_1(\boldsymbol{x}) \doteq I(\boldsymbol{x}(t), t)$ and $t + dt$ the time when the second image is taken, i.e. $I_2(\boldsymbol{x}) \doteq I(\boldsymbol{x}(t + dt), t + dt)$. The notation "$dt$" suggests an infinitesimal increment of time (and hence motion). Also to associate the displacement $\Delta\boldsymbol{x}$ with the notion of *velocity* in the infinitesimal case, we write $\Delta\boldsymbol{x} \doteq \boldsymbol{u}\, dt$ for a (velocity) vector $\boldsymbol{u} \in \mathbb{R}^2$. Thus, $h(\boldsymbol{x}(t)) = \boldsymbol{x}(t + dt) = \boldsymbol{x}(t) + \boldsymbol{u}\, dt$. With this notation, equation (4.9) can be rewritten as

$$I(\boldsymbol{x}(t), t) = I(\boldsymbol{x}(t) + \boldsymbol{u}\, dt, t + dt). \tag{4.10}$$

Applying Taylor series expansion around $\boldsymbol{x}(t)$ to the right-hand side and neglecting higher-order terms we obtain

$$\boxed{\nabla I(\boldsymbol{x}(t), t)^T \boldsymbol{u} + I_t(\boldsymbol{x}(t), t) = 0,} \tag{4.11}$$

where

$$\nabla I(\boldsymbol{x}, t) \doteq \begin{bmatrix} I_x(\boldsymbol{x}, t) \\ I_y(\boldsymbol{x}, t) \end{bmatrix} = \begin{bmatrix} \frac{\partial I}{\partial x}(\boldsymbol{x}, t) \\ \frac{\partial I}{\partial y}(\boldsymbol{x}, t) \end{bmatrix} \in \mathbb{R}^2, \quad I_t(\boldsymbol{x}, t) \doteq \frac{\partial I}{\partial t}(\boldsymbol{x}, t) \in \mathbb{R}, \tag{4.12}$$

where $\nabla I$ and $I_t$ are the spatial and temporal derivatives of $I(\boldsymbol{x}, t)$, respectively. The spatial derivative $\nabla I$ is often called the *image gradient*.[7] We will discuss how to compute these derivatives from discretized images in Appendix 4.A of this chapter. If $\boldsymbol{x}(t) = [x(t), y(t)]^T$ is the trajectory of the image of a point moving across the image plane as time $t$ changes, $I(\boldsymbol{x}(t), t)$ should remain constant. Thus,

[7] Be aware that, strictly speaking, the gradient of a function is a covector and should be represented as a row vector. But in this book we define it to be a column vector (see Appendix C), to be consistent with all the other vectors.

another way of deriving the above equation is in terms of the total derivative of $I(\boldsymbol{x}(t), t) = I(x(t), y(t), t)$ with respect to time,

$$\frac{dI(x(t), y(t), t)}{dt} = 0, \tag{4.13}$$

which yields

$$\boxed{\frac{\partial I}{\partial x}\frac{dx}{dt} + \frac{\partial I}{\partial y}\frac{dy}{dt} + \frac{\partial I}{\partial t} = 0.} \tag{4.14}$$

This equation is identical to (4.11) once we notice that $\boldsymbol{u} \doteq [u_x, u_y]^T = [\frac{dx}{dt}, \frac{dy}{dt}]^T \in \mathbb{R}^2$. We also call this equation the *brightness constancy constraint*. It is the continuous version of (4.4) for the simplest translational model. Depending on where the constraint is evaluated, this equation can be used to compute what is called *optical flow*, or to *track photometric features* in a sequence of moving images.

When we fix our attention at a particular image location $\bar{\boldsymbol{x}}$ and use (4.14) to compute the velocity of "particles flowing" through that pixel, $\boldsymbol{u}(\bar{\boldsymbol{x}}, t)$ is called optical flow. When the attention is on a particular particle $\boldsymbol{x}(t)$ instead, and (4.14) is computed at the location $\boldsymbol{x}(t)$ as it moves through the image domain, we refer to the computation of $\boldsymbol{u}(\boldsymbol{x}(t), t)$ as *feature tracking*. Optical flow and feature tracking are obviously related by $\boldsymbol{x}(t+dt) = \boldsymbol{x}(t) + \boldsymbol{u}(\boldsymbol{x}(t), t)dt$. The only difference, at the conceptual level, is where the vector $\boldsymbol{u}(\boldsymbol{x}, t)$ is computed: in optical flow it is computed at a fixed location in the image, whereas in feature tracking it is computed at the point $\boldsymbol{x}(t)$.

Before we delve into the study of optical flow and feature tracking, notice that (4.11), if computed at each point, provides only one equation for two unknowns $(u_x, u_y)$. This is the aperture problem we have hinted at earlier.

*The aperture problem*

We start by rewriting equation (4.11) in a more compact form as

$$\boxed{\nabla I^T \boldsymbol{u} + I_t = 0.} \tag{4.15}$$

For simplicity we omit "$t$" from $(x(t), y(t))$ in $I(x(t), y(t), t)$ and write only $I(x, y, t)$, or $I(\boldsymbol{x}, t)$.

The brightness constancy constraint captures the relationship between the image velocity $\boldsymbol{u}$ of an image point $\boldsymbol{x}$ and spatial and temporal derivatives $\nabla I, I_t$, which are directly measurable from images. As we have already noticed, the equation provides a single constraint for two unknowns in $\boldsymbol{u} = [u_x, u_y]^T$. From the linear-algebraic point of view there are infinitely many solutions $\boldsymbol{u}$ that satisfy this equation. All we can compute is the projection of the actual optical flow vector in the direction of the image gradient $\nabla I$. This component is also referred to as *normal flow* and can be thought of as a minimum norm vector $\boldsymbol{u}_n \in \mathbb{R}^2$ that satisfies the brightness constancy constraint. It is given by a projection of the true

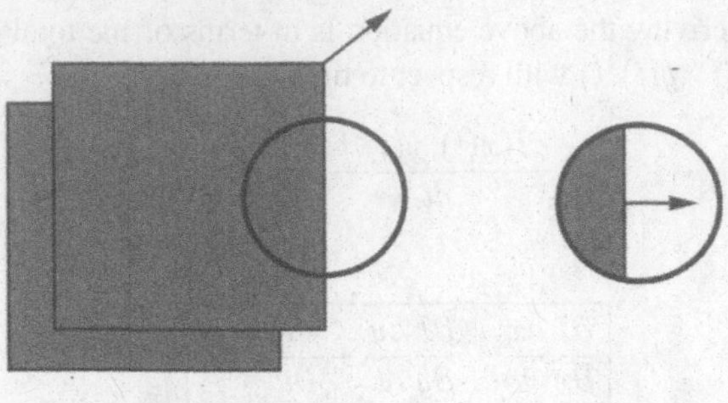

Figure 4.4. In spite of the fact that the square moves diagonally between two consecutive frames, only horizontal motion can be observed through the aperture.

motion vector $\boldsymbol{u}$ onto the gradient direction and is given by

$$\boldsymbol{u}_n \doteq \frac{\nabla I^T \boldsymbol{u}}{\|\nabla I\|} \frac{\nabla I}{\|\nabla I\|} = -\frac{I_t}{\|\nabla I\|} \frac{\nabla I}{\|\nabla I\|}. \tag{4.16}$$

This observation is a consequence of the *aperture problem* and can be easily visualized. For example, consider viewing the square in Figure 4.4 through a small aperture. In spite of the fact that the square moves diagonally between the two consecutive frames, only horizontal motion can be observed through the aperture, and nothing can be said about motion along the direction of the edge.

It is only when the brightness constancy constraint is applied to each point $\tilde{\boldsymbol{x}}$ in a region $W(\boldsymbol{x})$ that contains "sufficient texture," and the motion $\boldsymbol{u}$ is assumed to be constant in the region, that the equations provide enough constraints on $\boldsymbol{u}$. This constancy assumption enables us to integrate the constraints for all points in the region $W(\boldsymbol{x})$ and seek the best image velocity consistent with all the point constraints. In order to account for the effect of noise in the model (4.6), optical flow computation is often formulated as a minimization of the following quadratic error function based on the gradient constraint:

$$E_b(\boldsymbol{u}) = \sum_{W(\boldsymbol{x})} \left[\nabla I^T(\tilde{\boldsymbol{x}}, t)\boldsymbol{u}(\boldsymbol{x}) + I_t(\tilde{\boldsymbol{x}}, t)\right]^2, \tag{4.17}$$

where the subscript "$b$" indicates brightness constancy. To obtain a linear least-squares estimate of $\boldsymbol{u}(\boldsymbol{x})$ at each image location, we compute the derivative with respect to $\boldsymbol{u}$ of the error function $E_b(\boldsymbol{u})$:

$$\begin{aligned} \nabla E_b(\boldsymbol{u}) &= 2 \sum_{W(\boldsymbol{x})} \nabla I(\nabla I^T \boldsymbol{u} + I_t) \\ &= 2 \sum_{W(\boldsymbol{x})} \left( \begin{bmatrix} I_x^2 & I_x I_y \\ I_x I_y & I_y^2 \end{bmatrix} \boldsymbol{u} + \begin{bmatrix} I_x I_t \\ I_y I_t \end{bmatrix} \right). \end{aligned}$$

For $\boldsymbol{u}$ that minimizes $E_b$, it is necessary that $\nabla E_b(\boldsymbol{u}) = 0$. This yields

$$\begin{bmatrix} \sum I_x^2 & \sum I_x I_y \\ \sum I_x I_y & \sum I_y^2 \end{bmatrix} \boldsymbol{u} + \begin{bmatrix} \sum I_x I_t \\ \sum I_y I_t \end{bmatrix} = 0, \tag{4.18}$$

or, in matrix form,

$$Gu + b = 0. \tag{4.19}$$

Solving this equation (if $G$ is invertible) gives the least-squares estimate of image velocity

$$\boxed{u = -G^{-1}b.} \tag{4.20}$$

Note, however, that the matrix $G$ is not guaranteed to be invertible. If the intensity variation in a local image window varies only along one dimension (e.g., $I_x = 0$ or $I_y = 0$) or vanishes ($I_x = 0$ and $I_y = 0$), then $G$ is *not* invertible. These singularities have been previously mentioned as the *aperture* and *blank wall problem*, respectively. Based on these observations we see that it is the local properties of image irradiance in the window $W(x)$ that determine whether the problem is ill posed.

Since it seems that the correspondence problem can be solved, under the brightness constancy assumption, for points $x$ where $G(x)$ is invertible, it is convenient to *define* such points as "feature points" at least according to the quadratic criterion above. As we will see shortly, this definition is also consistent with other criteria.

*The "sum of squared differences" (SSD) criterion*

Let us now go back to the simplest translational deformation model

$$h(\tilde{x}) = \tilde{x} + \Delta x, \quad \forall\, \tilde{x} \in W(x). \tag{4.21}$$

In order to track a feature point $x$ by computing its image displacement $\Delta x$, we can seek the location $x + \Delta x$ on the image at time $t + dt$ whose window is "most similar" to the window $W(x)$. A common way of measuring similarity is by using the "sum of squared differences" (SSD) criterion. The SSD approach considers an image window $W$ centered at a location $(x, y)$ at time $t$ and other candidate locations $(x + dx, y + dy)$ in the image at time $t + dt$, where the point could have moved between two frames. The goal is to find a displacement $\Delta x = (dx, dy)$ at a location in the image $(x, y)$ that minimizes the SSD criterion

$$E_t(dx, dy) \doteq \sum_{W(x,y)} [I(x + dx, y + dy, t + dt) - I(x, y, t)]^2, \tag{4.22}$$

where the subscript "$t$" indicates the translational deformation model. Comparing this with the error function (4.17), an advantage of the SSD criterion is that in principle we no longer need to compute derivatives of $I(x, y, t)$, although one can easily show that $u\, dt = (-G^{-1}b)\, dt$ is the first-order approximation of the displacement $\Delta x = (dx, dy)$. We leave this as an exercise to the reader (see Exercise 4.4). One alternative for computing the displacement is to evaluate the function at each location and choose the one that gives the minimum error. This formulation is due to [Lucas and Kanade, 1981] and was originally proposed in the context of stereo algorithms and was later refined by [Tomasi and Kanade, 1992] in a more general feature-tracking context.

In Algorithm 4.1 we summarize a basic algorithm for feature tracking or optical flow; a more effective version of this algorithm that involves a multi-resolution representation and subpixel refinement is described in Chapter 11 (Algorithm 11.2).

---

**Algorithm 4.1 (Basic feature tracking and optical flow).**

Given an image $I(\boldsymbol{x})$ at time $t$, set a window $W$ of fixed size, use the filters given in Appendix 4.A to compute the image gradient $(I_x, I_y)$, and compute $G(\boldsymbol{x}) \doteq \begin{bmatrix} \sum I_x^2 & \sum I_x I_y \\ \sum I_x I_y & \sum I_y^2 \end{bmatrix}$ at every pixel $\boldsymbol{x}$. Then, either

- (feature tracking) select a number of point features by choosing $\boldsymbol{x}_1, \boldsymbol{x}_2, \ldots$ such that $G(\boldsymbol{x}_i)$ is invertible, or
- (optical flow) select $\boldsymbol{x}_i$ to be on a fixed grid.

An invertibility test of $G$ that is more robust to the effects of noise will be described in Algorithm 4.2.

- Compute $\boldsymbol{b}(\boldsymbol{x}, t) \doteq \begin{bmatrix} \sum I_x I_t \\ \sum I_y I_t \end{bmatrix}$.
- If $G(\boldsymbol{x})$ is invertible (which is guaranteed for point features), compute the displacement $\boldsymbol{u}(\boldsymbol{x}, t)$ from equation (4.20). If $G(\boldsymbol{x})$ is not invertible, return $\boldsymbol{u}(\boldsymbol{x}, t) = 0$.

The displacement of the pixel $\boldsymbol{x}$ at time $t$ is therefore given by $\boldsymbol{u}(\boldsymbol{x}, t) = -G(\boldsymbol{x})^{-1}\boldsymbol{b}(\boldsymbol{x}, t)$ wherever $G(\boldsymbol{x})$ is invertible.

- (Feature tracking) at time $t + 1$, repeat the operation at $\boldsymbol{x} + \boldsymbol{u}(\boldsymbol{x}, t)$.
- (Optical flow) at time $t + 1$, repeat the operation at $\boldsymbol{x}$.

---

### 4.3.2 *Large baseline: affine model and normalized cross-correlation*

The small-baseline tracking algorithm presented in the previous section results in very efficient and fast implementations. However, when features are tracked over an extended time period, the estimation error resulting from matching templates between two adjacent frames accumulates in time. This leads to eventually losing track of the originally selected features. To avoid this problem, instead of matching image regions between adjacent frames, one could match image regions between the initial frame, say $I_1$, and the current frame, say $I_2$. On the other hand, the deformation of the image regions between the first frame and the current frame can no longer be modeled by a simple translational model. Instead, a commonly adopted model is that of affine deformation of image regions that support point features, $I_1(\tilde{x}) = I_2(h(\tilde{x}))$, where the function $h$ has the form

$$h(\tilde{x}) = A\tilde{x} + d = \begin{bmatrix} a_1 & a_2 \\ a_3 & a_4 \end{bmatrix} \tilde{x} + \begin{bmatrix} d_1 \\ d_2 \end{bmatrix}, \quad \forall \tilde{x} \in W(x). \tag{4.23}$$

As in the pure translation model (4.9), we can formulate the brightness constancy constraint with this more general six-parameter affine model for the two images:

$$I_1(\tilde{x}) = I_2(A\tilde{x} + d), \quad \forall\, \tilde{x} \in W(x). \tag{4.24}$$

Enforcing the above assumption over a region of the image, we can estimate the unknown affine parameters $A$ and $d$ by integrating the above constraint for all the points in the region $W(x)$,

$$E_a(A, d) \doteq \sum_{W(x)} \left[I_2(A\tilde{x} + d) - I_1(\tilde{x})\right]^2, \tag{4.25}$$

where the subscript "$a$" indicates the affine deformation model. By approximating the function $I_2(A\tilde{x} + d)$ to first order around the point $A_0 = I_{2\times 2}, d_0 = 0_{2\times 1}$,

$$I_2(A\tilde{x} + d) \approx I_2(\tilde{x}) + \nabla I_2^T(\tilde{x})[(A - A_0)\tilde{x} + d],$$

the above minimization problem can be solved using linear least-squares, yielding estimates of the unknown parameters $A \in \mathbb{R}^{2\times 2}$ and $d \in \mathbb{R}^2$ directly from measurements of spatial and temporal gradients of the image. In Exercise 4.5 we walk the reader through the steps necessary to implement such a tracking algorithm. In Chapter 11, we will combine this affine model with contrast compensation to derive a practical feature-tracking algorithm that works for a moderate baseline.

*Normalized cross-correlation (NCC) criterion*

In the previous sections we used the SSD as a cost function for template matching. Although the SSD allows for a linear least-squares solution in the unknowns, there are also some drawbacks to this choice. For example, the SSD is not invariant to scalings and shifts in image intensities, often caused by changing lighting conditions over time. For the purpose of template matching, a better choice is *normalized cross-correlation*. Given two nonuniform image regions $I_1(\tilde{x})$ and $I_2(h(\tilde{x}))$, with $\tilde{x} \in W(x)$ and $N = |W(x)|$ (the number of pixels in the window), the normalized cross-correlation (NCC) is defined as

$$\mathrm{NCC}(h) = \frac{\sum_{W(x)} \left(I_1(\tilde{x}) - \bar{I}_1\right)\left(I_2(h(\tilde{x})) - \bar{I}_2\right)}{\sqrt{\sum_{W(x)} (I_1(\tilde{x}) - \bar{I}_1)^2 \sum_{W(x)} (I_2(h(\tilde{x})) - \bar{I}_2)^2}}, \tag{4.26}$$

where $\bar{I}_1, \bar{I}_2$ are the mean intensities:

$$\begin{aligned} \bar{I}_1 &= \tfrac{1}{N} \textstyle\sum_{W(x)} I_1(\tilde{x}), \\ \bar{I}_2 &= \tfrac{1}{N} \textstyle\sum_{W(x)} I_2(h(\tilde{x})). \end{aligned}$$

The normalized cross-correlation value always ranges between $-1$ and $+1$, irrespective of the size of the window. When the normalized cross-correlation is 1, the two image regions match perfectly. In particular, in the case of the affine model the normalized cross-correlation becomes

$$\mathrm{NCC}(A, \boldsymbol{d}) = \frac{\sum_{W(\boldsymbol{x})} \left(I_1(\tilde{\boldsymbol{x}}) - \bar{I}_1\right)\left(I_2(A\tilde{\boldsymbol{x}} + \boldsymbol{d}) - \bar{I}_2\right)}{\sqrt{\sum_{W(\boldsymbol{x})}(I_1(\tilde{\boldsymbol{x}}) - \bar{I}_1)^2 \sum_{W(\boldsymbol{x})}(I_2(A\tilde{\boldsymbol{x}} + \boldsymbol{d}) - \bar{I}_2)^2}}. \tag{4.27}$$

So, we look for $(\hat{A}, \hat{\boldsymbol{d}}) = \arg\max_{A,\boldsymbol{d}} \mathrm{NCC}(A, \boldsymbol{d})$. In Chapter 11, we will combine NCC with robust statistics techniques to derive a practical algorithm that can match features between two images with a large baseline.

### 4.3.3 Point feature selection

In previous sections we have seen how to compute the translational or affine deformation of a photometric feature, and we have distinguished the case where the computation is performed at a fixed set of locations (optical flow) from the case where point features are tracked over time (feature tracking). One issue we have not addressed in this second case is how to initially select the points to be tracked. However, we have hinted on various occasions at the possibility of selecting as "feature points" the locations that allow us to solve the correspondence problem easily. In this section we make this more precise by giving a numerical algorithm to select such features.

As the reader may have noticed, the description of any of those feature points relies on knowing the gradient of the image. Hence, before we can give any numerical algorithm for feature selection, the reader needs to know how to compute the image gradient $\nabla I = [I_x, I_y]^T$ in an accurate and robust way. The description of how to compute the gradient of a discretized image is in Appendix 4.A.

The solution to the tracking or correspondence problem for the case of pure translation relied on inverting the matrix $G$ made of the spatial gradients of the image (4.20). For $G$ to be invertible, the region must have nontrivial gradients along two independent directions, resembling therefore a "corner" structure, as shown in Figure 4.5. Alternatively, if we regard the corner as the "intersection"

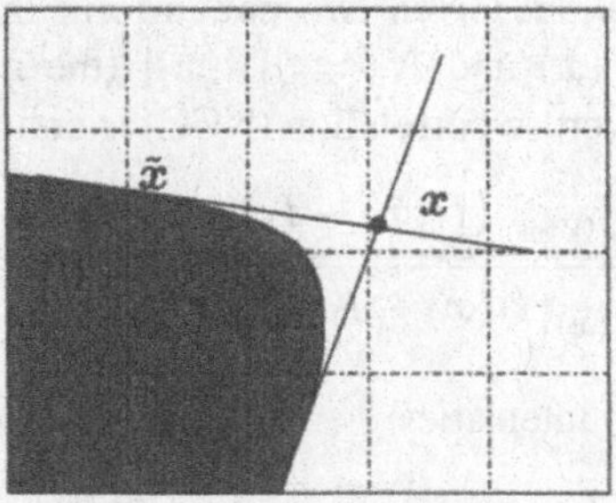


Figure 4.5. A corner feature $\boldsymbol{x}$ is the virtual intersection of local edges (within a window).

of all the edges inside the window, then the existence of at least a corner point

$\boldsymbol{x} = [x, y]^T$ means that over the window $W(\boldsymbol{x})$, the following minimization has a solution:

$$\min_{\boldsymbol{x}} \; E_c(\boldsymbol{x}) \doteq \sum_{\tilde{\boldsymbol{x}} \in W(\boldsymbol{x})} \left[\nabla I^T(\tilde{\boldsymbol{x}})(\tilde{\boldsymbol{x}} - \boldsymbol{x})\right]^2, \tag{4.28}$$

where $\nabla I(\tilde{\boldsymbol{x}})$ is the gradient calculated at $\tilde{\boldsymbol{x}} = [\tilde{x}, \tilde{y}]^T \in W(\boldsymbol{x})$. It is then easy to check that the existence of a local minimum for this error function is equivalent to the summation of the outer product of the gradients, i.e.

$$G(\boldsymbol{x}) = \sum_{\tilde{\boldsymbol{x}} \in W(\boldsymbol{x})} \nabla I(\tilde{\boldsymbol{x}}) \nabla I^T(\tilde{\boldsymbol{x}}) = \begin{bmatrix} \sum I_x^2 & \sum I_x I_y \\ \sum I_x I_y & \sum I_y^2 \end{bmatrix} \in \mathbb{R}^{2\times 2}, \tag{4.29}$$

being nonsingular. If $\sigma_2$, the smallest singular value of $G$, is above a specified threshold $\tau$, then $G$ is invertible, (4.20) can be solved, and therefore, we say that the point $\boldsymbol{x}$ is a feature point. If both singular values of $G$ are close to zero, the feature window has almost constant brightness. If only one of the singular values is close to zero, the brightness varies mostly along a single direction. In both cases, the point cannot be localized or matched in another image. This leads to a simple algorithm to extract point (or corner) features; see Algorithm 4.2.

---

**Algorithm 4.2 (Corner detector).**

---

Given an image $I(x, y)$, follow the steps to detect whether a given pixel $(x, y)$ is a corner feature:

- set a threshold $\tau \in \mathbb{R}$ and a window $W$ of fixed size, and compute the image gradient $(I_x, I_y)$ using the filters given in Appendix 4.A;
- at all pixels in the window $W$ around $(x, y)$ compute the matrix
$$G = \begin{bmatrix} \sum I_x^2 & \sum I_x I_y \\ \sum I_x I_y & \sum I_y^2 \end{bmatrix}; \tag{4.30}$$
- if the smallest singular value $\sigma_2(G)$ is bigger than the prefixed threshold $\tau$, then mark the pixel as a feature (or corner) point.

---

Although we have used the word "corner," the reader should observe that the test above guarantees only that the irradiance function $I$ is "changing enough" in two independent directions within the window of interest. Another way in which this can happen is for the window to contain "sufficient texture," causing enough variation along at least two independent directions.

A variation to the above algorithm is the well-known Harris corner detector [Harris and Stephens, 1988]. The main idea is to threshold the quantity

$$C(G) = \det(G) + k \times \text{trace}^2(G), \tag{4.31}$$

where $k \in \mathbb{R}$ is a (usually small) scalar, and different choices of $k$ may result in favoring gradient variation in one or more than one direction, or maybe both. To see this, let the two eigenvalues (which in this case coincide with the singular

Figure 4.6. An example of the response of the Harris feature detector using $5\times 5$ integration windows and parameter $k = 0.04$. Some apparent corners around the boundary of the image are not detected due to the size of window chosen.

values) of $G$ be $\sigma_1, \sigma_2$. Then

$$C(G) = \sigma_1\sigma_2 + k(\sigma_1 + \sigma_2)^2 = (1 + 2k)\sigma_1\sigma_2 + k(\sigma_1^2 + \sigma_2^2). \tag{4.32}$$

Note that if $k$ is large and either one of the eigenvalues is large, so will be $C(G)$. That is, features with significant gradient variation in at least one direction will likely pass a threshold. If $k$ is small, then both eigenvalues need to be big enough to make $C(G)$ pass the threshold. In this case, only the corner feature is favored. Simple thresholding operations often do not yield satisfactory results, which lead to a detection of too many corners, which are not well localized. Partial improvements can be obtained by searching for the local minima in the regions, where the response of the detector is high. Alternatively, more sophisticated techniques can be used, which utilize contour (or edge) detection techniques and indeed search for the high curvature points of the detected contours [Wuescher and Boyer, 1991]. In Chapter 11 we will explore further details that are crucial in implementing an effective feature detection and selection algorithm.

## 4.4 Tracking line features

As we will see in future chapters, besides point features, line (or edge) features, which typically correspond to boundaries of homogeneous regions, also provide important geometric information about the 3-D structure of objects in the scene. In this section, we study how to extract and track such features.

### 4.4.1 Edge features and edge detection

As mentioned above, when the matrix $G$ in (4.29) has both singular values close to zero, it corresponds to a textureless "blank wall." When one of the singular values is large and the other one is close to zero, the brightness varies mostly along a single direction. But that does not imply a sudden change of brightness value in the direction of the gradient. For example, an image of a shaded marble sphere does vary in brightness, but the variation is smooth, and therefore the entire surface is better interpreted as one smooth region instead of one with edges everywhere. Thus, by "an edge" in an image, we typically refer to a place where there is a distinctive "peak" in the gradient. Of course, the notion of a "peak" depends on the resolution of the image and the size of the window chosen. What appears as smooth shading on a small patch in a high-resolution image may appear as a sharp discontinuity on a large patch in a subsampled image.

We therefore label a pixel $\boldsymbol{x}$ as an "edge feature" only if the gradient norm $\|\nabla I\|$ reaches a local maximum compared to its neighboring pixels. This simple idea results in the well-known Canny edge-detection algorithm [Canny, 1986].

**Algorithm 4.3 (Canny edge detector).**

Given an image $I(x, y)$, follow the steps to detect whether a given pixel $(x, y)$ is on an edge

- set a threshold $\tau > 0$ and standard deviation $\sigma > 0$ for the Gaussian function $g_\sigma$ used to derive the filter (see Appendix 4.A for details);
- compute the gradient vector $\nabla I = [I_x, I_y]^T$ (see Appendix 4.A);
- if $\|\nabla I(x, y)\|^2 = \nabla I^T \nabla I$ is a local maximum along the gradient and larger than the prefixed threshold $\tau$, then mark it as an edge pixel.

Figure 4.7 demonstrates edges detected by the Canny edge detector on a gray-level image.

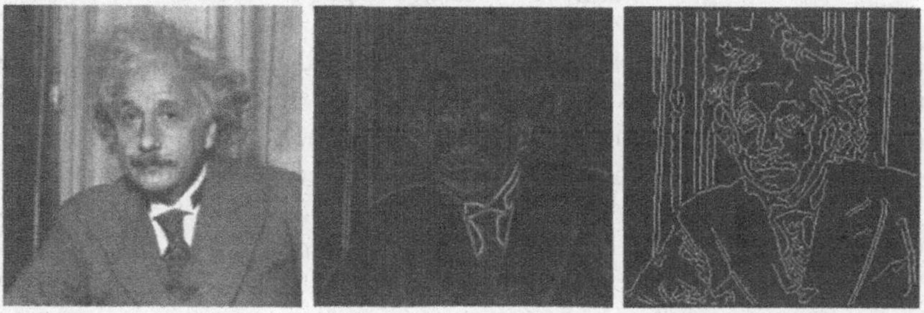

Figure 4.7. Original image, gradient magnitude, and detected edge pixels of an image of Einstein.

### 4.4.2 *Composition of edge elements: line fitting*

In order to compensate for the effects of digitization and thresholding that destroy the continuity of the gradient magnitude function $\|\nabla I\|$, the edge-detection stage is often followed by a *connected component analysis*, which enables us to group neighboring pixels with common gradient orientation to form a connected contour or more specifically a candidate line $\boldsymbol{\ell}$. The connected component algorithm can be found in most image processing or computer vision textbooks, and we refer the reader to [Gonzalez and Woods, 1992]. Using results from the connected component analysis, the *line fitting stage* typically involves the computation of the Hough or Radon transform, followed by a peak detection in the parameter space. Both of these techniques are well established in image processing and the algorithms are available as a part of standard image processing toolboxes (see Exercise 4.9).

Alternatively, a conceptually simpler way to obtain line feature candidates is by directly fitting lines to the segments obtained by connected component analysis. Each connected component $C^k$ is a list of edge pixels $\{(x_i, y_i)\}_{i=1}^n$, which are connected and grouped based on their gradient orientation, forming a line support region, say $W(\boldsymbol{\ell})$. The line parameters can then be directly computed from the eigenvalues $\lambda_1, \lambda_2$ and eigenvectors $v_1, v_2$ of the matrix $D^k$ associated with the line support region:

$$D^k \doteq \begin{bmatrix} \sum_i \tilde{x}_i^2 & \sum_i \tilde{x}_i \tilde{y}_i \\ \sum_i \tilde{x}_i \tilde{y}_i & \sum_i \tilde{y}_i^2 \end{bmatrix} \in \mathbb{R}^{2\times 2}, \tag{4.33}$$

where $\tilde{x} = x_i - \bar{x}$ and $\tilde{y} = y_i - \bar{y}$ are the mean-corrected pixel coordinates of every pixel $(x_i, y_i)$ in the connected component, and $\bar{x} = \frac{1}{n}\sum_i x_i$ and $\bar{y} = \frac{1}{n}\sum_i y_i$ are the means. In the case of an ideal line, one of the eigenvalues should be zero. The quality of the line fit is characterized by the ratio of the two eigenvalues $\frac{\lambda_1}{\lambda_2}$ (with $\lambda_1 > \lambda_2$) of $D^k$.

On the 2-D image plane, any point $(x, y)$ on a line must satisfy an equation of the form

$$\sin(\theta)x - \cos(\theta)y = \rho. \tag{4.34}$$

Geometrically, $\theta$ is the angle between the line $\boldsymbol{\ell}$ and the $x$-axis, and $\rho$ is the distance from the origin to the line $\boldsymbol{\ell}$ (Figure 4.8). In this notation, the unit eigenvector $v_1$ (corresponding to the larger eigenvalue $\lambda_1$) is of the form $v_1 = [\cos(\theta), \sin(\theta)]^T$. Then, parameters of the line $\boldsymbol{\ell} : (\rho, \theta)$ are determined from $v_1$ as

$$\begin{aligned} \theta &= \arctan(v_1(2)/v_1(1)), & (4.35)\\ \rho &= \bar{x}\sin(\theta) - \bar{y}\cos(\theta), & (4.36) \end{aligned}$$

where $(\bar{x}, \bar{y})$ is the midpoint of the line segment. We leave the derivation of these formulae to the reader as an exercise (see Exercise 4.7).

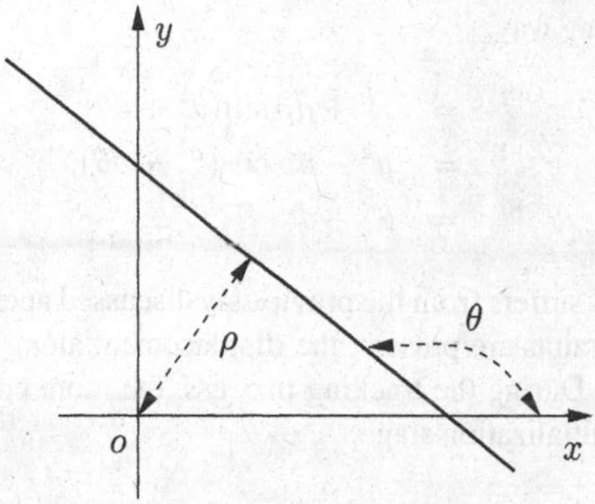


Figure 4.8. Parameterization of a line in 2-D.

Figure 4.9. Edge detection and line fitting results.

### 4.4.3 *Tracking and matching line segments*

The techniques for associating line features across multiple frames depend, as in the point feature case, on the baseline between the views. The simplest image-based line tracking technique starts with associating a window support region $W(\ell)$, containing the edge pixels forming a line support region.[8] The selected window is first transformed to a canonical image coordinate frame, making the line orientation vertical. At sample points $(x_i, y_i)$ along the line support region, the displacement $d\rho$ in the direction perpendicular to the line is computed. Once this has been done for some number of sample points, the parameters of the new line segment can be obtained, giving rise to the change of the line orientation by $d\theta$. The remaining points can then be updated using the computed parameters $d\rho$

[8]The size of the region can vary depending on whether extended lines are being tracked or just small pieces of connected contours.

and $d\theta$ in the following way:

$$x^{k+1} = x^k + d\rho \sin(\theta^k + d\theta), \tag{4.37}$$

$$y^{k+1} = y^k - d\rho \cos(\theta^k + d\theta), \tag{4.38}$$

$$\theta^{k+1} = \theta^k + d\theta. \tag{4.39}$$

Note that this method suffers from the previously discussed aperture problem. Unless additional constraints are present, the displacement along the edge direction cannot be measured. During the tracking process, the more costly line detection is done only in the initialization stage.

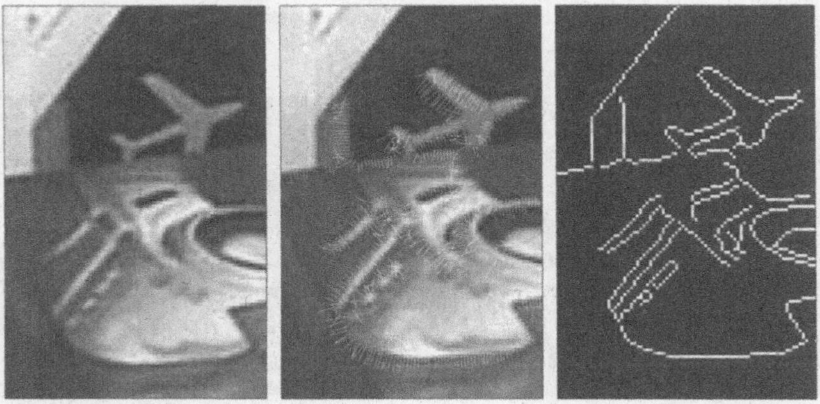

Figure 4.10. Edge tracking by computing the normal displacement of the edge between adjacent frames.

In case of line matching across wide baselines, the support regions $W(\boldsymbol{\ell})$ associated with the candidate line features subtend the entire extent of the line. Due to the fact that the line support regions automatically contain orientation information, standard window matching criteria (such as SSD and NCC), introduced in Section 4.3, can be used.

## 4.5 Summary

This chapter describes the crucial step of going from measurements of light intensity to geometric primitives. The notions of "point feature" and "line feature" are introduced, and basic algorithms for feature detection, tracking, and matching are described. Further refinements of these algorithms, e.g., affine tracking, subpixel iterations, and multiscale implementation, are explored in the exercises; practical issues associated with their implementation will be discussed in Chapter 11.

## 4.6 Exercises

**Exercise 4.1 (Motion model).** Consider measuring image motion $h(\mathbf{x})$ and noticing that $h(\mathbf{x}) = \mathbf{x} + \Delta\mathbf{x}, \ \forall\, \mathbf{x} \in \Omega$; i.e. each point on the image translates by the same amount $\Delta\mathbf{x}$. What particular motion $(R, T)$ and 3-D structure $\mathbf{X}$ must the scene undergo to satisfy this model?

**Exercise 4.2** Repeat the exercise above for an affine motion model, $h(\mathbf{x}) = A\mathbf{x} + \mathbf{d}$.

**Exercise 4.3** Repeat the exercise above for a general linear motion model, $h(\mathbf{x}) = H\mathbf{x}$, in homogeneous coordinates.

**Exercise 4.4 (LLS approximation of translational flow).** Consider the problem of finding a displacement $(dx, dy)$ at a location in the image $(x, y)$ that minimizes the SSD criterion

$$\mathrm{SSD}(dx, dy) \doteq \sum_{W(x,y)} [I(x + dx, y + dy, t + dt) - I(x, y, t))]^2.$$

If we approximate the function $I(x + dx, y + dy, t + dt)$ up to the first order term of its Taylor expansion,

$$I(x + dx, y + dy, t + dt) \approx I(x, y, t) + I_t(x, y, t)dt + \nabla I^T(x, y, t)[dx, dy]^T,$$

we can find a solution to the above minimization problem. Explain under what conditions a unique solution to $(dx, dy)$ exists. Compare the solution to the optical flow solution $\mathbf{u} = -G^{-1}\mathbf{b}$.

**Exercise 4.5 (LLS approximation of affine flow).** To obtain an approximate solution to $(A, \mathbf{d})$ that minimizes the function

$$E_a(A, \mathbf{d}) \doteq \sum_{W(\mathbf{x})} [I(A\tilde{\mathbf{x}} + \mathbf{d}, t + dt) - I(\tilde{\mathbf{x}}, t)]^2, \tag{4.40}$$

follow the steps outlined below:

- Approximate the function $I(A\mathbf{x} + \mathbf{d}, t + dt)$ to first order as
$$I(A\tilde{\mathbf{x}} + \mathbf{d}, t + dt) \approx I(\tilde{\mathbf{x}}, t + dt) + \nabla I^T(\tilde{\mathbf{x}}, t + dt)[(A - I_{2\times 2})\tilde{\mathbf{x}} + \mathbf{d}].$$
- Consider the matrix $D = A - I_{2\times 2} \in \mathbb{R}^{2\times 2}$ and vector $\mathbf{d} \in \mathbb{R}^2$ as new unknowns. Collect the unknowns $(D, \mathbf{d})$ into the vector $\mathbf{y} = [d_{11}, d_{12}, d_{21}, d_{22}, d_1, d_2]^T$ and set $I_t = I(\tilde{\mathbf{x}}, t + dt) - I(\tilde{\mathbf{x}}, t)$.
- Compute the derivative of the objective function $E_a(D, \mathbf{d})$ with respect to the unknowns and set it to zero. Show that the resulting estimate of $\mathbf{y} \in \mathbb{R}^6$ is equivalent to the solution of the following systems of linear equations,
$$\sum_{W(\mathbf{x})} \begin{bmatrix} G_1 & G_2 \\ G_2^T & G_3 \end{bmatrix} \mathbf{y} = \sum_{W(\mathbf{x})} \mathbf{b}, \quad \text{where} \quad G_3 \doteq \begin{bmatrix} I_x^2 & I_x I_y \\ I_x I_y & I_y^2 \end{bmatrix},$$
$\mathbf{b} \doteq [xI_tI_x, xI_tI_y, yI_tI_x, yI_tI_y, I_tI_x, I_tI_y]^T$, and $G_1, G_2$ are
$$G_1 \doteq \begin{bmatrix} x^2I_x^2 & x^2I_xI_y & xyI_x^2 & xyI_xI_y \\ x^2I_xI_y & x^2I_y^2 & xyI_xI_y & xyI_y^2 \\ xyI_x^2 & xyI_xI_y & y^2I_x^2 & y^2I_xI_y \\ xyI_xI_y & xyI_y^2 & y^2I_xI_y & y^2I_y^2 \end{bmatrix}, \quad G_2 \doteq \begin{bmatrix} xI_xI_y & xI_x^2 \\ xI_y^2 & xI_xI_y \\ yI_xI_y & yI_x^2 \\ yI_y^2 & yI_xI_y \end{bmatrix}.$$

- Write down the linear least-squares estimate of $y$ and discuss under what condition the solution is well-defined.

**Exercise 4.6 (Eigenvalues of the sum of outer products).** Given a set of vectors $u_1, u_2, \ldots, u_m \in \mathbb{R}^n$, prove that all eigenvalues of the matrix

$$G = \sum_{i=1}^{m} u_i u_i^T \in \mathbb{R}^{n \times n} \tag{4.41}$$

are nonnegative. This shows that the eigenvalues of $G$ are the same as the singular values of $G$. (Note: you may take it for granted that all the eigenvalues are real, since $G$ is a real symmetric matrix.)

**Exercise 4.7** Suppose $\{(x_i, y_i)\}_{i=1}^n$ are coordinates of $n$ sample points from a straight line in $\mathbb{R}^2$. Show that the matrix $D$ defined in (4.33) has rank 1. What is the geometric interpretation of the two eigenvectors $v_1, v_2$ of $D$ in terms of the line? Since every line in $\mathbb{R}^2$ can be expressed in terms of an equation, $ax + by + c = 0$, derive an expression for the parameters $a, b, c$ in terms of $v_1$ and $v_2$.

**Exercise 4.8 (Programming: implementation of the corner detector).** Implement a version of the corner detector using Matlab. Mark the most distinctive, say 20 to 50, feature points in a given image.

After you are done, you may try to play with it. Here are some suggestions:

- Identify and compare practical methods to choose the threshold $\tau$ or other parameters. Evaluate the choice by altering the levels of brightness, saturation, and contrast in the image.

- In practice, you may want to select only one pixel around a corner feature. Devise a method to choose the "best" pixel (or subpixel location) within a window, instead of every pixel above the threshold.

- Devise some quality measure for feature points and sort them according to that measure. Note that the threshold has only "soft" control on the number of features selected. With such a quality measure, however, you can specify any number of features you want.

**Exercise 4.9 (Programming: implementation of the line feature detector).** Implement a version of the line feature detector using Matlab. Select the line segments whose length exceeds the predefined threshold $l$. Here are some guidelines on how to proceed:

1. Run the edge-detection algorithm implemented by the function `BW = egde(I,param)` in Matlab.

   - Experiment with different choices of thresholds. Alternatively, you can implement individual steps of the Canny edge detector. Visualize both gradient magnitude $M = \sqrt{I_x^2 + I_y^2}$ and gradient orientation $\Theta = \text{atan2}(I_y, I_x)$.
   - Run the connected component algorithm `L = bwlabel(BW)` in Matlab and group the pixels with similar gradient orientations as described in Section 4.4.
   - Estimate the line parameters of each linear connected group based on equations (4.35) and (4.36), and visualize the results.

2. On the same image, experiment with the function L = `radon(I,` $\theta$`)` and suggest how to use the function to detect line segments in the image. Discuss the advantages and disadvantages of these two methods.

**Exercise 4.10 (Programming: subpixel iteration).** Both the linear and the affine model for point feature tracking can be refined by subpixel iterations as well as by using multiscale deformation models that allow handling larger deformations. In order to achieve subpixel accuracy, implement the following iteration:

- $\delta^0 = -G^{-1}e^0$,
- $\delta^{i+1} = -G^{-1}e^i$,
- $d^{i+1} \leftarrow d^i + \delta^{i+1}$,

where we define following quantities based on equation (4.19),

- $e^0 \doteq b$,
- $e^{i+1} \doteq \sum_{W(x)} \nabla I(\tilde{x})\big(I(\tilde{x} + d^i, t + dt) - I(\tilde{x}, t)\big)$.

At each step, $\tilde{x} + d^i$ is in general not on the pixel grid, so it is necessary to interpolate the brightness values to obtain image intensity at that location.

**Exercise 4.11 (Programming: multiscale implementation).** One problem common to all differential techniques is that they fail if the displacement across frames is bigger than a few pixels. One possible way to overcome this inconvenience is to use a coarse-to-fine strategy:

- Build a pyramid of images by smoothing and subsampling the original images (see, for instance, [Burt and Adelson, 1983]).
- Select features at the desired level of definition and then propagate the selection up the pyramid.
- Track the features at the coarser level.
- Propagate the displacement to finer resolutions and use that displacement as an initial step for the subpixel iteration described in the previous section.

## 4.A Computing image gradients

Let us neglect for the moment the discrete nature of digital images. Conceptually, the image gradient $\nabla I(x,y) = [I_x(x,y), I_y(x,y)]^T \in \mathbb{R}^2$ is defined by a vector whose individual components are given by the two partial derivatives

$$I_x(x,y) = \frac{\partial I}{\partial x}(x,y), \quad I_y(x,y) = \frac{\partial I}{\partial y}(x,y). \tag{4.42}$$

In order to simplify the notation, we will omit the argument $(x,y)$ and simply write $\nabla I = [I_x, I_y]^T$. While the notion of derivative is well defined for smooth functions, additional steps need to be taken in computing the derivatives of digital images.

*Sampling of a continuous signal*

The starting point of this development lies in the relationship between continuous and sampled discrete signals and the theory of sampling and reconstruction [Oppenheim et al., 1999]. Let us assume that we have a sampled version of a continuous signal $f(x), x \in \mathbb{R}$, denoted by

$$f[x] = f(xT), \quad x \in \mathbb{Z}, \tag{4.43}$$

where $f[x]$ is the value of the continuous function $f(x)$ sampled at integer values of $x$ with $T$ being the sampling period of the signal (Figure 4.11). We will adopt the notation of discretely sampled signals with the argument in square brackets.

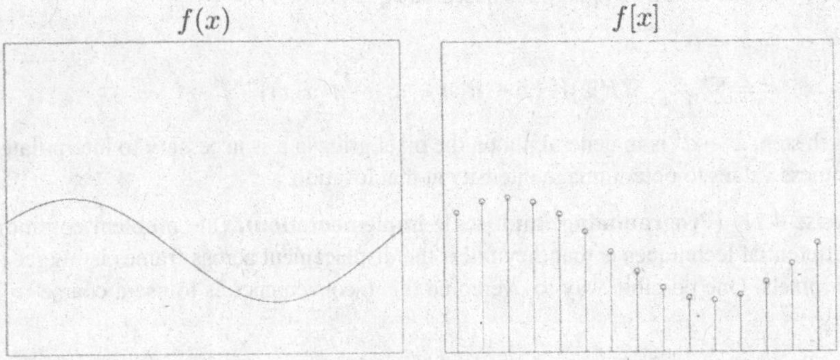


Figure 4.11. Continuous signal $f(x)$ and its discrete sampled version $f[x]$.

Consider a continuous signal $f(x)$ and its Fourier transform $F(\omega)$. The well-known *Nyquist sampling theorem* states that if the continuous signal $f(x)$ is band-limited, i.e. $|F(\omega)| = 0$ for all $\omega > \omega_n$, it can be reconstructed exactly from a set of discrete samples, provided that the sampling frequency $\omega_s > 2\omega_n$; $\omega_n$ is called the Nyquist frequency. The relationship between the sampling period and the sampling frequency is $\omega_s = \frac{2\pi}{T}$. Once the above relationship is satisfied, the original signal $f(x)$ can be reconstructed by multiplication of its sampled signal $f[x]$ in the frequency domain with an ideal reconstruction filter, denoted by $h(x)$, whose Fourier transform $H(\omega)$ is 1 between the frequencies $-\pi/T$ and $\pi/T$, and 0 elsewhere. That is, the reconstruction filter $h(x)$ is a sync function:

$$h(x) = \frac{\sin(\pi x/T)}{\pi x/T}, \quad x \in \mathbb{R}. \tag{4.44}$$

A multiplication in the frequency domain corresponds to a convolution in the spatial domain. Therefore,

$$f(x) = f[x] * h(x), \quad x \in \mathbb{R}, \tag{4.45}$$

as long as $\omega_n(f) < \frac{\pi}{T}$.

*Derivative of a sampled signal*

Knowing the relationship between the sampled function $f[x]$ and its continuous version $f(x)$, one can approach the computation of the derivative of the sampled function by first computing the derivative of the continuous function $f(x)$ and then sampling the result. We will outline this process for 1-D signals and then describe how to carry out the computation for 2-D images. Applying the derivative operator to both sides of equation (4.45) yields

$$D\{f(x)\} = D\{f[x] * h(x)\}. \tag{4.46}$$

Expressing the right hand side in terms of the convolution, and using the fact that both the derivative and the convolution are linear operators, we can bring the derivative operator inside of the convolution and write

$$\begin{aligned} D\{f(x)\} &= D\left\{\sum_{k=-\infty}^{k=\infty} f[k]h(x-k)\right\} \\ &= \sum_{k=-\infty}^{k=\infty} f[k]D\{h(x-k)\} = f[x] * D\{h(x)\}. \end{aligned}$$

Notice that the derivative operation is being applied only to continuous entities. Once the derivative of the continuous function has been computed, all we need to do is to sample the result. Denoting the sampling operator as $S\{\cdot\}$ and $D\{f(x)\}$ as $f'(x)$ we have

$$S\{f'(x)\} = S\{f[x] * D\{h(x)\}\} = f[x] * S\{h'(x)\} = f[x] * h'[x]. \tag{4.47}$$

Hence in an ideal situation the derivative of the sampled function can be computed as a convolution of the sampled signal with the sampled derivative of an ideal sync $h'(x)$ (Figure 4.12), where

$$h'(x) = \frac{(\pi^2 x/T^2)\cos(\pi x/T) - \pi/T \sin(\pi x/T)}{(\pi x/T)^2}, \quad x \in \mathbb{R}.$$

Note that in general the value of the function $f'[x]$ receives contribution from all samples of $h'[x]$. However, since the extent of $h[x]$ is infinite and the functions falls off very slowly far away from the origin, the convolution is not practically feasible and simple truncating would yield undesirable artifacts. In practice the computation of derivatives is accomplished by convolving the signal with a finite filter. In case of 1-D signals, a commonly used approximation to the ideal sync and its derivative is a Gaussian and its derivative, respectively, defined as

$$g(x) = \frac{1}{\sqrt{2\pi}\sigma} e^{\frac{-x^2}{2\sigma^2}}, \quad g'(x) = -\frac{x}{\sigma^2\sqrt{2\pi}\sigma} e^{\frac{-x^2}{2\sigma^2}}. \tag{4.48}$$

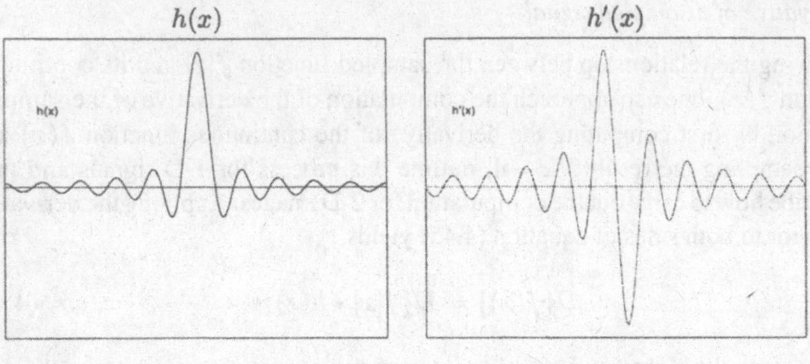


Figure 4.12. Ideal sync function and its derivative.

Note that the Gaussian, like the sync, extends to infinity, and therefore it needs to be truncated.[9] The derivative of a 1-D signal can then be simply computed by convolution with a finite-size filter, which is obtained by sampling and truncating the continuous derivative of the Gaussian. The number of samples $w$ needed is typically related to the variance $\sigma$. An adequate relationship between the two is $w = 5\sigma$, imposing the fact that the window subtends 98.76% of the area under the curve. In such a case the convolution becomes

$$f'[x] = f[x] * g'[x] = \sum_{k=-\frac{w}{2}}^{k=\frac{w}{2}} f[k]g'[x-k], \quad x, k \in \mathbb{Z}. \tag{4.49}$$

Examples of the Gaussian filter and its derivative are shown in Figure 4.13.

*Image gradient*

In order to compute the derivative of a 2-D signal defined by equation (4.42) we have to revisit the relationship between the continuous and sampled versions of the signal

$$I(x, y) = I[x, y] * h(x, y), \quad x, y \in \mathbb{R}, \tag{4.50}$$

where

$$h(x, y) = \frac{\sin(\pi x/T)\sin(\pi y/T)}{\pi^2 xy/T^2}, \quad x, y \in \mathbb{R}, \tag{4.51}$$

is a 2-D ideal sync. Notice that this function is separable, namely $h(x, y) = h(x)h(y)$. Without loss of generality consider the derivative with respect to $x$. Applying again the derivative operator to both sides we obtain

$$D_x\{I(x, y)\} = D_x\{I[x, y] * h(x, y)\}. \tag{4.52}$$

[9]Nevertheless, the value of a Gaussian function drops to zero exponentially and much faster than a sync, although its Fourier transform, also a Gaussian function, is not an ideal band-pass filter like the sync.

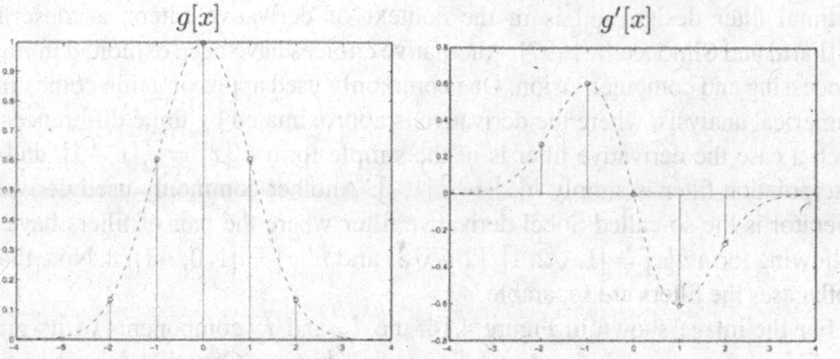


Figure 4.13. Examples of a 1-D five-tap Gaussian filter and its derivative, sampled from a continuous Gaussian function with a variance $\sigma = 1$. The numerical values of the samples are $g[x] = [0.1353, 0.6065, 1.0000, 0.6065, 0.1353]$, and $g'[x] = [0.2707, 0.6065, 0, -0.6065, -0.2707]$, respectively.

Since an ideal sync is separable we can write

$$D_x\{I(x,y)\} = I[x,y] * D_x\{h(x,y)\} = I[x,y] * D_x\{h(x)\} * h(y). \quad (4.53)$$

At last sampling the result we obtain the expression for $I_x$ component of the image gradient

$$\begin{aligned} S\{D_x\{I(x,y)\}\} &= S\{I[x,y] * D_x\{h(x) * h(y)\}\}, \\ I_x[x,y] &= I[x,y] * h'[x] * h[y]. \end{aligned}$$

Similarly the partial derivative $I_y$ is given by

$$I_y[x,y] = I[x,y] * h[x] * h'[y]. \quad (4.54)$$

Note that when computing the partial image derivatives, the image is convolved in one direction with the derivative filter and in the other direction with the interpolation filter. By the same argument as in the 1-D case, we approximate the ideal sync function with a Gaussian function, which falls off faster, and we sample from it and truncate it to obtain a finite-size filter. The computation of image derivatives is then accomplished as a pair of 1-D convolutions with filters obtained by sampling the continuous Gaussian function and its derivative, as shown in Figure 4.13. The image gradient at the pixel $[x,y]^T \in \mathbb{Z}^2$ is then given by

$$I_x[x,y] = I[x,y] * g'[x] * g[y] = \sum_{k=-\frac{w}{2}}^{\frac{w}{2}} \sum_{l=-\frac{w}{2}}^{\frac{w}{2}} I[k,l]g'[x-k]g[y-l],$$

$$I_y[x,y] = I[x,y] * g[x] * g'[y] = \sum_{k=-\frac{w}{2}}^{\frac{w}{2}} \sum_{l=-\frac{w}{2}}^{\frac{w}{2}} I[k,l]g[x-k]g'[y-l].$$

Recall that our choice is only an approximation to the ideal derivative filter. More systematic means of designing such approximations is a subject of optimal filter design and is in the context of derivative filters, as described in [Farid and Simoncelli, 1997]. Alternative choices have been exploited in image processing and computer vision. One commonly used approximation comes from numerical analysis, where the derivative is approximated by finite differences. In such a case the derivative filter is of the simple form $h'[x] = \frac{1}{2}[1, -1]$, and the interpolation filter is simply $h[x] = \frac{1}{2}[1, 1]$. Another commonly used derivative operator is the so-called Sobel derivative filter where the pair of filters have the following form $h[x] = [1, \sqrt{2}, 1]/(2 + \sqrt{2})$ and $h'[x] = [1, 0, -1]/3$. Note that in both cases the filters are separable.

For the image shown in Figure 4.14, the $I_x$ and $I_y$ components of its gradient $\nabla I$ were computed via convolution with a five-tap Gaussian derivative filter shown in Figure 4.13.

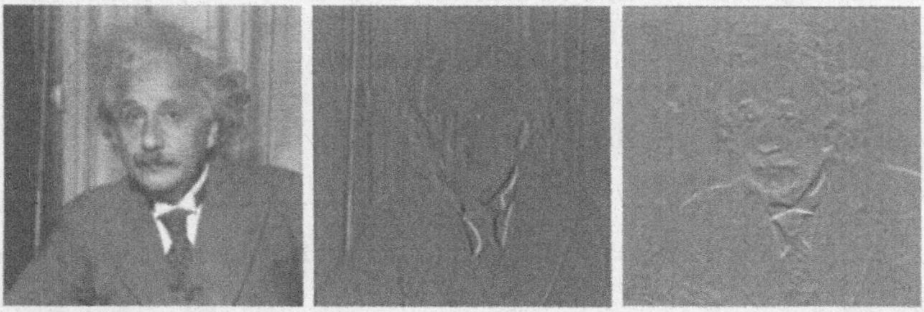

Figure 4.14. Left: original image. Middle and right: Horizontal component $I_x$ and vertical component $I_y$ of the image gradient $\nabla I$.

### *Image smoothing*

In many instances due to the presence of noise in the image formation process, it is often desirable to smooth the image in order to suppress the high frequency component. For this purpose the Gaussian filter is again suitable choice. The image smoothing is then simply accomplished by two 1-D convolutions with the Gaussian. Convolution with the Gaussian can be done efficiently due to the fact that the Gaussian is again separable. The smoothed image then becomes

$$\tilde{I}(x, y) = I(x, y) * g(x, y) = I(x, y) * g(x) * g(y). \tag{4.55}$$

The same expression written in terms of convolution with the filter size $w$ is

$$\tilde{I}[x, y] = I[x, y] * g[x, y] = \sum_{k=-\frac{w}{2}}^{\frac{w}{2}} \sum_{l=-\frac{w}{2}}^{\frac{w}{2}} I[k, l] g[x - k] g[y - l]. \tag{4.56}$$

Figure 4.15 demonstrates the effect of smoothing a noisy image via convolution with a Gaussian.

Figure 4.15. Left: the image "Lena" corrupted by white noise. Right: the corrupted image smoothed by convolution with a 2-D Gaussian.

## Further readings

*Extraction of corners, edges, and contours*

Gradient-based edge detectors like the Canny edge detector [Canny, 1986] and the Harris corner detector [Harris and Stephens, 1988] introduced in this chapter are widely available publicly, for instance, [Meer and Georgescu, www]. Further studies on the extraction of edge elements can be found in the work of [Casadei and Mitter, 1998, Parent and Zucker, 1989, Medioni et al., 2000]. Since lines are special curves with a constant curvature zero, they can also be extracted using curve extraction techniques based on the *constant curvature criterion* [Wuescher and Boyer, 1991]. Besides gradient-based edge detection methods, edges as boundaries of homogeneous regions can also be extracted using the *active contour methods* [Kass et al., 1987, Cohen, 1991, Menet et al., 1990]. One advantage of active contour is its robustness in generating continuous boundaries, but it typically involves solving partial differential equations; see [Kichenassamy et al., 1996, Sapiro, 2001, Osher and Sethian, 1988], and [Chan and Vese, 1999]. In time-critical applications, such as robot navigation, the gradient-based edge detection methods are more commonly used. The Hough transformation is another popular tool, which in principle enables one to extract any type of geometric primitives, including corners, edges, and curves. But its limitation is that one usually needs to specify the size of the primitive a priori.

*Feature tracking and optical flow*

Image motion (either feature tracking or optical flow) refers to the motion of brightness patterns on the image. It is only under restrictive assumptions on the photometry of the scene, which we discussed in Appendix 3.A, that such image motion is actually related to the motion of the scene. For instance, one can imagine a painted marble sphere rotating in space, and a static spherical mirror where the ambient light is moved to match the image motion of the first sphere. The distinction between motion field (the motion of the projection of points in the scene onto the image) and optical flow (the motion of brightness patterns on the image) has been elucidated by [Verri and Poggio, 1989].

The feature-tracking schemes given in this chapter mainly follow the work of [Lucas and Kanade, 1981, Tomasi and Kanade, 1992]. The affine flow tracking method was due to [Shi and Tomasi, 1994]. Multiscale estimation methods of global affine flow fields have been introduced by [Bergen et al., 1992]. The use robust estimation techniques in the context of optical flow computation have been proposed by [Black and Anandan, 1993]. Feature extracting and tracking, as we have described it, is an intrinsically local operation in space and time. Therefore, it is extremely difficult to maintain tracking of a feature over extended lengths of time. Typically, a feature point becomes occluded or changes its appearance up to the point of not passing the SSD test. This does not mean that we cannot integrate motion information over time. In fact, it is possible that even if individual features appear and disappear, their motion is consistent with a global 3-D interpretation, as we will see in later parts of the book. Alternative to feature tracking include deformable contours [Blake and Isard, 1998], learning-based approaches [Yacoob and Davis, 1998] and optical flow [Verri and Poggio, 1989, Weickert et al., 1998, Nagel, 1987]. Computation of qualitative ego-motion from normal flow was addressed in [Fermüller and Aloimonos, 1995].

# Part II

# Geometry of Two Views

# Chapter 5

# Reconstruction from Two Calibrated Views

*We see because we move; we move because we see.*
– James J. Gibson, *The Perception of the Visual World*

In this chapter we begin unveiling the basic geometry that relates images of points to their 3-D position. We start with the simplest case of two calibrated cameras, and describe an algorithm, first proposed by the British psychologist H.C. Longuet-Higgins in 1981, to reconstruct the relative pose (i.e. position and orientation) of the cameras as well as the locations of the points in space from their projection onto the two images.

It has been long known in photogrammetry that the coordinates of the projection of a point and the two camera optical centers form a triangle (Figure 5.1), a fact that can be written as an algebraic constraint involving the camera poses and image coordinates but *not* the 3-D position of the points. Given enough points, therefore, this constraint can be used to solve for the camera poses. Once those are known, the 3-D position of the points can be obtained easily by triangulation. The interesting feature of the constraint is that although it is nonlinear in the unknown camera poses, it can be solved by two linear steps in closed form. Therefore, in the absence of any noise or uncertainty, given two images taken from calibrated cameras, one can in principle recover camera pose and position of the points in space with a few steps of simple linear algebra.

While we have not yet indicated how to calibrate the cameras (which we will do in Chapter 6), this chapter serves to introduce the basic building blocks of the geometry of two views, known as "epipolar geometry." The simple algorithms to be introduced in this chapter, although merely conceptual,[1] allow us to introduce the basic ideas that will be revisited in later chapters of the book to derive more powerful algorithms that can deal with uncertainty in the measurements as well as with uncalibrated cameras.

## 5.1 Epipolar geometry

Consider two images of the same scene taken from two distinct vantage points. If we assume that the camera is *calibrated*, as described in Chapter 3 (the calibration matrix $K$ is the identity), the homogeneous image coordinates $\boldsymbol{x}$ and the spatial coordinates $\boldsymbol{X}$ of a point $p$, with respect to the camera frame, are related by[2]

$$\lambda \boldsymbol{x} = \Pi_0 \boldsymbol{X}, \tag{5.1}$$

where $\Pi_0 = [I, 0]$. That is, the image $\boldsymbol{x}$ differs from the actual 3-D coordinates of the point by an unknown (depth) scale $\lambda \in \mathbb{R}_+$. For simplicity, we will assume that the scene is *static* (that is, there are no moving objects) and that the position of corresponding feature points across images is available, for instance from one of the algorithms described in Chapter 4. If we call $\boldsymbol{x}_1, \boldsymbol{x}_2$ the corresponding points in two views, they will then be related by a precise geometric relationship that we describe in this section.

### *5.1.1 The epipolar constraint and the essential matrix*

Following Chapter 3, an orthonormal reference frame is associated with each camera, with its origin $o$ at the optical center and the $z$-axis aligned with the optical axis. The relationship between the 3-D coordinates of a point in the inertial "world" coordinate frame and the camera frame can be expressed by a rigid-body transformation. Without loss of generality, we can assume the world frame to be one of the cameras, while the other is positioned and oriented according to a Euclidean transformation $g = (R, T) \in SE(3)$. If we call the 3-D coordinates of a point $p$ relative to the two camera frames $\boldsymbol{X}_1 \in \mathbb{R}^3$ and $\boldsymbol{X}_2 \in \mathbb{R}^3$, they are related by a rigid-body transformation in the following way:

$$\boldsymbol{X}_2 = R\boldsymbol{X}_1 + T.$$

Now let $\boldsymbol{x}_1, \boldsymbol{x}_2 \in \mathbb{R}^3$ be the homogeneous coordinates of the projection of *the same* point $p$ in the two image planes. Since $\boldsymbol{X}_i = \lambda_i \boldsymbol{x}_i, i = 1, 2$, this equation

[1] They are not suitable for real images, which are typically corrupted by noise. In Section 5.2.3 of this chapter, we show how to modify them so as to minimize the effect of noise and obtain an optimal solution.

[2] We remind the reader that we do not distinguish between ordinary and homogeneous coordinates; in the former case $\boldsymbol{x} \in \mathbb{R}^2$, whereas in the latter $\boldsymbol{x} \in \mathbb{R}^3$ with the last component being 1. Similarly, $\boldsymbol{X} \in \mathbb{R}^3$ or $\boldsymbol{X} \in \mathbb{R}^4$ depends on whether ordinary or homogeneous coordinates are used.

can be written in terms of the image coordinates $\boldsymbol{x}_i$ and the depths $\lambda_i$ as

$$\lambda_2 \boldsymbol{x}_2 = R\lambda_1 \boldsymbol{x}_1 + T.$$

In order to eliminate the depths $\lambda_i$ in the preceding equation, premultiply both sides by $\widehat{T}$ to obtain

$$\lambda_2 \widehat{T} \boldsymbol{x}_2 = \widehat{T} R \lambda_1 \boldsymbol{x}_1.$$

Since the vector $\widehat{T}\boldsymbol{x}_2 = T \times \boldsymbol{x}_2$ is perpendicular to the vector $\boldsymbol{x}_2$, the inner product $\langle \boldsymbol{x}_2, \widehat{T}\boldsymbol{x}_2 \rangle = \boldsymbol{x}_2{}^T \widehat{T} \boldsymbol{x}_2$ is zero. Premultiplying the previous equation by $\boldsymbol{x}_2^T$ yields that the quantity $\boldsymbol{x}_2^T \widehat{T} R \lambda_1 \boldsymbol{x}_1$ is zero. Since $\lambda_1 > 0$, we have proven the following result:

**Theorem 5.1 (Epipolar constraint).** *Consider two images $\boldsymbol{x}_1, \boldsymbol{x}_2$ of the same point $p$ from two camera positions with relative pose $(R, T)$, where $R \in SO(3)$ is the relative orientation and $T \in \mathbb{R}^3$ is the relative position. Then $\boldsymbol{x}_1, \boldsymbol{x}_2$ satisfy*

$$\langle \boldsymbol{x}_2, T \times R\boldsymbol{x}_1 \rangle = 0, \quad \text{or} \quad \boxed{\boldsymbol{x}_2^T \widehat{T} R \boldsymbol{x}_1 = 0.} \tag{5.2}$$

The matrix

$$E \doteq \widehat{T} R \quad \in \mathbb{R}^{3 \times 3}$$

in the epipolar constraint equation (5.2) is called the *essential matrix*. It encodes the relative pose between the two cameras. The epipolar constraint (5.2) is also called the *essential constraint*. Since the epipolar constraint is bilinear in each of its arguments $\boldsymbol{x}_1$ and $\boldsymbol{x}_2$, it is also called the *bilinear constraint*. We will revisit this bilinear nature in later chapters.

In addition to the preceding algebraic derivation, this constraint follows immediately from geometric considerations, as illustrated in Figure 5.1. The vector connecting the first camera center $o_1$ and the point $p$, the vector connecting $o_2$

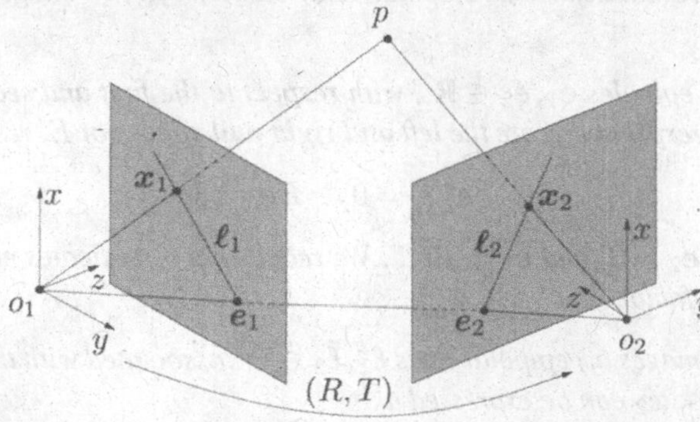


Figure 5.1. Two projections $\boldsymbol{x}_1, \boldsymbol{x}_2 \in \mathbb{R}^3$ of a 3-D point $p$ from two vantage points. The Euclidean transformation between the two cameras is given by $(R, T) \in SE(3)$. The intersections of the line $(o_1, o_2)$ with each image plane are called *epipoles* and denoted by $\boldsymbol{e}_1$ and $\boldsymbol{e}_2$. The lines $\boldsymbol{\ell}_1, \boldsymbol{\ell}_2$ are called *epipolar lines*, which are the intersection of the plane $(o_1, o_2, p)$ with the two image planes.

and $p$, and the vector connecting the two optical centers $o_1$ and $o_2$ clearly form a triangle. Therefore, the three vectors lie on the same plane. Their triple product,[3] which measures the volume of the parallelepiped determined by the three vectors, is therefore zero. This is true for the coordinates of the points $\boldsymbol{X}_i,\ i = 1, 2$, as well as for the homogeneous coordinates of their projection $\boldsymbol{x}_i,\ i = 1, 2$, since $\boldsymbol{X}_i$ and $\boldsymbol{x}_i$ (as vectors) differ only be a scalar factor. The constraint (5.2) is just the triple product written in the second camera frame; $R\boldsymbol{x}_1$ is simply the direction of the vector $\overrightarrow{o_1 p}$, and $T$ is the vector $\overrightarrow{o_2 o_1}$ with respect to the second camera frame. The translation $T$ between the two camera centers $o_1$ and $o_2$ is also called the *baseline*.

Associated with this picture, we define the following set of geometric entities, which will facilitate our future study:

**Definition 5.2 (Epipolar geometric entities).**

1. *The plane $(o_1, o_2, p)$ determined by the two centers of projection $o_1, o_2$ and the point $p$ is called an* epipolar plane *associated with the camera configuration and point $p$. There is one epipolar plane for each point $p$.*

2. *The projection $\mathbf{e}_1(\mathbf{e}_2)$ of one camera center onto the image plane of the other camera frame is called an* epipole. *Note that the projection may occur outside the physical boundary of the imaging sensor.*

3. *The intersection of the epipolar plane of $p$ with one image plane is a line $\boldsymbol{\ell}_1(\boldsymbol{\ell}_2)$, which is called the* epipolar line *of $p$. We usually use the normal vector $\boldsymbol{\ell}_1(\boldsymbol{\ell}_2)$ to the epipolar plane to denote this line.*[4]

From the definitions, we immediately have the following relations among epipoles, epipolar lines, and image points:

**Proposition 5.3 (Properties of epipoles and epipolar lines).** *Given an essential matrix $E = \widehat{T}R$ that defines an epipolar relation between two images $\boldsymbol{x}_1, \boldsymbol{x}_2$, we have:*

1. *The two epipoles $\mathbf{e}_1, \mathbf{e}_2 \in \mathbb{R}^3$, with respect to the first and second camera frames, respectively, are the left and right null spaces of $E$, respectively:*

$$\mathbf{e}_2^T E = 0, \quad E\mathbf{e}_1 = 0. \tag{5.3}$$

*That is, $\mathbf{e}_2 \sim T$ and $\mathbf{e}_1 \sim R^T T$. We recall that $\sim$ indicates equality up to a scalar factor.*

2. *The (coimages of) epipolar lines $\boldsymbol{\ell}_1, \boldsymbol{\ell}_2 \in \mathbb{R}^3$ associated with the two image points $\boldsymbol{x}_1, \boldsymbol{x}_2$ can be expressed as*

$$\boldsymbol{\ell}_2 \sim E\boldsymbol{x}_1, \quad \boldsymbol{\ell}_1 \sim E^T\boldsymbol{x}_2 \quad \in \mathbb{R}^3, \tag{5.4}$$

---

[3] As we have seen in Chapter 2, the triple product of three vectors is the inner product of one with the cross product of the other two.

[4] Hence the vector $\boldsymbol{\ell}_1(\boldsymbol{\ell}_2)$ is in fact the coimage of the epipolar line.

*where $\ell_1, \ell_2$ are in fact the normal vectors to the epipolar plane expressed with respect to the two camera frames, respectively.*

3. *In each image, both the image point and the epipole lie on the epipolar line*

$$\ell_i^T e_i = 0, \quad \ell_i^T x_i = 0, \quad i = 1, 2. \tag{5.5}$$

The proof is simple, and we leave it to the reader as an exercise. Figure 5.2 illustrates the relationships among 3-D points, images, epipolar lines, and epipoles.

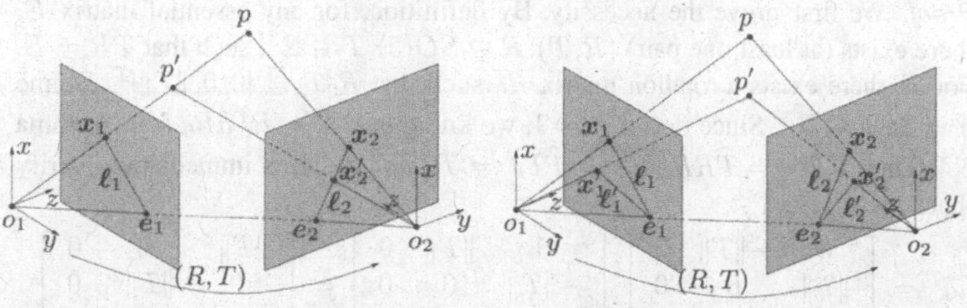


Figure 5.2. Left: the essential matrix $E$ associated with the epipolar constraint maps an image point $x_1$ in the first image to an epipolar line $\ell_2 = Ex_1$ in the second image; the precise location of its corresponding image ($x_2$ or $x_2'$) depends on where the 3-D point ($p$ or $p'$) lies on the ray $(o_1, x_1)$; Right: When $(o_1, o_2, p)$ and $(o_1, o_2, p')$ are two different planes, they intersect at the two image planes at two pairs of epipolar lines $(\ell_1, \ell_2)$ and $(\ell_1', \ell_2')$, respectively, and these epipolar lines always pass through the pair of epipoles $(e_1, e_2)$.

### 5.1.2 *Elementary properties of the essential matrix*

The matrix $E = \widehat{T}R \in \mathbb{R}^{3\times3}$ in equation (5.2) contains information about the relative position $T$ and orientation $R \in SO(3)$ between the two cameras. Matrices of this form belong to a very special set of matrices in $\mathbb{R}^{3\times3}$ called the *essential space* and denoted by $\mathcal{E}$:

$$\mathcal{E} \doteq \left\{ \widehat{T}R \,\middle|\, R \in SO(3), T \in \mathbb{R}^3 \right\} \quad \subset \mathbb{R}^{3\times3}.$$

Before we study the structure of essential matrices, we introduce a useful lemma from linear algebra.

**Lemma 5.4 (The hat operator).** *For a vector $T \in \mathbb{R}^3$ and a matrix $K \in \mathbb{R}^{3\times3}$, if $\det(K) = +1$ and $T' = KT$, then $\widehat{T} = K^T\widehat{T'}K$.*

*Proof.* Since both $K^T\widehat{(\cdot)}K$ and $\widehat{K^{-1}(\cdot)}$ are linear maps from $\mathbb{R}^3$ to $\mathbb{R}^{3\times3}$, one may directly verify that these two linear maps agree on the basis vectors $[1,0,0]^T$, $[0,1,0]^T$, and $[0,0,1]^T$ (using the fact that $\det(K) = 1$). □

The following theorem, due to [Huang and Faugeras, 1989], captures the algebraic structure of essential matrices in terms of their singular value decomposition (see Appendix A for a review on the SVD):

**Theorem 5.5 (Characterization of the essential matrix).** *A nonzero matrix* $E \in \mathbb{R}^{3\times 3}$ *is an essential matrix* if and only if $E$ *has a singular value decomposition (SVD)* $E = U\Sigma V^T$ *with*

$$\Sigma = diag\{\sigma, \sigma, 0\}$$

*for some* $\sigma \in \mathbb{R}_+$ *and* $U, V \in SO(3)$.

*Proof.* We first prove the necessity. By definition, for any essential matrix $E$, there exists (at least one pair) $(R, T)$, $R \in SO(3)$, $T \in \mathbb{R}^3$, such that $\widehat{T}R = E$. For $T$, there exists a rotation matrix $R_0$ such that $R_0 T = [0, 0, \|T\|]^T$. Define $a = R_0 T \in \mathbb{R}^3$. Since $\det(R_0) = 1$, we know that $\widehat{T} = R_0^T \widehat{a} R_0$ from Lemma 5.4. Then $EE^T = \widehat{T}RR^T\widehat{T}^T = \widehat{T}\widehat{T}^T = R_0^T \widehat{a}\widehat{a}^T R_0$. It is immediate to verify that

$$\widehat{a}\widehat{a}^T = \begin{bmatrix} 0 & -\|T\| & 0 \\ \|T\| & 0 & 0 \\ 0 & 0 & 0 \end{bmatrix} \begin{bmatrix} 0 & \|T\| & 0 \\ -\|T\| & 0 & 0 \\ 0 & 0 & 0 \end{bmatrix} = \begin{bmatrix} \|T\|^2 & 0 & 0 \\ 0 & \|T\|^2 & 0 \\ 0 & 0 & 0 \end{bmatrix}.$$

So, the singular values of the essential matrix $E = \widehat{T}R$ are $(\|T\|, \|T\|, 0)$. However, in the standard SVD of $E = U\Sigma V^T$, $U$ and $V$ are only orthonormal, and their determinants can be $\pm 1$.[5] We still need to prove that $U, V \in SO(3)$ (i.e. they have determinant $+1$) to establish the theorem. We already have $E = \widehat{T}R = R_0^T \widehat{a} R_0 R$. Let $R_Z(\theta)$ be the matrix that represents a rotation around the $Z$-axis by an angle of $\theta$ radians; i.e. $R_Z(\theta) \doteq e^{\widehat{e}_3 \theta}$ with $e_3 = [0, 0, 1]^T \in \mathbb{R}^3$. Then

$$R_Z\left(+\frac{\pi}{2}\right) = \begin{bmatrix} 0 & -1 & 0 \\ 1 & 0 & 0 \\ 0 & 0 & 1 \end{bmatrix}.$$

Then $\widehat{a} = R_Z(+\frac{\pi}{2}) R_Z^T(+\frac{\pi}{2})\widehat{a} = R_Z(+\frac{\pi}{2})\ \mathrm{diag}\{\|T\|, \|T\|, 0\}$. Therefore,

$$E = \widehat{T}R = R_0^T R_Z\left(+\frac{\pi}{2}\right)\ \mathrm{diag}\{\|T\|, \|T\|, 0\} R_0 R.$$

So, in the SVD of $E = U\Sigma V^T$, we may choose $U = R_0^T R_Z(+\frac{\pi}{2})$ and $V^T = R_0 R$. Since we have constructed both $U$ and $V$ as products of matrices in $SO(3)$, they are in $SO(3)$, too; that is, both $U$ and $V$ are rotation matrices.

We now prove sufficiency. If a given matrix $E \in \mathbb{R}^{3\times 3}$ has SVD $E = U\Sigma V^T$ with $U, V \in SO(3)$ and $\Sigma = \mathrm{diag}\{\sigma, \sigma, 0\}$, define $(R_1, T_1) \in SE(3)$ and $(R_2, T_2) \in SE(3)$ to be

$$\begin{cases} (\widehat{T}_1, R_1) &= (UR_Z(+\frac{\pi}{2})\Sigma U^T, UR_Z^T(+\frac{\pi}{2})V^T), \\ (\widehat{T}_2, R_2) &= (UR_Z(-\frac{\pi}{2})\Sigma U^T, UR_Z^T(-\frac{\pi}{2})V^T). \end{cases} \tag{5.6}$$

[5]Interested readers can verify this using the Matlab routine: SVD.

It is now easy to verify that $\widehat{T}_1 R_1 = \widehat{T}_2 R_2 = E$. Thus, $E$ is an essential matrix. □

Given a rotation matrix $R \in SO(3)$ and a translation vector $T \in \mathbb{R}^3$, it is immediate to construct an essential matrix $E = \widehat{T}R$. The inverse problem, that is how to retrieve $T$ and $R$ from a given essential matrix $E$, is less obvious. In the sufficiency proof for the above theorem, we have used the SVD to construct two solutions for $(R, T)$. Are these the only solutions? Before we can answer this question in the upcoming Theorem 5.7, we need the following lemma.

**Lemma 5.6.** *Consider an arbitrary nonzero skew-symmetric matrix $\widehat{T} \in so(3)$ with $T \in \mathbb{R}^3$. If for a rotation matrix $R \in SO(3)$, $\widehat{T}R$ is also a skew-symmetric matrix, then $R = I$ or $R = e^{\widehat{u}\pi}$, where $u = \frac{T}{\|T\|}$. Further, $\widehat{T}e^{\widehat{u}\pi} = -\widehat{T}$.*

*Proof.* Without loss of generality, we assume that $T$ is of unit length. Since $\widehat{T}R$ is also a skew-symmetric matrix, $(\widehat{T}R)^T = -\widehat{T}R$. This equation gives

$$R\widehat{T}R = \widehat{T}. \tag{5.7}$$

Since $R$ is a rotation matrix, there exist $\omega \in \mathbb{R}^3$, $\|\omega\| = 1$, and $\theta \in \mathbb{R}$ such that $R = e^{\widehat{\omega}\theta}$. If $\theta = 0$ the lemma is proved. Hence consider the case $\theta \neq 0$. Then, (5.7) is rewritten as $e^{\widehat{\omega}\theta}\widehat{T}e^{\widehat{\omega}\theta} = \widehat{T}$. Applying this equation to $\omega$, we get $e^{\widehat{\omega}\theta}\widehat{T}e^{\widehat{\omega}\theta}\omega = \widehat{T}\omega$. Since $e^{\widehat{\omega}\theta}\omega = \omega$, we obtain $e^{\widehat{\omega}\theta}\widehat{T}\omega = \widehat{T}\omega$. Since $\omega$ is the only eigenvector associated with the eigenvalue 1 of the matrix $e^{\widehat{\omega}\theta}$, and $\widehat{T}\omega$ is orthogonal to $\omega$, $\widehat{T}\omega$ has to be zero. Thus, $\omega$ is equal to either $\frac{T}{\|T\|}$ or $-\frac{T}{\|T\|}$; i.e. $\omega = \pm u$. Then $R$ has the form $e^{\widehat{\omega}\theta}$, which commutes with $\widehat{T}$. Thus from (5.7), we get

$$e^{2\widehat{\omega}\theta}\widehat{T} = \widehat{T}. \tag{5.8}$$

According to *Rodrigues' formula* (2.16) from Chapter 2, we have

$$e^{2\widehat{\omega}\theta} = I + \widehat{\omega}\sin(2\theta) + \widehat{\omega}^2(1 - \cos(2\theta)),$$

and (5.8) yields

$$\widehat{\omega}^2 \sin(2\theta) + \widehat{\omega}^3(1 - \cos(2\theta)) = 0.$$

Since $\widehat{\omega}^2$ and $\widehat{\omega}^3$ are linearly independent (we leave this as an exercise to the reader), we have $\sin(2\theta) = 1 - \cos(2\theta) = 0$. That is, $\theta$ is equal to $2k\pi$ or $2k\pi + \pi$, $k \in \mathbb{Z}$. Therefore, $R$ is equal to $I$ or $e^{\widehat{\omega}\pi}$. Now if $\omega = u = \frac{T}{\|T\|}$, then it is direct from the geometric meaning of the rotation $e^{\widehat{\omega}\pi}$ that $e^{\widehat{\omega}\pi}\widehat{T} = -\widehat{T}$. On the other hand, if $\omega = -u = -\frac{T}{\|T\|}$, then it follows that $e^{\widehat{\omega}}\widehat{T} = -\widehat{T}$. Thus, in any case the conclusions of the lemma follow. □

The following theorem shows exactly how many rotation and translation pairs $(R, T)$ one can extract from an essential matrix, and the solutions are given in closed form by equation (5.9).

**Theorem 5.7 (Pose recovery from the essential matrix).** *There exist exactly two relative poses $(R, T)$ with $R \in SO(3)$ and $T \in \mathbb{R}^3$ corresponding to a nonzero essential matrix $E \in \mathcal{E}$.*

*Proof.* Assume that $(R_1, T_1) \in SE(3)$ and $(R_2, T_2) \in SE(3)$ are both solutions for the equation $\widehat{T}R = E$. Then we have $\widehat{T}_1 R_1 = \widehat{T}_2 R_2$. This yields $\widehat{T}_1 = \widehat{T}_2 R_2 R_1^T$. Since $\widehat{T}_1, \widehat{T}_2$ are both skew-symmetric matrices and $R_2 R_1^T$ is a rotation matrix, from the preceding lemma, we have that either $(R_2, T_2) = (R_1, T_1)$ or $(R_2, T_2) = (e^{\widehat{u}_1 \pi} R_1, -T_1)$ with $u_1 = T_1/\|T_1\|$. Therefore, given an essential matrix $E$ there are exactly *two* pairs of $(R, T)$ such that $\widehat{T}R = E$. Further, if $E$ has the SVD: $E = U\Sigma V^T$ with $U, V \in SO(3)$, the following formulae give the two distinct solutions (recall that $R_Z(\theta) \doteq e^{\widehat{e}_3 \theta}$ with $e_3 = [0, 0, 1]^T \in \mathbb{R}^3$)

$$\boxed{\begin{array}{rcl} (\widehat{T}_1, R_1) & = & (UR_Z(+\frac{\pi}{2})\Sigma U^T, UR_Z^T(+\frac{\pi}{2})V^T), \\ (\widehat{T}_2, R_2) & = & (UR_Z(-\frac{\pi}{2})\Sigma U^T, UR_Z^T(-\frac{\pi}{2})V^T). \end{array}} \tag{5.9}$$

□

**Example 5.8 (Two solutions to an essential matrix).** It is immediate to verify that $\widehat{e_3} R_Z\left(+\frac{\pi}{2}\right) = \widehat{-e_3} R_Z\left(-\frac{\pi}{2}\right)$, since

$$\begin{bmatrix} 0 & 1 & 0 \\ -1 & 0 & 0 \\ 0 & 0 & 0 \end{bmatrix} \begin{bmatrix} 0 & -1 & 0 \\ 1 & 0 & 0 \\ 0 & 0 & 1 \end{bmatrix} = \begin{bmatrix} 0 & -1 & 0 \\ 1 & 0 & 0 \\ 0 & 0 & 0 \end{bmatrix} \begin{bmatrix} 0 & 1 & 0 \\ -1 & 0 & 0 \\ 0 & 0 & 1 \end{bmatrix}.$$

These two solutions together are usually referred to as a "twisted pair," due to the manner in which the two solutions are related geometrically, as illustrated in Figure 5.3. A physically correct solution can be chosen by enforcing that the reconstructed points be visible, i.e. they have positive depth. We discuss this issue further in Exercise 5.11. ■

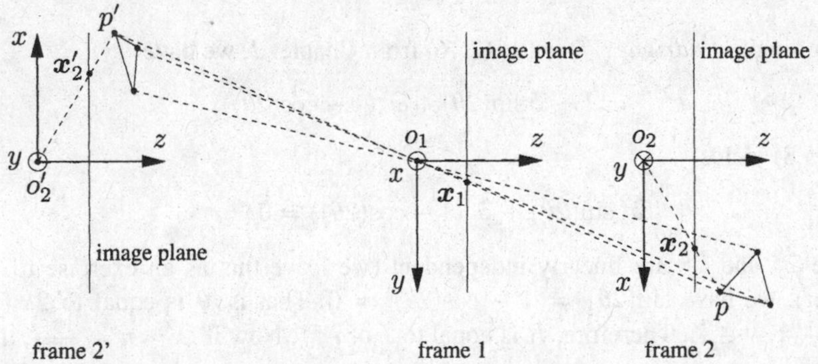


Figure 5.3. Two pairs of camera frames, i.e. $(1, 2)$ and $(1, 2')$, generate the same essential matrix. The frame 2 and frame $2'$ differ by a translation and a $180°$ rotation (a twist) around the $z$-axis, and the two pose pairs give rise to the same image coordinates. For the same set of image pairs $\boldsymbol{x}_1$ and $\boldsymbol{x}_2 = \boldsymbol{x}_2'$, the recovered structures $p$ and $p'$ might be different. Notice that with respect to the camera frame 1, the point $p'$ has a negative depth.

## 5.2 Basic reconstruction algorithms

In the previous section, we have seen that images of corresponding points are related by the epipolar constraint, which involves the unknown relative pose between the cameras. Therefore, given a number of corresponding points, we could use the epipolar constraints to try to recover camera pose. In this section, we show a simple closed-form solution to this problem. It consists of two steps: First a matrix $E$ is recovered from a number of epipolar constraints; then relative translation and orientation are extracted from $E$. However, since the matrix $E$ recovered using correspondence data in the epipolar constraint may not be an essential matrix, it needs to be projected into the space of essential matrices prior to extraction of the relative pose of the cameras using equation (5.9).

Although the linear algorithm that we propose here is suboptimal when the measurements are corrupted by noise, it is important for illustrating the geometric structure of the space of essential matrices. We leave the more practical issues with noise and optimality to Section 5.2.3.

### *5.2.1 The eight-point linear algorithm*

Let $E = \widehat{T}R$ be the essential matrix associated with the epipolar constraint (5.2). The entries of this $3 \times 3$ matrix are denoted by

$$E = \begin{bmatrix} e_{11} & e_{12} & e_{13} \\ e_{21} & e_{22} & e_{23} \\ e_{31} & e_{32} & e_{33} \end{bmatrix} \in \mathbb{R}^{3\times 3} \tag{5.10}$$

and stacked into a vector $E^s \in \mathbb{R}^9$, which is typically referred to as the *stacked version* of the matrix $E$ (Appendix A.1.3):

$$E^s \doteq [e_{11}, e_{21}, e_{31}, e_{12}, e_{22}, e_{32}, e_{13}, e_{23}, e_{33}]^T \quad \in \mathbb{R}^9.$$

The inverse operation from $E^s$ to its matrix version is then called *unstacking*. We further denote the *Kronecker product* $\otimes$ (also see Appendix A.1.3) of two vectors $\boldsymbol{x}_1$ and $\boldsymbol{x}_2$ by

$$\boldsymbol{a} \doteq \boldsymbol{x}_1 \otimes \boldsymbol{x}_2. \tag{5.11}$$

Or, more specifically, if $\boldsymbol{x}_1 = [x_1, y_1, z_1]^T \in \mathbb{R}^3$ and $\boldsymbol{x}_2 = [x_2, y_2, z_2]^T \in \mathbb{R}^3$, then

$$\boldsymbol{a} = [x_1x_2, x_1y_2, x_1z_2, y_1x_2, y_1y_2, y_1z_2, z_1x_2, z_1y_2, z_1z_2]^T \quad \in \mathbb{R}^9. \tag{5.12}$$

Since the epipolar constraint $\boldsymbol{x}_2^T E \boldsymbol{x}_1 = 0$ is linear in the entries of $E$, using the above notation we can rewrite it as the inner product of $\boldsymbol{a}$ and $E^s$:

$$\boxed{\boldsymbol{a}^T E^s = 0.}$$

This is just another way of writing equation (5.2) that emphasizes the linear dependence of the epipolar constraint on the elements of the essential matrix. Now, given a set of corresponding image points $(\boldsymbol{x}_1^j, \boldsymbol{x}_2^j)$, $j = 1, 2, \ldots, n$, define a matrix $\chi \in \mathbb{R}^{n\times 9}$ associated with these measurements to be

$$\chi \doteq [\boldsymbol{a}^1, \boldsymbol{a}^2, \ldots, \boldsymbol{a}^n]^T, \tag{5.13}$$

where the $j$th row $\boldsymbol{a}^j$ is the Kronecker product of each pair $(\boldsymbol{x}_1^j, \boldsymbol{x}_2^j)$ using (5.12). In the absence of noise, the vector $E^s$ satisfies

$$\chi E^s = 0. \tag{5.14}$$

This linear equation may now be solved for the vector $E^s$. For the solution to be unique (up to a scalar factor, ruling out the trivial solution $E^s = 0$), the rank of the matrix $\chi \in \mathbb{R}^{9\times n}$ needs to be exactly eight. This should be the case given $n \geq 8$ "ideal" corresponding points, as shown in Figure 5.4. In general, however, since correspondences may be prone to errors, there may be no solution to (5.14). In such a case, one can choose the $E^s$ that minimizes the least-squares error function $\|\chi E^s\|^2$. This is achieved by choosing $E^s$ to be the eigenvector of $\chi^T\chi$ that corresponds to its smallest eigenvalue, as we show in Appendix A. We would also like to draw attention to the case when the rank of $\chi$ is less than eight even for number of points greater than nine. In this instance there are multiple solutions to (5.14). This happens when the feature points are not in "general position," for example when they all lie on a plane. We will specifically deal with the planar case in the next section.

Figure 5.4. Eight pairs of corresponding image points in two views of the Tai-He palace in the Forbidden City, Beijing, China (photos courtesy of Jie Zhang).

However, even in the absence of noise, for a vector $E^s$ to be the solution of our problem, it is not sufficient that it be in the null space of $\chi$. In fact, it has to satisfy an additional constraint, that its matrix form $E$ belong to the space of essential matrices. Enforcing this structure in the determination of the null space of $\chi$ is difficult. Therefore, as a first cut, we estimate the null space of $\chi$, *ignoring the internal structure of essential matrix*, obtaining a matrix, say $F$, that probably does not belong to the essential space $\mathcal{E}$, and then "orthogonally" project the matrix thus obtained onto the essential space. This process is illustrated in Figure 5.5. The following theorem says precisely what this projection is.

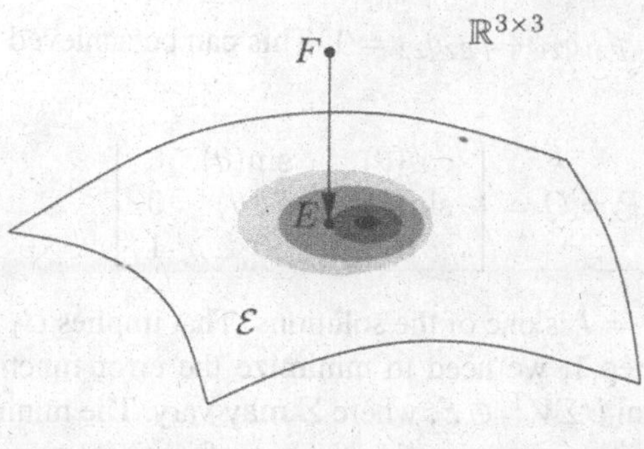


Figure 5.5. Among all points in the essential space $\mathcal{E} \subset \mathbb{R}^{3\times 3}$, $E$ has the shortest Frobenius distance to $F$. However, the least-square error may not be the smallest for so-obtained $E$ among all points in $\mathcal{E}$.

**Theorem 5.9 (Projection onto the essential space).** *Given a real matrix $F \in \mathbb{R}^{3\times 3}$ with SVD $F = U\text{diag}\{\lambda_1, \lambda_2, \lambda_3\}V^T$ with $U, V \in SO(3)$, $\lambda_1 \geq \lambda_2 \geq \lambda_3$, then the essential matrix $E \in \mathcal{E}$ that minimizes the error $\|E - F\|_f^2$ is given by $E = U\text{diag}\{\sigma, \sigma, 0\}V^T$ with $\sigma = (\lambda_1 + \lambda_2)/2$. The subscript "$f$" indicates the Frobenius norm of a matrix. This is the square norm of the sum of the squares of all the entries of the matrix (see Appendix A).*

*Proof.* For any fixed matrix $\Sigma = \text{diag}\{\sigma, \sigma, 0\}$, we define a subset $\mathcal{E}_\Sigma$ of the essential space $\mathcal{E}$ to be the set of all essential matrices with SVD of the form $U_1\Sigma V_1^T$, $U_1, V_1 \in SO(3)$. To simplify the notation, define $\Sigma_\lambda = \text{diag}\{\lambda_1, \lambda_2, \lambda_3\}$. We now prove the theorem in two steps:

*Step 1:* We prove that for a fixed $\Sigma$, the essential matrix $E \in \mathcal{E}_\Sigma$ that minimizes the error $\|E - F\|_f^2$ has a solution $E = U\Sigma V^T$ (not necessarily unique). Since $E \in \mathcal{E}_\Sigma$ has the form $E = U_1\Sigma V_1^T$, we get

$$\|E - F\|_f^2 = \|U_1\Sigma V_1^T - U\Sigma_\lambda V^T\|_f^2 = \|\Sigma_\lambda - U^T U_1 \Sigma V_1^T V\|_f^2.$$

Define $P = U^T U_1, Q = V^T V_1 \in SO(3)$, which have the form

$$P = \begin{bmatrix} p_{11} & p_{12} & p_{13} \\ p_{21} & p_{22} & p_{23} \\ p_{31} & p_{32} & p_{33} \end{bmatrix}, \quad Q = \begin{bmatrix} q_{11} & q_{12} & q_{13} \\ q_{21} & q_{22} & q_{23} \\ q_{31} & q_{32} & q_{33} \end{bmatrix}. \tag{5.15}$$

Then

$$\begin{aligned} \|E - F\|_f^2 &= \|\Sigma_\lambda - U^T U_1 \Sigma V_1^T V\|_f^2 \\ &= \text{trace}(\Sigma_\lambda^2) - 2\text{trace}(P\Sigma Q^T \Sigma_\lambda) + \text{trace}(\Sigma^2). \end{aligned}$$

Expanding the second term, using $\Sigma = \text{diag}\{\sigma, \sigma, 0\}$ and the notation $p_{ij}, q_{ij}$ for the entries of $P, Q$, we have

$$\text{trace}(P\Sigma Q^T \Sigma_\lambda) = \sigma\big(\lambda_1(p_{11}q_{11} + p_{12}q_{12}) + \lambda_2(p_{21}q_{21} + p_{22}q_{22})\big).$$

Since $P, Q$ are rotation matrices, $p_{11}q_{11} + p_{12}q_{12} \leq 1$ and $p_{21}q_{21} + p_{22}q_{22} \leq 1$. Since $\Sigma, \Sigma_\lambda$ are fixed and $\lambda_1, \lambda_2 \geq 0$, the error $\|E - F\|_f^2$ is minimized when

$p_{11}q_{11} + p_{12}q_{12} = p_{21}q_{21} + p_{22}q_{22} = 1$. This can be achieved when $P, Q$ are of the general form

$$P = Q = \begin{bmatrix} \cos(\theta) & -\sin(\theta) & 0 \\ \sin(\theta) & \cos(\theta) & 0 \\ 0 & 0 & 1 \end{bmatrix}.$$

Obviously, $P = Q = I$ is one of the solutions. That implies $U_1 = U, V_1 = V$.

*Step 2:* From Step 1, we need to minimize the error function only over the matrices of the form $U\Sigma V^T \in \mathcal{E}$, where $\Sigma$ may vary. The minimization problem is then converted to one of minimizing the error function

$$\|E - F\|_f^2 = (\lambda_1 - \sigma)^2 + (\lambda_2 - \sigma)^2 + (\lambda_3 - 0)^2.$$

Clearly, the $\sigma$ that minimizes this error function is given by $\sigma = (\lambda_1 + \lambda_2)/2$. □

As we have already pointed out, the epipolar constraint allows us to recover the essential matrix only up to a scalar factor (since the epipolar constraint (5.2) is homogeneous in $E$, it is not modified by multiplying it by any nonzero constant). A typical choice to fix this ambiguity is to assume a unit translation, that is, $\|T\| = \|E\| = 1$. We call the resulting essential matrix *normalized*.

**Remark 5.10.** *The reader may have noticed that the above theorem relies on a special assumption that in the SVD of $E$ both matrices $U$ and $V$ are rotation matrices in $SO(3)$. This is not always true when $E$ is estimated from noisy data. In fact, standard SVD routines do* not *guarantee that the computed $U$ and $V$ have positive determinant. The problem can be easily resolved once one notices that the sign of the essential matrix $E$ is also arbitrary (even after normalization). The above projection can operate either on $+E$ or $-E$. We leave it as an exercise to the reader that one of the (noisy) matrices $\pm E$ will always have an SVD that satisfies the conditions of Theorem 5.9.*

According to Theorem 5.7, each normalized essential matrix $E$ gives two possible poses $(R, T)$. So from $\pm E$, we can recover the pose up to four solutions. In fact, three of the solutions can be eliminated by imposing the positive depth constraint. We leave the details to the reader as an exercise (see Exercise 5.11).

The overall algorithm, which is due to [Longuet-Higgins, 1981], can then be summarized as Algorithm 5.1.

To account for the possible sign change $\pm E$, in the last step of the algorithm, the "+" and "−" signs in the equations for $R$ and $T$ should be arbitrarily combined so that all four solutions can be obtained.

**Example 5.11 (A numerical example).** Suppose that

$$R = \begin{bmatrix} \cos(\pi/4) & 0 & \sin(\pi/4) \\ 0 & 1 & 0 \\ -\sin(\pi/4) & 0 & \cos(\pi/4) \end{bmatrix} = \begin{bmatrix} \frac{\sqrt{2}}{2} & 0 & \frac{\sqrt{2}}{2} \\ 0 & 1 & 0 \\ -\frac{\sqrt{2}}{2} & 0 & \frac{\sqrt{2}}{2} \end{bmatrix}, \quad T = \begin{bmatrix} 2 \\ 0 \\ 0 \end{bmatrix}.$$

**Algorithm 5.1 (The eight-point algorithm).**

For a given set of image correspondences $(x_1^j, x_2^j)$, $j = 1, 2, \ldots, n$ $(n \geq 8)$, this algorithm recovers $(R, T) \in SE(3)$, which satisfy

$$x_2^{jT} \widehat{T} R x_1^j = 0, \quad j = 1, 2, \ldots, n.$$

1. **Compute a first approximation of the essential matrix**
   Construct $\chi = [a^1, a^2, \ldots, a^n]^T \in \mathbb{R}^{n \times 9}$ from correspondences $x_1^j$ and $x_2^j$ as in (5.12), namely,
   $$a^j = x_1^j \otimes x_2^j \quad \in \mathbb{R}^9.$$
   Find the vector $E^s \in \mathbb{R}^9$ of unit length such that $\|\chi E^s\|$ is minimized as follows: compute the SVD of $\chi = U_\chi \Sigma_\chi V_\chi^T$ and define $E^s$ to be the ninth column of $V_\chi$. Unstack the nine elements of $E^s$ into a square $3 \times 3$ matrix $E$ as in (5.10). Note that this matrix will in general *not* be in the essential space.
2. **Project onto the essential space**
   Compute the singular value decomposition of the matrix $E$ recovered from data to be
   $$E = U \mathrm{diag}\{\sigma_1, \sigma_2, \sigma_3\} V^T,$$
   where $\sigma_1 \geq \sigma_2 \geq \sigma_3 \geq 0$ and $U, V \in SO(3)$. In general, since $E$ may not be an essential matrix, $\sigma_1 \neq \sigma_2$ and $\sigma_3 \neq 0$. But its projection onto the normalized essential space is $U \Sigma V^T$, where $\Sigma = \mathrm{diag}\{1, 1, 0\}$.
3. **Recover the displacement from the essential matrix**
   We now need only $U$ and $V$ to extract $R$ and $T$ from the essential matrix as
   $$R = U R_Z^T\left(\pm\frac{\pi}{2}\right) V^T, \quad \widehat{T} = U R_Z\left(\pm\frac{\pi}{2}\right) \Sigma U^T.$$
   where $R_Z^T\left(\pm\frac{\pi}{2}\right) \doteq \begin{bmatrix} 0 & \pm 1 & 0 \\ \mp 1 & 0 & 0 \\ 0 & 0 & 1 \end{bmatrix}$.

Then the essential matrix is

$$E = \widehat{T} R = \begin{bmatrix} 0 & 0 & 0 \\ \sqrt{2} & 0 & -\sqrt{2} \\ 0 & 2 & 0 \end{bmatrix}.$$

Since $\|T\| = 2$, the $E$ obtained here is not normalized. It is also easy to see this from its SVD,

$$E = U \Sigma V^T \doteq \begin{bmatrix} 0 & 0 & -1 \\ -1 & 0 & 0 \\ 0 & 1 & 0 \end{bmatrix} \begin{bmatrix} 2 & 0 & 0 \\ 0 & 2 & 0 \\ 0 & 0 & 0 \end{bmatrix} \begin{bmatrix} -\frac{\sqrt{2}}{2} & 0 & -\frac{\sqrt{2}}{2} \\ 0 & 1 & 0 \\ \frac{\sqrt{2}}{2} & 0 & -\frac{\sqrt{2}}{2} \end{bmatrix}^T,$$

where the nonzero singular values are 2 instead of 1. Normalizing $E$ is equivalent to replacing the above $\Sigma$ by

$$\Sigma = \mathrm{diag}\{1, 1, 0\}.$$

It is then easy to compute the four possible decompositions $(R, \widehat{T})$ for $E$:

1. $UR_Z^T\left(\frac{\pi}{2}\right)V^T = \begin{bmatrix} \frac{\sqrt{2}}{2} & 0 & \frac{\sqrt{2}}{2} \\ 0 & -1 & 0 \\ \frac{\sqrt{2}}{2} & 0 & -\frac{\sqrt{2}}{2} \end{bmatrix}, \quad UR_Z\left(\frac{\pi}{2}\right)\Sigma U^T = \begin{bmatrix} 0 & 0 & 0 \\ 0 & 0 & 1 \\ 0 & -1 & 0 \end{bmatrix};$

2. $UR_Z^T\left(\frac{\pi}{2}\right)V^T = \begin{bmatrix} \frac{\sqrt{2}}{2} & 0 & \frac{\sqrt{2}}{2} \\ 0 & -1 & 0 \\ \frac{\sqrt{2}}{2} & 0 & -\frac{\sqrt{2}}{2} \end{bmatrix}, \quad UR_Z\left(-\frac{\pi}{2}\right)\Sigma U^T = \begin{bmatrix} 0 & 0 & 0 \\ 0 & 0 & -1 \\ 0 & 1 & 0 \end{bmatrix};$

3. $UR_Z^T\left(-\frac{\pi}{2}\right)V^T = \begin{bmatrix} \frac{\sqrt{2}}{2} & 0 & \frac{\sqrt{2}}{2} \\ 0 & 1 & 0 \\ -\frac{\sqrt{2}}{2} & 0 & \frac{\sqrt{2}}{2} \end{bmatrix}, \quad UR_Z\left(-\frac{\pi}{2}\right)\Sigma U^T = \begin{bmatrix} 0 & 0 & 0 \\ 0 & 0 & -1 \\ 0 & 1 & 0 \end{bmatrix};$

4. $UR_Z^T\left(-\frac{\pi}{2}\right)V^T = \begin{bmatrix} \frac{\sqrt{2}}{2} & 0 & \frac{\sqrt{2}}{2} \\ 0 & 1 & 0 \\ -\frac{\sqrt{2}}{2} & 0 & \frac{\sqrt{2}}{2} \end{bmatrix}, \quad UR_Z\left(\frac{\pi}{2}\right)\Sigma U^T = \begin{bmatrix} 0 & 0 & 0 \\ 0 & 0 & 1 \\ 0 & -1 & 0 \end{bmatrix}.$

Clearly, the third solution is exactly the original motion $(R, \widehat{T})$ except that the translation $T$ is recovered up to a scalar factor (i.e. it is normalized to unit norm). ■

Despite its simplicity, the above algorithm, when used in practice, suffers from some shortcomings that are discussed below.

*Number of points*

The number of points, eight, assumed by the algorithm, is mostly for convenience and simplicity of presentation. In fact, the matrix $E$ (as a function of $(R, T)$) has only a total of five degrees of freedom: three for rotation and two for translation (up to a scalar factor). By utilizing some additional algebraic properties of $E$, we may reduce the necessary number of points. For instance, knowing $\det(E) = 0$, we may relax the condition $\text{rank}(\chi) = 8$ to $\text{rank}(\chi) = 7$, and get two solutions $E_1^s$ and $E_2^s \in \mathbb{R}^9$ from the null space of $\chi$. Nevertheless, there is usually only one $\alpha \in \mathbb{R}$ such that

$$\det(E_1 + \alpha E_2) = 0.$$

Therefore, seven points is all we need to have a relatively simpler algorithm. As shown in Exercise 5.13, in fact, a linear algorithm exists for only six points if more complicated algebraic properties of the essential matrix are used. Hence, it should not be a surprise, as shown by [Kruppa, 1913], that one needs only five points in general position to recover $(R, T)$. It can be shown that there are up to ten (possibly complex) solutions, though the solutions are not obtainable in closed form. Furthermore, for many special motions, one needs only up to four points to determine the associated essential matrix. For instance, planar motions (Exercise 5.6) and motions induced from symmetry (Chapter 10) have this nice property.

*Number of solutions and positive depth constraint*

Since both $E$ and $-E$ satisfy the same set of epipolar constraints, they in general give rise to $2 \times 2 = 4$ possible solutions for $(R, T)$. However, this does not pose a problem, because only one of the solutions guarantees that the depths of all the 3-D points reconstructed are *positive* with respect to both camera frames. That is, in general, three out of the four solutions will be physically impossible and hence may be discarded (see Exercise 5.11).

*Structure requirement: general position*

In order for the above algorithm to work properly, the condition that the given eight points be in "general position" is very important. It can be easily shown that if these points form certain degenerate configurations, called critical surfaces, the algorithm will fail (see Exercise 5.14). A case of some practical importance occurs when all the points happen to lie on the same 2-D plane in $\mathbb{R}^3$. We will discuss the geometry for the planar case in Section 5.3, and also later within the context of multiple-view geometry (Chapter 9).

*Motion requirement: sufficient parallax*

In the derivation of the epipolar constraint we have implicitly assumed that $E \neq 0$, which allowed us to derive the eight-point algorithm where the essential matrix is normalized to $\|E\| = 1$. Due to the structure of the essential matrix, $E = 0 \Leftrightarrow T = 0$. Therefore, the eight-point algorithm requires that the translation (or baseline) $T \neq 0$. The translation $T$ induces parallax in the image plane. In practice, due to noise, the algorithm will likely return an answer even when there is no translation. However, in this case the estimated direction of translation will be meaningless. Therefore, one needs to exercise caution to make sure that there is "sufficient parallax" for the algorithm to be well conditioned. It has been observed experimentally that even for purely rotational motion, i.e. $T = 0$, the "spurious" translation created by noise in the image measurements is sufficient for the eight-point algorithm to return a correct estimate of $R$.

*Infinitesimal viewpoint change*

It is often the case in applications that the two views described in this chapter are taken by a moving camera rather than by two static cameras. The derivation of the epipolar constraint and the associated eight-point algorithm does not change, as long as the two vantage points are distinct. In the limit that the two viewpoints come infinitesimally close, the epipolar constraint takes a related but different form called the continuous epipolar constraint, which we will study in Section 5.4. The continuous case is typically of more significance for applications in robot vision, where one is often interested in recovering the linear and angular velocities of the camera.

*Multiple motion hypotheses*

In the case of multiple moving objects in the scene, image points may no longer satisfy the same epipolar constraint. For example, if we know that there are two independent moving objects with motions, say $(R^1, T^1)$ and $(R^2, T^2)$, then the two images $(\boldsymbol{x}_1, \boldsymbol{x}_2)$ of a point $p$ on one of these objects should satisfy instead the equation

$$(\boldsymbol{x}_2^T E^1 \boldsymbol{x}_1)(\boldsymbol{x}_2^T E^2 \boldsymbol{x}_1) = 0, \tag{5.16}$$

corresponding to the fact that the point $p$ moves according to either motion 1 or motion 2. Here $E^1 = \widehat{T^1} R^1$ and $E^2 = \widehat{T^2} R^2$. As we will see, from this equation it is still possible to recover $E^1$ and $E^2$ if enough points are visible on either object. Generalizing to more than two independent motions requires some attention; we will study the multiple-motion problem in Chapter 7.

### 5.2.2 Euclidean constraints and structure reconstruction

The eight-point algorithm just described uses as input a set of eight or more point correspondences and returns the relative pose (rotation and translation) between the two cameras up to an arbitrary scale $\gamma \in \mathbb{R}^+$. Without loss of generality, we may assume this scale to be $\gamma = 1$, which is equivalent to scaling translation to unit length. Relative pose and point correspondences can then be used to retrieve the position of the points in 3-D by recovering their depths relative to each camera frame.

Consider the basic rigid-body equation, where the pose $(R, T)$ has been recovered, with the translation $T$ defined up to the scale $\gamma$. In terms of the images and the depths, it is given by

$$\lambda_2^j \boldsymbol{x}_2^j = \lambda_1^j R \boldsymbol{x}_1^j + \gamma T, \quad j = 1, 2, \dots, n. \tag{5.17}$$

Notice that since $(R, T)$ are known, the equations given by (5.17) are linear in both the structural scale $\lambda$'s and the motion scale $\gamma$'s, and therefore they can be easily solved. For each point, $\lambda_1, \lambda_2$ are its depths with respect to the first and second camera frames, respectively. One of them is therefore redundant; for instance, if $\lambda_1$ is known, $\lambda_2$ is simply a function of $(R, T)$. Hence we can eliminate, say, $\lambda_2$ from the above equation by multiplying both sides by $\widehat{\boldsymbol{x}_2}$, which yields

$$\lambda_1^j \widehat{\boldsymbol{x}_2^j} R \boldsymbol{x}_1^j + \gamma \widehat{\boldsymbol{x}_2^j} T = 0, \quad j = 1, 2, \dots, n. \tag{5.18}$$

This is equivalent to solving the linear equation

$$M^j \bar{\lambda}^j \doteq \begin{bmatrix} \widehat{\boldsymbol{x}_2^j} R \boldsymbol{x}_1^j, & \widehat{\boldsymbol{x}_2^j} T \end{bmatrix} \begin{bmatrix} \lambda_1^j \\ \gamma \end{bmatrix} = 0, \tag{5.19}$$

where $M^j = \begin{bmatrix} \widehat{\boldsymbol{x}_2^j} R \boldsymbol{x}_1^j, & \widehat{\boldsymbol{x}_2^j} T \end{bmatrix} \in \mathbb{R}^{3 \times 2}$ and $\bar{\lambda}^j = [\lambda_1^j, \gamma]^T \in \mathbb{R}^2$, for $j = 1, 2, \dots, n$. In order to have a unique solution, the matrix $M^j$ needs to be of rank 1. This is not the case only when $\widehat{x_2}T = 0$, i.e. when the point $p$ lies on the line connecting the two optical centers $o_1$ and $o_2$.

Notice that all the $n$ equations above share the same $\gamma$; we define a vector $\vec{\lambda} = [\lambda_1^1, \lambda_1^2, \ldots, \lambda_1^n, \gamma]^T \in \mathbb{R}^{n+1}$ and a matrix $M \in \mathbb{R}^{3n\times(n+1)}$ as

$$M \doteq \begin{bmatrix} \widehat{x_2^1}Rx_1^1 & 0 & 0 & 0 & 0 & \widehat{x_2^1}T \\ 0 & \widehat{x_2^2}Rx_1^2 & 0 & 0 & 0 & \widehat{x_2^2}T \\ 0 & 0 & \ddots & 0 & 0 & \vdots \\ 0 & 0 & 0 & \widehat{x_2^{n-1}}Rx_1^{n-1} & 0 & \widehat{x_2^{n-1}}T \\ 0 & 0 & 0 & 0 & \widehat{x_2^n}Rx_1^n & \widehat{x_2^n}T \end{bmatrix}. \quad (5.20)$$

Then the equation

$$M\vec{\lambda} = 0 \quad (5.21)$$

determines all the unknown depths *up to a single universal scale*. The linear least-squares estimate of $\vec{\lambda}$ is simply the eigenvector of $M^TM$ that corresponds to its smallest eigenvalue. Note that this scale ambiguity is intrinsic, since without any prior knowledge about the scene and camera motion, one cannot disambiguate whether the camera moved twice the distance while looking at a scene twice larger but two times further away.

### 5.2.3 Optimal pose and structure

The eight-point algorithm given in the previous section assumes that *exact* point correspondences are given. In the presence of noise in image correspondences, we have suggested possible ways of estimating the essential matrix by solving a least-squares problem followed by a projection onto the essential space. But in practice, this will not be satisfying in at least two respects:

1. There is no guarantee that the estimated pose $(R, T)$, is as close as possible to the true solution.
2. Even if we were to accept such an $(R, T)$, a noisy image pair, say $(\tilde{x}_1, \tilde{x}_2)$, would not necessarily give rise to a consistent 3-D reconstruction, as shown in Figure 5.6.

At this stage of development, we do not want to bring in all the technical details associated with optimal estimation, since they would bury the geometric intuition. We will therefore discuss only the key ideas, and leave the technical details to Appendix 5.A as well as Chapter 11, where we will address more practical issues.

#### *Choice of optimization objectives*

Recall from Chapter 3 that a calibrated camera can be described as a plane perpendicular to the $z$-axis at a distance 1 from the origin; therefore, the coordinates of image points $x_1$ and $x_2$ are of the form $[x, y, 1]^T \in \mathbb{R}^3$. In practice, we cannot

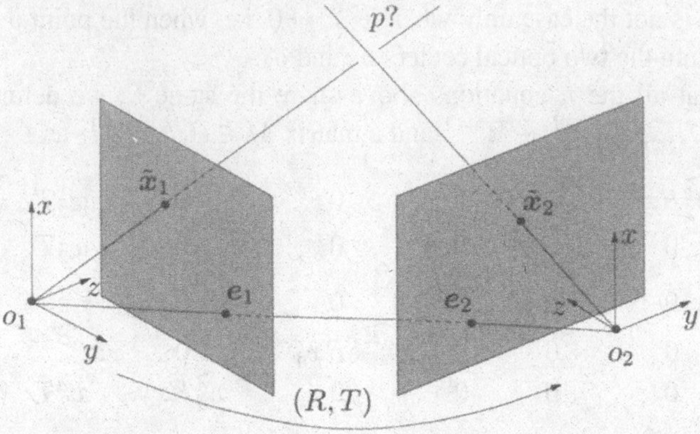


Figure 5.6. Rays extended from a noisy image pair $\tilde{x}_1, \tilde{x}_2 \in \mathbb{R}^3$ do not intersect at any point $p$ in 3-D if they do not satisfy the epipolar constraint precisely.

measure the actual coordinates but only their noisy versions, say

$$\tilde{x}_1^j = x_1^j + w_1^j, \quad \tilde{x}_2^j = x_2^j + w_2^j, \quad j = 1, 2, \ldots, n, \tag{5.22}$$

where $x_1^j$ and $x_2^j$ denote the "ideal" image coordinates and $w_1^j = [w_{11}^j, w_{12}^j, 0]^T$ and $w_2^j = [w_{21}^j, w_{22}^j, 0]^T$ are localization errors in the correspondence. Notice that it is the (unknown) ideal image coordinates $(x_1^j, x_2^j)$ that satisfy the epipolar constraint $x_2^{jT}\widehat{T}Rx_1^j = 0$, and *not* the (measured) noisy ones $(\tilde{x}_1^j, \tilde{x}_2^j)$. One could think of the ideal coordinates as a "model," and $w_i^j$ as the discrepancy between the model and the measurements: $\tilde{x}_i^j = x_i^j + w_i^j$. Therefore, in general, we seek the parameters $(x, R, T)$ that minimize the discrepancy between the model and the data, i.e. $w_i^j$. In order to do so, we first need to decide how to *evaluate* the discrepancy, which determines the choice of optimization objective.

Unfortunately, there is no "correct," uncontroversial, universally accepted objective function, and the choice of discrepancy measure is part of the design process, since it depends on what assumptions are made on the *residuals* $w_i^j$. Different assumptions result in different choices of discrepancy measures, which eventually result in different "optimal" solutions $(x^*, R^*, T^*)$.

For instance, one may assume that $w = \{w_i^j\}$ are samples from a distribution that depends on the unknown *parameters* $(x, R, T)$, which are considered deterministic but unknown. In this case, based on the model generating the data, one can derive an expression of the *likelihood function* $p(w|x, R, T)$ and choose to maximize it (or, more conveniently, its logarithm) with respect to the unknown parameters. Then the "optimal solution," in the sense of *maximum likelihood*, is given by

$$(x^*, R^*, T^*) = \arg\max \phi_{ML}(x, R, T) \doteq \sum_{i,j} \log p((\tilde{x}_i^j - x_i^j)|x, R, T).$$

Naturally, different likelihood functions can result in very different optimal solutions. Indeed, there is no guarantee that the maximum is unique, since $p$ can be multimodal, and therefore there may be several choices of parameters that achieve the maximum. Constructing the likelihood function for the location of point features from first principles, starting from the noise characteristics of the photosensitive elements of the sensor, is difficult because of the many nonlinear steps involved in feature detection and tracking. Therefore, it is common to assume that the likelihood belongs to a family of density functions, the most popular choice being the normal (or Gaussian) distribution.

Sometimes, however, one may have reasons to believe that $(\boldsymbol{x}, R, T)$ are not just unknown parameters that can take any value. Instead, even before any measurement is gathered, one can say that some values are more probable than others, a fact that can be described by a joint *a priori* probability density (or *prior*) $p(\boldsymbol{x}, R, T)$. For instance, for a robot navigating on a flat surface, rotation about the horizontal axis may be very improbable, as would translation along the vertical axis. When combined with the likelihood function, the prior can be used to determine the *a posteriori density* , or *posterior* $p(\boldsymbol{x}, R, T|\{\tilde{\boldsymbol{x}}_i^j\})$ using Bayes rule. In this case, one may seek the maximum of the posterior *given* the value of the measurements. This is the *maximum a posteriori* estimate

$$(\boldsymbol{x}^*, R^*, T^*) = \arg\max \phi_{MAP}(\boldsymbol{x}, R, T) \doteq p(\boldsymbol{x}, R, T|\{\tilde{\boldsymbol{x}}_i^j\}).$$

Although this choice has several advantages, in our case it requires defining a probability density on the space of camera poses $SO(3) \times \mathbb{S}^2$, which has a nontrivial geometric structure. This is well beyond the scope of this book, and we will therefore not discuss this criterion further here.

In what follows, we will take a more minimalistic approach to optimality, and simply assume that $\{w_i^j\}$ are unknown values ("errors," or "residuals") whose norms need to be minimized. In this case, we do not postulate any probabilistic description, and we simply seek $(\boldsymbol{x}^*, R^*, T^*) = \arg\min \phi(\boldsymbol{x}, R, T)$, where $\phi$ is, for instance, the squared 2-norm:

$$\phi(\boldsymbol{x}, R, T) \doteq \sum_j \|w_1^j\|_2^2 + \|w_2^j\|_2^2 = \sum_j \|\tilde{\boldsymbol{x}}_1^j - \boldsymbol{x}_1^j\|_2^2 + \|\tilde{\boldsymbol{x}}_2^j - \boldsymbol{x}_2^j\|_2^2.$$

This corresponds to a *least-squares estimator*. Since $\boldsymbol{x}_1^j$ and $\boldsymbol{x}_2^j$ are the recovered 3-D points projected back onto the image planes, the above criterion is often called the "reprojection error."

However, the unknowns for the above minimization problem are not completely free; for example, they need to satisfy the epipolar constraint $\boldsymbol{x}_2^T \widehat{T} R \boldsymbol{x}_1 = 0$. Hence, with the choice of the least-squares criterion, we can pose the problem of reconstruction as a constrained optimization: given $\tilde{\boldsymbol{x}}_i^j, i = 1, 2,\ j = 1, 2, \ldots, n$, minimize

$$\phi(\boldsymbol{x}, R, T) \doteq \sum_{j=1}^{n} \sum_{i=1}^{2} \|\tilde{\boldsymbol{x}}_i^j - \boldsymbol{x}_i^j\|_2^2 \tag{5.23}$$

subject to

$$\boldsymbol{x}_2^{jT} \widehat{T} R \boldsymbol{x}_1^j = 0, \quad \boldsymbol{x}_1^{jT} e_3 = 1, \quad \boldsymbol{x}_2^{jT} e_3 = 1, \quad j = 1, 2, \ldots, n. \tag{5.24}$$

Using Lagrange multipliers (Appendix C), we can convert this constrained optimization problem to an unconstrained one. Details on how to carry out the optimization are outlined in Appendix 5.A.

**Remark 5.12 (Equivalence to bundle adjustment).** *The reader may have noticed that the depth parameters $\lambda_i$, despite being unknown, are missing from the optimization problem of equation (5.24). This is not an oversight: indeed, the depth parameters play the role of Lagrange multipliers in the constrained optimization problem described above, and therefore they enter the optimization problem indirectly. Alternatively, one can write the optimization problem in unconstrained form:*

$$\sum_{j=1}^{n} \left\| \tilde{\boldsymbol{x}}_1^j - \pi_1(\boldsymbol{X}^j) \right\|_2^2 + \left\| \tilde{\boldsymbol{x}}_2^j - \pi_2(\boldsymbol{X}^j) \right\|_2^2, \tag{5.25}$$

*where $\pi_1$ and $\pi_2$ denote the projection of a point $\boldsymbol{X}$ in space onto the first and second images, respectively. If we choose the first camera frame as the reference, then the above expression can be simplified to*[6]

$$\phi(\boldsymbol{x}_1, R, T, \lambda) = \sum_{j=1}^{n} \left\| \tilde{\boldsymbol{x}}_1^j - \boldsymbol{x}_1^j \right\|_2^2 + \left\| \tilde{\boldsymbol{x}}_2^j - \pi(R\lambda_1^j \boldsymbol{x}_1^j + T) \right\|_2^2. \tag{5.26}$$

*Minimizing the above expression with respect to the unknowns $(R, T, \boldsymbol{x}_1, \lambda)$ is known in the literature as* bundle adjustment. *Bundle adjustment and the constrained optimization described above are simply two different ways to parameterize the same optimization objective. As we will see in Appendix 5.A, the constrained form better highlights the geometric structure of the problem, and serves as a guide to develop effective approximations.*

In the remainder of this section, we limit ourselves to describing a simplified cost functional that approximates the reprojection error resulting in simpler optimization algorithms, while retaining a strong geometric interpretation. In this approximation, the unknown $\boldsymbol{x}$ is approximated by the measured $\tilde{\boldsymbol{x}}$, so that the cost function $\phi$ depends only on camera pose $(R, T)$ (see Appendix 5.A for more details):

$$\phi(R, T) \doteq \sum_{j=1}^{n} \frac{(\tilde{\boldsymbol{x}}_2^{jT} \widehat{T} R \tilde{\boldsymbol{x}}_1^j)^2}{\|\widehat{e}_3 \widehat{T} R \tilde{\boldsymbol{x}}_1^j\|^2} + \frac{(\tilde{\boldsymbol{x}}_2^{jT} \widehat{T} R \tilde{\boldsymbol{x}}_1^j)^2}{\|\tilde{\boldsymbol{x}}_2^{jT} \widehat{T} R \widehat{e}_3^T\|^2}. \tag{5.27}$$

Geometrically, this expression can be interpreted as distances from the image points $\tilde{\boldsymbol{x}}_1^j$ and $\tilde{\boldsymbol{x}}_2^j$ to corresponding epipolar lines in the two image planes, respectively, as shown in Figure 5.7. For instance, the reader can verify as an exercise

[6]Here we use $\pi$ to denote the standard planar projection introduced in Chapter 3: $[X, Y, Z]^T \mapsto [X/Z, Y/Z, 1]^T$.

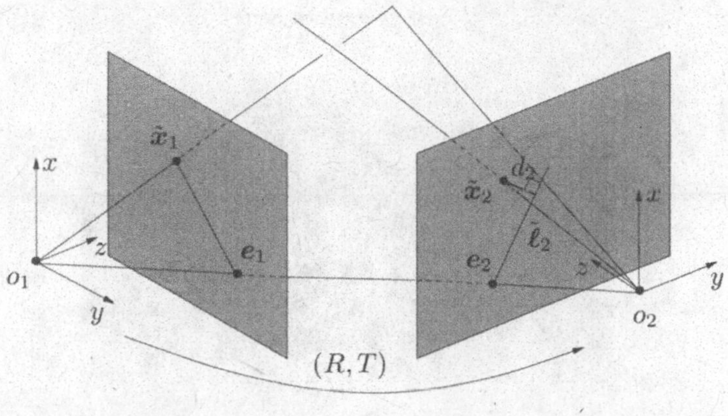


Figure 5.7. Two noisy image points $\tilde{x}_1, \tilde{x}_2 \in \mathbb{R}^3$. Here $\tilde{\ell}_2$ is an epipolar line that is the intersection of the second image plane with the epipolar plane. The distance $d_2$ is the geometric distance between the second image point $\tilde{x}_2$ and the epipolar line. Symmetrically, one can define a similar geometric distance $d_1$ in the first image plane.

(Exercise 5.12) that following the notation in the figure, we have

$$d_2^2 = \frac{(\tilde{x}_2^T \widehat{T} R \tilde{x}_1)^2}{\|\widehat{e}_3 \widehat{T} R \tilde{x}_1\|^2}.$$

In the presence of noise, minimizing the above objective function, although more difficult, improves the results of the linear eight-point algorithm.

**Example 5.13 (Comparison with the linear algorithm).** Figure 5.8 demonstrates the effect of the optimization: numerical simulations were run for both the linear eight-point algorithm and the nonlinear optimization. Values of the objective function $\phi(R, T)$ at different $T$ are plotted (with $R$ fixed at the ground truth); "+" denotes the true translation $T$, "$*$" is the estimated $T$ from the linear eight-point algorithm, and "$\circ$" is the estimated $T$ by upgrading the linear algorithm result with the optimization. ■

*Structure triangulation*

If we were given the optimal estimate of camera pose $(R, T)$, obtained, for instance, from Algorithm 5.5 in Appendix 5.A, we can find a pair of images $(x_1^*, x_2^*)$ that satisfy the epipolar constraint $x_2^T \widehat{T} R x_1 = 0$ and minimize the (reprojection) error

$$\phi(x) = \|\tilde{x}_1 - x_1\|^2 + \|\tilde{x}_2 - x_2\|^2. \tag{5.28}$$

This is called the *triangulation problem.* The key to its solution is to find what exactly the reprojection error depends on, which can be more easily explained geometrically by Figure 5.9. As we see from the figure, the value of the reprojection error depends only on the position of the epipolar plane $P$: when the plane $P$ rotates around the baseline $(o_1, o_2)$, the image pair $(x_1, x_2)$, which minimizes the distance $\|\tilde{x}_1 - x_1\|^2 + \|\tilde{x}_2 - x_2\|^2$, changes accordingly, and so does the error. To

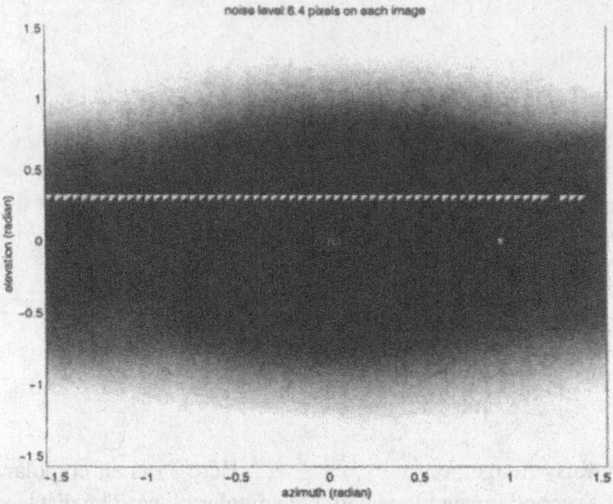


Figure 5.8. Improvement by nonlinear optimization. A two-dimensional projection of the five-dimensional residual function $\phi(R, T)$ is shown in greyscale. The residual corresponds to the two-dimensional function $\phi(\hat{R}, T)$ with rotation fixed at the true value. The location of the solution found by the linear algorithm is shown as "$*$," and it can be seen that it is quite far from the true minimum (darkest point in the center of the image, marked by "+").The solution obtained by nonlinear optimization is marked by "$\circ$," which shows a significant improvement.

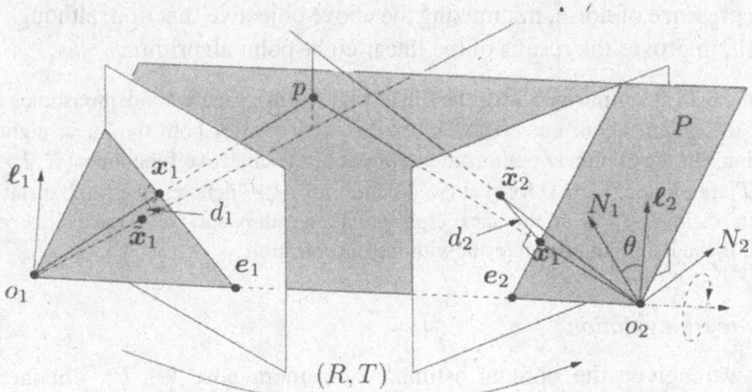


Figure 5.9. For a fixed epipolar plane $P$, the pair of images $(\boldsymbol{x}_1, \boldsymbol{x}_2)$ that minimize the reprojection error $d_1^2 + d_2^2$ must be points on the two epipolar lines and closest to $\tilde{\boldsymbol{x}}_1, \tilde{\boldsymbol{x}}_2$, respectively. Hence the reprojection error is a function only of the position of the epipolar plane $P$.

parameterize the position of the epipolar plane, let $(e_2, N_1, N_2)$ be an orthonormal basis in the second camera frame. Then $P$ is determined by its normal vector $\ell_2$ (with respect to the second camera frame), which in turn is determined by the angle $\theta$ between $\ell_2$ and $N_1$ (Figure 5.9). Hence the reprojection error $\phi$ should be a function that depends only on $\theta$. There is typically only one $\theta^*$ that minimizes the error $\phi(\theta)$. Once it is found, the corresponding image pair $(\boldsymbol{x}_1^*, \boldsymbol{x}_2^*)$ and 3-D point $p$ are determined. Details of the related algorithm can be found in Appendix 5.A.

## 5.3 Planar scenes and homography

In order for the eight-point algorithm to give a unique solution (up to a scalar factor) for the camera motion, it is crucial that the feature points in 3-D be in general position. When the points happen to form certain degenerate configurations, the solution might no longer be unique. Exercise 5.14 explains why this may occur when all the feature points happen to lie on certain 2-D surfaces, called critical surfaces.[7] Many of these critical surfaces occur rarely in practice, and their importance is limited. However, 2-D planes, which happen to be a special case of critical surfaces, are ubiquitous in man-made environments and in aerial imaging.

Therefore, if one applies the eight-point algorithm to images of points all lying on the same 2-D plane, the algorithm will fail to provide a unique solution (as we will soon see why). On the other hand, in many applications, a scene can indeed be approximately planar (e.g., the landing pad for a helicopter) or piecewise planar (e.g., the corridors inside a building). We therefore devote this section to this special but important case.

### *5.3.1 Planar homography*

Let us consider two images of points $p$ on a 2-D plane $P$ in 3-D space. For simplicity, we will assume throughout the section that the optical center of the camera never passes through the plane.

Now suppose that two images $(\boldsymbol{x}_1, \boldsymbol{x}_2)$ are given for a point $p \in P$ with respect to two camera frames. Let the coordinate transformation between the two frames be

$$\boldsymbol{X}_2 = R\boldsymbol{X}_1 + T, \tag{5.29}$$

where $\boldsymbol{X}_1, \boldsymbol{X}_2$ are the coordinates of $p$ relative to camera frames 1 and 2, respectively. As we have already seen, the two images $\boldsymbol{x}_1, \boldsymbol{x}_2$ of $p$ satisfy the epipolar constraint

$$\boldsymbol{x}_2^T E \boldsymbol{x}_1 = \boldsymbol{x}_2^T \widehat{T} R \boldsymbol{x}_1 = 0.$$

However, for points on the same plane $P$, their images will share an extra constraint that makes the epipolar constraint alone no longer sufficient.

[7] In general, such critical surfaces can be described by certain quadratic equations in the $X, Y, Z$ coordinates of the point, hence are often referred to as quadratic surfaces.

Let $N = [n_1, n_2, n_2]^T \in \mathbb{S}^2$ be the unit normal vector of the plane $P$ with respect to the first camera frame, and let $d > 0$ denote the distance from the plane $P$ to the optical center of the first camera. Then we have

$$N^T \boldsymbol{X}_1 = n_1 X + n_2 Y + n_3 Z = d \quad \Leftrightarrow \quad \frac{1}{d} N^T \boldsymbol{X}_1 = 1, \quad \forall \boldsymbol{X}_1 \in P. \tag{5.30}$$

Substituting equation (5.30) into equation (5.29) gives

$$\boldsymbol{X}_2 = R\boldsymbol{X}_1 + T = R\boldsymbol{X}_1 + T\frac{1}{d}N^T\boldsymbol{X}_1 = \left(R + \frac{1}{d}TN^T\right)\boldsymbol{X}_1. \tag{5.31}$$

We call the matrix

$$H \doteq R + \frac{1}{d}TN^T \quad \in \mathbb{R}^{3\times 3} \tag{5.32}$$

the *(planar) homography matrix*, since it denotes a linear transformation from $\boldsymbol{X}_1 \in \mathbb{R}^3$ to $\boldsymbol{X}_2 \in \mathbb{R}^3$ as

$$\boldsymbol{X}_2 = H\boldsymbol{X}_1.$$

Note that the matrix $H$ depends on the motion parameters $\{R, T\}$ as well as the structure parameters $\{N, d\}$ of the plane $P$. Due to the inherent scale ambiguity in the term $\frac{1}{d}T$ in equation (5.32), one can at most expect to recover from $H$ the ratio of the translation $T$ scaled by the distance $d$. From

$$\lambda_1 \boldsymbol{x}_1 = \boldsymbol{X}_1, \quad \lambda_2 \boldsymbol{x}_2 = \boldsymbol{X}_2, \quad \boldsymbol{X}_2 = H\boldsymbol{X}_1, \tag{5.33}$$

we have

$$\lambda_2 \boldsymbol{x}_2 = H\lambda_1 \boldsymbol{x}_1 \quad \Leftrightarrow \quad \boldsymbol{x}_2 \sim H\boldsymbol{x}_1, \tag{5.34}$$

where we recall that $\sim$ indicates equality up to a scalar factor. Often, the equation

$$\boxed{\boldsymbol{x}_2 \sim H\boldsymbol{x}_1} \tag{5.35}$$

itself is referred to as a *(planar) homography* mapping induced by a plane $P$. Despite the scale ambiguity, as illustrated in Figure 5.10, $H$ introduces a special map between points in the first image and those in the second in the following sense:

1. For any point $\boldsymbol{x}_1$ in the first image that is the image of some point, say $p$ on the plane $P$, its corresponding second image $\boldsymbol{x}_2$ is uniquely determined as $\boldsymbol{x}_2 \sim H\boldsymbol{x}_1$, since for any other point, say $\boldsymbol{x}_2'$, on the same epipolar line $\boldsymbol{\ell}_2 \sim E\boldsymbol{x}_1 \in \mathbb{R}^3$, the ray $o_2\boldsymbol{x}_2'$ will intersect the ray $o_1\boldsymbol{x}_1$ at a point $p'$ out of the plane.

2. On the other hand, if $\boldsymbol{x}_1$ is the image of some point, say $p'$, not on the plane $P$, then $\boldsymbol{x}_2 \sim H\boldsymbol{x}_1$ is only a point that is on the same epipolar line $\boldsymbol{\ell}_2 \sim E\boldsymbol{x}_1$ as its actual corresponding image $\boldsymbol{x}_2'$. That is, $\boldsymbol{\ell}_2^T\boldsymbol{x}_2 = \boldsymbol{\ell}_2^T\boldsymbol{x}_2' = 0$.

We hence have the following result:

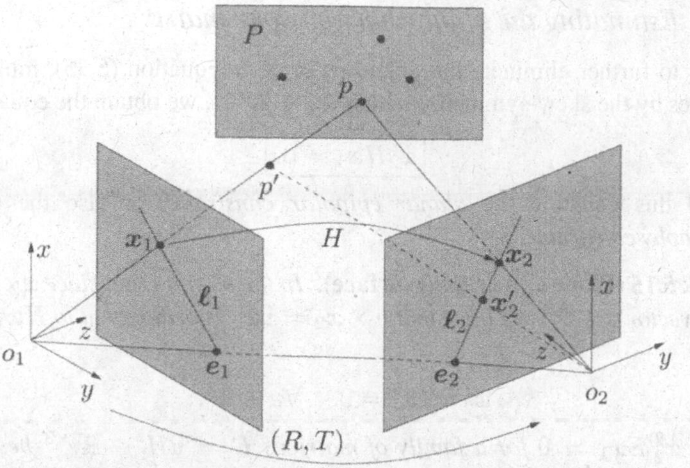


Figure 5.10. Two images $x_1, x_2 \in \mathbb{R}^3$ of a 3-D point $p$ on a plane $P$. They are related by a homography $H$ that is induced by the plane.

**Proposition 5.14 (Homography for epipolar lines).** *Given a homography $H$ (induced by plane $P$ in 3-D) between two images, for any pair of corresponding images $(x_1, x_2)$ of a 3-D point $p$ that is not necessarily on $P$, the associated epipolar lines are*

$$\boxed{\ell_2 \sim \widehat{x_2} H x_1, \quad \ell_1 \sim H^T \ell_2.} \tag{5.36}$$

*Proof.* If $p$ is not on $P$, the first equation is true from point 2 in above discussion. Note that for points on the plane $P$, $x_2 = Hx_1$ implies $\widehat{x_2}Hx_1 = 0$, and the first equation is still true as long as we adopt the convention that $v \sim 0$, $\forall v \in \mathbb{R}^3$. The second equation is easily proven using the definition of a line $\ell^T x = 0$. □

This property of the homography allows one to compute epipolar lines without knowing the essential matrix. We will explore further the relationships between the essential matrix and the planar homography in Section 5.3.4.

In addition to the fact that the homography matrix $H$ encodes information about the camera motion and the scene structure, knowing it directly facilitates establishing correspondence between points in the first and the second images. As we will see soon, $H$ can be computed in general from a small number of corresponding image pairs. Once $H$ is known, correspondence between images of other points on the same plane can then be fully established, since the corresponding location $x_2$ for an image point $x_1$ is simply $Hx_1$. Proposition 5.14 suggests that correspondence between images of points not on the plane can also be established, since $H$ contains information about the epipolar lines.

### 5.3.2 Estimating the planar homography matrix

In order to further eliminate the unknown scale in equation (5.35), multiplying both sides by the skew-symmetric matrix $\widehat{\boldsymbol{x}_2} \in \mathbb{R}^{3\times3}$, we obtain the equation

$$\boxed{\widehat{\boldsymbol{x}_2} H \boldsymbol{x}_1 = 0.} \tag{5.37}$$

We call this equation the *planar epipolar constraint*, or also the *(planar) homography constraint.*

**Remark 5.15 (Plane as a critical surface).** *In the planar case, since $\boldsymbol{x}_2 \sim H\boldsymbol{x}_1$, for any vector $u \in \mathbb{R}^3$, we have that $u \times \boldsymbol{x}_2 = \widehat{u}\boldsymbol{x}_2$ is orthogonal to $H\boldsymbol{x}_1$. Hence we have*

$$\boldsymbol{x}_2^T \widehat{u} H \boldsymbol{x}_1 = 0, \quad \forall u \in \mathbb{R}^3.$$

*That is, $\boldsymbol{x}_2^T E \boldsymbol{x}_1 = 0$ for a family of matrices $E = \widehat{u}H \in \mathbb{R}^{3\times3}$ besides the essential matrix $E = \widehat{T}R$. This explains why the eight-point algorithm does not apply to feature points from a planar scene.*

**Example 5.16 (Homography from a pure rotation).** The homographic relation $\boldsymbol{x}_2 \sim H\boldsymbol{x}_1$ also shows up when the camera is purely rotating, i.e. $\boldsymbol{X}_2 = R\boldsymbol{X}_1$. In this case, the homography matrix $H$ becomes $H = R$, since $T = 0$. Consequently, we have the constraint

$$\widehat{\boldsymbol{x}_2} R \boldsymbol{x}_1 = 0.$$

One may view this as a special planar scene case, since without translation, information about the depth of the scene is completely lost in the images, and one might as well interpret the scene to be planar (e.g., all the points lie on a plane infinitely far away). As the distance of the plane $d$ goes to infinity, $\lim_{d\to\infty} H = R$.

The homography from purely rotational motion can be used to construct image mosaics of the type shown in Figure 5.11. For additional references on how to construct panoramic mosaics the reader can refer to [Szeliski and Shum, 1997, Sawhney and Kumar, 1999], where the latter includes compensation for radial distortion. ■

Figure 5.11. Mosaic from the rotational homography.

Since equation (5.37) is *linear* in $H$, by stacking the entries of $H$ as a vector,

$$H^s \doteq [H_{11}, H_{21}, H_{31}, H_{12}, H_{22}, H_{32}, H_{13}, H_{23}, H_{33}]^T \quad \in \mathbb{R}^9, \tag{5.38}$$

we may rewrite equation (5.37) as

$$\boldsymbol{a}^T H^s = 0,$$

where the matrix $\boldsymbol{a} \doteq \boldsymbol{x}_1 \otimes \widehat{\boldsymbol{x}_2} \in \mathbb{R}^{9\times 3}$ is the Kronecker product of $\widehat{\boldsymbol{x}_2}$ and $\boldsymbol{x}_1$ (see Appendix A.1.3).

Since the matrix $\widehat{\boldsymbol{x}_2}$ is only of rank 2, so is the matrix $\boldsymbol{a}$. Thus, even though the equation $\widehat{\boldsymbol{x}_2}H\boldsymbol{x}_1 = 0$ has three rows, it only imposes two independent constraints on $H$. With this notation, given $n$ pairs of images $\{(\boldsymbol{x}_1^j, \boldsymbol{x}_2^j)\}_{j=1}^n$ from points on the same plane $P$, by defining $\chi \doteq [\boldsymbol{a}^1, \boldsymbol{a}^2, \ldots, \boldsymbol{a}^n]^T \in \mathbb{R}^{3n\times 9}$, we may combine all the equations (5.37) for all the image pairs and rewrite them as

$$\chi H^s = 0. \tag{5.39}$$

In order to solve uniquely (up to a scalar factor) for $H^s$, we must have $\operatorname{rank}(\chi) = 8$. Since each pair of image points gives two constraints, we expect that at least four point correspondences would be necessary for a unique estimate of $H$. We leave the proof of the following statement as an exercise to the reader.

**Proposition 5.17 (Four-point homography).** *We have $rank(\chi) = 8$ if and only if there exists a set of four points (out of the $n$) such that no three of them are collinear; i.e. they are in a general configuration in the plane.*

Thus, if there are more than four image correspondences of which no three in each image are collinear, we may apply standard linear least-squares estimation to find $\min \|\chi H^s\|^2$ to recover $H$ up to a scalar factor. That is, we are able to recover $H$ of the form

$$H_L \doteq \lambda H = \lambda\left(R + \frac{1}{d}TN^T\right) \in \mathbb{R}^{3\times 3} \tag{5.40}$$

for some (unknown) scalar factor $\lambda$.

Knowing $H_L$, the next thing is obviously to determine the scalar factor $\lambda$ by taking into account the structure of $H$.

**Lemma 5.18 (Normalization of the planar homography).** *For a matrix of the form $H_L = \lambda\left(R + \frac{1}{d}TN^T\right)$, we have*

$$|\lambda| = \sigma_2(H_L), \tag{5.41}$$

*where $\sigma_2(H_L) \in \mathbb{R}$ is the second largest singular value of $H_L$.*

*Proof.* Let $u = \frac{1}{d}R^T T \in \mathbb{R}^3$. Then we have

$$H_L^T H_L = \lambda^2(I + uN^T + Nu^T + \|u\|^2 NN^T).$$

Obviously, the vector $u \times N = \widehat{u}N \in \mathbb{R}^3$, which is orthogonal to both $u$ and $N$, is an eigenvector and $H_L^T H_L(\widehat{u}N) = \lambda^2(\widehat{u}N)$. Hence $|\lambda|$ is a singular value of $H_L$. We only have to show that it is the second largest. Let $v = \|u\|N$, $w = u/\|u\| \in \mathbb{R}^3$. We have

$$Q = uN^T + Nu^T + \|u\|^2 NN^T = (w+v)(w+v)^T - ww^T.$$

The matrix $Q$ has a positive, a negative, and a zero eigenvalue, except that when $u \sim N$, $Q$ will have two repeated zero eigenvalues. In any case, $H_L^T H_L$ has $\lambda^2$ as its second-largest eigenvalue. □

Then, if $\{\sigma_1, \sigma_2, \sigma_3\}$ are the singular values of $H_L$ recovered from linear least-squares estimation, we set a new

$$H = H_L/\sigma_2(H_L).$$

This recovers $H$ up to the form $H = \pm\left(R + \frac{1}{d}TN^T\right)$. To get the correct sign, we may use $\lambda_2^j \boldsymbol{x}_2^j = H\lambda_1^j \boldsymbol{x}_1^j$ and the fact that $\lambda_1^j, \lambda_2^j > 0$ to impose the positive depth constraint

$$(\boldsymbol{x}_2^j)^T H \boldsymbol{x}_1^j > 0, \quad \forall j = 1, 2, \ldots, n.$$

Thus, if the points $\{p\}_{j=1}^n$ are in general configuration on the plane, then the matrix $H = \left(R + \frac{1}{d}TN^T\right)$ can be uniquely determined from the image pair.

### 5.3.3 *Decomposing the planar homography matrix*

After we have recovered $H$ of the form $H = \left(R + \frac{1}{d}TN^T\right)$, we now study how to decompose such a matrix into its motion and structure parameters, namely $\{R, \frac{T}{d}, N\}$.

**Theorem 5.19 (Decomposition of the planar homography matrix).** *Given a matrix $H = \left(R + \frac{1}{d}TN^T\right)$, there are at most two physically possible solutions for a decomposition into parameters $\left\{R, \frac{1}{d}T, N\right\}$ given in Table 5.1.*

*Proof.* First notice that $H$ preserves the length of any vector orthogonal to $N$, i.e. if $N \perp a$ for some $a \in \mathbb{R}^3$, we have $\|Ha\|^2 = \|Ra\|^2 = \|a\|^2$. Also, if we know the plane spanned by the vectors that are orthogonal to $N$, we then know $N$ itself. Let us first recover the vector $N$ based on this knowledge.

The symmetric matrix $H^T H$ will have three eigenvalues $\sigma_1^2 \geq \sigma_2^2 \geq \sigma_3^2 \geq 0$, and from Lemma 5.18 we know that $\sigma_2 = 1$. Since $H^T H$ is symmetric, it can be diagonalized by an orthogonal matrix $V \in SO(3)$ such that

$$H^T H = V \Sigma V^T, \tag{5.42}$$

where $\Sigma = \text{diag}\{\sigma_1^2, \sigma_2^2, \sigma_3^2\}$. If $[v_1, v_2, v_3]$ are the three column vectors of $V$, we have

$$H^T H v_1 = \sigma_1^2 v_1, \quad H^T H v_2 = v_2, \quad H^T H v_3 = \sigma_3^2 v_3. \tag{5.43}$$

Hence $v_2$ is orthogonal to both $N$ and $T$, and its length is preserved under the map $H$. Also, it is easy to check that the length of two other unit-length vectors defined as

$$u_1 \doteq \frac{\sqrt{1 - \sigma_3^2}\, v_1 + \sqrt{\sigma_1^2 - 1}\, v_3}{\sqrt{\sigma_1^2 - \sigma_3^2}}, \quad u_2 \doteq \frac{\sqrt{1 - \sigma_3^2}\, v_1 - \sqrt{\sigma_1^2 - 1}\, v_3}{\sqrt{\sigma_1^2 - \sigma_3^2}} \tag{5.44}$$

is also preserved under the map $H$. Furthermore, it is easy to verify that $H$ preserves the length of any vectors inside each of the two subspaces

$$S_1 = \text{span}\{v_2, u_1\}, \quad S_2 = \text{span}\{v_2, u_2\}. \tag{5.45}$$

Since $v_2$ is orthogonal to $u_1$ and $u_2$, $\widehat{v_2}u_1$ is a unit normal vector to $S_1$, and $\widehat{v_2}u_2$ a unit normal vector to $S_2$. Then $\{v_2, u_1, \widehat{v_2}u_1\}$ and $\{v_2, u_2, \widehat{v_2}u_2\}$ form two sets of orthonormal bases for $\mathbb{R}^3$. Notice that we have

$$Rv_2 = Hv_2, \quad Ru_i = Hu_i, \quad R(\widehat{v_2}u_i) = \widehat{Hv_2}Hu_i$$

if $N$ is the normal to the subspace $S_i$, $i = 1, 2$, as shown in Figure 5.12.

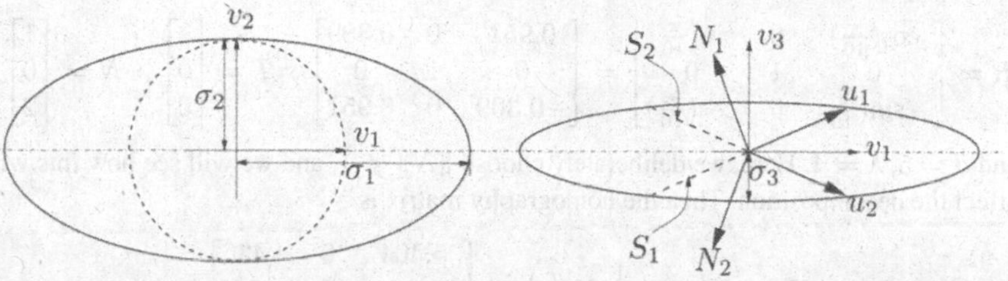


Figure 5.12. In terms of singular vectors $(v_1, v_2, v_3)$ and singular values $(\sigma_1, \sigma_2, \sigma_3)$ of the matrix $H$, there are two candidate subspaces $S_1$ and $S_2$ on which the vectors' length is preserved by the homography matrix $H$.

Define the matrices

$$U_1 = [v_2, u_1, \widehat{v_2}u_1], \quad W_1 = [Hv_2, Hu_1, \widehat{Hv_2}Hu_1];$$
$$U_2 = [v_2, u_2, \widehat{v_2}u_2], \quad W_2 = [Hv_2, Hu_2, \widehat{Hv_2}Hu_2].$$

We then have

$$RU_1 = W_1, \quad RU_2 = W_2.$$

This suggests that each subspace $S_1$, or $S_2$ may give rise to a solution to the decomposition. By taking into account the extra sign ambiguity in the term $\frac{1}{d}TN^T$, we then obtain four solutions for decomposing $H = R + \frac{1}{d}TN^T$ to $\{R, \frac{1}{d}T, N\}$. They are given in Table 5.1.

| | | | | | |
|---|---|---|---|---|---|
| Solution 1 | $R_1 = W_1U_1^T$ <br> $N_1 = \widehat{v_2}u_1$ <br> $\frac{1}{d}T_1 = (H - R_1)N_1$ | Solution 3 | $R_3 = R_1$ <br> $N_3 = -N_1$ <br> $\frac{1}{d}T_3 = -\frac{1}{d}T_1$ |
| Solution 2 | $R_2 = W_2U_2^T$ <br> $N_2 = \widehat{v_2}u_2$ <br> $\frac{1}{d}T_2 = (H - R_2)N_2$ | Solution 4 | $R_4 = R_2$ <br> $N_4 = -N_2$ <br> $\frac{1}{d}T_4 = -\frac{1}{d}T_2$ |

Table 5.1. Four solutions for the planar homography decomposition, only two of which satisfy the positive depth constraint.

In order to reduce the number of physically possible solutions, we may impose the positive depth constraint (Exercise 5.11); since the camera can see only points that are in front of it, we must have $N^Te_3 = n_3 > 0$. Suppose that solution 1 is the true one; this constraint will then eliminate solution 3 as being physically impossible. Similarly, one of solutions 2 or 4 will be eliminated. For the case that $T \sim N$, we have $\sigma_3^2 = 0$ in the above proof. Hence $u_1 = u_2$, and solutions 1 and 2 are equivalent. Imposing the positive depth constraint leads to a unique solution for all motion and structure parameters. □

**Example 5.20 (A numerical example).** Suppose that

$$R = \begin{bmatrix} \cos(\frac{\pi}{10}) & 0 & \sin(\frac{\pi}{10}) \\ 0 & 1 & 0 \\ -\sin(\frac{\pi}{10}) & 0 & \cos(\frac{\pi}{10}) \end{bmatrix} = \begin{bmatrix} 0.951 & 0 & 0.309 \\ 0 & 1 & 0 \\ -0.309 & 0 & 0.951 \end{bmatrix}, \quad T = \begin{bmatrix} 2 \\ 0 \\ 0 \end{bmatrix}, \quad N = \begin{bmatrix} 1 \\ 0 \\ 2 \end{bmatrix},$$

and $d = 5, \lambda = 4$. Here, we deliberately choose $\|N\| \neq 1$, and we will see how this will affect the decomposition. Then the homography matrix is

$$H_L = \lambda \left( R + \frac{1}{d} T N^T \right) = \begin{bmatrix} 5.404 & 0 & 4.436 \\ 0 & 4 & 0 \\ -1.236 & 0 & 3.804 \end{bmatrix}.$$

The singular values of $H_L$ are $\{7.197, 4.000, 3.619\}$. The middle one is exactly the scale $\lambda$. Hence for the normalized homography matrix $H_L/4 \rightarrow H$, the matrix $H^T H$ has the SVD[8]

$$V \Sigma V^T \doteq \begin{bmatrix} 0.675 & 0 & -0.738 \\ 0 & 1 & 0 \\ 0.738 & 0 & 0.675 \end{bmatrix} \begin{bmatrix} 3.237 & 0 & 0 \\ 0 & 1 & 0 \\ 0 & 0 & 0.819 \end{bmatrix} \begin{bmatrix} 0.675 & 0 & -0.738 \\ 0 & 1 & 0 \\ 0.738 & 0 & 0.675 \end{bmatrix}^T.$$

Then the two vectors $u_1$ and $u_2$ are given by

$$u_1 = [-0.525, 0, 0.851]^T; \quad u_2 = [0.894, 0, -0.447]^T.$$

The four solutions to the decomposition are

$$R_1 = \begin{bmatrix} 0.704 & 0 & 0.710 \\ 0 & 1 & 0 \\ -0.710 & 0 & 0.704 \end{bmatrix}, \quad N_1 = \begin{bmatrix} 0.851 \\ 0 \\ 0.525 \end{bmatrix}, \quad \frac{1}{d} T_1 = \begin{bmatrix} 0.760 \\ 0 \\ 0.471 \end{bmatrix};$$

$$R_2 = \begin{bmatrix} 0.951 & 0 & 0.309 \\ 0 & 1 & 0 \\ -0.309 & 0 & 0.951 \end{bmatrix}, \quad N_2 = \begin{bmatrix} -0.447 \\ 0 \\ -0.894 \end{bmatrix}, \quad \frac{1}{d} T_2 = \begin{bmatrix} -0.894 \\ 0 \\ 0 \end{bmatrix};$$

$$R_3 = \begin{bmatrix} 0.704 & 0 & 0.710 \\ 0 & 1 & 0 \\ -0.710 & 0 & 0.704 \end{bmatrix}, \quad N_3 = \begin{bmatrix} -0.851 \\ 0 \\ -0.525 \end{bmatrix}, \quad \frac{1}{d} T_3 = \begin{bmatrix} -0.760 \\ 0 \\ -0.471 \end{bmatrix};$$

$$R_4 = \begin{bmatrix} 0.951 & 0 & 0.309 \\ 0 & 1 & 0 \\ -0.309 & 0 & 0.951 \end{bmatrix}, \quad N_4 = \begin{bmatrix} 0.447 \\ 0 \\ 0.894 \end{bmatrix}, \quad \frac{1}{d} T_4 = \begin{bmatrix} 0.894 \\ 0 \\ 0 \end{bmatrix}.$$

Obviously, the fourth solution is the correct one: The original $\|N\| \neq 1$, and $N$ is recovered up to a scalar factor (with its length normalized to 1), and hence in the solution we should expect $\frac{1}{d} T_4 = \frac{\|N\|}{d} T$. Notice that the first solution also satisfies $N_1^T e_3 > 0$,

[8]The Matlab routine SVD does not always guarantee that $V \in SO(3)$. When using the routine, if one finds that $\det(V) = -1$, replace both $V$'s by $-V$.

which indicates a plane in front of the camera. Hence it corresponds to another physically possible solution (from the decomposition). ■

We will investigate the geometric relation between the remaining two physically possible solutions in the exercises (see Exercise 5.19). We conclude this section by presenting the following four-point Algorithm 5.2 for motion estimation from a planar scene. Examples of use of this algorithm on real images are shown in Figure 5.13.

---

**Algorithm 5.2 (The four-point algorithm for a planar scene).**

---

For a given set of image pairs $(\boldsymbol{x}_1^j, \boldsymbol{x}_2^j)$, $j = 1, 2, \ldots, n$ $(n \geq 4)$, of points on a plane $N^T \boldsymbol{X} = d$, this algorithm finds $\{R, \frac{1}{d}T, N\}$ that solves

$$\widehat{\boldsymbol{x}_2^j}^T \left( R + \frac{1}{d} T N^T \right) \boldsymbol{x}_1^j = 0, \quad j = 1, 2, \ldots, n.$$

1. **Compute a first approximation of the homography matrix**
   Construct $\chi = [\boldsymbol{a}^1, \boldsymbol{a}^2, \ldots, \boldsymbol{a}^n]^T \in \mathbb{R}^{3n \times 9}$ from correspondences $\boldsymbol{x}_1^j$ and $\boldsymbol{x}_2^j$, where $\boldsymbol{a}^j = \boldsymbol{x}_1^j \otimes \widehat{\boldsymbol{x}_2^j} \in \mathbb{R}^{9 \times 3}$. Find the vector $H_L^s \in \mathbb{R}^9$ of unit length that solves
   $$\chi H_L^s = 0$$
   as follows: compute the SVD of $\chi = U_\chi \Sigma_\chi V_\chi^T$ and define $H_L^s$ to be the ninth column of $V_\chi$. Unstack the nine elements of $H_L^s$ into a square $3 \times 3$ matrix $H_L$.
2. **Normalization of the homography matrix**
   Compute the singular values $\{\sigma_1, \sigma_2, \sigma_3\}$ of the matrix $H_L$ and normalize it as
   $$H = H_L / \sigma_2.$$
   Correct the sign of $H$ according to sign $\left((\boldsymbol{x}_2^j)^T H \boldsymbol{x}_1^j\right)$ for $j = 1, 2, \ldots, n$.
3. **Decomposition of the homography matrix**
   Compute the singular value decomposition of
   $$H^T H = V \Sigma V^T$$
   and compute the four solutions for a decomposition $\{R, \frac{1}{d}T, N\}$ as in the proof of Theorem 5.19. Select the two physically possible ones by imposing the positive depth constraint $N^T e_3 > 0$.

---

## 5.3.4 *Relationships between the homography and the essential matrix*

In practice, especially when the scene is piecewise planar, we often need to compute the essential matrix $E$ with a given homography $H$ computed from some four points known to be planar; or in the opposite situation, the essential matrix $E$ may have been already estimated using points in general position, and we then want to compute the homography for a particular (usually smaller) set of coplanar

Figure 5.13. Homography between the left and middle images is determined by the building facade on the top, and the ground plane on the bottom. The right image is the warped image overlayed on the first image based on the estimated homography $H$. Note that all points on the reference plane are aligned, whereas points outside the reference plane are offset by an amount that is proportional to their distance from the reference plane.

points. We hence need to understand the relationship between the essential matrix $E$ and the homography $H$.

**Theorem 5.21 (Relationships between the homography and essential matrix).** *For a matrix $E = \widehat{T}R$ and a matrix $H = R + Tu^T$ for some nonsingular $R \in \mathbb{R}^{3\times3}$, $T, u \in \mathbb{R}^3$, with $\|T\| = 1$, we have:*

1. $E = \widehat{T}H$;
2. $H^T E + E^T H = 0$;
3. $H = \widehat{T}^T E + Tv^T$, *for some* $v \in \mathbb{R}^3$.

*Proof.* The proof of item 1 is easy, since $\widehat{T}T = 0$. For item 2, notice that $H^T E = (R + Tu^T)^T \widehat{T}R = R^T \widehat{T}R$ is a skew-symmetric matrix, and hence $H^T E = -E^T H$. For item 3, notice that

$$\widehat{T}H = \widehat{T}R = \widehat{T}\widehat{T}^T\widehat{T}R = \widehat{T}\widehat{T}^T E,$$

since $\widehat{T}\widehat{T}^T v = (I - TT^T)v$ represents an orthogonal projection of $v$ onto the subspace (a plane) orthogonal to $T$ (see Exercise 5.3). Therefore, $\widehat{T}(H - \widehat{T}^T E) = 0$. That is, all the columns of $H - \widehat{T}^T E$ are parallel to $T$, and hence we have $H - \widehat{T}^T E = Tv^T$ for some $v \in \mathbb{R}^3$. □

Notice that neither the statement nor the proof of the theorem assumes that $R$ is a rotation matrix. Hence, the results will also be applicable to the case in which the camera is not calibrated, which will be discussed in the next chapter.

This theorem directly implies two useful corollaries stated below that allow us to easily compute $E$ from $H$ as well as $H$ from $E$ with minimum extra information from images.[9] The first corollary is a direct consequence of the above theorem and Proposition 5.14:

**Corollary 5.22 (From homography to the essential matrix).** *Given a homography $H$ and two pairs of images $(\boldsymbol{x}_1^i, \boldsymbol{x}_2^i), i = 1, 2$, of two points not on the plane $P$ from which $H$ is induced, we have*

$$E = \widehat{T}H, \tag{5.46}$$

*where $T \sim \widehat{\boldsymbol{\ell}_2^1}\boldsymbol{\ell}_2^2$ and $\|T\| = 1$.*

*Proof.* According to Proposition 5.14, $\boldsymbol{\ell}_2^i$ is the epipolar line $\boldsymbol{\ell}_2^i \sim \widehat{\boldsymbol{x}_2^i} H\boldsymbol{x}_1^i$, $i = 1, 2$. Both epipolar lines $\boldsymbol{\ell}_2^1, \boldsymbol{\ell}_2^2$ pass through the epipole $\boldsymbol{e}_2 \sim T$. This can be illustrated by Figure 5.14. □

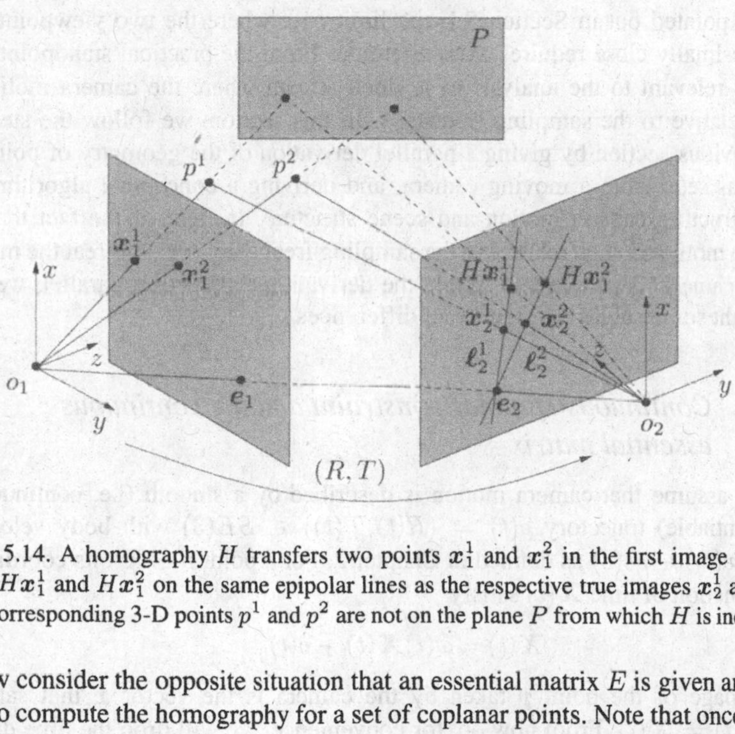


Figure 5.14. A homography $H$ transfers two points $\boldsymbol{x}_1^1$ and $\boldsymbol{x}_1^2$ in the first image to two points $H\boldsymbol{x}_1^1$ and $H\boldsymbol{x}_1^2$ on the same epipolar lines as the respective true images $\boldsymbol{x}_2^1$ and $\boldsymbol{x}_2^2$ if the corresponding 3-D points $p^1$ and $p^2$ are not on the plane $P$ from which $H$ is induced.

Now consider the opposite situation that an essential matrix $E$ is given and we want to compute the homography for a set of coplanar points. Note that once $E$ is known, the vector $T$ is also known (up to a scalar factor) as the left null space of $E$. We may typically choose $T$ to be of unit length.

[9] Although in principle, to compute $E$ from $H$, one does not need any extra information but only has to decompose $H$ and find $R$ and $T$ using Theorem 5.19, the corollary will allow us to bypass that by much simpler techniques, which, unlike Theorem 5.19, will also be applicable to the uncalibrated case.

**Corollary 5.23 (From essential matrix to homography).** *Given an essential matrix $E$ and three pairs of images $(\boldsymbol{x}_1^i, \boldsymbol{x}_2^i), i = 1,2,3$, of three points in 3-D, the homography $H$ induced by the plane specified by the three points then is*

$$H = \widehat{T}^T E + Tv^T, \tag{5.47}$$

*where $v = [v_1, v_2, v_3]^T \in \mathbb{R}^3$ solves the system of three linear equations*

$$\widehat{\boldsymbol{x}_2^i}(\widehat{T}^T E + Tv^T)\boldsymbol{x}_1^i = 0, \quad i = 1,2,3. \tag{5.48}$$

*Proof.* We leave the proof to the reader as an exercise. □

## 5.4 Continuous motion case[10]

As we pointed out in Section 5.1, the limit case where the two viewpoints are infinitesimally close requires extra attention. From the practical standpoint, this case is relevant to the analysis of a video stream where the camera motion is slow relative to the sampling frequency. In this section, we follow the steps of the previous section by giving a parallel derivation of the geometry of points in space as seen from a moving camera, and deriving a conceptual algorithm for reconstructing camera motion and scene structure. In light of the fact that the camera motion is slow relative to the sampling frequency, we will treat the motion of the camera as continuous. While the derivations proceed in parallel, we will highlight some subtle but significant differences.

### 5.4.1 *Continuous epipolar constraint and the continuous essential matrix*

Let us assume that camera motion is described by a smooth (i.e. continuously differentiable) trajectory $g(t) = (R(t), T(t)) \in SE(3)$ with body velocities $(\omega(t), v(t)) \in se(3)$ as defined in Chapter 2. For a point $p \in \mathbb{R}^3$, its coordinates as a function of time $\boldsymbol{X}(t)$ satisfy

$$\dot{\boldsymbol{X}}(t) = \widehat{\omega}(t)\boldsymbol{X}(t) + v(t). \tag{5.49}$$

The image of the point $p$ taken by the camera is the vector $\boldsymbol{x}$ that satisfies $\lambda(t)\boldsymbol{x}(t) = \boldsymbol{X}(t)$. From now on, for convenience, we will drop the time dependency from the notation. Denote the velocity of the image point $\boldsymbol{x}$ by $\boldsymbol{u} \doteq \dot{\boldsymbol{x}} \in \mathbb{R}^3$. The velocity $\boldsymbol{u}$ is also called *image motion field*, which under the brightness constancy assumption discussed in Chapter 4 can be approximated by the *optical*

[10] This section can be skipped without loss of continuity if the reader is not interested in the continuous-motion case.

*flow*. To obtain an explicit expression for $\boldsymbol{u}$, we notice that

$$\boldsymbol{X} = \lambda \boldsymbol{x}, \quad \dot{\boldsymbol{X}} = \dot{\lambda}\boldsymbol{x} + \lambda\dot{\boldsymbol{x}}.$$

Substituting this into equation (5.49), we obtain

$$\dot{\boldsymbol{x}} = \widehat{\omega}\boldsymbol{x} + \frac{1}{\lambda}v - \frac{\dot{\lambda}}{\lambda}\boldsymbol{x}. \tag{5.50}$$

Then the image velocity $\boldsymbol{u} = \dot{\boldsymbol{x}}$ depends not only on the camera motion but also on the depth scale $\lambda$ of the point. For the planar perspective projection and the spherical perspective projection, the expression for $\boldsymbol{u}$ will be slightly different. We leave the detail to the reader as an exercise (see Exercise 5.20).

To eliminate the depth scale $\lambda$, consider now the inner product of the vectors in (5.50) with the vector $(v \times \boldsymbol{x})$. We obtain

$$\dot{\boldsymbol{x}}^T \widehat{v}\boldsymbol{x} = \boldsymbol{x}^T \widehat{\omega}^T \widehat{v}\boldsymbol{x}.$$

We can rewrite the above equation in an equivalent way:

$$\boxed{\boldsymbol{u}^T \widehat{v}\boldsymbol{x} + \boldsymbol{x}^T \widehat{\omega}\widehat{v}\boldsymbol{x} = 0.} \tag{5.51}$$

This constraint plays the same role for the case of continuous-time images as the epipolar constraint for two discrete image, in the sense that it does not depend on the position of the point in space, but only on its projection and the motion parameters. We call it the *continuous epipolar constraint*.

Before proceeding with an analysis of equation (5.51), we state a lemma that will become useful in the remainder of this section.

**Lemma 5.24.** *Consider the matrices $M_1, M_2 \in \mathbb{R}^{3\times 3}$. Then $\boldsymbol{x}^T M_1 \boldsymbol{x} = \boldsymbol{x}^T M_2 \boldsymbol{x}$ for all $\boldsymbol{x} \in \mathbb{R}^3$ if and only if $M_1 - M_2$ is a skew-symmetric matrix, i.e. $M_1 - M_2 \in so(3)$.*

We leave the proof of this lemma as an exercise. Following the lemma, for any skew-symmetric matrix $M \in \mathbb{R}^{3\times 3}$, $\boldsymbol{x}^T M \boldsymbol{x} = 0$. Since $\frac{1}{2}(\widehat{\omega}\widehat{v} - \widehat{v}\widehat{\omega})$ is a skew-symmetric matrix, $\boldsymbol{x}^T \frac{1}{2}(\widehat{\omega}\widehat{v} - \widehat{v}\widehat{\omega})\boldsymbol{x} = 0$. If we define the *symmetric epipolar component* to be the matrix

$$s \doteq \frac{1}{2}(\widehat{\omega}\widehat{v} + \widehat{v}\widehat{\omega}) \quad \in \mathbb{R}^{3\times 3},$$

then we have that

$$\boldsymbol{x}^T s \boldsymbol{x} = \boldsymbol{x}^T \widehat{\omega}\widehat{v}\boldsymbol{x},$$

so that the continuous epipolar constraint may be rewritten as

$$\boldsymbol{u}^T \widehat{v}\boldsymbol{x} + \boldsymbol{x}^T s \boldsymbol{x} = 0. \tag{5.52}$$

This equation shows that for the matrix $\widehat{\omega}\widehat{v}$, only its symmetric component $s = \frac{1}{2}(\widehat{\omega}\widehat{v} + \widehat{v}\widehat{\omega})$ can be recovered from the epipolar equation (5.51) or equivalently

(5.52).[11] This structure is substantially different from that of the discrete case, and it cannot be derived by a first-order approximation of the essential matrix $\widehat{T}R$. In fact, a naive discretization of the discrete epipolar equation may lead to a constraint involving only a matrix of the form $\widehat{v}\widehat{\omega}$, whereas in the true continuous case we have to deal with only its symmetric component $s = \frac{1}{2}(\widehat{\omega}\widehat{v} + \widehat{v}\widehat{\omega})$ plus another term as given in (5.52). The set of matrices of interest in the case of continuous motions is thus the space of $6 \times 3$ matrices of the form

$$\mathcal{E}' \doteq \left\{ \begin{bmatrix} \widehat{v} \\ \frac{1}{2}(\widehat{\omega}\widehat{v} + \widehat{v}\widehat{\omega}) \end{bmatrix} \middle| \, \omega, v \in \mathbb{R}^3 \right\} \quad \subset \mathbb{R}^{6\times 3},$$

which we call the *continuous essential space*. A matrix in this space is called a *continuous essential matrix*. Note that the continuous epipolar constraint (5.52) is homogeneous in the linear velocity $v$. Thus $v$ may be recovered only up to a constant scalar factor. Consequently, in motion recovery, we will concern ourselves with matrices belonging to the *normalized continuous essential space* with $v$ scaled to unit norm:

$$\mathcal{E}'_1 = \left\{ \begin{bmatrix} \widehat{v} \\ \frac{1}{2}(\widehat{\omega}\widehat{v} + \widehat{v}\widehat{\omega}) \end{bmatrix} \middle| \, \omega \in \mathbb{R}^3, v \in \mathbb{S}^2 \right\} \quad \subset \mathbb{R}^{6\times 3}.$$

## 5.4.2 *Properties of the continuous essential matrix*

The skew-symmetric part of a continuous essential matrix simply corresponds to the velocity $v$. The characterization of the (normalized) essential matrix focuses only on the symmetric matrix part $s = \frac{1}{2}(\widehat{\omega}\widehat{v} + \widehat{v}\widehat{\omega})$. We call the space of all the matrices of this form the *symmetric epipolar space*

$$\mathcal{S} \doteq \left\{ \frac{1}{2}(\widehat{\omega}\widehat{v} + \widehat{v}\widehat{\omega}) \middle| \, \omega \in \mathbb{R}^3, v \in \mathbb{S}^2 \right\} \subset \mathbb{R}^{3\times 3}.$$

The motion estimation problem is now reduced to that of *recovering the velocity* $(\omega, v)$ *with* $\omega \in \mathbb{R}^3$ *and* $v \in \mathbb{S}^2$ *from a given symmetric epipolar component* $s$.

The characterization of symmetric epipolar components depends on a characterization of matrices of the form $\widehat{\omega}\widehat{v} \in \mathbb{R}^{3\times 3}$, which is given in the following lemma. Of use in the lemma is the matrix $R_Y(\theta)$ defined to be the rotation around the $Y$-axis by an angle $\theta \in \mathbb{R}$, i.e. $R_Y(\theta) = e^{\widehat{e}_2\theta}$ with $e_2 = [0, 1, 0]^T \in \mathbb{R}^3$.

**Lemma 5.25.** *A matrix* $Q \in \mathbb{R}^{3\times 3}$ *has the form* $Q = \widehat{\omega}\widehat{v}$ *with* $\omega \in \mathbb{R}^3$, $v \in \mathbb{S}^2$ *if and only if*

$$Q = -VR_Y(\theta)diag\{\lambda, \lambda\cos(\theta), 0\}V^T \qquad (5.53)$$

[11] This redundancy is the reason why different forms of the continuous epipolar constraint exist in the literature [Zhuang and Haralick, 1984, Ponce and Genc, 1998, Vieville and Faugeras, 1995, Maybank, 1993, Brooks et al., 1997], and accordingly, various approaches have been proposed to recover $\omega$ and $v$ (see [Tian et al., 1996]).

*for some rotation matrix* $V \in SO(3)$, *the positive scalar* $\lambda = \|\omega\|$, *and* $\cos(\theta) = \omega^T v/\lambda$.

*Proof.* We first prove the necessity. The proof follows from the geometric meaning of $\widehat{\omega}\widehat{v}$ multiplied by any vector $q \in \mathbb{R}^3$:

$$\widehat{\omega}\widehat{v}q = \omega \times (v \times q).$$

Let $b \in \mathbb{S}^2$ be the unit vector perpendicular to both $\omega$ and $v$. That is, $b = \frac{v \times \omega}{\|v \times \omega\|}$. (If $v \times \omega = 0$, $b$ is not uniquely defined. In this case, $\omega, v$ are parallel, and the rest of the proof follows if one picks any vector $b$ orthogonal to $v$ and $\omega$.) Then $\omega = \lambda \exp(\widehat{b}\theta)v$ (according to this definition, $\theta$ is the angle between $\omega$ and $v$, and $0 \leq \theta \leq \pi$). It is easy to check that if the matrix $V$ is defined to be

$$V = \left(e^{\widehat{b}\frac{\pi}{2}}v, b, v\right),$$

then $Q$ has the given form (5.53).

We now prove the sufficiency. Given a matrix $Q$ that can be decomposed into the form (5.53), define the orthogonal matrix $U = -VR_Y(\theta) \in O(3)$. (Recall that $O(3)$ represents the space of all orthogonal matrices of determinant $\pm 1$.) Let the two skew-symmetric matrices $\widehat{\omega}$ and $\widehat{v}$ be given by

$$\widehat{\omega} = UR_Z\left(\pm\frac{\pi}{2}\right)\Sigma_\lambda U^T, \quad \widehat{v} = VR_Z\left(\pm\frac{\pi}{2}\right)\Sigma_1 V^T, \tag{5.54}$$

where $\Sigma_\lambda = \text{diag}\{\lambda, \lambda, 0\}$ and $\Sigma_1 = \text{diag}\{1, 1, 0\}$. Then

$$\begin{aligned}
\widehat{\omega}\widehat{v} &= UR_Z\left(\pm\frac{\pi}{2}\right)\Sigma_\lambda U^T VR_Z\left(\pm\frac{\pi}{2}\right)\Sigma_1 V^T \\
&= UR_Z\left(\pm\frac{\pi}{2}\right)\Sigma_\lambda(-R_Y^T(\theta))R_Z\left(\pm\frac{\pi}{2}\right)\Sigma_1 V^T \\
&= U\text{diag}\{\lambda, \lambda\cos(\theta), 0\}V^T \\
&= Q.
\end{aligned} \tag{5.55}$$

Since $\omega$ and $v$ have to be, respectively, the left and the right zero eigenvectors of $Q$, the reconstruction given in (5.54) is unique up to a sign. □

Based on the above lemma, the following theorem reveals the structure of the symmetric epipolar component.

**Theorem 5.26 (Characterization of the symmetric epipolar component).** *A real symmetric matrix* $s \in \mathbb{R}^{3\times 3}$ *is a symmetric epipolar component if and only if* $s$ *can be diagonalized as* $s = V\Sigma V^T$ *with* $V \in SO(3)$ *and*

$$\Sigma = diag\{\sigma_1, \sigma_2, \sigma_3\}$$

*with* $\sigma_1 \geq 0, \sigma_3 \leq 0$, *and* $\sigma_2 = \sigma_1 + \sigma_3$.

*Proof.* We first prove the necessity. Suppose $s$ is a symmetric epipolar component. Thus there exist $\omega \in \mathbb{R}^3, v \in \mathbb{S}^2$ such that $s = \frac{1}{2}(\widehat{\omega}\widehat{v} + \widehat{v}\widehat{\omega})$. Since $s$ is a symmetric matrix, it is diagonalizable, all its eigenvalues are real, and all the eigenvectors are orthogonal to each other. It then suffices to check that its eigenvalues satisfy the given conditions.

Let the unit vector $b$, the rotation matrix $V$, $\theta$, and $\lambda$ be the same as in the proof of Lemma 5.25. According to the lemma, we have

$$\widehat{\omega}\widehat{v} = -VR_Y(\theta)\text{diag}\{\lambda, \lambda\cos(\theta), 0\}V^T.$$

Since $(\widehat{\omega}\widehat{v})^T = \widehat{v}\widehat{\omega}$, we have

$$s = \frac{1}{2}V\left(-R_Y(\theta)\text{diag}\{\lambda, \lambda\cos(\theta), 0\} - \text{diag}\{\lambda, \lambda\cos(\theta), 0\}R_Y^T(\theta)\right)V^T.$$

Define the matrix $D(\lambda, \theta) \in \mathbb{R}^{3\times 3}$ to be

$$\begin{aligned} D(\lambda, \theta) &= -R_Y(\theta)\text{diag}\{\lambda, \lambda\cos(\theta), 0\} - \text{diag}\{\lambda, \lambda\cos(\theta), 0\}R_Y^T(\theta) \\ &= \lambda \begin{bmatrix} -2\cos(\theta) & 0 & \sin(\theta) \\ 0 & -2\cos(\theta) & 0 \\ \sin(\theta) & 0 & 0 \end{bmatrix}. \end{aligned}$$

Directly calculating its eigenvalues and eigenvectors, we obtain that $D(\lambda, \theta)$ is equal to

$$R_Y\left(\frac{\theta - \pi}{2}\right)\text{diag}\left\{\lambda(1 - \cos(\theta)), -2\lambda\cos(\theta), \lambda(-1 - \cos(\theta))\right\}R_Y^T\left(\frac{\theta - \pi}{2}\right). \tag{5.56}$$

Thus $s = \frac{1}{2}VD(\lambda, \theta)V^T$ has eigenvalues

$$\left\{\frac{1}{2}\lambda(1 - \cos(\theta)), \quad -\lambda\cos(\theta), \quad \frac{1}{2}\lambda(-1 - \cos(\theta))\right\}, \tag{5.57}$$

which satisfy the given conditions.

We now prove the sufficiency. Given $s = V_1\text{diag}\{\sigma_1, \sigma_2, \sigma_3\}V_1^T$ with $\sigma_1 \geq 0, \sigma_3 \leq 0, \sigma_2 = \sigma_1 + \sigma_3$, and $V_1^T \in SO(3)$, these three eigenvalues uniquely determine $\lambda, \theta \in \mathbb{R}$ such that the $\sigma_i$'s have the form given in (5.57):

$$\begin{aligned} \lambda &= \sigma_1 - \sigma_3, & \lambda &\geq 0, \\ \theta &= \arccos(-\sigma_2/\lambda), & \theta &\in [0, \pi]. \end{aligned}$$

Define a matrix $V \in SO(3)$ to be $V = V_1R_Y^T\left(\frac{\theta}{2} - \frac{\pi}{2}\right)$. Then $s = \frac{1}{2}VD(\lambda, \theta)V^T$. According to Lemma 5.25, there exist vectors $v \in \mathbb{S}^2$ and $\omega \in \mathbb{R}^3$ such that

$$\widehat{\omega}\widehat{v} = -VR_Y(\theta)\text{diag}\{\lambda, \lambda\cos(\theta), 0\}V^T.$$

Therefore, $\frac{1}{2}(\widehat{\omega}\widehat{v} + \widehat{v}\widehat{\omega}) = \frac{1}{2}VD(\lambda, \theta)V^T = s$. □

Figure 5.15 gives a geometric interpretation of the three eigenvectors of the symmetric epipolar component $s$ for the case in which both $\omega, v$ are of unit length. The constructive proof given above is important since it gives an explicit decomposition of the symmetric epipolar component $s$, which will be studied in more detail next.

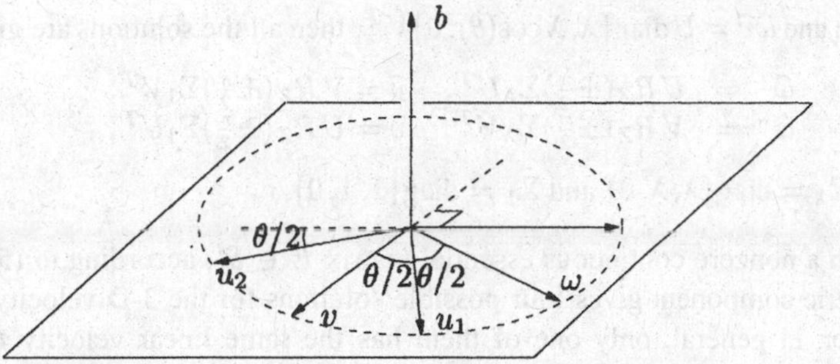


Figure 5.15. Vectors $u_1, u_2, b$ are the three eigenvectors of a symmetric epipolar component $\frac{1}{2}(\widehat{\omega}\widehat{v}+\widehat{v}\widehat{\omega})$. In particular, $b$ is the normal vector to the plane spanned by $\omega$ and $v$, and $u_1, u_2$ are both in this plane. The vector $u_1$ is the average of $\omega$ and $v$, and $u_2$ is orthogonal to both $b$ and $u_1$.

Following the proof of Theorem 5.26, if we already know the eigenvector decomposition of a symmetric epipolar component $s$, we certainly can find at least one solution $(\omega, v)$ such that $s = \frac{1}{2}(\widehat{\omega}\widehat{v} + \widehat{v}\widehat{\omega})$. We now discuss uniqueness, i.e. how many solutions exist for $s = \frac{1}{2}(\widehat{\omega}\widehat{v} + \widehat{v}\widehat{\omega})$.

**Theorem 5.27 (Velocity recovery from the symmetric epipolar component).** *There exist exactly four 3-D velocities $(\omega, v)$ with $\omega \in \mathbb{R}^3$ and $v \in \mathbb{S}^2$ corresponding to a nonzero $s \in \mathcal{S}$.*

*Proof.* Suppose $(\omega_1, v_1)$ and $(\omega_2, v_2)$ are both solutions for $s = \frac{1}{2}(\widehat{\omega}\widehat{v} + \widehat{v}\widehat{\omega})$. Then we have

$$\widehat{v}_1\widehat{\omega}_1 + \widehat{\omega}_1\widehat{v}_1 = \widehat{v}_2\widehat{\omega}_2 + \widehat{\omega}_2\widehat{v}_2. \tag{5.58}$$

From Lemma 5.25, we may write

$$\begin{aligned} \widehat{\omega}_1\widehat{v}_1 &= -V_1 R_Y(\theta_1)\text{diag}\{\lambda_1, \lambda_1\cos(\theta_1), 0\}V_1^T, \\ \widehat{\omega}_2\widehat{v}_2 &= -V_2 R_Y(\theta_2)\text{diag}\{\lambda_2, \lambda_2\cos(\theta_2), 0\}V_2^T. \end{aligned} \tag{5.59}$$

Let $W = V_1^T V_2 \in SO(3)$. Then from (5.58),

$$D(\lambda_1, \theta_1) = W D(\lambda_2, \theta_2) W^T. \tag{5.60}$$

Since both sides of (5.60) have the same eigenvalues, according to (5.56), we have

$$\lambda_1 = \lambda_2, \quad \theta_2 = \theta_1.$$

We can then denote both $\theta_1$ and $\theta_2$ by $\theta$. It is immediate to check that the only possible rotation matrix $W$ that satisfies (5.60) is given by $I_{3\times 3}$,

$$\begin{bmatrix} -\cos(\theta) & 0 & \sin(\theta) \\ 0 & -1 & 0 \\ \sin(\theta) & 0 & \cos(\theta) \end{bmatrix}, \quad \text{or} \quad \begin{bmatrix} \cos(\theta) & 0 & -\sin(\theta) \\ 0 & -1 & 0 \\ -\sin(\theta) & 0 & -\cos(\theta) \end{bmatrix}.$$

From the geometric meaning of $V_1$ and $V_2$, all the cases give either $\widehat{\omega}_1\widehat{v}_1 = \widehat{\omega}_2\widehat{v}_2$ or $\widehat{\omega}_1\widehat{v}_1 = \widehat{v}_2\widehat{\omega}_2$. Thus, according to the proof of Lemma 5.25, if $(\omega, v)$ is one solution and $\widehat{\omega}\widehat{v} = U\text{diag}\{\lambda, \lambda\cos(\theta), 0\}V^T$, then all the solutions are given by

$$\begin{aligned} \widehat{\omega} &= UR_Z(\pm\tfrac{\pi}{2})\Sigma_\lambda U^T, \quad \widehat{v} = VR_Z(\pm\tfrac{\pi}{2})\Sigma_1 V^T, \\ \widehat{\omega} &= VR_Z(\pm\tfrac{\pi}{2})\Sigma_\lambda V^T, \quad \widehat{v} = UR_Z(\pm\tfrac{\pi}{2})\Sigma_1 U^T, \end{aligned} \tag{5.61}$$

where $\Sigma_\lambda = \text{diag}\{\lambda, \lambda, 0\}$ and $\Sigma_1 = \text{diag}\{1, 1, 0\}$. □

Given a nonzero continuous essential matrix $E \in \mathcal{E}'$, according to (5.61), its symmetric component gives four possible solutions for the 3-D velocity $(\omega, v)$. However, in general, only one of them has the same linear velocity $v$ as the skew-symmetric part of $E$. Hence, compared to the discrete case, where there are two 3-D motions $(R, T)$ associated with an essential matrix, the velocity $(\omega, v)$ corresponding to a continuous essential matrix is unique. This is because in the continuous case, the *twisted-pair ambiguity*, which occurs in the discrete case and is caused by a 180° rotation of the camera around the translation direction, see Example 5.8, is now avoided.

### 5.4.3 *The eight-point linear algorithm*

Based on the preceding study of the continuous essential matrix, this section describes an algorithm to recover the 3-D velocity of the camera from a set of (possibly noisy) optical flow measurements.

Let $E = \begin{bmatrix} \widehat{v} \\ s \end{bmatrix} \in \mathcal{E}'_1$ with $s = \frac{1}{2}(\widehat{\omega}\widehat{v} + \widehat{v}\widehat{\omega})$ be the essential matrix associated with the continuous epipolar constraint (5.52). Since the submatrix $\widehat{v}$ is skew-symmetric and $s$ is symmetric, they have the following form

$$\widehat{v} = \begin{bmatrix} 0 & -v_3 & v_2 \\ v_3 & 0 & -v_1 \\ -v_2 & v_1 & 0 \end{bmatrix}, \quad s = \begin{bmatrix} s_1 & s_2 & s_3 \\ s_2 & s_4 & s_5 \\ s_3 & s_5 & s_6 \end{bmatrix}. \tag{5.62}$$

Define the continuous version of the "stacked" vector $E^s \in \mathbb{R}^9$ to be

$$E^s \doteq [v_1, v_2, v_3, s_1, s_2, s_3, s_4, s_5, s_6]^T. \tag{5.63}$$

Define a vector $\boldsymbol{a} \in \mathbb{R}^9$ associated with the optical flow $(\boldsymbol{x}, \boldsymbol{u})$ with $\boldsymbol{x} = [x, y, z]^T \in \mathbb{R}^3$, $\boldsymbol{u} = [u_1, u_2, u_3]^T \in \mathbb{R}^3$ to be[12]

$$\boldsymbol{a} \doteq [u_3 y - u_2 z, u_1 z - u_3 x, u_2 x - u_1 y, x^2, 2xy, 2xz, y^2, 2yz, z^2]^T. \tag{5.64}$$

The continuous epipolar constraint (5.52) can be then rewritten as

$$\boldsymbol{a}^T E^s = 0.$$

Given a set of (possibly noisy) optical flow vectors $(\boldsymbol{x}^j, \boldsymbol{u}^j)$, $j = 1, 2, \ldots, n$, generated by the same motion, define a matrix $\chi \in \mathbb{R}^{n\times 9}$ associated with these

[12]For a planar perspective projection, $z = 1$ and $u_3 = 0$; thus the expression for $\boldsymbol{a}$ can be simplified.

measurements to be

$$\chi \doteq [\boldsymbol{a}^1, \boldsymbol{a}^2, \ldots, \boldsymbol{a}^n]^T, \tag{5.65}$$

where $\boldsymbol{a}^j$ are defined for each pair $(\boldsymbol{x}^j, \boldsymbol{u}^j)$ using (5.64). In the absence of noise, the vector $E^s$ has to satisfy

$$\chi E^s = 0. \tag{5.66}$$

In order for this equation to have a unique solution for $E^s$, the rank of the matrix $\chi$ has to be eight. Thus, *for this algorithm, the optical flow vectors of at least eight points are needed to recover the 3-D velocity, i.e.* $n \geq 8$, although the minimum number of optical flow vectors needed for a finite number of solutions is actually five, as discussed by [Maybank, 1993].

When the measurements are noisy, there may be no solution to $\chi E^s = 0$. As in the discrete case, one may approximate the solution by minimizing the least-squares error function $\|\chi E^s\|^2$.

Since the vector $E^s$ is recovered from noisy measurements, the symmetric part $s$ of $E$ directly recovered from unstacking $E^s$ is not necessarily a symmetric epipolar component. Thus one cannot directly use the previously derived results for symmetric epipolar components to recover the 3-D velocity. In analogy to the discrete case, we can project the symmetric matrix $s$ onto the space of symmetric epipolar components.

**Theorem 5.28 (Projection onto the symmetric epipolar space).** *If a real symmetric matrix $F \in \mathbb{R}^{3\times 3}$ is diagonalized as $F = V\mathrm{diag}\{\lambda_1, \lambda_2, \lambda_3\}V^T$ with $V \in SO(3)$, $\lambda_1 \geq 0, \lambda_3 \leq 0$, and $\lambda_1 \geq \lambda_2 \geq \lambda_3$, then the symmetric epipolar component $E \in \mathcal{S}$ that minimizes the error $\|E - F\|_f^2$ is given by $E = V\mathrm{diag}\{\sigma_1, \sigma_2, \sigma_2\}V^T$ with*

$$\sigma_1 = \frac{2\lambda_1 + \lambda_2 - \lambda_3}{3}, \quad \sigma_2 = \frac{\lambda_1 + 2\lambda_2 + \lambda_3}{3}, \quad \sigma_3 = \frac{2\lambda_3 + \lambda_2 - \lambda_1}{3}. \tag{5.67}$$

*Proof.* Define $\mathcal{S}_\Sigma$ to be the subspace of $\mathcal{S}$ whose elements have the same eigenvalues $\Sigma = \mathrm{diag}\{\sigma_1, \sigma_2, \sigma_3\}$. Thus every matrix $E \in \mathcal{S}_\Sigma$ has the form $E = V_1\Sigma V_1^T$ for some $V_1 \in SO(3)$. To simplify the notation, define $\Sigma_\lambda = \mathrm{diag}\{\lambda_1, \lambda_2, \lambda_3\}$. We now prove this theorem in two steps.

*Step 1:* We prove that the matrix $E \in \mathcal{S}_\Sigma$ that minimizes the error $\|E - F\|_f^2$ is given by $E = V\Sigma V^T$. Since $E \in \mathcal{S}_\Sigma$ has the form $E = V_1\Sigma V_1^T$, we get

$$\|E - F\|_f^2 = \|V_1\Sigma V_1^T - V\Sigma_\lambda V^T\|_f^2 = \|\Sigma_\lambda - V^T V_1 \Sigma V_1^T V\|_f^2.$$

Define $W = V^T V_1 \in SO(3)$ and denote its entries by

$$W = \begin{bmatrix} w_1 & w_2 & w_3 \\ w_4 & w_5 & w_6 \\ w_7 & w_8 & w_9 \end{bmatrix}. \tag{5.68}$$

Then

$$\begin{aligned}\|E-F\|_f^2 &= \|\Sigma_\lambda - W\Sigma W^T\|_f^2 \\ &= \text{trace}(\Sigma_\lambda^2) - 2\text{trace}(W\Sigma W^T\Sigma_\lambda) + \text{trace}(\Sigma^2). \quad (5.69)\end{aligned}$$

Substituting (5.68) into the second term, and using the fact that $\sigma_2 = \sigma_1 + \sigma_3$ and $W$ is a rotation matrix, we get

$$\begin{aligned}\text{trace}(W\Sigma W^T\Sigma_\lambda) &= \sigma_1(\lambda_1(1-w_3^2) + \lambda_2(1-w_6^2) + \lambda_3(1-w_9^2)) \\ &\quad + \sigma_3(\lambda_1(1-w_1^2) + \lambda_2(1-w_4^2) + \lambda_3(1-w_7^2)).\end{aligned}$$

Minimizing $\|E-F\|_f^2$ is equivalent to maximizing trace$(W\Sigma W^T\Sigma_\lambda)$. From the above equation, trace$(W\Sigma W^T\Sigma_\lambda)$ is maximized if and only if $w_3 = w_6 = 0$, $w_9^2 = 1$, $w_4 = w_7 = 0$, and $w_1^2 = 1$. Since $W$ is a rotation matrix, we also have $w_2 = w_8 = 0$, and $w_5^2 = 1$. All possible $W$ give a unique matrix in $S_\Sigma$ that minimizes $\|E-F\|_f^2$: $E = V\Sigma V^T$.

*Step 2:* From step one, we need only to minimize the error function over the matrices that have the form $V\Sigma V^T \in S$. The optimization problem is then converted to one of minimizing the error function

$$\|E-F\|_f^2 = (\lambda_1 - \sigma_1)^2 + (\lambda_2 - \sigma_2)^2 + (\lambda_3 - \sigma_3)^2$$

subject to the constraint

$$\sigma_2 = \sigma_1 + \sigma_3.$$

The formulae (5.67) for $\sigma_1, \sigma_2, \sigma_3$ are directly obtained from solving this minimization problem. □

**Remark 5.29.** *In the preceding theorem, for a symmetric matrix $F$ that does not satisfy the conditions $\lambda_1 \geq 0$ and $\lambda_3 \leq 0$, one chooses $\lambda_1' = \max\{\lambda_1, 0\}$ and $\lambda_3' = \min\{\lambda_3, 0\}$ prior to applying the above theorem.*

Finally, we outline an eigenvalue-decomposition algorithm, Algorithm 5.3, for estimating 3-D velocity from optical flows of eight points, which serves as a continuous counterpart of the eight-point algorithm given in Section 5.2.

**Remark 5.30.** *Since both $E, -E \in \mathcal{E}_1'$ satisfy the same set of continuous epipolar constraints, both $(\omega, \pm v)$ are possible solutions for the given set of optical flow vectors. However, as in the discrete case, one can get rid of the ambiguous solution by enforcing the positive depth constraint (Exercise 5.11).*

In situations where the motion of the camera is partially constrained, the above linear algorithm can be further simplified. The following example illustrates such a scenario.

**Example 5.31 (Constrained motion estimation).** This example shows how to utilize constraints on motion to be estimated in order to simplify the proposed linear motion estimation algorithm in the continuous case. Let $g(t) \in SE(3)$ represent the position and orientation of an aircraft relative to the spatial frame; the inputs $\omega_1, \omega_2, \omega_3 \in \mathbb{R}$ stand for

**Algorithm 5.3 (The continuous eight-point algorithm).**

For a given set of images and optical flow vectors $(\boldsymbol{x}^j, \boldsymbol{u}^j)$, $j = 1, 2, \ldots, n$, this algorithm finds $(\omega, v) \in SE(3)$ that solves

$$\boldsymbol{u}^{jT}\widehat{v}\boldsymbol{x}^j + \boldsymbol{x}^{jT}\widehat{\omega}\widehat{v}\boldsymbol{x}^j = 0, \quad j = 1, 2, \ldots, n.$$

1. **Estimate the essential vector**
Define a matrix $\chi \in \mathbb{R}^{n\times 9}$ whose $j$th row is constructed from $\boldsymbol{x}^j$ and $\boldsymbol{u}^j$ as in (5.64). Use the SVD to find the vector $E^s \in \mathbb{R}^9$ such that $\chi E^s = 0$: $\chi = U_\chi \Sigma_\chi V_\chi^T$ and $E^s = V_\chi(:, 9)$. Recover the vector $v_0 \in \mathbb{S}^2$ from the first three entries of $E^s$ and a symmetric matrix $s \in \mathbb{R}^{3\times 3}$ from the remaining six entries as in (5.63). Multiply $E^s$ with a scalar such that the vector $v_0$ becomes of unit norm.

2. **Recover the symmetric epipolar component**
Find the eigenvalue decomposition of the symmetric matrix $s$:
$$s = V_1 \mathrm{diag}\{\lambda_1, \lambda_2, \lambda_3\} V_1^T,$$
with $\lambda_1 \geq \lambda_2 \geq \lambda_3$. Project the symmetric matrix $s$ onto the symmetric epipolar space $\mathcal{S}$. We then have the new $s = V_1 \mathrm{diag}\{\sigma_1, \sigma_2, \sigma_3\} V_1^T$ with
$$\sigma_1 = \frac{2\lambda_1 + \lambda_2 - \lambda_3}{3}, \quad \sigma_2 = \frac{\lambda_1 + 2\lambda_2 + \lambda_3}{3}, \quad \sigma_3 = \frac{2\lambda_3 + \lambda_2 - \lambda_1}{3}.$$

3. **Recover the velocity from the symmetric epipolar component**
Define
$$\begin{aligned} \lambda &= \sigma_1 - \sigma_3, \quad \lambda \geq 0, \\ \theta &= \arccos(-\sigma_2/\lambda), \quad \theta \in [0, \pi]. \end{aligned}$$
Let $V = V_1 R_Y^T\left(\frac{\theta}{2} - \frac{\pi}{2}\right) \in SO(3)$ and $U = -V R_Y(\theta) \in O(3)$. Then the four possible 3-D velocities corresponding to the matrix $s$ are given by
$$\begin{aligned} \widehat{\omega} &= U R_Z(\pm\tfrac{\pi}{2}) \Sigma_\lambda U^T, \quad \widehat{v} = V R_Z(\pm\tfrac{\pi}{2}) \Sigma_1 V^T, \\ \widehat{\omega} &= V R_Z(\pm\tfrac{\pi}{2}) \Sigma_\lambda V^T, \quad \widehat{v} = U R_Z(\pm\tfrac{\pi}{2}) \Sigma_1 U^T, \end{aligned}$$
where $\Sigma_\lambda = \mathrm{diag}\{\lambda, \lambda, 0\}$ and $\Sigma_1 = \mathrm{diag}\{1, 1, 0\}$.

4. **Recover velocity from the continuous essential matrix**
From the four velocities recovered from the matrix $s$ in step 3, choose the pair $(\omega^*, v^*)$ that satisfies
$$v^{*T} v_0 = \max_i \{v_i^T v_0\}.$$
Then the estimated 3-D velocity $(\omega, v)$ with $\omega \in \mathbb{R}^3$ and $v \in \mathbb{S}^2$ is given by
$$\omega = \omega^*, \quad v = v_0.$$

---

the rates of the rotation about the axes of the aircraft, and $v_1 \in \mathbb{R}$ is the velocity of the aircraft. Using the standard homogeneous representation for $g$ (see Chapter 2), the kinematic equations of the aircraft motion are given by

$$\dot{g} = \begin{bmatrix} 0 & -\omega_3 & \omega_2 & v_1 \\ \omega_3 & 0 & -\omega_1 & 0 \\ -\omega_2 & \omega_1 & 0 & 0 \\ 0 & 0 & 0 & 0 \end{bmatrix} g,$$

where $\omega_1$ stands for pitch rate, $\omega_2$ for roll rate, $\omega_3$ for yaw rate, and $v_1$ for the velocity of the aircraft. Then the 3-D velocity $(\omega, v)$ in the continuous epipolar constraint (5.52) has the form $\omega = [\omega_1, \omega_2, \omega_3]^T$, $v = [v_1, 0, 0]^T$. For Algorithm 5.3, we have extra constraints on the symmetric matrix $s = \frac{1}{2}(\widehat{\omega}\widehat{v} + \widehat{v}\widehat{\omega})$: $s_1 = s_5 = 0$ and $s_4 = s_6$. Then there are only four different essential parameters left to determine, and we can redefine the motion parameter vector $E^s \in \mathbb{R}^4$ to be $E^s \doteq [v_1, s_2, s_3, s_4]^T$. Then the measurement vector $\boldsymbol{a} \in \mathbb{R}^4$ is given by $\boldsymbol{a} = [u_3 y - u_2 z, 2xy, 2xz, y^2 + z^2]^T$. The continuous epipolar constraint can then be rewritten as

$$\boldsymbol{a}^T E^s = 0.$$

If we define the matrix $\chi$ from $\boldsymbol{a}$ as in (5.65), the matrix $\chi^T\chi$ is a $4 \times 4$ matrix rather than a $9 \times 9$. For estimating the velocity $(\omega, v)$, the dimension of the problem is then reduced from nine to four. In this special case, the minimum number of optical flow measurements needed to guarantee a unique solution of $E^s$ is reduced to four instead of eight. Furthermore, the symmetric matrix $s$ recovered from $E^s$ is automatically in the space $\mathcal{S}$, and the remaining steps of the algorithm can thus be dramatically simplified. From this simplified algorithm, the angular velocity $\omega = [\omega_1, \omega_2, \omega_3]^T$ can be fully recovered from the images. The velocity information can then be used for controlling the aircraft. ■

As in the discrete case, the linear algorithm proposed above is not optimal, since it does not enforce the structure of the parameter space during the minimization. Therefore, the recovered velocity does not necessarily minimize the originally chosen error function $\|\chi E^s(\omega, v)\|^2$ on the space $\mathcal{E}_1'$.

Additionally, as in the discrete case, we have to assume that translation is not zero. If the motion is purely rotational, then one can prove that there are infinitely many solutions to the epipolar constraint-related equations. We leave this as an exercise to the reader.

### *5.4.4 Euclidean constraints and structure reconstruction*

As in the discrete case, the purpose of exploiting Euclidean constraints is to reconstruct the scales of the motion and structure. From the above linear algorithm, we know that we can recover the linear velocity $v$ only up to an arbitrary scalar factor. Without loss of generality, we may assume that the velocity of the camera motion to be $(\omega, \eta v)$ with $\|v\| = 1$ and $\eta \in \mathbb{R}$. By now, only the scale factor $\eta$ is unknown. Substituting $\boldsymbol{X}(t) = \lambda(t)\boldsymbol{x}(t)$ into the equation

$$\dot{\boldsymbol{X}}(t) = \widehat{\omega}\boldsymbol{X}(t) + \eta v(t),$$

we obtain for the image $\boldsymbol{x}^j$ of each point $p^j \in \mathbb{E}^3$, $j = 1, 2, \ldots, n$,

$$\dot{\lambda}^j \boldsymbol{x}^j + \lambda^j \dot{\boldsymbol{x}}^j = \widehat{\omega}(\lambda^j \boldsymbol{x}^j) + \eta v \;\Leftrightarrow\; \dot{\lambda}^j \boldsymbol{x}^j + \lambda^j(\dot{\boldsymbol{x}}^j - \widehat{\omega}\boldsymbol{x}^j) - \eta v = 0. \quad (5.70)$$

As one may expect, in the continuous case, the scale information is then encoded in $\lambda, \dot{\lambda}$ for the location of the 3-D point, and $\eta \in \mathbb{R}^+$ for the linear velocity $v$. Knowing $\boldsymbol{x}, \dot{\boldsymbol{x}}, \omega$, and $v$, we see that these constraints are all linear in $\lambda^j, \dot{\lambda}^j, 1 \leq j \leq n$, and $\eta$. Also, if $\boldsymbol{x}^j, 1 \leq j \leq n$ are linearly independent of $v$, i.e. the feature points do not line up with the direction of translation, it can be shown that these linear constraints are not degenerate; hence the unknown scales are determined up to a universal scalar factor. We may then arrange all the unknown scalars into a single vector $\vec{\lambda}$:

$$\vec{\lambda} = [\lambda^1, \lambda^2 \ldots, \lambda^n, \dot{\lambda}^1, \dot{\lambda}^2, \ldots, \dot{\lambda}^n, \eta]^T \quad \in \mathbb{R}^{2n+1}.$$

For $n$ optical flow vectors, $\vec{\lambda}$ is a $(2n+1)$-dimensional vector. (5.70) gives $3n$ (scalar) linear equations. The problem of solving $\vec{\lambda}$ from (5.70) is usually over determined. It is easy to check that in the absence of noise the set of equations given by (5.70) uniquely determines $\vec{\lambda}$ if the configuration is noncritical. We can therefore write all the equations in the matrix form

$$M\vec{\lambda} = 0,$$

with $M \in \mathbb{R}^{3n \times (2n+1)}$ a matrix depending on $\omega, v$, and $\{(\boldsymbol{x}^j, \dot{\boldsymbol{x}}^j)\}_{j=1}^n$. Then, in the presence of noise, the linear least-squares estimate of $\vec{\lambda}$ is simply the eigenvector of $M^T M$ corresponding to the smallest eigenvalue.

Notice that the time derivative of the scales $\{\dot{\lambda}^j\}_{j=1}^n$ can also be estimated. Suppose we have done the above recovery for a time interval, say $(t_0, t_f)$. Then we have the estimate $\vec{\lambda}(t)$ as a function of time $t$. But $\vec{\lambda}(t)$ at each time $t$ is determined only up to an arbitrary scalar factor. Hence $\rho(t)\vec{\lambda}(t)$ is also a valid estimation for any positive function $\rho(t)$ defined on $(t_0, t_f)$. However, since $\rho(t)$ is multiplied by both $\lambda(t)$ and $\dot{\lambda}(t)$, their ratio

$$r(t) = \dot{\lambda}(t)/\lambda(t)$$

is independent of the choice of $\rho(t)$. Notice that $\frac{d}{dt}(\ln \lambda) = \dot{\lambda}/\lambda$. Let the logarithm of the structural scale $\lambda$ be $y = \ln \lambda$. Then a time-consistent estimation $\lambda(t)$ needs to satisfy the following ordinary differential equation, which we call the *dynamical scale ODE*

$$\dot{y}(t) = r(t).$$

Given $y(t_0) = y_0 = \ln(\lambda(t_0))$, we solve this ODE and obtain $y(t)$ for $t \in [t_0, t_f]$. Then we can recover a consistent scale $\lambda(t)$ given by

$$\lambda(t) = \exp(y(t)).$$

Hence (structure and motion) scales estimated at different time instances now are all relative to the same scale at time $t_0$. Therefore, in the continuous case, we are also able to recover all the scales as functions of time up to a universal scalar factor. The reader must be aware that the above scheme is only *conceptual*. In reality, the ratio function $r(t)$ would never be available for *every* time instant in $[t_0, t_f]$.

*Universal scale ambiguity*

In both the discrete and continuous cases, in principle, the proposed schemes can reconstruct both the Euclidean structure and motion up to a universal scalar factor. This ambiguity is intrinsic, since one can scale the entire world up or down with a scaling factor while all the images obtained remain the same. In all the algorithms proposed above, this factor is fixed (rather arbitrarily, in fact) by imposing the translation scale to be 1. In practice, this scale and its unit can also be chosen to be directly related to some known length, size, distance, or motion of an object in space.

### 5.4.5 Continuous homography for a planar scene

In this section, we consider the continuous version of the case that we have studied in Section 5.3, where all the feature points of interest are lying on a plane $P$. Planar scenes are a degenerate case for the discrete epipolar constraint, and also for the continuous case. Recall that in the continuous scenario, instead of having image pairs, we measure the image point $\boldsymbol{x}$ and its optical flow $\boldsymbol{u} = \dot{\boldsymbol{x}}$. Other assumptions are the same as in Section 5.3.

Suppose the camera undergoes a rigid-body motion with body angular and linear velocities $\omega, v$. Then the time derivative of coordinates $\boldsymbol{X} \in \mathbb{R}^3$ of a point $p$ (with respect to the camera frame) satisfies[13]

$$\dot{\boldsymbol{X}} = \widehat{\omega}\boldsymbol{X} + v. \tag{5.71}$$

Let $N \in \mathbb{R}^3$ be the surface normal to $P$ (with respect to the camera frame) at time $t$. Then, if $d(t) > 0$ is the distance from the optical center of the camera to the plane $P$ at time $t$, then

$$N^T\boldsymbol{X} = d \quad \Leftrightarrow \quad \frac{1}{d}N^T\boldsymbol{X} = 1, \quad \forall \boldsymbol{X} \in P. \tag{5.72}$$

Substituting equation (5.72) into equation (5.71) yields the relation

$$\dot{\boldsymbol{X}} = \widehat{\omega}\boldsymbol{X} + v = \widehat{\omega}\boldsymbol{X} + v\frac{1}{d}N^T\boldsymbol{X} = \left(\widehat{\omega} + \frac{1}{d}vN^T\right)\boldsymbol{X}. \tag{5.73}$$

As in the discrete case, we call the matrix

$$H \doteq \left(\widehat{\omega} + \frac{1}{d}vN^T\right) \in \mathbb{R}^{3\times 3} \tag{5.74}$$

the *continuous homography matrix*. For simplicity, here we use the same symbol $H$ to denote it, and it really is a continuous (or infinitesimal) version of the (discrete) homography matrix $H = R + \frac{1}{d}TN^T$ studied in Section 5.3.

[13]Here, as in previous cases, we assume implicitly that time dependency of $\boldsymbol{X}$ on $t$ is smooth so that we can take derivatives whenever necessary. However, for simplicity, we drop the dependency of $\boldsymbol{X}$ on $t$ in the notation $\boldsymbol{X}(t)$.

Note that the matrix $H$ depends both on the continuous motion parameters $\{\omega, v\}$ and structure parameters $\{N, d\}$ that we wish to recover. As in the discrete case, there is an inherent scale ambiguity in the term $\frac{1}{d}v$ in equation (5.74). Thus, in general, knowing $H$, one can recover only the ratio of the camera translational velocity scaled by the distance to the plane.

From the relation

$$\lambda \boldsymbol{x} = \boldsymbol{X}, \quad \dot{\lambda}\boldsymbol{x} + \lambda \boldsymbol{u} = \dot{\boldsymbol{X}}, \quad \dot{\boldsymbol{X}} = H\boldsymbol{X}, \tag{5.75}$$

we have

$$\boxed{\boldsymbol{u} = H\boldsymbol{x} - \frac{\dot{\lambda}}{\lambda}\boldsymbol{x}.} \tag{5.76}$$

This is indeed the continuous version of the planar homography.

### 5.4.6 Estimating the continuous homography matrix

In order to further eliminate the depth scale $\lambda$ in equation (5.76), multiplying both sides by the skew-symmetric matrix $\widehat{\boldsymbol{x}} \in \mathbb{R}^{3\times 3}$, we obtain the equation

$$\boxed{\widehat{\boldsymbol{x}}H\boldsymbol{x} = \widehat{\boldsymbol{x}}\boldsymbol{u}.} \tag{5.77}$$

We may call this the *continuous homography constraint* or the *continuous planar epipolar constraint* as a continuous version of the discrete case.

Since this constraint is linear in $H$, by stacking the entries of $H$ as

$$H^s \doteq [H_{11}, H_{21}, H_{31}, H_{12}, H_{22}, H_{32}, H_{13}, H_{23}, H_{33}]^T \quad \in \mathbb{R}^9,$$

we may rewrite (5.77) as

$$\boldsymbol{a}^T H^s = \widehat{\boldsymbol{x}}\boldsymbol{u},$$

where $\boldsymbol{a} \in \mathbb{R}^{9\times 3}$ is the Kronecker product $\boldsymbol{x} \otimes \widehat{\boldsymbol{x}}$. However, since the skew-symmetric matrix $\widehat{\boldsymbol{x}}$ is only of rank 2, the equation imposes only two constraints on the entries of $H$. Given a set of $n$ image point and velocity pairs $\{(\boldsymbol{x}^j, \boldsymbol{u}^j)\}_{j=1}^n$ of points on the plane, we may stack all equations $\boldsymbol{a}^{jT}H^s = \widehat{\boldsymbol{x}^j}\boldsymbol{u}^j, j = 1, 2, \ldots, n$, into a single equation

$$\chi H^s = B, \tag{5.78}$$

where $\chi \doteq \left[\boldsymbol{a}^1, \ldots, \boldsymbol{a}^n\right]^T \in \mathbb{R}^{3n\times 9}$ and $B \doteq \left[(\widehat{\boldsymbol{x}^1}\boldsymbol{u}^1)^T, \ldots, (\widehat{\boldsymbol{x}^j}\boldsymbol{u}^j)^T\right]^T \in \mathbb{R}^{3n}$.

In order to solve uniquely (up to a scalar factor) for $H^s$, we must have rank$(\chi) = 8$. Since each pair of image points gives two constraints, we expect that at least four optical flow pairs would be necessary for a unique estimate of $H$ (up to a scalar factor). In analogy with the discrete case, we have the following statement, the proof of which we leave to the reader as a linear-algebra exercise.

**Proposition 5.32 (Four-point continuous homography).** *We have $rank(\mathcal{X}) = 8$ if and only if there exists a set of four points (out of the $n$) such that any three of them are not collinear; i.e. they are in general configuration in the plane.*

Then, if optical flow at more than four points in general configuration in the plane is given, using linear least-squares techniques, equation (5.78) can be used to recover $H^s$ up to one dimension, since $\mathcal{X}$ has a one-dimensional null space. That is, we can recover $H_L = H - \xi H_K$, where $H_L$ corresponds to the minimum-norm linear least-squares estimate of $H$ solving $\min \|\mathcal{X}H^s - B\|^2$, and $H_K$ corresponds to a vector in $\text{null}(\mathcal{X})$ and $\xi \in \mathbb{R}$ is an unknown scalar factor.

By inspection of equation (5.77) one can see that $H_K = I$, since $\widehat{\boldsymbol{x}}I\boldsymbol{x} = \widehat{\boldsymbol{x}}\boldsymbol{x} = 0$. Then we have

$$H = H_L + \xi I. \tag{5.79}$$

Thus, in order to recover $H$, we need only to identify the unknown $\xi$. So far, we have not considered the special structure of the matrix $H$. Next, we give constraints imposed by the structure of $H$ that will allow us to identify $\xi$, and thus uniquely recover $H$.

**Lemma 5.33.** *Suppose $u, v \in \mathbb{R}^3$, and $\|u\|^2 = \|v\|^2 = \alpha$. If $u \neq v$, the matrix $D = uv^T + vu^T \in \mathbb{R}^{3\times 3}$ has eigenvalues $\{\lambda_1, 0, \lambda_3\}$, where $\lambda_1 > 0$, and $\lambda_3 < 0$. If $u = \pm v$, the matrix $D$ has eigenvalues $\{\pm 2\alpha, 0, 0\}$.*

*Proof.* Let $\beta = u^T v$. If $u \neq \pm v$, we have $-\alpha < \beta < \alpha$. We can solve the eigenvalues and eigenvectors of $D$ by

$$\begin{aligned} D(u+v) &= (\beta+\alpha)(u+v), \\ D(u \times v) &= 0, \\ D(u-v) &= (\beta-\alpha)(u-v). \end{aligned}$$

Clearly, $\lambda_1 = (\beta+\alpha) > 0$ and $\lambda_3 = \beta - \alpha < 0$. It is easy to check the conditions on $D$ when $u = \pm v$. □

**Lemma 5.34 (Normalization of the continuous homography matrix).** *Given the $H_L$ part of a continuous planar homography matrix of the form $H = H_L + \xi I$, we have*

$$\xi = -\frac{1}{2}\gamma_2(H_L + H_L^T), \tag{5.80}$$

*where $\gamma_2(H_L + H_L^T) \in \mathbb{R}$ is the second-largest eigenvalue of $H_L + H_L^T$.*

*Proof.* In this proof, we will work with sorted eigenvalues; that is, if $\{\lambda_1, \lambda_2, \lambda_3\}$ are eigenvalues of some matrix, then $\lambda_1 \geq \lambda_2 \geq \lambda_3$. If the points are not in general configuration, then $\text{rank}(\mathcal{X}) < 7$, and the problem is under constrained. Now suppose the points are in general configuration. Then by least-squares estimation we may recover $H_L = H - \xi I$ for some unknown $\xi \in \mathbb{R}$. By Lemma 5.33, $H + H^T = \frac{1}{d}vN^T + \frac{1}{d}Nv^T$ has eigenvalues $\{\lambda_1, \lambda_2, \lambda_3\}$, where $\lambda_1 \geq 0$, $\lambda_2 = 0$, and $\lambda_3 \leq 0$. So compute the eigenvalues of $H_L + H_L^T$ and denote them by $\{\gamma_1, \gamma_2, \gamma_3\}$. Since we have $H = H_L + \xi I$, then $\lambda_i = \gamma_i + 2\xi$, for $i = 1, 2, 3$. Since we must have $\lambda_2 = 0$, we have $\xi = -\frac{1}{2}\gamma_2$. □

Therefore, knowing $H_L$, we can fully recover the continuous homography matrix as $H = H_L - \frac{1}{2}\gamma_2 I$.

### 5.4.7 *Decomposing the continuous homography matrix*

We now address the task of decomposing the recovered $H = \widehat{\omega} + \frac{1}{d}vN^T$ into its motion and structure parameters $\{\omega, \frac{v}{d}, N\}$. The following constructive proof provides an algebraic technique for the recovery of motion and structure parameters.

**Theorem 5.35 (Decomposition of continuous homography matrix).** *Given a matrix $H \in \mathbb{R}^{3\times 3}$ in the form $H = \widehat{\omega} + \frac{1}{d}vN^T$, one can recover the motion and structure parameters $\{\widehat{\omega}, \frac{1}{d}v, N\}$ up to at most two physically possible solutions. There is a unique solution if $v = 0$, $v \times N = 0$, or $e_3^T v = 0$, where $e_3 = [0, 0, 1]^T$ is the optical axis.*

*Proof.* Compute the eigenvalue/eigenvector pairs of $H + H^T$ and denote them by $\{\lambda_i, u_i\}$, $i = 1, 2, 3$. If $\lambda_i = 0$ for $i = 1, 2, 3$, then we have $v = 0$ and $\widehat{\omega} = H$. In this case we cannot recover the normal of the plane $N$. Otherwise, if $\lambda_1 > 0$, and $\lambda_3 < 0$, then we have $v \times N \neq 0$. Let $\alpha = \|v/d\| > 0$, let $\tilde{v} = v/\sqrt{\alpha}$ and $\tilde{N} = \sqrt{\alpha}N$, and let $\beta = \tilde{v}^T\tilde{N}$. According to Lemma 5.33, the eigenvalue/eigenvector pairs of $H + H^T$ are given by

$$\begin{aligned} \lambda_1 &= \beta + \alpha > 0, & u_1 &= \tfrac{1}{\|\tilde{v}+\tilde{N}\|}(\tilde{v} + \tilde{N}), \\ \lambda_3 &= \beta - \alpha < 0, & u_3 &= \pm\tfrac{1}{\|\tilde{v}-\tilde{N}\|}(\tilde{v} - \tilde{N}). \end{aligned} \tag{5.81}$$

Then $\alpha = \frac{1}{2}(\lambda_1 - \lambda_3)$. It is easy to check that $\|\tilde{v} + \tilde{N}\|^2 = 2\lambda_1$, $\|\tilde{v} - \tilde{N}\|^2 = -2\lambda_3$. Together with (5.81), we have two solutions (due to the two possible signs for $u_3$):

$$\begin{aligned} \tilde{v}_1 &= \tfrac{1}{2}\left(\sqrt{2\lambda_1}\, u_1 + \sqrt{-2\lambda_3}\, u_3\right), & \tilde{v}_2 &= \tfrac{1}{2}\left(\sqrt{2\lambda_1}\, u_1 - \sqrt{-2\lambda_3}\, u_3\right), \\ \tilde{N}_1 &= \tfrac{1}{2}\left(\sqrt{2\lambda_1}\, u_1 - \sqrt{-2\lambda_3}\, u_3\right), & \tilde{N}_2 &= \tfrac{1}{2}\left(\sqrt{2\lambda_1}\, u_1 + \sqrt{-2\lambda_3}\, u_3\right), \\ \widehat{\omega}_1 &= H - \tilde{v}_1\tilde{N}_1^T, & \widehat{\omega}_2 &= H - \tilde{v}_2\tilde{N}_2^T. \end{aligned}$$

In the presence of noise, the estimate of $\widehat{\omega} = H - \tilde{v}\tilde{N}^T$ is not necessarily an element in $so(3)$. In algorithms, one may take its skew-symmetric part,

$$\widehat{\omega} = \frac{1}{2}\left((H - \tilde{v}\tilde{N}^T) - (H - \tilde{v}\tilde{N}^T)^T\right).$$

There is another sign ambiguity, since $(-\tilde{v})(-\tilde{N})^T = \tilde{v}\tilde{N}^T$. This sign ambiguity leads to a total of four possible solutions for decomposing $H$ back to $\{\widehat{\omega}, \frac{1}{d}v, N\}$ given in Table 5.2.

| | | | |
|---|---|---|---|
| Solution 1 | $\frac{1}{d}v_1 = \sqrt{\alpha}\tilde{v}_1$<br>$N_1 = \frac{1}{\sqrt{\alpha}}\tilde{N}_1$<br>$\widehat{\omega}_1 = H - \tilde{v}_1\tilde{N}_1^T$ | Solution 3 | $\frac{1}{d}v_3 = -\frac{1}{d}v_1$<br>$N_3 = -N_1$<br>$\widehat{\omega}_3 = \widehat{\omega}_1$ |
| Solution 2 | $\frac{1}{d}v_2 = \sqrt{\alpha}\tilde{v}_2$<br>$N_2 = \frac{1}{\sqrt{\alpha}}\tilde{N}_2$<br>$\widehat{\omega}_2 = H - \tilde{v}_2\tilde{N}_2^T$ | Solution 4 | $\frac{1}{d}v_4 = -\frac{1}{d}v_2$<br>$v_4 = -N_2$<br>$\widehat{\omega}_4 = \widehat{\omega}_2$ |

Table 5.2. Four solutions for continuous planar homography decomposition. Here $\alpha$ is computed as before as $\alpha = \frac{1}{2}(\lambda_1 - \lambda_3)$.

In order to reduce the number of physically possible solutions, we impose the positive depth constraint: since the camera can only see points that are in front of it, we must have $N^T e_3 > 0$. Therefore, if solution 1 is the correct one, this constraint will eliminate solution 3 as being physically impossible. If $v^T e_3 \neq 0$, one of solutions 2 or 4 will be eliminated, whereas if $v^T e_3 = 0$, both solutions 2 and 4 will be eliminated. For the case that $v \times N = 0$, it is easy to see that solutions 1 and 2 are equivalent, and that imposing the positive depth constraint leads to a unique solution. □

Despite the fact that as in the discrete case, there is a close relationship between the continuous epipolar constraint and continuous homography, we will not develop the details here. Basic intuition and necessary technical tools have already been established in this chapter, and at this point interested readers may finish that part of the story with ease, or more broadly, apply these techniques to solve other special problems that one may encounter in real-world applications.

We summarize Sections 5.4.6 and 5.4.7 by presenting the continuous four-point Algorithm 5.4 for motion estimation from a planar scene.

## 5.5 Summary

Given corresponding points in two images $(\boldsymbol{x}_1, \boldsymbol{x}_2)$ of a point $p$, or, in continuous time, optical flow $(\boldsymbol{u}, \boldsymbol{x})$, we summarize the constraints and relations between the image data and the unknown motion parameters in Table 5.3.

Despite the similarity between the discrete and the continuous case, one must be aware that there are indeed important subtle differences between these two cases, since the differentiation with respect to time $t$ changes the algebraic relation between image data and unknown motion parameters.

In the presence of noise, the motion recovery problem in general becomes a problem of minimizing a cost function associated with statistical optimality or geometric error criteria subject to the above constraints. Once the camera motion is recovered, an overall 3-D reconstruction of both the camera motion and scene structure can be obtained up to a global scaling factor.

**Algorithm 5.4 (The continuous four-point algorithm for a planar scene).**

For a given set of optical flow vectors $(\boldsymbol{u}^j, \boldsymbol{x}^j)$, $j = 1, 2, \ldots, n$ $(n \geq 4)$, of points on a plane $N^T \boldsymbol{X} = d$, this algorithm finds $\{\widehat{\omega}, \frac{1}{d}v, N\}$ that solves

$$\widehat{\boldsymbol{x}^j}^T \left( \widehat{\omega} + \frac{1}{d} v N^T \right) \boldsymbol{x}^j = \widehat{\boldsymbol{x}^j} \boldsymbol{u}^j, \quad j = 1, 2, \ldots, n.$$

1. **Compute a first approximation of the continuous homography matrix**
   Construct the matrix $\chi = [\boldsymbol{a}^1, \boldsymbol{a}^2, \ldots, \boldsymbol{a}^n]^T \in \mathbb{R}^{3n \times 9}$, $B = [\boldsymbol{b}^{1T}, \boldsymbol{b}^{2T}, \ldots, \boldsymbol{b}^{nT}]^T \in \mathbb{R}^{3n}$ from the optical flow $(\boldsymbol{u}^j, \boldsymbol{x}^j)$, where $\boldsymbol{a}^j = \boldsymbol{x}^j \otimes \widehat{\boldsymbol{x}^j} \in \mathbb{R}^{9 \times 3}$ and $\boldsymbol{b} = \widehat{\boldsymbol{x}}\boldsymbol{u} \in \mathbb{R}^3$. Find the vector $H_L^s \in \mathbb{R}^9$ as
   $$H_L^s = \chi^\dagger B,$$
   where $\chi^\dagger \in \mathbb{R}^{9 \times 3n}$ is the pseudo-inverse of $\chi$. Unstack $H_L^s$ to obtain the $3 \times 3$ matrix $H_L$.
2. **Normalization of the continuous homography matrix**
   Compute the eigenvalue values $\{\gamma_1, \gamma_2, \gamma_3\}$ of the matrix $H_L^T + H_L$ and normalize it as
   $$H = H_L - \frac{1}{2}\gamma_2 I.$$
3. **Decomposition of the continuous homography matrix**
   Compute the eigenvalue decomposition of
   $$H^T + H = U \Lambda U^T$$
   and compute the four solutions for a decomposition $\{\widehat{\omega}, \frac{1}{d}v, N\}$ as in the proof of Theorem 5.35. Select the two physically possible ones by imposing the positive depth constraint $N^T e_3 > 0$.

## 5.6 Exercises

**Exercise 5.1 (Linear equation).** Solve $x \in \mathbb{R}^n$ from the linear equation

$$Ax = b,$$

where $A \in \mathbb{R}^{m \times n}$ and $b \in \mathbb{R}^m$. In terms of conditions on the matrix $A$ and vector $b$, describe when a solution exists and when it is unique. In case the solution is not unique, describe the entire solution set.

**Exercise 5.2 (Properties of skew-symmetric matrices).**

1. Prove Lemma 5.4.
2. Prove Lemma 5.24.

**Exercise 5.3 (Skew-symmetric matrix continued).** Given a vector $T \in \mathbb{R}^3$ with unit length, i.e. $\|T\| = 1$, show that:

1. The identity holds: $\widehat{T}^T\widehat{T} = \widehat{T}\widehat{T}^T = I - TT^T$ (note that the superscript $^T$ stands for matrix transpose).

| | Epipolar constraint | (Planar) homography |
|---|---|---|
| Discrete motion | $\boldsymbol{x}_2^T \widehat{T} R \boldsymbol{x}_1 = 0$ | $\widehat{\boldsymbol{x}_2}(R + \frac{1}{d} T N^T)\boldsymbol{x}_1 = 0$ |
| Matrices | $E = \widehat{T} R$ | $H = R + \frac{1}{d} T N^T$ |
| Relation | $\exists v \in \mathbb{R}^3,\ H = \widehat{T}^T E + T v^T$ | |
| Continuous motion | $\boldsymbol{x}^T \widehat{\omega} \widehat{v} \boldsymbol{x} + \boldsymbol{u}^T \widehat{v} \boldsymbol{x} = 0$ | $\widehat{\boldsymbol{x}}(\widehat{\omega} + \frac{1}{d} v N^T)\boldsymbol{x} = \widehat{\boldsymbol{u}} \boldsymbol{x}$ |
| Matrices | $E = \begin{bmatrix} \frac{1}{2}(\widehat{\omega}\widehat{v} + \widehat{v}\widehat{\omega}) \\ \widehat{v} \end{bmatrix}$ | $H = \widehat{\omega} + \frac{1}{d} v N^T$ |
| Linear algorithms | 8 points | 4 points |
| Decomposition | 1 solution | 2 solutions |

Table 5.3. Here the number of points is required by corresponding linear algorithms, and we count only the number of physically possible solutions from corresponding decomposition algorithms *after* applying the positive depth constraint.

2. Explain the effect of multiplying a vector $u \in \mathbb{R}^3$ by the matrix $P = I - TT^T$. Show that $P^n = P$ for any integer $n$.
3. Show that $\widehat{T}^T \widehat{T} \widehat{T} = \widehat{T} \widehat{T}^T \widehat{T} = \widehat{T}$. Explain geometrically why this is true.
4. How do the above statements need to be changed if the vector $T$ is not of unit length?

**Exercise 5.4 (A rank condition for the epipolar constraint).** Show that $\boldsymbol{x}_2^T \widehat{T} R \boldsymbol{x}_1 = 0$ if and only if

$$\operatorname{rank}\,[\widehat{\boldsymbol{x}_2} R \boldsymbol{x}_1,\ \widehat{\boldsymbol{x}_2} T] \leq 1.$$

**Exercise 5.5 (Parallel epipolar lines).** Explain under what conditions the family of epipolar lines in at least one of the image planes will be parallel to each other. Where is the corresponding epipole (in terms of its homogeneous coordinates)?

**Exercise 5.6 (Essential matrix for planar motion).** Suppose we know that the camera always moves on a plane, say the $XY$ plane. Show that:

1. The essential matrix $E = \widehat{T} R$ is of the special form

$$E = \begin{bmatrix} 0 & 0 & a \\ 0 & 0 & b \\ c & d & 0 \end{bmatrix}, \quad a, b, c, d \in \mathbb{R}. \tag{5.82}$$

2. Without using the SVD-based decomposition introduced in this chapter, find a solution to $(R, T)$ in terms of $a, b, c, d$.

**Exercise 5.7 (Rectified essential matrix).** Suppose that using the linear algorithm, you obtain an essential matrix $E$ of the form

$$E = \begin{bmatrix} 0 & 0 & 0 \\ 0 & 0 & a \\ 0 & -a & 0 \end{bmatrix}, \quad a \in \mathbb{R}. \tag{5.83}$$

What type of motion $(R, T)$ does the camera undergo? How many solutions exist exactly?

**Exercise 5.8 (Triangulation).** Given two images $\boldsymbol{x}_1, \boldsymbol{x}_2$ of a point $p$ together with the relative camera motion $(R, T)$, $\boldsymbol{X}_2 = R\boldsymbol{X}_1 + T$:

1. express the depth of $p$ with respect to the first image, i.e. $\lambda_1$ in terms of $\boldsymbol{x}_1, \boldsymbol{x}_2$, and $(R, T)$;
2. express the depth of $p$ with respect to the second image, i.e. $\lambda_2$ in terms of $\boldsymbol{x}_1, \boldsymbol{x}_2$, and $(R, T)$.

**Exercise 5.9 (Rotational motion).** Assume that the camera undergoes pure rotational motion; i.e. it rotates around its center. Let $R \in SO(3)$ be the rotation of the camera and $\omega \in so(3)$ be the angular velocity. Show that in this case, we have:

1. discrete case: $\boldsymbol{x}_2^T \widehat{T} R \boldsymbol{x}_1 \equiv 0, \quad \forall T \in \mathbb{R}^3$;
2. continuous case: $\boldsymbol{x}^T \widehat{\omega} \widehat{v} \boldsymbol{x} + \boldsymbol{u}^T \widehat{v} \boldsymbol{x} \equiv 0, \quad \forall v \in \mathbb{R}^3$.

**Exercise 5.10 (Projection onto $O(3)$).** Given an arbitrary $3 \times 3$ matrix $M \in \mathbb{R}^{3\times 3}$ with positive singular values, find the orthogonal matrix $R \in O(3)$ such that the error $\|R - M\|_f^2$ is minimized. Is the solution unique? Note: Here we allow $\det(R) = \pm 1$.

**Exercise 5.11 (Four motions related to an epipolar constraint).** Suppose $E = \widehat{T}R$ is a solution to the epipolar constraint $\boldsymbol{x}_2^T E \boldsymbol{x}_1 = 0$. Then $-E$ is also an essential matrix, which obviously satisfies the same epipolar constraint (for given corresponding images).

1. Explain geometrically how these four motions are related. [Hint: Consider a pure translation case. If $R$ is a rotation about $T$ by an angle $\pi$, then $\widehat{T}R = -\widehat{T}$, which is in fact the *twisted pair* ambiguity.]
2. Show that in general, for three out of the four solutions, the equation $\lambda_2 \boldsymbol{x}_2 = \lambda_1 R \boldsymbol{x}_1 + T$ will yield either negative $\lambda_1$ or negative $\lambda_2$ or both. Hence only one solution satisfies the positive depth constraint.

**Exercise 5.12 (Geometric distance to an epipolar line).** Given two image points $\boldsymbol{x}_1, \tilde{\boldsymbol{x}}_2$ with respect to camera frames with their relative motion $(R, T)$, show that the geometric distance $d_2$ defined in Figure 5.7 is given by the formula

$$d_2^2 = \frac{(\tilde{\boldsymbol{x}}_2^T \widehat{T} R \boldsymbol{x}_1)^2}{\|\widehat{e}_3 \widehat{T} R \boldsymbol{x}_1\|^2},$$

where $e_3 = [0, 0, 1]^T \in \mathbb{R}^3$.

**Exercise 5.13 (A six-point algorithm).** In this exercise, we show how to use some of the (algebraic) structure of the essential matrix to reduce the number of matched pairs of points from 8 to 6.

1. Show that if a matrix $E$ is an essential matrix, then it satisfies the identity
$$EE^T E = \frac{1}{2}\text{trace}(EE^T)E.$$
2. Show that the dimension of the space of matrices $\{F\} \subset \mathbb{R}^{3\times 3}$ that satisfy the epipolar constraints
$$(\boldsymbol{x}_2^j)^T F \boldsymbol{x}_1^j = 0, \quad j = 1, 2, \ldots, 6,$$

is three. Hence the essential matrix $E$ can be expressed as a linear combination $E = \alpha_1 F_1 + \alpha_2 F_2 + \alpha_3 F_3$ for some linearly independent matrices $F_1, F_2, F_3$ that satisfy the above equations.

3. To further determine the coefficients $\alpha_1, \alpha_2, \alpha_3$, show that the identity in (a) gives nine scalar equations linearly in the nine unknowns $\{\alpha_1^i \alpha_2^j \alpha_3^k\}$, $i + j + k = 3$, $0 \le i, j, k \le 3$. (Why nine?) Hence, the essential matrix $E$ can be determined from six pairs of matched points.

**Exercise 5.14 (Critical surfaces).** To have a unique solution (up to a scalar factor), it is very important for the points considered in the above six-point or eight-point algorithms to be in general position. If a (dense) set of points whose images allow at least two distinct essential matrices, we say that they are "critical," Let $\boldsymbol{X} \in \mathbb{R}^3$ be coordinates of such a point and $(R, T)$ be the motion of a camera. Let $\boldsymbol{x}_1 \sim \boldsymbol{X}$ and $\boldsymbol{x}_2 \sim (R\boldsymbol{X} + T)$ be two images of the point.

1. Show that if
$$(R\boldsymbol{X} + T)^T \widehat{T'} R' \boldsymbol{X} = 0,$$
then
$$\boldsymbol{x}_2^T \widehat{T} R \boldsymbol{x}_1 = 0, \quad \boldsymbol{x}_2^T \widehat{T'} R' \boldsymbol{x}_1 = 0.$$

2. Show that for points $\boldsymbol{X} \in \mathbb{R}^3$ that satisfy the equation $(R\boldsymbol{X} + T)^T \widehat{T'} R' \boldsymbol{X} = 0$, their homogeneous coordinates $\bar{\boldsymbol{X}} = [X, 1]^T \in \mathbb{R}^4$ satisfy the quadratic equation
$$\bar{\boldsymbol{X}}^T \begin{bmatrix} R^T \widehat{T'} R' + R'^T \widehat{T'}^T R & R'^T \widehat{T'}^T T \\ T^T \widehat{T'} R' & 0 \end{bmatrix} \bar{\boldsymbol{X}} = 0.$$
This quadratic surface is denoted by $C_1 \subset \mathbb{R}^3$ and is called a *critical surface*. So no matter how many points one chooses on such a surface, their two corresponding images always satisfy epipolar constraints for at least two different essential matrices.

3. Symmetrically, points defined by the equation $(R'\boldsymbol{X} + T')^T \widehat{T} R \boldsymbol{X} = 0$ will have similar properties. This gives another quadratic surface,
$$C_2 : \bar{\boldsymbol{X}}^T \begin{bmatrix} R'^T \widehat{T} R + R^T \widehat{T}^T R' & R^T \widehat{T}^T T' \\ T'^T \widehat{T} R & 0 \end{bmatrix} \bar{\boldsymbol{X}} = 0.$$
Argue that a set of points on the surface $C_1$ observed from two vantage points related by $(R, T)$ could be interpreted as a corresponding set of points on the surface $C_2$ observed from two vantage points related by $(R', T')$.

**Exercise 5.15 (Estimation of the homography).** We say that two images are related by a *homography* if the homogeneous coordinates of the two images $\boldsymbol{x}_1, \boldsymbol{x}_2$ of every point satisfy
$$\boldsymbol{x}_2 \sim H \boldsymbol{x}_1$$
for some nonsingular matrix $H \in \mathbb{R}^{3 \times 3}$. Show that in general one needs four pairs of $(\boldsymbol{x}_1, \boldsymbol{x}_2)$ to determine the matrix $H$ (up to a scalar factor).

**Exercise 5.16** Under a homography $H \in \mathbb{R}^{3\times3}$ from $\mathbb{R}^2$ to $\mathbb{R}^2$, a standard unit square with the homogeneous coordinates for the four corners

$$(0,0,1),\ (1,0,1),\ (1,1,1),\ (0,1,1)$$

is mapped to

$$(6,5,1),\ (4,3,1),\ (6,4.5,1),\ (10,8,1),$$

respectively. Determine the matrix $H$ with its last entry $H_{33}$ normalized to 1.

**Exercise 5.17 (Epipolar line homography from an essential matrix).** From the geometric interpretation of epipolar lines in Figure 5.2, we know that there is a one-to-one map between the family of epipolar lines $\{\boldsymbol{\ell}_1\}$ in the first image plane (through the epipole $e_1$) and the family of epipolar lines $\{\boldsymbol{\ell}_2\}$ in the second. Suppose that the essential matrix $E$ is known. Show that this map is in fact a homography. That is, there exists a nonsingular matrix $H \in \mathbb{R}^{3\times3}$ such that

$$\boldsymbol{\ell}_2 \sim H\boldsymbol{\ell}_1$$

for any pair of corresponding epipolar lines $(\boldsymbol{\ell}_1, \boldsymbol{\ell}_2)$. Find an explicit form for $H$ in terms of $E$.

**Exercise 5.18 (Homography with respect to the second camera frame).** In the chapter, we have learned that for a transformation $\boldsymbol{X}_2 = R\boldsymbol{X}_1 + T$ on a plane $N^T\boldsymbol{X}_1 = 1$ (expressed in the first camera frame), we have a homography $H = R + TN^T$ such that $\boldsymbol{x}_2 \sim H\boldsymbol{x}_1$ relates the two images of the plane.

1. Now switch roles of the first and the second camera frames and show that the new homography matrix becomes

$$\tilde{H} = \left(R^T + \frac{-R^T T}{1 + N^T R^T T} N^T R^T\right). \tag{5.84}$$

2. What is the relationship between $H$ and $\tilde{H}$? Provide a formal proof to your answer. Explain why this should be expected.

**Exercise 5.19 (Two physically possible solutions for the homography decomposition).** Let us study in the nature of the two physically possible solutions for the homography decomposition. Without loss of generality, suppose that the true homography matrix is $H = I + ab^T$ with $\|a\| = 1$.

1. Show that $R' = -I + 2aa^T$ is a rotation matrix.
2. Show that $H' = R' + (-a)(b + 2a)^T$ is equal to $-H$.
3. Since $(H')^T H' = H^T H$, conclude that both $\{I, a, b\}$ and $\{R', -a, (b + 2a)\}$ are solutions from the homography decomposition of $H$.
4. Argue that, under certain conditions on the relationship between $a$ and $b$, the second solution is also physically possible.
5. What is the geometric relationship between these two solutions? Draw a figure to illustrate your answer.

**Exercise 5.20 (Various expressions for the image motion field).** In the continuous-motion case, suppose that the camera motion is $(\omega, v)$, and $\boldsymbol{u} = \dot{\boldsymbol{x}}$ is the velocity of the image $\boldsymbol{x}$ of a point $\boldsymbol{X} = [X, Y, Z]^T$ in space. Show that:

1. For a spherical perspective projection; i.e. $\lambda = \|\boldsymbol{X}\|$, we have
$$\boldsymbol{u} = -\widehat{\boldsymbol{x}}\omega + \frac{1}{\lambda}\widehat{\boldsymbol{x}}^2 v. \tag{5.85}$$

2. For a planar perspective projection; i.e. $\lambda = Z$, we have
$$\boldsymbol{u} = (-\widehat{\boldsymbol{x}} + \boldsymbol{x}e_3^T\widehat{\boldsymbol{x}})\omega + \frac{1}{\lambda}(I - \boldsymbol{x}e_3^T)v, \tag{5.86}$$
or in coordinates,
$$\begin{bmatrix}\dot{x}\\ \dot{y}\end{bmatrix} = \begin{bmatrix}-xy & x^2 & -y\\ -(1+y^2) & xy & x\end{bmatrix}\omega + \frac{1}{\lambda}\begin{bmatrix}1 & 0 & -x\\ 0 & 1 & -y\end{bmatrix}v. \tag{5.87}$$

3. Show that in the planar perspective case, equation (5.76) is equivalent to
$$\boldsymbol{u} = (I - \boldsymbol{x}e_3^T)H\boldsymbol{x}. \tag{5.88}$$
From this equation, discuss under what conditions the motion field for a planar scene is an affine function of the image coordinates; i.e.
$$\boldsymbol{u} = A\boldsymbol{x}, \tag{5.89}$$
where $A$ is a constant $3 \times 3$ affine matrix that does not depend on the image point $\boldsymbol{x}$.

**Exercise 5.21 (Programming: implementation of (discrete) eight-point algorithm).** Implement a version of the three-step pose estimation algorithm for two views. Your Matlab code should be responsible for

- Initialization: Generate a set of $n$ ($\geq 8$) 3-D points; generate a rigid-body motion $(R, T)$ between two camera frames and project (the coordinates of) the points (relative to the camera frame) onto the image plane correctly. Here you may assume that the focal length is 1. This step will give you corresponding images as input to the algorithm.

- Motion Recovery: using the corresponding images and the algorithm to compute the motion $(\tilde{R}, \tilde{T})$ and compare it to the ground truth $(R, T)$.

After you get the correct answer from the above steps, here are a few suggestions for you to try with the algorithm (or improve it):

- A more realistic way to generate these 3-D points is to make sure that they are all indeed "in front of" the image plane before and after the camera moves.

- Systematically add some noise to the projected images and see how the algorithm responds. Try different camera motions and different layouts of the points in 3-D.

- Finally, to make the algorithm fail, take all the 3-D points from some plane in front of the camera. Run the program and see what you get (especially with some noise on the images).

**Exercise 5.22 (Programming: implementation of the continuous eight-point algorithm).** Implement a version of the four-step velocity estimation algorithm for optical flow.

- Initialization: Choose a set of $n$ ($\geq 8$) 3-D points and a rigid-body velocity $(\omega, v)$. Correctly obtain the image $\boldsymbol{x}$ and compute the image velocity $\boldsymbol{u} = \dot{\boldsymbol{x}}$. You need to figure out how to compute $\boldsymbol{u}$ from $(\omega, v)$ and $\boldsymbol{X}$. Here you may assume that the focal length is 1. This step will give you images and their velocities as input to the algorithm.
- Motion Recovery: Use the algorithm to compute the motion $(\tilde{\omega}, \tilde{v})$ and compare it to the ground truth $(\omega, v)$.

# 5.A Optimization subject to the epipolar constraint

In this appendix, we will study the problem of minimizing the reprojection error (5.23) subject to the fact that the underlying unknowns must satisfy the epipolar constraint. This yields an optimal estimate, in the sense of least-squares, of camera motion between the two views.

*Constraint elimination by Lagrange multipliers*

Our goal here is, given $\tilde{\boldsymbol{x}}_i^j, i = 1, 2, \; j = 1, 2, \ldots, n$, to find

$$(\boldsymbol{x}^*, R^*, T^*) = \arg\min \phi(\boldsymbol{x}, R, T) \doteq \sum_{j=1}^{n} \sum_{i=1}^{2} \|\tilde{\boldsymbol{x}}_i^j - \boldsymbol{x}_i^j\|_2^2$$

subject to

$$\boldsymbol{x}_2^{jT} \widehat{T} R \boldsymbol{x}_1^j = 0, \quad \boldsymbol{x}_1^{jT} e_3 = 1, \quad \boldsymbol{x}_2^{jT} e_3 = 1, \quad j = 1, 2, \ldots, n. \tag{5.90}$$

Using Lagrange multipliers (Appendix C) $\lambda^j, \gamma^j, \eta^j$, we can convert the above minimization problem to an unconstrained minimization problem over $R \in SO(3), T \in \mathbb{S}^2, \boldsymbol{x}_1^j, \boldsymbol{x}_2^j, \lambda^j, \gamma^j, \eta^j$. Consider the Lagrangian function associated with this constrained optimization problem

$$\min \sum_{j=1}^{n} \|\tilde{\boldsymbol{x}}_1^j - \boldsymbol{x}_1^j\|^2 + \|\tilde{\boldsymbol{x}}_2^j - \boldsymbol{x}_2^j\|^2 + \lambda^j \boldsymbol{x}_2^{jT} \widehat{T} R \boldsymbol{x}_1^j + \gamma^j (\boldsymbol{x}_1^{jT} e_3 - 1) + \eta^j (\boldsymbol{x}_2^{jT} e_3 - 1). \tag{5.91}$$

A necessary condition for the existence of a minimum is $\nabla L = 0$, where the derivative is taken with respect to $\boldsymbol{x}_1^j, \boldsymbol{x}_2^j, \lambda^j, \gamma^j, \eta^j$. Setting the derivative with respect to the Lagrange multipliers $\lambda^j, \gamma^j, \eta^j$ to zero returns the equality constraints, and setting the derivative with respect to $\boldsymbol{x}_1^j, \boldsymbol{x}_2^j$ to zero yields

$$2(\tilde{\boldsymbol{x}}_1^j - \boldsymbol{x}_1^j) + \lambda^j R^T \widehat{T}^T \boldsymbol{x}_2^j + \gamma^j e_3 = 0,$$
$$2(\tilde{\boldsymbol{x}}_2^j - \boldsymbol{x}_2^j) + \lambda^j \widehat{T} R \boldsymbol{x}_1^j + \eta^j e_3 = 0.$$

Simplifying these equations by premultiplying both by the matrix $\widehat{e}_3^T \widehat{e}_3$, we obtain

$$\begin{aligned} \boldsymbol{x}_1^j &= \tilde{\boldsymbol{x}}_1^j - \tfrac{1}{2} \lambda^j \widehat{e}_3^T \widehat{e}_3 R^T \widehat{T}^T \boldsymbol{x}_2^j, \\ \boldsymbol{x}_2^j &= \tilde{\boldsymbol{x}}_2^j - \tfrac{1}{2} \lambda^j \widehat{e}_3^T \widehat{e}_3 \widehat{T} R \boldsymbol{x}_1^j. \end{aligned} \tag{5.92}$$

Together with $\boldsymbol{x}_2^{jT}\widehat{T}R\boldsymbol{x}_1^j = 0$, we may solve for the Lagrange multipliers $\lambda^j$ in different expressions,[22]

$$\lambda^j = \frac{2(\boldsymbol{x}_2^{jT}\widehat{T}R\tilde{\boldsymbol{x}}_1^j + \tilde{\boldsymbol{x}}_2^{jT}\widehat{T}R\boldsymbol{x}_1^j)}{\boldsymbol{x}_1^{jT}R^T\widehat{T}^T\widehat{e}_3^T\widehat{e}_3\widehat{T}R\boldsymbol{x}_1^j + \boldsymbol{x}_2^{jT}\widehat{T}R\widehat{e}_3^T\widehat{e}_3R^T\widehat{T}^T\boldsymbol{x}_2^j} \tag{5.93}$$

or

$$\lambda^j = \frac{2\boldsymbol{x}_2^{jT}\widehat{T}R\tilde{\boldsymbol{x}}_1^j}{\boldsymbol{x}_1^{jT}R^T\widehat{T}^T\widehat{e}_3^T\widehat{e}_3\widehat{T}R\boldsymbol{x}_1^j} = \frac{2\tilde{\boldsymbol{x}}_2^{jT}\widehat{T}R\boldsymbol{x}_1^j}{\boldsymbol{x}_2^{jT}\widehat{T}R\widehat{e}_3^T\widehat{e}_3R^T\widehat{T}^T\boldsymbol{x}_2^j}. \tag{5.94}$$

Substituting (5.92) and (5.93) into the least-squares cost function of equation (5.91), we obtain

$$\phi(\boldsymbol{x}, R, T) = \sum_{j=1}^{n} \frac{(\boldsymbol{x}_2^{jT}\widehat{T}R\tilde{\boldsymbol{x}}_1^j + \tilde{\boldsymbol{x}}_2^{jT}\widehat{T}R\boldsymbol{x}_1^j)^2}{\|\widehat{e}_3\widehat{T}R\boldsymbol{x}_1^j\|^2 + \|\boldsymbol{x}_2^{jT}\widehat{T}R\widehat{e}_3^T\|^2}. \tag{5.95}$$

If one uses instead (5.92) and (5.94), one gets

$$\phi(\boldsymbol{x}, R, T) = \sum_{j=1}^{n} \frac{(\tilde{\boldsymbol{x}}_2^{jT}\widehat{T}R\boldsymbol{x}_1^j)^2}{\|\widehat{e}_3\widehat{T}R\boldsymbol{x}_1^j\|^2} + \frac{(\boldsymbol{x}_2^{jT}\widehat{T}R\tilde{\boldsymbol{x}}_1^j)^2}{\|\boldsymbol{x}_2^{jT}\widehat{T}R\widehat{e}_3^T\|^2}. \tag{5.96}$$

These expressions for $\phi$ can finally be minimized with respect to $(R, T)$ as well as $\boldsymbol{x} = \{\boldsymbol{x}_i^j\}$. In doing so, however, one has to make sure that the unknowns are constrained so that $R \in SO(3)$ and $T \in \mathbb{S}^2$ are explicitly enforced. In Appendix C we discuss methods for minimizing a function with unknowns in spaces like $SO(3) \times \mathbb{S}^2$, that can be used to minimize $\phi(\boldsymbol{x}, R, T)$ once $\boldsymbol{x}$ is known. Since $\boldsymbol{x}$ is *not* known, one can set up an *alternating minimization scheme where an initial approximation of* $\boldsymbol{x}$ *is used to estimate an approximation of* $(R, T)$*, which is used, in turn, to update the estimates of* $\boldsymbol{x}$. It can be shown that each such iteration decreases the cost function, and therefore convergence to a local extremum is guaranteed, since the cost function is bounded below by zero. The overall process is described in Algorithm 5.5. As we mentioned before, this is equivalent to the so-called *bundle adjustment* for the two-view case, that is the direct minimization of the reprojection error with respect to all unknowns. Equivalence is intended in the sense that, at the optimum, the two solutions coincide.

*Structure triangulation*

In step 3 of Algorithm 5.5, for each pair of images $(\tilde{\boldsymbol{x}}_1, \tilde{\boldsymbol{x}}_2)$ and a fixed $(R, T)$, $\boldsymbol{x}_1$ and $\boldsymbol{x}_2$ can be computed by minimizing the same reprojection error function $\phi(\boldsymbol{x}) = \|\tilde{\boldsymbol{x}}_1 - \boldsymbol{x}_1\|^2 + \|\tilde{\boldsymbol{x}}_2 - \boldsymbol{x}_2\|^2$ for each pair of image points. Assuming that the notation is the same as in Figure 5.9, let $\boldsymbol{\ell}_2 \in \mathbb{R}^3$ be the normal vector (of unit length) to the epipolar plane spanned by $(\boldsymbol{x}_2, \boldsymbol{e}_2)$.[23] Given such an $\boldsymbol{\ell}_2$, $\boldsymbol{x}_1$ and $\boldsymbol{x}_2$

[22] Since we have multiple equations to solve for one unknown $\lambda^j$, the redundancy gives rise to different expressions depending on which equation in (5.92) is used.

[23] $\boldsymbol{\ell}_2$ can also be interpreted as the coimage of the epipolar line in the second image, but here we do not use that interpretation.

---

**Algorithm 5.5 (Optimal triangulation).**

1. **Initialization**
   Initialize $\boldsymbol{x}_1$ and $\boldsymbol{x}_2$ as $\tilde{\boldsymbol{x}}_1$ and $\tilde{\boldsymbol{x}}_2$, respectively. Also initialize $(R, T)$ with the pose initialized by the solution from the eight-point linear algorithm.
2. **Pose estimation**
   For $\boldsymbol{x}_1$ and $\boldsymbol{x}_2$ computed from the previous step, update $(R, T)$ by minimizing the reprojection error $\phi(\boldsymbol{x}, R, T)$ given in its unconstrained form (5.95) or (5.96).
3. **Structure triangulation**
   For each image pair $(\tilde{\boldsymbol{x}}_1, \tilde{\boldsymbol{x}}_2)$ and $(R, T)$ computed from the previous step, solve for $\boldsymbol{x}_1$ and $\boldsymbol{x}_2$ that minimize the reprojection error $\phi(\boldsymbol{x}) = \|\boldsymbol{x}_1 - \tilde{\boldsymbol{x}}_1\|^2 + \|\boldsymbol{x}_2 - \tilde{\boldsymbol{x}}_2\|^2$.
4. Return to step 2 until the decrement in the value of $\phi$ is below a threshold.

---

are determined by

$$\boldsymbol{x}_1(\boldsymbol{\ell}_1) = \frac{\widehat{e}_3 \boldsymbol{\ell}_1 \boldsymbol{\ell}_1^T \widehat{e}_3^T \tilde{\boldsymbol{x}}_1 + \widehat{\boldsymbol{\ell}_1}^T \widehat{\boldsymbol{\ell}_1} e_3}{e_3^T \widehat{\boldsymbol{\ell}_1}^T \widehat{\boldsymbol{\ell}_1} e_3}, \quad \boldsymbol{x}_2(\boldsymbol{\ell}_2) = \frac{\widehat{e}_3 \boldsymbol{\ell}_2 \boldsymbol{\ell}_2^T \widehat{e}_3^T \tilde{\boldsymbol{x}}_2 + \widehat{\boldsymbol{\ell}_2}^T \widehat{\boldsymbol{\ell}_2} e_3}{e_3^T \widehat{\boldsymbol{\ell}_2}^T \widehat{\boldsymbol{\ell}_2} e_3},$$

where $\boldsymbol{\ell}_1 = R^T \boldsymbol{\ell}_2 \in \mathbb{R}^3$. Then the distance can be explicitly expressed as

$$\|\tilde{\boldsymbol{x}}_2 - \boldsymbol{x}_2\|^2 + \|\tilde{\boldsymbol{x}}_1 - \boldsymbol{x}_1\|^2 = \|\tilde{\boldsymbol{x}}_2\|^2 + \frac{\boldsymbol{\ell}_2^T A \boldsymbol{\ell}_2}{\boldsymbol{\ell}_2^T B \boldsymbol{\ell}_2} + \|\tilde{\boldsymbol{x}}_1\|^2 + \frac{\boldsymbol{\ell}_1^T C \boldsymbol{\ell}_1}{\boldsymbol{\ell}_1^T D \boldsymbol{\ell}_1},$$

where $A, B, C, D \in \mathbb{R}^{3\times3}$ are defined as functions of $(\tilde{\boldsymbol{x}}_1, \tilde{\boldsymbol{x}}_2)$:

$$\begin{aligned} A &= I - (\widehat{e}_3 \tilde{\boldsymbol{x}}_2 \tilde{\boldsymbol{x}}_2^T \widehat{e}_3^T + \widehat{\tilde{\boldsymbol{x}}_2} \widehat{e}_3 + \widehat{e}_3 \widehat{\tilde{\boldsymbol{x}}_2}), \quad B = \widehat{e}_3^T \widehat{e}_3, \\ C &= I - (\widehat{e}_3 \tilde{\boldsymbol{x}}_1 \tilde{\boldsymbol{x}}_1^T \widehat{e}_3^T + \widehat{\tilde{\boldsymbol{x}}_1} \widehat{e}_3 + \widehat{e}_3 \widehat{\tilde{\boldsymbol{x}}_1}), \quad D = \widehat{e}_3^T \widehat{e}_3. \end{aligned} \tag{5.97}$$

Then the problem of finding the optimal $\boldsymbol{x}_1^*$ and $\boldsymbol{x}_2^*$ becomes a problem of finding the normal vector $\boldsymbol{\ell}_2^*$ that minimizes the function of a sum of two *singular Rayleigh quotients*:

$$\min_{\boldsymbol{\ell}_2^T T = 0, \boldsymbol{\ell}_2^T \boldsymbol{\ell}_2 = 1} V(\boldsymbol{\ell}_2) = \frac{\boldsymbol{\ell}_2^T A \boldsymbol{\ell}_2}{\boldsymbol{\ell}_2^T B \boldsymbol{\ell}_2} + \frac{\boldsymbol{\ell}_2^T R C R^T \boldsymbol{\ell}_2}{\boldsymbol{\ell}_2^T R D R^T \boldsymbol{\ell}_2}. \tag{5.98}$$

This is an optimization problem on the unit circle $\mathbb{S}^1$ in the plane orthogonal to the (epipole) vector $e_2 (\sim T)$.[24] If $N_1, N_2 \in \mathbb{R}^3$ are vectors such that $(e_2, N_1, N_2)$ form an orthonormal basis of $\mathbb{R}^3$ in the second camera frame, then $\boldsymbol{\ell}_2 = \cos(\theta) N_1 + \sin(\theta) N_2$ with $\theta \in \mathbb{R}$. We need only to find $\theta^*$ that minimizes the function $V(\boldsymbol{\ell}_2(\theta))$. From the geometric interpretation of the optimal solution, we also know that the global minimum $\theta^*$ should lie between two values: $\theta_1$ and $\theta_2$ such that $\boldsymbol{\ell}_2(\theta_1)$ and $\boldsymbol{\ell}_2(\theta_2)$ correspond to normal vectors of the two planes

---

[24] Therefore, geometrically, motion and structure recovery from $n$ pairs of image correspondences is really an optimization problem on the space $SO(3) \times \mathbb{S}^2 \times \mathbb{T}^n$, where $\mathbb{T}^n$ is an $n$-torus, i.e. an $n$-fold product of $\mathbb{S}^1$.

spanned by $(\tilde{x}_2, e_2)$ and $(R\tilde{x}_1, e_2)$, respectively.[25] The problem now becomes a simple bounded minimization problem for a scalar function (in $\theta$) and can be efficiently solved using standard optimization routines (such as "`fmin`" in Matlab or Newton's algorithm, described in Appendix C).

## Historical notes

The origins of epipolar geometry can be dated back as early as the mid nineteenth century and appeared in the work of Hesse on studying the two-view geometry using seven points (see [Maybank and Faugeras, 1992] and references therein). Kruppa proved in 1913 that five points in general position are all one needs to solve the two-view problem up to a finite number of solutions [Kruppa, 1913]. Kruppa's proof was later improved in the work of [Demazure, 1988] where the actual number of solutions was proven, with a simpler proof given later by [Heyden and Sparr, 1999]. A constructive proof can be found in [Philip, 1996], and in particular, a linear algorithm is provided if there are six matched points, from which Exercise 5.13 was constructed. A more efficient five-point algorithm that enables real-time implementation has been recently implemented by [Nistér, 2003].

*The eight-point and four-point algorithms*

To our knowledge, the epipolar constraint first appeared in [Thompson, 1959]. The (discrete) eight-point linear algorithm introduced in this chapter is due to the work of [Longuet-Higgins, 1981] and [Huang and Faugeras, 1989], which sparked a wide interest in the structure from motion problem in computer vision and led to the development of numerous linear and nonlinear algorithms for motion estimation from two views. Early work on these subjects can be found in the books or manuscripts of [Faugeras, 1993, Kanatani, 1993b, Maybank, 1993, Weng et al., 1993b]. An improvement of the eight-point algorithm based on normalizing image coordinates was later given by [Hartley, 1997]. [Soatto et al., 1996] studied further the dynamical aspect of epipolar geometry and designed a Kalman filter on the manifold of essential matrices for dynamical motion estimation. We will study Kalman-filter-based approaches in Chapter 12.

The homography (discrete or continuous) between two images of a planar scene has been extensively studied and used in the computer vision literature. Early results on this subject can be found in [Subbarao and Waxman, 1985, Waxman and Ullman, 1985, Kanatani, 1985, Longuet-Higgins, 1986]. The four-point algorithm based on decomposing the homography matrix was first given by [Faugeras and Lustman, 1988]. A thorough discussion on the homography and the relationships between the two physically possible solutions in Theorem 5.19 can be found in [Weng et al., 1993b] and references therein. This chapter is a very

[25] If $\tilde{x}_1, \tilde{x}_2$ already satisfy the epipolar constraint, these two planes coincide.

concise summary and supplement to these early results in computer vision. In Chapter 9 we will see how the epipolar constraint and homography can be unified into a single type of constraint.

*Critical surfaces*

Regarding the criticality or ambiguity of the two-view geometry mentioned before (such as the critical surfaces), the interested reader may find more details in [Adiv, 1985, Longuet-Higgins, 1988, Maybank, 1993, Soatto and Brockett, 1998] or the book of [Faugeras and Luong, 2001]. More discussions on the criticality and degeneracy in camera calibration and multiple-view reconstruction can be found in later chapters.

*Objective functions for estimating epipolar geometry*

Many objective functions have been used in the computer vision literature for estimating the two-view epipolar geometry, such as "epipolar improvement" [Weng et al., 1993a], "normalized epipolar constraint" [Weng et al., 1993a, Luong and Faugeras, 1996, Zhang, 1998c], "minimizing the reprojection error" [Weng et al., 1993a], and "triangulation" [Hartley and Sturm, 1997]. The method presented in this chapter follows that of [Ma et al., 2001b].

As discussed in Section 5.A, there is no closed-form solution to an optimal motion and structure recovery problem if the reprojection error is chosen to be the objective since the problem involves solving algebraic equations of order six [Hartley and Sturm, 1997, Ma et al., 2001b]. The solution is typically found through iterative numerical schemes such as the ones described in Appendix C. It has, however, been shown by [Oliensis, 2001] that if one chooses to minimize the angle (not distance) between the measured $\tilde{\boldsymbol{x}}$ and recovered $\boldsymbol{x}$, a closed-form solution is available. Hence, solvability of a reconstruction problem does depend on the choice of objective function. In the multiple-view setting, minimizing reprojection error corresponds to a nonlinear optimization procedure [Spetsakis and Aloimonos, 1988], often referred to as "bundle adjustment," which we will discuss in Chapter 11.

*The continuous motion case*

The search for the continuous counterpart of the eight-point algorithm has produced many different versions in the computer vision literature due to its subtle difference from the discrete case. To our knowledge, the first algorithm was proposed in 1984 by [Zhuang and Haralick, 1984] with a simplified version given in [Zhuang et al., 1988]; and a first-order algorithm was given by [Waxman et al., 1987]. Other algorithms solved for rotation and translation separately using either numerical optimization techniques [Bruss and Horn, 1983] or linear subspace methods [Heeger and Jepson, 1992, Jepson and Heeger, 1993]. [Kanatani, 1993a] proposed a linear algorithm reformulating Zhuang's approach in terms of essential parameters and twisted flow. See [Tian et al., 1996] for some experimental comparisons of these methods, while analytical results on the sensitivity of two-view geometry can be found in [Daniilidis and Nagel, 1990, Spetsakis, 1994, Daniilidis and Spetsakis, 1997] and estimation bias study in the work of [Heeger and Jepson, 1992, Kanatani, 1993b]. [Fermüller et al., 1997] has further shown that the distortion induced on the structure from errors in the motion estimates is governed by the so-called Cremona transformation. The parallel development of the continuous eight-point algorithm presented in this chapter follows that of [Ma et al., 2000a], where the interested reader may also find a more detailed account of related bibliography and history. Besides the linear methods, a study of the (nonlinear) optimal solutions to the continuous motion case was given in [Chiuso et al., 2000].

# Chapter 6

# Reconstruction from Two Uncalibrated Views

*The real voyage of discovery consists not in seeking new landscapes, but in having new eyes.*

– Marcel Proust

In Chapter 3 we have seen that the projection of a point in space with coordinates $\boldsymbol{X}$ onto the image plane has (homogeneous) coordinates $\boldsymbol{x}'$ that satisfy the equation (3.19)

$$\lambda \boldsymbol{x}' = K\Pi_0 g\boldsymbol{X} = K[R,\ T]\boldsymbol{X}, \tag{6.1}$$

where $\Pi_0 = [I,\ 0] \in \mathbb{R}^{3\times 4}$, and $g \in SE(3)$ is the pose of the camera in the (chosen) world reference frame. In the equation above, the matrix $K$, which was defined in equation (3.14) as

$$K = \begin{bmatrix} fs_x & s_\theta & o_x \\ 0 & fs_y & o_y \\ 0 & 0 & 1 \end{bmatrix} \in \mathbb{R}^{3\times 3}, \tag{6.2}$$

describes "intrinsic" properties of the camera, such as the position of the optical center $(o_x, o_y)$, the size of the pixel $(s_x, s_y)$, its skew factor $s_\theta$, and the focal length $f$. The matrix $K$ is called the *intrinsic parameter matrix*, or simply calibration matrix, and it maps metric coordinates (units of meters) into image coordinates (units of pixels). In what follows, we denote pixel coordinates with a prime superscript $\boldsymbol{x}'$, whereas metric coordinates are indicated simply by $\boldsymbol{x}(= K^{-1}\boldsymbol{x}')$, following the convention used in Chapter 3. The rigid-body motion $g = (R, T)$ represents the "extrinsic" properties of the camera, namely, its position and orientation relative to a chosen world reference frame. The parameters $g$ are therefore called *extrinsic calibration parameters*.

In Chapter 5 we have seen that in the calibrated case, the intrinsic calibration matrix is the identity, $K = I$, and two views of a sufficient number of points are enough to recover the camera pose and the position of the points in space up to a global scale factor. If the intrinsic calibration matrix is *not* the identity, i.e. $K \neq I$, but it is nevertheless known, the problem can be easily rephrased so that results for the calibrated case can still be applied. In fact, just multiply equation (6.1) on both sides by $K^{-1}$ (from equation (6.2) notice that $K$ is always invertible), and let $\boldsymbol{x} \doteq K^{-1}\boldsymbol{x}'$. Then we get back to the calibrated case $\lambda \boldsymbol{x} = \Pi_0 g \boldsymbol{X}$. Hence the knowledge of matrix $K$ is crucial for recovery of the true 3-D Euclidean structure of the scene.

Unfortunately, the intrinsic calibration matrix $K$ is usually not known. This chapter explores different options available for estimating the calibration matrix $K$, or recovering spatial properties of the scene despite the lack of knowledge of $K$.

### *Taxonomy of calibration or uncalibrated reconstruction procedures*

The procedure of inferring the calibration matrix $K$ is called (intrinsic) *camera calibration*. If the user has access to the camera and has a known object available, this procedure is conceptually simple, and several calibration software distributions are available. Most of them require the known object to have a regular appearance and a simple geometry, e.g., a planar checkerboard pattern. In this case, the known object is often called a *calibration rig*. However, the calibration procedure can be applied to most objects that have a number of distinct points with known position relative to some reference frame. We address two such standard procedures in Section 6.5.2 for a 3-D pattern serving as a calibration rig and Section 6.5.3 for a planar pattern, respectively.

In many practical situations one does not have access to the camera, and a collection of images is all that is available. In this case, calibration with a rig is obviously not possible. However, one may still have partial information available, either on the scene or on the camera. For instance, many man-made objects contain planar surfaces, parallel lines, right angles, and symmetric structures, which provide strong constraints on $K$. In Section 6.5.1 we will show how one can perform calibration in the presence of *partial scene knowledge*. Naturally, one has to be aware that enforcing such knowledge can lead to gross errors if the assumptions are not satisfied. For instance, if one were to assume that the edges of the building in Figure 6.1 were parallel, the resulting errors in calibration would affect the reconstruction.

In addition to prior assumptions on the scene, one may have partial knowledge available about the camera. For instance, one may know that each image has been taken with the same camera, and therefore the calibration matrix $K$ for each view is the same. Another common scenario is one where the camera is available for precalibration, but some of the camera parameters may change during the shoot-

Figure 6.1. Frank O. Gehry's "Ginger and Fred" building in Prague. Photo courtesy of Don Barker.

ing of a particular sequence. For instance, the focal length can change as a result of zooming or focusing, while the other parameters, such as the skew factor or the size of the pixel, remain constant and known. In such a case, one can obtain constraints on $K$ and the images that under suitable assumptions allow the recovery of the calibration matrix. In the presence of *partial camera knowledge*, given a number of views, one can recover the calibration matrix and therefore the Euclidean 3-D scene structure and camera motion. Since these techniques typically involve more than two views, they will be studied in Chapter 8. However, in Section 6.4.5 we will preview the basic ideas as they pertain to uncalibrated reconstruction.

The least-constrained scenario occurs when one has images of a scene taken with different unknown cameras, and no knowledge of the scene is available. In this case, one cannot recover the camera calibration and the physically correct metric 3-D model of the scene. However, one can still recover *some* information of the scene, namely, a distorted version of the original Euclidean 3-D structure, also called *projective reconstruction*. We will discuss this in detail in Section 6.3.

## *Organization of this chapter*

The taxonomy of calibration procedures described above goes from full knowledge of the scene (calibration with a rig) to a complete lack of knowledge about the scene and arbitrary cameras (projective reconstruction). In this chapter, we will follow the reverse order and describe first the geometry of two uncalibrated views in the absence of any prior knowledge. In Sections 6.3 and 6.4 we will discuss how to resolve the projective ambiguity and upgrade the distorted 3-D structure to Euclidean. In Section 6.4.5, we preview how to recover calibration given projection matrices up to a projective transformation using the *absolute quadric constraints*, which readily connects two-view geometry to multiple-view geometry (to be studied in Chapter 8). In Section 6.5, we describe some simple procedures to calibrate a camera with a known object. Finally, in Section 6.6 we study a direct approach to camera autocalibration based on the pairwise relationships between views captured by *Kruppa's equations*.

Before we begin this program, at the beginning of the chapter we point out a useful analogy by showing that an uncalibrated camera moving in Euclidean space is equivalent to a calibrated camera moving in a "distorted" space, governed by a different way of measuring distances and angles (Section 6.1), and we show how the framework of epipolar geometry studied in the last chapter is modified in the presence of uncalibrated cameras (Section 6.2).

## 6.1 Uncalibrated camera or distorted space?

In the standard Euclidean space, the canonical inner product between two vectors is given by $\langle u, v\rangle \doteq u^T v$.[1] We will show that working with an uncalibrated camera in a Euclidean space is equivalent to working with a calibrated camera in a "distorted" space, where the inner product between two vectors is given by $\langle u, v\rangle_S = u^T S v$ for some symmetric and positive definite matrix $S$. As a consequence, all reconstruction algorithms described in the previous chapter for the calibrated case, if rewritten in terms of this new inner product, yield a reconstruction of camera pose and scene structure in the distorted space. Only in the particular case where $S = I$ is the reconstruction physically correct, corresponding to the true Euclidean structure.

To understand the geometry associated with an uncalibrated camera, consider a linear map $\psi$, represented by a matrix $K$ that transforms spatial coordinates $\boldsymbol{X}$ as follows

$$\psi : \mathbb{R}^3 \to \mathbb{R}^3; \quad \boldsymbol{X} \mapsto \boldsymbol{X}' = K\boldsymbol{X}.$$

For instance, in our case, $K$ can be the calibration matrix that maps metric coordinates into pixels. The map $\psi$ induces a transformation of the inner product as follows

$$\langle \psi^{-1}(u), \psi^{-1}(v)\rangle = u^T K^{-T} K^{-1} v \doteq \langle u, v\rangle_{K^{-T}K^{-1}}, \quad \forall u, v \in \mathbb{R}^3. \tag{6.3}$$

Therefore, if one wants to write the inner product between two vectors, but only their pixel coordinates $u, v$ are available, one has to weigh the inner product by a matrix as indicated in equation (6.3):

$$\langle u, v\rangle_S = u^T S v, \quad \text{where} \quad S = K^{-T} K^{-1}. \tag{6.4}$$

[1] Definitions and properties of inner products can be found in Appendix A.

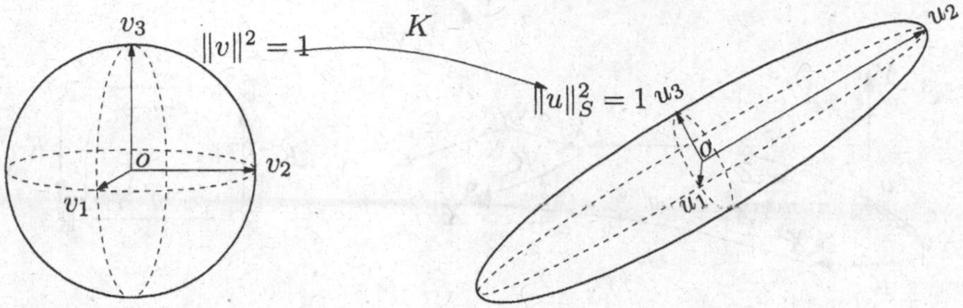


Figure 6.2. Effect of the matrix $K$ as a map $K : v \mapsto u = Kv$, where points on the sphere $\|v\|^2 = 1$ are mapped to points on an ellipsoid $\|u\|_S^2 = 1$ (a "unit sphere" under the metric $S$). The principal axes of the ellipsoid are the eigenvalues of $S$.

This is the inner product between two vectors, but expressed in terms of their pixel coordinates. The matrix $S$ is called the *metric* of the space. The distortion of the space induced by $S$ alters both the length of the vectors as well as the angles between them according to the modified definition of the inner product. Hence, under this metric, the length of a vector $u$ is measured as $\|u\|_S = \sqrt{\langle u, u\rangle_S}$. Figure 6.2 illustrates the effect of $S$ on the space. A unit sphere in the distorted space looks like an ellipsoid. Once transformed back to the Euclidean space it looks like the familiar sphere. In order to complete the picture, however, we need to understand how a camera moves in the distorted space.

Rigid-body motions, as we know from Chapter 2, must preserve distances and angles. But angles are now expressed in terms of the new metric $S$. So what does a rigid motion look like in the distorted space? The Euclidean coordinates $\boldsymbol{X}$ of a moving point $p$ at time $t$ are given by the familiar Euclidean transformation

$$\boldsymbol{X} = R\boldsymbol{X}_0 + T. \tag{6.5}$$

The coordinate transformation in the uncalibrated camera coordinates $\boldsymbol{X}'$ is then given by

$$K\boldsymbol{X} = KR\boldsymbol{X}_0 + KT \quad \Leftrightarrow \quad \boldsymbol{X}' = KRK^{-1}\boldsymbol{X}_0' + T', \tag{6.6}$$

where $\boldsymbol{X}' = K\boldsymbol{X}$ and $T' = KT$. Therefore, the transformation mapping $\boldsymbol{X}_0'$ to $\boldsymbol{X}'$ can be written in homogeneous coordinates as

$$G' = \left\{ g' = \begin{bmatrix} KRK^{-1} & T' \\ 0 & 1 \end{bmatrix} \middle| \, T' \in \mathbb{R}^3, R \in SO(3) \right\} \quad \subset \mathbb{R}^{4\times 4}. \tag{6.7}$$

These transformations form a matrix group (Appendix A), which is called the *conjugate* of the Euclidean group $G = SE(3)$. The relation between a Euclidean motion $g$ and its conjugate $g'$ is illustrated in Figure 6.3. Applying this to the image formation model (6.1), we get

$$\lambda \boldsymbol{x}' = K\Pi_0 g\boldsymbol{X}_0 = KR\boldsymbol{X}_0 + KT = KRK^{-1}K\boldsymbol{X}_0 + KT = \Pi_0 g'\boldsymbol{X}_0'.$$

Note that the above relationship is similar to the calibrated case, but it relates uncalibrated quantities from the distorted space $\boldsymbol{X}_0', \boldsymbol{x}'$ via the conjugate motion

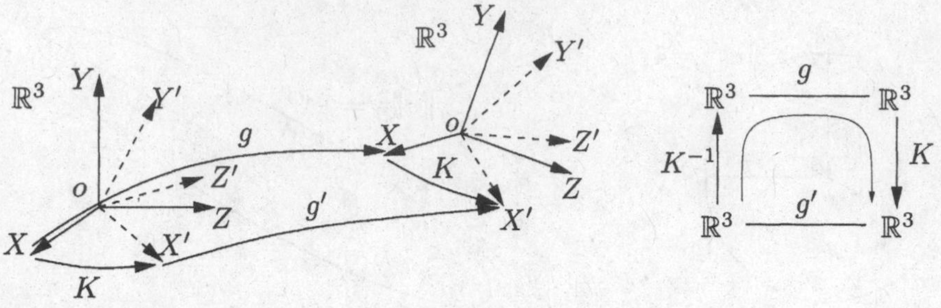


Figure 6.3. A rigid-body motion of a Euclidean coordinate frame $(X, Y, Z)$ is given by $g = (R, T) \in SE(3)$. With respect to the uncalibrated coordinate frame $(X', Y', Z')$, the transformation is $g' = (KRK^{-1}, KT)$.

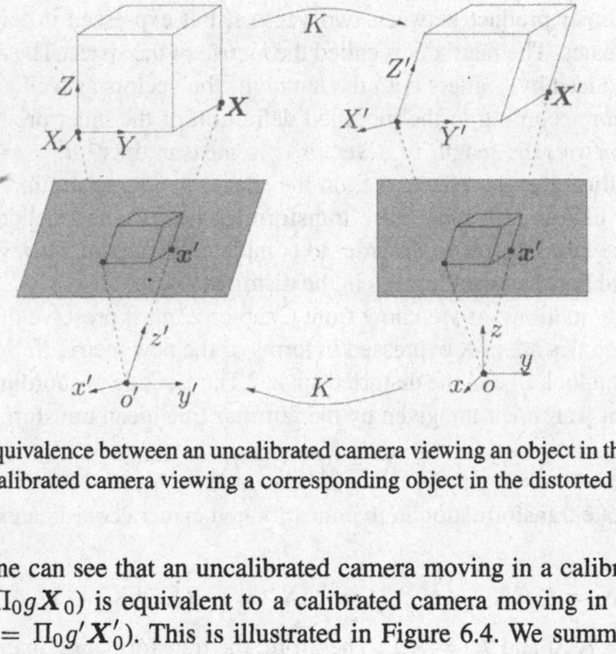


Figure 6.4. Equivalence between an uncalibrated camera viewing an object in the Euclidean space and a calibrated camera viewing a corresponding object in the distorted space.

$g'$. Hence one can see that an uncalibrated camera moving in a calibrated space ($\lambda \boldsymbol{x}' = K\Pi_0 g \boldsymbol{X}_0$) is equivalent to a calibrated camera moving in a distorted space ($\lambda \boldsymbol{x}' = \Pi_0 g' \boldsymbol{X}'_0$). This is illustrated in Figure 6.4. We summarize these observations in the following remark.

**Remark 6.1 (Uncalibrated camera and distorted space).** *An uncalibrated camera with calibration matrix $K$ viewing points in a calibrated (Euclidean) world moving with $(R, T)$ is equivalent to a calibrated camera viewing points in a distorted space governed by an inner product $\langle u, v\rangle_S \doteq u^T S v$, moving with $(KRK^{-1}, KT)$. Furthermore, $S = K^{-T}K^{-1}$.*

In the case of a row vector, say $u^T$, the map $\psi$ will transform it to another row vector $\psi(u^T) = u^T K$. Correspondingly, this induces an inner product on row vectors

$$\langle u^T, v^T\rangle_S = \langle \psi^{-1}(u^T), \psi^{-1}(v^T)\rangle = u^T S^{-1} v, \quad \text{where} \quad S^{-1} = KK^T.$$

The above apparently innocuous remark will be the key to understanding the geometry of uncalibrated views, and to deriving methods for camera calibration in later sections. Since the standard inner product $\langle u, v\rangle = u^T v$ is a fundamental invariant under Euclidean transformations, it will be invariant too in the distorted space, but now written in terms of the new inner product of either column vectors or row vectors. As we will soon see in this chapter, all constraints that will allow us to derive information about the camera calibration (i.e. the matrix $K$, $S$, or, $S^{-1}$) will be essentially in the form of the new inner product.

## 6.2 Uncalibrated epipolar geometry

In this section we study epipolar geometry for uncalibrated cameras or, equivalently, two-view geometry in distorted spaces. In particular, we will derive the epipolar constraint in terms of uncalibrated image coordinates, and we will see how the structure of the essential matrix is modified by the calibration matrix. For simplicity, we will assume that the same camera has captured both images, so that $K_1 = K_2 = K$. Extension to different cameras is straightforward, although it involves more elaborate notation.

### *6.2.1 The fundamental matrix*

The epipolar constraint (5.2) derived in Section 5.1.1 for calibrated cameras can be extended to uncalibrated cameras in a straightforward manner. The constraint expresses the fact that the volume identified by the three vectors $\boldsymbol{x}_1, \boldsymbol{x}_2$, and $T$ is zero, or in other words, the three vectors are coplanar. Such a volume is given by the triple product of the three vectors, written relative to the same reference frame. The triple product of three vector is the inner product of one with the cross product of the other two (see Chapter 2). In the "distorted space," which we have described in the previous section, the three vectors are given by $\boldsymbol{x}_2', T' = KT$, and $KR\boldsymbol{x}_1 = KRK^{-1}\boldsymbol{x}_1'$, and their triple product is

$$\boxed{\boldsymbol{x}_2'^T \widehat{T'} KRK^{-1}\boldsymbol{x}_1' = 0,} \tag{6.8}$$

as the reader can easily verify. Equation (6.8) is the uncalibrated version of the epipolar constraint studied in Chapter 5.

An alternative way of deriving the epipolar constraint for uncalibrated cameras is by direct substitution of $\boldsymbol{x} = K^{-1}\boldsymbol{x}'$ into the epipolar constraint

$$\boldsymbol{x}_2^T \widehat{T} R \boldsymbol{x}_1 = 0 \quad \Leftrightarrow \quad \boldsymbol{x}_2'^T \underbrace{K^{-T}\widehat{T}RK^{-1}}_{F} \boldsymbol{x}_1' = 0. \tag{6.9}$$

The matrix $F$ defined in the equation above is called the *fundamental matrix*:

$$F \doteq K^{-T}\widehat{T}RK^{-1} \in \mathbb{R}^{3\times 3}. \tag{6.10}$$

When $K = I$, the fundamental matrix $F$ is identical to the essential matrix $E = \widehat{T}R$ that we studied in Chapter 5.

Yet another derivation can be obtained by following the same algebraic procedure we proposed for the calibrated case, i.e. by direct elimination of the unknown depth scales $\lambda_1, \lambda_2$ from the rigid-body motion equation

$$\lambda_2 \boldsymbol{x}_2 = R\lambda_1 \boldsymbol{x}_1 + T,$$

where $\lambda \boldsymbol{x} = \boldsymbol{X}$. Multiplying both sides by the calibration matrix $K$, we see that

$$\lambda_2 K\boldsymbol{x}_2 = KR\lambda_1 \boldsymbol{x}_1 + KT \quad \Leftrightarrow \quad \lambda_2 \boldsymbol{x}_2' = KRK^{-1}\lambda_1 \boldsymbol{x}_1' + T', \tag{6.11}$$

where $\boldsymbol{x}' = K\boldsymbol{x}$ and $T' = KT$. In order to eliminate the unknown depth, we multiply both sides of (6.11) by $T' \times \boldsymbol{x}_2' = \widehat{T'}\boldsymbol{x}_2'$. This yields the same constraint as equation (6.8).

At this point the question arises as to the relationship between the different versions of the epipolar constraints in equations (6.8) and (6.9). To reconcile these two expressions, we recall that for $T \in \mathbb{R}^3$ and a nonsingular matrix $K$, Lemma 5.4 states that $K^{-T}\widehat{T}K^{-1} = \widehat{KT}$ if $\det(K) = +1$. Under the same condition we have that $F = K^{-T}\widehat{T}RK^{-1} = K^{-T}\widehat{T}K^{-1}KRK^{-1} = \widehat{T'}KRK^{-1}$. We collect these equalities as

$$\boxed{F = K^{-T}\widehat{T}RK^{-1} = \widehat{T}'KRK^{-1}} \tag{6.12}$$

if $\det(K) = +1$. In case $\det(K) \neq 1$, one can simply scale all the matrices by a factor. In any case, we have $K^{-T}\widehat{T}RK^{-1} \sim \widehat{T'}KRK^{-1}$. Thus, without loss of generality, we will always assume $\det(K) = 1$. Note that if desired, one can always divide $K$ by its last entry $k_{33}$ to convert it back to the form (6.2) that carries the physical interpretation of $K$ in terms of optical center, focal length, and skew factor of the pixels. We will see in later sections that the second form of $F$ in equation (6.12) is more convenient to use for camera calibration. Before getting further into the theory of calibration, we take a closer look at some of the properties of the fundamental matrix.

### 6.2.2 *Properties of the fundamental matrix*

The fundamental matrix maps, or "transfers," a point $\boldsymbol{x}_1'$ in the first view to a vector $\boldsymbol{\ell}_2 \doteq F\boldsymbol{x}_1' \in \mathbb{R}^3$ in the second view via

$$\boldsymbol{x}_2'^T F \boldsymbol{x}_1' = \boldsymbol{x}_2'^T \boldsymbol{\ell}_2 = 0.$$

In fact, the vector $\boldsymbol{\ell}_2$ (in the coordinate frame of the second camera) defines implicitly a line in the image plane as the collection of image points $\{\boldsymbol{x}_2'\}$ that satisfy the equation

$$\boldsymbol{\ell}_2^T \boldsymbol{x}_2' = 0.$$

Strictly speaking, the vector $\boldsymbol{\ell}_2$ should be called the "coimage" of the line (Section 3.3.4). By abuse of notation, we will refer to this vector as representing the line it-

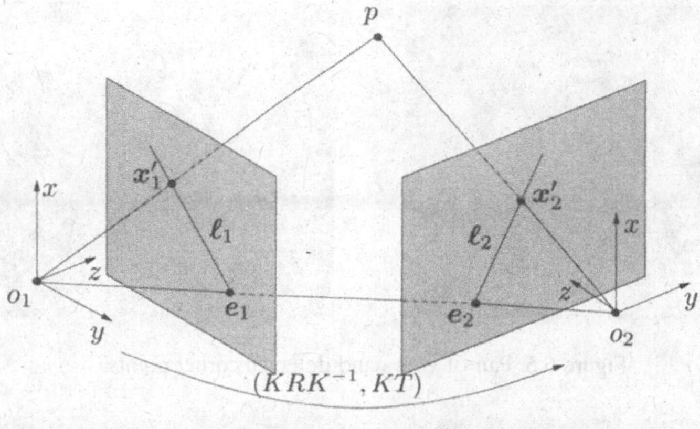


Figure 6.5. Two projections $\boldsymbol{x}_1', \boldsymbol{x}_2' \in \mathbb{R}^3$ of a 3-D point $p$ from two vantage points, $o_1$ and $o_2$. The transformation between the two vantage points is given by $(KRK^{-1}, KT)$. The epipoles are the projections of the center of one camera onto the image plane of the other, and the epipolar lines corresponding to each point are determined by the intersection of the image plane with the epipolar plane, formed by the point itself and the two camera centers.

self. Similarly, we may interpret the equation $\boldsymbol{\ell}_1 \doteq F^T \boldsymbol{x}_2' \in \mathbb{R}^3$ as $F$ transferring a point in the second image to a line in the first. These lines are called *epipolar lines*, illustrated in Figure 6.5. Geometrically, each image point, together with the two camera centers, identifies the *epipolar plane*; this plane intersects the other image plane in the epipolar line. So, each point in one image plane determines an epipolar line in the other image plane on which its corresponding point must lie, and vice versa.

**Lemma 6.2 (Epipolar matching lemma).** *Two image points $\boldsymbol{x}_1', \boldsymbol{x}_2'$ correspond to a single point in space if and only if $\boldsymbol{x}_1'$ is on the epipolar line $\boldsymbol{\ell}_1 = F^T \boldsymbol{x}_2'$ or equivalently, $\boldsymbol{x}_2'$ is on the epipolar line $\boldsymbol{\ell}_2 = F\boldsymbol{x}_1'$.*

Despite its simplicity, this lemma is rather useful in establishing correspondence across images. In fact, knowing the fundamental matrix $F$ allows us to restrict the search for corresponding points on the epipolar line only, rather than on the entire image, as shown in Figure 6.7 for some of the points detected in Figure 6.6. This fact will be further exploited in Chapter 11 for feature matching.

The epipole, denoted by $\boldsymbol{e}$, is the point where the *baseline* (the line joining the two camera centers $o_1, o_2$) intersects the image plane in each view, as shown in Figure 6.5. It can be computed as the null space of the fundamental matrix $F$ (or $F^T$). Let $\boldsymbol{e}_i \in \mathbb{R}^3, i = 1, 2$, be the epipole with respect to the first and second views respectively (as shown in Figure 6.5). One can verify that

$$\boldsymbol{e}_2^T F = 0, \quad F\boldsymbol{e}_1 = 0,$$

and therefore $\boldsymbol{e}_2 = KT = T'$ and $\boldsymbol{e}_1 = KR^T T$. We leave it as an exercise for the reader to verify that all epipolar lines defined as above must pass through the epipole in each image. For instance, we must have $\boldsymbol{e}_2^T \boldsymbol{\ell}_2 = 0$ regardless of which point in the first image is transferred to $\boldsymbol{\ell}_2$.

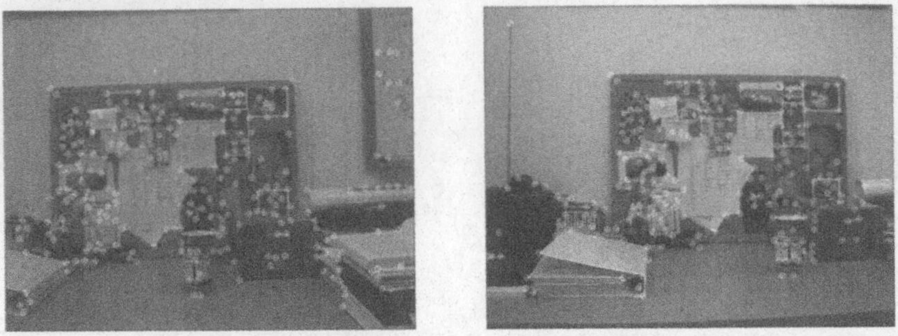

Figure 6.6. Pair of views and detected corner points.

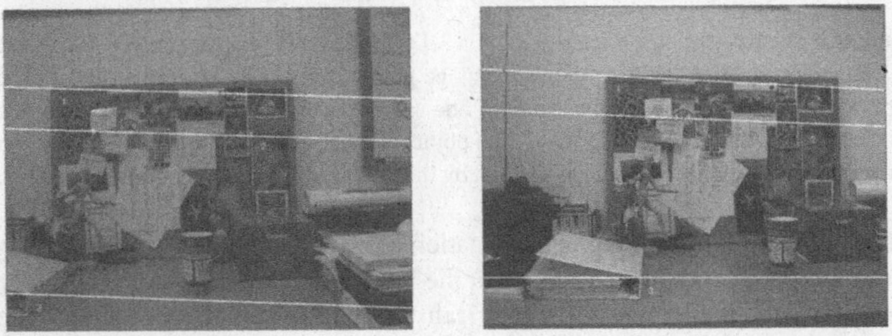

Figure 6.7. Epipolar geometry between two views of the scene. Note the corresponding points in two views lie on the associated epipolar lines. The epipoles for this pair of views lie outside of the image plane.

Since the fundamental matrix $F$ is the product of a skew-symmetric matrix $\widehat{T'}$ of rank 2 and a matrix $KRK^{-1} \in \mathbb{R}^{3\times 3}$ of rank 3, it must have rank 2. Hence $F$ can be characterized in terms of the singular value decomposition (SVD) $F = U\Sigma V^T$, with

$$\Sigma = \text{diag}\{\sigma_1, \sigma_2, 0\} \tag{6.13}$$

for some $\sigma_1, \sigma_2 \in \mathbb{R}_+$. In contrast to the case of an essential matrix, where the two nonzero singular values were equal, here we have $\sigma_1 \geq \sigma_2$. Indeed, any matrix of rank 2 can be a fundamental matrix, in the sense that one can always find two projection matrices whose epipolar geometry is governed by the given matrix.

The fundamental matrix can be estimated from a collection of eight or more corresponding points in two images. The algorithm, which we outline in Appendix 6.A, is essentially identical to the eight-point algorithm we described for the calibrated case. However, while in the calibrated case the essential matrix provided all the information necessary to recover camera pose and hence enabled the recovery of the 3-D Euclidean structure, in the uncalibrated case we cannot simply "unravel" $R$ and $T$ from the fundamental matrix. A simple way to see this is by using a counting argument: $F$ has at most eight free parameters (nine elements defined up to a scale factor), but it is composed of products of the matrix $K$ (five degrees of freedom), the matrix $R$ (three degrees of freedom) and $T$ (two degrees of freedom: three elements defined up to a scalar factor). Therefore, from the eight degrees of freedom in $F$ we cannot possibly recover the ten degrees of freedom in $K$, $R$, and $T$.

Thus, we cannot just "extract" the relative camera pose[2] $(KRK^{-1}, T')$ from $F$:

$$F = \widehat{T'}KRK^{-1} \quad \mapsto \quad \Pi = [KRK^{-1}, T']. \tag{6.14}$$

Note that there are in fact infinitely many matrices $\Pi$ that correspond to the same $F$. This can be observed directly from the epipolar constraint by noticing that

$$\boldsymbol{x}_2'^T \widehat{T'} KRK^{-1} \boldsymbol{x}_1' = \boldsymbol{x}_2'^T \widehat{T'}(KRK^{-1} + T'v^T)\boldsymbol{x}_1' = \boldsymbol{x}_2'^T \widehat{T'} R' \boldsymbol{x}_1' = 0 \tag{6.15}$$

for an arbitrary vector $v \in \mathbb{R}^3$, since $\widehat{T'}(T'v^T) = (\widehat{T'}T')v^T = 0$.

Hence if one were able to untangle the uncalibrated camera pose from a fundamental matrix $F = \widehat{T'}KRK^{-1}$, the result would, in general, be given by

$$F = \widehat{T'}KRK^{-1} \quad \mapsto \quad \Pi = [KRK^{-1} + T'v^T, \ v_4 T'] \tag{6.16}$$

for some value of[3] $v = [v_1, v_2, v_3] \in \mathbb{R}^3$, $v_4 \in \mathbb{R}$. Note that this four-parameter family of ambiguous decompositions was not present in the calibrated case, since the matrix $R'$ there had to be a rotation. This ambiguity will play an essential role in the stratified reconstruction approach to be introduced in Section 6.4. In particular, in Section 6.4.2 we will show how to extract a "canonical" choice of $\Pi$ from a fundamental matrix $F$.

## 6.3 Ambiguities and constraints in image formation

The geometric model of image formation (6.1) is composed of three matrix products, involving the unknown extrinsic parameters $g$, the projection matrix $\Pi_0$, and the unknown intrinsic parameter matrix $K$. Each multiplication of unknowns conceals a potential ambiguity, since the product can be interleaved by multiplication by an invertible matrix and its inverse. For instance, if $M = BC$ is an equality between matrices of suitable dimensions, then $M = BC = (BH^{-1})(HC) \doteq \tilde{B}\tilde{C}$ for any invertible matrix $H$, and therefore, one cannot distinguish the pair $(B, C)$ from $(\tilde{B}, \tilde{C})$ from measurements of $M$. However, if the factors are constrained to have a particular structure, then one may be able to recover the "true" unknowns $B$ and $C$. For instance, if we require that $B$ be upper triangular and $C$ be

---

[2]Here we again assume without loss of generality that the first camera frame is chosen as the reference, i.e. $\Pi_1 = [I, 0]$.

[3]The presence of $v_4$ is necessary because the true scale of the vector $T'$ in $F = \widehat{T'}KRK^{-1}$ is unknown. When we compute a normalized left zero eigenvector of $F$, we often choose $\|v_4 T'\| = 1$.

a rotation matrix, then $B$ and $C$ can be recovered uniquely from $M$ (via the QR decomposition; see Appendix A).

In the specific case of image formation, the product involves four matrices $(K, \Pi_0, g, \boldsymbol{X})$,

$$\lambda \boldsymbol{x}' = \Pi \boldsymbol{X} = K \Pi_0 g \boldsymbol{X}. \tag{6.17}$$

If we write each potential ambiguity explicitly, we have

$$\lambda \boldsymbol{x}' = \Pi \boldsymbol{X} = K \Pi_0 g \boldsymbol{X} = \underbrace{K R_0^{-1} R_0 \Pi_0 H^{-1}}_{\tilde{\Pi}} \underbrace{H g g_w^{-1} g_w \boldsymbol{X}}_{\tilde{\boldsymbol{X}}} \tag{6.18}$$

for some $R_0 \in GL(3)$, $H \in GL(4)$, and $g_w \in SE(3)$. Here $H$ is a general linear transformation of the homogeneous coordinates $\boldsymbol{X}$; such a matrix, which is introduced in Appendix A, is called a *projective matrix*, or *homography*.[4]

In this section we will show that ambiguities in $R_0$ and $g_w$ carry no consequence, in the sense that they can be fixed by the user with an arbitrary choice of Euclidean coordinate frames. On the other hand, the ambiguity in the projection matrix $\Pi$, caused by the matrix $H$, leads to a "distortion" of the space where $\boldsymbol{X}$ lives, and hence of the reconstruction. The process of "rectifying" the space is equivalent to identifying the metric of the space (Section 6.1), which in turn is equivalent to calibrating the camera.

### *6.3.1 Structure of the intrinsic parameter matrix*

In Chapter 3 we have derived the calibration matrix $K$ from a very simple geometric model of image formation, and we have concluded that $K$ is upper triangular and given by equation (6.2). However, in principle, $K$ could be an arbitrary invertible matrix (that transforms the Euclidean space to the "distorted" space). In this section we show that there is no loss of generality in assuming that $K$ is upper triangular, and therefore we give a geometric justification for this choice, which also highlights the structure of the ambiguity.

So, let us assume that $K$ is a general invertible $3 \times 3$ matrix. Since it is subject to an arbitrary scalar factor, we can normalize it by imposing that its determinant be $+1$. Invertible matrices with determinant $+1$ constitute the special linear group, $SL(3)$. Therefore, we will assume that $K \in SL(3)$. If we consider equation (6.1),

$$\lambda \boldsymbol{x}' = K \Pi_0 g \boldsymbol{X} = K R_0^{-1} R_0 [R, \; T] \boldsymbol{X} \doteq \tilde{K} [\tilde{R}, \tilde{T}] \boldsymbol{X} = \tilde{K} \Pi_0 \tilde{g} \boldsymbol{X}, \tag{6.19}$$

it is clear that for $\tilde{R} \doteq R_0 R$ to be a rotation matrix, we must have $R_0 \in SO(3)$ and $R_0^{-1} = R_0^T$. Therefore, we conclude that from measurements of $\boldsymbol{x}'$ we cannot distinguish $K$ from $\tilde{K} = K R_0^T$ and $g = [R, T]$ from $\tilde{g} = [R_0 R, R_0 T]$ (see

[4]Notice that $H$ here is a $4 \times 4$ nonsingular matrix. It is in fact an element in the general linear group $GL(4)$, also called a 3-D homography; hence the letter "$H$" is used to represent it. In Chapter 5, for the case of planar scenes, we have encountered a 2-D homography matrix $H$ that is a $3 \times 3$ matrix. This should generate no confusion, since the dimension of $H$ will always be clear from the context.

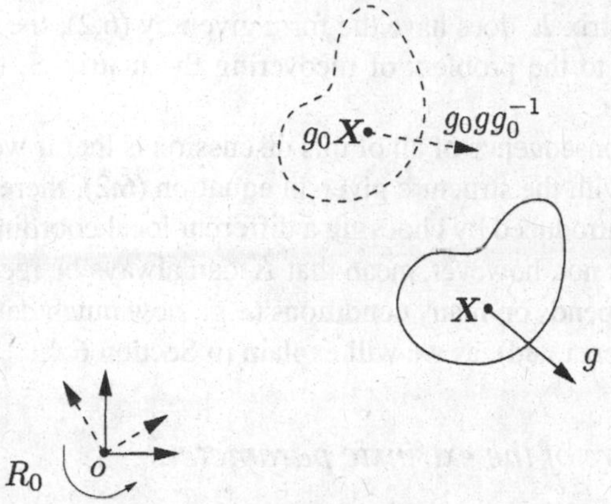


Figure 6.8. The effect of $R_0$ is best viewed by fixing the camera and assuming that the world is moving (relative to the camera). For the choice of $\tilde{K} = KR_0^T$, the world, including the moving object, is simply rotated by $R_0$ around the camera center $o$, a conjugation of a rigid-body motion $g_0 = (R_0, 0)$ in terms of the 3-D coordinates. It is thus essentially impossible to distinguish the two sets of images obtained by a camera with calibration matrix $K$ and one with $\tilde{K} = KR_0^T$.

Figure 6.8). We now show that invertible matrices with determinant $+1$, defined up to an arbitrary rotation, are an equivalence class, and that we can choose as a representative element for each equivalence class an upper triangular matrix.

In fact, let us consider an arbitrary matrix $K \in SL(3)$. The matrix $K$ has a QR decomposition $K = QR$, where $Q$ is upper triangular and $R$ is a rotation matrix (see Appendix A). Therefore, the set of nonsingular matrices $K$ defined up to an arbitrary rotation $R$ is equivalent to the set of upper-triangular matrices with unit determinant.

To summarize this discussion, without knowing the camera motion and scene structure, the camera calibration matrix $K \in SL(3)$ can be recovered only up to an equivalence class[5] $\tilde{K} \in SL(3)/SO(3)$. Notice that if $\tilde{K} = KR_0^T$, then $\tilde{K}^{-T}\tilde{K}^{-1} = K^{-T}K^{-1}$. That is, $\tilde{K}$ and $K$ induce the same inner product on the uncalibrated space, which should be comforting in light of the claim we made in Section 6.1 that recovering the calibration matrix $K$ is equivalent to recovering the metric $S = K^{-T}K^{-1}$, and therefore we would not want a different choice for $K$ to alter the metric.

To complete the picture, we note that the equation $S = K^{-T}K^{-1}$ gives a finite-to-one correspondence between upper-triangular matrices and the set of all $3 \times 3$ symmetric matrices with determinant $+1$ according to the *Cholesky factorization* of the matrix $S$ (Appendix A). Usually, only one of the upper triangular matrices corresponding to a given $S$ has the physical interpretation as the intrinsic parameters of a camera (e.g., all its diagonal entries need to be positive). Thus, if

[5]If more than two views are considered, the derivation is more involved, but the structure of the ambiguity remains the same.

the calibration matrix $K$ does have the form given by (6.2), the calibration problem is equivalent to the problem of recovering the matrix $S$, the metric of the uncalibrated space.

The practical consequence of all of this discussion is that if we restrict $K$ to be upper triangular with the structure given in equation (6.2), there is essentially no extra ambiguity introduced by choosing a different local coordinate frame for the camera. This does not, however, mean that $K$ can always be recovered! Whether this is possible depends on many conditions (e.g., how much data $\boldsymbol{x}'$ is available, how it has been generated), as we will explain in Section 6.4.

### 6.3.2 *Structure of the extrinsic parameters*

The coordinates $\boldsymbol{X}$ are expressed relative to *some* reference frame, which we call the "world" reference frame; $g$ transforms them to the camera coordinates via $g\boldsymbol{X}$. However, changing the world reference frame, and modifying the transformation to the camera frame accordingly, has no effect on the image points:

$$g\boldsymbol{X} = (gg_w^{-1})(g_w\boldsymbol{X}) \tag{6.20}$$

for any $g_w \in SE(3)$. Since the world reference frame is arbitrary, we can choose it at will. A common choice is to have the world reference coincide with one of the cameras, so that $gg_w^{-1} = I$, the identity transformation. In any case, the recovered $g_w\boldsymbol{X}$ will differ from the original coordinates $\boldsymbol{X}$ by a Euclidean transformation $g_w$, as illustrated in Figure 6.9.

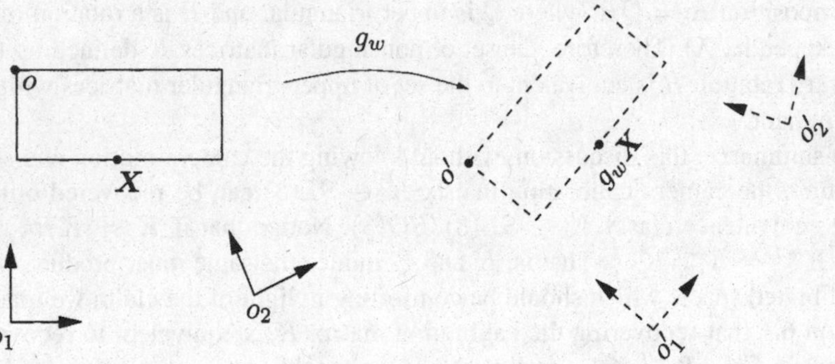


Figure 6.9. The effect of $g_w$ is as if the world, including the object and the camera, were transformed (i.e. rotated and translated) by a Euclidean transformation. It is impossible to distinguish the two sets of images obtained under these two scenarios.

### 6.3.3 *Structure of the projection matrix*

If we collect the intrinsic and extrinsic parameters into the projection matrix $\Pi = [KR, KT]$, we can rewrite the image formation process as

$$\lambda\boldsymbol{x}' = \Pi\boldsymbol{X} = (\Pi H^{-1})(H\boldsymbol{X}) \doteq \tilde{\Pi}\tilde{\boldsymbol{X}} \tag{6.21}$$

for any nonsingular $4 \times 4$ matrix $H$. Therefore, one cannot distinguish the camera $\Pi$ imaging the real world $\boldsymbol{X}$ from the camera $\tilde{\Pi}$ imaging a "distorted" world $\tilde{\boldsymbol{X}}$.

This statement, however, is not entirely accurate, because the matrix $H$, in this special context, cannot be arbitrary. We know at least that for it to correspond to a valid imaging process, $\tilde{\Pi}H$ must have the same structure as $\Pi$, namely, its left $3 \times 3$ block has to be the product of an upper-triangular matrix and a rotation. As we will see in Section 6.4, this structure will provide useful constraints to infer the intrinsic parameters $K$. If we further assume that the first camera frame is the reference, the ambiguity associated with $g_w$ will be resolved, and the choice will yield additional constraints on the form of $H$.

## 6.4 Stratified reconstruction

Reconstructing the full camera calibration and the Euclidean representation of the 3-D scene from uncalibrated views is a complex process, whose success relies on the solution of difficult nonlinear equations as well as on other factors over which the user often has little control. As described in the previous section, the ambiguities associated with $K$ and $g_w$ are rather harmless and can be fixed easily. In the following we will focus our attention on the ambiguity characterized by the transformation $H$ and discuss a strategy on how to resolve it. It turns out that even in the presence of the ambiguity $H$ it is relatively simple to obtain "some" 3-D reconstruction, not the true (Euclidean) reconstruction, but one that differs from it by a geometric transformation. The richer the class of transformations, the easier the reconstruction, the larger the ambiguity. This motivates dividing the approach into steps, where the reconstruction is first computed up to a general transformation, and then successively refined until one possibly obtains a Euclidean reconstruction. This procedure is appealing because in some applications a full-fledged Euclidean reconstruction is not necessary, for instance in visual servoing or in image-based rendering, and therefore one can forgo the later and more complex stages of reconstruction. If, however, Euclidean reconstruction is the final goal, one should be aware that computing it in one fell swoop gives better results. We will outline such a technique in Section 6.4.5.

We start by describing the hierarchy of transformations from projective, to affine to Euclidean, and then illustrate how to compute the reconstruction at each level of the hierarchy (see Appendix A for details about Euclidean, affine, and projective transformations). This process is illustrated in Figure 6.10.

### 6.4.1 *Geometric stratification*

Let us go back to equation (6.1), written for two views:

$$\lambda_i \boldsymbol{x}'_i = K_i \Pi_0 g_{ie} \boldsymbol{X}_e, \quad i = 1, 2. \tag{6.22}$$

To explicitly distinguish the Euclidean structure $\boldsymbol{X}_e$ from structures to be defined for other stages in the stratification hierarchy, we use the subscript "$e$" to indicate

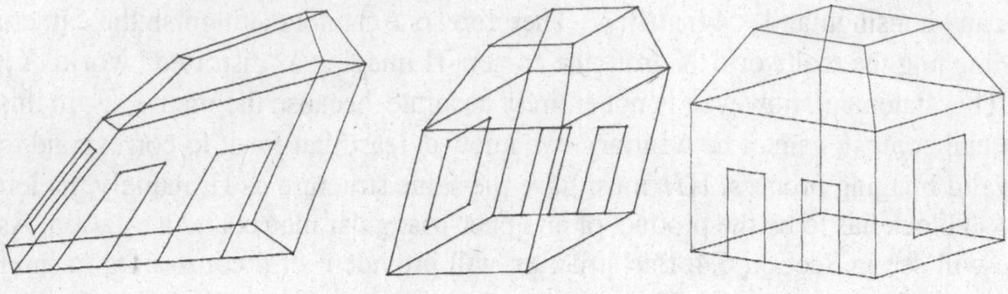

Figure 6.10. Illustration of the stratified approach: projective structure $\boldsymbol{X}_p$, affine structure $\boldsymbol{X}_a$, and Euclidean structure $\boldsymbol{X}_e$ obtained in different stages of reconstruction.

"Euclidean"; $g_{ie}, i = 1, 2$ denotes the (Euclidean) pose of the $i$th camera frame relative to the world reference frame. For generality, we allow the calibration matrix $K$ to be different in each view, so we have $K_1$ and $K_2$. The corresponding projection matrix then is

$$\Pi_{ie} \doteq K_i \Pi_0 g_{ie}, \quad i = 1, 2.$$

From Section 6.3.2 we know that the choice of the world reference frame is arbitrary. Therefore, we choose it in a way that allows $g_{ie}$ to acquire a particularly simple form in one of the views. For instance, we can require that $g_{1e}$ be the identity $g_{1e} = (I, 0)$, so that equation (6.22) for the first camera becomes

$$\lambda_1 \boldsymbol{x}_1' = K_1 \Pi_0 \boldsymbol{X}_e. \tag{6.23}$$

This corresponds to choosing the world frame to coincide with the Euclidean reference frame whose origin is the same as the first camera center. In particular, $\Pi_{1e} = K_1[I, 0]$.

Even after we do so, however, there are still ambiguities in the previous equation. In fact, as we have seen in Section 6.3.3, one can always choose an arbitrary $4 \times 4$ invertible matrix $H \in GL(4)$ such that

$$K_1 \Pi_0 \boldsymbol{X}_e = K_1 \Pi_0 H^{-1} H \boldsymbol{X}_e = \Pi_{1p} \boldsymbol{X}_p, \tag{6.24}$$

and similarly for the second camera, where

$$\Pi_{ip} \doteq \Pi_{ie} H^{-1} \quad \text{and} \quad \boldsymbol{X}_p \doteq H \boldsymbol{X}_e, \quad i = 1, 2. \tag{6.25}$$

Since $H$ indicates an arbitrary linear transformation of the homogeneous coordinates, called a projective transformation, we use the subscript "$p$" to indicate that the reconstruction is up to a "projective" transformation.

Again, among all possible $H$, we choose one that makes the projection model simple. For instance, we can require that $\Pi_{1p}$ be equal to $\Pi_0 = [I, \ 0]$, which corresponds to setting the projective reference frame to the first view. Note that unless $K_1 = I$, in general we cannot find a Euclidean transformation $g_{1e}$ to make $\Pi_{1e}$ equal to the standard camera. However, we can always find a general invertible matrix $H \in GL(4)$ that achieves this goal.

If we decompose $H^{-1}$ into blocks $H^{-1} = \begin{bmatrix} G & b \\ v^T & v_4 \end{bmatrix}$, where $G \in \mathbb{R}^{3\times3}$, $v, b \in \mathbb{R}^3$, and $v_4 \in \mathbb{R}$, then

$$\Pi_{1p} = \Pi_{1e} H^{-1} = K_1[I,\ 0]H^{-1} = [I,\ 0]$$

yields $G = K_1^{-1}$ and $b = 0$. Therefore, we have eliminated some of the degrees of freedom in $H$, and we are left with an arbitrary transformation of the general form

$$H^{-1} = \begin{bmatrix} K_1^{-1} & 0 \\ v^T & v_4 \end{bmatrix} \in \mathbb{R}^{4\times4}. \tag{6.26}$$

For readers familiar with the terminology of projective geometry, $H$ corresponds to a choice of projective basis in the first view. The matrix $H^{-1}$ can be decomposed into the product of two invertible matrices,

$$H^{-1} = \begin{bmatrix} K_1^{-1} & 0 \\ v^T & v_4 \end{bmatrix} = \begin{bmatrix} K_1^{-1} & 0 \\ 0 & 1 \end{bmatrix} \begin{bmatrix} I & 0 \\ v^T & v_4 \end{bmatrix} \doteq H_a^{-1} H_p^{-1}, \tag{6.27}$$

where the first matrix on the right-hand side is in fact an *affine* transformation of the coordinates (since its last row is $[0,\ 0,\ 0,\ 1]$), while the second matrix still represents a projective transformation.

With these two choices of reference frames, the image formation model for two views becomes

$$\begin{aligned} \lambda_1 \boldsymbol{x}_1' &= \Pi_0 \boldsymbol{X}_p, \\ \lambda_2 \boldsymbol{x}_2' &= \Pi_{2p} \boldsymbol{X}_p = K_2 \Pi_0 g_{2e} H_a^{-1} H_p^{-1} \boldsymbol{X}_p, \end{aligned}$$

where

$$\boldsymbol{X}_p = H_p \overbrace{H_a \underbrace{g_e \boldsymbol{X}}_{\boldsymbol{X}_e}}^{\boldsymbol{X}_a}. \tag{6.28}$$

Here, $\boldsymbol{X}$ represents the "true" 3-D structure, $\boldsymbol{X}_e$ differs from it by a Euclidean transformation (a rigid motion) $g_e = (R_e, T_e)$, $\boldsymbol{X}_a$ differs from it by a more general affine transformation $H_a g_e$, and finally, $\boldsymbol{X}_p$ differs from it by a general linear (projective) transformation $H_p H_a g_e$. Correspondingly, for $i = 1, 2$ we define projection matrices associated with

- a Euclidean (calibrated) camera: $\Pi_{ie} \doteq K_i \Pi_0 g_{ie}$,
- an affine (weakly calibrated) camera: $\Pi_{ia} \doteq K_i \Pi_0 g_{ie} H_a^{-1}$,
- a projective (uncalibrated) camera: $\Pi_{ip} \doteq K_i \Pi_0 g_{ie} H_a^{-1} H_p^{-1}$.

This is summarized in Table 6.1, where for simplicity, we have assumed that the camera parameters are not changing; i.e. $K_1 = K_2 = K$ and $g_{2e} = (R, T)$. Notice that in the table, the projective camera $\Pi_{2p}$ is of the same form as obtained from the decomposition of a fundamental matrix in equation (6.16). Also, it is

| | Camera projection | 3-D structure |
|---|---|---|
| Euclid. | $\Pi_{1e} = [K, 0],\ \Pi_{2e} = [KR,\ KT]$ | $\boldsymbol{X}_e = g_e \boldsymbol{X} = \begin{bmatrix} R_e & T_e \\ 0 & 1 \end{bmatrix} \boldsymbol{X}$ |
| Affine | $\Pi_{2a} = [KRK^{-1},\ KT]$ | $\boldsymbol{X}_a = H_a \boldsymbol{X}_e = \begin{bmatrix} K & 0 \\ 0 & 1 \end{bmatrix} \boldsymbol{X}_e$ |
| Project. | $\Pi_{2p} = [KRK^{-1} + KTv^T, v_4 KT]$ | $\boldsymbol{X}_p = H_p \boldsymbol{X}_a = \begin{bmatrix} I & 0 \\ -v^T v_4^{-1} & v_4^{-1} \end{bmatrix} \boldsymbol{X}_a$ |

Table 6.1. Relationships between three types of reconstruction (assuming $K_1 = K_2 = K$). For both affine and projective cases, the projection matrix with respect to the first view is the same $\Pi_{1a} = \Pi_{1p} = [I, 0]$, due to choice of the (projective) coordinate frame.

useful to notice that both $H_p$ and $H_p^{-1}$ are of the same form: their first three rows are $[I,\ 0]$.

In the next three subsections we will show how to obtain a projective reconstruction, and how to update that first to affine, and finally to a full-fledged Euclidean reconstruction. Since this chapter covers only the two-view case, we do not give an exhaustive account of all the possible techniques for reconstruction. Instead, we report a few examples here, and refer the reader to Part III, where we treat the more general multiple-view case.

### 6.4.2 *Projective reconstruction*

In this section we address the problem of reconstructing the projective structure $\boldsymbol{X}_p$ and the projective cameras $\Pi_{ip}$ given a number of point correspondences between two views $\{(\boldsymbol{x}_1', \boldsymbol{x}_2')\}$. The rationale is simple: from point correspondences we retrieve the fundamental matrix $F$, from the fundamental matrix $F$ we retrieve the projective camera $\Pi_{2p}$ (since $\Pi_{1p} = [I,\ 0]$ already), and finally, we triangulate to obtain the projective structure $\boldsymbol{X}_p$.

From the properties studied in Section 6.2.2, the decomposition of a fundamental matrix into the calibration and camera pose is not unique. Indeed, it corresponds to a particular choice of the free parameters $v$ and $v_4$ (6.16). Fortunately, these are the only possible ambiguities. The following theorem guarantees that all projection matrices $\Pi_{2p}$ that give rise to the same fundamental matrix must be related by a transformation of the form $H_p$.

**Theorem 6.3 (Projective reconstruction).** $(\Pi_{1p}, \Pi_{2p})$ *and* $(\Pi_{1p}, \tilde{\Pi}_{2p})$ *are two pairs of projection matrices that yield the same fundamental matrix* $F$ *if and only if there exists a nonsingular transformation matrix* $H_p$ *such that* $\tilde{\Pi}_{2p} \sim \Pi_{2p} H_p^{-1}$, *or equivalently,* $\Pi_{2p} \sim \tilde{\Pi}_{2p} H_p$.

*Proof.* Recall that $\Pi_{1p} = [I, 0]$ is fixed by the choice of the projective reference frame. Set $\Pi_{2p} = [C, c]$ with $C \in \mathbb{R}^{3\times 3}, c \in \mathbb{R}^3$, and similarly, set $\tilde{\Pi}_{2p} = [B, b]$.

Since they both give rise to the same fundamental matrix (which is defined up to a scalar factor), we have

$$\widehat{c}C \sim \widehat{b}B. \tag{6.29}$$

Since $c$ and $b$ span the left null spaces of the left-hand side and the right-hand side respectively, they are linearly dependent: $c \sim b$. Consequently, we have $C \sim B + bv^T$ for some $v \in \mathbb{R}^3$. Hence we have

$$[C, c] \sim [B, b] \begin{bmatrix} I & 0 \\ v^T & v_4 \end{bmatrix} \tag{6.30}$$

and $\Pi_{2p} \sim \tilde{\Pi}_{2p} H_p$ for an $H_p$ of the form defined above. □

The next step will be to choose a particular set of cameras for a given fundamental matrix. Finally, we will be able to recover the projective structure.

*Canonical decomposition of the fundamental matrix*

The previous discussion shows that there is a four-parameter family of transformations $H_p$ parameterized by $v = [v_1, v_2, v_3]^T$ and $v_4$ that gives rise to exactly the same relationship among pairs of views. While any choice of $H_p$ will result in the same fundamental matrix, different choices of $H_p$ will result in different reconstructions of the 3-D coordinates $\boldsymbol{X}_p$, which presents a problem. In other words, while the map from two projection matrices to a fundamental matrix is determined by

$$\boxed{\Pi_{1p} = [I,\ 0],\ \Pi_{2p} = [B,\ b]} \quad \mapsto \quad F = \widehat{b}B, \tag{6.31}$$

the inverse map from a fundamental matrix to two projection matrices is one-to-many, because of the freedom in the choice of $H_p$, as we have seen in the proof of Theorem 6.3.

One way to resolve this problem is to fix a particular choice of $H_p$, so that each $F$ maps to only one particular pair $(\Pi_{1p}, \Pi_{2p})$. For this choice to be called *canonical*, $(\Pi_{1p}, \Pi_{2p})$ must depend only on $F$ and not on $(v_1, v_2, v_3, v_4)$. As we have seen in Exercise 5.3, if $\|T'\| = 1$, we have the identity $\widehat{T'}\widehat{T'}^T\widehat{T'} = \widehat{T'}$. Therefore, the following choice of projection matrices $(\Pi_{1p}, \Pi_{2p})$ results in the fundamental matrix

$$F \quad \mapsto \quad \boxed{\Pi_{1p} = [I,\ 0],\ \Pi_{2p} = [(\widehat{T'})^T F,\ T']} \tag{6.32}$$

for $(T')^T F = 0$ and $\|T'\| = 1$. Note that the projection matrices defined above depend only on $F$, since the vector $T'$, the epipole in the second view, can be directly computed as the left null space of $F$. Hence, this decomposition can be obtained directly from the fundamental matrix. This is called the *canonical decomposition* of a fundamental matrix into two camera projection matrices.

**Example 6.4 (A numerical example).** Suppose that

$$K = \begin{bmatrix} 500 & 0 & 250 \\ 0 & 500 & 250 \\ 0 & 0 & 1 \end{bmatrix}, \quad R = \begin{bmatrix} 1 & 0 & 0 \\ 0 & 1 & 0 \\ 0 & 0 & 1 \end{bmatrix}, \quad T = \begin{bmatrix} 0 \\ 1 \\ 2 \end{bmatrix}.$$

Then the fundamental matrix is $F = \widehat{KT}$, and the (normalized) epipole is $T' = KT/\|KT\|$,

$$F = \begin{bmatrix} 0 & -2 & 1000 \\ 2 & 0 & -500 \\ -1000 & 500 & 0 \end{bmatrix}, \quad T' = \begin{bmatrix} 0.4472 \\ 0.8944 \\ 0.0018 \end{bmatrix}.$$

The canonical decomposition of $F$ gives $\Pi_{1p} = [I,\ 0]$, and

$$\Pi_{2p} = [(\widehat{T'})^T F,\ T'] = \begin{bmatrix} 894.4 & -447.2 & -0.9 & 0.4472 \\ -447.2 & 223.6 & -1.8 & 0.8944 \\ -0.9 & -1.8 & 1118.0 & 0.0018 \end{bmatrix}.$$

By direct computation, we can check that this pair of projection matrices gives the same fundamental matrix

$$\widehat{T'}(\widehat{T'})^T F = \begin{bmatrix} 0 & -2 & 1000 \\ 2 & 0 & -500 \\ -1000 & 500 & 0 \end{bmatrix} = F.$$

However, we know that another pair of projection matrices $\Pi_{1a} = [I,\ 0]$ and

$$\Pi_{2a} = [KRK^{-1},\ KT] = \begin{bmatrix} 1 & 0 & 0 & 500 \\ 0 & 1 & 0 & 1000 \\ 0 & 0 & 1 & 2 \end{bmatrix}$$

also gives the same fundamental matrix $F$. Therefore, we should have $\Pi_{ip} \sim \Pi_{ia} H_p^{-1}, i = 1, 2$, for some $H_p \in \mathbb{R}^{4\times 4}$. One may easily verify that indeed,

$$\begin{bmatrix} 894.4 & -447.2 & -0.9 & 0.4472 \\ -447.2 & 223.6 & -1.8 & 0.8944 \\ -0.9 & -1.8 & 1118.0 & 0.0018 \end{bmatrix} \sim \begin{bmatrix} 1 & 0 & 0 & 500 \\ 0 & 1 & 0 & 1000 \\ 0 & 0 & 1 & 2 \end{bmatrix} \begin{bmatrix} I & 0 \\ -\frac{(T')^T}{\|KT\|} & \frac{1}{\|KT\|^2} \end{bmatrix},$$

since the right-hand side multiplied by $\|KT\| = 1118$ is exactly the left-hand side. From this example, we notice that numerically, the entries of $H_p^{-1}$ are rather unbalanced, differing by orders of magnitude! Therefore, numerical accuracy may have a significant effect in computing the uncalibrated epipolar geometry and the subsequent reconstruction. In practice, this could severely degrade the quality of reconstruction. We will show in Appendix 6.A how to resolve the numerical issue by properly normalizing the image coordinates. ■

Since the first three columns $(\widehat{T'})^T F$ of $\Pi_{2p}$ in the canonical decomposition form a singular matrix, this choice will always differ from $KRK^{-1}$ in the affine $\Pi_{2a}$, which we know to be nonsingular. In practice, if partial knowledge of $(v_1, v_2, v_3, v_4)$ or the calibration $K$ is available, one should not choose the canonical decomposition, since a better decomposition may become available (see Exercise 6.10).

*Projective reconstruction with respect to the canonical decomposition*

The canonical projection matrices $(\Pi_{1p}, \Pi_{2p})$ now relate the given pair of uncalibrated images $\boldsymbol{x}_1'$ and $\boldsymbol{x}_2'$ to the unknown structure $\boldsymbol{X}_p$, which is related to the true Euclidean structure by a projective transformation:

$$\begin{aligned} \lambda_1 \boldsymbol{x}_1' &= \Pi_{1p}\boldsymbol{X}_p &= [I, 0]\boldsymbol{X}_p, \\ \lambda_2 \boldsymbol{x}_2' &= \Pi_{2p}\boldsymbol{X}_p &= [(\widehat{T'})^T F, T']\boldsymbol{X}_p. \end{aligned} \tag{6.33}$$

This equation is linear in all the unknowns. Therefore, it enables us to recover $\boldsymbol{X}_p$ in a straightforward manner. Eliminating the unknown scales $\lambda_1$ and $\lambda_2$ by multiplying both sides of equation by $\widehat{\boldsymbol{x}_1'}$ and $\widehat{\boldsymbol{x}_2'}$ in (6.33) we obtain $\widehat{\boldsymbol{x}_i'}\Pi_{ip}\boldsymbol{X}_p = 0, i = 1, 2$, which gives three scalar equations per image, only two of which are linearly independent. Thus, two corresponding image points of the form $\boldsymbol{x}_1' = [x_1, y_1, 1]^T$ and $\boldsymbol{x}_2' = [x_2, y_2, 1]^T$ yield four linearly independent constraints on $\boldsymbol{X}_p$:

$$\begin{aligned} (x_1\pi_1^{3T})\boldsymbol{X}_p = \pi_1^{1T}\boldsymbol{X}_p, \qquad (y_1\pi_1^{3T})\boldsymbol{X}_p = \pi_1^{2T}\boldsymbol{X}_p, \\ (x_2\pi_2^{3T})\boldsymbol{X}_p = \pi_2^{1T}\boldsymbol{X}_p, \qquad (y_2\pi_2^{3T})\boldsymbol{X}_p = \pi_2^{2T}\boldsymbol{X}_p, \end{aligned}$$

where $\Pi_{1p} = [\pi_1^1, \pi_1^2, \pi_1^3]^T$ and $\Pi_{2p} = [\pi_2^1, \pi_2^2, \pi_2^3]^T$ are the projection matrices written in terms of their three row vectors. The projective structure can then be recovered as the least-squares solution of a system of linear equations

$$M\boldsymbol{X}_p = 0.$$

This can be done, for instance, using the SVD (Appendix A). As in the calibrated case, this linear reconstruction is suboptimal in the presence of noise.

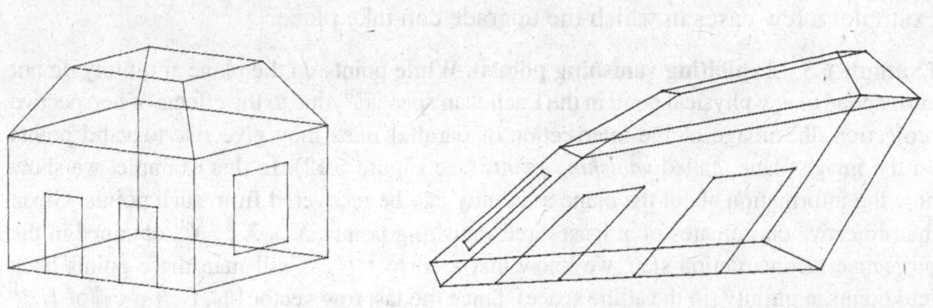

Figure 6.11. The true Euclidean structure $\boldsymbol{X}_e$ of the scene and the projective structure $\boldsymbol{X}_p = H\boldsymbol{X}_e$ obtained by the algorithm described above.

In order to "upgrade" the projective structure $\boldsymbol{X}_p$ to the Euclidean structure $\boldsymbol{X}_e$, we have to recover the unknown transformation $H^{-1} = H_a^{-1}H_p^{-1} \in \mathbb{R}^{4\times 4}$, such that

$$\boldsymbol{X}_e = H_a^{-1}\boldsymbol{X}_a = H_a^{-1}H_p^{-1}\boldsymbol{X}_p. \tag{6.34}$$

This can be done in two steps: first recover the affine structure $\boldsymbol{X}_a$ by identifying the matrix $H_p^{-1}$, and then recover the Euclidean structure $\boldsymbol{X}_e$ by identifying the matrix $H_a^{-1}$ (or equivalently $K$).[6]

### 6.4.3 *Affine reconstruction*

As we outlined in the previous section, in case of general motion and in the absence of any knowledge on the scene or the camera, all we can compute is the fundamental matrix $F$ and the corresponding projective structure $\boldsymbol{X}_p$. In order to upgrade the projective structure to affine, we need to find the transformation

$$H_p^{-1} = \begin{bmatrix} I & 0 \\ v^T & v_4 \end{bmatrix} \in \mathbb{R}^{4\times 4} \tag{6.35}$$

such that $\boldsymbol{X}_a = H_p^{-1}\boldsymbol{X}_p$ and $\Pi_{ia} = \Pi_{ip}H_p, i = 1, 2$. Note that $H_p^{-1}$ has the special property of mapping all points $\boldsymbol{X}_p$ that satisfy the equation $[v^T, v_4]\boldsymbol{X}_p = 0$ to points with homogeneous coordinates $\boldsymbol{X}_a = [X, Y, Z, 0]^T$, with the last coordinate equal to 0. Such points in the affine space correspond to points infinitely far away from the center of the chosen world coordinate frame (for the affine space). The locus of such points $\{\boldsymbol{X}_p\}$ is called the *plane at infinity* in the projective space, denoted by $P_\infty$ (see Appendix 3.B). It is characterized by the vector

$$\pi_\infty^T \doteq [v^T, v_4] = [v_1, v_2, v_3, v_4] \quad \in \mathbb{R}^4$$

as $\boldsymbol{X}_p \in P_\infty$ if and only if $\pi_\infty^T\boldsymbol{X}_p = 0$. Note that such $\pi_\infty^T$ is defined only up to an arbitrary scalar factor.[7] Hence, what really matters is the ratio between $v$ and $v_4$. A standard choice for $\pi_\infty^T$ is simply $[v^T/v_4, 1]$ when $v_4 \neq 0$.[8] In any case, finding this plane is equivalent to finding the vector $v$ and $v_4$, and hence equivalent to finding the projective-to-affine upgrade matrix $H_p^{-1}$ (or $H_p$). Below we give as examples a few cases in which the upgrade can take place.

**Example 6.5 (Exploiting vanishing points).** While points on the plane at infinity do not correspond to any physical point in the Euclidean space $\mathbb{E}^3$, due to the effects of perspective projection, the image of the intersection of parallel lines may give rise to valid points on the image plane, called *vanishing points* (see Figure 6.12). In this example, we show how the information about the plane at infinity can be recovered from such points. Given the projective coordinates of at least three vanishing points $\boldsymbol{X}_p^1, \boldsymbol{X}_p^2, \boldsymbol{X}_p^3$ obtained in the projective reconstruction step, we know that a correct $H_p^{-1}$ will map these points back into points at infinity (in the affine space). Since the last row vector $[v_1, v_2, v_3, v_4]$ of $H_p^{-1}$ corresponds to a plane in the projective space where all points at infinity lie, the vanishing

---

[6]The reader should be aware that this separation is only for convenience: if the matrix $H = H_pH_a$ or $K$ can be estimated directly, as we will see in some methods given later, these two steps can be bypassed.

[7]Any other vector differing from $\pi_\infty^T$ by a nonzero scalar factor would define exactly the same plane as $\pi_\infty^T$.

[8]In our applications, $v_4$ is never zero, since the world origin is the first camera center, which is always a physical point at finite distance.

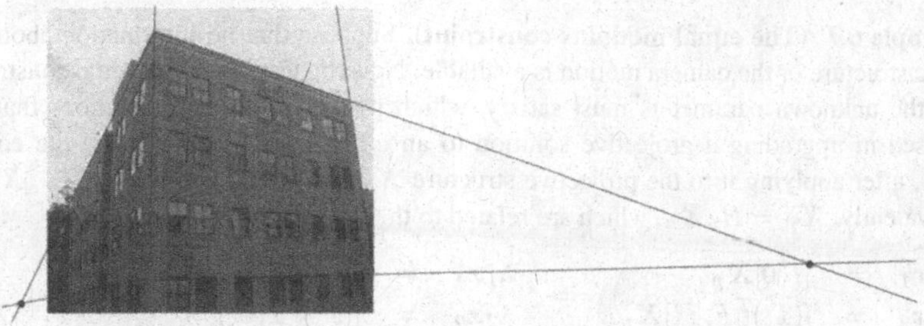

Figure 6.12. An example of parallel lines and the associated vanishing points.

points must satisfy

$$[v_1, v_2, v_3, v_4]\boldsymbol{X}_p^j = 0, \quad j = 1, 2, 3.$$

Then three vanishing points in general are sufficient to determine the vector $[v_1, v_2, v_3, v_4]$ (up to scale). Once this vector has been determined, the projective upgrade $H_p^{-1}$ is defined, and the projective structure $\boldsymbol{X}_p$ can be upgraded to the affine structure $\boldsymbol{X}_a$,

$$\boldsymbol{X}_a = \begin{bmatrix} I & 0 \\ v^T & v_4 \end{bmatrix} \boldsymbol{X}_p. \tag{6.36}$$

Therefore, the projective upgrade can be computed when images of at least three vanishing points are available. ■

**Example 6.6 (Direct affine reconstruction from pure translation).** For simplicity, let us assume $K_1 = K_2 = K$ in this example. Then two views related by a pure translation can be described as

$$\lambda_2 K \boldsymbol{x}_2 = \lambda_1 K \boldsymbol{x}_1 + KT,$$

or in pixel coordinates

$$\lambda_2 \boldsymbol{x}_2' = \lambda_1 \boldsymbol{x}_1' + T', \tag{6.37}$$

where $T' = KT$ and $\boldsymbol{x}' = K\boldsymbol{x}$. We notice that the vector $T'$ can be computed directly from the epipolar constraint

$$\boldsymbol{x}_2'^T \widehat{T'} \boldsymbol{x}_1' = 0.$$

Once $T'$ is computed, the unknown scales can be computed directly from equation (6.37) as

$$\lambda_1 = -\frac{(\widehat{\boldsymbol{x}_2'}\boldsymbol{x}_1')^T \widehat{\boldsymbol{x}_2'} T'}{\|\widehat{\boldsymbol{x}_2'}\boldsymbol{x}_1'\|^2}, \quad \lambda_2 = \frac{(\widehat{\boldsymbol{x}_1'}\boldsymbol{x}_2')^T \widehat{\boldsymbol{x}_1'} T'}{\|\widehat{\boldsymbol{x}_1'}\boldsymbol{x}_2'\|^2}.$$

One can easily verify that the structure $\boldsymbol{X}' = \lambda \boldsymbol{x}'$ obtained in this manner is indeed related to the Euclidean structure $\boldsymbol{X}_e$ by an unknown affine transformation of the form

$$H_a = \begin{bmatrix} K & 0 \\ 0 & 1 \end{bmatrix}. \tag{6.38}$$

Therefore, in the case of pure translation, the step of recovering a projective structure can be easily bypassed, and we may directly obtain an affine reconstruction from two views. ■

**Example 6.7 (The equal modulus constraint).** Suppose that no information about the scene structure or the camera motion is available. Nevertheless, there are some constraints that the unknown parameters must satisfy, which provide additional equations that can be used in upgrading a projective solution to an affine one. Had we found the correct $H_p^{-1}$, after applying it to the projective structure $\boldsymbol{X}_p$, we would get $\boldsymbol{X}_a = H_p^{-1}\boldsymbol{X}_p$ or, equivalently, $\boldsymbol{X}_p = H_p\boldsymbol{X}_a$, which are related to the images $\boldsymbol{x}'_1, \boldsymbol{x}'_2$ by

$$\begin{array}{rcl} \lambda_1\boldsymbol{x}'_1 & = & [I,0]\boldsymbol{X}_p, \\ \lambda_2\boldsymbol{x}'_2 & = & [(\widehat{T'})^TF, T']\boldsymbol{X}_p. \end{array} \quad \Rightarrow \quad \begin{array}{rcl} \lambda_1\boldsymbol{x}'_1 & = & [I,0]\boldsymbol{X}_a, \\ \lambda_2\boldsymbol{x}'_2 & = & [(\widehat{T'})^TF - T'v^Tv_4^{-1}, T'v_4^{-1}]\boldsymbol{X}_a. \end{array}$$

Without loss of generality, we assume $v_4 = 1$ (otherwise, simply divide $v$ by $v_4$). Then the first three columns of the second projection matrix $\Pi_{2a}$ become $(\widehat{T'})^TF - T'v^T$. If we know that $K_1 = K_2 = K$ is constant, the matrix $(\widehat{T'})^TF - T'v^T$ must be related to the matrix $KRK^{-1}$ by

$$(\widehat{T'})^TF - T'v^T \sim KRK^{-1}.$$

Hence the matrix $(\widehat{T'})^TF - T'v^T$ must have eigenvalues of the form

$$\sigma\left((\widehat{T'})^TF - T'v^T\right) = \{\alpha, \alpha e^{i\theta}, \alpha e^{-i\theta}\}, \quad \alpha, \theta \in \mathbb{R},\ i \doteq \sqrt{-1}.$$

Since the eigenvalues have equal modulus, it is easy to verify that coefficients of the characteristic polynomial of $(\widehat{T'})^TF - T'v^T$,

$$\det\left(\lambda I - ((\widehat{T'})^TF - T'v^T)\right) = \lambda^3 + a_2\lambda^2 + a_1\lambda + a_0, \tag{6.39}$$

must satisfy

$$a_0a_2^3 = a_1^3. \tag{6.40}$$

Note that $(a_1, a_2, a_3)$ are all linear functions of the entries of the unknown vector $v$. Showing this is left as an exercise to the reader (see Exercise 6.13). This equation is called the *modulus constraint*. Note that this constraint depends only on the images, and hence it is intrinsic. But this constraint gives only one (fourth order polynomial) equation in the three unknowns $v_1, v_2, v_3$ in $v$. So in general, from only two images it is impossible to uniquely determine $v$ and obtain an affine reconstruction, unless more images are available or some extra information about the camera calibration is known. For instance, one can assume that the optical center is known and that the skew factor is zero. In Section 6.4.5 and Chapter 8, we will introduce additional intrinsic constraints for multiple views and show that in that case, a full reconstruction of the Euclidean structure becomes possible. ■

### 6.4.4 *Euclidean reconstruction*

For simplicity, in this subsection, we assume that $K_1 = K_2 = K$ is constant. The upgrade from the affine structure $\boldsymbol{X}_a$ to Euclidean $\boldsymbol{X}_e$ requires knowledge of the metric $S = K^{-T}K^{-1}$ (or equivalently, $S^{-1} = KK^T$) of the uncalibrated space. Suppose we have two views of an affine structure obtained by one of the methods in the previous section,

$$\begin{array}{rcccl} \lambda_1\boldsymbol{x}'_1 & = & \Pi_{1a}\boldsymbol{X}_a & = & [I,\ 0]\boldsymbol{X}_a, \\ \lambda_2\boldsymbol{x}'_2 & = & \Pi_{2a}\boldsymbol{X}_a & = & [KRK^{-1},\ KT]\boldsymbol{X}_a. \end{array}$$

Hence the first $3 \times 3$ submatrix of the projection matrix $\Pi_{2a}$ must be $C \doteq KRK^{-1} \in \mathbb{R}^{3\times3}$.[9] Since our ultimate goal is to recover $S^{-1}$, then given $C$ we can ask how much information it reveals about $S^{-1}$. It is easy to verify that the relationship between the two is

$$S^{-1} - CS^{-1}C^T = 0. \tag{6.41}$$

This is a special type of equation, called a *Lyapunov equation*. In order for $S^{-1}$ to be a solution of the above matrix equation, it has to be in the symmetric real kernel,[10] denoted by SRker, of the *Lyapunov map*

$$L : \mathbb{C}^{3\times3} \to \mathbb{C}^{3\times3}; \quad X \mapsto X - CXC^T. \tag{6.42}$$

**Lemma 6.8 (Kernel of a Lyapunov map).** *Given a rotation matrix $R$ not of the form $e^{\widehat{u}k\pi}$ for some $k \in \mathbb{Z}$ and some $u \in \mathbb{R}^3$ of unit length, the symmetric real kernel associated with the Lyapunov map $L$, SRker($L$), is two-dimensional. Otherwise, the symmetric real kernel is four-dimensional if $k$ is odd and six-dimensional if $k$ is even.*

The proof of this lemma follows directly from the properties of Lyapunov maps and can be found in Appendix A. Interested readers can prove it as part of Exercises 6.1 and 6.2. As for how many matrices of the form $C \doteq KRK^{-1}$ with $R \in SO(3)$ can uniquely determine the metric $S^{-1}$, we have the following theorem as a result of Lemma 6.8:

**Theorem 6.9 (Calibration from two rotational motions).** *Given two matrices $C_i = KR_iK^{-1} \in KSO(3)K^{-1}, i = 1, 2$, where $R_i = e^{\widehat{u}_i\theta_i}$ with $\|u_i\| = 1$ and $\theta_i$'s not equal to $k\pi, k \in \mathbb{Z}$, then SRker($L_1$) $\cap$ SRker($L_2$) is one-dimensional if and only if $u_1$ and $u_2$ are linearly independent.*

*Proof.* Necessity is straightforward: if two rotation matrices $R_1$ and $R_2$ have the same axis, they have the same eigenvectors hence SRker($L_1$) $=$ SRker($L_2$), where $L_i : X \mapsto X - C_iXC_i^T, i = 1, 2$. We now only need to prove sufficiency. We may assume that $u_1$ and $u_2$ are the two rotation axes of $R_1$ and $R_2$, respectively, and are linearly independent (i.e. distinct). Since the rotation angles of both $R_1$ and $R_2$ are not $k\pi$, both SRker($L_1$) and SRker($L_2$) are two-dimensional according to Lemma 6.8. Since $u_1$ and $u_2$ are linearly independent, the matrices $Ku_1u_1^TK^T$ and $Ku_2u_2^TK^T$ are linearly independent and are in SRker($L_1$) and SRker($L_2$), respectively. Thus SRker($L_1$) is not fully contained in SRker($L_2$) because it already contains both $Ku_2u_2^TK^T$ and $S^{-1} = KK^T$ and could not possibly contain another independent $Ku_1u_1^TK^T$. Hence, the intersection SRker($L_1$) $\cap$ SRker($L_2$) is of dimension at most 1. Then $X = S^{-1}$ is the only solution in the intersection. □

---

[9]The first three columns of $\Pi_{2a}$ might differ by a scalar factor from $KRK^{-1}$, but this is no problem, since we know that the determinant of $KRK^{-1}$ has to be 1.

[10]That is, $S^{-1}$ is a real and symmetric matrix that solves equation (6.41). The definition of the kernel of a linear map can be found in Appendix A.

According to this theorem, we need two matrices of the form $C = KRK^{-1}$ with independent rotations $R$ in order to determine uniquely the matrix $S^{-1} = KK^T$, and hence the camera calibration matrix $K$. Thus, if only a pair of affine projection matrices is given, we can recover the calibration only up to a one-parameter family, corresponding to the two-dimensional kernel of one Lyapunov map. If two pairs of affine projection matrices are given and if the associated rotations are around different axes, we can in general fully calibrate the camera and hence upgrade the affine structure to a Euclidean one.

The previous lemma reveals that information about the camera calibration $K$ is encoded only in the rotational part of the motion and captured in the matrix $C = KRK^{-1}$. This observation can be exploited for the purpose of calibration using purely rotational motions.

**Example 6.10 (Direct calibration from pure rotation).** In case two images are related by a pure rotation, the corresponding image pairs $(\boldsymbol{x}_1', \boldsymbol{x}_2')$ satisfy

$$\lambda_2 \boldsymbol{x}_2' = KRK^{-1} \lambda_1 \boldsymbol{x}_1', \tag{6.43}$$

for some scalars $\lambda_1, \lambda_2 \in \mathbb{R}_+$. Then the two images are related by a homography

$$\boldsymbol{x}_2' \sim KRK^{-1} \boldsymbol{x}_1' \quad \Leftrightarrow \quad \widehat{\boldsymbol{x}_2'} KRK^{-1} \boldsymbol{x}_1' = 0. \tag{6.44}$$

From the last linear equation, in general, four pairs of corresponding image points uniquely determine the matrix $C \doteq KRK^{-1} \in \mathbb{R}^{3\times 3}$. By rotating the camera about a different axis, we obtain another matrix of the same form, and together the camera calibration $K$ can be uniquely determined. This method was proposed by [Hartley, 1994b], and further analysis of this technique can be found in [de Agapito et al., 1999]. ■

### 6.4.5 *Direct stratification from multiple views (preview)*

In general, even if a number of views of the same scene are given, but the calibration of the camera changes in each view, it is impossible to recover the correct (Euclidean) structure of the scene, camera motion, and camera calibration altogether. In the absence of additional information, a projective reconstruction is all we can do.

When the camera does not change between views or only some of the parameters change (as it is often the case), given sufficiently many views taken from sufficiently different vantage points, the Euclidean structure and the camera motion can be recovered together with the camera calibration. Moreover, there exist methods that allow us to directly solve for the matrix $H = H_p H_a$ without going through all the stratification steps. Here we give only a brief preview of such methods, since we will address them again in Chapter 8.

To this end, consider the structure of the projection matrix highlighted in Section 6.4.1:

$$\Pi_{ie} \boldsymbol{X}_e = \Pi_{ie} H^{-1} H \boldsymbol{X}_e \doteq \Pi_{ip} \boldsymbol{X}_p, \quad i = 1, 2, \ldots, m, \tag{6.45}$$

for some $H \in GL(4)$. The techniques for obtaining the projective reconstruction $\boldsymbol{X}_p = H\boldsymbol{X}_e$ consistent with all the views and the associated projection matrices

$\Pi_{ip}$ will be explained in full detail in Chapter 8 (Section 8.5), when certain rank conditions associated with multiple images are introduced. Here, we assume that $\Pi_{ip}$ and $\boldsymbol{X}_p$ are already available. By choosing the first frame as the reference $\Pi_{1e} = [K_1, 0]$ and $\Pi_{1p} = [I, 0]$ in order for the transformation $H$ (defined up to a scalar factor) to preserve this choice of reference frame, it has to assume the general form of (6.26) introduced in the earlier section. Therefore, if we define $B_i \in \mathbb{R}^{3\times3}, b_i \in \mathbb{R}^3$ to be the blocks of $\Pi_{ip} \doteq [B_i, b_i]$, and take into account the fact that equation (6.45) has to be satisfied for all $\boldsymbol{X}_e$, we have $\Pi_{ip}H \sim \Pi_{ie}$. Isolating the left $3 \times 3$ block of this equation, we obtain the homogeneous equality (up to a scalar factor)

$$(B_i - b_i v^T)K_1 \sim K_i R_i, \tag{6.46}$$

where we have used the fact that $\Pi_{ie} = [K_i R_i,\ K_i T_i]$. Multiplying each side of the equation above by its transpose, and recalling that $S_i^{-1} = K_i K_i^T$, we get

$$(B_i - b_i v^T)S_1^{-1}(B_i - b_i v^T)^T \sim S_i^{-1}. \tag{6.47}$$

Note that in case $v$ is known, the above equation provides a similar type of constraint as equation (6.41). In order to relate more explicitly the unknowns $S_i^{-1}$ and $v$, we can rewrite (6.47) in terms of homogeneous coordinates and obtain the following relationship in terms of the projection matrices $\Pi_{ip}$:

$$\boxed{\Pi_{ip} Q \Pi_{ip}^T \sim S_i^{-1},} \tag{6.48}$$

where

$$Q \doteq H \begin{bmatrix} I_{3\times3} & 0 \\ 0^T & 0 \end{bmatrix} H^T \quad \in \mathbb{R}^{4\times4} \tag{6.49}$$

is a $4 \times 4$ symmetric positive semi-definite matrix of rank 3. Equation (6.48) is called the *absolute quadric constraint*.[11] If the calibration matrix is the same in all views, $S_1^{-1} = S_i^{-1} = S^{-1}$, after eliminating the unknown scale and taking into account the symmetry of the matrices, the above equation in general gives five third-order polynomial equations in the unknowns $S^{-1}$ and $v$. Iterative nonlinear minimization techniques are typically employed in order to solve for the eight unknowns (six for the $3 \times 3$ symmetric matrix $S$ up to a scalar factor and three for the vector $v$).

Although equations (6.47) and (6.48) are exactly the same, the second form is particularly useful when some partial knowledge of the intrinsic camera parameters $K$ is available; for instance, the camera skew factor $s_\theta$ is zero or the coordinates of the principal point $(o_x, o_y)$ are known. The most commonly encountered assumptions about the intrinsic camera parameters and the type of

[11] The name for these constraints comes from projective-geometric interpretation of the matrices $S^{-1}$ and $Q$ as quadratic conics in the projective spaces $\mathbb{P}^3$ and $\mathbb{P}^4$, respectively.

constraints they yield on

$$S^{-1} = \begin{bmatrix} s_{11} & s_{12} & s_{13} \\ s_{12} & s_{22} & s_{23} \\ s_{13} & s_{23} & s_{33} \end{bmatrix} = \begin{bmatrix} \alpha_x^2 + s_\theta^2 + o_x^2 & s_\theta \alpha_y + o_x o_y & o_x \\ s_\theta \alpha_y + o_x o_y & \alpha_y^2 + o_y^2 & o_y \\ o_x & o_y & 1 \end{bmatrix} \quad (6.50)$$

are summarized in Table 6.2. The utility of some of these constraints in scenarios

| Partial camera knowledge | Constraints on $S^{-1}$ | type of constraints |
|---|---|---|
| zero skew | $s_{12}s_{33} = s_{13}s_{23}$ | quadratic |
| principal point known | $s_{13} = s_{23} = 0$ | linear |
| zero skew and principal point known | $s_{12} = s_{13} = s_{23} = 0$ | linear |
| known aspect ratio $r = \frac{\alpha_x}{\alpha_y}$, zero skew, principal point known | $r^2 s_{11} = s_{22}$, $s_{12} = s_{13} = s_{23} = 0$ | linear |

Table 6.2. Commonly encountered assumptions about the intrinsic camera parameters $K$ and the constraints they yield on $S^{-1}$, in addition to the absolute quadric constraints.

commonly encountered in practice will be covered in more detail in Chapter 11.

Alternatively, if all $S_i^{-1}$'s are identical, given a sufficient number of views, $Q$ and $S^{-1}$ can be solved simultaneously as long as the motion of the camera is "rich enough."[12] Since the number of views needed for this method is often greater than two, we will revisit this formulation in Chapter 8 for the general case. If all cameras are different, one can never garner enough data to solve for all $S_i^{-1}$'s from equation (6.48).

## 6.5 Calibration with scene knowledge

In this section, we discuss additional techniques for camera calibration by exploiting scene knowledge. We first discuss, in Section 6.5.1, how to exploit orthogonality and parallelism constraints between vectors. In Section 6.5.2, we discuss how to use complete (Euclidean) knowledge of an object in the scene if available. More complete exposition of the constraints provided by partial scene knowledge follows more naturally from a multiple-view formulation[13] and will be covered in Chapter 10.

[12] For many restricted camera motion sequences, the camera parameters typically cannot be uniquely determined by any autocalibration schemes; see Exercise 6.16 and Section 8.5.

[13] Even for a single image.

### 6.5.1 *Partial scene knowledge*

In man-made environments, information such as *orthogonality* and *parallelism* among line features (e.g., edges of buildings, streets) can be safely assumed. For instance, Figure 6.13 shows three sets of lines that are image of lines in space that are likely pairwise parallel, with each set orthogonal to the other two.

Figure 6.13. Three sets of parallel lines: solid, dashed, and dotted. These sets are mutually orthogonal to each other. Image size: 400 × 300 pixels.

We say "likely" because there is no way to verify, based on images alone, whether such information is true or not. The reader should be aware that from images we can only *assume* that lines are parallel or perpendicular, and provide algorithms that exploit such assumptions. If the assumptions turn out to be violated (as for instance, Figure 6.1), the resulting calibration will be wrong, and we have no way of verifying that. We now demonstrate how such assumptions, when satisfied, can be used to calibrate the camera. A set of parallel lines intersect at the same point infinitely far. The projection of this point onto the image plane is called a vanishing point. If the two vectors $\boldsymbol{\ell}^1, \boldsymbol{\ell}^2 \in \mathbb{R}^3$ represent the (co)images of the two parallel lines,[14] then the corresponding vanishing point is given by

$$\boldsymbol{v} \sim \boldsymbol{\ell}^1 \times \boldsymbol{\ell}^2 = \widehat{\boldsymbol{\ell}^1}\boldsymbol{\ell}^2, \tag{6.51}$$

where $\sim$ indicates homogeneous equality up to a scalar factor. Since for an image such as that in Figure 6.14 the three sets of lines are mutually orthogonal, we may assume that, by a proper choice of the world coordinate frame, their 3-D directions coincide with the three principal directions: $e_1 = [1,0,0]^T$, $e_2 = [0,1,0]^T$, $e_3 = [0,0,1]^T$. In the image, the vanishing points corresponding to the three sets of parallel lines are respectively

$$\boldsymbol{v}_1 = KRe_1, \quad \boldsymbol{v}_2 = KRe_2, \quad \boldsymbol{v}_3 = KRe_3.$$

[14]We recall that we use the convention of using superscripts to enumerate different features in the scene and subscripts to enumerate different images of the same feature.

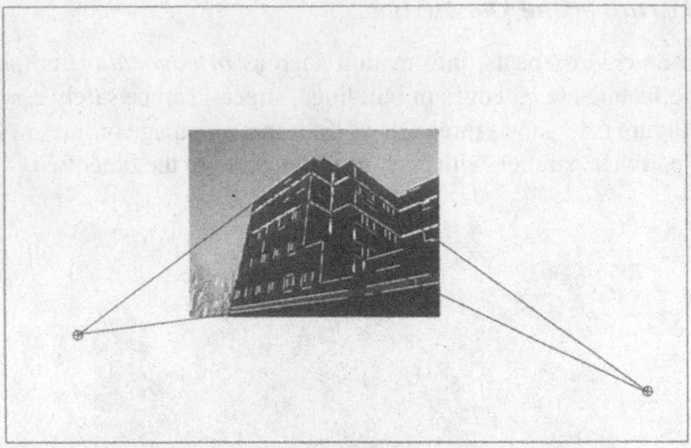

Figure 6.14. An example of two vanishing points associated with two out of the three sets of parallel lines in Figure 6.13.

Note that the coordinates of the vanishing points depend only on rotation and internal parameters of the camera, but not on translation. The orthogonality relations among $e_1, e_2, e_3$ readily provide constraints on the calibration matrix $K$. In particular, we have

$$\boldsymbol{v}_i^T S \boldsymbol{v}_j = \boldsymbol{v}_i^T K^{-T} K^{-1} \boldsymbol{v}_j = e_i^T R^T R e_j = e_i^T e_j = 0, \quad i \neq j,$$

where again $S = K^{-T}K^{-1}$ is the symmetric matrix associated with the uncalibrated camera

$$S = K^{-T}K^{-1} = \begin{bmatrix} s_{11} & s_{12} & s_{13} \\ s_{12} & s_{22} & s_{23} \\ s_{13} & s_{23} & s_{33} \end{bmatrix} \in \mathbb{R}^{3\times 3}.$$

When three vanishing points are detected, they provide three independent constraints on the matrix $S$:

$$\boldsymbol{v}_1^T S \boldsymbol{v}_2 = 0, \quad \boldsymbol{v}_1^T S \boldsymbol{v}_3 = 0, \quad \boldsymbol{v}_2^T S \boldsymbol{v}_3 = 0.$$

In general, the symmetric matrix $S$ has five degrees of freedom. Without additional constraints we can recover $S$ only up to a two-parameter family of solutions from the three linear equations above. With the assumption of zero skew ($fs_\theta = 0$) and known aspect ratio ($fs_x = fs_y$), one may obtain a unique solution for $K$ from a single image as above. As an example, the camera calibration for the image in Figure 6.13 is

$$K = \begin{bmatrix} fs_x & fs_\theta & o_x \\ 0 & fs_y & o_y \\ 0 & 0 & 1 \end{bmatrix} = \begin{bmatrix} 409.33 & 0 & 177.46 \\ 0 & 409.33 & 165.75 \\ 0 & 0 & 1 \end{bmatrix} \in \mathbb{R}^{3\times 3}. \quad (6.52)$$

A more detailed study of the degeneracies in exploiting these types of constraints can be found in [Liebowitz and Zisserman, 1999].

### 6.5.2 Calibration with a rig

Calibration with a rig is the method of choice for camera calibration when one has access to the camera and can place a known object in the scene.

Under these conditions, one can use an object with a distinct number of points, whose coordinates relative to some reference frame are known with high accuracy, as a *calibration rig*. Notice that a calibration rig could be an actual object manufactured primarily for the purpose of camera calibration (Figure 6.15), or simply an object in the scene with known geometry, for instance a golf ball with dots painted on it, or the rim of a car wheel whose alignment needs to be computed from a collection of cameras.

Figure 6.15. An example of a calibration rig for laboratory use: a checkerboard-textured cube.

Let $\boldsymbol{X} = [X, Y, Z, 1]^T$ be the coordinates of a point $p$ on the rig. Its image has the pixel coordinates $\boldsymbol{x}' = [x', y', 1]^T$ that satisfy equation (6.1). If we let $\pi_1, \pi_2, \pi_3 \in \mathbb{R}^4$ be the three row vectors of the projection matrix $\Pi = K\Pi_0 g \in \mathbb{R}^{3\times 4}$, then equation (6.1) can be written for each point $p^i$ on the rig as

$$\lambda^i \begin{bmatrix} x'^i \\ y'^i \\ 1 \end{bmatrix} = \begin{bmatrix} \pi_1^T \\ \pi_2^T \\ \pi_3^T \end{bmatrix} \begin{bmatrix} X^i \\ Y^i \\ Z^i \\ 1 \end{bmatrix}. \tag{6.53}$$

From the third row we get $\lambda^i = \pi_3^T \boldsymbol{X}^i$. Hence for each point we obtain the following two equations

$$\begin{aligned} x'^i(\pi_3^T \boldsymbol{X}^i) &= \pi_1^T \boldsymbol{X}^i, \\ y'^i(\pi_3^T \boldsymbol{X}^i) &= \pi_2^T \boldsymbol{X}^i. \end{aligned}$$

Unlike previous formulations, here $X^i, Y^i$, and $Z^i$ are known, and so are $x'^i, y'^i$. We can therefore stack all the unknown entries of $\Pi$ into a vector and rewrite the equations above as a system of linear equations

$$M\Pi^s = 0,$$

where $\Pi^s$ is a stacked projection matrix $\Pi$,

$$\Pi^s = [\pi_{11}, \pi_{21}, \pi_{31}, \pi_{12}, \pi_{22}, \pi_{32}, \pi_{13}, \pi_{23}, \pi_{33}, \pi_{14}, \pi_{24}, \pi_{34}]^T \quad \in \mathbb{R}^{12},$$

and rows of $M$ are functions of $(x'^i, y'^i)$ and $(X^i, Y^i, Z^i)$. A linear (suboptimal) estimate of $\Pi^s$ can then be obtained by minimizing the least-squares criterion

$$\min \|M\Pi^s\|^2 \quad \text{subject to} \quad \|\Pi^s\|^2 = 1 \tag{6.54}$$

without taking into account the structure of the unknown vector $\Pi^s$. This can be accomplished, as usual, using the SVD. If we denote the (nonlinear) projection map from $\boldsymbol{X} = [X, Y, Z]^T$ to $\boldsymbol{x}' = [x', y']^T$ by

$$\boldsymbol{x}' \doteq h(\Pi \boldsymbol{X}),$$

where $h(\boldsymbol{X}) = [X,\ Y]/Z$, then the estimate can be further refined by a nonlinear minimization of the objective function

$$\min_{\Pi} \sum_i \left\|\boldsymbol{x}'^i - h(\Pi \boldsymbol{X}^i)\right\|^2.$$

After obtaining an estimate of the projection matrix

$$\Pi = K[R \,|\, T] = [KR \,|\, KT],$$

we can factor the first $3 \times 3$ submatrix into the calibration matrix $K \in \mathbb{R}^{3\times 3}$ (in its upper triangular form) and rotation matrix $R \in SO(3)$ using a routine QR decomposition

$$KR = \begin{bmatrix} \pi_{11} & \pi_{12} & \pi_{13} \\ \pi_{21} & \pi_{22} & \pi_{23} \\ \pi_{31} & \pi_{32} & \pi_{33} \end{bmatrix},$$

which yields an estimate of the intrinsic parameters in the calibration matrix $K$ and that of the rotational component of the extrinsic parameters. Estimating translation completes the calibration procedure:

$$T = K^{-1} \begin{bmatrix} \pi_{14} \\ \pi_{24} \\ \pi_{34} \end{bmatrix}.$$

This procedure requires that the points in space have known coordinates $\{\boldsymbol{X}^i\}$, and that they be in *general position*, i.e. that they do not lie on a set of measure zero, for instance a plane. However, the simplest and most common calibration rigs available are indeed planar checkerboard patterns! Therefore, we study the case of a planar calibration rig in detail below.

### *6.5.3 Calibration with a planar pattern*

Although the approach just described requires only one image of a known object, it does not return a unique and well-conditioned estimate of the calibration if the object is planar. Since nonplanar calibration rigs are not easy to manufacture, a more commonly adopted approach consists in capturing several images of a known planar object, such as a checkerboard like the one shown in Figure 6.16.

Figure 6.16. Two images of the checkerboard for camera calibration. The resolution of the images is 640 × 480.

Since we are free to choose the world reference frame, we choose it aligned with the board so that points on it have coordinates of the special form $\boldsymbol{X} = [X, Y, 0, 1]^T$. Notice that the center of the world frame needs to be *on* the board, and the $Z$-axis of the world frame is the normal vector. Then, with respect to the camera coordinate frame, the image $\boldsymbol{x}'$ of a point $\boldsymbol{X}$ on the board is given by equation (6.1), which for the given choice of coordinate frames simplifies to

$$\lambda \begin{bmatrix} x' \\ y' \\ 1 \end{bmatrix} = K[r_1, r_2, T] \begin{bmatrix} X \\ Y \\ 1 \end{bmatrix}, \tag{6.55}$$

where $r_1, r_2 \in \mathbb{R}^3$ are the first and second columns of the rotation matrix $R$. Notice that the matrix

$$H \doteq K[r_1, r_2, T] \quad \in \mathbb{R}^{3\times3} \tag{6.56}$$

is a linear transformation of the homogeneous coordinates $[X, Y, 1]^T$ to the homogeneous coordinates $\boldsymbol{x}' = [x', y', 1]^T$; i.e. it is a homography between the checkerboard plane and the image plane.

Applying the common trick of multiplying both sides by the skew-symmetric matrix $\widehat{\boldsymbol{x}'}$ in order to eliminate $\lambda$ (recall that $\widehat{\boldsymbol{x}'}\boldsymbol{x}' = 0$) yields

$$\widehat{\boldsymbol{x}'} H[X, Y, 1]^T = 0. \tag{6.57}$$

Notice that in the above equation, we know both $\boldsymbol{x}'$ (measured from the image) and $[X, Y, 1]^T$ (given from knowledge of the checkerboard). Hence $H$ can be solved (up to a scalar factor) linearly from such equations if sufficiently many points on the checkerboard are given. We know from the previous chapter that at least four images of such points are needed in order to solve for the homography $H$ up to a scalar factor.

Once we know $H$, we observe that its first two columns are simply $[h_1, h_2] \sim K[r_1, r_2]$. This is equivalent to $K^{-1}[h_1, h_2] \sim [r_1, r_2]$. Since $r_1, r_2$ are orthonormal vectors, we obtain two equations that the calibration matrix $K$ has to satisfy:

$$h_1^T K^{-T} K^{-1} h_2 = 0, \quad h_1^T K^{-T} K^{-1} h_1 = h_2^T K^{-T} K^{-1} h_2. \tag{6.58}$$

These equations are quadratic in the entries of $K$. However, if we are willing to neglect the structure of $S = K^{-T}K^{-1} \in \mathbb{R}^{3\times 3}$, we can just recover it linearly from the equations above, as usual, using the SVD. Once $S$ is known, $K$ can be retrieved, in principle,[15] using the Cholesky factorization (see Appendix A).

From each image, we obtain two such (linear) equations in $S = K^{-T}K^{-1}$. The calibration matrix $K$ has five unknowns $fs_x, fs_y, fs_\theta, o_x, o_y$; so does $S$. We then need at least three images to determine them, since each image produces two equations. However, often the skew term $fs_\theta$ is small compared to the other parameters, and therefore may be assumed to be zero. In that case, one has only four parameters in $K$ (or $S$) to determine: $fs_x, fs_y, o_x, o_y$. Another way to see this is that in the zero-skew case, there is an extra linear constraint that the matrix $S$ needs to satisfy:

$$e_1^T S e_2 = 0.$$

As an example, for the two images given in Figure 6.16, the calibration result given by the above scheme is

$$K = \begin{bmatrix} fs_x & fs_\theta & o_x \\ 0 & fs_y & o_y \\ 0 & 0 & 1 \end{bmatrix} = \begin{bmatrix} 769.942 & 0 & 319.562 \\ 0 & 769.086 & 244.162 \\ 0 & 0 & 1 \end{bmatrix} \in \mathbb{R}^{3\times 3}.$$

The method we have outlined above is merely conceptual, in the sense that it will not perform satisfactorily in the presence of noise in the measurements of $\boldsymbol{x}'$ and $\boldsymbol{X}$. In practice, some refinement of the solution based on nonlinear optimization techniques is necessary, and if needed, such techniques can also simultaneously estimate radial distortion.

Several free software packages are publicly available that perform camera calibration using a planar rig. For instance, Intel's OpenCV software library (`http://sourceforge.net/projects/opencvlibrary/`) provides code that is easy to use, accurate, and well documented.

## 6.6 Dinner with Kruppa

This last section is dedicated to exploring intrinsic constraints among calibration parameters that date back to the work of Kruppa in 1913. While these were historically the first constraints to be exploited for autocalibration, their use has become less widespread due to the difficulty in solving the strongly nonlinear equations. In this section we will derive the equations and outline their basic properties. We

[15] We say "in principle" because in practice, noise may prevent the matrix $S$ recovered from data from being positive definite.

will explore further properties and examples in the appendix and in the exercises at the end of the chapter.

In Section 6.2.1 we introduced the fundamental matrix $F = \widehat{T'}KRK^{-1}$, which is the uncalibrated counterpart of the essential matrix from Chapter 5. Since we are interested in the recovery of the unknown matrix $K$, let us see whether there are additional relationships between $K$ and $F$ that can be exploited for that purpose, without any additional knowledge of the scene.

In Section 6.1, we anticipated that all constraints on the camera calibration are related to the new inner product in the distorted space,

$$\langle u, v\rangle_S = u^T S v, \quad \langle u^T, v^T\rangle_S = u^T S^{-1} v \tag{6.59}$$

for column vectors and row vectors, respectively. We have seen that this is indeed the case with the calibration methods using the Lyapunov equation, the absolute quadric constraint, the vanishing point, and the planar rig. There is another important equality that can be derived from this fact: the fundamental matrix $F = \widehat{T'}KRK^{-1}$ and $S^{-1} = KK^T$ satisfy the following equation:

$$FS^{-1}F^T = \widehat{T'}S^{-1}\widehat{T'}^T, \tag{6.60}$$

which is obviously in the form of the inner product for row vectors. We call this equation the *normalized matrix Kruppa's equation*. In general, however, the fundamental matrix $F$ is determined only up to a scalar factor. So, there is a scale factor $\lambda \in \mathbb{R}$ to be taken into account, i.e. $F = \lambda\widehat{T'}KRK^{-1}$ if we assume $\|T'\| = 1$. We then have the *matrix Kruppa's equation*

$$\boxed{FS^{-1}F^T = \lambda^2\widehat{T'}S^{-1}\widehat{T'}^T.} \tag{6.61}$$

It seems that this matrix equation gives nine scalar constraints in terms of the unknowns $\lambda$ and $S^{-1}$. However, since both sides of the equation are symmetric matrices and have some additional special structure, after $\lambda$ is eliminated from the equations, only two of them impose algebraically independent constraints on $S^{-1}$, which we prove in Proposition 6.11 in Appendix 6.B. These constraints are of order two in the unknown entries of $S^{-1}$. A detailed study of additional algebraic properties of Kruppa's equations are given in Appendix 6.B at the end of this chapter.

Kruppa's equations are not the only intrinsic constraints that will allow us to recover information about the camera calibration and hence the true Euclidean metric from uncalibrated images. It can be shown that there exist ambiguous solutions for $K$ that satisfy Kruppa's equations but do not give rise to a valid Euclidean reconstruction (Appendix 6.B.2). Thus, to obtain a complete picture of camera calibration, we must study carefully, during each stage of the inversion of the image formation process, what ambiguous solutions could be introduced and how to eliminate them. As we have seen before, both the absolute quadric constraints and the modulus constraints serve the same purpose. We conclude this chapter by comparing them with Kruppa's equations in Table 6.3.

| | Kruppa's equations | Modulus constraint | Absolute quadric constraint |
|---|---|---|---|
| Known | $F$ | $F$ | $\Pi_{ip} = \Pi_i H^{-1}$ |
| Unknowns | $S^{-1} = KK^T$ | $v = [v_1, v_2, v_3]^T$ | $S^{-1}$ and $v$ |
| # of equations | 2 | 1 | 5 |
| Orders | 2nd order | 4th order | 3rd order |

Table 6.3. Three types of intrinsic constraints from a pair of images, assuming that the calibration matrix $K$, hence $S^{-1} = KK^T$, is constant. The number of equations are counted under generic conditions.

## 6.7 Summary

In this chapter we have studied the geometry of pairs of images when the calibration matrix of the camera is unknown. The epipolar geometry that we have studied for the calibrated case in the previous chapter can be extended to the uncalibrated case, and Table 6.4 gives a brief comparison.

| | Calibrated case | Uncalibrated case |
|---|---|---|
| Image point | $\boldsymbol{x}$ | $\boldsymbol{x}' = K\boldsymbol{x}$ |
| Camera (motion) | $g = (R, T)$ | $g' = (KRK^{-1}, KT)$ |
| Epipolar constraint | $\boldsymbol{x}_2^T E \boldsymbol{x}_1 = 0$ | $(\boldsymbol{x}_2')^T F \boldsymbol{x}_1' = 0$ |
| Fundamental matrix | $E = \widehat{T}R$ | $F = \widehat{T'}KRK^{-1}, T' = KT$ |
| Epipoles | $E\boldsymbol{e}_1 = 0,\ \boldsymbol{e}_2^T E = 0$ | $F\boldsymbol{e}_1 = 0,\ \boldsymbol{e}_2^T F = 0$ |
| Epipolar lines | $\boldsymbol{\ell}_1 = E^T \boldsymbol{x}_2,\ \boldsymbol{\ell}_2 = E\boldsymbol{x}_1$ | $\boldsymbol{\ell}_1 = F^T \boldsymbol{x}_2',\ \boldsymbol{\ell}_2 = F\boldsymbol{x}_1'$ |
| Decomposition | $E \mapsto [R, T]$ | $F \mapsto [(\widehat{T'})^T F, T']$ |
| Reconstruction | Euclidean: $\boldsymbol{X}_e$ | Projective: $\boldsymbol{X}_p = H\boldsymbol{X}_e$ |

Table 6.4. Comparison of calibrated and uncalibrated two-view geometry (assuming that the calibration matrix $K_1 = K_2 = K$ is the same for both images).

The ultimate task confronting us in the uncalibrated case is to retrieve the unknown camera calibration, the three-dimensional structure, and camera motion. There are many different types of information that can be exploited to this end. Table 6.5 summarizes and compares the methods introduced in this chapter, and further discussed in later chapters in a multiple-view setting.

## 6.8 Exercises

**Exercise 6.1 (Lyapunov maps).** Let $\{\lambda\}_{i=1}^n$ be the (distinct) eigenvalues of the matrix $C \in \mathbb{C}^{n \times n}$. Find *all* the $n^2$ eigenvalues and eigenvectors of the following linear maps:

| | Euclidean | Affine | Projective |
|---|---|---|---|
| Structure | $\boldsymbol{X}_e = g_e \boldsymbol{X}$ | $\boldsymbol{X}_a = H_a \boldsymbol{X}_e$ | $\boldsymbol{X}_p = H_p \boldsymbol{X}_a$ |
| Transformation | $g_e = \begin{bmatrix} R & T \\ 0 & 1 \end{bmatrix}$ | $H_a = \begin{bmatrix} K & 0 \\ 0 & 1 \end{bmatrix}$ | $H_p = \begin{bmatrix} I & 0 \\ -v^T v_4^{-1} & v_4^{-1} \end{bmatrix}$ |
| Projection | $\Pi_e = [KR, KT]$ | $\Pi_a = \Pi_e H_a^{-1}$ | $\Pi_p = \Pi_a H_p^{-1}$ |
| 3-step upgrade | $\boldsymbol{X}_e \leftarrow \boldsymbol{X}_a$ | $\boldsymbol{X}_a \leftarrow \boldsymbol{X}_p$ | $\boldsymbol{X}_p \leftarrow \{\boldsymbol{x}_1', \boldsymbol{x}_2'\}$ |
| Info. needed | Calibration $K$ | Plane at infinity $\pi_\infty^T = [v^T, v_4]$ | Fundamental matrix $F$ |
| Methods | Lyapunov eqn. | Vanishing points | Canonical decomposition |
| | Pure rotation | Pure translation | |
| | Kruppa's eqn. | Modulus constraint | |
| 2-step upgrade | $\boldsymbol{X}_e \leftarrow \boldsymbol{X}_p$ | | $\boldsymbol{X}_p \leftarrow \{\boldsymbol{x}_i'\}_{i=1}^m$ |
| Info. needed | Calibration $K$ and $\pi_\infty^T = [v^T, v_4]$ | | Multiple-view matrix* |
| Methods | Absolute quadric constraint | | Rank conditions* |
| 1-step upgrade | $\{\boldsymbol{x}_i\}_{i=1}^m \leftarrow \{\boldsymbol{x}_i'\}_{i=1}^m$ | | |
| Info. needed | Calibration $K$ | | |
| Methods | Orthogonality & parallelism, symmetry* or calibration rig | | |

Table 6.5. Summary of methods for reconstruction under an uncalibrated camera. (*: Topics to be studied systematically in Chapters of Part III.)

1. $L_1 : \mathbb{C}^{n\times n} \to \mathbb{C}^{n\times n}; \quad X \mapsto C^T X - XC.$
2. $L_2 : \mathbb{C}^{n\times n} \to \mathbb{C}^{n\times n}; \quad X \mapsto C^T XC - X.$

(Hint: construct the eigenvectors of $L_1$ and $L_2$ from left eigenvectors $\{u_i\}_{i=1}^n$ and right eigenvectors $\{v_j\}_{j=1}^n$ of $C$.)

**Exercise 6.2 (Lyapunov equations).** Using results from the above exercise, show that the set of $X$ that satisfies the equations

$$C^T XC - X = 0, \quad \text{where} \quad C = KRK^{-1}, \tag{6.62}$$

and

$$C^T X - XC = 0, \quad \text{where} \quad C = K\widehat{\omega}K^{-1}, \tag{6.63}$$

is a two-dimensional subspace for general $R \in SO(3)$ and $\widehat{\omega} \in so(3)$.

**Exercise 6.3 (Invariant conic under a rotation).** For the equation

$$S - RSR^T = 0, \quad R \in SO(3), \quad S^T = S \in \mathbb{R}^{3\times 3},$$

prove Lemma 6.8 for the case in which the rotation angle is between 0 and $\pi$. What are the two typical solutions for $S$ in terms of eigenvectors of the rotation matrix $R$? Since a symmetric real matrix $S$ is usually used to represent a conic $x^T Sx = 1$, an $S$ that satisfies the above equation obviously gives a conic that is invariant under the rotation $R$.

**Exercise 6.4 (Canonical decomposition of the fundamental matrix).** Given a fundamental matrix $F = \widehat{T'}KRK^{-1}$, show that the coordinate transformation $\boldsymbol{X}_2 = (\widehat{T'})^T F\boldsymbol{X}_1 + T'$ also gives rise to the same fundamental matrix.

**Exercise 6.5** Construct a rotation matrix $R \in SO(3)$ such that $RT = e_3 = [0,0,1]^T$ for a given unit vector $T = [T_1, T_2, T_3]^T$. Is such $R$ unique? How many can you find? Using this fact, show that for any $T \in \mathbb{R}^3$, we can always find $R_1, R_2, R_3 \in SO(3)$ such that

$$R_1\widehat{T}R_1^T = \widehat{e}_1, \quad R_2\widehat{T}R_2^T = \widehat{e}_2, \quad R_3\widehat{T}R_3^T = \widehat{e}_3. \tag{6.64}$$

**Exercise 6.6 (A few identities associated with a fundamental matrix).** Given a fundamental matrix $F = \widehat{T'}KRK^{-1}$ with $K \in SL(3)$, $R \in SO(3)$ and $T' = KT \in \mathbb{R}^3$ of unit length, besides the well-known Kruppa's equations (6.61), we have the following identities:

1. $F^T\widehat{T'}F = \widehat{KR^TT}$.
2. $KRK^{-1} - (\widehat{T'})^T F = T'v^T$ for some $v \in \mathbb{R}^3$.

(Note that the first identity is used in the proof of Theorem 6.15.)

**Exercise 6.7 (Planar motion).** First convince yourself that for a fundamental matrix $F$, in general, we have rank$(F + F^T) = 3$. However, show that if $F$ is a fundamental matrix associated with a planar motion $(R,T) \in SE(2)$, then

$$\text{rank}(F + F^T) = 2. \tag{6.65}$$

**Exercise 6.8 (Pure translational motion).** Derive the fundamental matrix and the associated Kruppa's equation associated with a pure translational motion. Can you extract information about the camera calibration from such motions? Explain why.

**Exercise 6.9 (Time-varying calibration matrix).** Derive the fundamental matrix $F$ and the associated Kruppa's equation for the case in which the calibration matrices for the first and second views are different, say $K_1$ and $K_2$. What are the left and right null spaces of $F$ (i.e. the epipoles)?

**Exercise 6.10 (Improvement of the canonical decomposition).** Even if the fundamental matrix happens to be an essential matrix $F = \widehat{T}R$ with $\|T\| = 1$, the canonical decomposition does not give the correct decomposition to the pair $(R,T)$. Instead, we have $(U,T)$ with $U = \widehat{T}^T\widehat{T}R$ from the canonical decomposition.

Knowing that there exists a vector $v \in \mathbb{R}^3$ such that $R = U + Tv^T$ allows us to upgrade $U$ to $R$ by further imposing the fact that $U + Tv^T$ needs to be a rotation matrix.

1. Using the fact that $R^TR = I$ for any rotation matrix $R$, show that to recover the correct $v$, we in general need to solve an equation of the form
$$uv^T + vu^T + vv^T = W, \tag{6.66}$$
with the vector $u \in \mathbb{R}^3$ and the rank-2 matrix $W \in \mathbb{R}^{3\times 3}$ known. What are $u$ and $W$ in terms of $U$ and $T$?
2. Find a solution to the above problem. The solution you find is then a second solution to the problem of decomposing an essential matrix (different from the one given in the previous chapter).

3. In practice, $F$ could be only approximately an essential matrix, which is often the case when we have a good guess on the camera calibration. Then the above problem becomes how to find the optimal $v$ such that $\|uv^T + vu^T + vv^T - W\|_f^2$ is minimized, where $W$ is not necessarily a rank-2 matrix. Find a solution to this problem. (Hint: use the SVD of the matrix $W$ and apply the fixed rank approximation theorem in Appendix A.)

The resulting decomposition $(U + Tv^T, T)$ can be considered an improved version of the canonical decomposition for the same fundamental matrix $F$.

**Exercise 6.11 (Critical surface continued).** Analogously to the calibrated case (see Exercise 5.14), derive the expression for a (critical) surface that allows two fundamental matrices. In particular, show that planes are critical surfaces in this sense.

**Exercise 6.12 (Uncalibrated planar homography and the fundamental matrix).** In this exercise, we generalize the study in Section 5.3 on planar scenes to the uncalibrated case.

1. Show that two uncalibrated images of feature points on a 3-D plane $N^T \boldsymbol{X}_1' = d$ satisfy the homography
$$\widehat{\boldsymbol{x}_2'} H \boldsymbol{x}_1' = 0, \quad \text{where} \quad H = KRK^{-1} + \frac{1}{d} T' N^T \in \mathbb{R}^{3\times 3}. \tag{6.67}$$

2. Generalize the results in Theorem 5.21 and Corollaries 5.22 and 5.23 to relationships between the (uncalibrated) homography matrix and the fundamental matrix.

**Exercise 6.13** Show that if $\phi(\lambda) = \lambda^3 + a_2\lambda^2 + a_1\lambda + a_0$ is the characteristic polynomial of a matrix $M + uv^T$ for $M \in \mathbb{R}^{3\times 3}, u, v \in \mathbb{R}^3$, then the coefficients $a_0, a_1, a_2$ are all linear functions in $v$. This fact was used in characterizing the modulus constraint in this chapter.

**Exercise 6.14 (Autocalibration for focal length).** Suppose the calibration matrix $K$ is of the diagonal form
$$K = \begin{bmatrix} f_x & 0 & 0 \\ 0 & f_y & 0 \\ 0 & 0 & 1 \end{bmatrix}. \tag{6.68}$$

This corresponds to a practical situation when the center of the image is known and the skew factor is zero.

1. Write down Kruppa's equations for this special case.

2. Discuss when a unique solution for $f_x$ and $f_y$ can be obtained from such equations (i.e. what is the minimum number of image pairs needed and what are the requirements for the relative motion?).

**Exercise 6.15 (Other forms of Kruppa's equations).** Given a fundamental matrix $F = \lambda \widehat{T'} KRK^{-1}$ with $T' = KT$ and $\|T'\| = 1$:

1. Show that $F^T \widehat{T'} F = \lambda^2 \widehat{KR^T T}$. Now define the vector $\alpha = \lambda^2 KR^T T = \lambda^2 KR^T K^{-1} T'$ and conclude that it can be directly computed from the given $F$.

2. Show that the following equation for $S = K^{-T} K^{-1}$ must hold:
$$\alpha^T S \alpha = \lambda^4 T'^T S T'. \tag{6.69}$$

3. Notice that this is a constraint on $S$, unlike Kruppa's equations, which are constraints on $S^{-1}$. Combining equation (6.69) with Kruppa's equations given in (6.90), we have

$$\lambda^2 = \frac{d_1^T S^{-1} d_1}{k_1^T S^{-1} k_1} = \frac{d_2^T S^{-1} d_2}{k_2^T S^{-1} k_2} = \frac{d_1^T S^{-1} d_2}{k_1^T S^{-1} k_2} = \sqrt{\frac{\alpha^T S \alpha}{T'^T S T'}}. \tag{6.70}$$

Is the last equation algebraically independent of the two Kruppa's equations? Verify this numerically (e.g., in Matlab) or symbolically (e.g., in Mathematica).

**Exercise 6.16 (Reconstruction up to subgroups).** Assume that neither the camera calibration $K \in SL(3)$ nor the scene structure $\boldsymbol{X}$ is known, but we do know that the camera motion is subject to:

1. A planar motion $SE(2)$; i.e. the camera always moves on a plane;
2. A free rotation $SO(3)$; i.e. the camera rotates only around its center;
3. A free translation $\mathbb{R}^3$.

Describe the extent to which we are able to recover the scene structure and the camera calibration for each type of motion (subgroup).

**Exercise 6.17 (Autocalibration from continuous motion).** The continuous-motion case of camera autocalibration is quite different from the discrete case, and this exercise shows how.

1. Prove that in the uncalibrated case, if the linear velocity $v$ is nonzero, the continuous epipolar constraint is

$$(\dot{\boldsymbol{x}}')^T K^{-T} \widehat{v} K^{-1} \boldsymbol{x}' + (\boldsymbol{x}')^T K^{-T} \widehat{\omega} \widehat{v} K^{-1} \boldsymbol{x}' = 0. \tag{6.71}$$

2. Show that the above equation is equivalent to

$$(\dot{\boldsymbol{x}}')^T \widehat{v'} \boldsymbol{x}' + (\boldsymbol{x}')^T \frac{1}{2} (\widehat{\omega'} S^{-1} \widehat{v'} + \widehat{v'} S^{-1} \widehat{\omega'}) \boldsymbol{x}' = 0 \tag{6.72}$$

for $v' = Kv, \omega' = K\omega$.

3. Using the above equation, conclude that only the eigenvalues of $S^{-1} = KK^T$, i.e. the singular values of $K$, can be recovered from the continuous epipolar constraint.
4. However, show that if $v = 0$ and there is only rotational motion, the epipolar constraint becomes

$$\widehat{\boldsymbol{x}'} \dot{\boldsymbol{x}}' = \widehat{\boldsymbol{x}'} K \widehat{\omega} K^{-1} \boldsymbol{x}'. \tag{6.73}$$

From this equation, show that, as in the discrete case, two linearly independent rotations $\omega_1, \omega_2$ can uniquely determine the camera calibration $K$.

**Exercise 6.18 (Programming: camera calibration with a known object).** Generate 3-D coordinates $\boldsymbol{X}$ of $n$ points in the world coordinate frame and their projections in the first camera frame in terms of pixel coordinates $\boldsymbol{x}'$. When generating $\boldsymbol{x}'$ assume a relative displacement between the world coordinate frame and the camera frame $(R, T) \in SE(3)$ and a calibration matrix $K$ having the following form

$$K = \begin{bmatrix} 500 & 0.5 & 250 \\ 0 & 500 & 250 \\ 0 & 0 & 1 \end{bmatrix}.$$

Write a Matlab function `[R,T,K] = calibration2Dto3D(X,x)` that computes the relative displacement $(R,T) \in SE(3)$ and the intrinsic camera parameter matrix $K \in SL(3)$ used to generate $\boldsymbol{x}'$.

**Exercise 6.19 (Programming: projective reconstruction).** Generate 3-D coordinates $\boldsymbol{X}$ of $n$ points in the coordinate frame of the first camera and second camera and their image projections $\boldsymbol{x}_1$ and $\boldsymbol{x}_2$ in the two views. Assume a general displacement between the two views $(R,T) \in SE(3)$ and the intrinsic calibration matrix $K$ is the same as in the exercise above.

1. Write a Matlab function `F = points2F(x1, x2)` for computing the fundamental matrix $F$ relating the two views.

2. Compute the canonical decomposition $R_c, T_c$ of $F$ and the associated projective structure $\boldsymbol{X}_p$. The function should be called `[Xp] = points2Xp(x1, x2, Rc, Tc)`.

## 6.A From images to fundamental matrices

As in the calibrated case, the fundamental matrix can be recovered from the measurements using linear techniques. For any fundamental matrix $F$ with entries

$$F = \begin{bmatrix} f_1 & f_4 & f_7 \\ f_2 & f_5 & f_8 \\ f_3 & f_6 & f_9 \end{bmatrix} \quad \in \mathbb{R}^{3\times 3}, \tag{6.74}$$

we stack it into a vector $F^s \in \mathbb{R}^9$ as we did for the essential matrix,

$$F^s \doteq [f_1, f_2, f_3, f_4, f_5, f_6, f_7, f_8, f_9]^T \quad \in \mathbb{R}^9.$$

Since the epipolar constraint $\boldsymbol{x}_2'^T F \boldsymbol{x}_1' = 0$ is linear in the entries of $F$, we can rewrite it as

$$\boldsymbol{a}^T F^s = 0, \tag{6.75}$$

where $\boldsymbol{a} \doteq \boldsymbol{x}_1' \otimes \boldsymbol{x}_2'$ is given explicitly by

$$\boldsymbol{a} = [x_1'x_2', x_1'y_2', x_1'z_2', y_1'x_2', y_1'y_2', y_1'z_2', z_1'x_2', z_1'y_2', z_1'z_2']^T \quad \in \mathbb{R}^9. \tag{6.76}$$

Given a set of $n \geq 8$ corresponding points, form the matrix of measurements $\chi \doteq [\boldsymbol{a}^1, \boldsymbol{a}^2, \ldots, \boldsymbol{a}^n]^T$; $F^s$ can be obtained as the minimizing solution of the least-squares objective function $\|\chi F^s\|^2$. Such a solution corresponds to the eigenvector associated with the smallest eigenvalue of $\chi^T\chi$, and can be found conveniently using the SVD (just compute the SVD of $\chi$ and choose $F^s$ to be the right singular vector associated with the smallest singular value, i.e. the last column of $V$, as in Appendix A). The recovery of the fundamental matrix is summarized in Algorithm 6.1.

**Algorithm 6.1 (The eight-point algorithm for the fundamental matrix).**

For a given set of image correspondences $(\boldsymbol{x}_1'^j, \boldsymbol{x}_2'^j)$, $j = 1, 2, \ldots, n$ $(n \geq 8)$, this algorithm finds the fundamental matrix $F$ that minimizes in a least-squares sense the epipolar constraint

$$(\boldsymbol{x}_2'^j)^T F \boldsymbol{x}_1'^j = 0, \quad j = 1, 2, \ldots, n.$$

1. **Compute a first approximation of the fundamental matrix**
   Construct the $\chi \in \mathbb{R}^{n \times 9}$ from transformed correspondences $\boldsymbol{x}_1'^j$ and $\boldsymbol{x}_2'^j$ as in (6.76), namely,
   $$\chi^j = \boldsymbol{x}_1'^j \otimes \boldsymbol{x}_2'^j \quad \in \mathbb{R}^9.$$
   Find the vector $F^s \in \mathbb{R}^9$ of unit length such that $\|\chi F^s\|$ is minimized as follows: Compute the SVD of $\chi = U_\chi \Sigma_\chi V_\chi^T$ and define $F^s$ to be the ninth column of $V_\chi$. Unstack the nine elements of $F^s$ into a square $3 \times 3$ matrix $F$. Note that this matrix will in general *not* be a fundamental matrix.

2. **Impose the rank constraint and recover the fundamental matrix**
   Compute the singular value decomposition of the matrix $F = U\mathrm{diag}\{\sigma_1, \sigma_2, \sigma_3\}V^T$. Impose the rank-2 constraint by setting $\sigma_3 = 0$ and set the fundamental matrix to be
   $$F = U\mathrm{diag}\{\sigma_1, \sigma_2, 0\}V^T.$$

*Normalization of image coordinates*

Since image coordinates $\boldsymbol{x}_1'$ and $\boldsymbol{x}_2'$ are measured in pixels, the individual entries of the matrix $\chi$ can vary by two orders of magnitude (e.g., between 0 and 512), which affects the conditioning of the matrix $\chi$ (see Appendix A). Errors in the values $x'$ and $y'$ will introduce uneven errors in the recovered entries of $F^s$ and hence $F$. Since we know how to handle linear transformations in the epipolar framework, we can use that to our advantage and choose the transformation that "balances" the coordinates. This can be done, for instance, by transforming the points $\{\boldsymbol{x}_1'^j\}_{j=1}^n$ by an affine matrix $H_1 \in \mathbb{R}^{3\times3}$ so that the resulting points $\{H_1\boldsymbol{x}_1'^j\}_{j=1}^n$ have zero mean and unit variance. This can be accomplished by transforming the pixel coordinates $\boldsymbol{x}_1'$ into the "normalized" coordinates $\tilde{\boldsymbol{x}}_i$ via

$$\tilde{\boldsymbol{x}}_i \doteq H_i \boldsymbol{x}_i' = \begin{bmatrix} 1/\sigma_{x_i} & 0 & -\mu_{x_i}/\sigma_{x_i} \\ 0 & 1/\sigma_{y_i} & -\mu_{y_i}/\sigma_{y_i} \\ 0 & 0 & 1 \end{bmatrix} \begin{bmatrix} x_i' \\ y_i' \\ 1 \end{bmatrix}, \tag{6.77}$$

where $\mu_{x_i}$ is the average (or mean) and $\sigma_{x_i}$ is the standard deviation of the set of $x$-coordinates $\{(x_i')^j\}_{j=1}^n$ in the $i$th image $i = 1, 2$,

$$\mu_{x_i} \doteq \frac{1}{n}\sum_{j=1}^n (x_i')^j, \quad \sigma_{x_i} \doteq \sqrt{\frac{1}{n}\sum_{j=1}^n [(x_i')^j - \mu_{x_i}]^2}; \tag{6.78}$$

and $\mu_{y_i}$ and $\sigma_{y_i}$ are defined similarly. After the transformation, the normalized coordinates are $\tilde{x}_1 = H_1 x_1'$ and $\tilde{x}_2 = H_2 x_2'$, and the epipolar constraint becomes

$$x_2'^T F x_1' = \tilde{x}_2^T \underbrace{H_2^{-T} F H_1^{-1}}_{\tilde{F}} \tilde{x}_1 = 0. \tag{6.79}$$

Now one can use the above eight-point linear algorithm with the new image pairs $(\tilde{x}_1, \tilde{x}_2)$ to estimate the matrix $\tilde{F} \doteq H_2^{-T} F H_1^{-1}$ first and then recover $F$ as $F = H_2^T \tilde{F} H_1$.

Notice that this normalization has little effect in the calibrated case, since metric coordinates are expressed relative to the optical center, which is often close to the center of the image.

*Number of points*

The number of points (eight or more) assumed to be given in the algorithm above is mostly for convenience. Technically, by utilizing additional algebraic properties of $F$, we may reduce the number of points necessary. For instance, knowing $\det(F) = 0$, we may relax the condition $\text{rank}(\mathcal{X}) = 8$ to $\text{rank}(\mathcal{X}) = 7$, and get two solutions $F_1^s$ and $F_2^s \in \mathbb{R}^9$ from the 2-D null space of $\mathcal{X}$. Nevertheless, there is usually only one $\alpha \in \mathbb{R}$ such that

$$\det(F_1 + \alpha F_2) = 0.$$

Once $\alpha$ is found from solving the above equation, one has $F = F_1 + \alpha F_2$. Therefore, seven points are all that one needs to have a relatively simple algorithm for estimating the fundamental matrix $F$.

The $\det(F) = 0$ constraint has also been used in explicit parametrization of the fundamental matrix assuming that one of the columns can be written as a linear combination of the other two

$$F = \begin{bmatrix} f_1 & f_4 & \alpha f_1 + \beta f_4 \\ f_2 & f_5 & \alpha f_2 + \beta f_5 \\ f_3 & f_6 & \alpha f_3 + \beta f_6 \end{bmatrix}. \tag{6.80}$$

By assuming that one of the columns can be written as a linear combination of the other two, we have introduced a singularity, when the two columns we choose as a basis are linearly dependent. Geometrically this corresponds to the second epipole being at infinity. In fact, in the parametrization above, the coefficients $\alpha, \beta$ are related to the homogeneous coordinates of the epipole in the second view $e_2 = [\alpha, \beta, -1]^T$, since $Fe_2 = 0$. This singularity can be detected and eliminated by choosing an alternative basis. Additional symmetric parametrizations of the fundamental matrix in terms of both epipoles has been suggested by [Luong and Vieville, 1994], and will be illustrated in section 11.3.1 in Chapter 11.

*Improvement by nonlinear optimization*

Similarly as in the calibrated case, the eight-point algorithm is suboptimal in the presence of noise in image correspondences. Ideally, under the same assumptions as in Chapter 5, the optimal estimate of the fundamental matrix $F$ and the associated projective structure can be obtained by minimizing the reprojection error

$$\phi(\boldsymbol{x}, F) \doteq \sum_{j=1}^{n} \|\tilde{\boldsymbol{x}'}_1^j - \boldsymbol{x}'^j_1\|_2^2 + \|\tilde{\boldsymbol{x}'}_2^j - \boldsymbol{x}'^j_2\|_2^2, \tag{6.81}$$

where $\tilde{\boldsymbol{x}'}_1^j, \tilde{\boldsymbol{x}'}_2^j$ are noisy measurements, and $\boldsymbol{x}'^j_1, \boldsymbol{x}'^j_2$ are the ideal corresponding points that satisfy the epipolar constraint $\boldsymbol{x}'^{jT}_2 F \boldsymbol{x}'^j_2 = 0$.

To obtain an approximation of the reprojection error, we assume the same discrepancy model between the ideal correspondences and the measurements as in the calibrated case (Chapter 5), $\tilde{\boldsymbol{x}'}_i^j = \boldsymbol{x}'^j_i + w_i^j$, with $w_1^j$ and $w_2^j$ of i.i.d. normal distribution $\mathcal{N}(0, \sigma^2)$. Substituting this model into the epipolar constraint, we obtain

$$\tilde{\boldsymbol{x}'}_2^{jT} F \tilde{\boldsymbol{x}'}_1^j = w_2^{jT} F \tilde{\boldsymbol{x}'}_1^j + \tilde{\boldsymbol{x}'}_2^{jT} F w_1 + w_2^{jT} F w_1^j. \tag{6.82}$$

Since the image coordinates $\tilde{\boldsymbol{x}'}_1^j$ and $\tilde{\boldsymbol{x}'}_2^j$ are usually of magnitude larger than $w_1^j$ and $w_2^j$ the last term in the equation above can be omitted. An approximate cost function that only takes into account the first-order effects of the noise is given by

$$\phi(F) = \sum_{j=1}^{n} \frac{\left(\tilde{\boldsymbol{x}'}_2^{jT} F \tilde{\boldsymbol{x}'}_1^j\right)^2}{\|\widehat{e}_3 F \tilde{\boldsymbol{x}'}_1^j\|^2 + \|\tilde{\boldsymbol{x}'}_2^{jT} F \widehat{e}_3\|^2}, \tag{6.83}$$

which is the uncalibrated approximation to the expression (5.95) derived in Section 5.A. This simplified cost function has been shown in [Sampson, 1982] to be a first order approximation of the reprojection error, called the *Sampson distance*. The solution to the minimization of $\phi(F)$ calls for nonlinear optimization techniques discussed in Appendix C. More detailed discussion on different choices of objective functions and parametrizations of the fundamental matrix can be found in [Torr and Murray, 1997, Torr and Zisserman, 2000].

The above simplified objective function is also commonly used in the context of linear, iteratively reweighted method for estimation of the fundamental matrix [Torr and Murray, 1997]. In such case the denominator is considered to be a weight in the weighted linear least squares estimation problem. In the weight computation stage, the value of $F$ from the previous iteration is used.

### *Uncalibrated homography for a planar scene*

Just as in the calibrated case, the epipolar constraint will not be sufficient to describe relationships between two images of a planar scene (Section 5.3), due to the fact that a plane is a critical surface for the epipolar constraint (Exercise 5.14). Nevertheless, it is straightforward to generalize the results in Section 5.3 on planar homography to the uncalibrated case. We leave this to the reader as an exercise (Exercise 6.12).

## 6.B Properties of Kruppa's equations

In this appendix, we study the properties of Kruppa's equations introduced in Section 6.6. Historically, Kruppa's equations were introduced to the computer vision community for camera autocalibration (or self-calibration). Despite interesting algebraic properties, calibration algorithms based on Kruppa's equations are often ill-conditioned. The detailed study provided below will help the reader understand the reason for this difficulty. To complete the picture, through Section 6.B.2, we also hope to reveal the connection between Kruppa's equations and the stratified reconstruction approach introduced in the chapter.

**Proposition 6.11 (Kruppa's equations from the fundamental matrix).** *After eliminating $\lambda$ from the matrix Kruppa's equation $FS^{-1}F^T = \lambda^2 \widehat{T'} S^{-1} (\widehat{T'})^T$, one obtains at most two algebraically independent equations in the unknown entries of $S^{-1}$.*

*Proof.* Since the fundamental matrix $F = \widehat{T'}KRK^{-1}$ can be recovered only up to a scalar factor, let us assume, for the moment, that $\|T'\| = \|KT\| = 1$. Let $R_0 \in SO(3)$ be a rotation matrix such that

$$e_3 = R_0 T', \tag{6.84}$$

where $e_3 = [0,\ 0,\ 1]^T$. Such a matrix always exists, since $SO(3)$ preserves the norm of vectors. Indeed, there are infinitely many $R_0$ that satisfy this equation. Consider now the matrix $D \doteq R_0 F$. A convenient expression for it is given by

$$D \doteq R_0 F = \widehat{e}_3 R_0 K R K^{-1} \doteq \begin{bmatrix} d_1^T \\ d_2^T \\ 0 \end{bmatrix} \in \mathbb{R}^{3\times 3}. \tag{6.85}$$

The zero in the last row comes from our choice of $R_0$ as follows:

$$\begin{aligned} D &= R_0 \widehat{T'} K R K^{-1} = R_0 \widehat{T'} R_0^T R_0 K R K^{-1} \\ &= \widehat{R_0 T'} R_0 K R K^{-1} = \widehat{e}_3 R_0 (K R K^{-1}). \end{aligned}$$

Now we see that since $e_3$ is in the (left) null space of $D$, its last row must be zero. The remaining two rows are given by

$$\begin{cases} d_1^T = (-e_2^T R_0)(KRK^{-1}), \\ d_2^T = (e_1^T R_0)(KRK^{-1}). \end{cases} \tag{6.86}$$

Now the crucial observation comes from the fact that the row vectors $d_1^T$ and $d_2^T$ are obtained from the row vectors

$$k_1^T \doteq -e_2^T R_0, \quad k_2^T \doteq e_1^T R_0 \quad \in \mathbb{R}^3,$$

respectively, through a transformation $KRK^{-1}$, which is nothing but a rotation $R$ expressed in the distorted space. Since inner product is preserved under rotation in a Euclidean space, the inner product on the row vectors

$$\langle u^T, v^T \rangle_S = u^T K K^T v = u^T S^{-1} v \tag{6.87}$$

must be preserved under rotation in the distorted space:

$$\begin{cases} \langle d_1^T, d_2^T \rangle_S = \langle k_1^T, k_2^T \rangle_S, \\ \langle d_1^T, d_1^T \rangle_S = \langle k_1^T, k_1^T \rangle_S, \\ \langle d_2^T, d_2^T \rangle_S = \langle k_2^T, k_2^T \rangle_S. \end{cases} \tag{6.88}$$

So far, we have relied on the assumption that $\|T'\| = 1$. In general, that is not the case. Thus, from the definition (6.85) of the matrix $D$, its row vectors $d_1, d_2$ must be scaled accordingly by the norm of $T'$. Taking this into account, we can write the conservation of the inner product as

$$\begin{cases} d_1^T S^{-1} d_2 = \lambda^2 k_1^T S^{-1} k_2, \\ d_1^T S^{-1} d_1 = \lambda^2 k_1^T S^{-1} k_1, \\ d_2^T S^{-1} d_2 = \lambda^2 k_2^T S^{-1} k_2, \end{cases} \tag{6.89}$$

where $\lambda \doteq \|T'\|$. In order to eliminate the dependency on $\lambda$, we can consider ratios of inner products. We obtain, therefore two independent constraints on the six independent components of the symmetric matrix $S^{-1} = KK^T$, which are traditionally called *Kruppa's equations*:

$$\boxed{\frac{\langle d_1^T, d_2^T \rangle_S}{\langle k_1^T, k_2^T \rangle_S} = \frac{\langle d_1^T, d_1^T \rangle_S}{\langle k_1^T, k_1^T \rangle_S} = \frac{\langle d_2^T, d_2^T \rangle_S}{\langle k_2^T, k_2^T \rangle_S}.} \tag{6.90}$$

In general, these two equations are algebraically independent. □

**Remark 6.12.** *One must be aware that solving Kruppa's equations for camera calibration is not equivalent to the camera autocalibration problem in the sense that there may exist solutions of Kruppa's equations that are not solutions of a "valid" calibration. It is possible that even if a sufficient set of images is given,*[16] *the associated Kruppa's equations do not give enough constraints to solve for the calibration matrix $K$. See Section 6.B.2 for a more detailed account of these issues.*

Given a sufficient number of images, we can take pairs of views and obtain Kruppa's equations as above. Each pair supposedly gives two equations, hence, in principle, at least three images (hence three pairs of images) are needed in order to solve for all the five unknown parameters in the calibration matrix $K$. Unfortunately, there is no closed-form solution known for Kruppa's equations, so numerical methods must be deployed to look for an answer. In general, however, since Kruppa's equations (6.90) are nonlinear in $S^{-1}$, numerical solutions

[16]Here we are deliberately vague on the word "sufficient," which simply suggests that information about the calibration matrix is fully encoded in the images.

are often problematic [Bougnoux, 1998, Luong and Faugeras, 1997]. Nevertheless, under many special camera motions, Kruppa's equations can be converted to linear equations that are much easier to solve. In the next section we explore these special but important cases.

### 6.B.1 Linearly independent Kruppa's equations under special motions

Note that the use of Kruppa's equations for a particular camera motion assumes that one has access to the camera and can move it at will (or that one is guaranteed that the motion satisfies the requirements). This is tantamount to having control over the imaging process, at which point one might as well employ a calibration rig and follow the procedures outlined in Section 6.5.2. In other words, the techniques we describe in this section cannot be applied when only images are given, with no additional information on the camera motion. Nevertheless, there are practical situations where these techniques come of use. For instance, pure rotation about the optical center can be approximated when the camera undergoes a pan-tilt motion and objects in the scene are "far away enough." The screw motion (when the axis of rotation and the direction of translation are parallel, defined in Chapter 2) and motions on a plane (when the axis of rotation is perpendicular to the direction of translation) are also important cases to be covered in this section, especially the latter, since most autonomous robots move on a plane.

Given a fundamental matrix $F = \widehat{T'}KRK^{-1}$ with $T'$ of unit length, the normalized matrix Kruppa's equation (6.60) can be rewritten as

$$\widehat{T'}(S^{-1} - KRK^{-1}S^{-1}K^{-T}R^TK^T)\widehat{T'}^T = 0. \tag{6.91}$$

According to this form, if we define $C = KRK^{-1}$ and two linear maps, a Lyapunov map $\sigma : \mathbb{R}^{3\times3} \to \mathbb{R}^{3\times3}; X \mapsto X - CXC^T$, and another linear map $\tau : \mathbb{R}^{3\times3} \to \mathbb{R}^{3\times3}; Y \mapsto \widehat{T'}Y\widehat{T'}^T$, then the solution $S^{-1}$ of equation (6.91) is exactly the (symmetric and real) kernel of the composition map

$$\tau \circ \sigma : \quad \mathbb{R}^{3\times3} \xrightarrow{\sigma} \mathbb{R}^{3\times3} \xrightarrow{\tau} \mathbb{R}^{3\times3}, \tag{6.92}$$

i.e. the symmetric and real solution of $\tau(\sigma(X)) = 0$. We call the composite map $\kappa$, and its symmetric real kernel $\mathrm{SRker}(\kappa)$. This interpretation of Kruppa's equations decomposes the effects of rotation and translation: if there is no translation, i.e. $T = 0$, then the map $\tau$ is the identity; if the translation is nonzero, the kernel is enlarged due to the composition with the map $\tau$. In general, the kernel of $\sigma$ is only two-dimensional, following Lemma 6.8.

*Pure rotation*

As we have discussed as an example in Section 6.4.4, a simple way to calibrate a camera is to rotate it about two different axes. For each rotation, we can get a matrix of the form $C = KRK^{-1}$, and hence $S^{-1} = KK^T$ is a solution to the

Lyapunov equation

$$S^{-1} - CS^{-1}C^T = 0. \tag{6.93}$$

This matrix equation in general gives three linearly independent equations in the (unknown) entries of $S^{-1}$. Thus the autocalibration problem in this case becomes linear, and a unique solution is usually guaranteed if the two rotation axes are different (Theorem 6.9). However, in practice, one usually does not know exactly where the center of a camera is, and "rotating" the camera does not necessarily guarantee that the rotation is about the exact optical center. In a general case with translational motion, however, the symmetric real kernel of the composition map $\kappa = \tau \circ \sigma$ is usually three-dimensional; hence the conditions for a unique calibration are much more complicated. Furthermore, as we will see soon, the dimension of the kernel is not always three. In many cases of practical importance, this kernel may become four-dimensional instead, which in fact corresponds to certain degeneracy of Kruppa's equations. Solutions for Kruppa's equations are even more complicated due to the unknown scale $\lambda$. Nevertheless, the following lemma shows that the conditions on uniqueness depend only on the camera motion, which will to some extent simplify our analysis.

**Lemma 6.13.** *Given a fundamental matrix $F = \widehat{T'}KRK^{-1}$ with $T' = KT$, a symmetric matrix $X \in \mathbb{R}^{3\times 3}$ is a solution of $FXF^T = \lambda^2 \widehat{T'}X\widehat{T'}^T$ if and only if $Y = K^{-1}XK^{-T}$ is a solution of $EYE^T = \lambda^2 \widehat{T}Y\widehat{T}^T$ with $E = \widehat{T}R$.*

The proof of this lemma consists in simple algebraic manipulations. More important, however, is its interpretation: given a set of fundamental matrices $F_i = \widehat{T_i'}KR_iK^{-1}$ with $T_i' = KT_i, i = 1, 2, \ldots, m$, there is a one-to-one correspondence between the set of solutions of the equations

$$F_iXF_i^T = \lambda_i^2 \widehat{T_i'}X\widehat{T_i'}^T, \quad i = 1, 2, \ldots, m,$$

and the set of solutions of the equations

$$E_iYE_i^T = \lambda_i^2 \widehat{T}_iY\widehat{T}_i^T, \quad i = 1, 2, \ldots, m,$$

where $E_i = \widehat{T}_iR_i$ are essential matrices associated with the given fundamental matrices. Note that these essential matrices are determined only by the motion of the camera. Therefore, we conclude:

> *Conditions for the uniqueness of the solution of Kruppa's equations depend only on the camera motion.*

Our next task is then to study *how* the solutions of Kruppa's equations depend on the camera motion and under what conditions the solutions can be simplified.

*Translation perpendicular or parallel to the rotation axis*

From the derivation of Kruppa's equations (6.90) or (6.61), we observe that the reason why they are nonlinear is that we do not usually know the scale $\lambda$. It is then helpful to know under what conditions the scale factor can be inferred easily. Here we will study two special cases for which we are able to derive $\lambda$ directly from $F$. The fundamental matrix can then be *normalized*, and we can therefore solve for the camera calibration parameters from the normalized matrix Kruppa's equations, which are linear! These two cases are those in which the rotation axis is *parallel* or *perpendicular* to the direction of translation. That is, if the motion is represented by $(R, T) \in SE(3)$ and the unit vector $\omega \in \mathbb{R}^3$ is the axis of $R \in SO(3)$, the two cases are:

1. $\omega$ is parallel to $T$ (i.e. the screw motion), and
2. $\omega$ is perpendicular to $T$ (e.g., the planar motion).

As we will see below, these two cases are of great theoretical importance: not only does the algorithm becomes linear, but it also reveals certain subtleties in Kruppa's equations and explains when the nonlinear Kruppa's equations might become ill-conditioned.

Although motions with translation parallel or perpendicular to rotation are only a zero-measure subset of $SE(3)$, they are very commonly encountered in applications: many image sequences are taken by moving the camera around an object with an *orbital motion* as shown in Figure 6.17, in which case the rotation axis and translation direction are perpendicular to each other.

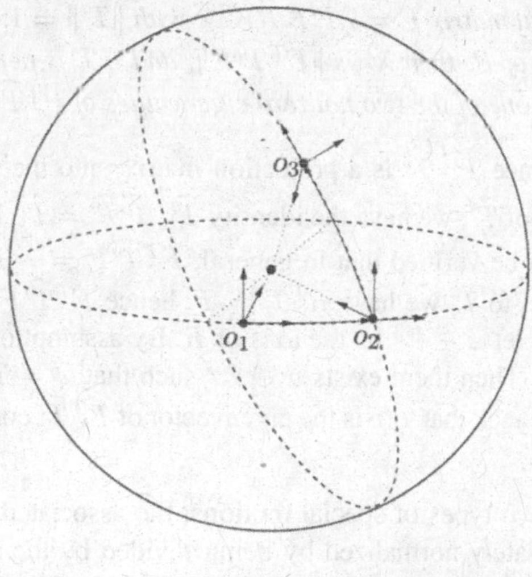


Figure 6.17. Two consecutive orbital motions with independent rotations: the camera optical axis is always pointing to the center of the globe, where the object resides.

**Theorem 6.14 (Normalization of Kruppa's equations).** *Consider an unnormalized fundamental matrix* $F = \widehat{T'}KRK^{-1}$, *where* $R = e^{\widehat{\omega}\theta}$, $\theta \in (0, \pi)$, *and the axis* $\omega \in \mathbb{R}^3$ *is parallel or perpendicular to* $T = K^{-1}T'$. *Let* $\mathbf{e} = T'/\|T'\| \in \mathbb{R}^3$. *Then if* $\lambda \in \mathbb{R}$ *and a positive definite matrix* $S$ *are a solution to the matrix Kruppa's equation* $FS^{-1}F^T = \lambda^2\widehat{\mathbf{e}}S^{-1}\widehat{\mathbf{e}}^T$, *we must have* $\lambda^2 = \|T'\|^2$.

*Proof.* Lemma 6.13 implies that we need to prove only the essential case: if $\gamma \in \mathbb{R}$ and a positive definite matrix $Y$ is a solution to the matrix Kruppa's equation: $\widehat{T}RYR^T\widehat{T}^T = \gamma^2\widehat{T}Y\widehat{T}^T$ associated with the essential matrix $\widehat{T}R$, then we must have $\gamma^2 = 1$. In other words, $Y$ is automatically a solution of the normalized matrix Kruppa's equation $\widehat{T}RYR^T\widehat{T}^T = \widehat{T}Y\widehat{T}^T$.

Without loss of generality, we assume $\|T\| = 1$. For the parallel case, let $x \in \mathbb{R}^3$ be a vector of unit length in the plane spanned by the column vectors of $\widehat{T}$. All such $x$ lie on a unit circle. There exists $x_0 \in \mathbb{R}^3$ on the circle such that $x_0^T Y x_0$ is maximum. We then have $x_0^T RYR^T x_0 = \gamma^2 x_0^T Y x_0$, and hence $\gamma^2 \leq 1$. Similarly, if we pick $x_0$ such that $x_0^T Y x_0$ is minimum, we have $\gamma^2 \geq 1$. Therefore, $\gamma^2 = 1$. For the perpendicular case, since the columns of $\widehat{T}$ span the subspace that is perpendicular to the vector $T$, the eigenvector $\omega$ of $R$ is in this subspace. Thus we have $\omega^T RYR^T\omega = \gamma^2\omega^T Y\omega \Rightarrow \omega^T Y\omega = \gamma^2\omega^T Y\omega$. Hence $\gamma^2 = 1$ if $Y$ is positive definite. □

This theorem claims that for the two types of special motions considered here, there is no solution for $\lambda$ in Kruppa's equation (6.61) besides the true scale of the fundamental matrix. Therefore, we can decompose the problem into finding $\lambda$ first and then solving for $S$ or $S^{-1}$. The following theorem allows us to directly compute the scale $\lambda$ in the two special cases for a given fundamental matrix.

**Theorem 6.15 (Normalization of the fundamental matrix).** *Given an unnormalized fundamental matrix $F = \lambda\widehat{T'}KRK^{-1}$ with $\|T'\| = 1$, if $T = K^{-1}T'$ is parallel to the axis of $R$, then $\lambda^2$ is $\|F^T\widehat{T'}F\|$, and if $T$ is perpendicular to the axis of $R$, then $\lambda$ is one of the two nonzero eigenvalues of $F^T\widehat{T'}$.*

*Proof.* Note that since $\widehat{T'}^T\widehat{T'}$ is a projection matrix onto the plane spanned by the column vectors of $\widehat{T'}$, we have the identity $\widehat{T'}^T\widehat{T'}\widehat{T'} = \widehat{T'}$. First we prove the parallel case. It can be verified that in general, $F^T\widehat{T'}F = \lambda^2\widehat{KR^TT}$. Since the axis of $R$ is parallel to $T$, we have $R^TT = T$, hence $F^T\widehat{T'}F = \lambda^2\widehat{T'}$. For the perpendicular case, let $\omega \in \mathbb{R}^3$ be the axis of $R$. By assumption, $T = K^{-1}T'$ is perpendicular to $\omega$. Then there exists $v \in \mathbb{R}^3$ such that $\omega = \widehat{T}K^{-1}v$. It is then straightforward to check that $\widehat{T'}v$ is the eigenvector of $F^T\widehat{T'}$ corresponding to the eigenvalue $\lambda$. □

Then for these two types of special motions, the associated fundamental matrix can be immediately normalized by being divided by the scale $\lambda$. Once the fundamental matrices are normalized, the problem of finding the calibration matrix $S^{-1}$ from the normalized matrix Kruppa's equations (6.60) becomes a simple *linear* one. A normalized matrix Kruppa's equation in general imposes *three* linearly independent constraints on the unknown calibration matrix given by (6.88). However, this is no longer the case for the special motions that we are considering here.

**Theorem 6.16 (Degeneracy of the normalized Kruppa's equations).** *Consider a camera motion $(R, T) \in SE(3)$, where $R = e^{\widehat{\omega}\theta}$ has the angle $\theta \in (0, \pi)$.*

*If the axis $\omega \in \mathbb{R}^3$ is parallel or perpendicular to $T$, then the normalized matrix Kruppa's equation $\widehat{T}RYR^T\widehat{T}^T = \widehat{T}Y\widehat{T}^T$ imposes only two linearly independent constraints on the symmetric matrix $Y$.*

*Proof.* For the parallel case, by restricting $Y$ to the plane spanned by the column vectors of $\widehat{T}$, it yields a symmetric matrix $\tilde{Y}$ in $\mathbb{R}^{2\times 2}$. The rotation matrix $R \in SO(3)$ restricted to this plane is a rotation $\tilde{R} \in SO(2)$. The normalized matrix Kruppa's equation is then equivalent to $\tilde{Y} - \tilde{R}\tilde{Y}\tilde{R}^T = 0$. Since $0 < \theta < \pi$, this equation imposes exactly two constraints on the three-dimensional space of $2 \times 2$ symmetric real matrices. The identity $I_{2\times 2}$ is the only solution. Hence, the normalized Kruppa's equation imposes exactly two linearly independent constraints on $Y$. For the perpendicular case, since $\omega$ is in the plane spanned by the column vectors of $\widehat{T}$, there exists $v \in \mathbb{R}^3$ such that $[\omega, v]$ form an orthonormal basis of the plane. Then the normalized matrix Kruppa's equation is equivalent to

$$\widehat{T}RYR^T\widehat{T}^T = \widehat{T}Y\widehat{T}^T \quad \Leftrightarrow \quad [\omega, v]^T RYR^T[\omega, v] = [\omega, v]^T Y [\omega, v].$$

Since $R^T\omega = \omega$, the above matrix equation is equivalent to the two equations

$$v^T RY\omega = v^T Y\omega, \quad v^T RYR^T v = v^T Y v. \tag{6.94}$$

These are the only two constraints on $Y$ imposed by the normalized Kruppa's equation. □

According to this theorem, although we can normalize the fundamental matrix when the rotation axis and the direction of translation are parallel or perpendicular, we get only two independent constraints from the resulting (normalized) Kruppa's equations; i.e. one of the three linear equations in (6.88) depends on the other two. So, degeneracy does occur with normalized Kruppa's equations. Hence, for these motions, we still need at least three such fundamental matrices to uniquely determine the unknown calibration.

**Example 6.17 (A numerical example).** Suppose that

$$K = \begin{bmatrix} 4 & 0 & 2 \\ 0 & 4 & 2 \\ 0 & 0 & 1 \end{bmatrix}, \quad R = \begin{bmatrix} \cos(\pi/4) & 0 & \sin(\pi/4) \\ 0 & 1 & 0 \\ -\sin(\pi/4) & 0 & \cos(\pi/4) \end{bmatrix}, \quad T = \begin{bmatrix} 2 \\ 0 \\ 0 \end{bmatrix}.$$

Notice that here the translation and the rotation axes are orthogonal ($R \perp T$). Then the (normalized) epipole is $T' = KT/\|KT\| = [1, 0, 0]^T$, and suppose that the (unnormalized) fundamental matrix is $F = \widehat{KT}KRK^{-1}$

$$F = \begin{bmatrix} 0 & 0 & 0 \\ \sqrt{2} & 0 & -6\sqrt{2} \\ -2\sqrt{2} & 8 & 12\sqrt{2} - 16 \end{bmatrix}, \quad T' = \begin{bmatrix} 1 \\ 0 \\ 0 \end{bmatrix}.$$

Note that the scale difference between $KT$ and the normalized epipole $T'$ is $\lambda = \|KT\| = 8$. The eigenvalues of the matrix $F^T\widehat{T'}$ are $\{0, 8, 8.4853\}$. The middle one is exactly the missing scale. It is then easy to verify that $S^{-1} = KK^T$ is a solution to the normalized

Kruppa's equations, because

$$FS^{-1}F^T = \begin{bmatrix} 0 & 0 & 0 \\ 0 & 64 & -128 \\ 0 & -128 & 1280 \end{bmatrix} = \lambda^2 \widehat{T'} S^{-1} \widehat{T'}.$$

Denote the entries of the symmetric matrix by

$$S^{-1} = \begin{bmatrix} s_1 & s_2 & s_3 \\ s_2 & s_4 & s_5 \\ s_3 & s_5 & s_6 \end{bmatrix}.$$

The (unnormalized) Kruppa's equation $FS^{-1}F^T = \lambda^2 \widehat{T'} S^{-1} \widehat{T'}^T$ gives rise to

$$\begin{aligned} \lambda^2 s_6 &= 2s_1 - 24s_3 + 72s_6, & (i) \\ -\lambda^2 s_5 &= -4s_1 + 8\sqrt{2}s_2 + (48 - 16\sqrt{2})s_3 - 48\sqrt{2}s_5 + (-144 + 96\sqrt{2})s_6, & (ii) \\ \lambda^2 s_4 &= 8s_1 - 32\sqrt{2}s_2 - (96 - 64\sqrt{2})s_3 + 64s_4 + (-256 + 192\sqrt{2})s_5 \\ &\quad +(544 - 384\sqrt{2})s_6. & (iii) \end{aligned}$$

We know here that $\lambda^2 = 64$ from the eigenvalues of $F^T \widehat{T'}$. Substituting it in the above equations, we only get two linearly independent equations in $s_1, s_2, \ldots, s_6$, since

$$-4 \cdot [(i) + (ii)] = (iii).$$

Therefore, after normalization, we only get two linearly independent constraints on the entries of $S^{-1}$. This is consistent with Theorem 6.16. ■

Under these special motions, what happens to the unnormalized Kruppa's equations? If we do not normalize the fundamental matrix and directly use the unnormalized Kruppa's equations (6.90) to solve for calibration, the two nonlinear equations in (6.90) also might still be algebraically independent but numerically ill conditioned due to the linear dependency around the true solution (i.e. when $\lambda$ is around the true scale). Hence, normalizing Kruppa's equations under such special motions becomes crucial for obtaining numerically reliable solutions to the camera calibration.

**Remark 6.18 (Rotation by 180°).** *Theorem 6.16 does not cover the case in which the rotation angle $\theta$ is $\pi$ radians (i.e. 180°). However, if one allows the rotation to be $\pi$, the solutions of the normalized Kruppa's equations are even more troublesome. For example, $\widehat{T} e^{\widehat{\omega}\pi} = -\widehat{T}$ if $\omega$ is of unit length and parallel to $T$. Therefore, if $R = e^{\widehat{\omega}\pi}$, the corresponding Kruppa's equations are completely degenerate, and they impose no constraints on the calibration matrix.*

**Remark 6.19 (Number of solutions for the perpendicular case).** *Although Theorem 6.15 claims that for the perpendicular case $\lambda$ is one of the two nonzero eigenvalues of $F^T \widehat{T'}$, unfortunately, there is no way to tell which one is the correct one. Simulations show that it could be either the larger or smaller one. Therefore, in a numerical algorithm, for $m \geq 3$ given fundamental matrices, one needs to consider all possible $2^m$ combinations. According to Theorem 6.14, in the noise-free case, only one of the solutions can be positive definite, which corresponds to the true calibration.*

We summarize in Table 6.6 facts about the special motions studied in this subsection.

| Cases | Type of constraints | # of constraints on $S^{-1}$ |
|---|---|---|
| $T = 0$ | Lyapunov equation (linear) | 3 |
| $R \perp T$ | Normalized Kruppa (linear) | 2 |
| $R \parallel T$ | Normalized Kruppa (linear) | 2 |
| Others | Unnormalized Kruppa (nonlinear) | 2 |

Table 6.6. Maximum numbers of independent constraints on the camera calibration. Here we assume that the rotation angle $\theta$ satisfies $0 < \theta < \pi$.

## 6.B.2 *Cheirality constraints*

Solutions that satisfy Kruppa's equations do not necessarily give rise to physically viable reconstruction, in the sense that all the reconstructed points should be at finite depth (in front of the camera), and that certain ordering constraints among points should be preserved (i.e. which points are in front of which others). As we will show in this subsection, these ambiguous solutions to Kruppa's equations are closely related to the stratified reconstruction approach through the notion of plane at infinity. Such ordering constraints, which are not present in Kruppa's equations, are called *cheirality constraints*, named by [Hartley, 1998a]. The following theorem reveals interesting relationships among Kruppa's equations, Lyapunov equations, plane at infinity, and cheirality.

**Theorem 6.20 (Kruppa's equations and cheirality).** *Consider a camera with calibration matrix $I$ and motion $(R, T)$. If $T \neq 0$, among all the solutions $Y = K^{-1}K^{-T}$ of Kruppa's equation $EYE^T = \lambda^2 \widehat{T} Y \widehat{T}^T$ associated with $E = \widehat{T}R$, only those that guarantee that $KRK^{-1} \in SO(3)$ provide a* valid Euclidean reconstruction *of both camera motion and scene structure in the sense that any other solution pushes some plane $P \subset \mathbb{R}^3$ to infinity, and feature points on different sides of the plane $P$ have different signs of recovered depth.*

*Proof.* The images $\boldsymbol{x}_1, \boldsymbol{x}_2$ of any point $p \in \mathbb{R}^3$ satisfy the coordinate transformation

$$\lambda_2 \boldsymbol{x}_2 = \lambda_1 R \boldsymbol{x}_1 + T.$$

If there exists $Y = K^{-1}K^{-T}$ such that $EYE^T = \lambda^2 \widehat{T} Y \widehat{T}^T$ for some $\lambda \in \mathbb{R}$, then the matrix $F = K^{-T}EK^{-1} = \widehat{T'}KRK^{-1}$ is also an essential matrix with $T' = KT$; that is, there exists $\tilde{R} \in SO(3)$ such that $F = \widehat{T'}\tilde{R}$. Under the new calibration $K$, the coordinate transformation is

$$\lambda_2 K \boldsymbol{x}_2 = \lambda_1 KRK^{-1}(K\boldsymbol{x}_1) + T'.$$

Since $F = \widehat{T'}\tilde{R} = \widehat{T'}KRK^{-1}$, we have $KRK^{-1} = \tilde{R} + T'v^T$ for some $v \in \mathbb{R}^3$. Then the above equation becomes $\lambda_2 K\boldsymbol{x}_2 = \lambda_1 \tilde{R}(K\boldsymbol{x}_1) + \lambda_1 T'v^T(K\boldsymbol{x}_1) + T'$. Let $\beta = \lambda_1 v^T(K\boldsymbol{x}_1) \in \mathbb{R}$. We can further rewrite the equation as

$$\lambda_2 K\boldsymbol{x}_2 = \lambda_1 \tilde{R}K\boldsymbol{x}_1 + (\beta + 1)T'. \tag{6.95}$$

Nevertheless, with respect to the solution $K$, the reconstructed images $K\boldsymbol{x}_1, K\boldsymbol{x}_2$ and $(\tilde{R}, T')$ must also satisfy

$$\gamma_2 K\boldsymbol{x}_2 = \gamma_1 \tilde{R}K\boldsymbol{x}_1 + T' \tag{6.96}$$

for some scale factors $\gamma_1, \gamma_2 \in \mathbb{R}$. Now we prove by contradiction that $v \neq 0$ cannot occur in a valid Euclidean reconstruction. Suppose that $v \neq 0$, and we define the plane $P = \{\boldsymbol{X} \in \mathbb{R}^3 | v^T\boldsymbol{X} = -1\}$. Then for any $\boldsymbol{X} = \lambda_1 K\boldsymbol{x}_1 \in P$, we have $\beta = -1$. Hence, from (6.95), $K\boldsymbol{x}_1$ and $K\boldsymbol{x}_2$ satisfy $\lambda_2 K\boldsymbol{x}_2 = \lambda_1 \tilde{R}K\boldsymbol{x}_1$. Since $K\boldsymbol{x}_1$ and $K\boldsymbol{x}_2$ also satisfy (6.96) and $T' \neq 0$, both $\gamma_1$ and $\gamma_2$ in (6.96) must be $\infty$. That is, the plane $P$ is "pushed" to infinity by the solution $K$. For points not on the plane $P$, we have $\beta + 1 \neq 0$. Comparing the two equations (6.95) and (6.96), we get $\gamma_i = \lambda_i/(\beta + 1), i = 1, 2$. Then, for a point on the far side of the plane $P$, i.e. $\beta + 1 < 0$, the recovered depth scale $\gamma$ is negative; for a point on the near side of $P$, i.e. $\beta + 1 > 0$, the recovered depth scale $\gamma$ is positive. Thus, we must have that $v = 0$. □

For the matrix $C = KRK^{-1}$ to be a rotation matrix, it must satisfy the Lyapunov equation $CS^{-1}C^T = S^{-1}$. It is known from Theorem 6.9 that in general, two such Lyapunov equations uniquely determine $S^{-1}$. Thus, as a consequence of Theorem 6.20,

> *A camera calibration can be uniquely determined by two independent rotations regardless of translation if the image of every point in the world is available.*

An intuitive reason for this is provided in Figure 6.18. The theorem above then resolves the apparent discrepancy between Kruppa's equations and the necessary and sufficient condition for a unique calibration: Kruppa's equations do *not* provide sufficient conditions for a valid calibration, i.e. ones that result in a valid Euclidean reconstruction of both the camera motion and scene structure. However, the results given in Theorem 6.20 are somewhat difficult to harness in algorithms. For example, in order to exclude invalid solutions, one needs feature points on or beyond the plane $P$, which in theory could be anywhere in $\mathbb{R}^3$.[17]

**Remark 6.21 (Quasi-affine reconstruction).** *According to the above theorem, if only finitely many feature points are measured, a solution of the calibration matrix $K$ that may allow a valid Euclidean reconstruction should induce a plane $P$ not cutting through the convex hull spanned by all the feature points and camera centers. Such a reconstruction is called quasi-affine in [Hartley, 1998a].*

---

[17]Basically, such constraints give *inequality* (rather than equality) constraints on the possible solutions of camera calibration.

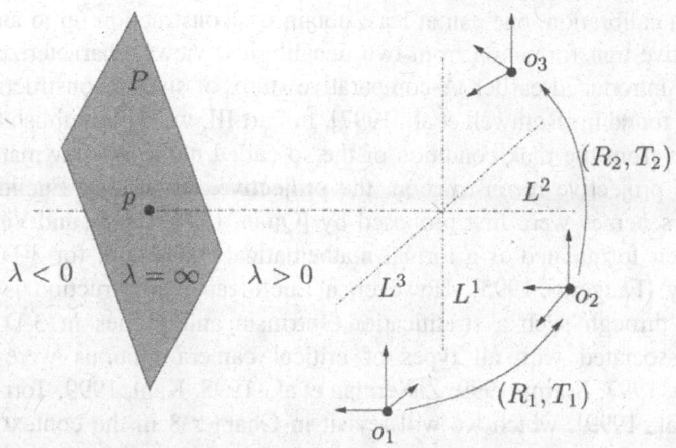


Figure 6.18. A camera undergoes two motions $(R_1, T_1)$ and $(R_2, T_2)$ observing a rig consisting of three straight lines $L^1, L^2, L^3$. Then camera calibration is uniquely determined as long as $R_1$ and $R_2$ have independent rotation axes and rotation angles in $(0, \pi)$, regardless of $T_1, T_2$. This is because for any invalid solution $K$, the associated plane $P$ must intersect the three lines at some point, say $p$ (see the proof of Theorem 6.20). Then the reconstructed depth $\lambda$ of point $p$ with respect to the solution $K$ would be infinite (points beyond the plane $P$ would have negative recovered depth). This gives us a criterion to exclude such invalid solutions.

# Historical notes

The original formulation of Kruppa's equations for camera autocalibration was due to [Maybank and Faugeras, 1992]. Based on that, many camera calibration algorithms were developed [Faugeras et al., 1992, Luong and Faugeras, 1996, Zeller and Faugeras, 1996, Ponce et al., 1994], as were many algorithms for robustly and accurately estimating fundamental matrices [Boufama and Mohr, 1995, Chai and Ma, 1998, Zhang et al., 1995, Torr and Murray, 1997, Zhang, 1998a, Zhang, 1998c, Tang et al., 1999]. A general discussion on estimation in the presence of bilinear constraints can be found in [Konderink and van Doorn, 1997]. However, the numerical stability and algebraic degeneracy of Kruppa's equation limits its use in camera calibration in many practical situations, as pointed out in [Bougnoux, 1998, Ma et al., 2000b]. Many recent autocalibration techniques are based on other intrinsic constraints: the absolute quadric constraints [Heyden and Åström, 1996, Triggs, 1997] or the modulus constraints [Pollefeys and Gool, 1999].

### *Stratification*

Following the work of Koenderink and Van Doorn on affine projection in 1980s, [Quan and Mohr, 1991] showed how to obtain a projective shape (from some reference points), and the idea of affine stratification was initially formulated. [Faugeras, 1992, Hartley et al., 1992] formally showed that, without knowing the camera calibration, one can at least obtain a reconstruction up to an (unknown) projective transformation from two uncalibrated views, characterized by the $H$ matrix introduced earlier. A comparative study of such reconstruction methods can be found in [Rothwell et al., 1997]. In Part III, we will establish that this fact follows from the rank condition of the so-called multiple-view matrix. Starting with a projective reconstruction, the projective, affine, and Euclidean stratification schemes were first proposed by [Quan, 1993, Luong and Vieville, 1994] and later formulated as a formal mathematical framework for 3-D reconstruction by [Faugeras, 1995]. However, a Euclidean reconstruction is not always viable through such a stratification. Intrinsic ambiguities in 3-D reconstruction associated with all types of critical camera motions were studied by [Sturm, 1997, Sturm, 1999, Zisserman et al., 1998, Kahl, 1999, Torr et al., 1999, Ma et al., 1999], which we will revisit in Chapter 8 in the context of multiple images.

*Camera knowledge*

Autocalibration with time-varying camera intrinsic parameters were studied by [Heyden and Åström, 1997, Enciso and Vieville, 1997, Pollefeys et al., 1998]. [Pollefeys et al., 1996] gave a simpler solution to the case in which only the focal length is varying, which is also the case studied by [Rousso and Shilat, 1998]. In practice, one often has certain knowledge on the camera calibration. Different situations then lead to many different calibration or autocalibration techniques. For instance, with a reasonable guess on the camera calibration and motion, [Beardsley et al., 1997] showed that one can actually do better than a projective reconstruction with the so-called quasi-Euclidean reconstruction. Camera calibration under various special motions (e.g., rotation, screw motion, planar motion) was studied by [Hartley, 1994b, Ma et al., 2000b]. [Seo and Hong, 1999] studied further a hybrid case with a rotating and zooming camera.

*Scene knowledge*

As we have seen in this chapter, (partial) knowledge of the scene enables simpler calibration schemes for the camera, such as a calibration rig [Tsai, 1986a], a planar calibration rig [Zhang, 1998b], planar scenes [Triggs, 1998], orthogonality [Svedberg and Carlsson, 1999], and parallelism, i.e. vanishing points. Although we did not mention in the chapter how to detect and compute vanishing points automatically from images, there are numerous mature techniques to perform this task: [Quan and Mohr, 1989, Collins and Weiss, 1990, Caprile and Torre, 1990, Lutton et al., 1994, Shufelt, 1999, Kosecká and Zhang, 2002]. A simple algorithm based on polynomial factorization will be given in the exercises of Chapter 7. Later, in Chapter 10, we will give a unified study of all types of scene knowledge based on the more general notion of "symmetry." Although such knowledge typically enables calibration from a single view, a principled explanation has to be postponed until we have understood thoroughly the geometry of multiple views.

*Stereo rigs*

Although the two-view geometry studied in this chapter and the previous one governs that of a stereo rig, the latter setup often results in special and simpler calibration and reconstruction techniques that we do not cover in detail in this book. Early work on the geometry of stereo rigs can be found in [Yakimovsky and Cunningham, 1978, Weng et al., 1992a, Zisserman et al., 1995, Devernay and Faugeras, 1996] and [Zhang et al., 1996]. More detailed study on stereo rigs can be found in [Horaud and Csurka, 1998, Csurka et al., 1998]. Autocalibration of a stereo rig under planar motion can be found in [Brooks et al., 1996], pure translation in [Ruf et al., 1998], pure rotation in [Ruf and Horaud, 1999a], and articulated motion in [Ruf and Horaud, 1999b].

*The continuous-motion case*

In the continuous-motion case, in contrast to the discrete case, only two camera parameters can be recovered from the continuous epipolar constraint (see Exercise 6.17). The interested reader may refer to [Brooks et al., 1997, Brodsky et al., 1998] for possible ways of conducting calibration in this case.

# Chapter 7
# Estimation of Multiple Motions from Two Views

*The elegance of a mathematical theorem is directly proportional to the number of independent ideas one can see in the theorem and inversely proportional to the effort it takes to see them.*

– George Pólya, *Mathematical Discovery, 1981*

So far we have been concerned with single rigid-body motions. Consequently, the algorithms described can be applied to a camera moving within a static scene, or to a single rigid object moving relative to a camera. In practice, this assumption is rather restrictive: interaction with real-world scenes requires negotiating physical space with multiple objects. In this chapter we consider the case of scenes populated with multiple *rigid* objects moving independently.

For simplicity, we restrict our attention to *two views*, leaving the multiple-view case to Part III. Also, in the interest of generality, we will assume that camera calibration is not available, and therefore we deal with fundamental matrices rather than essential matrices. However, the results we obtain can be easily specialized to the case of calibrated cameras.

Given a number of independently moving objects, our goal in this chapter is threefold. We want to study (a) how many objects are in the scene, (b) the motion and shape of each object, and (c) which point belongs to which object. As we will see, much of what we have learned so far can be generalized to the case of multiple rigid bodies.

## 7.1 Multibody epipolar constraint and the fundamental matrix

From previous chapters, we know that two (homogeneous) images $\boldsymbol{x}_1, \boldsymbol{x}_2 \in \mathbb{R}^3$ of a point $p$ in 3-D space undergoing a rigid-body motion $(R, T)$ (with $T \neq 0$) satisfy the *epipolar constraint*

$$\boldsymbol{x}_2^T F \boldsymbol{x}_1 = 0, \tag{7.1}$$

where $F = \widehat{T}R \in \mathbb{R}^{3\times3}$ is the fundamental matrix (or the essential matrix in the calibrated case).[1] For a simpler notation, in this chapter we will drop the prime superscript "$'$" from the image $\boldsymbol{x}$ even if it is uncalibrated. This will not cause any confusion, since we are not studying calibration here, and all results will be based on the epipolar constraint only, which does not discriminate between calibrated and uncalibrated cameras.

To generalize this notion of epipolar constraint for a single motion to multiple motions, let us first take a look at a simple example:

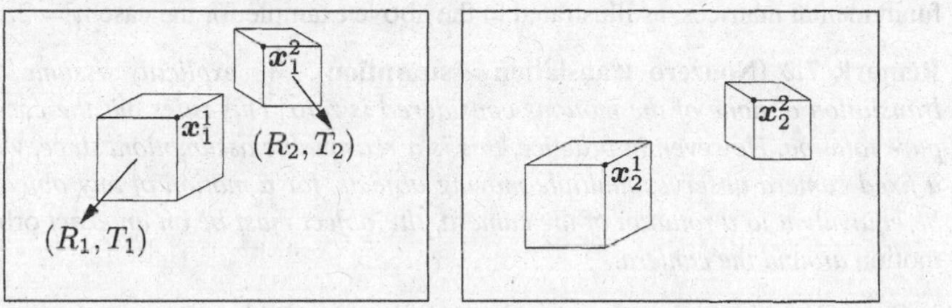


Figure 7.1. Two views of two independently moving objects.

**Example 7.1 (Two rigid-body motions).** Imagine the simplest scenario in which there are two independently moving objects in the scene as shown in Figure 7.1. Each image pair $(\boldsymbol{x}_1^1, \boldsymbol{x}_2^1)$ or $(\boldsymbol{x}_1^2, \boldsymbol{x}_2^2)$ satisfies the equation

$$\left(\boldsymbol{x}_2^T F_1 \boldsymbol{x}_1\right)\left(\boldsymbol{x}_2^T F_2 \boldsymbol{x}_1\right) = 0$$

for $F_1 = \widehat{T_1}R_1$ and $F_2 = \widehat{T_2}R_2$. This equation is no longer bilinear but rather biquadratic in the two images $\boldsymbol{x}_1$ and $\boldsymbol{x}_2$ of any point $p$ on one of these objects. However, we can still imagine that if enough image pairs $(\boldsymbol{x}_1, \boldsymbol{x}_2)$ are given, without our knowing which object or motion each pair belongs to, some information about the two fundamental matrices $F_1$ and $F_2$ can still be found from such equations. ■

In general, given an image pair $(\boldsymbol{x}_1, \boldsymbol{x}_2)$ of a point on the $i$th moving object (say there are $n$ of them), the image pair and the fundamental matrix $F_i \in \mathbb{R}^{3\times3}$

[1] In the uncalibrated case, simply replace $R$ by $R' = K_2 R K_1^{-1}$ and $T$ by $T' = K_2 T$. Such a change of symbols, however, carries no significance for the rest of this chapter, since we will mostly work at the level of the fundamental matrix.

associated with the $i$th motion satisfy the epipolar constraint

$$\boldsymbol{x}_2^T F_i \boldsymbol{x}_1 = 0. \tag{7.2}$$

Even if we do not know which object or motion the image pair $(\boldsymbol{x}_1, \boldsymbol{x}_2)$ belongs to, the equation $\boldsymbol{x}_2^T F_i \boldsymbol{x}_1 = 0$ holds for some (unknown) fundamental matrix $F_i \in \mathbb{R}^{3\times3}$. Thus, the following constraint must be satisfied by any image pair $(\boldsymbol{x}_1, \boldsymbol{x}_2)$, regardless of the object that the 3-D point associated with the image pair belongs to

$$\boxed{f(\boldsymbol{x}_1, \boldsymbol{x}_2) \doteq \prod_{k=1}^{n} \left(\boldsymbol{x}_2^T F_i \boldsymbol{x}_1\right) = 0.} \tag{7.3}$$

We call this constraint the *multibody epipolar constraint*, since it is a natural generalization of the epipolar constraint valid for $n = 1$. The main difference is that the multibody epipolar constraint is defined for an arbitrary number of objects, which is typically unknown. Furthermore, even if $n$ is known, the algebraic structure of the constraint is neither bilinear in the image points nor linear in the fundamental matrices, as illustrated in the above example for the case $n = 2$.

**Remark 7.2 (Nonzero translation assumption).** *We explicitly assume that translation of none of the motions considered is zero. This rules out the case of pure rotation. However, in practice, this is a reasonable assumption, since, when a fixed camera observes multiple moving objects, for a motion of any object to be equivalent to a rotation of the camera, the object must be on an exact* orbital motion *around the camera.*

Our goal in this chapter is to tackle the following problem:

---

**Problem 7.1 (Multibody structure from motion).**

---

Given a collection of image pairs $\{(\boldsymbol{x}_1^j, \boldsymbol{x}_2^j)\}_{j=1}^N$ corresponding to an unknown number of independent and rigidly moving objects, estimate the number of independent motions $n$, the fundamental matrices $\{F_i\}_{i=1}^n$, and the object to which each image pair belongs.

---

Before we delve into analyzing this problem, let us first try to understand the difficulties associated with this problem:

1. **Number of motions.** We know only that $(\boldsymbol{x}_1, \boldsymbol{x}_2)$ satisfies the equation $\prod_{i=1}^n (\boldsymbol{x}_2^T F_i \boldsymbol{x}_1) = 0$ for some $n$. We typically do not even know what $n$, the number of independent motions, is. All we know here is the type of equations that such image pairs satisfy, which is the only information that can be used to determine $n$, as we will show in the next section.

2. **Feature association.** If we know which image pairs belong to the same motion, then for each motion the problem reduces to a classic two-view case that we have studied extensively in Chapters 5 and 6. But here we do not know such an association in the first place.

3. **Nonlinearity.** Even if $n$ is known and a sufficient number $N$ of image pairs is given, immediately estimating $F_i$ from the equation $f(\boldsymbol{x}_1, \boldsymbol{x}_2) = \prod_{i=1}^n (\boldsymbol{x}_1^T F_i \boldsymbol{x}_2) = 0$ is a difficult nonlinear problem. As we will discuss in more detail below, what we can estimate linearly are the coefficients of $f(\boldsymbol{x}_1, \boldsymbol{x}_2)$ treated as a homogeneous polynomial of degree $n$ in six variables $(x_1, y_1, z_1, x_2, y_2, z_2)$: the entrics of $\boldsymbol{x}_1, \boldsymbol{x}_2$, respectively.
4. **Polynomial factorization.** Once we have obtained the homogeneous polynomial $f(\boldsymbol{x}_1, \boldsymbol{x}_2)$ of degree $n$, the remaining question is how to use it to group image pairs according to their motions or, equivalently, retrieve each individual fundamental matrix $F_i$. This often requires factoring the polynomial $f$, which is typically not a simple algebraic problem.

In this chapter, We will seek a linear solution to this problem that has a seemingly nonlinear nature. A standard technique in algebra to render a nonlinear problem linear is to "embed" it into a higher-dimensional space. But to do this properly, we must first understand what type of algebraic equation $f(\boldsymbol{x}_1, \boldsymbol{x}_2) = 0$ is. Again, let us start with the simplest case, two images of two motions:

**Example 7.3 (Two-body epipolar constraint).** From the previous example, we know that the polynomial $f$ in the case of $n = 2$ is simply

$$f(\boldsymbol{x}_1, \boldsymbol{x}_2) = \left(\boldsymbol{x}_2^T F_1 \boldsymbol{x}_1\right)\left(\boldsymbol{x}_2^T F_2 \boldsymbol{x}_1\right) = 0. \tag{7.4}$$

Then $f(\boldsymbol{x})$ can be viewed as a homogeneous polynomial of degree 4 in the entries $(x_1, y_1, z_1)$ of $\boldsymbol{x}_1$ and the entries $(x_2, y_2, z_2)$ of $\boldsymbol{x}_2$. However, due to the repetition of $\boldsymbol{x}_1$ and $\boldsymbol{x}_2$, there are only 36 monomials $m_i$ in $f$ that are generated by the six variables. That is, $f$ can be written as a linear combination

$$f = \sum_{i=1}^{36} a_i m_i,$$

where the $a_i$'s are coefficients (dependent on the entries of $F_1$ and $F_2$), and $m_i$ are monomials sorted in the *degree-lexicographic* order:

$$\begin{array}{llllll} x_1^2x_2^2, & x_1^2x_2y_2, & x_1^2x_2z_2, & x_1^2y_2^2, & x_1^2y_2z_2, & x_1^2z_2^2, \\ x_1y_1x_2^2, & x_1y_1x_2y_2, & x_1y_1x_2z_2, & x_1y_1y_2^2, & x_1y_1y_2z_2, & x_1y_1z_2^2, \\ x_1z_1x_2^2, & x_1z_1x_2y_2, & x_1z_1x_2z_2, & x_1z_1y_2^2, & x_1z_1y_2z_2, & x_1z_1z_2^2, \\ y_1^2x_2^2, & y_1^2x_2y_2, & y_1^2x_2z_2, & y_1^2y_2^2, & y_1^2y_2z_2, & y_1^2z_2^2, \\ y_1z_1x_2^2, & y_1z_1x_2y_2, & y_1z_1x_2z_2, & y_1z_1y_2^2, & y_1z_1y_2z_2, & y_1z_1z_2^2, \\ z_1^2x_2^2, & z_1^2x_2y_2, & z_1^2x_2z_2, & z_1^2y_2^2, & z_1^2y_2z_2, & z_1^2z_2^2. \end{array} \tag{7.5}$$

One may view these 36 monomials as a "basis" in the space $\mathbb{R}^{36}$. Given randomly chosen values for $x_1, y_1, z_1, x_2, y_2, z_2$, the vector $m = [m_1, m_2, \ldots, m_{36}]^T \in \mathbb{R}^{36}$ will span the entire space unless these values come from corresponding image pairs, in which case they satisfy the equation

$$\sum_{i=1}^{36} a_i m_i = 0. \tag{7.6}$$

In other words, the vector $m$ lies in a 35-dimensional subspace, and its normal is given by the vector $a = [a_1, a_2, \ldots, a_{36}]^T \in \mathbb{R}^{36}$. Hence we may expect that given sufficiently many image pairs, we may solve linearly for the coefficients $a_i$ from the above equation. Information about $F_1$ and $F_2$ is then encoded (not so trivially) in these coefficients. ■

To formalize the idea in the above example, we need to introduce two maps that will help us systematically "embed" our problem into higher dimensional spaces and hence render it linear.

**Definition 7.4 (Veronese map).** *For any $k$ and $n$, we define the* Veronese map *of degree $n$,*

$$\nu_n : [x_1, x_2, \ldots, x_k]^T \mapsto [\ldots, \boldsymbol{x}^{\boldsymbol{n}}, \ldots]^T, \tag{7.7}$$

*where $\boldsymbol{x}^{\boldsymbol{n}} = x_1^{n_1} x_2^{n_2} \cdots x_k^{n_k}$ ranges over all monomials of degree $n_1 + n_2 + \cdots + n_k = n$ in the variables $x_1, x_2, \ldots, x_k$, sorted in the degree-lexicographic order.*

We leave it as an exercise to the reader to show that the number of monomials of degree $n$ in $k$ variables is the binomial coefficient

$$\binom{n+k-1}{n} = \binom{n+k-1}{k-1}.$$

In this chapter, we will repeatedly deal with the case $k = 3$, and therefore, for convenience, we define the number

$$C_n \doteq \binom{n+3-1}{3-1} = \frac{(n+2)(n+1)}{2}. \tag{7.8}$$

**Example 7.5** The Veronese map of degree $n$ for an image $\boldsymbol{x} = [x, y, z]^T$ is the vector

$$\nu_n(\boldsymbol{x}) = [x^n, x^{n-1}y, x^{n-1}z, \ldots, z^n]^T \in \mathbb{R}^{C_n}. \tag{7.9}$$

In the case $n = 2$, we have

$$\nu_2(\boldsymbol{x}) = [x^2, xy, xz, y^2, yz, z^2]^T \quad \in \mathbb{R}^6. \tag{7.10}$$

■

**Example 7.6** The Veronese map of degree 2 for $u = [1, 2, 3]^T \in \mathbb{R}^3$ is

$$\nu_2(u) = [1 \cdot 1, 1 \cdot 2, 1 \cdot 3, 2 \cdot 2, 2 \cdot 3, 3 \cdot 3]^T = [1, 2, 3, 4, 6, 9]^T \quad \in \mathbb{R}^6.$$

■

For any two vectors of dimension $m$ and $n$, respectively, we can define their Kronecker product (Appendix A) as

$$[x_1, x_2, \ldots, x_m]^T \otimes [y_1, y_2, \ldots, y_n]^T \doteq [\ldots, x_i y_j, \ldots]^T \quad \in \mathbb{R}^{mn}, \tag{7.11}$$

where the coordinates in the target space $\mathbb{R}^{mn}$ range over all pairwise products of coordinates $x_i$ and $y_j$ in the lexicographic order.

**Example 7.7** Given $u = [1, 2, 3]^T \in \mathbb{R}^3$ and $v = [10, 15]^T \in \mathbb{R}^2$, we have

$$u \otimes v = [1 \cdot 10, 1 \cdot 15, 2 \cdot 10, 2 \cdot 15, 3 \cdot 10, 3 \cdot 15]^T = [10, 15, 20, 30, 30, 45]^T \quad \in \mathbb{R}^6.$$

We leave to the reader to compute $u \otimes u$ and hence verify that $u \otimes u \neq \nu_2(u)$. ■

**Example 7.8** Given two images $\boldsymbol{x}_1 = [x_1, y_1, z_1]^T$ and $\boldsymbol{x}_2 = [x_2, y_2, z_2]^T$, the Kronecker product of their Veronese maps $\nu_2(\boldsymbol{x}_1) \in \mathbb{R}^6$ and $\nu_2(\boldsymbol{x}_2) \in \mathbb{R}^6$ is a vector in $\mathbb{R}^{36}$ whose entries are exactly the monomials given in (7.5). That is,

$$\nu_2(\boldsymbol{x}_1) \otimes \nu_2(\boldsymbol{x}_2) = [m_1, m_2, \ldots, m_{36}]^T \quad \in \mathbb{R}^{36}, \tag{7.12}$$

where the $m_i$'s are monomials given in the degree-lexicographic order as in (7.5). ■

Using the notation introduced, we then can rewrite the multibody epipolar constraint $f(\boldsymbol{x}_1, \boldsymbol{x}_2) = \prod_{i=1}^n (\boldsymbol{x}_2^T F_i \boldsymbol{x}_1) = 0$ in a form that is better suited for further computation.

**Lemma 7.9 (Bilinear form and multibody fundamental matrix).** *Using the Veronese map, the multibody epipolar constraint $f(\boldsymbol{x}_1, \boldsymbol{x}_2) = \prod_{i=1}^n (\boldsymbol{x}_2^T F_i \boldsymbol{x}_1) = 0$ can be rewritten as a bilinear form:*

$$\boxed{f(\boldsymbol{x}_1, \boldsymbol{x}_2) = \nu_n(\boldsymbol{x}_2)^T F \nu_n(\boldsymbol{x}_1) = 0,} \tag{7.13}$$

*where $F$ is an $C_n \times C_n$ matrix as a symmetric function of $F_1, F_2 \ldots, F_n$. We call it the* multibody fundamental matrix.

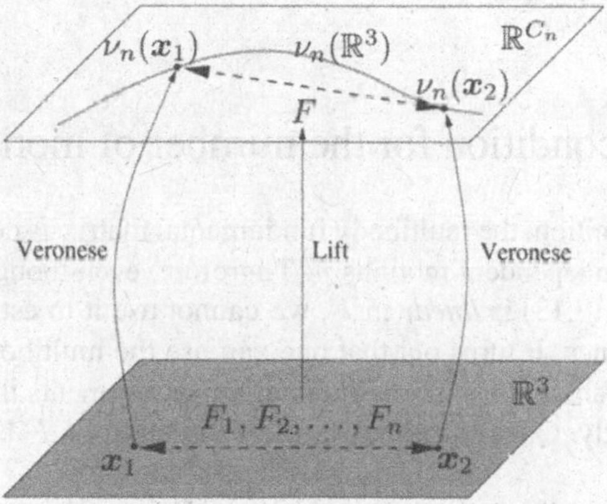


Figure 7.2. If a pair of images $(\boldsymbol{x}_1, \boldsymbol{x}_2)$ in $\mathbb{R}^3$ satisfy an epipolar constraint for one of the $F_1, F_2, \ldots,$ or $F_n$, then their images under the Veronese map must satisfy a bilinear multibody epipolar constraint in $\mathbb{R}^{C_n}$. Note that $\nu_n(\mathbb{R}^3)$ is a (three-dimensional) surface, the so-called Veronese surface, in the space $\mathbb{R}^{C_n}$.

*Proof.* Let $\boldsymbol{\ell}_i = F_i \boldsymbol{x}_1 \in \mathbb{R}^3$, for $i = 1, 2, \ldots, n$, be the epipolar lines of $\boldsymbol{x}_1$ associated with the $n$ motions. Then the multibody epipolar constraint $f(\boldsymbol{x}_1, \boldsymbol{x}_2) = \prod_{i=1}^n \boldsymbol{x}_2^T \boldsymbol{\ell}_i$ is a homogeneous polynomial of degree $n$ in $\boldsymbol{x}_2 = [x_2, y_2, z_2]^T$; i.e.

$$f(\boldsymbol{x}_1, \boldsymbol{x}_2) = \sum a_{n_1,n_2,n_3} x_2^{n_1} y_2^{n_2} z_2^{n_3} \doteq \sum a_{\boldsymbol{n}} \boldsymbol{x}_2^{\boldsymbol{n}} = \nu_n(\boldsymbol{x}_2)^T \boldsymbol{a},$$

where $\boldsymbol{a} \in \mathbb{R}^{C_n}$ is the vector of coefficients. From the properties of polynomial multiplication, each $a_{\boldsymbol{n}}$ is a symmetric multilinear function of $(\boldsymbol{\ell}_1, \boldsymbol{\ell}_2, ..., \boldsymbol{\ell}_n)$, i.e. it is linear in each $\boldsymbol{\ell}_i$ and $a_{\boldsymbol{n}}(\boldsymbol{\ell}_1, \boldsymbol{\ell}_2, ..., \boldsymbol{\ell}_n) = a_{\boldsymbol{n}}(\boldsymbol{\ell}_{\sigma(1)}, \boldsymbol{\ell}_{\sigma(2)}, ..., \boldsymbol{\ell}_{\sigma(n)})$ for all

$\sigma \in \mathfrak{S}_n$, where $\mathfrak{S}_n$ is the permutation group of $n$ elements. Since each $\boldsymbol{\ell}_i$ is linear in $\boldsymbol{x}_1$, each $a_n$ is indeed a homogeneous polynomial of degree $n$ in $\boldsymbol{x}_1$, i.e. $a_n = \boldsymbol{f}_n^T \nu_n(\boldsymbol{x}_1)$, where each entry of $\boldsymbol{f}_n \in \mathbb{R}^{C_n}$ is a symmetric multilinear function of the entries of the $F_i$'s. Letting

$$F \doteq [\boldsymbol{f}_{n,0,0}\,,\ \boldsymbol{f}_{n-1,1,0}\,,\ \cdots\,,\ \boldsymbol{f}_{0,0,n}]^T \in \mathbb{R}^{C_n \times C_n},$$

we obtain

$$f(\boldsymbol{x}_1, \boldsymbol{x}_2) = \nu_n(\boldsymbol{x}_2)^T F \nu_n(\boldsymbol{x}_1) = 0.$$

This process is illustrated in Figure 7.2. □

Equation (7.13) resembles the bilinear form of the epipolar constraint for a single rigid-body motion studied in Chapter 5. For that reason, we will interchangeably refer to both equations (7.3) and (7.13) as the *multibody epipolar constraint*. A generalized notion of the "epipole" will be introduced later, when we study how information on the epipole associated with each fundamental matrix $F_i$ is encoded in the multibody fundamental matrix $F$. For now, let us first study how to obtain the matrix $F$.

## 7.2 A rank condition for the number of motions

Notice that, by definition, the multibody fundamental matrix $F$ depends explicitly on the number of independent motions $n$. Therefore, even though the multibody epipolar constraint (7.13) is *linear* in $F$, we cannot use it to estimate $F$ without knowing $n$ in advance. It turns out that one can use the multibody epipolar constraint to derive a rank constraint on the image measurements that allows one to compute $n$ explicitly. Once $n$ is known, the estimation of $F$ becomes a *linear* problem.

To estimate the multibody fundamental matrix $F$, in analogy to the eight-point algorithm, we may rewrite the above bilinear form in a linear form using the Kronecker product as

$$\boxed{f(\boldsymbol{x}_1, \boldsymbol{x}_2) = [\nu_n(\boldsymbol{x}_2) \otimes \nu_n(\boldsymbol{x}_1)]^T F^s = 0,} \tag{7.14}$$

where $F^s \in \mathbb{R}^{C_n^2}$ is the stacked version of the matrix $F$.

**Remark 7.10 (A different embedding).** *Readers may have noticed that there is another way to "convert" the multibody epipolar equation into a linear form by switching the order of the Veronese map and the Kronecker product in the above process:*

$$f(\boldsymbol{x}_1, \boldsymbol{x}_2) = \prod_{i=1}^{n} (\boldsymbol{x}_2^T F_i \boldsymbol{x}_1) = \prod_{i=1}^{n} [(\boldsymbol{x}_2 \otimes \boldsymbol{x}_1)^T F_i^s] = \nu_n(\boldsymbol{x}_2 \otimes \boldsymbol{x}_1)^T F^*, \tag{7.15}$$

*where $F^*$ is a vector dependent on $F_1, F_2, \ldots, F_n$. However, since $\boldsymbol{x}_2 \otimes \boldsymbol{x}_1 \in \mathbb{R}^9$, it seems that $\nu_n(\boldsymbol{x}_2 \otimes \boldsymbol{x}_1)$ produces $\binom{n+8}{8}$ terms of monomials. In general,*

$$\binom{n+8}{8} > C_n^2.$$

*Nevertheless, one may easily verify from the simple case $n = 2$ that the $\binom{2+8}{8} = 45$ monomials generated from $\nu_2(\boldsymbol{x}_2 \otimes \boldsymbol{x}_1)$ are not all independent. In fact there are only $C_2^2 = 36$ of them, which are the same ones given by $\nu_2(\boldsymbol{x}_2) \otimes \nu_2(\boldsymbol{x}_1)$. Hence, not all coefficients of $F^*$ can be estimated independently.*

Now if multiple, say $N$, image pairs $\{(\boldsymbol{x}_1^j, \boldsymbol{x}_2^j)\}_{j=1}^N$ are given for points on the $n$ rigid bodies, $F^s$ must satisfy the system of linear equations $A_n F^s = 0$, where

$$A_n \doteq \begin{bmatrix} \left(\nu_n(\boldsymbol{x}_2^1) \otimes \nu_n(\boldsymbol{x}_1^1)\right)^T \\ \left(\nu_n(\boldsymbol{x}_2^2) \otimes \nu_n(\boldsymbol{x}_1^2)\right)^T \\ \vdots \\ \left(\nu_n(\boldsymbol{x}_2^N) \otimes \nu_n(\boldsymbol{x}_1^N)\right)^T \end{bmatrix} \in \mathbb{R}^{N \times C_n^2}. \tag{7.16}$$

In other words, the vector $F^s$ is in the (right) null space of the matrix $A_n$. In order to determine $F^s$ uniquely (up to a scalar factor), the matrix $A_n$ must have rank exactly

$$\operatorname{rank}(A_n) = C_n^2 - 1. \tag{7.17}$$

For this to be true, the conditions of the following lemma must hold.

**Lemma 7.11.** *The number of image pairs $N$ needed to solve linearly the multibody fundamental matrix $F$ from the above equations is*

$$\boxed{N \geq C_n^2 - 1 = \left[\frac{(n+2)(n+1)}{2}\right]^2 - 1.} \tag{7.18}$$

The above rank condition on the matrix $A_n$ in fact gives us an effective criterion to determine the number of independent motions $n$ from the given set of image pairs.

**Theorem 7.12 (Rank condition for the number of independent motions).** *Let $\{\boldsymbol{x}_1^j, \boldsymbol{x}_2^j\}_{j=1}^N$ be a collection of image pairs corresponding to 3-D points in general configuration and undergoing an unknown number $n$ of distinct rigid-body motions with nonzero translation. Let $A_i \in \mathbb{R}^{N \times C_i^2}$ be the matrix defined in (7.16), but computed using the Veronese map $\nu_i$ of degree $i \geq 1$. Then, if the number of image pairs is big enough ($N \geq C_n^2 - 1$ when $n$ is known) and at least eight* *points correspond to each motion, we have*

$$rank(A_i) \begin{cases} > C_i^2 - 1, & \text{if } i < n, \\ = C_i^2 - 1, & \text{if } i = n, \\ < C_i^2 - 1, & \text{if } i > n. \end{cases} \tag{7.19}$$

*Therefore, the number of independent motions $n$ is given by*

$$\boxed{n \doteq \min\{i : \text{rank}(A_i) = C_i^2 - 1\}.} \tag{7.20}$$

*Proof.* Since each fundamental matrix $F_i$ has rank 2, the polynomial $f_i = \boldsymbol{x}_2^T F_i \boldsymbol{x}_1$ is irreducible over the real field $\mathbb{R}$. Let $Z_i$ be the set of $(\boldsymbol{x}_1, \boldsymbol{x}_2)$ that satisfy $\boldsymbol{x}_2^T F_i \boldsymbol{x}_1 = 0$. Then due to the irreducibility of $f_i$, any polynomial $g$ in $\boldsymbol{x}_1$ and $\boldsymbol{x}_2$ that vanishes on the entire set $Z_i$ must be of the form $g = f_i h$, where $h$ is some polynomial. Hence if $F_1, F_2, \ldots, F_n$ are distinct, a polynomial that vanishes on the set $\cup_{i=1}^n Z_i$ must be of the form $g = f_1 f_2 \cdots f_n h$ for some $h$. Therefore, the only polynomial of *minimal* degree that vanishes on the same set is

$$f = f_1 f_2 \cdots f_n = \left(\boldsymbol{x}_2^T F_1 \boldsymbol{x}_1\right)\left(\boldsymbol{x}_2^T F_2 \boldsymbol{x}_1\right) \cdots \left(\boldsymbol{x}_2^T F_n \boldsymbol{x}_1\right). \tag{7.21}$$

Since the entries of $\nu_n(\boldsymbol{x}_2) \otimes \nu_n(\boldsymbol{x}_1)$ are exactly the independent monomials of $f$ (as we will show below), this implies that if the number of data points per motion is at least eight and $N \geq C_n^2 - 1$, then:

1. There is no polynomial of degree $i < n$ whose coefficients are in the null space of $A_i$, i.e. rank$(A_i) > C_i^2 - 1$ for $i < n$;
2. There is a unique polynomial of degree $n$, namely $f$, with coefficients in the null space of $A_n$, i.e. rank$(A_n) = C_n^2 - 1$;
3. There is more than one polynomial of degree $i > n$ (one for each independent choice of the $(i-n)$-degree polynomial $h$) with coefficients in the null space of $A_i$, i.e. rank$(A_i) < C_i^2 - 1$ for $i > n$.

The rest of the proof consists in showing that the entries of $\nu_n(\boldsymbol{x}_2) \otimes \nu_n(\boldsymbol{x}_1)$ are exactly the independent monomials in the polynomial $f$, which we do by induction. Since the claim is obvious for $n = 1$, we assume that it is true for $n$ and prove it for $n + 1$. Let $\boldsymbol{x}_1 = (x_1, y_1, z_1)$ and $\boldsymbol{x}_2 = (x_2, y_2, z_2)$. Then the entries of $\nu_n(\boldsymbol{x}_2) \otimes \nu_n(\boldsymbol{x}_1)$ are of the form $(x_2^{m_1} y_2^{m_2} z_2^{m_3})(x_1^{n_1} y_1^{n_2} z_1^{n_3})$ with $m_1 + m_2 + m_3 = n_1 + n_2 + n_3 = n$, while the entries of $\boldsymbol{x}_2 \otimes \boldsymbol{x}_1$ are of the form $(x_2^{i_1} y_2^{i_2} z_2^{i_3})(x_1^{j_1} y_1^{j_2} z_1^{j_3})$ with $i_1 + i_2 + i_3 = j_1 + j_2 + j_3 = 1$. Thus a basis for the product of these monomials is given by the entries of $\nu_{n+1}(\boldsymbol{x}_2) \otimes \nu_{n+1}(\boldsymbol{x}_1)$. □

The significance of this theorem is that the number of independent motions can now be determined incrementally using Algorithm 7.1. Once the number $n$ of motions is found, the multibody fundamental matrix $F$ is simply the 1-D null space of the corresponding matrix $A_n$. Nevertheless, in order for Algorithm 7.1 to work properly, the number of image pairs needed must be at least $N \geq C_n^2 - 1$.

**Algorithm 7.1 (The number of independent motions and the multibody fundamental matrix).**

Given a collection of pairs of points $\{x_1^j, x_2^j\}_{j=1}^N$ undergoing an unknown number of different rigid-body motions,

1. Set $i = 1$.
2. Compute the matrix
$$A_i = \begin{bmatrix} (\nu_i(x_2^1) \otimes \nu_i(x_1^1))^T \\ (\nu_i(x_2^2) \otimes \nu_i(x_1^2))^T \\ \vdots \\ (\nu_i(x_2^N) \otimes \nu_i(x_1^N))^T \end{bmatrix} \in \mathbb{R}^{N \times C_i^2}. \tag{7.22}$$
3. If $\text{rank}(A_i) = C_i^2 = \left[\frac{(i+2)(i+1)}{2}\right]^2$, then set $i = i + 1$ and go back to step 2.
4. Now we have $\text{rank}(A_i) = C_i^2 - 1$. The number $n$ of independent motions is then the current $i$, and the only eigenvector corresponding to the zero eigenvalue of the current matrix $A_i$ gives the stacked version $F^s$ of the multibody fundamental matrix $F$.

For $n = 1, 2, 3, 4$, the minimum $N$ is $8, 35, 99, 225$, respectively. When $n$ is large, $N$ grows approximately in the order of $O(n^4)$, a price to pay for trying to solve the Problem 7.1 linearly. Nevertheless, we will discuss many variations of this general scheme in later sections or in exercises that will dramatically reduce the number of image points required (especially for large $n$) and render the scheme much more practical.

## 7.3 Geometric properties of the multibody fundamental matrix

In this section, we study the relationships between the multibody fundamental matrix $F$ and the epipoles $e_1, e_2, \ldots, e_n$ associated with the fundamental matrices $F_1, F_2, \ldots, F_n$. The relationships between epipoles and epipolar lines will be studied in the next section, where we will show how they can be computed from the multibody fundamental matrix $F$.

First of all, recall that the epipole $e_i$ associated with the $i$th motion in the second image is defined as the left null space of the rank-2 fundamental matrix $F_i$, that is,

$$e_i^T F_i \doteq 0. \tag{7.23}$$

Hence, the following polynomial (in $x$) is zero for every $e_i$, $i = 1, 2, \ldots, n$,

$$(e_i^T F_1 x)(e_i^T F_2 x) \cdots (e_i^T F_n x) = \nu_n(e_i)^T F \nu_n(x) = 0. \tag{7.24}$$

We call the vector $\nu_n(\boldsymbol{e}_i)$ the *embedded epipole* associated with the $i$th motion. Since $\nu_n(\boldsymbol{x})$ as a vector spans the entire space $\mathbb{R}^{C_n}$ when $\boldsymbol{x}$ ranges over $\mathbb{R}^3$ (or $\mathbb{R}^3$),[2] we have

$$\nu_n(\boldsymbol{e}_i)^T F = 0. \tag{7.25}$$

Therefore, the embedded epipoles $\{\nu_n(\boldsymbol{e}_i)\}_{i=1}^n$ lie in the left null space of $F$, while the epipoles $\{\boldsymbol{e}_i\}_{i=1}^n$ lie in the left null space of $\{F_i\}_{i=1}^n$. Hence, the rank of $F$ is bounded, depending on the number of *distinct* (pairwise linearly independent) epipoles as stated in Lemmas 7.13 and 7.14.

**Lemma 7.13 (Null space of $F$ when the epipoles are distinct).** *Let $F$ be the multibody fundamental matrix generated by the fundamental matrices $F_1, F_2, \ldots, F_n$ with pairwise linearly independent epipoles $\boldsymbol{e}_1, \boldsymbol{e}_2, \ldots, \boldsymbol{e}_n$. Then the (left) null space of $F \in \mathbb{R}^{C_n \times C_n}$ contains at least the $n$ linearly independent vectors*

$$\nu_n(\boldsymbol{e}_i) \in \mathbb{R}^{C_n}, \quad i = 1, 2, \ldots, n. \tag{7.26}$$

*Therefore, the rank of the multibody fundamental matrix $F$ is bounded by*

$$\boxed{rank(F) \leq (C_n - n).} \tag{7.27}$$

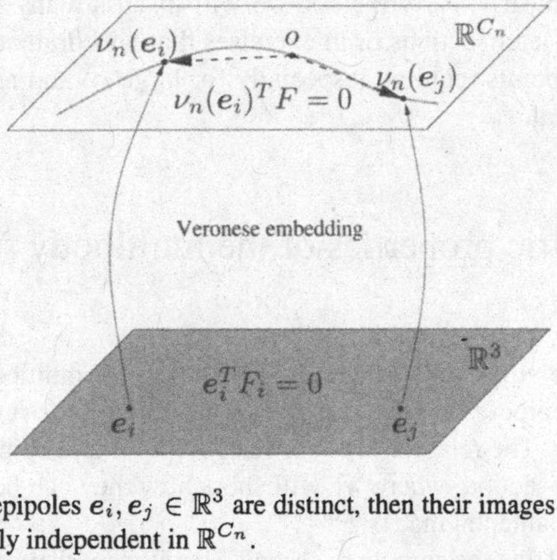


Figure 7.3. If two epipoles $\boldsymbol{e}_i, \boldsymbol{e}_j \in \mathbb{R}^3$ are distinct, then their images under the Veronese map must be linearly independent in $\mathbb{R}^{C_n}$.

*Proof.* We need to show only that if the $\boldsymbol{e}_i$'s are distinct, then the $\nu_n(\boldsymbol{e}_i)$'s are linearly independent. If we let $\boldsymbol{e}_i = [x_i, y_i, z_i]^T, i = 1, 2, \ldots, n$, then we need to

[2]This is simply because the $C_n$ monomials in $\nu_n(\boldsymbol{x})$ are linearly independent.

prove only that the rank of the matrix

$$U \doteq \begin{bmatrix} \nu_n(e_1)^T \\ \nu_n(e_2)^T \\ \vdots \\ \nu_n(e_n)^T \end{bmatrix} = \begin{bmatrix} x_1^n & x_1^{n-1}y_1 & x_1^{n-1}z_1 & \cdots & z_1^n \\ x_2^n & x_2^{n-1}y_2 & x_2^{n-1}z_2 & \cdots & z_2^n \\ \vdots & \vdots & \vdots & \ddots & \vdots \\ x_n^n & x_n^{n-1}y_n & x_n^{n-1}z_n & \cdots & z_n^n \end{bmatrix} \in \mathbb{R}^{n \times C_n} \quad (7.28)$$

is exactly $n$. Since the $e_i$'s are distinct, we can assume without loss of generality that $\{[x_i, z_i]\}_{i=1}^n$ are already distinct and that $z_i \neq 0$.[3] Then, after dividing the $i$th row of $U$ by $z_i^n$ and letting $t_i = x_i/z_i$, we can extract the following Vandermonde submatrix of $U$

$$V \doteq \begin{bmatrix} t_1^{n-1} & t_1^{n-2} & \cdots & 1 \\ t_2^{n-1} & t_2^{n-2} & \cdots & 1 \\ \vdots & \vdots & \ddots & \vdots \\ t_n^{n-1} & t_n^{n-2} & \cdots & 1 \end{bmatrix} \in \mathbb{R}^{n \times n}. \quad (7.29)$$

Since $\det(V) = \prod_{i<j}(t_i - t_j)$, the Vandermonde matrix $V$ has rank $n$ if and only if $t_1, t_2, \ldots, t_n$ are distinct. Hence $\text{rank}(U) = \text{rank}(V) = n$. Figure 7.3 illustrates the intuition for $n = 2$ distinct epipoles. □

Even though we know that the linearly independent vectors $\nu_n(e_i)$'s lie in the left null space of $F$, we do not know whether the $n$-dimensional subspace spanned by them will be exactly the left null space of $F$; i.e. we do not know whether $\text{rank}(F) = C_n - n$. Simulations confirm that this is true when all the epipoles are distinct but a rigorous proof remains an open problem (see Exercise 7.11).

Now, if one of the epipoles is repeated, then one would expect that the dimension of the null space of $F$ decreases. However, this is *not* the case: the null space of $F$ is actually enlarged by higher-order derivatives of the Veronese map, as stated in the following lemma.

**Lemma 7.14 (Null space of $F$ when one epipole is repeated).** *Let $F$ be the multibody fundamental matrix generated by the fundamental matrices $F_1, F_2, \ldots, F_n$ with epipoles $e_1, e_2, \ldots, e_n$. Let $e_1$ be repeated $k$ times, i.e. $e_1 = e_2 = \cdots = e_k$, and let the other $n - k$ epipoles be distinct. Then the rank of the multibody fundamental matrix $F$ is bounded by*

$$\textit{rank}(F) \leq C_n - C_{k-1} - (n - k). \quad (7.30)$$

[3] This assumption is not always satisfied, e.g., for $n = 3$ motions with epipoles along the $X$-axis, $Y$-axis, and $Z$-axis, respectively. However, as long as the $e_i$'s are distinct, one can always find a nonsingular linear transformation $e_i \mapsto Le_i$ on $\mathbb{R}^3$ that makes the assumption true. Furthermore, this linear transformation induces a linear transformation on the lifted space $\mathbb{R}^{C_n}$ that preserves the rank of the matrix $U$.

*Proof.* When $k = 2$, $\boldsymbol{e}_1 = \boldsymbol{e}_2$ is a repeated root of $\nu_n(\boldsymbol{x})^T F$ as a polynomial (matrix) in $\boldsymbol{x} = [x, y, z]^T$. Hence we have

$$\frac{\partial \nu_n(\boldsymbol{x})^T}{\partial x} F \bigg|_{\boldsymbol{x}=\boldsymbol{e}_1} = 0, \quad \frac{\partial \nu_n(\boldsymbol{x})^T}{\partial y} F \bigg|_{\boldsymbol{x}=\boldsymbol{e}_1} = 0, \quad \frac{\partial \nu_n(\boldsymbol{x})^T}{\partial z} F \bigg|_{\boldsymbol{x}=\boldsymbol{e}_1} = 0.$$

Notice that the Jacobian matrix of the Veronese map $D\nu_n(\boldsymbol{x})$ is of full rank for all $\boldsymbol{x} \in \mathbb{R}^3$, because

$$D\nu_n(\boldsymbol{x})^T D\nu_n(\boldsymbol{x}) \geq \boldsymbol{x}^T \boldsymbol{x}\, I_{3\times 3},$$

where the inequality is in the sense of positive semidefiniteness. Thus, the vectors $\frac{\partial \nu_n(\boldsymbol{e}_1)}{\partial x}, \frac{\partial \nu_n(\boldsymbol{e}_1)}{\partial y}, \frac{\partial \nu_n(\boldsymbol{e}_1)}{\partial z}$ are linearly independent, because they are the columns of $D\nu_n(\boldsymbol{e}_1)$, and $\boldsymbol{e}_1 \neq 0$. In addition, their span contains $\nu_n(\boldsymbol{e}_1)$, because

$$n\nu_n(\boldsymbol{x}) = D\nu_n(\boldsymbol{x})\boldsymbol{x} \doteq \left[\frac{\partial \nu_n(\boldsymbol{x})}{\partial x}, \frac{\partial \nu_n(\boldsymbol{x})}{\partial y}, \frac{\partial \nu_n(\boldsymbol{x})}{\partial z}\right] \boldsymbol{x}, \quad \forall \boldsymbol{x} \in \mathbb{R}^3. \tag{7.31}$$

Hence $\operatorname{rank}(F) \leq C_n - C_1 - (n-1) = C_n - 3 - (n-1)$. Now if $k > 2$, one should consider the $(k-1)$th order partial derivatives of $\nu_n(\boldsymbol{x})$ evaluated at $\boldsymbol{e}_1$. There is a total of $C_{k-1}$ such partial derivatives, which give rise to $C_{k-1}$ linearly independent vectors in the (left) null space of $F$. Similar to the case $k = 2$, one can show that the embedded epipole $\nu_n(\boldsymbol{e}_1)$ is in the span of these higher-order partial derivatives. □

**Example 7.15 (Two repeated epipoles).** In the two-body problem, if $F_1$ and $F_2$ have the same (left) epipole, i.e. $F_1 = \widehat{T}R_1$ and $F_2 = \widehat{T}R_2$, then the rank of the two-body fundamental matrix $F$ is $C_2 - C_1 - (2-2) = 6 - 3 = 3$ instead of $C_2 - 2 = 4$. ■

Since the null space of $F$ is enlarged by higher-order derivatives of the Veronese map evaluated at repeated epipoles, in order to identify the embedded epipoles $\nu_n(\boldsymbol{e}_i)$ from the left null space of $F$ we will need to exploit the algebraic structure of the Veronese map. Let us denote the image of the real projective space $\mathbb{RP}^3$ under the Veronese map of degree $n$ by $\nu_n(\mathbb{R}^3)$.[4] The following theorem establishes a key relationship between the null space of $F$ and the epipoles of each fundamental matrix.

**Theorem 7.16 (Veronese null space of the multibody fundamental matrix).** *The intersection of the left null space of the multibody fundamental matrix $F$, null($F$), with the Veronese surface $\nu_n(\mathbb{R}^3)$ is exactly*

$$\boxed{\textit{null}(F) \cap \nu_n(\mathbb{R}^3) = \{\nu_n(\boldsymbol{e}_i)\}_{i=1}^{n}.} \tag{7.32}$$

*Proof.* Let $\boldsymbol{x} \in \mathbb{R}^3$ be a vector whose Veronese map is in the left null space of $F$. We then have

$$\nu_n(\boldsymbol{x})^T F = 0 \quad \Leftrightarrow \quad \nu_n(\boldsymbol{x})^T F \nu_n(\boldsymbol{y}) = 0, \forall \boldsymbol{y} \in \mathbb{R}^3. \tag{7.33}$$

[4]This is the so-called (real) Veronese surface in algebraic geometry [Harris, 1992].

Since $F$ is a multibody fundamental matrix,

$$\nu_n(\boldsymbol{x})^T F \nu_n(\boldsymbol{y}) = \prod_{i=1}^{n} (\boldsymbol{x}^T F_i \boldsymbol{y}).$$

This means that for this $\boldsymbol{x}$,

$$\prod_{i=1}^{n} (\boldsymbol{x}^T F_i \boldsymbol{y}) = 0, \quad \forall \boldsymbol{y} \in \mathbb{R}^3. \tag{7.34}$$

If $\boldsymbol{x}^T F_i \neq 0$ for all $i = 1, 2, \ldots, n$, then the set of $\boldsymbol{y}$ that satisfy the above equation is simply the union of $n$ two-dimensional subspaces in $\mathbb{R}^3$, which will never fill the entire space $\mathbb{R}^3$. Hence we must have $\boldsymbol{x}^T F_i = 0$ for some $i$. Therefore, $\boldsymbol{x}$ is one of the epipoles. □

The significance of Theorem 7.16 is that in spite of the fact that repeated epipoles may enlarge the null space of $F$, and that we do not know whether the dimension of the null space equals $n$ for distinct epipoles, one may always find the epipoles exactly by intersecting the left null space of $F$ with the Veronese surface $\nu_n(\mathbb{R}^3)$, as illustrated in Figure 7.4.

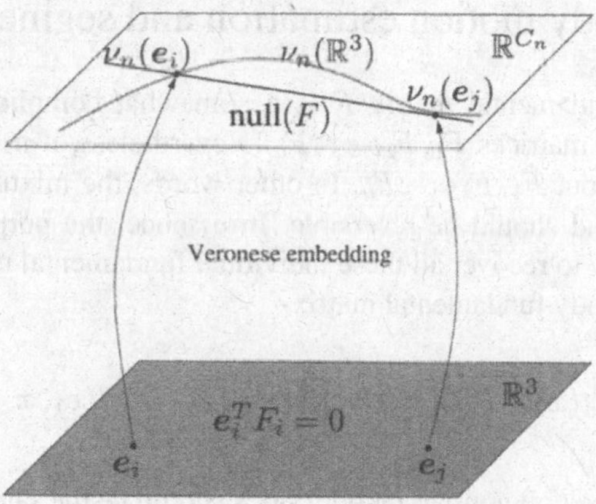


Figure 7.4. The intersection of $\nu_n(\mathbb{R}^3)$ and null($F$) is exactly $n$ points representing the Veronese map of the $n$ epipoles, repeated or not.

**Example 7.17 (Representation of a Veronese surface).** Consider the Veronese map $\nu_2 : [x, y, z]^T \in \mathbb{R}^3 \mapsto [x^2, xy, xz, y^2, yz, z^2]^T \in \mathbb{R}^6$. Given the image of the coordinates $X = [X_1, X_2, X_3, X_4, X_5, X_6]^T \in \mathbb{R}^6$, then a point $X \in \mathbb{R}^6$ is in the image of the Veronese map if and only if

$$\operatorname{rank} \begin{bmatrix} X_1 & X_2 & X_3 \\ X_2 & X_4 & X_5 \\ X_3 & X_5 & X_6 \end{bmatrix} = 1. \tag{7.35}$$

Hence the Veronese surface $\nu_2(\mathbb{R}^3)$, for instance, can be represented also as the locus of $2 \times 2$ minors of the above $3 \times 3$ matrix. ■

The question now is how to compute the intersection of null($F$) with $\nu_n(\mathbb{R}^3)$ in practice. One possible approach is to determine a vector $v \in \mathbb{R}^n$ such that $Bv \in \nu_n(\mathbb{R}^3)$, where $B$ is a matrix whose columns form a basis for the (left) null space of $F$. Finding $v$, hence the epipoles, is equivalent to solving for the roots of polynomials of degree $n$ in $n-1$ variables. Although this is feasible for $n = 2, 3$, it is computationally formidable for $n > 3$.

In the next section, we introduce a more systematic approach that combines the multibody epipolar geometry developed so far with a polynomial factorization technique given in Appendix 7.A at the end this chapter. In essence, we will show that the epipoles (and also the epipolar lines) can be computed by solving for the roots of a polynomial of degree $n$ in *one* variable plus one linear system in $n$ variables. Given the epipoles and the epipolar lines, the computation of individual fundamental matrices becomes a *linear* problem. Therefore, we will be able to reach the conclusion that there exists a *closed-form solution* to Problem 7.1 if and only if $n \leq 4$; and if $n > 4$, one has only to solve numerically a univariate polynomial equation of degree $n$.

## 7.4 Multibody motion estimation and segmentation

The multibody fundamental matrix $F$ is a somewhat complicated mixture of the $n$ fundamental matrices $F_1, F_2, \ldots, F_n$. Nevertheless, it in fact contains all the information about $F_1, F_2 \ldots, F_n$. In other words, the mixture does not lose any information and should be reversible. In essence, the purpose of this section is to show how to recover all these individual fundamental matrices from the (estimated) multibody fundamental matrix

$$f(\boldsymbol{x}_1, \boldsymbol{x}_2) = \nu_n(\boldsymbol{x}_2)^T F \nu_n(\boldsymbol{x}_1) = \prod_{i=1}^{n} (\boldsymbol{x}_2^T F_i \boldsymbol{x}_1), \quad \forall \boldsymbol{x}_1, \boldsymbol{x}_2 \in \mathbb{R}^3. \tag{7.36}$$

Notice that since the polynomials on both sides should be the same one, the above equality holds even if $\boldsymbol{x}_1, \boldsymbol{x}_2$ are not a corresponding image pair. Unfortunately, there is no known numerical algorithm that can effectively factor a polynomial such as the one on the left-hand side into the one on the right-hand side. In this book, we will hence adopt a more geometric approach to solve this problem.

In general, if there are $n$ independently moving objects in front of a camera, their image pairs will generate, say in the second image plane, $n$ groups of epipolar lines $\{\boldsymbol{\ell}_1\}$, $\{\boldsymbol{\ell}_2\}$, $\ldots$, $\{\boldsymbol{\ell}_n\}$ that intersect respectively at $n$ distinct epipoles $\boldsymbol{e}_1, \boldsymbol{e}_2, \ldots, \boldsymbol{e}_n$, as shown in Figure 7.5. Therefore, in order to determine which image pair belongs to which rigid-body motion, we may decompose the problem into two steps:

1. use the multibody fundamental matrix $F$ to compute for each image pair $(\boldsymbol{x}_1, \boldsymbol{x}_2)$ its epipolar line $\boldsymbol{\ell}$ (in the second image plane);

2. use the multibody fundamental matrix and all the epipolar lines to compute the $n$ epipoles of the $n$ motions (in the second image plane).

Once all the epipoles are found, grouping each image pair to an object is done by checking for which epipole $\boldsymbol{e}$ the epipolar line $\boldsymbol{\ell}$ of this image pair satisfies

$$\boldsymbol{e}^T\boldsymbol{\ell} = 0.$$

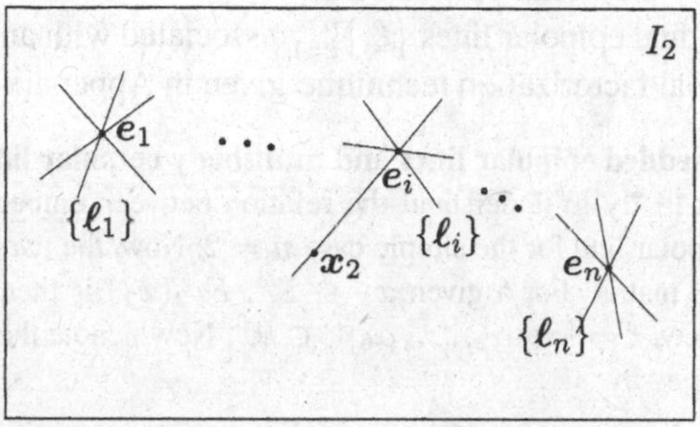


Figure 7.5. When $n$ objects move independently in front of a fixed camera, epipolar lines associated with image pairs from the objects intersect respectively at $n$ distinct epipoles. Again, epipolar lines and epipoles are in the second image.

Let $(\boldsymbol{x}_1, \boldsymbol{x}_2)$ be an image pair associated with the $i$th motion. The corresponding epipolar line in the second image is then $\boldsymbol{\ell}_i = F_i\boldsymbol{x}_1 \in \mathbb{R}^3$. Obviously, from $\boldsymbol{x}_2^T F_i \boldsymbol{x}_1 = 0$ and $\boldsymbol{e}_i^T F_i = 0$ we have the relations

$$\boxed{\boldsymbol{x}_2^T\boldsymbol{\ell}_i = 0, \quad \boldsymbol{e}_i^T\boldsymbol{\ell}_i = 0.} \tag{7.37}$$

Nevertheless, here the multibody fundamental matrix $F$ does not directly give the corresponding epipolar line $\boldsymbol{\ell}$ for each point in the first image, nor the $n$ epipoles $\boldsymbol{e}_1, \boldsymbol{e}_2, \ldots, \boldsymbol{e}_n$. To retrieve individual epipoles from a multibody fundamental matrix $F$, we need to exploit the relationship between the original epipoles $\boldsymbol{e}_i$'s and other geometric entities that can be generated from the multibody fundamental matrix $F$.

### 7.4.1 *Estimation of epipolar lines and epipoles*

Given a point $\boldsymbol{x}_1$ in the first image frame, the epipolar lines associated with it are defined as $\boldsymbol{\ell}_i \doteq F_i\boldsymbol{x}_1 \in \mathbb{R}^3$, $i = 1, 2, \ldots, n$. From the epipolar constraint, we know that one of these lines passes through the corresponding point in the second frame $\boldsymbol{x}_2$; i.e. there exists $i$ such that $\boldsymbol{x}_2^T\boldsymbol{\ell}_i = 0$. Let $F$ be the multibody fundamental matrix. We have that

$$f(\boldsymbol{x}_1, \boldsymbol{x}_2) = \nu_n(\boldsymbol{x}_2)^T F \nu_n(\boldsymbol{x}_1) = \prod_{i=1}^{n}(\boldsymbol{x}_2^T F_i \boldsymbol{x}_1) = \prod_{i=1}^{n}(\boldsymbol{x}_2^T\boldsymbol{\ell}_i), \tag{7.38}$$

from which we conclude that the vector $\tilde{\boldsymbol{\ell}} \doteq F\nu_n(\boldsymbol{x}_1) \in \mathbb{R}^{C_n}$ represents the coefficients of the homogeneous polynomial in $\boldsymbol{x}$:

$$\boxed{g(\boldsymbol{x}) \doteq (\boldsymbol{x}^T\boldsymbol{\ell}_1)(\boldsymbol{x}^T\boldsymbol{\ell}_2)\cdots(\boldsymbol{x}^T\boldsymbol{\ell}_n) = \nu_n(\boldsymbol{x})^T\tilde{\boldsymbol{\ell}}.} \tag{7.39}$$

We call the vector $\tilde{\boldsymbol{\ell}}$ the *multibody epipolar line* associated with $\boldsymbol{x}_1$. Notice that $\tilde{\boldsymbol{\ell}}$ is a vector representation of the symmetric tensor product of all the epipolar lines $\boldsymbol{\ell}_1, \boldsymbol{\ell}_2, \ldots, \boldsymbol{\ell}_n$, and it is in general *not* the embedded (through the Veronese map) $\nu_n(\boldsymbol{\ell}_i)$ of any particular epipolar line $\boldsymbol{\ell}_i, i = 1, 2, \ldots, n$. From $\tilde{\boldsymbol{\ell}}$, we can compute the individual epipolar lines $\{\boldsymbol{\ell}_i\}_{i=1}^n$ associated with any image point $\boldsymbol{x}_1$ using the polynomial factorization technique given in Appendix 7.A.

**Example 7.18 (Embedded epipolar lines and multibody epipolar line).** Before we proceed any further, let us try to understand the relation between embedded epipolar lines and a multibody epipolar line for the simple case $n = 2$. Now, the two-body fundamental matrix $F$ is a $6 \times 6$ matrix. For a given $\boldsymbol{x}_1 \in \mathbb{R}^3$, $F\nu_2(\boldsymbol{x}_1)$ is then a six-dimensional vector with entries, say, $\tilde{\boldsymbol{\ell}} = [\alpha_1, \alpha_2, \ldots, \alpha_6]^T \in \mathbb{R}^6$. Now denote the two epipolar lines $\boldsymbol{\ell}_1 = F_1\boldsymbol{x}_1, \boldsymbol{\ell}_2 = F_2\boldsymbol{x}_1$ by

$$\boldsymbol{\ell}_1 = [a_1, b_1, c_1]^T, \quad \boldsymbol{\ell}_2 = [a_2, b_2, c_2]^T \quad \in \mathbb{R}^3. \tag{7.40}$$

Then the following two homogeneous polynomials (in $x, y, z$) are equal:

$$\alpha_1x^2+\alpha_2xy+\alpha_3xz+\alpha_4y^2+\alpha_5yz+\alpha_6z^2 = (a_1x+b_1y+c_1z)(a_2x+b_2y+c_2z). \tag{7.41}$$

Since the $\alpha_i$'s are given by $F\nu_2(\boldsymbol{x}_1)$, one would expect to be able to retrieve $[a_1, b_1, c_1]^T$ and $[a_2, b_2, c_2]^T$ from this identity. We leave the detail to the reader as an exercise (see Exercise 7.3). But the reader should notice that the multibody epipolar line $\tilde{\boldsymbol{\ell}}$ and the two embedded ones $\nu_n(\boldsymbol{\ell}_1)$ and $\nu_n(\boldsymbol{\ell}_2)$ are three different vectors in $\mathbb{R}^6$. ■

**Comment 7.19 (Polynomial factorization).** *Mathematically, factoring the polynomial $g(\boldsymbol{x})$ in (7.39) is a much simpler problem than directly factoring the original polynomial $f(\boldsymbol{x}_1, \boldsymbol{x}_2)$ as in (7.36). There is yet no known algorithm for the latter problem except for some special cases (e.g., $n = 2$). But a general algorithm can be found for the former problem. See Appendix 7.A.*

**Example 7.20** Given a homogeneous polynomial of degree 2 in two variables $f(x, y) = \alpha_1x^2 + \alpha_2xy + \alpha_3y^2$, to factor it into the form $f(x, y) = (a_1x + b_1y)(a_2x + b_2y)$, we may divide $f(x, y)$ by $y^2$ and let $z = x/y$ and obtain $g(z) = \alpha_1z^2 + \alpha_2z + \alpha_3$. Solving the roots of this quadratic equation gives $g(z) = \alpha_1(z - z_1)(z - z_2)$. Since $g(z) = g(x/y)$, multiplying it now by $y^2$, we get $f(x, y) = \alpha_1(x - z_1y)(x - z_2y)$. As we see, the factorization is in general not unique: both $[a_1, b_1]^T$ and $[a_2, b_2]^T$ are determined up to an arbitrary scalar factor. A general algorithm for factoring a polynomial with three variables and of higher degree can be found in Appendix 7.A. ■

In essence, the multibody fundamental matrix $F$ allows us to "transfer" a point $\boldsymbol{x}_1$ in the first image to a set of epipolar lines in the second image. This is exactly the multibody version of the conventional "epipolar transfer" that maps a point in the first image to an epipolar line in the second image. The *multibody epipolar transfer* process can be described by the sequence of maps

$$\boldsymbol{x}_1 \overset{\text{Veronese}}{\longmapsto} \nu_n(\boldsymbol{x}_1) \overset{\text{epipolar transfer}}{\longmapsto} \mathcal{F}\nu_n(\boldsymbol{x}_1) \overset{\text{polynomial factorization}}{\longmapsto} \{\boldsymbol{\ell}_i\}_{i=1}^n,$$

which is illustrated geometrically in Figure 7.6.

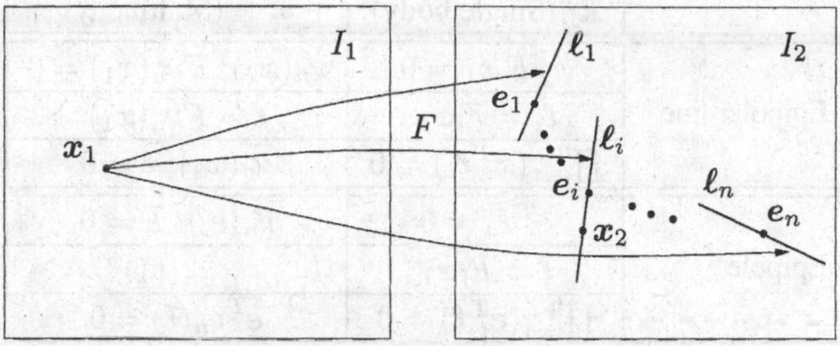


Figure 7.6. The multibody fundamental matrix $\mathcal{F}$ maps each point $\boldsymbol{x}_1$ in the first image to $n$ epipolar lines $\boldsymbol{\ell}_1, \boldsymbol{\ell}_2, \ldots, \boldsymbol{\ell}_n$ that pass through the $n$ epipoles $e_1, e_2, \ldots, e_n$, respectively. Furthermore, one of these epipolar lines passes through $\boldsymbol{x}_2$

Given a set of epipolar lines, we now describe how to compute the epipoles. Recall that the (left) epipole associated with each rank-2 fundamental matrix $F_i \in \mathbb{R}^{3\times3}$ is defined as the vector $e_i \in \mathbb{R}^3$ lying in the (left) null space of $F_i$; that is, $e_i$ satisfies $e_i^T F_i = 0$. Now let $\boldsymbol{\ell} \in \mathbb{R}^3$ be an arbitrary epipolar line associated with some image point in the first frame. Then there exists an $i$ such that $e_i^T \boldsymbol{\ell} = 0$. Therefore, every epipolar line $\boldsymbol{\ell}$ has to satisfy the polynomial constraint

$$\boxed{h(\boldsymbol{\ell}) \doteq (e_1^T \boldsymbol{\ell})(e_2^T \boldsymbol{\ell}) \cdots (e_n^T \boldsymbol{\ell}) = \tilde{e}^T \nu_n(\boldsymbol{\ell}) = 0,} \tag{7.42}$$

regardless of the motion with which it is associated. We call the vector $\tilde{e} \in \mathbb{R}^{C_n}$ the *multibody epipole* associated with the $n$ motions. As before, $\tilde{e}$ is a vector representation of the symmetric tensor product of the individual epipoles $e_1, e_2, \ldots, e_n$, and it is in general different from any of the embedded epipoles $\nu_n(e_i), i = 1, 2, \ldots, n$.

**Example 7.21 (Continued from Example 7.18.)** Denote the two epipoles $e_1, e_2$ of $F_1, F_2$, respectively, by

$$e_1 = [x_1, y_1, z_1]^T, \quad e_2 = [x_2, y_2, z_2]^T \quad \in \mathbb{R}^3, \tag{7.43}$$

and let $\boldsymbol{\ell} = [a, b, c]^T$. Then equation (7.42) implies that any epipolar line (e.g., $\boldsymbol{\ell}_1 = F_1\boldsymbol{x}_1$ or $\boldsymbol{\ell}_2 = F_2\boldsymbol{x}_1$) retrieved as described above satisfies the equation

$$(x_1a + y_1b + z_1c)(x_2a + y_2b + z_2c) = 0. \tag{7.44}$$

In other words,

$$x_1x_2a^2 + (x_1y_2 + y_1x_2)ab + (x_1z_2 + z_1x_2)ac + y_1y_2b^2 + (y_1z_2 + z_1y_2)bc + z_1z_2c^2 = 0. \tag{7.45}$$

As one may see, the information about the epipoles $e_1$ and $e_2$ is then encoded in the coefficients of the above homogeneous polynomial (in $a, b, c$), i.e. the multibody epipole

$$\tilde{e} = [x_1x_2, (x_1y_2 + y_1x_2), (x_1z_2 + z_1x_2), y_1y_2, (y_1z_2 + z_1y_2), z_1z_2]^T \in \mathbb{R}^6, \tag{7.46}$$

which can be linearly solved from (7.45) if sufficiently many epipolar lines are given. ■

We summarize the relationship between (multibody) epipolar lines and epipoles studied so far in Table 7.1.

| | $\mathbb{R}^3$ (Single body) | $\mathbb{R}^{C_n}$ (Multibody) |
|---|---|---|
| Epipolar line | $\boldsymbol{x}_2^T F_i \boldsymbol{x}_1 = 0$ | $\nu_n(\boldsymbol{x}_2)^T F \nu_n(\boldsymbol{x}_1) = 0$ |
| | $\boldsymbol{\ell}_i = F_i \boldsymbol{x}_1$ | $\tilde{\boldsymbol{\ell}} = F \nu_n(\boldsymbol{x}_1)$ |
| | $\prod_{i=1}^n (\boldsymbol{x}_2^T \boldsymbol{\ell}_i) = 0$ | $\nu_n(\boldsymbol{x}_2)^T \tilde{\boldsymbol{\ell}} = 0$ |
| Epipole | $\boldsymbol{e}_i^T F_i = 0$ | $\nu_n(\boldsymbol{e}_i)^T F = 0$ |
| | $\boldsymbol{\ell} = F_j \boldsymbol{x}_1$ | $\nu_n(\boldsymbol{\ell})$ |
| | $\prod_{i=1}^n (\boldsymbol{e}_i^T \boldsymbol{\ell}) = 0$ | $\tilde{\boldsymbol{e}}^T \nu_n(\boldsymbol{\ell}) = 0$ |

Table 7.1. Multibody epipolar line and epipole.

Now, given a collection $\{\boldsymbol{\ell}^j\}_{j=1}^m$ of $m \geq C_n - 1$ epipolar lines (which can be computed from the multibody epipolar transfer described before), we can obtain the multibody epipole $\tilde{\boldsymbol{e}} \in \mathbb{R}^{C_n}$ as the solution of the linear system

$$B_n \tilde{\boldsymbol{e}} \doteq \begin{bmatrix} \nu_n(\boldsymbol{\ell}^1)^T \\ \nu_n(\boldsymbol{\ell}^2)^T \\ \vdots \\ \nu_n(\boldsymbol{\ell}^m)^T \end{bmatrix} \tilde{\boldsymbol{e}} = 0. \tag{7.47}$$

In order for equation (7.47) to have a unique solution (up to a scalar factor), we will need to replace $n$ by the number of distinct epipoles $n_e$, as stated by the following proposition:

**Proposition 7.22 (Number of distinct epipoles).** *Assume that we are given a collection of epipolar lines $\{\boldsymbol{\ell}^j\}_{j=1}^m$ corresponding to 3-D points in general configuration undergoing $n$ distinct rigid-body motions with nonzero translation. Then, if the number of epipolar lines $m$ is at least $C_n - 1$, we have*

$$\mathit{rank}(B_i) \begin{cases} > C_i - 1, & \text{if } i < n_e, \\ = C_i - 1, & \text{if } i = n_e, \\ < C_i - 1, & \text{if } i > n_e. \end{cases} \tag{7.48}$$

*Therefore, the number of distinct epipoles $n_e \leq n$ is given by*

$$\boxed{n_e \doteq \min\{i : \mathit{rank}(B_i) = C_i - 1\}.} \tag{7.49}$$

*Proof.* Similar to the proof of Theorem 7.12. □

Once the number of distinct epipoles, $n_e$, has been computed, the vector $\tilde{\boldsymbol{e}} \in \mathbb{R}^{C_{n_e}}$ can be obtained from the linear system $B_{n_e} \tilde{\boldsymbol{e}} = 0$. Note that, typically, only $(C_n - 1)$ epipolar lines are needed to estimate the multibody epipole $\tilde{\boldsymbol{e}}$, a number much less than the total number $(C_n^2 - 1)$ of image pairs needed to estimate the multibody fundamental matrix $\mathcal{F}$. Once $\tilde{e}$ has been computed, the individual epipoles $\{e_i\}_{i=1}^{n_e}$ can be computed from $\tilde{e}$ using the factorization technique of Appendix 7.A.

### 7.4.2 *Recovery of individual fundamental matrices*

Given the epipolar lines and the epipoles, we now show how to recover each fundamental matrix $\{F_i\}_{i=1}^n$. To avoid degenerate cases, we assume that all the epipoles are distinct, i.e.

$$n_e = n.$$

Let $F_i = [\boldsymbol{f}_i^1\ \boldsymbol{f}_i^2\ \boldsymbol{f}_i^3] \in \mathbb{R}^{3\times 3}$ be the fundamental matrix associated with motion $i$, with columns $\boldsymbol{f}_i^1, \boldsymbol{f}_i^2, \boldsymbol{f}_i^3 \in \mathbb{R}^3$. We know from Section 7.4.1 that given $\boldsymbol{x}_1 = [x_1, y_1, z_1]^T \in \mathbb{R}^3$, the vector $\mathcal{F}\nu_n(\boldsymbol{x}_1) \in \mathbb{R}^{C_n}$ represents the coefficients of the following homogeneous polynomial in $\boldsymbol{x}$:

$$g(\boldsymbol{x}) = \left(\boldsymbol{x}^T(\boldsymbol{f}_1^1 x_1 + \boldsymbol{f}_1^2 y_1 + \boldsymbol{f}_1^3 z_1)\right)\cdots\left(\boldsymbol{x}^T(\boldsymbol{f}_n^1 x_1 + \boldsymbol{f}_n^2 y_1 + \boldsymbol{f}_n^3 z_1)\right).$$

Therefore, given the multibody fundamental matrix $\mathcal{F}$, one can estimate any linear combination of the columns of the fundamental matrix $F_i$ up to a scalar factor, i.e. we can obtain vectors $\boldsymbol{\ell}_i \in \mathbb{R}^3$ satisfying

$$\lambda_i \boldsymbol{\ell}_i \doteq (\boldsymbol{f}_i^1 x_1 + \boldsymbol{f}_i^2 y_1 + \boldsymbol{f}_i^3 z_1), \quad \lambda_i \in \mathbb{R},\ i = 1, 2, \ldots, n.$$

These vectors are nothing but the epipolar lines associated with the multibody epipolar line $\mathcal{F}\nu_n(\boldsymbol{x}_1)$, which can be computed using the polynomial factorization technique of Appendix 7.A. Notice that in particular we can obtain the three columns of $F_i$ up to scale by choosing $\boldsymbol{x}_1 = [1, 0, 0]^T$, $\boldsymbol{x}_1 = [0, 1, 0]^T$, and $\boldsymbol{x}_1 = [0, 0, 1]^T$. However:

1. We do not know the fundamental matrix to which the recovered epipolar lines belong;
2. The recovered epipolar lines, hence the columns of each $F_i$, are obtained up to a scalar factor only. Hence, we do not know the relative scales between the columns of the same fundamental matrix.

The first problem is easily solvable: if a recovered epipolar line $\boldsymbol{\ell} \in \mathbb{R}^3$ corresponds to a linear combination of columns of the fundamental matrix $F_i$, then it must be perpendicular to the previously computed epipole $e_i$; i.e. we must have $e_i^T \boldsymbol{\ell} = 0$. As for the second problem, for each $i$ let $\boldsymbol{\ell}_i^j$ be the epipolar line associated with $\boldsymbol{x}_1^j$ that is perpendicular to $e_i$, for $j = 1, 2, \ldots, m$. Since the $\boldsymbol{x}_1^j$'s can be chosen arbitrarily, we choose the first three to be $\boldsymbol{x}_1^1 = [1, 0, 0]^T$, $\boldsymbol{x}_1^2 = [0, 1, 0]^T$, and $\boldsymbol{x}_1^3 = [0, 0, 1]^T$ to form a simple basis. Then for every

$\boldsymbol{x}_1^j = [x_1^j, y_1^j, z_1^j]^T, j \geq 4$, there exist unknown scales $\lambda_i^j \in \mathbb{R}$ such that

$$\begin{aligned} \lambda_i^j \boldsymbol{\ell}_i^j &= \boldsymbol{f}_i^1 x_1^j \quad + \boldsymbol{f}_i^2 y_1^j \quad + \boldsymbol{f}_i^3 z_1^j \\ &= (\lambda_i^1 \boldsymbol{\ell}_i^1) x_1^j + (\lambda_i^2 \boldsymbol{\ell}_i^2) y_1^j + (\lambda_i^3 \boldsymbol{\ell}_i^3) z_1^j, \quad j \geq 4. \end{aligned}$$

Multiplying both sides by $\widehat{\boldsymbol{\ell}_i^j}$, we obtain

$$0 = \widehat{\boldsymbol{\ell}_i^j} \left( (\lambda_i^1 \boldsymbol{\ell}_i^1) x_1^j + (\lambda_i^2 \boldsymbol{\ell}_i^2) y_1^j + (\lambda_i^3 \boldsymbol{\ell}_i^3) z_1^j \right), \quad j \geq 4, \tag{7.50}$$

where $\lambda_i^1, \lambda_i^2, \lambda_i^3$ are the only unknowns. Therefore, the fundamental matrices are given by

$$\boxed{F_i = [\boldsymbol{f}_i^1 \ \boldsymbol{f}_i^2 \ \boldsymbol{f}_i^3] = [\lambda_i^1 \boldsymbol{\ell}_i^1 \ \lambda_i^2 \boldsymbol{\ell}_i^2 \ \lambda_i^3 \boldsymbol{\ell}_i^3],} \tag{7.51}$$

where $\lambda_i^1$, $\lambda_i^2$ and $\lambda_i^3$ can be obtained as the solution to the linear system

$$\begin{bmatrix} \widehat{\boldsymbol{\ell}_i^4} [x_1^4 \boldsymbol{\ell}_i^1 \ y_1^4 \boldsymbol{\ell}_i^2 \ z_1^4 \boldsymbol{\ell}_i^3] \\ \widehat{\boldsymbol{\ell}_i^5} [x_1^5 \boldsymbol{\ell}_i^1 \ y_1^5 \boldsymbol{\ell}_i^2 \ z_1^5 \boldsymbol{\ell}_i^3] \\ \vdots \\ \widehat{\boldsymbol{\ell}_i^m} [x_1^m \boldsymbol{\ell}_i^1 \ y_1^m \boldsymbol{\ell}_i^2 \ z_1^m \boldsymbol{\ell}_i^3] \end{bmatrix} \begin{bmatrix} \lambda_i^1 \\ \lambda_i^2 \\ \lambda_i^3 \end{bmatrix} = 0. \tag{7.52}$$

We have given a constructive proof for the following statement.

**Theorem 7.23 (Factorization of the multibody fundamental matrix).** *Let $\mathcal{F} \in \mathbb{R}^{C_n \times C_n}$ be the multibody fundamental matrix associated with fundamental matrices $\{F_i \in \mathbb{R}^{3\times 3}\}_{i=1}^n$. If the $n$ epipoles are distinct, then the matrices $\{F_i\}_{i=1}^n$ can be uniquely determined (up to a scalar factor).*

### 7.4.3 3-D motion segmentation

The 3-D motion segmentation problem refers to the problem of assigning each image pair $\{(\boldsymbol{x}_1^j, \boldsymbol{x}_2^j)\}_{j=1}^N$ to the motion it corresponds. This can be easily done from either the epipoles $\{\boldsymbol{e}_i\}_{i=1}^n$ and epipolar lines $\{\boldsymbol{\ell}^j\}_{j=1}^m$, or from the fundamental matrices $\{F_i\}_{i=1}^n$, as follows.

1. *Motion segmentation from the epipoles and epipolar lines*: Given an image pair $(\boldsymbol{x}_1, \boldsymbol{x}_2)$, the factorization of $\tilde{\boldsymbol{\ell}} = \mathcal{F} \nu_n(\boldsymbol{x}_1)$ gives $n$ epipolar lines. One of these lines, say $\boldsymbol{\ell}$, passes through $\boldsymbol{x}_2$; i.e. $\boldsymbol{\ell}^T \boldsymbol{x}_2 = 0$. The pair $(\boldsymbol{x}_1, \boldsymbol{x}_2)$ is assigned to the $i$th motion if $\boldsymbol{\ell}^T \boldsymbol{e}_i = 0$.

2. *Motion segmentation from the fundamental matrices*: The image pair $(\boldsymbol{x}_1, \boldsymbol{x}_2)$ is assigned to the $i$th motion if $\boldsymbol{x}_2^T F_i \boldsymbol{x}_1 = 0$.

Figure 7.7 illustrates how a particular image pair, say $(\boldsymbol{x}_1, \boldsymbol{x}_2)$, that belongs to the $i$th motion, $i = 1, 2, \ldots, n$, is successfully segmented.

In the presence of noise, $(\boldsymbol{x}_1, \boldsymbol{x}_2)$ is assigned to the motion $i$ that minimizes $(\boldsymbol{e}_i^T \boldsymbol{\ell})^2$ or $(\boldsymbol{x}_2^T F_i \boldsymbol{x}_1)^2$. In the scheme illustrated by the figure, the epipolar lines

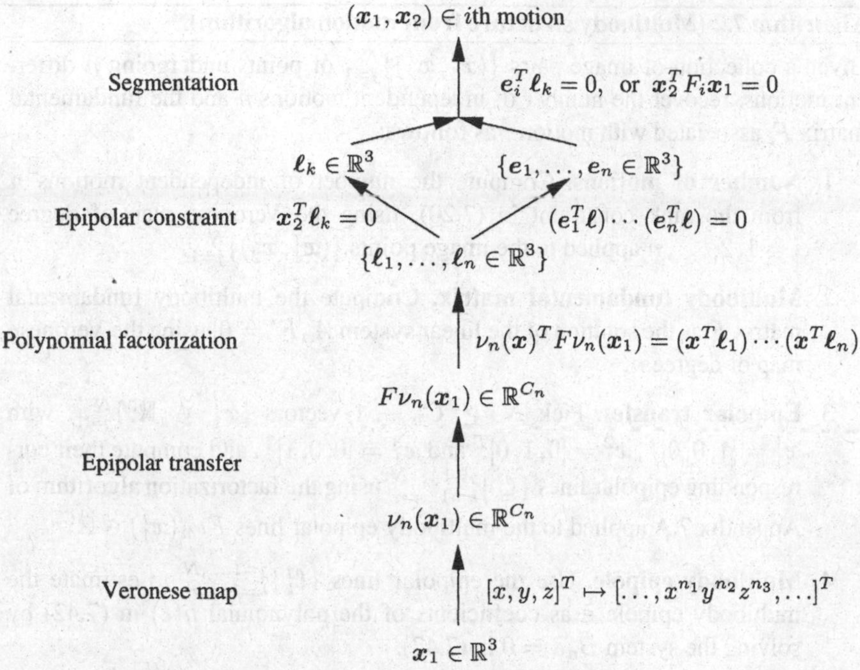


Figure 7.7. Transformation diagram associated with the segmentation of an image pair $(\boldsymbol{x}_1, \boldsymbol{x}_2)$ in the presence of $n$ motions.

$\{\boldsymbol{\ell}\}$ computed from polynomial factorization for all given image pairs can be used for three purposes:

1. Estimation of epipoles $\boldsymbol{e}_i, i = 1, 2, \ldots, n$;
2. Retrieval of individual fundamental matrices $F_i, i = 1, 2, \ldots, n$;
3. Segmentation of image pairs based on $\boldsymbol{e}_i^T \boldsymbol{\ell} = 0, i = 1, 2, \ldots, n$.

## 7.5 Multibody structure from motion

Algorithm 7.2 presents a complete algorithm for multibody motion estimation and segmentation from two perspective views.

One of the main drawbacks of Algorithm 7.2 is that it needs many image correspondences in order to compute the multibody fundamental matrix, which often makes it impractical for large $n$ (see Remark 7.24 below). In practice, one can significantly reduce the data requirements by incorporating partial knowledge about the motion or segmentation of the objects with minor changes to the general algorithm. We discuss a few such variations to Algorithm 7.2, assuming linear motions and constant-velocity motions in Exercises 7.6 and 7.8.

**Algorithm 7.2 (Multibody structure from motion algorithm).**

Given a collection of image pairs $\{(x_1^j, x_2^j)\}_{j=1}^N$ of points undergoing $n$ different motions, recover the number of independent motions $n$ and the fundamental matrix $F_i$ associated with motion $i$ as follows:

1. **Number of motions.** Compute the number of independent motions $n$ from the rank constraint in (7.20), using the Veronese map of degree $i = 1, 2, \ldots, n$ applied to the image points $\{(x_1^j, x_2^j)\}_{j=1}^N$.
2. **Multibody fundamental matrix.** Compute the multibody fundamental matrix $F$ as the solution of the linear system $A_n F^s = 0$, using the Veronese map of degree $n$.
3. **Epipolar transfer.** Pick $N \geq C_n - 1$ vectors $\{x_1^j \in \mathbb{R}^3\}_{j=1}^N$, with $x_1^1 = [1,0,0]^T$, $x_1^2 = [0,1,0]^T$ and $x_1^3 = [0,0,1]^T$, and compute their corresponding epipolar lines $\{\ell_k^j\}_{k=1,\ldots,n}^{j=1,\ldots,N}$ using the factorization algorithm of Appendix 7.A applied to the multibody epipolar lines $F\nu_n(x_1^j) \in \mathbb{R}^{C_n}$.
4. **Multibody epipole.** Use the epipolar lines $\{\ell_k^j\}_{k=1,\ldots,n}^{j=1,\ldots,N}$ to estimate the multibody epipole $\tilde{e}$ as coefficients of the polynomial $h(\ell)$ in (7.42) by solving the system $B_n\tilde{e} = 0$ in (7.47).
5. **Individual epipoles.** Use the polynomial factorization algorithm of Appendix 7.A to compute the individual epipoles $\{e_i\}_{i=1}^n$ from the multibody epipole $\tilde{e} \in \mathbb{R}^{C_n}$.
6. **Individual fundamental matrices.** For each $j$, choose $k(i)$ such that $e_i^T \ell_{k(i)}^j = 0$; i.e. assign each epipolar line to its motion. Then use equations (7.51) and (7.52) to obtain each fundamental matrix $F_i$ from the epipolar lines assigned to epipole $i$.
7. **Feature segmentation by motion.** Assign image pair $(x_1^j, x_2^j)$ to motion $i$ if $e_i^T \ell_{k(i)}^j = 0$ or if $(x_2^j)^T F_i x_1^j = 0$.

The only step of Algorithm 7.2 that requires $O(n^4)$ image pairs is the estimation of the multibody fundamental matrix $F$. Step 2 requires a large number of data points, because $F$ is estimated linearly without taking into account the rich internal (algebraic) structure of $F$ (e.g., $\text{rank}(F) \leq C_n - n$). Therefore, one should expect to be able to reduce the number of image pairs needed by considering constraints among entries of $F$, in the same spirit that the eight-point algorithm for $n = 1$ can be reduced to seven points if the algebraic property $\det(F) = 0$ is used.

**Remark 7.24 (Issues related to the algorithm).** *Despite its algebraic simplicity, there are many important and subtle issues related to Algorithm 7.2 that we have not addressed:*

1. Special motions and repeated epipoles. *Algorithm 7.2 works for distinct motions with nonzero translation. It does not work for some special motions, e.g., pure rotation or repeated epipoles. If two individual fundamental matrices share the same (left) epipoles, we cannot segment the epipolar lines as described in step 6 of Algorithm 7.2. In this case, one may consider the right epipoles (in the first image frame) instead, since it is extremely rare that two motions give rise to the same left and right epipoles.*[5]

2. Algebraic solvability. *The only nonlinear part of Algorithm 7.2 is to factor homogeneous polynomials of degree $n$ in step 3 and step 5. Therefore, the multibody structure from motion problem is algebraically solvable (i.e. there is a closed-form solution) if and only if the number of motions is $n \leq 4$. When $n \geq 5$, the above algorithm must rely on a numerical solution for the roots of those polynomials. See Proposition 7.30 in Appendix 7.A.*

3. Computational complexity. *In terms of data, Algorithm 7.2 requires $O(n^4)$ image pairs to estimate the multibody fundamental matrix $F$ associated with the $n$ motions. In terms of numerical computation, it needs to factor $O(n)$ polynomials*[6] *and hence solve for the roots of $O(n)$ univariate polynomials of degree $n$*[7]*. As for the remaining computation, which can be well approximated by the most costly step 1 and step 2, the complexity is $O(n^6)$.*

4. Statistical optimality. *Algorithm 7.2 gives a purely algebraic solution to the multibody structure from motion problem. Since the polynomial factorization in step 3 and step 5 is very robust to noise, one should pay attention to step 2, which is sensitive to noise, because it does not exploit the algebraic structure of the multibody fundamental matrix $F$. For nonlinear optimal solutions, please refer to [Vidal and Sastry, 2003].*

**Example 7.25 (A simple demonstration).** The proposed algorithm is tested below by segmenting a real image sequence with $n = 3$ moving objects: a truck, a car, and a box. Figure 7.8 shows two frames of the sequence with the tracked features superimposed. We track a total of $N = 173$ point features across the two views: 44 for the truck, 48 for the car, and 81 for the box. Figure 7.9 plots the segmentation of the image points obtained using Algorithm 7.2. Notice that the obtained segmentation has no mismatches.

■

[5] This happens only when the rotation axes of the two motions are equal to each other and parallel to the translation.

[6] One needs about $C_n - 1 \approx O(n^2)$ epipolar lines to compute the epipoles and fundamental matrices, which can be obtained from $O(n)$ polynomial factorizations, since each one generates $n$ epipolar lines. Hence it is not necessary to compute the epipolar lines for all $N = C_n^2 - 1 \approx O(n^4)$ image pairs in step 3.

[7] The numerical complexity of solving for the roots for an $n$th-order polynomial in one variable is polynomial in $n$ for a given error bound; see [Smale, 1997].

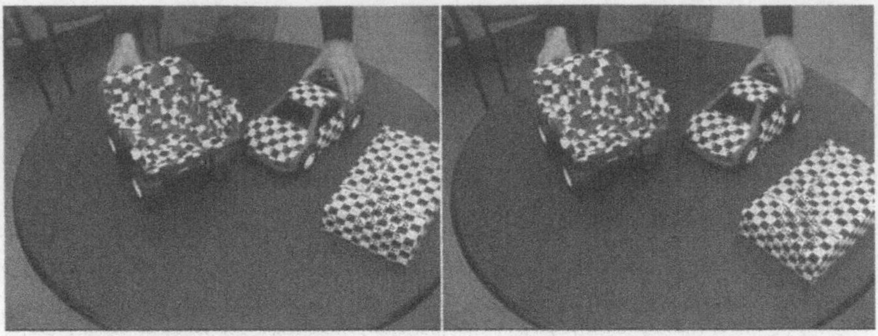

($a$) First image frame ($b$) Second image frame

Figure 7.8. A motion sequence with a truck, a car, and a box. Tracked features are marked as follows: "○" for the truck, "□" for the car, and "△" for the box.

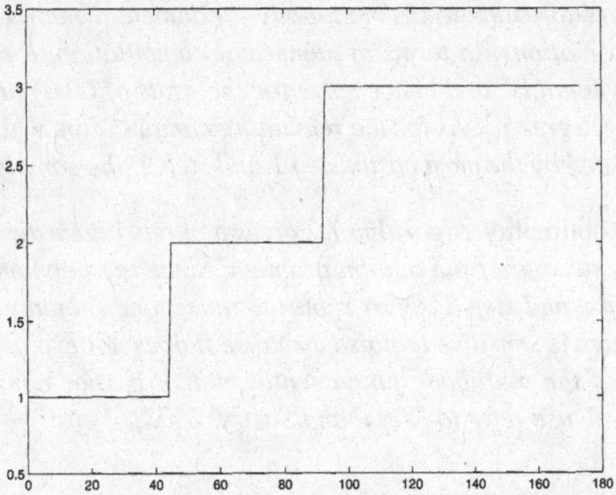


Figure 7.9. Motion segmentation results. Each image pair is assigned to the fundamental matrix for which the algebraic error is minimized. The first 44 points correspond to the truck, the next 48 to the car, and the last 81 to the box. The correct segmentation is obtained.

## 7.6 Summary

In this chapter we have shown how to solve the reconstruction problem when there are multiple independently moving objects in the scene. The solution is a generalization of the single-motion case by the use of the Veronese map, which renders a multilinear problem linear. Table 7.2 gives a comparison of geometric entities associated with two views of 1 rigid-body motion and $n$ rigid-body motions. So far, in the three chapters (5, 6, and 7) of Part II, we have studied two-view geometry for both discrete and continuous motions, both calibrated and uncalibrated cameras, both generic and planar scenes, and both single-body and

| Comparison of | two views of 1 body | two views of $n$ bodies |
|---|---|---|
| An image pair | $\boldsymbol{x}_1, \boldsymbol{x}_2 \in \mathbb{R}^3$ | $\nu_n(\boldsymbol{x}_1), \nu_n(\boldsymbol{x}_2) \in \mathbb{R}^{C_n}$ |
| Epipolar constraint | $\boldsymbol{x}_2^T F \boldsymbol{x}_1 = 0$ | $\nu_n(\boldsymbol{x}_2)^T \mathcal{F} \nu_n(\boldsymbol{x}_1) = 0$ |
| fundamental matrix | $F \in \mathbb{R}^{3\times 3}$ | $\mathcal{F} \in \mathbb{R}^{C_n \times C_n}$ |
| Linear recovery from $m$ image pairs | $\begin{bmatrix} \boldsymbol{x}_2^1 \otimes \boldsymbol{x}_1^1 \\ \boldsymbol{x}_2^2 \otimes \boldsymbol{x}_1^2 \\ \vdots \\ \boldsymbol{x}_2^m \otimes \boldsymbol{x}_1^m \end{bmatrix} F^s = 0$ | $\begin{bmatrix} \nu_n(\boldsymbol{x}_2^1) \otimes \nu_n(\boldsymbol{x}_1^1) \\ \nu_n(\boldsymbol{x}_2^2) \otimes \nu_n(\boldsymbol{x}_1^2) \\ \vdots \\ \nu_n(\boldsymbol{x}_2^m) \otimes \nu_n(\boldsymbol{x}_1^m) \end{bmatrix} \mathcal{F}^s = 0$ |
| Epipole | $\boldsymbol{e}^T F = 0$ | $\nu_n(\boldsymbol{e})^T \mathcal{F} = 0$ |
| Epipolar transfer | $\boldsymbol{\ell} = F\boldsymbol{x}_1 \in \mathbb{R}^3$ | $\tilde{\boldsymbol{\ell}} = \mathcal{F}\nu_n(\boldsymbol{x}_1) \in \mathbb{R}^{C_n}$ |
| Epipolar line & point | $\boldsymbol{x}_2^T \boldsymbol{\ell} = 0$ | $\nu_n(\boldsymbol{x}_2)^T \tilde{\boldsymbol{\ell}} = 0$ |
| Epipolar line & epipole | $\boldsymbol{e}^T \boldsymbol{\ell} = 0$ | $\tilde{\boldsymbol{e}}^T \nu_n(\boldsymbol{\ell}) = 0$ |

Table 7.2. Comparison between the geometry for two views of 1 rigid-body motion and that for $n$ rigid-body motions.

multibody motions. In the chapters of Part III, we will see how to generalize these results to the multiple-view setting.

## 7.7 Exercises

**Exercise 7.1 (0-D data segmentation).** Given a set of points $\{\boldsymbol{x}_i\}_{i=1}^N$ drawn from $n$ ($< N$) different unknown values $a_1, a_2, \ldots, a_n$ in $\mathbb{R}$, then each $\boldsymbol{x}$ satisfies the polynomial equation

$$p(\boldsymbol{x}) = (\boldsymbol{x} - a_1)(\boldsymbol{x} - a_2)\cdots(\boldsymbol{x} - a_n) = 0. \tag{7.53}$$

Show that:

1. We have that the rank of the matrix

$$A_m \doteq \begin{bmatrix} \boldsymbol{x}_1^m & \boldsymbol{x}_1^{m-1} & \cdots & 1 \\ \boldsymbol{x}_2^m & \boldsymbol{x}_2^{m-1} & \cdots & 1 \\ \vdots & \vdots & \ddots & \vdots \\ \boldsymbol{x}_N^m & \boldsymbol{x}_N^{m-1} & \cdots & 1 \end{bmatrix} \in \mathbb{R}^{N\times(m+1)} \tag{7.54}$$

is $m$ if and only if $m = n$. This rank condition determines the number $n$ of different values.

2. The coefficients of the polynomial $p(\boldsymbol{x})$ are simply the only vector (up to a scalar factor) in the null space of $A_n$. How does one recover $a_1, a_2, \ldots, a_n$ from $p(\boldsymbol{x})$?

**Exercise 7.2 (Monomials).** Show that the number of monomials of degree $n$ in $k$ variables is the binomial coefficient

$$\binom{n+k-1}{n} = \binom{n+k-1}{k-1}.$$

**Exercise 7.3** If you are told that a second-degree homogeneous polynomial

$$p = \alpha_1 x^2 + \alpha_2 xy + \alpha_3 xz + \alpha_4 y^2 + \alpha_5 yz + \alpha_6 z^2 \tag{7.55}$$

is factorizable into two linear factors as $p = (a_1 x + b_1 y + c_1 z)(a_2 x + b_2 y + c_2 z)$, find a solution to $(a_i, b_i, c_i), i = 1, 2$, in terms of $\alpha_1, \alpha_2, \dots, \alpha_6$. Find an example of a second homogeneous polynomial that is, however, not factorizable in this fashion.

**Exercise 7.4** Show that:

1. The set of vectors $\{\nu_2(\boldsymbol{x}) | \boldsymbol{x} \in \mathbb{R}^3\} \subset \mathbb{R}^6$ spans the entire space $\mathbb{R}^6$;
2. The set of vectors $\{\nu_2(\boldsymbol{x}_1) \otimes \nu_2(\boldsymbol{x}_2) | \boldsymbol{x}_1, \boldsymbol{x}_2 \in \mathbb{R}^3\} \subset \mathbb{R}^{36}$ spans the entire space $\mathbb{R}^{36}$.

That is, if $\boldsymbol{x}_1, \boldsymbol{x}_2$ are not two corresponding images, they do not necessarily lie in a 35-dimensional subspace that is orthogonal to $F^s$.

**Exercise 7.5 (Two moving objects).** In case that there are $n = 2$ moving objects with fundamental matrices $F_1, F_2 \in \mathbb{R}^{3\times3}$, write down entries of the associated two-body fundamental matrix $F \in \mathbb{R}^{6\times6}$ in terms of entries of $F_1$ and $F_2$. Do you see any obvious solution for recovering $F_1$ and $F_2$ from entries of $F$?

**Exercise 7.6 (Multiple linearly moving objects).** In many practical situations, the motion of the objects can be well approximated by a linear motion; i.e. there is only translation but no rotation. In this case, the epipolar constraint reduces to $\boldsymbol{x}_2^T \widehat{\boldsymbol{e}_i} \boldsymbol{x}_1 = 0$ or $\boldsymbol{e}_i^T \widehat{\boldsymbol{x}_2} \boldsymbol{x}_1 = 0$, where $\boldsymbol{e}_i \in \mathbb{R}^3$ represents the epipole associated with the $i$th motion $\boldsymbol{e}_i \sim T_i, i = 1, 2, \dots, n$. Therefore, the vector $\boldsymbol{\ell} = \widehat{\boldsymbol{x}_2} \boldsymbol{x}_1 \in \mathbb{R}^3$ is an epipolar line satisfying the equation

$$g(\boldsymbol{\ell}) = (\boldsymbol{e}_1^T \boldsymbol{\ell})(\boldsymbol{e}_2^T \boldsymbol{\ell}) \cdots (\boldsymbol{e}_n^T \boldsymbol{\ell}) = 0. \tag{7.56}$$

Therefore, given a set of image pairs $\{(\boldsymbol{x}_2^j, \boldsymbol{x}_1^j)\}_{j=1}^N$ of points undergoing $n$ *distinct* linear motions $\boldsymbol{e}_1, \boldsymbol{e}_2, \dots, \boldsymbol{e}_n \in \mathbb{R}^3$, one can use the set of epipolar lines $\boldsymbol{\ell}^j = \widehat{\boldsymbol{x}_2^j} \boldsymbol{x}_1^j, j = 1, 2, \dots, N$, to estimate the epipoles $\boldsymbol{e}_i$ using step 4 and step 5 of Algorithm 7.2. Notice that the epipoles are recovered directly using polynomial factorization *without* estimating the multibody fundamental matrix $F$ first. What is the minimal number of image correspondences needed to solve the general problem of linearly moving objects undergoing $n$ independent motions?

**Exercise 7.7 (Estimation of vanishing points).** We know from the previous chapters that images $\{\boldsymbol{\ell}\}$ of parallel lines in space intersect at the so-called vanishing points; see Figure 6.13. Suppose that there are in total three sets of parallel lines, and denote the three associated vanishing points by $\boldsymbol{v}_1, \boldsymbol{v}_2, \boldsymbol{v}_3 \in \mathbb{R}^3$. Then each image line satisfies the equation

$$(\boldsymbol{v}_1^T \boldsymbol{\ell})(\boldsymbol{v}_2^T \boldsymbol{\ell})(\boldsymbol{v}_3^T \boldsymbol{\ell}) = 0.$$

Based on this fact and the method introduced in this chapter, design an algorithm to estimate $\boldsymbol{v}_1, \boldsymbol{v}_2, \boldsymbol{v}_3$ from all the image lines (e.g., edges detected in Figure 6.13).

**Exercise 7.8 (Constant-velocity motions).** In case the motion of the objects in the scene changes slowly relative to the sampling rate, we may assume that for a number of image frames, say $m$, the motion of each object between consecutive pairs of images is the same. Hence *all* the feature points corresponding to the $m-1$ image pairs in between can be used to estimate the *same* multibody fundamental matrix. Suppose $m = 5$ and the number of independent motions is $n = 4$. How many feature points do we need to track between the consecutive image pairs in order to effectively segment the objects based on their motion?

**Exercise 7.9 (Segmenting planar motions).** We know from Exercise 5.6 that an essential matrix for a planar motion is of the form

$$E = \begin{bmatrix} 0 & 0 & a \\ 0 & 0 & b \\ c & d & 0 \end{bmatrix}, \quad a, b, c, d \in \mathbb{R}. \tag{7.57}$$

Given the special structure of $E$ (i.e. many zeros in it), derive a simplified algorithm for segmenting two planar motions (with two different essential matrices $E_1$ and $E_2$) from two calibrated images. Try to generalize your method to the case of $n$ motions.

**Exercise 7.10 (Segmenting affine flows).** Although the practicality of the general motion segmentation method given in this chapter is questionable, since it requires a large number of feature points from both views, the method can be made much more useful if we apply it to more restricted situations like the translational and planar motion cases. In this exercise, we explore another possibility.

From Chapter 4, we know that the optical flow $\boldsymbol{u} = \dot{\boldsymbol{x}} \in \mathbb{R}^3$ satisfies the brightness constancy constraint $\boldsymbol{y}^T \boldsymbol{u} = 0$ for $\boldsymbol{y} = [I_x, I_y, I_t]^T$. We further assume that the optical flow generated by different motions is piecewise affine; that is, $\boldsymbol{u} = A_i \boldsymbol{x}$ for some affine matrix

$$A_i = \begin{bmatrix} a_{11} & a_{12} & a_{13} \\ a_{21} & a_{22} & a_{23} \\ 0 & 0 & 1 \end{bmatrix} \in \mathbb{R}^{3\times 3}, \quad i = 1, 2, \ldots, n. \tag{7.58}$$

Thus, we have the following equation for every image point $\boldsymbol{x}$ and its image intensity gradient $\boldsymbol{y}$:

$$\boldsymbol{y}^T A_i \boldsymbol{x} = 0, \tag{7.59}$$

for some unknown affine matrix $A_i$, $i = 1, 2, \ldots, n$.

Unlike the fundamental matrix $F$, an affine matrix $A$ is always of rank 3. But $A$ has a special structure: its last row is always $[0, 0, 1]$. Using this fact, derive a similar scheme to the fundamental-matrix multibody segmentation for segmenting multiple affine flows. Justify why this is more feasible and practical than the feature-based method using the fundamental matrix.

**Exercise 7.11 (An open problem).** Under the same conditions of Lemma 7.13, prove or disprove

$$\operatorname{rank}(F) = C_n - n \tag{7.60}$$

under appropriate conditions.

## 7.A Homogeneous polynomial factorization

Let $\{\boldsymbol{\ell}_i\}_{i=1}^n$ be a collection of $n$ *distinct* vectors in $\mathbb{R}^3$ and let $p_n(\boldsymbol{x})$ be the homogeneous polynomial of degree $n$ in $\boldsymbol{x} = [x, y, z]^T \in \mathbb{R}^3$ given by

$$\begin{aligned} p_n(\boldsymbol{x}) = \boldsymbol{a}^T \nu_n(\boldsymbol{x}) &\doteq \sum a_{n_1,n_2,n_3}\, x^{n_1} y^{n_2} z^{n_3} = (\boldsymbol{\ell}_1^T \boldsymbol{x})(\boldsymbol{\ell}_2^T \boldsymbol{x}) \cdots (\boldsymbol{\ell}_n^T \boldsymbol{x}) \\ &= (\ell_{11}x + \ell_{12}y + \ell_{13}z)(\ell_{21}x + \ell_{22}y + \ell_{23}z) \cdots (\ell_{n1}x + \ell_{n2}y + \ell_{n3}z), \end{aligned} \quad (7.61)$$

where $\boldsymbol{a} \in \mathbb{R}^{C_n}$ is the vector of coefficients of the polynomial $p_n(\boldsymbol{x})$.

**Remark 7.26 (Symmetric multilinear tensor).** *Mathematically, the vector $\boldsymbol{a} \in \mathbb{R}^{C_n}$ is a vector representation for the symmetric tensor product of all the vectors $\boldsymbol{\ell}_1, \boldsymbol{\ell}_2, \ldots, \boldsymbol{\ell}_n \in \mathbb{R}^3$,*

$$Sym(\boldsymbol{\ell}_1 \otimes \boldsymbol{\ell}_2 \otimes \cdots \otimes \boldsymbol{\ell}_n) \doteq \sum_{\sigma \in \mathfrak{S}_n} \boldsymbol{\ell}_{\sigma(1)} \otimes \boldsymbol{\ell}_{\sigma(2)} \otimes \cdots \otimes \boldsymbol{\ell}_{\sigma(n)}, \quad (7.62)$$

*where $\mathfrak{S}_n$ is the permutation group of $n$ elements and $\otimes$ represents the tensor product of vectors.*

Given the vector of coefficients $\boldsymbol{a} \in \mathbb{R}^{C_n}$ of the polynomial $p_n(\boldsymbol{x})$, we would like to compute the set of vectors $\{\boldsymbol{\ell}_i\}_{i=1}^n$ up to a scalar factor. In our special context, one may view $\{\boldsymbol{\ell}_i\}_{i=1}^n$ as the epipolar lines, and $\boldsymbol{a}$ can then be interpreted as the multibody epipolar line $\tilde{\boldsymbol{\ell}}$. Our goal here is to compute these (individual) epipolar lines from the multibody one:

$$\boldsymbol{a} = \tilde{\boldsymbol{\ell}} \in \mathbb{R}^{C_n} \quad \mapsto \quad \{\boldsymbol{\ell}_1, \boldsymbol{\ell}_2, \ldots, \boldsymbol{\ell}_n \in \mathbb{R}^3\}. \quad (7.63)$$

To this end, we consider the last $n+1$ coefficients of $p_n(\boldsymbol{x})$, which define the following homogeneous polynomial of degree $n$ in $y$ and $z$:

$$\sum a_{0,n_2,n_3} y^{n_2} z^{n_3} = \prod_{i=1}^{n} (\ell_{i2} y + \ell_{i3} z). \quad (7.64)$$

Letting $w = y/z$, we have that

$$\prod_{i=1}^{n} (\ell_{i2} y + \ell_{i3} z) = 0 \quad \Leftrightarrow \quad \prod_{i=1}^{n} (\ell_{i2} w + \ell_{i3}) = 0,$$

and hence the $n$ roots of the univariate polynomial

$$q_n(w) = a_{0,n,0} w^n + a_{0,n-1,1} w^{n-1} + \cdots + a_{0,0,n} \quad (7.65)$$

are exactly $w_i = -\ell_{i3}/\ell_{i2}$, for $i = 1, 2, \ldots, n$. Therefore, after dividing $\boldsymbol{a}$ by $a_{0,n,0}$ (if nonzero), we obtain the last two entries of each $\boldsymbol{\ell}_i$ as

$$(\ell_{i2}\,,\ \ell_{i3}) = (1\,,\ -w_i), \quad i = 1, 2, \ldots, n. \quad (7.66)$$

If $\ell_{i2} = 0$ for some $i$, then some of the leading coefficients of $q_n(w)$ are zero, and we cannot proceed as before, because $q_n(w)$ has fewer than $n$ roots. More specifically, assume that the first $r \leq n$ coefficients of $q_n(w)$ are zero and divide $\boldsymbol{a}$ by the $(r+1)$st coefficient. In this case, we can choose $(\ell_{i2}, \ell_{i3}) = (0, 1)$, for

$i = 1, 2, \ldots, r$, and obtain $\{(\ell_{i2}, \ell_{i3})\}_{i=n-r+1}^{n}$ from the $n - r$ roots of $q_n(w)$ by using equation (7.66). Finally, if all the coefficients of $q_n(w)$ are equal to zero, we set $(\ell_{i2}, \ell_{i3}) = (0, 0)$, for all $i = 1, 2, \ldots, n$.[8]

**Remark 7.27 (Solvability of roots of a univariate polynomial).** *It is a known fact in abstract algebra that there is no closed-form solution for the roots of univariate polynomials of degree $n \geq 5$ [Abel, 1828, Galois, 1931]. Hence, there is no closed-form solution to homogeneous polynomial factorization for $n \geq 5$ either. Since one can always find the roots of a univariate polynomial numerically using efficient polynomial-time algorithms [Smale, 1997], we will consider this problem as solved.*

We are left with the computation of the coefficients of the variable $x$ of each factor of $p_n(\boldsymbol{x})$, i.e. $\{\ell_{i1}\}_{i=1}^{n}$. For that, we consider the $n$ coefficients $a_{1,n_2,n_2}$ of $p_n(\boldsymbol{x})$. We notice that these coefficients are *linear* functions of the unknowns $\{\ell_{i1}\}_{i=1}^{n}$, given that we already know $\{(\ell_{i2}, \ell_{i3})\}_{i=1}^{n}$. Therefore, we can solve for $\ell_{i1}$ from the linear system

$$\begin{bmatrix} \mathcal{V}_1 & \mathcal{V}_2 & \cdots & \mathcal{V}_n \end{bmatrix} \begin{bmatrix} \ell_{11} \\ \ell_{21} \\ \vdots \\ \ell_{n1} \end{bmatrix} = \begin{bmatrix} a_{1,n-1,0} \\ a_{1,n-2,1} \\ \vdots \\ a_{1,0,n-1} \end{bmatrix}, \tag{7.67}$$

where $\mathcal{V}_i \in \mathbb{R}^n$ are the coefficients of the following homogeneous polynomial of degree $n - 1$ in $y$ and $z$:

$$r_i(y, z) \doteq \prod_{k=1}^{i-1} (\ell_{k2} y + \ell_{k3} z) \cdot \prod_{k=i+1}^{n} (\ell_{k2} y + \ell_{k3} z). \tag{7.68}$$

In order for the linear system in (7.67) to have a unique solution, the column vectors $\{\mathcal{V}_i \in \mathbb{R}^n\}_{i=1}^{n}$ (in the matrix on the left-hand side) must be linearly independent. We leave to the reader as an exercise to prove the following proposition:

**Proposition 7.28.** *The vectors $\{\mathcal{V}_i \in \mathbb{R}^n\}_{i=1}^{n}$ are linearly independent if and only if the $n$ vectors $\{(\ell_{i2}, \ell_{i3})\}_{i=1}^{n}$ are pairwise linearly independent.*

This latter condition is always satisfied, except for some degenerate cases described in Example 7.29 below.

**Example 7.29 (Some degenerate cases).** There are essentially three cases in which the vectors $\{(\ell_{i2}, \ell_{i3})\}_{i=1}^{n}$ are not pairwise linearly independent:

1. The original polynomial $p_n(\boldsymbol{x})$ is such that the polynomial $q_n(w)$ has repeated roots, e.g., $p_n(\boldsymbol{x}) = (2x + y + 3z)(x + y + 3z)$.

[8] Under the assumption that $p_n(\boldsymbol{x})$ has distinct factors, this occurs only when $p_n(\boldsymbol{x}) = ax$ for some $a \in \mathbb{R}$.

2. The polynomial $q_n(w)$ associated with some factorizable $p_n(\boldsymbol{x})$, e.g., $p_n(\boldsymbol{x}) = (x+z)z$ as a polynomial in $x, y, z$, has more than one zero leading coefficient. In this case we have $(\ell_{i2}, \ell_{i3}) = (0, 1)$ for more than one $i$.
3. The original polynomial $p_n(\boldsymbol{x})$ is not factorizable. This happens, for example, when the vector of coefficients $\boldsymbol{a}$ is corrupted by noise. In this case the polynomial $q_n(w)$ may have complex roots, e.g., $p_n(\boldsymbol{x}) = x^2 + y^2 + yz + z^2$, and one could "project" these complex roots onto their real parts. This typically introduces repeated real roots in the resulting polynomial; e.g., after "projection," the above polynomial $p_n(\boldsymbol{x})$ is effectively converted to $x^2 + y^2 + yz + \frac{1}{4}z^2$.

■

In those degenerate cases, as long as the original polynomial $p_n(\boldsymbol{x})$ has $n$ distinct factors, one can always perform an invertible linear transformation

$$\boldsymbol{x} \mapsto L\boldsymbol{x}, \quad L \in \mathbb{R}^{3\times 3}, \tag{7.69}$$

that induces a linear transformation on the vector of coefficients $\boldsymbol{a} \mapsto T\boldsymbol{a}$, for some $T \in \mathbb{R}^{C_n \times C_n}$, such that the new vectors $\{(\ell_{i2}, \ell_{i3})\}_{i=1}^n$ are pairwise linearly independent. A typical choice for such $L$ is of the form

$$L \doteq \begin{bmatrix} 1 & t & t \\ 0 & 1 & t \\ 0 & 0 & 1 \end{bmatrix} \in \mathbb{R}^{3\times 3},$$

where $t \in \mathbb{R}$ can always be chosen so that the new polynomial $q_n(w)$ in (7.65) has distinct roots. Under this transformation, the polynomial $p_n(\boldsymbol{x})$ becomes

$$\begin{aligned} p_n'(\boldsymbol{x}) = p_n(L\boldsymbol{x}) &= \prod_{i=1}^n (\boldsymbol{\ell}_i^T L\boldsymbol{x}) \\ &= \prod_{i=1}^n \Bigg( \ell_{i1}x + \underbrace{[t\ell_{i1} + \ell_{i2}]}_{\ell_{i2}'(t)} y + \underbrace{[t\ell_{i1} + t\ell_{i2} + \ell_{i3}]}_{\ell_{i3}'(t)} z \Bigg). \end{aligned}$$

Therefore, the polynomial associated with $y$ and $z$ will have distinct roots for all $t \in \mathbb{R}$, except for the $t$'s, which are roots of the following second-order polynomial:

$$\ell_{r2}'(t)\ell_{s3}'(t) = \ell_{s2}'(t)\ell_{r3}'(t), \tag{7.70}$$

for some $r \neq s$, $1 \leq r, s \leq n$. Since there are a total of $n(n+1)/2$ such polynomials, each of them having at most two roots, we can choose $t$ arbitrarily, except for $n(n+1)$ values. Then the new polynomial $p_n'(\boldsymbol{x})$ satisfies the condition of Proposition 7.28 and can be factorized by the method given above. The recovered vectors are

$$\boldsymbol{\ell}_i' \doteq L^T \boldsymbol{\ell}_i \in \mathbb{R}^3, \quad i = 1, 2, \ldots, n, \tag{7.71}$$

which directly gives the original vectors $\boldsymbol{\ell}_i = L^{-T}\boldsymbol{\ell}_i'$ for $i = 1, 2, \ldots, n$.

We have thus proved the following proposition.

**Proposition 7.30.** *Factorization of a (factorizable) $n$th-order homogeneous polynomial is algebraically equivalent to solving the roots of a univariate polynomial of degree $n$ plus the solution to a linear system in $n$ variables. Hence the problem is algebraically solvable if and only if $n \leq 4$.*

The factorization technique discussed here has been used repeatedly in this chapter for the computation of the epipoles and epipolar lines associated with the multibody structure from motion problem.

# Further readings

The problem of motion segmentation falls into a more general class of multimodal pattern recognition problems. The algebraic geometric treatment adopted by this chapter for this class of problems follows the so-called *generalized principal component analysis* [Vidal et al., 2003].

The problem of multiple-motion estimation was first addressed in the work of [Shizawa and Mase, 1991] for the case in which there are transparent objects in the scene. The problem of two perspective views of two motions was studied by [Wolf and Shashua, 2001b]. The feasibility of a systematic solution to the most general problem ($n$ motions) was only recently addressed by the work of [Vidal et al., 2002b, Vidal et al., 2002c], which led to the material presented in this chapter. Regarding statistical optimality, nonlinear optimal algorithms have been developed recently, and they, as expected, have superior performance over the linear algorithm given in this chapter, but at the price of higher computational cost [Vidal and Sastry, 2003].

### *Various motion segmentation techniques*

Variations to the approach presented in this chapter for different camera models and special motions that are not covered in this book can be found in [Demirdjian and Horaud, 1992, Costeira and Kanade, 1995, Xu and Tsuji, 1996], [Torr, 1998, Avidan and Shashua, 2000, Han and Kanade, 2000, Kanatani, 2001, Shashua and Levin, 2001].

### *Linked (articulated) rigid-body motions*

The methods given in this chapter have their limitations, especially when the motion of multiple objects are not independent. For instance, in the case of a human movement, the motions of different limbs and joints are related in a systematic way, also referred to as articulated motions. Given the fundamental matrix associated with one motion, another one is very likely determined up to a one- or two-parameter family. Furthermore, the nature of available data in this case is also fundamentally different: typically, there are not so many distinctive features on each limb, and usually only the joints can be reliably distinguished and tracked. That prevents us from using the algorithms given in this chapter. The reader can consult the work of [Sinclair and Zesar, 1996, Sinclair et al., 1997,

Ruf and Horaud, 1999b] and [Sidenbladh et al., 2000, Bregler and Malik, 1998] for the study of linked rigid bodies. However, multiple-view geometry of linked rigid bodies remains largely an open topic.

# Part III

# Geometry of Multiple Views

# Chapter 8

# Multiple-View Geometry of Points and Lines

*An idea which can be used once is a trick. If it can be used more than once it becomes a method.*

– George Pólya and Gábor Szegö

In this chapter we study how the framework of epipolar geometry introduced in Part II generalizes to the case of multiple views. As we shall see, this entails studying constraints that corresponding points in different views must satisfy if they are the projection of the same point in space. Not only is this development crucial for understanding the geometry of multiple views but, as in the two-view case, these constraints may be used to derive algorithms for reconstructing camera configuration and, ultimately, the 3-D position of geometric primitives. The search for the $m$-view analogue of the epipolar constraint has been an active research area for almost two decades. It was realized early on in [Liu and Huang, 1986, Spetsakis and Aloimonos, 1987] that the relationship between three views of the same point or line can be characterized by the *trilinear constraints*. Consequently, the study of multiple-view geometry has involved multidimensional linear operators, also called tensors.

This chapter and the next will, however, take a *different* technical approach that involves only matrix linear algebra. In essence, we shall demonstrate that all the constraints between corresponding points or lines in $m$ views, including the epipolar constraint studied before, can be written as simple rank conditions on the *multiple-view matrices*. In order to put this approach into a historical context, in this chapter we shall also explore relationships between such matrix rank conditions and multilinear tensors.

A conceptual multiple-view factorization algorithm will be derived from the matrix rank conditions. This algorithm demonstrates the possibility of performing 3-D reconstruction by simultaneously utilizing all available views and features. In this chapter, we restrict our attention to individual point and line features. In the next chapter, we will show how to extend the same rank technique to study multiple images of points, lines, planes, as well as incidence relations among them. We leave issues related to the implementation of the algorithms to Part IV.

## 8.1 Basic notation for the (pre)image and coimage of points and lines

To set the stage, we recall the notation from Chapter 3 (Section 3.3.4). Consider a generic point $p \in \mathbb{E}^3$ in Euclidean space. The homogeneous coordinates of $p$ relative to a fixed world coordinate frame are denoted by $\boldsymbol{X} \doteq [X, Y, Z, 1]^T \in \mathbb{R}^4$. Then the (perspective) image $\boldsymbol{x}(t) \doteq [x(t), y(t), z(t)]^T \in \mathbb{R}^3$ of $p$, taken by a moving camera at time[1] $t$, satisfies the relationship

$$\lambda(t)\boldsymbol{x}(t) = K(t)\Pi_0 g(t)\boldsymbol{X}, \tag{8.1}$$

where $\lambda(t) \in \mathbb{R}_+$ is the (unknown) depth of the point $p$ relative to the camera frame, and $K(t) \in \mathbb{R}^{3\times 3}$ is the camera calibration matrix. Here we allow the intrinsic calibration parameters to change from frame to frame, since it involves no additional technical difficulty at this stage of analysis. Moreover, $\Pi_0 = [I, 0] \in \mathbb{R}^{3\times 4}$ is the standard (perspective) projection matrix, and $g(t) \in SE(3)$ is the coordinate transformation from the world frame to the camera frame at time $t$. In equation (8.1), $\boldsymbol{x}, \boldsymbol{X}$, and $g$ are all in *the homogeneous representation*. Suppose the transformation $g$ is specified by its rotation $R \in SO(3)$ and translation $T \in \mathbb{R}^3$. Then the homogeneous representation of $g$ is simply

$$g = \begin{bmatrix} R & T \\ 0 & 1 \end{bmatrix} \in \mathbb{R}^{4\times 4}. \tag{8.2}$$

Notice that equation (8.1) is equivalent to

$$\lambda(t)\boldsymbol{x}(t) = [K(t)R(t)\ K(t)T(t)]\boldsymbol{X}. \tag{8.3}$$

Now consider a point $p$ lying on a straight line $L \subset \mathbb{E}^3$, as shown in Figure 8.1. The line $L$ can be defined by a collection of points in $\mathbb{E}^3$ described (in homogeneous coordinates) as

$$L \doteq \{\boldsymbol{X} \mid \boldsymbol{X} = \boldsymbol{X}_o + \mu \boldsymbol{V},\ \mu \in \mathbb{R}\} \quad \subset \mathbb{R}^4, \tag{8.4}$$

[1] We remind the reader that $t$ is just an index of the view, and does not necessarily imply an order in which the views are captured; i.e. $t$ does not necessarily have to be interpreted as "time."

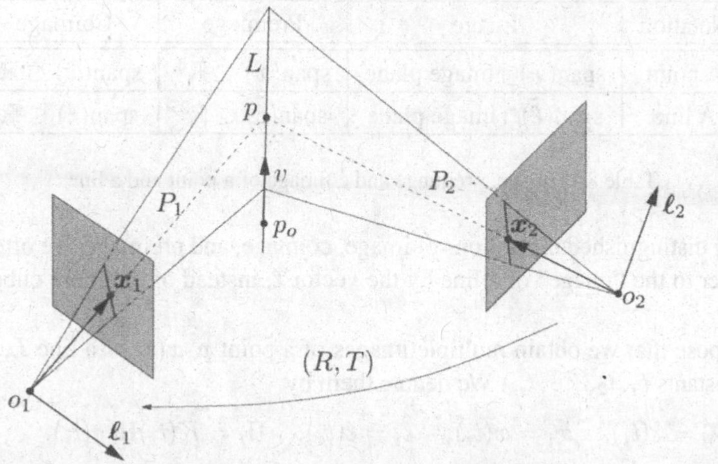


Figure 8.1. Images of a point $p$ on a line $L$: The preimages $P_1, P_2$ of the two image lines should intersect at the line $L$ in space; the preimages of the two image points $\boldsymbol{x}_1, \boldsymbol{x}_2$ should intersect at the point $p$ in space. Normal vectors $\boldsymbol{\ell}_1, \boldsymbol{\ell}_2$ to the planes indicate the two coimages of the line.

where $\boldsymbol{X}_o = [X_o, Y_o, Z_o, 1]^T \in \mathbb{R}^4$ are the coordinates of a "base point" $p_o$ on this line, and $\boldsymbol{V} = [V_1, V_2, V_3, 0]^T \in \mathbb{R}^4$ is a nonzero vector indicating the "direction" of the line.

Then the image of the line $L$ at time $t$ is simply the set of images $\{\boldsymbol{x}(t)\}$ of all points $\{p \in L\}$. It is clear that all such $\boldsymbol{x}(t)$ lie on the intersection of the plane $P$ in $\mathbb{R}^3$ passing through the center of projection with the image plane, as shown in Figure 8.1. Recall from Section 3.3.4 that we can use the normal vector $\boldsymbol{\ell}(t) = [a(t), b(t), c(t)]^T \in \mathbb{R}^3$ of this plane to denote the coimage of the line $L$. If $\boldsymbol{x}(t)$ is the image of a point $p$ on this line, then $\boldsymbol{\ell}(t)$ satisfies the orthogonality equation

$$\boldsymbol{\ell}(t)^T \boldsymbol{x}(t) = \boldsymbol{\ell}(t)^T K(t) \Pi_0 g(t) \boldsymbol{X} = 0. \tag{8.5}$$

Then the plane $P = \text{span}(\widehat{\boldsymbol{\ell}})$ is the preimage of the line $L$, defined in Section 3.3.4. This is illustrated in Figure 8.1. Similarly, if $\boldsymbol{x}$ is the image of a point $p$, its coimage is the plane orthogonal to $\boldsymbol{x}$ given by the span of the column vectors of the matrix $\widehat{\boldsymbol{x}}$. Recall from Section 3.3.4 that we have introduced the notation in Table 8.1 to represent the image, preimage, and coimage of points and lines. For simplicity, we will often use the vector $\boldsymbol{x}$ and the matrix $\widehat{\boldsymbol{\ell}}$, determined up to a scalar factor, to represent the preimage of a point and a line, respectively; and the matrix $\widehat{\boldsymbol{x}}$ and the vector $\boldsymbol{\ell}$, up to a scalar factor, to represent the coimage of a point and a line, respectively. Using this notation, for a line $L$ and a point $p \in L$, the relation between their (pre)image and coimage can be expressed in terms of the vectors $\boldsymbol{x}, \boldsymbol{\ell} \in \mathbb{R}^3$ as

$$\boldsymbol{\ell}^T \boldsymbol{x} = 0, \quad \widehat{\boldsymbol{x}}\boldsymbol{x} = 0, \quad \widehat{\boldsymbol{\ell}}\boldsymbol{\ell} = 0.$$

| Notation | Image | Preimage | Coimage |
|---|---|---|---|
| A point | span($\boldsymbol{x}$)$\cap$ image plane | span($\boldsymbol{x}$) $\subset \mathbb{R}^3$ | span($\widehat{\boldsymbol{x}}$) $\subset \mathbb{R}^3$ |
| A line | span($\widehat{\boldsymbol{\ell}}$)$\cap$ image plane | span($\widehat{\boldsymbol{\ell}}$) $\subset \mathbb{R}^3$ | span($\boldsymbol{\ell}$) $\subset \mathbb{R}^3$ |

Table 8.1. Image, preimage and coimage of a point and a line.

Having distinguished the notions of image, coimage, and preimage, we often simply refer to the "image" of a line by the vector $\boldsymbol{\ell}$, instead of the more cubersome $\widehat{\boldsymbol{\ell}}$.

Suppose that we obtain multiple images of a point $p$, $\boldsymbol{x}(t)$ or a line $L$, $\boldsymbol{\ell}(t)$ at time instants $t_1, t_2, \ldots, t_m$. We denote them by

$$\lambda_i \doteq \lambda(t_i), \quad \boldsymbol{x}_i \doteq \boldsymbol{x}(t_i), \quad \boldsymbol{\ell}_i \doteq \boldsymbol{\ell}(t_i), \quad \Pi_i \doteq K(t_i)\Pi_0 g(t_i). \tag{8.6}$$

The projection matrix $\Pi_i$ is then a $3 \times 4$ matrix that relates the $i$th image of a point $p$ to its world coordinates $\boldsymbol{X}$ by

$$\lambda_i \boldsymbol{x}_i = \Pi_i \boldsymbol{X} \tag{8.7}$$

and the $i$th coimage of a line $L$ to its world coordinates $(\boldsymbol{X}_o, \boldsymbol{V})$ by

$$\boldsymbol{\ell}_i^T \Pi_i \boldsymbol{X}_o = \boldsymbol{\ell}_i^T \Pi_i \boldsymbol{V} = 0 \tag{8.8}$$

for $i = 1, 2, \ldots, m$. For convenience, by abuse of notation, we often use $R_i \in \mathbb{R}^{3\times 3}$ to denote the first three columns of the projection matrix $\Pi_i$, and $T_i \in \mathbb{R}^3$ for the last column. That is,

$$\Pi_i \doteq [R_i, T_i] \in \mathbb{R}^{3\times 4}, \quad R_i \in \mathbb{R}^{3\times 3}, \quad T_i \in \mathbb{R}^3. \tag{8.9}$$

*Be aware that $(R_i, T_i)$ do not necessarily represent the actual camera rotation and translation unless the camera is fully calibrated, i.e. $K(t_i) \equiv I$.* In any case, the matrix $\Pi_i$ is of rank 3 and specifies how to project the (world) coordinates $\boldsymbol{X}$ of a point onto the image plane (with respect to the local camera frame).

Recall that in Section 3.3.4, we have formally defined a preimage as the set of points in $\mathbb{R}^3$ that give rise to the same image of a point or a line. This notion can be easily generalized to the multiple-view setting.

**Definition 8.1 (Preimage from multiple views).** *A preimage of multiple images of a point or a line is the (largest) set of 3-D points that give rise to the same set of multiple images of the point or the line.*

For example, in Figure 8.1, the preimage of the two image lines $\boldsymbol{\ell}_1, \boldsymbol{\ell}_2$ must be the intersection of the preimage of each image line, i.e. the intersection of the plane $P_1$ and the plane $P_2$. Obviously, from the figure, $L = P_1 \cap P_2$. Hence, the line $L$ in space is exactly the preimage of its two images. Equivalently, given multiple images of a point or a line, we can define their preimage to be the intersection

$$\begin{aligned}
\text{preimage}\,(\boldsymbol{x}_1, \ldots, \boldsymbol{x}_m) &= \text{preimage}\,(\boldsymbol{x}_1) \cap \cdots \cap \text{preimage}\,(\boldsymbol{x}_m), \\
\text{preimage}\,(\boldsymbol{\ell}_1, \ldots, \boldsymbol{\ell}_m) &= \text{preimage}\,(\boldsymbol{\ell}_1) \cap \cdots \cap \text{preimage}\,(\boldsymbol{\ell}_m).
\end{aligned}$$

By this definition, we can compute in principle the preimage for any set of image points or lines. For instance, the preimage of multiple image lines can be either an empty set, a point, a line, or a plane, depending on whether or not they come from the same line in space.

## 8.2 Preliminary rank conditions of multiple images

We first observe that in equations (8.7) and (8.8) the unknowns, $\lambda_i$, $\boldsymbol{X}$, $\boldsymbol{X}_o$, and $\boldsymbol{V}$, which encode the information about the location of the point $p$ or the line $L$, are not directly available from image measurements. As in the two-view case, in order to obtain intrinsic relationships between $\boldsymbol{x}$, $\boldsymbol{\ell}$, and $\Pi$ only, i.e. between the image measurements and the camera configuration, we need to eliminate these unknowns. There are many different but algebraically *equivalent* ways for eliminating these unknowns, which result in different kinds of constraints that have been studied in the computer vision literature. Here we will show a systematic way of eliminating the above unknowns that results in a complete set of conditions and provides a clear geometric characterization of all the constraints.

### *8.2.1 Point features*

Consider images of a 3-D point $\boldsymbol{X}$ seen in multiple views. We can rewrite equation (8.7) in matrix form in the following way

$$\begin{bmatrix} \boldsymbol{x}_1 & 0 & \cdots & 0 \\ 0 & \boldsymbol{x}_2 & \cdots & 0 \\ \vdots & \vdots & \ddots & \vdots \\ 0 & 0 & \cdots & \boldsymbol{x}_m \end{bmatrix} \begin{bmatrix} \lambda_1 \\ \lambda_2 \\ \vdots \\ \lambda_m \end{bmatrix} = \begin{bmatrix} \Pi_1 \\ \Pi_2 \\ \vdots \\ \Pi_m \end{bmatrix} \boldsymbol{X}. \tag{8.10}$$

This equation, after defining

$$\mathcal{I} \doteq \begin{bmatrix} \boldsymbol{x}_1 & 0 & \cdots & 0 \\ 0 & \boldsymbol{x}_2 & \cdots & 0 \\ \vdots & \vdots & \ddots & \vdots \\ 0 & 0 & \cdots & \boldsymbol{x}_m \end{bmatrix}, \quad \vec{\lambda} \doteq \begin{bmatrix} \lambda_1 \\ \lambda_2 \\ \vdots \\ \lambda_m \end{bmatrix}, \quad \Pi \doteq \begin{bmatrix} \Pi_1 \\ \Pi_2 \\ \vdots \\ \Pi_m \end{bmatrix}, \tag{8.11}$$

can be written as

$$\mathcal{I}\vec{\lambda} = \Pi \boldsymbol{X}. \tag{8.12}$$

We call $\vec{\lambda} \in \mathbb{R}^m$ the *depth scale vector*, and $\Pi \in \mathbb{R}^{3m \times 4}$ the *multiple-view projection matrix* associated with the *image matrix* $\mathcal{I} \in \mathbb{R}^{3m \times m}$. Note that with the exception of $\mathcal{I}$, everything else in this equation is unknown. Solving for the depth scales and the projection matrices directly from these equations is by no means straightforward. Hence, as in the two-view case, we decouple the recovery of the camera displacements $\Pi_i$ from the recovery of scene structure, $\lambda_i$, and $\boldsymbol{X}$.

Note that every column of the matrix $\mathcal{I} \in \mathbb{R}^{3m\times m}$ in equation (8.11) lies in a four-dimensional space spanned by columns of the matrix $\Pi \in \mathbb{R}^{3m\times 4}$. Hence in order for equation (8.12) to be satisfied, it is necessary for the columns of $\mathcal{I}$ and $\Pi$ to be linearly dependent. In other words, the matrix

$$N_p \doteq [\Pi, \mathcal{I}] = \begin{bmatrix} \Pi_1 & \boldsymbol{x}_1 & 0 & \cdots & 0 \\ \Pi_2 & 0 & \boldsymbol{x}_2 & \ddots & \vdots \\ \vdots & \vdots & \ddots & \ddots & 0 \\ \Pi_m & 0 & \cdots & 0 & \boldsymbol{x}_m \end{bmatrix} \in \mathbb{R}^{3m\times(m+4)} \qquad (8.13)$$

must have a nontrivial right null space; that is to say, its columns are dependent (note that for $m \geq 2, 3m \geq m+4$), and hence

$$\boxed{\text{rank}(N_p) \leq m+3.} \qquad (8.14)$$

In fact, from equation (8.12) it is immediate to see that the vector $u \doteq [\boldsymbol{X}^T, -\vec{\lambda}^T]^T \in \mathbb{R}^{m+4}$ is in the right null space of the matrix $N_p$, since $N_p u = 0$.

**Remark 8.2 (Positive depth).** *Even if rank$(N_p) = m+3$, there is no guarantee that $\vec{\lambda}$ in (8.10) will have positive entries.*[2] *In practice, if the point being observed is always in front of the camera, then $\mathcal{I}$, $\Pi$ and $\boldsymbol{X}$ in (8.10) will be such that the entries of $\vec{\lambda}$ are all positive. Since the solution to $N_p u = 0$ is unique if rank$(N_p) = m+3$ and $u \doteq [\boldsymbol{X}^T, -\vec{\lambda}^T]^T$ is a solution, then the last $m$ entries of $u$ have to be of the same sign.*

The above rank constraint can be expressed in a more compact way in terms of the coimages of the point. Let us define a matrix that "annihilates" $\mathcal{I}$,

$$\mathcal{I}^{\perp} \doteq \begin{bmatrix} \widehat{\boldsymbol{x}_1} & 0 & \cdots & 0 \\ 0 & \widehat{\boldsymbol{x}_2} & \cdots & 0 \\ \vdots & \vdots & \ddots & \vdots \\ 0 & 0 & \cdots & \widehat{\boldsymbol{x}_m} \end{bmatrix} \in \mathbb{R}^{3m\times 3m}. \qquad (8.15)$$

Since $\widehat{\boldsymbol{x}}_i \boldsymbol{x}_i = \boldsymbol{x}_i \times \boldsymbol{x}_i = 0$, we have

$$\mathcal{I}^{\perp}\mathcal{I} = 0. \qquad (8.16)$$

Therefore, premultiplying both sides of equation (8.12) by $\mathcal{I}^{\perp}$, we get

$$\mathcal{I}^{\perp}\Pi\boldsymbol{X} = 0. \qquad (8.17)$$

[2]In a homogeneous sense, this is not a problem, since all points on the same line through the camera center are equivalent. But it does matter if we want to choose appropriate reference frames such that the depth for the recovered 3-D point is physically meaningful, i.e. positive.

This means that the vector $\boldsymbol{X}$ is in the null space of the matrix

$$W_p \doteq \mathcal{I}^{\perp}\Pi = \begin{bmatrix} \widehat{\boldsymbol{x}_1}\Pi_1 \\ \widehat{\boldsymbol{x}_2}\Pi_2 \\ \vdots \\ \widehat{\boldsymbol{x}_m}\Pi_m \end{bmatrix} \in \mathbb{R}^{3m \times 4}, \tag{8.18}$$

and since the null space of $W_p$ is at least one-dimensional, we have

$$\text{rank}(W_p) \leq 3. \tag{8.19}$$

If all the images $\boldsymbol{x}_i$ are indeed from a single point (with coordinates $\boldsymbol{X}$), the matrix $W_p$ should have only a one-dimensional null space. Null spaces of both $W_p$ and $N_p$ then correspond to points that may have given rise to the same set of images. Note that individual rows of the matrix $W_p$ can simply be obtained by eliminating the unknown scales $\lambda_i$ from

$$\lambda_i \boldsymbol{x}_i = \Pi_i \boldsymbol{X} \Rightarrow \widehat{\boldsymbol{x}}_i \lambda_i \boldsymbol{x}_i = \widehat{\boldsymbol{x}}_i \Pi_i \boldsymbol{X} \Rightarrow 0 = \widehat{\boldsymbol{x}}_i \Pi_i \boldsymbol{X}. \tag{8.20}$$

We employed this style of elimination in deriving the epipolar constraint in the two-view case. Stating the rank conditions more precisely we have the following result:

**Lemma 8.3 (Images and coimages of a point).** *Given $m$ images $\boldsymbol{x}_i \in \mathbb{R}^3$, $i = 1, 2, \ldots, m$, of a point p with respect to $m$ camera frames defined by the projection matrices $\Pi_i = [R_i, T_i]$, we must have*

$$\boxed{\textit{rank}(W_p) = \textit{rank}(N_p) - m \leq 3.} \tag{8.21}$$

*Proof.* Note that the matrix

$$\begin{bmatrix} \mathcal{I}^{\perp} \\ \mathcal{I}^T \end{bmatrix} \in \mathbb{R}^{4m \times 3m}$$

has full rank $3m$. Multiplying it on the left by the matrix $N_p$ does not change the rank of $N_p$:

$$\begin{bmatrix} \mathcal{I}^{\perp} \\ \mathcal{I}^T \end{bmatrix} N_p = \begin{bmatrix} \mathcal{I}^{\perp} \\ \mathcal{I}^T \end{bmatrix} [\Pi, \mathcal{I}] = \begin{bmatrix} \mathcal{I}^{\perp}\Pi & 0 \\ \mathcal{I}^T\Pi & \mathcal{I}^T\mathcal{I} \end{bmatrix} \in \mathbb{R}^{4m \times (m+4)}.$$

Since the submatrix $\mathcal{I}^T\mathcal{I}$ has full rank $m$, the rank of the overall matrix on the right-hand side (= rank($N_p$)) is equal to rank($\mathcal{I}^{\perp}\Pi$) + $m$. Since $W_p = \mathcal{I}^{\perp}\Pi$, we conclude that

$$\text{rank}(W_p) = \text{rank}(N_p) - m. \tag{8.22}$$

□

This proof illustrates a useful algebraic technique (called the rank reduction lemma in Appendix A) for eliminating redundant rows and columns in a matrix, which we will use extensively in the rest of the book.

### 8.2.2 Line features

Consider now images of a single 3-D line seen in multiple views. A rank condition similar to the case of a point can be derived in terms of the coimages $\boldsymbol{\ell}_i$. From equation (8.8), the matrix

$$W_l \doteq \begin{bmatrix} \boldsymbol{\ell}_1^T \Pi_1 \\ \boldsymbol{\ell}_2^T \Pi_2 \\ \vdots \\ \boldsymbol{\ell}_m^T \Pi_m \end{bmatrix} \in \mathbb{R}^{m \times 4} \tag{8.23}$$

must satisfy the rank condition

$$\boxed{\text{rank}(W_l) \leq 2.} \tag{8.24}$$

The null space of $W_l$ is at least two-dimensional because from (8.8) the vectors $\boldsymbol{X}_o$ and $\boldsymbol{V}$ are both in the null space of the matrix $W_l$. In fact, any $\boldsymbol{X} \in \mathbb{R}^4$ in the null space of $W_l$ represents the homogeneous coordinates of some point lying on the line $L$, and vice versa. For the sake of completeness, we can also introduce the counterpart of $N_p$ for line features as

$$N_l \doteq \begin{bmatrix} \Pi_1 & \widehat{\boldsymbol{\ell}_1} & 0 & \cdots & 0 \\ \Pi_2 & 0 & \widehat{\boldsymbol{\ell}_2} & \ddots & \vdots \\ \vdots & \vdots & \ddots & \ddots & 0 \\ \Pi_m & 0 & \cdots & 0 & \widehat{\boldsymbol{\ell}_m} \end{bmatrix} \in \mathbb{R}^{3m \times (3m+4)}. \tag{8.25}$$

We leave to the reader as an exercise (see Exercise 8.4) to prove the following lemma.

**Lemma 8.4 (Images and coimages of a line).** *Given $m$ coimages $\boldsymbol{\ell}_i \in \mathbb{R}^3$ of a line $L$ with respect to $m$ camera frames (or projection matrices) $\Pi_i = [R_i, T_i]$, we must have*

$$\boxed{\mathit{rank}(W_l) = \mathit{rank}(N_l) - 2m \leq 2.} \tag{8.26}$$

The next three examples illustrate the geometric meaning of the rank conditions associated with the matrix $W$.

**Example 8.5 (Homogeneous representation of hyperplanes in $\mathbb{R}^3$).** To describe a hyperplane $\mathbb{R}^3$, a conventional way is to give a vector $\pi = [a, b, c, d]^T \in \mathbb{R}^4$ such that a point with homogeneous coordinates $\boldsymbol{X} = [X, Y, Z, 1]^T \in \mathbb{R}^4$ is on the plane if and only if

$$\pi^T \boldsymbol{X} = aX + bY + cZ + d = 0. \tag{8.27}$$

In general, we know that two planes intersect at a line and three intersect at a point. Hence the null spaces of the matrices

$$\Pi_l = \begin{bmatrix} (\pi^1)^T \\ (\pi^2)^T \end{bmatrix} \in \mathbb{R}^{2 \times 4}, \quad \Pi_p = \begin{bmatrix} (\pi^1)^T \\ (\pi^2)^T \\ (\pi^3)^T \end{bmatrix} \in \mathbb{R}^{3 \times 4} \tag{8.28}$$

represent a line and a point in $\mathbb{R}^3$, respectively (if $\text{rank}(\Pi_l) = 2$ and $\text{rank}(\Pi_p) = 3$). This is illustrated in Figure 8.2. Therefore, both the matrices $W_p$ and $W_l$ are nothing but representations of a point and a line in terms of a family of hyperplanes passing through them. Each row of $W_p$ and $W_l$ represents a hyperplane; any point in their null space is simply a point in the intersection of all the hyperplanes from the rows. The null space of $W_p$ and $W_l$ is nothing but the preimage of all the images of the point or the line (expressed in homogeneous coordinates with respect to the first camera frame). ■

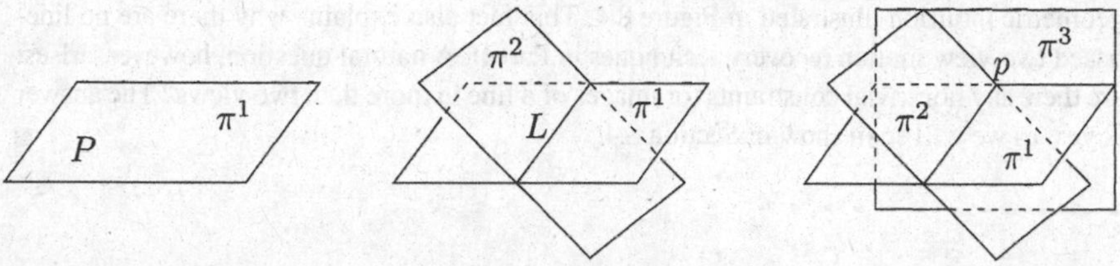


Figure 8.2. One homogeneous vector $\pi \in \mathbb{R}^4$ determines a hyperplane $P$; two hyperplanes in general determine a line $L$; three determine a point $p$.

**Example 8.6 (Geometric interpretation of $W_p$ for two views).** Consider the rows of the matrix $W_p \in \mathbb{R}^{3m\times 4}$: $m$ rows are trivially redundant due to the rank deficiency of $\widehat{x_i}$, and the remaining $2m$ rows each represent a hyperplane in space. Since $W_p X = 0$, all the planes must intersect at the same point $p$ (with coordinates $X$) in space. For $p$ to be the only intersection, we require $\text{rank}(W_p) = 3$. For $m = 2$ views, this puts an effective constraint on the four (independent) rows involved. This is illustrated in Figure 8.3. The reader may try to find out what happens if $\text{rank}(W_p) = 2$ (although an answer will be given in the next section). ■

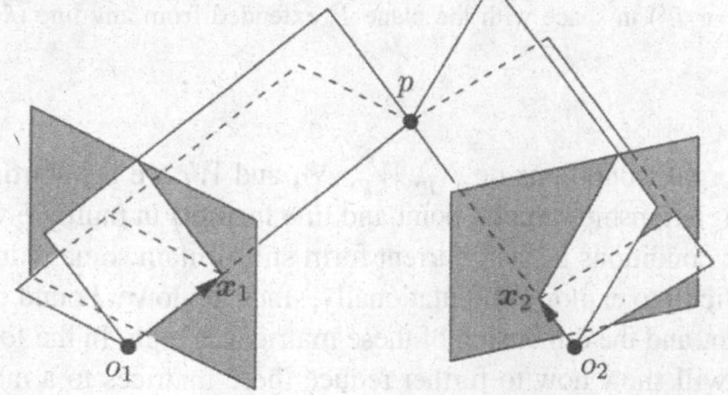


Figure 8.3. The two independent row vectors of $\widehat{x_1}\Pi_1$ represent two planes intersecting at the line $o_1 p$; together with the two planes from $\widehat{x_2}\Pi_2$, these four planes intersect at a point in space whose homogeneous coordinate vector $X$ is in the null space of the matrix $W_p$.

**Example 8.7 (Geometric interpretation of $W_l$ for two views).** For two (co)images $\ell_1, \ell_2$ of a line $L$, the rank condition (8.24) for the matrix $W_l$ becomes

$$\text{rank}(W_l) = \text{rank}\begin{bmatrix} \ell_1^T \Pi_1 \\ \ell_2^T \Pi_2 \end{bmatrix} \leq 2. \tag{8.29}$$

Here $W_l$ is a $2 \times 4$ matrix, that automatically has rank less than or equal to two. So in general, the preimage of two image lines always contains a line in space, and hence there is essentially no intrinsic constraint on two images of a line.[3] This is consistent with the geometric intuition illustrated in Figure 8.4. This fact also explains why there are no line-based two-view motion recovery techniques in Part II. A natural question, however, arises: are there any nontrivial constraints for images of a line in more than two views? The answer is yes, as we will soon show in Section 8.4. ■

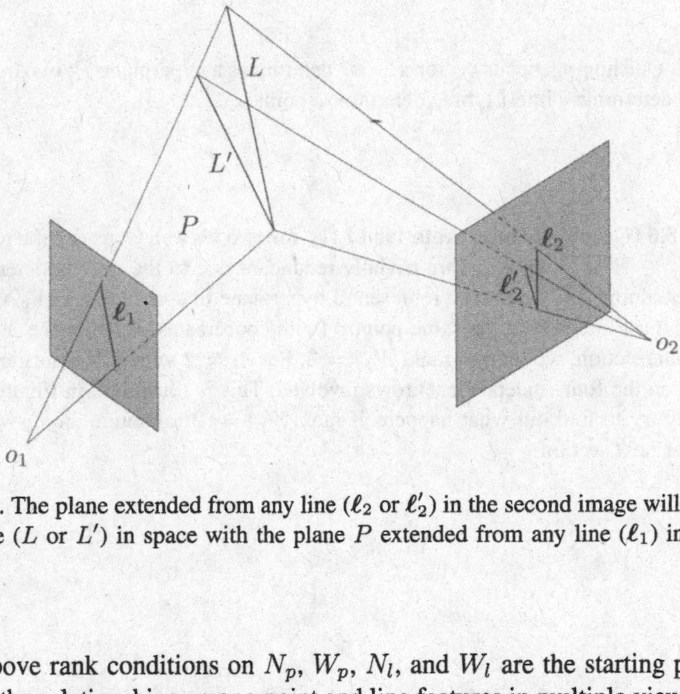


Figure 8.4. The plane extended from any line ($\ell_2$ or $\ell_2'$) in the second image will intersect at one line ($L$ or $L'$) in space with the plane $P$ extended from any line ($\ell_1$) in the first image.

The above rank conditions on $N_p$, $W_p$, $N_l$, and $W_l$ are the starting point for studying the relationships among point and line features in multiple views. However, the rank conditions in their current form still contain some redundancy and are rather difficult to exploit computationally, since the lower bound on their rank is not yet clear, and the dimension of these matrices is high. In the following two sections, we will show how to further reduce these matrices to a more compact form. As we will see, the reduction will simplify the study of the constraints that such matrix rank conditions impose upon multiple corresponding views, and will also lead to a useful algorithm for 3-D reconstruction.

[3]However, this is no longer the case if the lines are segments. See [Zhang, 1995].

## 8.3 Geometry of point features

We start this section with the case of point features and leave the line feature case to the next section.

### 8.3.1 *The multiple-view matrix of a point and its rank*

Without loss of generality, we may assume that the first (uncalibrated) camera frame is chosen to be the reference frame. That gives the projection matrices $\Pi_i, i = 1, 2, \ldots, m$, the form

$$\Pi_1 \doteq [I, 0],\ \Pi_2 = [R_2, T_2],\ \ldots,\ \Pi_m = [R_m, T_m] \quad \in \mathbb{R}^{3\times 4}, \tag{8.30}$$

where as before,[4] $R_i \in \mathbb{R}^{3\times 3}, i = 2, 3, \ldots, m$, collects the first three columns of $\Pi_i$, (therefore, it is not necessarily a rotation matrix unless the camera happens to be calibrated) and $T_i \in \mathbb{R}^3, i = 2, 3, \ldots, m$, is the fourth column of $\Pi_i$. Using this notation, the rank of the matrix $W_p$ will not change if we multiply it by a full-rank matrix $D_p \in \mathbb{R}^{4\times 5}$ of the following form:

$$W_p D_p = \begin{bmatrix} \widehat{\boldsymbol{x}_1}\Pi_1 \\ \widehat{\boldsymbol{x}_2}\Pi_2 \\ \widehat{\boldsymbol{x}_3}\Pi_3 \\ \vdots \\ \widehat{\boldsymbol{x}_m}\Pi_m \end{bmatrix} \begin{bmatrix} \widehat{\boldsymbol{x}_1} & \boldsymbol{x}_1 & 0 \\ 0 & 0 & 1 \end{bmatrix} = \begin{bmatrix} \widehat{\boldsymbol{x}_1}\widehat{\boldsymbol{x}_1} & 0 & 0 \\ \widehat{\boldsymbol{x}_2}R_2\widehat{\boldsymbol{x}_1} & \widehat{\boldsymbol{x}_2}R_2\boldsymbol{x}_1 & \widehat{\boldsymbol{x}_2}T_2 \\ \widehat{\boldsymbol{x}_3}R_3\widehat{\boldsymbol{x}_1} & \widehat{\boldsymbol{x}_3}R_3\boldsymbol{x}_1 & \widehat{\boldsymbol{x}_3}T_3 \\ \vdots & \vdots & \vdots \\ \widehat{\boldsymbol{x}_m}R_m\widehat{\boldsymbol{x}_1} & \widehat{\boldsymbol{x}_m}R_m\boldsymbol{x}_1 & \widehat{\boldsymbol{x}_m}T_m \end{bmatrix}.$$

Hence, $\text{rank}(W_p) \leq 3$ if and only if for the submatrix

$$M_p \doteq \begin{bmatrix} \widehat{\boldsymbol{x}_2}R_2\boldsymbol{x}_1 & \widehat{\boldsymbol{x}_2}T_2 \\ \widehat{\boldsymbol{x}_3}R_3\boldsymbol{x}_1 & \widehat{\boldsymbol{x}_3}T_3 \\ \vdots & \vdots \\ \widehat{\boldsymbol{x}_m}R_m\boldsymbol{x}_1 & \widehat{\boldsymbol{x}_m}T_m \end{bmatrix} \in \mathbb{R}^{3(m-1)\times 2} \tag{8.31}$$

of the above matrix has $\text{rank}(M_p) \leq 1$. We call $M_p$ the *multiple-view matrix* associated with a point feature $p$. Notice that $M_p$ involves both the image $\boldsymbol{x}_1$ and the coimages $\widehat{\boldsymbol{x}_2}, \widehat{\boldsymbol{x}_3}, \ldots, \widehat{\boldsymbol{x}_m}$ of the point $p$.

The above rank condition on the matrix $M_p$ can also be derived by manipulating the matrix $N_p$. We leave this as an exercise to the reader (Exercise 8.6). Using the fact stated in Lemma 8.3 and the observations above, we can summarize the relationship between the ranks of $N_p$, $W_p$, and $M_p$ in the following theorem.

**Theorem 8.8 (Rank conditions for point features).** *For multiple images of a point, the matrices $N_p$, $W_p$, and $M_p$ satisfy*

$$\boxed{\mathit{rank}(M_p) = \mathit{rank}(W_p) - 2 = \mathit{rank}(N_p) - (m+2) \leq 1.} \tag{8.32}$$

[4]Recall that here $R_i$ is not necessarily a rotation matrix unless the camera is calibrated.

**Comment 8.9 (Geometric information in $M_p$).** *Notice that $\widehat{\boldsymbol{x}}_i T_i$ is the normal vector to the epipolar plane formed between frames 1 and i, and so is the vector $\widehat{\boldsymbol{x}}_i R_i \boldsymbol{x}_1$. We further have $\lambda_1 \widehat{\boldsymbol{x}}_i R_i \boldsymbol{x}_1 + \widehat{\boldsymbol{x}}_i T_i = 0, i = 1, 2, \ldots, m$, which yields*

$$M_p \begin{bmatrix} \lambda_1 \\ 1 \end{bmatrix} = 0. \tag{8.33}$$

*So the coefficient that relates the two columns of $M_p$ is simply the depth of the point p in space with respect to the center of the first camera frame (the reference). Hence, the multiple-view matrix $M_p$ captures exactly the information about a point p in space that is missing from a single image ($\boldsymbol{x}_1$) but encoded in multiple images. This is shown in Figure 8.5.*

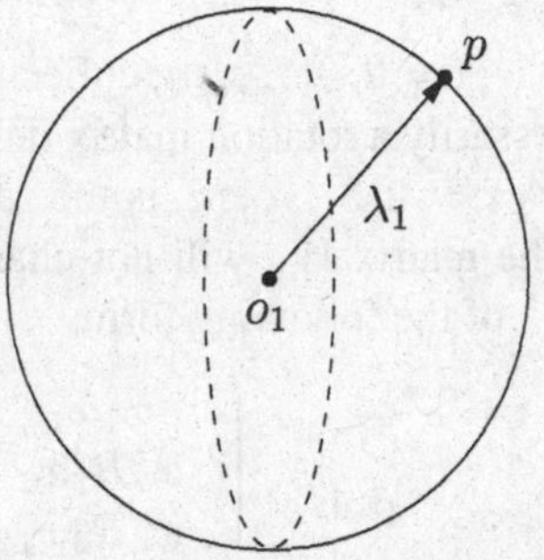


Figure 8.5. Geometric information in $M_p$: distance to the point $p$.

Due to the rank equality (8.32), $N_p$, $W_p$, and $M_p$ all give *equivalent* constraints that may arise among the $m$ views of a point. The matrix $M_p$ is obviously the most compact representation of these constraints. Notice that for $M_p$ to be of rank 1, it is necessary for the pair of vectors $\widehat{\boldsymbol{x}}_i T_i$, $\widehat{\boldsymbol{x}}_i R_i \boldsymbol{x}_1$ to be linearly dependent for each $i = 1, 2, \ldots, m$. This is exactly equivalent to the well-known bilinear epipolar constraints

$$\boldsymbol{x}_i^T \widehat{T}_i R_i \boldsymbol{x}_1 = 0 \tag{8.34}$$

between the $i$th and the first views (as we know from Exercise 5.4).

In order to see what extra constraints this rank condition implies, we need the following lemma.

**Lemma 8.10.** *Given any* nonzero *vectors $a_i, b_i \in \mathbb{R}^3, i = 1, 2, \ldots, n$, the matrix*

$$\begin{bmatrix} a_1 & b_1 \\ a_2 & b_2 \\ \vdots & \vdots \\ a_n & b_n \end{bmatrix} \in \mathbb{R}^{3n \times 2} \tag{8.35}$$

*is rank-deficient if and only if $a_i b_j^T - b_i a_j^T = 0$ for all $i, j = 1, 2, \ldots, n$.*

The lemma is simply a vector version of the fact that the determinant of any $2 \times 2$ minor of the given rank-deficient matrix is zero. The proof is left to the reader in Excercise 8.9. Applying this lemma, from the matrix $M_p$ that relates the first, $i$th, and $j$th views, we obtain the following equation:

$$\widehat{\boldsymbol{x}}_i R_i \boldsymbol{x}_1 (\widehat{\boldsymbol{x}_j} T_j)^T - \widehat{\boldsymbol{x}}_i T_i (\widehat{\boldsymbol{x}_j} R_j \boldsymbol{x}_1)^T = 0. \tag{8.36}$$

Notice that since $\widehat{u}^T = -\widehat{u}$, the above equation yields the *trilinear constraint*

$$\widehat{\boldsymbol{x}}_i (T_i \boldsymbol{x}_1^T R_j^T - R_i \boldsymbol{x}_1 T_j^T) \widehat{\boldsymbol{x}_j} = 0. \tag{8.37}$$

The above equation is a matrix equation, and it gives a total of $3 \times 3 = 9$ scalar (trilinear) equations, out of which one can show that only four are linearly independent (Exercise 8.24).

The bilinear and trilinear equations are related. As long as the corresponding entries in $\widehat{\boldsymbol{x}_j} T_j$ and $\widehat{\boldsymbol{x}_j} R_j \boldsymbol{x}_1$ are nonzero, each column of the above matrix equation implies that the two vectors $\widehat{\boldsymbol{x}}_i R_i \boldsymbol{x}_1$ and $\widehat{\boldsymbol{x}}_i T_i$ must be linearly dependent. Trilinear constraints (8.37) hence imply bilinear constraints (8.34), except for the special case in which $\widehat{\boldsymbol{x}_j} T_j = \widehat{\boldsymbol{x}_j} R_j \boldsymbol{x}_1 = 0$ for some view $j$. This case corresponds to a rare degenerate configuration in which the 3-D point $p$ lies on the line through the optical centers $o_1, o_j$, which we will discuss in more detail later.

Since $\text{rank}(M_p) \leq 1$ is equivalent to all the $2 \times 2$ minors of $M_p$ having zero determinant and all $2 \times 2$ minors of $M_p$ involving up to three images only, we can conclude that there are *no additional* linearly independent relationships among four views. Although we only proved it for the special case with $\Pi_1 = [I, \ 0]$, the general case differs only by a choice of reference frame.

So far, we have essentially given a proof for the following facts regarding the constraints among multiple images of a point feature.

**Theorem 8.11 (Linear relationships among multiple views of a point).** *For any given $m$ images of a point $p \in \mathbb{E}^3$ relative to $m$ camera frames, rank$(M_p) \leq 1$ yields the following:*

1. *Any algebraic constraint among the $m$ images can be reduced to only those involving two and three images at a time. Formulae of these bilinear and trilinear constraints are given by (8.34) and (8.37), respectively. There is no other irreducible relationship among point features involving more than three views.*

2. *Given $m$ images of a point, all trilinear constraints between any triplet of views algebraically imply all bilinear constraints between pairs of views, except for the degenerate case in which the point $p$ lies on the line through the optical centers $o_1, o_i$ for some $i$.*

From the discussion above, we see that the bilinear (epipolar) and trilinear constraints are certainly necessary for the rank of the matrix $M_p$ to be 1. But rigorously speaking, they are *not* sufficient. According to Lemma 8.10, for them to be equivalent to the rank condition, the vectors involved in the matrix $M_p$ need to be nonzero. This is not always true for certain degenerate cases, as mentioned above.

### 8.3.2 *Geometric interpretation of the rank condition*

This section explores more closely the relationships between the bilinear constraints, the trilinear constraints, the rank conditions, and their implication for the uniqueness of the preimage.

Given three vectors $\boldsymbol{x}_1, \boldsymbol{x}_2, \boldsymbol{x}_3 \in \mathbb{R}^3$, if they are indeed images of some 3-D point $p$ with respect to the three camera frames, as shown in Figure 8.6, they should automatically satisfy both the bilinear and trilinear constraints; i.e.

$$\begin{aligned} \text{bilinear:} \quad & \boldsymbol{x}_2^T \widehat{T_2} R_2 \boldsymbol{x}_1 = 0, \ \boldsymbol{x}_3^T \widehat{T_3} R_3 \boldsymbol{x}_1 = 0, \\ \text{trilinear:} \quad & \widehat{\boldsymbol{x}_2}(T_2 \boldsymbol{x}_1^T R_3^T - R_2 \boldsymbol{x}_1 T_3^T)\widehat{\boldsymbol{x}_3} = 0. \end{aligned}$$

Now consider the inverse problem: If the three vectors $\boldsymbol{x}_1, \boldsymbol{x}_2, \boldsymbol{x}_3$ satisfy the bilinear or trilinear constraints, are they necessarily images of some single point in space? If so, we refer to this single point (if it exists) as the *preimage* of the images $\boldsymbol{x}_1, \boldsymbol{x}_2$, and $\boldsymbol{x}_3$. The above question is answered by the following lemma.

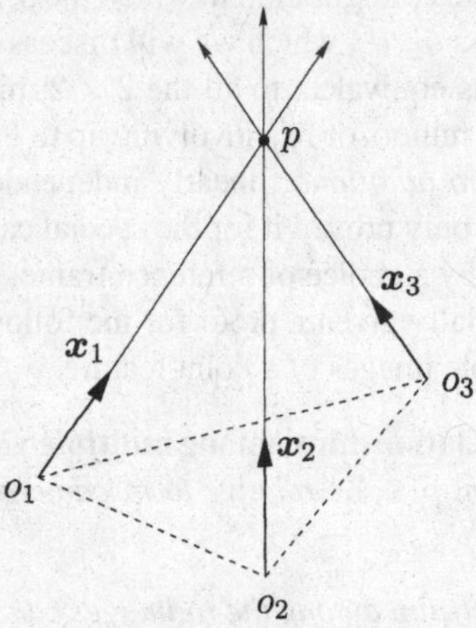


Figure 8.6. Three rays extended from the three images $\boldsymbol{x}_1, \boldsymbol{x}_2, \boldsymbol{x}_3$ intersect at one point $p$ in space, the preimage of $\boldsymbol{x}_1, \boldsymbol{x}_2, \boldsymbol{x}_3$.

**Lemma 8.12 (Properties of bilinear and trilinear constraints).** *Given three vectors $\boldsymbol{x}_1, \boldsymbol{x}_2, \boldsymbol{x}_3 \in \mathbb{R}^3$ and three camera frames with distinct optical centers, if the three images satisfy the epipolar constraint between each pair*[5]

$$\boldsymbol{x}_i^T \widehat{T_{ij}} R_{ij} \boldsymbol{x}_j = 0, \quad i, j = 1, 2, 3,$$

*a unique preimage $p$ is determined except for the case where the three lines associated to image points $\boldsymbol{x}_1, \boldsymbol{x}_2, \boldsymbol{x}_3$ are coplanar.*

[5]Here we use subscripts $ij$ to indicate that the transformation is from the $i$th to the $j$th frame. This convention will be used through the rest of this chapter.

*If these vectors satisfy all trilinear constraints*[6]

$$\widehat{x_j}(T_{ji}x_i^T R_{ki}^T - R_{ji}x_i T_{ki}^T)\widehat{x_k} = 0, \quad i,j,k = 1,2,3,$$

*then they determine a unique preimage $p \in \mathbb{E}^3$ except for when the three lines associated with the three images $x_1, x_2, x_3$ are collinear.*

*Proof.* A detailed proof for this lemma can be found in Appendix 8.A. □

Here we briefly summarize the geometric intuition behind the lemma. Geometrically, the three epipolar constraints simply imply that any pair of the three lines are coplanar. If all three lines are not coplanar, the intersection is uniquely determined, and so is the preimage. If all the lines do lie on the same plane, such a unique intersection is not always guaranteed. As shown in Figure 8.7, this may occur when the lines determined by the images lie on the plane spanned by the three optical centers $o_1, o_2, o_3$, the *trifocal plane*, or when the three optical centers lie on a straight line regardless of the image points, the *rectilinear motion*.

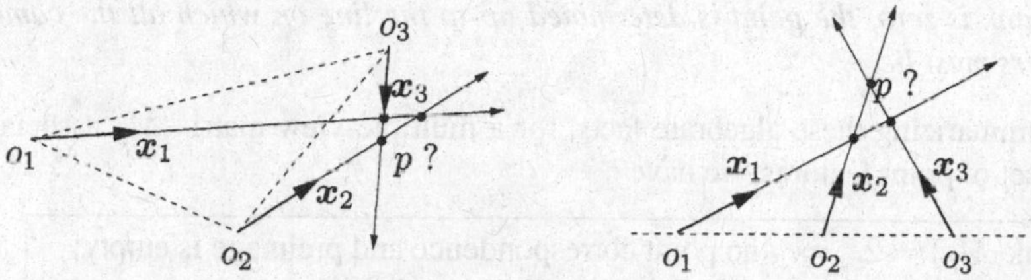


Figure 8.7. Two instances when the three lines determined by the three images $x_1, x_2, x_3$ lie on the same plane, in which case they may not necessarily intersect at a unique point $p$.

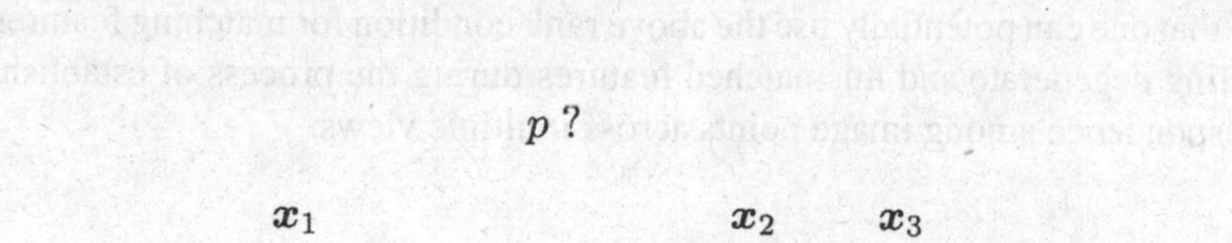

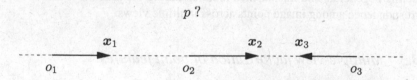

Figure 8.8. If the three images and the three optical centers lie on a straight line, any point on this line is a valid preimage that satisfies all the constraints.

The second case is more important and occurs frequently in practical applications (e.g., a car moving on the highway). In such a case, regardless of what 3-D feature point one chooses, epipolar constraints alone do not provide sufficient constraints to determine a unique preimage point from any given three images. Fortunately, the 3-D preimage is uniquely determined if the three images satisfy the trilinear constraint, except for one rare degenerate configuration in which the point

[6]Although there seem to be total of nine possible $(i, j, k)$, there are in fact only three different trilinear constraints due to the symmetry in the trilinear constraint equation.

$p$ in question lies on the line between collinear optical centers. This situation is demonstrated in Figure 8.8.

For more than three views, in order to check the uniqueness of the preimage, one needs to apply Lemma 8.12 to every pair or triplet of views. The possible number of combinations of degenerate cases often makes it very hard to draw any consistent conclusion. However, the two possible values for the rank of $M_p$ classify geometric configurations of the point relative to the $m$ camera frames into two and only two categories. If the rank is 1, then the relative location of the point $p$ is uniquely determined. If the rank is 0, i.e. $M_p = 0$, then $\widehat{\boldsymbol{x}}_i T_i = \widehat{\boldsymbol{x}}_i R_i \boldsymbol{x}_1 = 0$ for $i = 2, 3, \dots, m$. The only solution to the above equations is that all the camera centers $o_1, o_2, \dots, o_m$ lie on a line and the point $p$ can be anywhere on this line. Hence, in terms of the rank condition on the multiple-view matrix, Lemma 8.12 can be generalized to multiple views in a concise and unified way.

**Theorem 8.13 (Uniqueness of the preimage).** *Given $m$ vectors representing the $m$ images of a point in $m$ views, they correspond to the same point in the 3-D space if the rank of the $M_p$ matrix relative to any of the camera frames is one. If the rank is zero, the point is determined up to the line on which all the camera centers must lie.*

Summarizing these algebraic facts, for a multiple-view matrix $M_p$ associated to a set of point features, we have

$$
\boxed{\begin{array}{lcl}
\operatorname{rank}(M_p) = 2 & \Rightarrow & \text{no point correspondence and preimage is empty;} \\
\operatorname{rank}(M_p) = 1 & \Rightarrow & \text{point correspondence and preimage is unique;} \\
\operatorname{rank}(M_p) = 0 & \Rightarrow & \text{point correspondence and preimage is not unique.}
\end{array}}
$$

Note that one can potentially use the above rank condition for matching features or rejecting degenerate and mismatched features during the process of establishing correspondence among image points across multiple views.

### 8.3.3 *Multiple-view factorization of point features*

As we have demonstrated in the previous section, the rank condition of the multiple-view matrix $M_p$ captures all the constraints among multiple images simultaneously without specifying a particular set of pairs or triplets of frames. Ideally, one would like to formulate the entire problem of reconstructing 3-D motion and structure as the maximization of some global objective function subject to the rank condition. However, such an approach would rely on solving a costly nonlinear optimization problem. Instead, here we divide the overall problem into (linear) subproblems and show how the rank condition can be instrumental for estimating motion assuming known matched point features and estimating 3-D point coordinates given known motion. The complete conceptual algorithm that interleaves these steps will be given at the end of this section.

Suppose that $m$ images $\boldsymbol{x}_1^j, \boldsymbol{x}_2^j, \dots, \boldsymbol{x}_m^j$ of $n$ points $p^j$, $j = 1, 2, \dots, n$, are given, and we want to use them to estimate the unknown projection matrix $\Pi$.

This can be done in two steps. The rank condition of the matrix $M_p$ simply states that its two columns are linearly dependent, which can be written for the $j$th point $p^j$ in the form

$$\begin{bmatrix} \widehat{\boldsymbol{x}_2^j} R_2 \boldsymbol{x}_1^j \\ \widehat{\boldsymbol{x}_3^j} R_3 \boldsymbol{x}_1^j \\ \vdots \\ \widehat{\boldsymbol{x}_m^j} R_m \boldsymbol{x}_1^j \end{bmatrix} + \alpha^j \begin{bmatrix} \widehat{\boldsymbol{x}_2^j} T_2 \\ \widehat{\boldsymbol{x}_3^j} T_3 \\ \vdots \\ \widehat{\boldsymbol{x}_m^j} T_m \end{bmatrix} = 0 \quad \in \mathbb{R}^{3(m-1)\times 1}, \tag{8.38}$$

for proper $\alpha^j \in \mathbb{R}, j = 1, 2, \ldots, n$. Note that the individual rows of equation (8.38) can be obtained from $\lambda_i^j \boldsymbol{x}_i^j = \lambda_1^j R_i \boldsymbol{x}_1^j + T_i$ multiplying both sides by $\widehat{\boldsymbol{x}_i^j}$:

$$\widehat{\boldsymbol{x}_i^j} R_i \boldsymbol{x}_1^j + \widehat{\boldsymbol{x}_i^j} T_i / \lambda_1^j = 0. \tag{8.39}$$

Therefore, $\alpha^j = 1/\lambda_1^j$ can be interpreted as the inverse of the depth of point $p^j$ with respect to the first frame. Note that if we know $\alpha^j$, equation (8.39) is linear in $R_i$ and $T_i$. The estimation of camera motion is then equivalent to finding the (stacked) vectors $R_i^s = [r_{11}, r_{21}, r_{31}, r_{12}, r_{22}, r_{32}, r_{13}, r_{23}, r_{33}]^T \in \mathbb{R}^9$ and $T_i \in \mathbb{R}^3, i = 2, 3, \ldots, m$, such that

$$P_i \begin{bmatrix} R_i^s \\ T_i \end{bmatrix} = \begin{bmatrix} \boldsymbol{x}_1^{1T} \otimes \widehat{\boldsymbol{x}_i^1} & \alpha^1 \widehat{\boldsymbol{x}_i^1} \\ \boldsymbol{x}_1^{2T} \otimes \widehat{\boldsymbol{x}_i^2} & \alpha^2 \widehat{\boldsymbol{x}_i^2} \\ \vdots & \vdots \\ \boldsymbol{x}_1^{nT} \otimes \widehat{\boldsymbol{x}_i^n} & \alpha^n \widehat{\boldsymbol{x}_i^n} \end{bmatrix} \begin{bmatrix} R_i^s \\ T_i \end{bmatrix} = 0 \quad \in \mathbb{R}^{3n}, \tag{8.40}$$

where $\otimes$ is the *Kronecker product* between matrices (Appendix A). It can be shown that if the $\alpha^j$'s are known, the matrix $P_i \in \mathbb{R}^{3n\times 12}$ is of rank 11 if more than $n \geq 6$ points in general position are given. In that case, the null space of $P_i$ is unique up to a scale factor, and so is the projection matrix $\Pi_i = [R_i, T_i]$.

*Reconstruction from a calibrated camera*

To demonstrate the key ideas of the algorithm, we assume that the camera is calibrated; i.e. $K(t) = I$. We leave the uncalibrated case to Section 8.5. Then in the projection matrix $\Pi_i$, $R_i$ is a rotation matrix in $SO(3)$, and $T_i$ is a translation vector in $\mathbb{R}^3$. Let $\tilde{R}_i^s \in \mathbb{R}^9$ and $\tilde{T}_i \in \mathbb{R}^3$ be the (unique) solution of (8.40). In the presence of noise in feature measurements, $P_i$ typically has rank larger than 11; namely, it is of rank 12. In such a case, the solution is obtained as the eigenvector of $P_i$ associated with the smallest singular value. In order to guarantee that $\tilde{R}_i$ is indeed in $SO(3)$, we need to project and scale the obtained solution appropriately. Let $\tilde{R}_i = U_i S_i V_i^T$ be the SVD of $\tilde{R}_i$. Then the solution of (8.40) in $SO(3) \times \mathbb{R}^3$

is given by

$$R_i = \text{sign}(\det(U_i V_i^T)) U_i V_i^T \quad \in SO(3), \tag{8.41}$$

$$T_i = \frac{\text{sign}(\det(U_i V_i^T))}{\sqrt[3]{\det(S_i)}} \tilde{T}_i \quad \in \mathbb{R}^3. \tag{8.42}$$

The motion estimation step assumes that the initial $\alpha^j$'s are known. These initial estimates can be obtained from the first two images, using the *eight-point algorithm* for estimating the displacement $T_2 \in \mathbb{R}^3$ and $R_2 \in SO(3)$ (see Chapter 5), followed by estimation of the unknown scalars $\alpha^j$'s. Given the known motion, the equation given by the first row of equation (8.38) implies $\alpha^j \widehat{\boldsymbol{x}_2^j} T_2 = -\widehat{\boldsymbol{x}_2^j} R_2 \boldsymbol{x}_1^j$. The least-squares solution for the unknown scale factors is given by

$$\alpha^j = -\frac{\left(\widehat{\boldsymbol{x}_2^j} T_2\right)^T \widehat{\boldsymbol{x}_2^j} R_2 \boldsymbol{x}_1^j}{\left\|\widehat{\boldsymbol{x}_2^j} T_2\right\|^2}, \quad j = 1, 2, \ldots, n. \tag{8.43}$$

These initial values of $\alpha^j$ can therefore be used in equation (8.40) for recovery of $R_i$ and $T_i$ for $i = 3, 4, \ldots, m$. Recall that $T_2$ is recovered up to a scalar factor from the eight-point algorithm. We can then recover all the unknowns from (8.40) up to a single scale, since the $\alpha^j$'s were computed up to the same scale of $T_2$.

In the presence of noise, solving for the $\alpha^j$'s using only the first two frames may not necessarily be the best thing to do. Nevertheless, this arbitrary choice of $\alpha^j$'s allows us to compute all the motions $(R_i, T_i)$, $i = 2, 3, \ldots, m$. Given all the motions, the least-squares solution for the $\alpha^j$'s from (8.38) is given by

$$\alpha^j = -\frac{\sum_{i=2}^m \left(\widehat{\boldsymbol{x}_i^j} T_i\right)^T \widehat{\boldsymbol{x}_i^j} R_i \boldsymbol{x}_1^j}{\sum_{i=2}^m \left\|\widehat{\boldsymbol{x}_i^j} T_i\right\|^2}, \quad j = 1, 2, \ldots, n. \tag{8.44}$$

Note that these scalars $\alpha^j$ are the same as those in (8.43) if $m = 2$. One can then recompute the motion given these new $\alpha^j$'s and iterate these two steps until the difference between the reprojected image points $\{\tilde{\boldsymbol{x}}\}$ and the given ones $\{\boldsymbol{x}\}$, the so-called reprojection error, is small enough.

These observations suggest a natural iterative linear algorithm, Algorithm 8.1, for multiple-view motion and structure estimation.

The camera motion is then $[R_i, T_i]$, $i = 2, 3, \ldots, m$, and the structure of the points (with respect to the first camera frame) is given by the converged depth scalar $\lambda_1^j = 1/\alpha^j$, $j = 1, 2, \ldots, n$. The reason to set $\alpha_{k+1}^1 = 1$ is to fix the universal scale. It is equivalent to setting the distance of the first point to be the unit of measurement. There are a few notes for the proposed algorithm:

1. The iteration from step 4 is unnecessary if the data is perfect, i.e. noiseless. In this case, steps 1 to 3 give an exact $m$-view reconstruction based on the initialization from two views.

**Algorithm 8.1 (Factorization algorithm for multiple-view reconstruction).**

Given $m$ images $\boldsymbol{x}_1^j, \boldsymbol{x}_2^j, \ldots, \boldsymbol{x}_m^j$ of $n$ points $p^j$, $j = 1, 2, \ldots, n$, estimate the projection matrix $\Pi_i = [R_i, T_i]$, $i = 2, 3, \ldots, m$ and the 3-D structure as follows:

1. Initialization: $k = 0$
   (a) Compute $[R_2, T_2]$ using the eight-point algorithm for the first two views;
   (b) Compute $\alpha_k^j$ from (8.43);
   (c) Normalize $\alpha^j = \alpha_k^j / \alpha_k^1$ for $j = 1, 2, \ldots, n$.
2. Compute $[\tilde{R}_i, \tilde{T}_i]$ from (8.40) as the eigenvector associated with the smallest singular value of $P_i$, $i = 2, 3, \ldots, m$.
3. In the calibrated case, compute $[R_i, T_i]$ from (8.41) and (8.42); in the uncalibrated case, simply let $[R_i, T_i] = [\tilde{R}_i, \tilde{T}_i]$, for $i = 2, 3, \ldots, m$.
4. Compute the new $\alpha_{k+1}^j$ from (8.44) for $j = 1, 2, \ldots, n$. Normalize so that $\alpha^j = \alpha_{k+1}^j / \alpha_{k+1}^1$ and $T_i = \alpha_{k+1}^1 T_i$. Use the newly recovered $\alpha$'s and $[R, T]$'s to compute the reprojected images $\tilde{\boldsymbol{x}}$ for each point in all views.
5. If $\sum \|\boldsymbol{x} - \tilde{\boldsymbol{x}}\|^2 > \epsilon$, for a specified $\epsilon > 0$, then $k = k + 1$ and go to 2, else stop.

2. The algorithm makes use of all constraints simultaneously for motion and structure estimation. While the $[R_i, T_i]$'s seem to be estimated using pairs of views only, this is not exactly the case. The computation of the matrix $P_i$ depends on all the $\alpha^j$'s, each of which is in turn estimated from the $M_p^j$ matrix involving all views.
3. The algorithm can be used either in batch or in a recursive fashion: initialize with two views, recursively estimate camera motion, and automatically update scene structure when a new view becomes available.
4. In case a point is occluded in a particular image, the corresponding group of three rows in the $M_p$ matrix can be simply dropped without affecting the condition on its rank.

The above algorithm is only *conceptual* (and therefore naive in many ways). There are many possible ways to impose the rank condition, and each of them, although algebraically equivalent, can have significantly different numerical stability[7] and statistical properties. To make the situation even worse, under different conditions (e.g., long baseline or short baseline), correctly imposing such a rank condition requires different numerical techniques. This is even true for the standard eight-point algorithm in the two-view case (as we saw in Chapter 5). We will leave some of these practical issues to Part IV, and below we show a simple experiment demonstrating the peculiarities of the given algorithm.

**Example 8.14 (Reconstruction from images of a cube).** Figure 8.9 shows four images of a cube (on which the corresponding features are established manually), and Figure 8.10

[7] See, for example, Exercise 8.17.

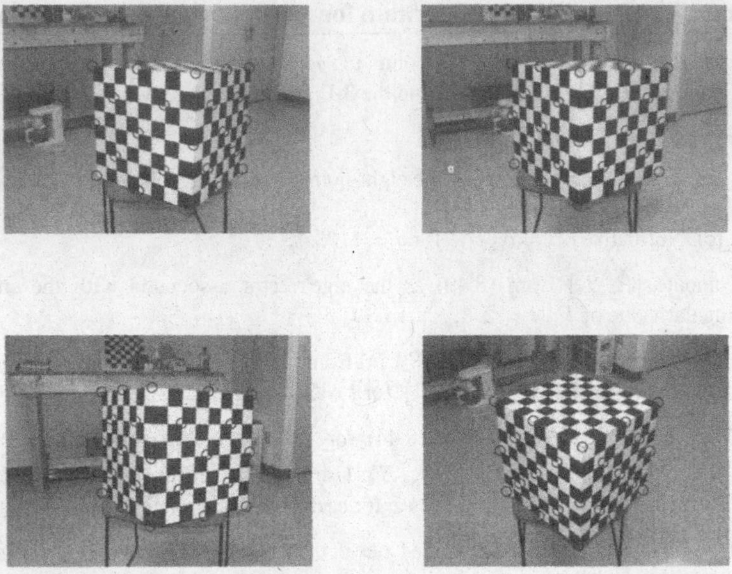

Figure 8.9. Four images of the cube and corresponding features. The camera is calibrated using techniques described in Chapter 6.

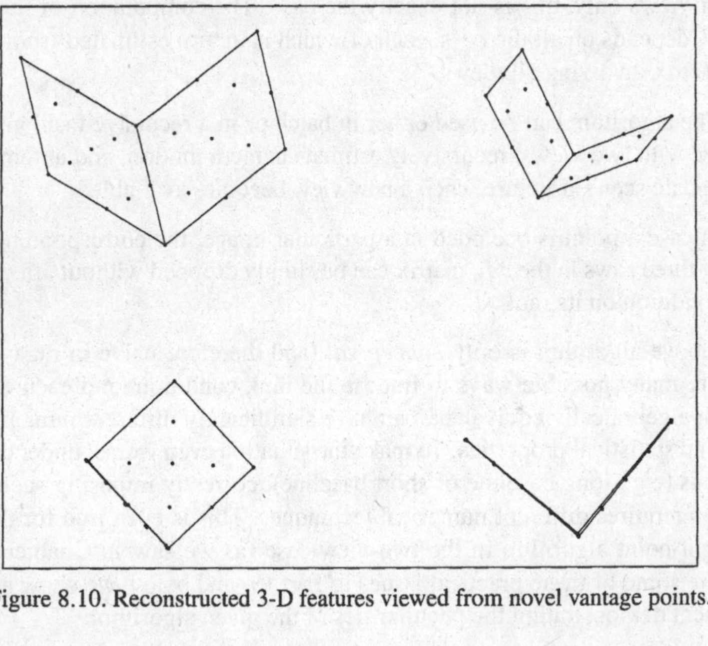

Figure 8.10. Reconstructed 3-D features viewed from novel vantage points.

shows the result of 3-D structure reconstructed from these images using Algorithm 8.1. The camera is roughly calibrated, and the error in the reconstructed right angles around each corner is within 1°. The choice of an image to be the reference view does not affect the reconstruction quality significantly. ■

## 8.4 Geometry of line features

The previous section focused on multiple-view relationships among point features. In this section, we follow a similar derivation for line features.

### 8.4.1 *The multiple-view matrix of a line and its rank*

According to Lemma 8.4, the matrix

$$W_l \doteq \begin{bmatrix} \boldsymbol{\ell}_1^T \Pi_1 \\ \boldsymbol{\ell}_2^T \Pi_2 \\ \vdots \\ \boldsymbol{\ell}_m^T \Pi_m \end{bmatrix} \in \mathbb{R}^{m \times 4} \tag{8.45}$$

associated with $m$ images of a line in space satisfies the rank condition

$$\operatorname{rank}(W_l) \leq 2. \tag{8.46}$$

Without loss of generality, we may assume that the first camera frame is chosen to be the reference frame, i.e. $\Pi_1 = [I, 0]$. Given this choice, the matrix $W_l$ and its associated rank condition can be expressed in a more compact form. Multiplying $W_l$ on the right by the full-rank matrix $D_l \in \mathbb{R}^{4\times5}$ does not change its rank:

$$W_l D_l = \begin{bmatrix} \boldsymbol{\ell}_1^T & 0 \\ \boldsymbol{\ell}_2^T R_2 & \boldsymbol{\ell}_2^T T_2 \\ \vdots & \vdots \\ \boldsymbol{\ell}_m^T R_m & \boldsymbol{\ell}_m^T T_m \end{bmatrix} \begin{bmatrix} \boldsymbol{\ell}_1 & \widehat{\boldsymbol{\ell}_1} & 0 \\ 0 & 0 & 1 \end{bmatrix} \in \mathbb{R}^{m\times5}. \tag{8.47}$$

This yields a new matrix

$$W_l' = W_l D_l = \begin{bmatrix} \boldsymbol{\ell}_1^T \boldsymbol{\ell}_1 & 0 & 0 \\ \boldsymbol{\ell}_2^T R_2 \boldsymbol{\ell}_1 & \boldsymbol{\ell}_2^T R_2 \widehat{\boldsymbol{\ell}_1} & \boldsymbol{\ell}_2^T T_2 \\ \vdots & \vdots & \vdots \\ \boldsymbol{\ell}_m^T R_m \boldsymbol{\ell}_1 & \boldsymbol{\ell}_m^T R_m \widehat{\boldsymbol{\ell}_1} & \boldsymbol{\ell}_m^T T_m \end{bmatrix} \in \mathbb{R}^{m\times5}. \tag{8.48}$$

Since $D_l$ is of full rank 4, the rank of $W_l'$ is the same as that of $W_l$:

$$\operatorname{rank}(W_l') = \operatorname{rank}(W_l) \leq 2.$$

For the matrix $W_l'$ to be of rank $\leq 2$, the submatrix

$$M_l \doteq \begin{bmatrix} \boldsymbol{\ell}_2^T R_2 \widehat{\boldsymbol{\ell}_1} & \boldsymbol{\ell}_2^T T_2 \\ \boldsymbol{\ell}_3^T R_3 \widehat{\boldsymbol{\ell}_1} & \boldsymbol{\ell}_3^T T_3 \\ \vdots & \vdots \\ \boldsymbol{\ell}_m^T R_m \widehat{\boldsymbol{\ell}_1} & \boldsymbol{\ell}_m^T T_m \end{bmatrix} \in \mathbb{R}^{(m-1)\times4} \tag{8.49}$$

of $W_l'$ must have rank of no more than one. We call the matrix $M_l$ the *multiple-view matrix* associated with a line feature $L$. Together with Lemma 8.4, we have proven the following theorem.

**Theorem 8.15 (Rank conditions for line features).** *For the two matrices $W_l$ and $M_l$ associated with $m$-views of a line, the following relationship holds:*

$$\boxed{\text{rank}(M_l) = \text{rank}(W_l) - 1 = \text{rank}(N_l) - (2m+1) \leq 1.} \tag{8.50}$$

*Therefore, rank$(W_l)$ is either 2 or 1, depending on whether rank$(M_l)$ is 1 or 0.*

For rank$(M_l) \leq 1$, it is necessary for every pair of row vectors of $M_l$ to be linearly dependent. Focusing on the first three columns of $M_l$, the linear dependency implies that $\boldsymbol{\ell}_i^T R_i \widehat{\boldsymbol{\ell}_1} \sim \boldsymbol{\ell}_j^T R_j \widehat{\boldsymbol{\ell}_1}$ for all $i, j$. This relation is equivalent to a trilinear equation

$$\boldsymbol{\ell}_i^T R_i \widehat{\boldsymbol{\ell}_1} R_j^T \boldsymbol{\ell}_j = 0. \tag{8.51}$$

Notice that this constraint involves only the camera orientation $R$. Now, taking into account the fourth column and considering the linear dependency between the $i$th and $j$th rows gives us another type of trilinear constraint,

$$\boldsymbol{\ell}_j^T T_j \boldsymbol{\ell}_i^T R_i \widehat{\boldsymbol{\ell}_1} - \boldsymbol{\ell}_i^T T_i \boldsymbol{\ell}_j^T R_j \widehat{\boldsymbol{\ell}_1} = 0, \tag{8.52}$$

that relates the first, $i$th and $j$th images. From Example 8.7, we know that nontrivial constraints on lines must involve at least three views. The above two equations confirm this fact.

Although the trilinear constraints (8.52) above are *necessary* for the rank of the matrix $M_l$ to be 1, they are *not sufficient*. For equation (8.52) to be sufficient and imply (8.51), it is required that the scalar $\boldsymbol{\ell}_i^T T_i$ in $M_l$ be nonzero. This is not true for certain degenerate cases such as the line $L$ being coplanar to the baseline $T_i$. But in any case, rank$(M_l) \leq 1$ if and only if all its $2 \times 2$ minors have zero determinant. Since all such minors involve only three images at a time, we can conclude that any constraint on lines is dependent on those involving only three images at a time.

Hence, we have essentially proven the following theorem.

**Theorem 8.16 (Constraints among multiple images of a line).** *Given $m$ views of a line $L$ in $\mathbb{E}^3$ in $m$ camera frames, the rank condition on the matrix $M_l$ implies that any algebraic constraints among these $m$ images can be reduced to only those involving three images at a time. These constraints include (8.51) and the trilinear constraint (8.52).*

### *8.4.2 Geometric interpretation of the rank condition*

From the above derivation, we know that given three vectors $\boldsymbol{\ell}_1, \boldsymbol{\ell}_2, \boldsymbol{\ell}_3 \in \mathbb{R}^3$ that are images of the same line $L$ in space, as shown in Figure 8.11, they satisfy the trilinear constraints

$$\boldsymbol{\ell}_2^T T_2 \boldsymbol{\ell}_3^T R_3 \widehat{\boldsymbol{\ell}_1} - \boldsymbol{\ell}_3^T T_3 \boldsymbol{\ell}_2^T R_2 \widehat{\boldsymbol{\ell}_1} = 0.$$

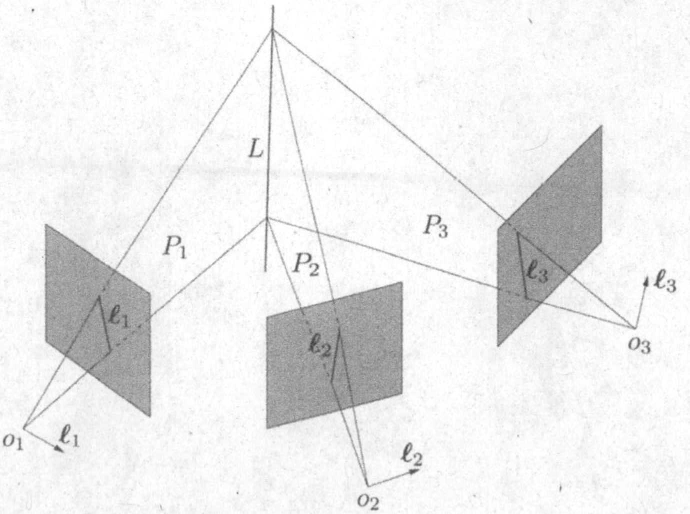


Figure 8.11. Three planes extended from the three images $\boldsymbol{\ell}_1, \boldsymbol{\ell}_2, \boldsymbol{\ell}_3$ intersect in one line $L$ in space, the preimage of $\boldsymbol{\ell}_1, \boldsymbol{\ell}_2, \boldsymbol{\ell}_3$.

As in the point case, we now ask the opposite question: If the three vectors $\boldsymbol{\ell}_1, \boldsymbol{\ell}_2, \boldsymbol{\ell}_3$ satisfy the trilinear constraints, are they necessarily images of a single line in space? The answer is given by the following lemma.

**Lemma 8.17 (Properties of trilinear constraints for lines).** *Given three camera frames with distinct optical centers and any three vectors $\boldsymbol{\ell}_1, \boldsymbol{\ell}_2, \boldsymbol{\ell}_3 \in \mathbb{R}^3$ that represent three image lines, if the three image lines satisfy the trilinear constraints*

$$\boldsymbol{\ell}_j^T T_{ji} \boldsymbol{\ell}_k^T R_{ki} \widehat{\boldsymbol{\ell}}_i - \boldsymbol{\ell}_k^T T_{ki} \boldsymbol{\ell}_j^T R_{ji} \widehat{\boldsymbol{\ell}}_i = 0, \quad i, j, k = 1, 2, 3,$$

*then their preimage $L$ is uniquely determined except for the case in which the preimage of every $\boldsymbol{\ell}_i$ is the same plane in space. This is the only degenerate case, and in this case, the matrix $M_l$ becomes zero.*

*Proof.* As shown in Figure 8.12, we denote the planes formed by the optical center $o_i$ of the $i$th frame and the image line $\boldsymbol{\ell}_i$ by $P_i$, and $\boldsymbol{\ell}_i \in \mathbb{R}^3$ is also the normal vector to $P_i$, $i = 1, 2, 3$. Denote the intersection line between $P_1$ and $P_2$ by $L_2$, and the intersection line between $P_1$ and $P_3$ by $L_3$. Geometrically, $-\boldsymbol{\ell}_i^T T_i = d_i$ is the distance from $o_1$ to the plane $P_i$, and $(\boldsymbol{\ell}_i^T R_i)^T = R_i^T \boldsymbol{\ell}_i$ is the unit normal vector of $P_i$ expressed in the first frame. Furthermore, $(\boldsymbol{\ell}_i^T R_i \widehat{\boldsymbol{\ell}}_1)^T$ is a vector parallel to $L_i$ with length $\sin(\theta_i)$, where $\theta_i \in [0, \pi]$ is the angle between the planes $P_1$ and $P_i$, $i = 2, 3$. Therefore, in the general case, the trilinear constraint first implies that $L_2$ is parallel to $L_3$, since $(\boldsymbol{\ell}_2^T R_2 \widehat{\boldsymbol{\ell}}_1)^T$ and $(\boldsymbol{\ell}_3^T R_3 \widehat{\boldsymbol{\ell}}_1)^T$ are linearly dependent. Secondly, the two terms in the trilinear constraints must have the same norm, which gives $d_2 \sin(\theta_3) = d_3 \sin(\theta_2)$. Since $d_2 / \sin(\theta_2)$ and $d_3 / \sin(\theta_2)$ are the distances from $o_1$ to $L_2$ and $L_3$ respectively, then $L_2$ must coincide with $L_3$, or in other words, the line $L$ in space is uniquely determined.

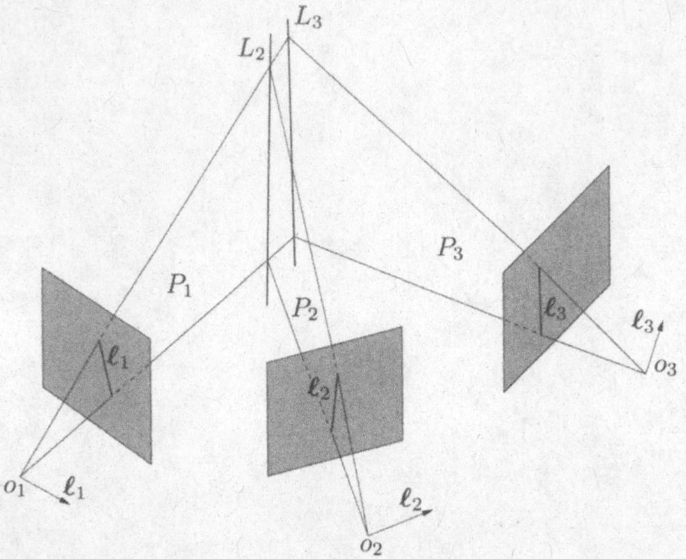


Figure 8.12. Three planes extended from the three images $\boldsymbol{\ell}_1, \boldsymbol{\ell}_2, \boldsymbol{\ell}_3$ intersect at lines $L_2$ and $L_3$, which should actually coincide.

The case in which $P_1$ coincides with only $P_2$ or only $P_3$ needs some attention. For example, if $P_1$ coincides with $P_2$ but not with $P_3$, then $d_2 = 0$ and $(\boldsymbol{\ell}_2^T R_2 \widehat{\boldsymbol{\ell}_1})^T = 0_{3\times 1}$. In this case the preimage $L$ is still uniquely determined as the intersection line between the two planes $P_1$ and $P_3$.

The only degeneracy occurs when $P_1$ coincides with both $P_2$ and $P_3$. In that case, $d_2 = d_3 = 0$ and $(\boldsymbol{\ell}_2^T R_2 \widehat{\boldsymbol{\ell}_1})^T = (\boldsymbol{\ell}_3^T R_3 \widehat{\boldsymbol{\ell}_1})^T = 0_{3\times 1}$. There is an infinite number of lines in $P_1 = P_2 = P_3$ that generate the same set of images $\boldsymbol{\ell}_1$, $\boldsymbol{\ell}_2$, and $\boldsymbol{\ell}_3$, as shown in Figure 8.13. □

**Comment 8.18 (Geometric information in $M_l$).** *First, recall that* $\mathbf{V} = [v^T, 0]^T$ *is the 3-D direction of the line. Without loss of generality, we assume* $\|v\| = 1$. *Notice from the proof of Lemma 8.17, that if* rank$(M_l) = 1$, *the first three entries in each row are exactly* $\sin(\theta_i)v^T$, *which is parallel to the direction of the line* $L$ *in space (with respect to the first view). If we normalize each row of* $M_l$ *by the norm of the first three entries, i.e. divide by* $\sin(\theta_i)$, *each row becomes* $[v^T, r]$, *where* $r = d_1/\sin(\theta_1) = \cdots = d_m/\sin(\theta_m)$ *is the distance from* $o_1$ *to the line* $L$. *Hence* $M_l$ *contains exactly the information of* $L$ *that is missing in the first image* $\boldsymbol{\ell}_1$. *Together with* $\boldsymbol{\ell}_1$, $v$ *and* $r$ *determine the 3-D location of* $L$, *as shown in Figure 8.14.*

For more than three views, in order to check the uniqueness of the preimage of the given image lines, one needs to apply the above lemma to every triplet of views. Since there are many possible combinations of degenerate cases, it is very hard to draw any consistent conclusion using trilinear relationships only.

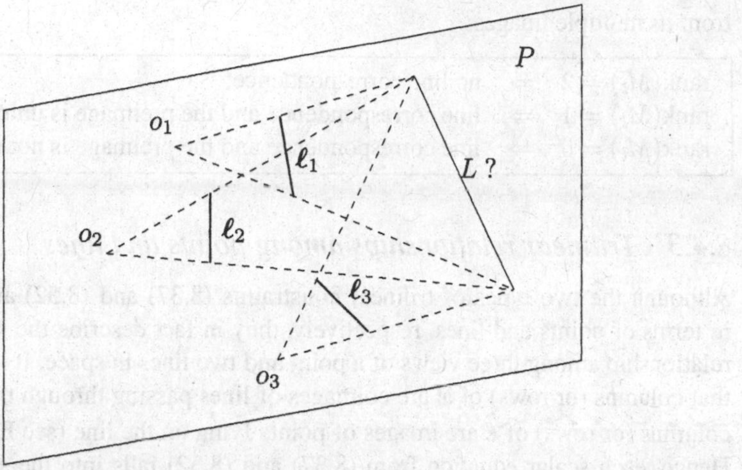


**Figure 8.13. If the three images and the three optical centers lie on a plane $P$, any line on this plane can be a valid preimage of $\ell_1, \ell_2, \ell_3$ that satisfies all the constraints.**

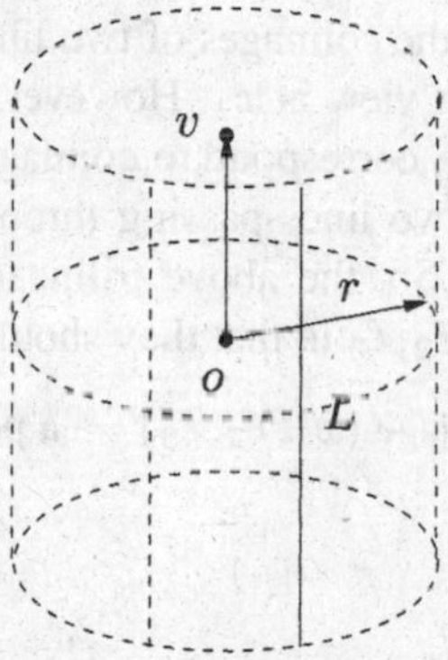


Figure 8.14. Geometric information in a matrix $M_l$: direction of and distance to the line $L$.

However, the previous lemma can be generalized to multiple views in terms of the rank condition on the multiple-view matrix.

**Theorem 8.19 (Uniqueness of the preimage).** *Given $m$ vectors in $\mathbb{R}^3$ representing images of lines $\ell_i$ with respect to $m$ camera frames, they correspond to the same line in space if the rank of the matrix $M_l$ relative to any of the camera frames is 1. If its rank is 0 (i.e. the matrix $M_l$ is zero), then the line is determined up to a plane on which all the camera centers must lie, as shown in Figure 8.13.*

The proof follows directly from Theorem 8.15, and the degenerate case can be easily derived from solving the equation $M_l = 0$. The case in which the line in space shares the same plane as all the centers of the camera is the only degenerate case in which one will not be able to determine the exact 3-D location of the line from its multiple images:

$$\boxed{\begin{array}{lcl} \text{rank}(M_l) = 2 & \Rightarrow & \text{no line correspondence;} \\ \text{rank}(M_l) = 1 & \Rightarrow & \text{line correspondence and the preimage is unique;} \\ \text{rank}(M_l) = 0 & \Rightarrow & \text{line correspondence and the preimage is not unique.} \end{array}}$$

### 8.4.3 *Trilinear relationships among points and lines*

Although the two types of trilinear constraints (8.37) and (8.52) are expressed in terms of points and lines, respectively, they in fact describe the same type of relationship among three views of a point and two lines in space. It is easy to see that columns (or rows) of $\widehat{\boldsymbol{x}}$ are coimages of lines passing through the point, and columns (or rows) of $\widehat{\boldsymbol{\ell}}$ are images of points lying on the line (see Exercise 8.1). Hence, each scalar equation from (8.37) and (8.52) falls into the same type of equation

$$\boxed{\boldsymbol{\ell}_2^T(T_2\boldsymbol{x}_1^T R_3^T - R_2\boldsymbol{x}_1 T_3^T)\boldsymbol{\ell}_3 = 0,} \tag{8.53}$$

where $\boldsymbol{\ell}_2, \boldsymbol{\ell}_3$ are respectively the coimages of two lines passing through the same point whose image in the first view is $\boldsymbol{x}_1$. However, for this equation to hold, it is not necessary that $\boldsymbol{\ell}_2$ and $\boldsymbol{\ell}_3$ correspond to coimages of the same line in space. They can be images of any two lines passing through the same point $p$, which is illustrated in Figure 8.15. So, the above trilinear equation (8.53) imposes a restriction on the images $\boldsymbol{x}_1, \boldsymbol{\ell}_2, \boldsymbol{\ell}_3$ in that they should satisfy

$$\text{preimage}\ (\boldsymbol{x}_1, \boldsymbol{\ell}_2, \boldsymbol{\ell}_3) = \text{a point.} \tag{8.54}$$

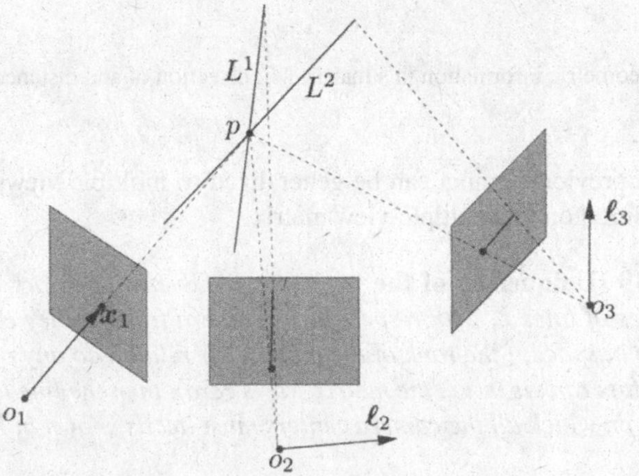


Figure 8.15. Images of two lines $L^1, L^2$ intersecting at a point $p$. Planes extended from the image lines might not intersect at the same line in space. But $\boldsymbol{x}_1, \boldsymbol{\ell}_2, \boldsymbol{\ell}_3$ satisfy the trilinear relationship.

**Example 8.20 (Tensorial representation of the trilinear constraint).** The above trilinear relationship (8.53) plays a similar role for three-view geometry to that of the epipolar constraint for two-view geometry. Here the fundamental matrix $F$ is replaced by the notion of *trifocal tensor*, traditionally denoted by $\mathcal{T}$. Like the fundamental matrix, the trifocal tensor depends (nonlinearly) on the motion parameters $R$ and $T$. The above trilinear relation can then be written formally as the *interior product* (or contraction) of the tensor $\mathcal{T}$ with the three vectors $\boldsymbol{x}_1, \boldsymbol{\ell}_2, \boldsymbol{\ell}_3$:

$$\mathcal{T}(\boldsymbol{x}_1, \boldsymbol{\ell}_2, \boldsymbol{\ell}_3) = \sum_{i,j,k=1}^{3,3,3} \mathcal{T}(i,j,k)\boldsymbol{x}_1(i)\boldsymbol{\ell}_2(j)\boldsymbol{\ell}_3(k) = 0. \tag{8.55}$$

Intuitively, the tensor $\mathcal{T}$ consists of $3 \times 3 \times 3$ entries, since the above trilinear relation has as many coefficients to be determined. Like the fundamental matrix, which has only seven free parameters,[8] there are in fact only 18 free parameters in $\mathcal{T}$. We will give a formal proof for this in Section 8.5.2 (Proposition 8.21). Nonetheless, if one is interested only in recovering $\mathcal{T}$ linearly from three views as we did for the fundamental matrix from two views, the internal dependency among the coefficients $\mathcal{T}(i,j,k)$'s can be ignored. The 27 coefficients can be recovered up to a scalar factor from 26 linear equations of the above type (8.53). We hence need at least 26 triplets $(\boldsymbol{x}_1, \boldsymbol{\ell}_2, \boldsymbol{\ell}_3)$ with correspondence relationships specified by Figure 8.15. We leave as an exercise for the reader to determine how many correspondences across three views are needed in order to have 26 linearly independent equations for solving $\mathcal{T}$ (see Exercise 8.24). Like the fundamental matrix, the trifocal tensor $\mathcal{T}$ also has a rich algebraic and geometric structure. In Exercise 8.23, we introduce a way of book-keeping its 27 entries that preserves nice geometric and algebraic properties. Since the trilinear constraints in most cases imply the epipolar constraints, we should expect that the fundamental matrix $F$ can be recovered from the trifocal tensor $\mathcal{T}$. This is indeed true. A step-by-step development for this fact is given in the exercises (see Exercise 8.25). ■

# 8.5 Uncalibrated factorization and stratification

We have studied in Chapter 6 how to achieve 3-D reconstruction in the case of an uncalibrated camera from two images. In this section, we will study uncalibrated reconstruction in the general multi-view setting based on the rank condition. In particular, we will show how the rank condition naturally reveals the relationships among projective (uncalibrated), affine (weakly calibrated), and Euclidean (calibrated) cameras and reconstructions. To clearly compare these cases, we will adopt the notation introduced in Chapter 6: we use $\boldsymbol{x}_i'$ now for the ($i$th) uncalibrated image, to be distinguished from the calibrated one $\boldsymbol{x}_i$. They are related by the equation $\boldsymbol{x}_i' = K\boldsymbol{x}$.

[8]Recall that the fundamental matrix $F \in \mathbb{R}^{3\times 3}$ is defined only up to a scalar factor and it further satisfies $\det(F) = 0$.

### 8.5.1 Equivalent multiple-view matrices

Recall that the relationship between an uncalibrated image $\boldsymbol{x}'$ and its 3-D counterpart in the uncalibrated setting is

$$\lambda_i \boldsymbol{x}_i' = \Pi_{ie}\boldsymbol{X}_e = \Pi_{ia}\boldsymbol{X}_a = \Pi_{ip}\boldsymbol{X}_p, \quad i = 1, 2, \ldots, m, \tag{8.56}$$

where, as defined in Chapter 6,

$$\begin{array}{lll} \Pi_{1e} = [K_1, 0], & \Pi_{ie} = [K_i R_i, K_i T_i], & i = 2, 3, \ldots, m, \\ \Pi_{1a} = [I,\ 0], & \Pi_{ia} = [K_i R_i K_1^{-1}, K_i T_i], & i = 2, 3, \ldots, m, \\ \Pi_{1p} = [I,\ 0], & \Pi_{ip} = [K_i R_i K_1^{-1} + K_i T_i v^T, v_4 K_i T_i], & i = 2, 3, \ldots, m, \end{array}$$

are the Euclidean, affine, and projective projection matrices with respect to the $m$ views, respectively. For generality, assume that the camera calibration matrix $K$ can be different for each view (in practice, only some of the entries of $K$ vary). To be more precise, a Euclidean structure $\boldsymbol{X}_e$, an affine structure $\boldsymbol{X}_a$, and a projective structure $\boldsymbol{X}_p$ generate the same set of $m$ images by projecting them through the following (multiple-view) projection matrices $\Pi_e$, $\Pi_a$, and $\Pi_p$, respectively:

$$\begin{bmatrix} K_1 & 0 \\ K_2R_2 & K_2T_2 \\ \vdots & \vdots \\ K_mR_m & K_mT_m \end{bmatrix}, \begin{bmatrix} I & 0 \\ K_2R_2K_1^{-1} & K_2T_2 \\ \vdots & \vdots \\ K_mR_mK_1^{-1} & K_mT_m \end{bmatrix}, \begin{bmatrix} I & 0 \\ K_2R_2K_1^{-1} + K_2T_2v^T & v_4K_2T_2 \\ \vdots & \vdots \\ K_mR_mK_1^{-1} + K_mT_mv^T & v_4K_mT_m \end{bmatrix}.$$

Recall from Chapter 6 that the relationships among these matrices are

$$\begin{bmatrix} K_1 & 0 \\ K_2R_2 & K_2T_2 \\ \vdots & \vdots \\ K_mR_m & K_mT_m \end{bmatrix} \begin{bmatrix} K_1^{-1} & 0 \\ 0 & 1 \end{bmatrix} = \begin{bmatrix} I & 0 \\ K_2R_2K_1^{-1} & K_2T_2 \\ \vdots & \vdots \\ K_mR_mK_1^{-1} & K_mT_m \end{bmatrix}$$

and

$$\begin{bmatrix} I & 0 \\ K_2R_2K_1^{-1} & K_2T_2 \\ \vdots & \vdots \\ K_mR_mK_1^{-1} & K_mT_m \end{bmatrix} \begin{bmatrix} I & 0 \\ v^T & v_4 \end{bmatrix} = \begin{bmatrix} I & 0 \\ K_2R_2K_1^{-1} + K_2T_2v^T & v_4K_2T_2 \\ \vdots & \vdots \\ K_mR_mK_1^{-1} + K_mT_mv^T & v_4K_mT_m \end{bmatrix}.$$

In the equation above, the matrix

$$H_p^{-1} = \begin{bmatrix} I & 0 \\ v^T & v_4 \end{bmatrix} \in \mathbb{R}^{4\times 4} \tag{8.57}$$

is the only type of linear transformation that preserves our choice of the first (affine or projective) camera frame to be the reference frame; i.e.

$$[I,\ 0]H_p^{-1} = [I,\ 0]. \tag{8.58}$$

Since with $\Pi_{1a} = \Pi_{1p} = [I,\ 0]$, both $\Pi_a$ and $\Pi_p$ now conform to the choice for the projection matrices in (8.30), they give rise to two multiple-view matrices with equal rank

$$\begin{bmatrix} \widehat{\boldsymbol{x}_2'} K_2 R_2 K_1^{-1} \boldsymbol{x}_1' & \widehat{\boldsymbol{x}_2'} K_2 T_2 \\ \widehat{\boldsymbol{x}_3'} K_3 R_3 K_1^{-1} \boldsymbol{x}_1' & \widehat{\boldsymbol{x}_3'} K_3 T_3 \\ \vdots & \vdots \\ \widehat{\boldsymbol{x}_m'} K_m R_m K_1^{-1} \boldsymbol{x}_1' & \widehat{\boldsymbol{x}_m'} K_3 T_m \end{bmatrix}, \begin{bmatrix} \widehat{\boldsymbol{x}_2'} (K_2 R_2 K_1^{-1} + K_2 T_2 v^T) \boldsymbol{x}_1' & \widehat{\boldsymbol{x}_2'} v_4 K_2 T_2 \\ \widehat{\boldsymbol{x}_3'} (K_3 R_3 K_1^{-1} + K_3 T_3 v^T) \boldsymbol{x}_1' & \widehat{\boldsymbol{x}_3'} v_4 K_3 T_3 \\ \vdots & \vdots \\ \widehat{\boldsymbol{x}_m'} (K_m R_m K_1^{-1} + K_m T_m v^T) \boldsymbol{x}_1' & \widehat{\boldsymbol{x}_m'} v_4 K_m T_m \end{bmatrix}.$$

Notice that the columns of the second matrix are simply linear combinations of those of the first matrix, and vice versa. These two matrices must have the same rank (see Exercise 8.20). However, the null spaces of these two matrices are usually different (Exercise 8.19). Similar results hold for the case of line features.

Thus, for the same set of uncalibrated images $\{\boldsymbol{x}'\}$, the projection matrix $\Pi$ that allows $\text{rank}(M) = 1$ is not unique. The remaining question is whether the projection matrices $\Pi_p$ or $\Pi_a$ (differing by a choice in $v, v_4$) are the only ones, or whether there exist other projection matrices that also satisfy the rank condition for the same set of images.

## 8.5.2 *Rank-based uncalibrated factorization*

In the first step of Algorithm 8.1, one needs to initialize $\alpha$ with $\lambda_1$ from any two views. Without calibration, the relationship between, say, the first two views is characterized by the fundamental matrix $F = \widehat{K_2 T_2} K_2 R_2 K_1^{-1}$. The difficulty with the uncalibrated case is that the fundamental matrix $F$ cannot be directly decomposed into the (affine) form $[K_2 R_2 K_1^{-1}, K_2 T]$. Thus, as we have contended in Section 6.4.2, we may first choose the canonical decomposition of $F$ and recover a projective solution of the form

$$F \quad \mapsto \quad \Pi_{2p} = [K_2 R_2 K_1^{-1} + K_2 T_2 v^T, v_4 K_2 T_2]. \tag{8.59}$$

Although $\Pi_{2p}$ gives rise to the same epipolar relationship between the first two views, the recovered 3-D structure depends on the actual choice of $v \in \mathbb{R}^3$ and $v_4 \in \mathbb{R}$. Different choices yield different values of $\alpha$, or equivalently, $\lambda_1$ (see Exercise 8.19).

But once a decomposition of $F$ is chosen for $\Pi_{2p}$, all the other projection matrices $\Pi_{3p}, \Pi_{4p}, \ldots, \Pi_{mp}$ are uniquely determined (up to a scale factor) from equation (8.40). Only in the projective case can we no longer normalize[9] the recovered $\Pi_{ip}$'s using $\det(R_i) = 1$. That is, the projection matrix $\Pi_p = (\Pi_{1p}, \Pi_{2p}, \ldots, \Pi_{mp})$ directly recovered from the first two steps of Algorithm 8.1

[9] We need to determine the projection matrices only up to a scalar factor, say $\gamma \in \mathbb{R}$. However, one must be aware that, once the scales on the image coordinates and 3-D coordinates are set, e.g., the last entry of $\boldsymbol{x}$ or $\boldsymbol{X}$ is set to 1, the scale of each projection matrix is in fact absorbed into $\lambda$ with respect to each view. It is easier to see this from the matrix $W_p$: although each row can be arbitrarily scaled, its null space does not change, hence preserves the same homogeneous coordinates of a 3-D point.

in the uncalibrated case must be of the following general form

$$\Pi_p = \begin{bmatrix} \Pi_{1p} \\ \Pi_{2p} \\ \vdots \\ \Pi_{mp} \end{bmatrix} = \begin{bmatrix} I & 0 \\ \gamma_2(K_2R_2K_1^{-1} + K_2T_2v^T) & \gamma_2 v_4 K_2 T_2 \\ \vdots & \vdots \\ \gamma_m(K_mR_mK_1^{-1} + K_mT_mv^T) & \gamma_m v_4 K_m T_m \end{bmatrix}, \quad (8.60)$$

for some unknown $v \in \mathbb{R}^3$, $v_4$, and $\gamma_i \in \mathbb{R}$, $i = 2, 3, \ldots, m$. Therefore, among the total of $12(m-1)$ parameters of the $m-1$ matrices $\Pi_{2p}, \Pi_{3p}, \ldots, \Pi_{mp} \in \mathbb{R}^{3\times 4}$, only $11(m-1)-4 = 11m-15$ can be determined due to the arbitrary choice of $v$ and $v_4$ (which have four parameters) and the $(m-1)$ scalars $\gamma_2, \gamma_3, \ldots, \gamma_m$. To summarize the above discussion, we have in fact proven the following statement.

**Proposition 8.21 (Total number of degrees of freedom).** *From $m$ images alone (in the absence of any extra information about calibration or motion), we can recover at most $11m - 15$ free parameters for the relative configuration of the $m$ camera frames.*

This in fact also explains why the fundamental matrix for two views should depend on only $11\times 2-15 = 7$ parameters[10] and the trifocal tensor for three views should depend on $11\times 3-15 = 18$ parameters. In particular, the three fundamental matrices for three pairs of views, say $F_{12}, F_{23}, F_{13} \in \mathbb{R}^{3\times 3}$, together should also have only 18 degrees of freedom. We already know that each has seven degrees of freedom. Hence the three fundamental matrices must be related, i.e. they should satisfy at least three more algebraically independent equations. We will leave the development of these equations to the reader as an exercise (see Exercise 8.21).

### *8.5.3 Direct stratification by the absolute quadric constraint*

So far, we have studied the type of solutions offered by the factorization algorithm in the uncalibrated case. That is, in the absence of any extra information about camera calibration or motion, using the image measurements only, we can directly recover a projection matrix $\Pi_p$ with

$$\Pi_{ip} = \left[\gamma_i(K_iR_iK_1^{-1} + K_iT_iv^T),\ \gamma_i v_4 K_i T_i\right]^T, \quad i = 2, 3, \ldots, m,$$

and its associated projective structure $\boldsymbol{X}_p$. In order to obtain a Euclidean reconstruction $\boldsymbol{X}_e$, we need to further eliminate the ambiguity introduced by the transformation $H_p$, the scale $\gamma$, and the camera calibration $K$. We may achieve this by accomplishing the following upgrade of the projection matrix,

$$\begin{bmatrix} I & 0 \\ \gamma_2(K_2R_2K_1^{-1} + K_2T_2v^T) & \gamma_2 v_4 K_2 T_2 \\ \vdots & \vdots \\ \gamma_m(K_mR_mK_1^{-1} + K_mT_mv^T) & \gamma_m v_4 K_m T_m \end{bmatrix} \mapsto \begin{bmatrix} K_1 & 0 \\ \gamma_2 K_2 R_2 & \gamma_2 K_2 T_2 \\ \vdots & \vdots \\ \gamma_m K_m R_m & \gamma_m K_m T_m \end{bmatrix},$$

[10] The nine entries of $F$ only have seven degrees of freedom, because $F$ is determined up to a scalar factor and $F$ is of rank 2, implying $\det(F) = 0$.

since as we leave to the reader to verify, the Euclidean structure $X_e$ can be obtained as the null space of the matrix

$$W_p = \begin{bmatrix} \widehat{x'_1}K_1 & 0 \\ \gamma_2\widehat{x'_2}K_2R_2 & \gamma_2\widehat{x'_2}K_2T_2 \\ \vdots & \vdots \\ \gamma_m\widehat{x'_m}K_mR_m & \gamma_m\widehat{x'_m}K_mT_m \end{bmatrix}. \tag{8.61}$$

To achieve the upgrade, we need to find the transformation matrix

$$H^{-1} \doteq \begin{bmatrix} K_1^{-1} & 0 \\ v^T & v_4 \end{bmatrix} = \begin{bmatrix} K_1^{-1} & 0 \\ 0 & 1 \end{bmatrix}\begin{bmatrix} I & 0 \\ v^T & v_4 \end{bmatrix} \quad \in \mathbb{R}^{4\times 4} \tag{8.62}$$

that relates the two matrices in the following way:

$$\begin{bmatrix} K_1 & 0 \\ \gamma_2K_2R_2 & \gamma_2K_2T_2 \\ \vdots & \vdots \\ \gamma_mK_mR_m & \gamma_mK_mT_m \end{bmatrix} H^{-1} = \begin{bmatrix} I & 0 \\ \gamma_2(K_2R_2K_1^{-1} + K_2T_2v^T) & \gamma_2v_4K_2T_2 \\ \vdots & \vdots \\ \gamma_m(K_mR_mK_1^{-1} + K_mT_mv^T) & \gamma_mv_4K_mT_m \end{bmatrix}.$$

In order to establish some additional constraints on $H^{-1}$, let us first have a look at the rows associated with the projection matrices

$$\begin{aligned} \Pi_{ie} &= [\gamma_iK_iR_i,\ \gamma_iK_iT_i], \\ \Pi_{ip} &= [\gamma_i(K_iR_iK_1^{-1} + K_iT_iv^T),\ \gamma_iv_4K_iT_i], \end{aligned}$$

for $i = 2, 3, \ldots, m$. Evidently, each pair of matrices are related by

$$\Pi_{ie}H^{-1} = \Pi_{ip} \quad \Leftrightarrow \quad \Pi_{ie} = \Pi_{ip}H, \quad i = 2, 3, \ldots, m. \tag{8.63}$$

Now, as in Chapter 6, define a matrix $Q$ to be

$$Q \doteq H \begin{bmatrix} I_{3\times 3} & 0 \\ 0^T & 0 \end{bmatrix} H^T \quad \in \mathbb{R}^{4\times 4}. \tag{8.64}$$

In general we know only that $Q$ is a rank-3 positive semi-definite symmetric matrix. Since $\Pi_{ie} = \Pi_{ip}H$ and $K_iR_i(K_iR_i)^T = K_iK_i^T = S_i^{-1}$, we directly have the following relationship:

$$\boxed{\Pi_{ip}Q(\Pi_{ip})^T = \gamma_i^2S_i^{-1}, \quad i = 2, 3, \ldots, m.} \tag{8.65}$$

These constraints are the *absolute quadric constraints* that we have seen in Chapter 6. Note that the $\Pi_{ip}$'s can be obtained directly from the factorization, but $Q \in \mathbb{R}^{4\times 4}$ and $S_i^{-1} \in \mathbb{R}^{3\times 3}$ are unknown matrices, and $\gamma_i \in \mathbb{R}$ are unknown scalars. As we have anticipated in Chapter 6 (e.g., Exercise 6.16), even in the case in which the camera calibration is constant $S_i^{-1} = S^{-1} = KK^T$, we cannot always expect to find a unique solution from such constraints unless the camera motions are rich enough. We call a sequence of motions $\{(R_i, T_i)\}$ *critical* if it leads to multiple solutions in the camera calibration $S^{-1}$ from its associated absolute quadric constraints.

**Theorem 8.22 (Critical motions for the absolute quadric constraints).** *A motion sequence or set is critical for camera calibration if and only if it belongs to one of the following four cases:*[11]

1. *A motion sequence whose relative rotations between views are generated by arbitrary rotations around a fixed axis (Figure 8.16 a). Formally, this is a motion subgroup $SO(2) \times \mathbb{R}^3$.*
2. *An orbital motion sequence (Figure 8.16 b), i.e. the circular group $\mathbb{S}^1$.*
3. *Eight camera poses that respect the symmetry of some (3-D) rectangular parallelepiped (Figure 8.16 c).*
4. *Four camera positions that respect the symmetry of some (2-D) rectangle and at each position the camera may rotate around the line connecting the centers of the camera and the rectangle (Figure 8.16 d).*

*These cases are illustrated in Figure 8.16.*

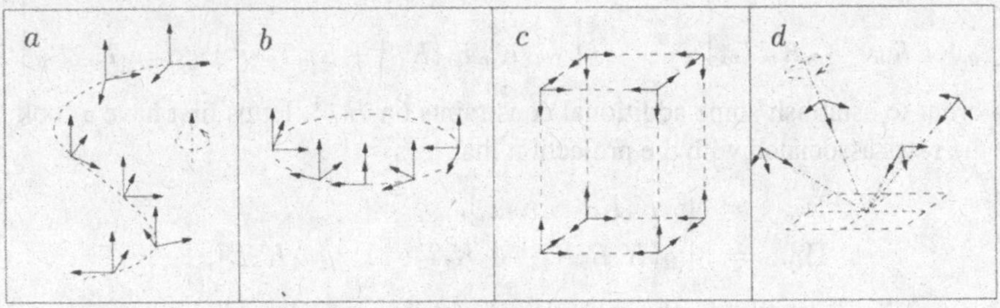


Figure 8.16. Four cases of critical motion sequences.

*Proof.* A complete proof consists of a careful classification of all possible relationships between the quadric given by $Q$ and the conic by $S^{-1}$. The interested reader can see [Sturm, 1997] for the details. □

Note that solving the absolute quadric constraints essentially provides a solution to the problem of camera calibration, anticipated in Chapter 6. Additional partial knowledge of the camera calibration, motion, and structure may give rise to alternative techniques for computing the camera calibration and the Euclidean structure. Discussions of such practical issues will be given in Chapters 10 and 11.

## 8.6 Summary

The intrinsic constraints among multiple images of a point or a line can be expressed in terms of rank conditions on the matrix $N$, $W$, or $M$. The relationship

[11] Here, for simplicity, we ignore that some trivial rotations by 180° can be added to each case.

among these rank conditions is summarized in Table 8.2. These rank conditions capture the relationships among corresponding geometric primitives in multiple images. They give rise to natural factorization-based algorithms for multiview recovery of 3-D structure and camera motion.

| Rank conditions | (Pre)image | Coimage | Jointly |
|---|---|---|---|
| Point | $\text{rank}(N_p) \leq m+3$ | $\text{rank}(W_p) \leq 3$ | $\text{rank}(M_p) \leq 1$ |
| Line | $\text{rank}(N_l) \leq 2m+2$ | $\text{rank}(W_l) \leq 2$ | $\text{rank}(M_l) \leq 1$ |

Table 8.2. Equivalent rank conditions on matrices in image, in coimage, and in image and coimage jointly.

Given $m$ images $\boldsymbol{x}_1, \boldsymbol{x}_2, \ldots, \boldsymbol{x}_m \in \mathbb{R}^3$ of a point $p$ and $m$ coimages $\boldsymbol{\ell}_1, \boldsymbol{\ell}_2, \ldots, \boldsymbol{\ell}_m \in \mathbb{R}^3$ of a line $L$ with respect to $m$ camera frames specified by $\Pi_1 = [I, 0]$, $\Pi_2 = [R_2, T_2]$, $\ldots$, $\Pi_m = [R_m, T_m] \in \mathbb{R}^{3\times 4}$, we summarize in Table 8.3 the results of this chapter on the comparison between points and lines.

## 8.7 Exercises

**Exercise 8.1 (Image and coimage of points and lines).** Suppose $p^1, p^2$ are two points on the line $L$, and $L^1, L^2$ are two lines intersecting at the point $p$. Let $\boldsymbol{x}, \boldsymbol{x}^1, \boldsymbol{x}^2$ be the images of the points $p, p^1, p^2$, respectively, and let $\boldsymbol{\ell}, \boldsymbol{\ell}^1, \boldsymbol{\ell}^2$ be the coimages of the lines $L, L^1, L^2$, respectively.

1. Show that

$$\boldsymbol{\ell} \sim \widehat{\boldsymbol{x}^1}\boldsymbol{x}^2, \quad \boldsymbol{x} \sim \widehat{\boldsymbol{\ell}^1}\boldsymbol{\ell}^2.$$

2. Show that for some $r, s, u, v \in \mathbb{R}^3$,

$$\boldsymbol{\ell}^1 = \widehat{\boldsymbol{x}}u, \quad \boldsymbol{\ell}^2 = \widehat{\boldsymbol{x}}v, \quad \boldsymbol{x}^1 = \widehat{\boldsymbol{\ell}}r, \quad \boldsymbol{x}^2 = \widehat{\boldsymbol{\ell}}s.$$

3. Draw a picture and convince yourself of the above relationships.

**Exercise 8.2 (Uncalibrated coimage and image).** We know that an uncalibrated image $\boldsymbol{x}'$ of a point $p$ is related to its calibrated counterpart by $\boldsymbol{x}' = K\boldsymbol{x}$.

1. Derive the relationship between the uncalibrated coimage of $p$ and its calibrated coimage.

2. Derive the relationship between an uncalibrated coimage of a line $L$ and its calibrated counterpart.

3. Derive the relationship between an uncalibrated (pre)image of a line $L$ and its calibrated counterpart.

Notice the difference in the effect of the calibration matrix $K$ on a (pre)image and on a coimage.

| Point feature | Line feature |
|---|---|
| $M_p = \begin{bmatrix} \widehat{\boldsymbol{x}_2} R_2 \boldsymbol{x}_1 & \widehat{\boldsymbol{x}_2} T_2 \\ \widehat{\boldsymbol{x}_3} R_3 \boldsymbol{x}_1 & \widehat{\boldsymbol{x}_3} T_3 \\ \vdots & \vdots \\ \widehat{\boldsymbol{x}_m} R_m \boldsymbol{x}_1 & \widehat{\boldsymbol{x}_m} T_m \end{bmatrix}$ | $M_l = \begin{bmatrix} \boldsymbol{\ell}_2^T R_2 \widehat{\boldsymbol{\ell}_1} & \boldsymbol{\ell}_2^T T_2 \\ \boldsymbol{\ell}_3^T R_3 \widehat{\boldsymbol{\ell}_1} & \boldsymbol{\ell}_3^T T_3 \\ \vdots & \vdots \\ \boldsymbol{\ell}_m^T R_m \widehat{\boldsymbol{\ell}_1} & \boldsymbol{\ell}_m^T T_m \end{bmatrix}$ |
| $\text{rank}(M_p) = 1$ | $\text{rank}(M_l) = 1$ |
| 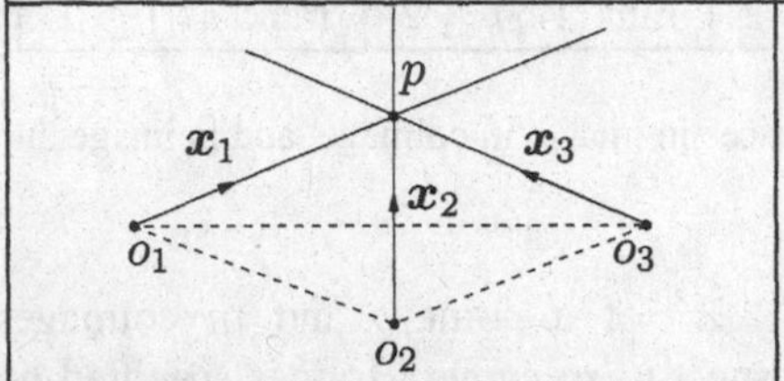  | 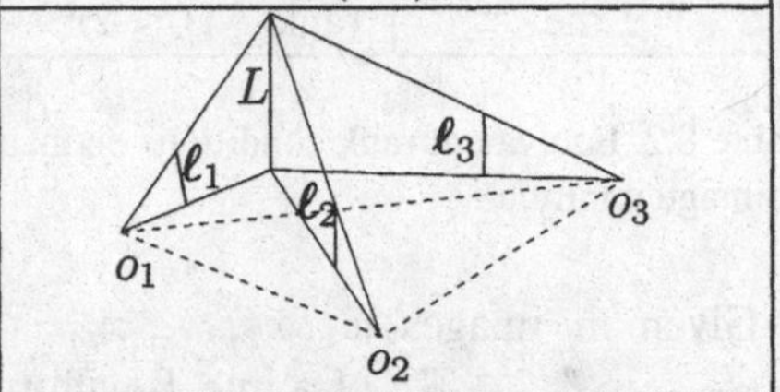  |
| $\text{rank}(M_p) = 0$<br>all images are colinear<br>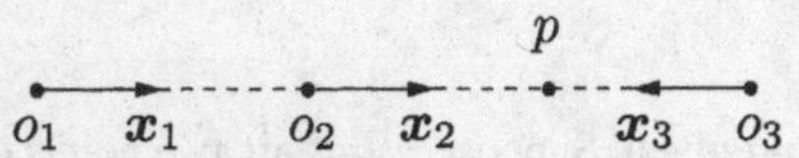  | $\text{rank}(M_l) = 0$<br>**all images are coplanar**<br>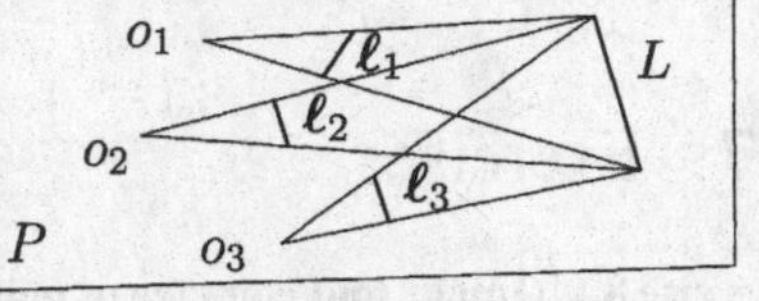  |
| Constraints from $\text{rank}(M_p) \leq 1$ | Constraints from $\text{rank}(M_l) \leq 1$ |
| $\boldsymbol{x}_i^T \widehat{T}_i R_i \boldsymbol{x}_1 = 0$ | $\boldsymbol{\ell}_i^T R_i \widehat{\boldsymbol{\ell}}_1 R_j^T \boldsymbol{\ell}_j = 0$ |
| $\widehat{\boldsymbol{x}}_i (T_i \boldsymbol{x}_1^T R_j^T - R_i \boldsymbol{x}_1 T_j^T) \widehat{\boldsymbol{x}_j} = 0$ | $\boldsymbol{\ell}_j^T (T_j \boldsymbol{\ell}_i^T R_i - R_j \boldsymbol{\ell}_i^T T_i) \widehat{\boldsymbol{\ell}}_1 = 0$ |
| 3-D information encoded in $M_p$ | 3-D information encoded in $M_l$ |
| 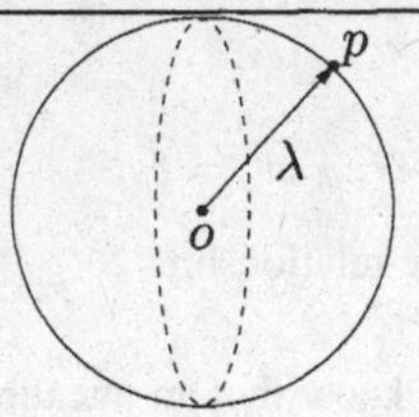  | 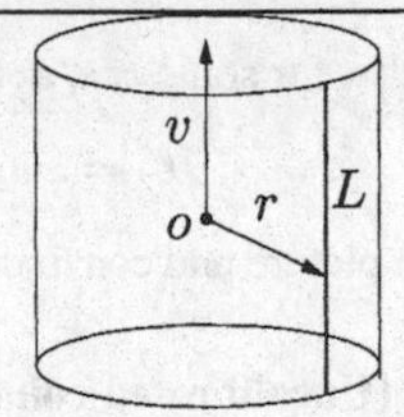  |

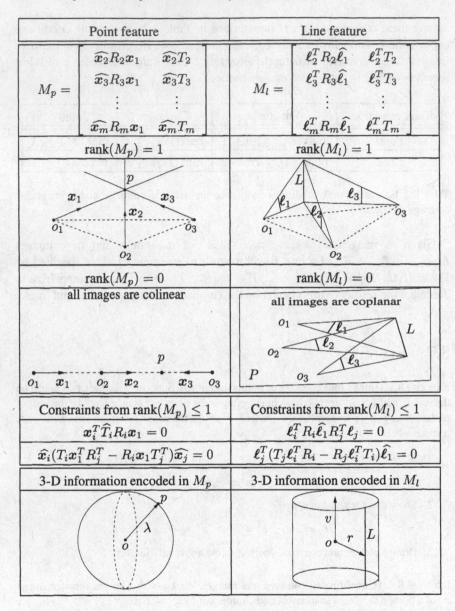

Table 8.3. Comparison between multiple images of a point and a line.

**Exercise 8.3 (A plane in $\mathbb{R}^3$).** Other than the homogeneous representation given in this chapter, describe a plane in $\mathbb{R}^3$ in terms of a base point (on it) and vectors that indicate its span. Find the relationship between such a representation and the homogeneous one for the same plane.

**Exercise 8.4 (Relationship between the ranks of $W_l$ and $N_l$).** Prove Lemma 8.4, relating the rank of the matrix $W_l$ to that of the matrix $N_l$ by

$$\text{rank}(N_l) = \text{rank}(W_l) + 2m.$$

Characterize the null space of the matrix $N_l$ given $\text{rank}(W_l) = 2$.

**Exercise 8.5 (Two-view geometry from $W_p$).** Given two images $\boldsymbol{x}_1, \boldsymbol{x}_2$ with respect to two camera frames specified by $\Pi_1, \Pi_2$, then:

1. Derive the epipolar constraint from the rank condition

$$\text{rank}(W_p) = \text{rank}\begin{bmatrix} \widehat{\boldsymbol{x}_1}\Pi_1 \\ \widehat{\boldsymbol{x}_2}\Pi_2 \end{bmatrix} \leq 3.$$

2. Assuming that the translation between the first and second camera frame is nonzero, show that

$$\text{rank}(W_p) = 3$$

except for a special case. What is this special case? This property means that the (homogeneous) coordinates of the 3-D point are uniquely determined by the null space of $W_p$ when translation is present. (Hint: to simplify the proof, one can consider the first camera frame to be the reference frame.)

This can be considered as a triangulation procedure for lines.

**Exercise 8.6 (Three-view geometry from $N_p$).** Given three images $\boldsymbol{x}_1, \boldsymbol{x}_2, \boldsymbol{x}_3$ with respect to three camera frames specified by $\Pi_1, \Pi_2, \Pi_3$ (with the first camera frame being the reference), show that

$$\text{rank}(N_p) \leq 6 \quad \Rightarrow \quad \text{rank}(M_p) \leq 1.$$

**Exercise 8.7 (Multilinear function).** A function $f(\cdot,\cdot) : \mathbb{R}^n \times \mathbb{R}^n \rightarrow \mathbb{R}$ is called *bilinear* if $f(x, y)$ is linear in $x \in \mathbb{R}^n$ if $y \in \mathbb{R}^n$ is fixed, and vice versa. That is,

$$\begin{aligned} f(\alpha x_1 + \beta x_2, y) &= \alpha f(x_1, y) + \beta f(x_2, y), \\ f(x, \alpha y_1 + \beta y_2) &= \alpha f(x, y_1) + \beta f(x, y_2), \end{aligned}$$

for any $x, x_1, x_2, y, y_1, y_2 \in \mathbb{R}^n$ and $\alpha, \beta \in \mathbb{R}$. Show that any bilinear function $f(x, y)$ can be written as

$$f(x, y) = x^T M y$$

for some matrix $M \in \mathbb{R}^{n \times n}$. Notice that in the epipolar constraint equation

$$\boldsymbol{x}_2^T F \boldsymbol{x}_1 = 0,$$

the left-hand side is a bilinear form in $\boldsymbol{x}_1, \boldsymbol{x}_2 \in \mathbb{R}^3$. Hence, the epipolar constraint is also referred to as the *bilinear constraint*. Similarly, we can define the *trilinear* or *quadrilinear* functions. In the most general case, we can define a multilinear function $f(x_1, \ldots, x_i, \ldots, x_m)$ that is linear in each $x_i \in \mathbb{R}^n$ if all other $x_j$'s with $j \neq i$ are fixed. Such a function is called *m-linear*. Another name for such a multilinear object is *tensor*. A two-tensor (i.e. a bilinear function) can be represented by a matrix. How can a higher-order tensor be represented?

**Exercise 8.8 (Minors of a matrix).** Suppose $M$ is an $m \times n$ matrix with $m \geq n$. Then $M$ is a "rectangular" matrix with $m$ rows and $n$ columns. We can pick $n$ (say $i_1, i_2, \ldots, i_n$) arbitrary distinct rows of $M$ and form an $n \times n$ submatrix of $M$. If we denote such an $n \times n$ submatrix by $M_{i_1,i_2,\ldots,i_n}$, its determinant $\det(M_{i_1,i_2,\ldots,i_n})$ is called a *minor* of $M$. Prove that the $n$ column vectors are linearly dependent if and only if all the minors of $M$ are identically zero, i.e.

$$\det(M_{i_1,i_2,\ldots,i_n}) = 0, \quad \forall i_1, i_2, \ldots, i_n \in \{1, 2, \ldots, m\}.$$

How many different minors can you possibly obtain?

**Exercise 8.9 (Linear dependency of two vectors).** Given any four *nonzero* vectors $a_1, a_2, \ldots, a_n, b_1, b_2, \ldots, b_n \in \mathbb{R}^3$, the matrix

$$\begin{bmatrix} a_1 & b_1 \\ a_2 & b_2 \\ \vdots & \vdots \\ a_n & b_n \end{bmatrix} \in \mathbb{R}^{3n \times 2} \tag{8.66}$$

is rank-deficient if and only if $a_i b_j^T - b_i a_j^T = 0$ for all $i, j = 1, 2, \ldots, n$. Explain what happens if some of the vectors are zero.

**Exercise 8.10 (Rectilinear motion and points).** Use the rank condition of the $M_p$ matrix to show that if the camera is calibrated and only translating on a straight line, then the relative translational scale between frames cannot be recovered from bilinear constraints, but it can be recovered from trilinear constraints. (Hint: use $R_i = I$ to simplify the constraints.)

**Exercise 8.11 (Pure rotational motion and points).** The rank condition on the matrix $M_p$ is so fundamental that it works for the pure rotational motion case too. Show that if there is no translation, the rank condition on the matrix $M_p$ is equivalent to the constraint that we may get in the case of pure rotational motion,

$$\widehat{\boldsymbol{x}_i} R_i \boldsymbol{x}_1 = 0. \tag{8.67}$$

(Hint: you need to convince yourself that in this case the rank of $N_p$ is no more than $m+2$.)

**Exercise 8.12 (Points on the plane at infinity).** Show that the multiple-view matrix $M_p$ associated with $m$ images of a point $p$ on the plane at infinity satisfies the rank condition $\text{rank}(M_p) \leq 1$. (Hint: the homogeneous coordinates for a point $p$ at infinity are of the form $\boldsymbol{X} = [X, Y, Z, 0]^T \in \mathbb{R}^4$.)

**Exercise 8.13 (Degenerate configurations).** Show that the only solution corresponding to the equation

$$\text{rank}(M_p) = 0$$

is that all the camera centers $o_1, o_2, \ldots, o_m$ lie on a line and the point $p$ can be anywhere on this line.

**Exercise 8.14 (Triangulation for line features).** Given two (co)images of a line $\boldsymbol{\ell}_1, \boldsymbol{\ell}_2 \in \mathbb{R}^3$ and the relative motion $(R, T)$ of the camera between the two vantage points, what is the 3-D location of the line with respect to each camera frame? Express the direction of the line and its distance to the center of the camera in terms of $\boldsymbol{\ell}_1, \boldsymbol{\ell}_2$, and $(R, T)$. Under what conditions are such a distance and direction not uniquely determined?

**Exercise 8.15 (Rectilinear motion and lines).** Suppose the camera (center) is moving on a straight line $L_o$. What is the multiple-view matrix $M$ associated with images of any line $L$ that is coplanar to $L_o$? Describe the set of all such lines $\{L\}$. What information about the camera motion $(R, T)$ are we able to get from images of such lines? Explain why.

**Exercise 8.16 (Purely rotational motion and lines).** Derive the constraints among multiple images of a line taken by a camera that rotates around its center. How many lines (in general position) are needed to determine linearly the rotation from two views? Explain.

**Exercise 8.17 (Numerical sensitivity of the rank condition).** Although the rank condition is a well-defined algebraic criterion, one must be cautious about using it in numerical algorithms. A matrix rank is invariant under arbitrary scaling of its rows or columns, but scaling does affect the sensitivity of the matrix to perturbation. Consider the two rank-1 matrices

$$M_1 = \begin{bmatrix} 0 & \epsilon \\ 0 & 1 \end{bmatrix}, \quad M_2 = \begin{bmatrix} 0 & 1 \\ 0 & 1 \end{bmatrix},$$

where $\epsilon$ is a small positive number. Perturb both by adding a vector $[\epsilon, 0]$ to their first rows. Compute the angles between the two row vectors of each matrix before and after the perturbation. Explain why for degenerate configurations (i.e. $M = 0$), the rank condition might be rather sensitive to noise.

**Exercise 8.18 (Invariant property of bilinear and trilinear relationships).** Suppose two sets of projection matrices

$$\Pi = \begin{bmatrix} I & 0 \\ R_2 & T_2 \\ R_3 & T_3 \end{bmatrix}, \quad \Pi' = \begin{bmatrix} I & 0 \\ R_2' & T_2' \\ R_3' & T_3' \end{bmatrix} \quad \in \mathbb{R}^{9\times 4}$$

are related by a common transformation $H_p$ of the form

$$H_p = \begin{bmatrix} I & 0 \\ v^T & v_4 \end{bmatrix} \quad \in \mathbb{R}^{4\times 4}.$$

That is, $[R_i, T_i]H_p \sim [R_i', T_i']$ are equal up to scale. Show that

1. $\Pi$ and $\Pi'$ give the same fundamental matrices (up to a scale factor).

2. $\Pi$ and $\Pi'$ give the same trifocal tensor (up to a scale factor).

(Hint: One can first show that the above claims are true when the equality $[R_i, T_i]H_p = [R_i', T_i']$ is exact. Then argue that the same two or three-view relationships hold even if one scales each projection matrix $\Pi_i'$ arbitrarily.)

**Exercise 8.19 (Change of depth).** Let two images $\boldsymbol{x}_1, \boldsymbol{x}_2$ of a point with respect to two camera frames be given. The projection matrix is $[R, T]$. Hence $\lambda_1 \widehat{\boldsymbol{x}_2} R \boldsymbol{x}_1 + \widehat{\boldsymbol{x}_2} T = 0$ for a depth scale $\lambda_1 \in \mathbb{R}$. Now suppose we have recovered the fundamental matrix $F = \widehat{T}R$. Since we do not know the calibration of the camera, we may recover the projection matrix only up to the decomposition

$$[R', T'] \doteq [R + Tv^T, v_4 T], \quad v = [v_1, v_2, v_3]^T \in \mathbb{R}^3, v_4 \in \mathbb{R}.$$

Find the corresponding "depth" $\lambda_1'$ for the image pair $(\boldsymbol{x}_1, \boldsymbol{x}_2)$ with respect to the new projection matrix.

**Exercise 8.20 (Invariant property of the rank condition).** We show that the rank conditions developed in this chapter are invariant under an arbitrary transformation $H_p$ of the form

$$H_p = \begin{bmatrix} I & 0 \\ v^T & v_4 \end{bmatrix} \in \mathbb{R}^{4\times 4}.$$

1. First derive the rank condition $\text{rank}(M_p) \leq 1$ from $\text{rank}(W_p) \leq 3$.
2. Suppose

$$\begin{bmatrix} R_2 & T_2 \\ R_3 & T_3 \\ \vdots & \vdots \\ R_m & T_m \end{bmatrix} H_p = \begin{bmatrix} R_2' & T_2' \\ R_3' & T_3' \\ \vdots & \vdots \\ R_m' & T_m' \end{bmatrix} \tag{8.68}$$

and let

$$M_p \doteq \begin{bmatrix} \widehat{x_2}R_2x_1 & \widehat{x_2}T_2 \\ \widehat{x_3}R_3x_1 & \widehat{x_3}T_3 \\ \vdots & \vdots \\ \widehat{x_m}R_mx_1 & \widehat{x_m}T_m \end{bmatrix}, \quad M_p' \doteq \begin{bmatrix} \widehat{x_2}R_2'x_1 & \widehat{x_2}T_2' \\ \widehat{x_3}R_3'x_1 & \widehat{x_3}T_3' \\ \vdots & \vdots \\ \widehat{x_m}R_m'x_1 & \widehat{x_m}T_m' \end{bmatrix}.$$

Show that

$$\text{rank}(M_p) \leq 1 \quad \Leftrightarrow \quad \text{rank}(M_p') \leq 1.$$

3. Based on the results from the previous exercise, explain how the null spaces of these two multiple-view matrices are related.

**Exercise 8.21 (Fundamental matrices among three views).** Denote by $[R_{ij}, T_{ij}]$ the relative coordinate transformation from the $j$th view to the $i$th such that

$$X_i = R_{ij}X_j + T_{ij}, \quad X_i, X_j \in \mathbb{R}^3.$$

Then the associated fundamental matrix is denoted by $F_{ij} = \widehat{T_{ij}}R_{ij}$. Now consider the three fundamental matrices $F_{21}, F_{31}$, and $F_{32}$.

1. Denote the image of the $j$th camera center $o_j$ in the $i$th view by $e_{ij} \in \mathbb{R}^3$. Show that

$$e_{ij} \sim T_{ij}, \quad e_{ji} \sim R_{ij}^T e_{ij}$$

where equality is in the homogeneous sense. This is illustrated in Figure 8.17.

2. Find the relationship between $F_{ij}$ and $F_{ji}$ and conclude that there are essentially three "independent" fundamental matrices as listed above. Verify that it is trivial that

$$e_{ij}^T F_{ij} = 0, \quad F_{ij}e_{ji} = 0.$$

3. Further, verify the nontrivial relation

$$e_{ik}^T F_{ij} e_{jk} = 0.$$

Conclude that the three fundamental matrices listed above are further related via the three equations

$$e_{23}^T F_{21} e_{13} = 0, \quad e_{32}^T F_{31} e_{12} = 0, \quad e_{31}^T F_{32} e_{21} = 0.$$

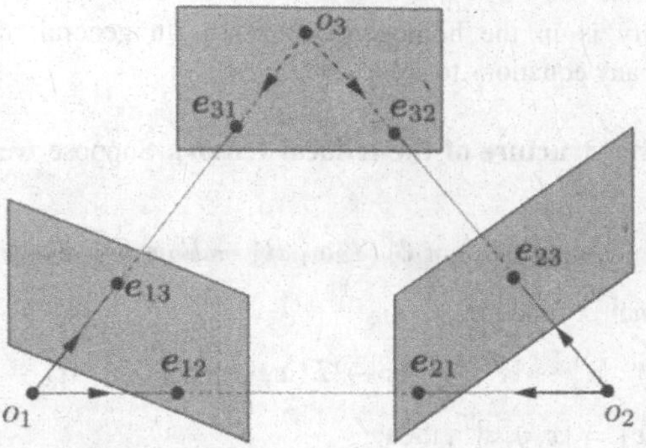


Figure 8.17. Images of camera centers (i.e. epipoles) among three views: the image of the $j$th camera center in the $i$th view is denoted by $e_{ij}$. There are therefore six in all.

Since $e_{jk}$ and $e_{ik}$ are essentially the (left) null space of $F_{jk}$ and $F_{ik}$, the above relation in fact imposes three algebraically independent constraints among the three matrices $F_{ij}, F_{jk}$, and $F_{ik}$.

**Exercise 8.22 (Three-view geometry from fundamental matrices between pairs).** Suppose we are given the three fundamental matrices $F_{21}, F_{31}$, and $F_{32}$ between three *uncalibrated*[12] views, and we want to recover a "consistent" set of projection matrices $\Pi_1, \Pi_2, \Pi_3$ such that they give back the same fundamental matrices. Without loss of generality we may assume $\Pi_1 = [I, 0]$, and $\Pi_2, \Pi_3$ need to be determined. The following steps provide guidelines on how to achieve this goal:

1. Compute the epipoles $T_{21}, T_{31}$ as the (normalized) left zero eigenvectors of $F_{21}, F_{31}$, respectively. However, verify (analytically or by simulation) that in general the straightforward canonical decomposition
$$\Pi_2 = [(\widehat{T_{21}})^T F_{21}, T_{21}], \quad \Pi_3 = [(\widehat{T_{31}})^T F_{31}, T_{31}]$$
*does not* necessarily give the same fundamental matrix $F_{32}$ between the second and third views.

2. Hence, instead of choosing the canonical decomposition, we need to modify at least one of the projection matrices, say
$$\Pi_3 \doteq [(\widehat{T_{31}})^T F_{31} + T_{31}v^T, T_{31}],$$
where $v$ needs to be determined such that $\Pi_3$ together with $\Pi_2$ gives the correct fundamental matrix $F_{32}$. Denote the pseudo-inverse of $\Pi_2$ by $\Pi_2^\dagger \in \mathbb{R}^{4\times 3}$ such that $\Pi_2\Pi_2^\dagger = I_{3\times 3}$. Verify that
$$\Pi_2[\Pi_2^\dagger, u] = [I, 0]$$
for some $u \in \mathbb{R}^4$. What is $u$? Find its expression in terms of the given data.

3. Now let $[R(v), T(v)] \doteq [\Pi_3\Pi_2^\dagger, \Pi_3 u]$. Hence $v$ has to be chosen in such a way that
$$\widehat{T(v)}R(v) \sim F_{32},$$

[12] The calibrated case follows easily due to the relatively simple structure of the essential matrix.

where equality is in the homogeneous sense. In general, this equality gives sufficiently many equations to solve for $v$.

**Exercise 8.23 (Basic structure of the trifocal tensor).** Suppose we know the trifocal tensor $\mathcal{T}$

$$\mathcal{T}(\boldsymbol{x}_1, \boldsymbol{\ell}_2, \boldsymbol{\ell}_3) = \boldsymbol{\ell}_2^T (T_2 \boldsymbol{x}_1^T R_3^T - R_2 \boldsymbol{x}_1 T_3^T) \boldsymbol{\ell}_3.$$

1. We define a matrix function

$$G(\boldsymbol{x}) : \mathbb{R}^3 \to \mathbb{R}^{3\times 3}; \quad \boldsymbol{x} \mapsto \mathcal{T}(\boldsymbol{x}, \cdot, \cdot) = (T_2 \boldsymbol{x}^T R_3^T - R_2 \boldsymbol{x} T_3^T).$$

   Show that if $\boldsymbol{x}_1 = [x, y, z]^T$, then

$$\mathcal{T}(\boldsymbol{x}_1, \boldsymbol{\ell}_2, \boldsymbol{\ell}_3) = \boldsymbol{\ell}_2^T G(\boldsymbol{x}_1) \boldsymbol{\ell}_3 = \boldsymbol{\ell}_2^T [x G(e_1) + y G(e_2) + z G(e_3)] \boldsymbol{\ell}_3.$$

2. Similarly, write down the corresponding matrix forms for

$$H(\boldsymbol{\ell}) = \mathcal{T}(\cdot, \boldsymbol{\ell}, \cdot) \quad \text{and} \quad H'(\boldsymbol{\ell}) = \mathcal{T}(\cdot, \cdot, \boldsymbol{\ell}).$$

**Exercise 8.24 (Linear estimation of the trifocal tensor).** In this exercise we show how to compute the trifocal tensor $\mathcal{T}$:

1. Show that from equation (8.37), three images $\boldsymbol{x}_1, \boldsymbol{x}_2, \boldsymbol{x}_3$ of a point $p$ in space give rise to four linearly independent constraints of the type (8.55).
2. Show that from equation (8.52), three coimages $\boldsymbol{\ell}_1, \boldsymbol{\ell}_2, \boldsymbol{\ell}_3$ of a line $L$ in space give rise to two linearly independent constraints of the type (8.55).
3. Conclude that in general, one needs $n$ points and $m$ lines across three views with $4n + 2m \geq 26$ to recover the trifocal tensor $\mathcal{T}$ linearly up to a scale factor.

**Exercise 8.25 (Fundamental matrices from a trifocal tensor).** According to its definition, the trifocal tensor $\mathcal{T}$ is simply a $3 \times 3 \times 3$ array of numbers whose "contraction" (multiplication by a vector) with a triplet of corresponding point and lines $(\boldsymbol{x}_1, \boldsymbol{\ell}_2, \boldsymbol{\ell}_3)$, as shown in Figure 8.15, is equal to

$$\mathcal{T}(\boldsymbol{x}_1, \boldsymbol{\ell}_2, \boldsymbol{\ell}_3) = \boldsymbol{\ell}_2^T (T_2 \boldsymbol{x}_1^T R_3^T - R_2 \boldsymbol{x}_1 T_3^T) \boldsymbol{\ell}_3 = 0.$$

Follow the steps to show that the fundamental matrix $F_{21} = \widehat{T_2} R_2$ and $F_{31} = \widehat{T_3} R_3$ can be recovered from decomposing the 27 entries of $\mathcal{T}(\cdot, \cdot, \cdot)$:

1. If we contract $\mathcal{T}$ with only a vector $\boldsymbol{x}_1$, we get a $3 \times 3$ matrix

$$G(\boldsymbol{x}_1) \doteq \mathcal{T}(\boldsymbol{x}_1, \cdot, \cdot) = T_2 \boldsymbol{x}_1^T R_3^T - R_2 \boldsymbol{x}_1 T_3^T \quad \in \mathbb{R}^{3\times 3}.$$

   Show that:

   (a) The left null space of $G(\boldsymbol{x}_1)$ is $k_l(\boldsymbol{x}_1) \sim \widehat{T_2} R_2 \boldsymbol{x}_1 \in \mathbb{R}^3$.

   (b) The right null space of $G(\boldsymbol{x}_1)$ is $k_r(\boldsymbol{x}_1) \sim \widehat{T_3} R_3 \boldsymbol{x}_1 \in \mathbb{R}^3$.

2. Note that $k_l(\boldsymbol{x}_1)$ is orthogonal to $T_2$. Recall that $e_1, e_2, e_3 \in \mathbb{R}^3$ stand for the standard basis vectors as columns of the identity matrix $I$. Show that $T_2' \doteq k_l(e_1) \times k_l(e_2) \sim T_2$. Similarly, show that $T_3' \doteq k_r(e_1) \times k_r(e_2) \sim T_3$. Note that the recovered $T_2', T_3'$ may not be of the same scale as $T_2, T_3$ originally used in the definition of $\mathcal{T}$. Conclude that the epipoles $T_2$ and $T_3$ can be determined from the trifocal tensor.

3. Now show that

$$\widehat{T_2'}G(\boldsymbol{x}_1) = -\widehat{T_2'}R_2\boldsymbol{x}_1T_3^T; \quad G(\boldsymbol{x}_1)\widehat{T_3'}^T = T_2\boldsymbol{x}_1^TR_3^T\widehat{T_3'}^T.$$

4. We then can finally conclude that $F_{21} = \widehat{T_2}R_2$ and $F_{31} = \widehat{T_3}R_3$ can be determined (up to a scale factor) as

$$\begin{aligned} F_{12} &\sim \widehat{T_2'}[G(e_1)T_3', G(e_2)T_3', G(e_3)T_3'], \\ F_{13} &\sim \widehat{T_3'}[G(e_1)^TT_2', G(e_2)^TT_2', G(e_3)^TT_2'], \end{aligned}$$

where $T_2', T_3'$ are computed from the second step.

In fact, $\mathcal{T}$ also implies the third fundamental matrix $F_{32}$, but it is not easy to express it explicitly. To prove this is true, we will, however, explore an alternative approach in the next exercise.

**Exercise 8.26 (Three-view geometry from trifocal tensors).** Consider three *uncalibrated views*. Under the same conditions as Exercise 8.25 above, suppose we have obtained the trifocal tensor $\mathcal{T}$ as

$$\mathcal{T}(\boldsymbol{x}_1, \boldsymbol{\ell}_2, \boldsymbol{\ell}_3) = \boldsymbol{\ell}_2^T(T_2\boldsymbol{x}_1^TR_3^T - R_2\boldsymbol{x}_1T_3^T)\boldsymbol{\ell}_3.$$

Our goal here is to recover a set of projection matrices $\Pi_1, \Pi_2, \Pi_3$ that would give back the same trifocal tensor. Follow the steps below to achieve this goal. First, suppose we have followed the steps in Exercise 8.25 and computed the epipoles $T_2' = \alpha_2T_2, T_3' = \alpha_3T_3$ from $\mathcal{T}$.[13]

1. Show that for the matrix $G(\boldsymbol{x}_1) = (T_2\boldsymbol{x}_1^TR_3^T - R_2\boldsymbol{x}_1T_3^T)$ defined in Exercise 8.25, we have

$$\begin{aligned} R_2' &\doteq -[G(e_1)T_3', G(e_2)T_3', G(e_3)T_3'] = R_2T_3^TT_3' - T_2(R_3^TT_3')^T, \\ R_3' &\doteq [G(e_1)^TT_2', G(e_2)^TT_2', G(e_3)^TT_2'] = R_3T_2^TT_2' - T_3(R_2^TT_2')^T. \end{aligned}$$

2. In general, we choose both $\|T_2'\| = \|T_3'\| = 1$. Suppose we define a new trifocal tensor using $[R_2', T_2']$ and $[R_3', T_3']$ computed above:

$$\mathcal{T}'(\boldsymbol{x}_1, \boldsymbol{\ell}_2, \boldsymbol{\ell}_3) = \boldsymbol{\ell}_2^T(T_2'\boldsymbol{x}_1^TR_3'^T - R_2'\boldsymbol{x}_1T_3'^T)\boldsymbol{\ell}_3.$$

Then what is the expression for $G'(\boldsymbol{x}_1) = (T_2'\boldsymbol{x}_1^TR_3'^T - R_2'\boldsymbol{x}_1T_3'^T)$ in terms of $R_i, T_i$ and the $\alpha_i$'s? Is $G'(\boldsymbol{x}_1)$ the same (up to a scale factor) as $G(\boldsymbol{x}_1)$ in general?

3. Verify that in general, we *do not* have that for some

$$H_p = \begin{bmatrix} I & 0 \\ v^T & v_4 \end{bmatrix} \in \mathbb{R}^{4\times 4},$$

the equality

$$[R_i \quad T_i]\,H_p \sim [R_i' \quad T_i'], \quad \text{for } i = 2, 3,$$

holds up to a scalar factor. (A numerical counterexample will do.)

[13] Here we use $\alpha_2, \alpha_3$ to explicitly denote the possible scale difference between the recovered $T_i'$'s and the original $T_i$'s, although they are equal in the homogeneous sense.

4. Now define $R_3'' \doteq -\widehat{T_3'}^2 R_3'$. Show that there is a transformation $H_p$ as above such that

$$[R_2 \quad T_2]\, H_p \sim [R_2' \quad T_2'], \quad [R_3 \quad T_3]\, H_p \sim [R_3'' \quad T_3'],$$

with equality up to a scale factor. (Hint: Notice that $I - T_3' T_3'^T = -\widehat{T_3'}^2$ when $\|T_3'\| = 1$, and consider in the matrix $H_p$ to choose $v = -\alpha_3 R_3^T T_3'$. You need to figure out what $v_4$ needs to be.)

5. Now verify that the resulting projection matrices do give the same trifocal tensor $\mathcal{T}$ (up to a scale factor). (Hint: No computation is actually needed.)

**Exercise 8.27 (Continuous-motion rank condition).** Suppose the camera is moving continuously and the projection matrix is $\Pi(t) = [R(t), T(t)] \in \mathbb{R}^{3\times 4}$. We assume $\Pi(0) = [I, 0]$ at time $t = 0$. For simplicity, we may further assume that the camera is calibrated, hence $R(t) \in SO(3)$. Then the image $\boldsymbol{x}(t) \in \mathbb{R}^3$ of a point $p$ satisfies $\widehat{\boldsymbol{x}(t)}\Pi(t)\boldsymbol{X} = 0, \forall t$, where $\boldsymbol{X} \in \mathbb{R}^4$ gives the (homogeneous) coordinates of $p$.

1. Show that at time $t = 0$, we have

$$\operatorname{rank}\begin{bmatrix} \widehat{\dot{\boldsymbol{x}}}\boldsymbol{x} + \widehat{\boldsymbol{x}}\widehat{\omega}\boldsymbol{x} & \widehat{\boldsymbol{x}}v \\ \widehat{\ddot{\boldsymbol{x}}}\boldsymbol{x} + \widehat{\dot{\boldsymbol{x}}}\widehat{\omega}\boldsymbol{x} + \widehat{\boldsymbol{x}}\widehat{\dot{\omega}}\boldsymbol{x} & \widehat{\dot{\boldsymbol{x}}}v + \widehat{\boldsymbol{x}}\dot{v} \end{bmatrix} \leq 1,$$

where $\omega, \dot{\omega} \in \mathbb{R}^3$ are the angular velocity and acceleration, respectively, and $v, \dot{v} \in \mathbb{R}^3$ are the linear ones.

2. Derive from the linear dependency of the first three rows of the above matrix the continuous-time epipolar constraint that we studied in Chapter 5:

$$\dot{\boldsymbol{x}}^T \widehat{v}\boldsymbol{x} + \boldsymbol{x}^T \widehat{\omega}\widehat{v}\boldsymbol{x} = 0.$$

3. What is the null space of the above two-column matrix? More specifically, suppose $[\lambda_1, \lambda_2]^T$ is in the null space of the above matrix. What is the ratio $\lambda_1/\lambda_2$?

**Exercise 8.28 (A simulation exercise for the rank condition: point case).** This exercise gives you hands-on experience with the reconstruction algorithm based on the rank condition:

1. In Matlab, generate, say, five images of a point with respect to five camera positions of your choice. Verify the rank of the associated matrices $N_p$, $W_p$ and the multiple-view matrix $M_p$. Perturb one of the images, and check the rank of $M_p$ again. Verify in simulation the condition when the rank of the matrix $M_p$ goes to zero.

2. Now take the images for six points (in general position) and verify the rank of the matrix $P_i$ defined in Section 8.3.3. Find a case in which for six (different) points, the rank of the matrix $P_i$ is less than 11.

3. Combined with the eight-point algorithm you coded before, verify in simulation the multiple-view algorithm given in this chapter (in the noise-free case).

**Exercise 8.29 (Multiple-view factorization for the line case).** Follow a similar development of Section 8.3.3 and construct a multiple-view factorization algorithm for the line case using the rank condition on $M_l$. In particular, answer the following questions:

1. How many lines (in general position) are needed?
2. How is the algorithm initialized?
3. How is structural information, i.e. the distance and orientation of a line, updated in each iteration?

The overall structure of the resulting algorithm should be exactly the same as the algorithm given for the point case, although the pure line case is rarely used in practice.

## 8.A Proof for the properties of bilinear and trilinear constraints

**Lemma 8.12 (Properties of bilinear and trilinear constraints).** *Given three vectors $\boldsymbol{x}_1, \boldsymbol{x}_2, \boldsymbol{x}_3 \in \mathbb{R}^3$ and three camera frames with distinct optical centers, if the three images satisfy epipolar constraints between every pair,*

$$\boldsymbol{x}_i^T \widehat{T_{ij}} R_{ij} \boldsymbol{x}_j = 0, \quad i, j = 1, 2, 3,$$

*a unique preimage p is determined except when the three vectors associated with image points $\boldsymbol{x}_1, \boldsymbol{x}_2, \boldsymbol{x}_3$ are coplanar.*

*If these vectors satisfy all trilinear constraints*

$$\widehat{\boldsymbol{x}_j}(T_{ji}\boldsymbol{x}_i^T R_{ki}^T - R_{ji}\boldsymbol{x}_i T_{ki}^T)\widehat{\boldsymbol{x}_k} = 0, \quad i, j, k = 1, 2, 3$$

*then they determine a unique preimage $p \in \mathbb{E}^3$ except when the three lines associated with the three images $\boldsymbol{x}_1, \boldsymbol{x}_2, \boldsymbol{x}_3$ are collinear.*

*Proof.* (Sketch) Let us first study whether bilinear constraints are sufficient to determine a unique preimage in space. For the given three vectors $\boldsymbol{x}_1, \boldsymbol{x}_2, \boldsymbol{x}_3$, suppose that they satisfy three epipolar constraints

$$\boldsymbol{x}_2^T F_{21} \boldsymbol{x}_1 = 0, \quad \boldsymbol{x}_3^T F_{31} \boldsymbol{x}_1 = 0, \quad \boldsymbol{x}_3^T F_{32} \boldsymbol{x}_2 = 0, \tag{8.69}$$

with $F_{ij} = \widehat{T_{ij}} R_{ij}$ the fundamental matrix between the $i$th and $j$th images. Note that each image (as a point on the image plane) and the corresponding optical center uniquely determine a line in space that passes through them. This gives us a total of three lines. Geometrically, the three epipolar constraints simply imply that each pair of the three lines are coplanar. So when do three pairs of coplanar lines intersect at exactly one point in space? If these three lines are not coplanar, the intersection is uniquely determined, and so is the preimage. If all of them do lie on the same plane, such a unique intersection is not always guaranteed. As shown in Figure 8.7, this may occur when the lines determined by the images lie on the plane spanned by the three optical centers $o_1, o_2, o_3$, the so-called *trifocal plane*, or when the three optical centers lie on a straight line regardless of the images.

The first case is of less practical importance, since 3-D points generically do not lie on the trifocal plane. The second case is more important: regardless of what 3-D feature points one chooses, epipolar constraints alone do not provide sufficient constraints to determine a unique 3-D point from any given three image vectors. In such a case, extra constraints need to be imposed on the three images in order to obtain a unique preimage. Would trilinear constraints suffice to salvage the situation? The answer is yes, and let us show why. Given any three vectors $\boldsymbol{x}_1, \boldsymbol{x}_2, \boldsymbol{x}_3 \in \mathbb{R}^3$, suppose they satisfy the trilinear constraint equation

$$\widehat{\boldsymbol{x}_2}(T_2\boldsymbol{x}_1^T R_3^T - R_2\boldsymbol{x}_1 T_3^T)\widehat{\boldsymbol{x}_3} = 0.$$

In order to detetmine $\boldsymbol{x}_3$ uniquely (up to a scale factor) from this equation, we need the matrix

$$\widehat{\boldsymbol{x}_2}(T_2\boldsymbol{x}_1^T R_3^T - R_2\boldsymbol{x}_1 T_3^T) \quad \in \mathbb{R}^{3\times 3}$$

to be of rank 1. The only case in which $\boldsymbol{x}_3$ is undetermined is that in which this matrix is of rank 0; that is,

$$\widehat{\boldsymbol{x}_2}(T_2\boldsymbol{x}_1^T R_3^T - R_2\boldsymbol{x}_1 T_3^T) = 0.$$

That is,

$$\text{range}(T_2\boldsymbol{x}_1^T R_3^T - R_2\boldsymbol{x}_1 T_3^T) \subset \text{span}\{\boldsymbol{x}_2\}. \tag{8.70}$$

If $T_3$ and $R_3\boldsymbol{x}_1$ are linearly independent, then (8.70) holds if and only if the vectors $R_2\boldsymbol{x}_1, T_2, \boldsymbol{x}_2$ are linearly dependent. This condition simply means that the line associated with the first image $\boldsymbol{x}_1$ coincides with the baseline determined by the optical centers $o_1, o_2$.[14] If $T_3$ and $R_3\boldsymbol{x}_1$ are linearly dependent, $\boldsymbol{x}_3$ is determined, since $R_3\boldsymbol{x}_1$ lies on the line determined by the optical centers $o_1, o_3$. Hence, we have shown that $\boldsymbol{x}_3$ cannot be uniquely determined from $\boldsymbol{x}_1, \boldsymbol{x}_2$ by the trilinear constraint if and only if

$$\widehat{T_2}R_2\boldsymbol{x}_1 = 0 \quad \text{and} \quad \widehat{T_2}\boldsymbol{x}_2 = 0. \tag{8.71}$$

Due to the symmetry of the trilinear constraint equation, $\boldsymbol{x}_2$ is not uniquely determined from $\boldsymbol{x}_1, \boldsymbol{x}_3$ by the trilinear constraint if and only if

$$\widehat{T_3}R_3\boldsymbol{x}_1 = 0 \quad \text{and} \quad \widehat{T_3}\boldsymbol{x}_3 = 0. \tag{8.72}$$

We still need to show that these three images indeed determine a unique preimage in space if either one of the images can be determined from the other two by the trilinear constraint. Without loss of generality, suppose it is $\boldsymbol{x}_3$ that can be uniquely determined from $\boldsymbol{x}_1$ and $\boldsymbol{x}_2$. Simply take the intersection $p' \in \mathbb{E}^3$ of the two lines associated with the first two images and project it back to the third image plane; such an intersection exists, since the two images satisfy the epipolar constraint. If these two lines are parallel, we take the intersection to be on the plane at infinity. Call this image $\boldsymbol{x}_3'$. Then $\boldsymbol{x}_3'$ automatically satisfies the trilinear constraint. Hence, $\boldsymbol{x}_3' \sim \boldsymbol{x}_3$ due to its uniqueness. Therefore, $p'$ is the 3-D point $p$

[14] In other words, the preimage point $p$ lies on the baseline between the first and second camera frames.

where all the three lines intersect in the first place. As we have argued before, the trilinear constraint (8.37) actually implies the bilinear constraint (8.34). Therefore, the 3-D preimage $p$ is uniquely determined if either $\boldsymbol{x}_3$ can be determined from $\boldsymbol{x}_1, \boldsymbol{x}_2$, or $\boldsymbol{x}_2$ can be determined from $\boldsymbol{x}_1, \boldsymbol{x}_3$. Figure 8.8 shows the only case in which the trilinear constraint may become degenerate. □

## Further references

### *Multilinear constraints*

After a great deal of work on the epipolar geometry of points, described in earlier chapters, the trilinear relationships were extended to the calibrated and uncalibrated case gradually and mostly independently for images of lines and later for points in many different forms. The relationships among three images of line features were first pointed out by [Spetsakis and Aloimonos, 1987, Liu and Huang, 1986]. The trilinear constraints among three images of points and lines were also studied in [Spetsakis and Aloimonos, 1990a] accompanied by algorithms. These relationships among three images of points were reformulated in the uncalibrated setting by [Shashua, 1994]; [Hartley, 1994a] soon pointed out its equivalence to the line case. [Quan, 1994, Quan, 1995] gave a closed-form solution to the six-point three-view (projective) reconstruction problem using the trifocal tensors. [Triggs, 1995] formulated bilinear, trilinear, and quadrilinear constraints (which will be introduced in the next chapter) among two, three, and four images, respectively, using a tensorial notation. [Faugeras and Mourrain, 1995] proved the dependence of the quadrilinear constraints on the trilinear and bilinear ones. A formal study of the relationships among these constraints based on polynomial rings can be found in [Heyden and Åström, 1997]. This line of work is now summarized in the books [Hartley and Zisserman, 2000, Faugeras and Luong, 2001].

There is a vast amount of literature studying the properties of the trilinear and quadrilinear constraints as well as associated estimation, calibration, and reconstruction algorithms. For three-view-based methods, please refer to [Hartley, 1994a, Quan, 1995, Armstrong et al., 1996, Torr and Zisserman, 1997, Faugeras and Papadopoulos, 1998, Papadopoulo and Faugeras, 1998], and that of [Avidan and Shashua, 1998, Canterakis, 2000]. For four-view based methods, see [Enciso and Vieville, 1997, Heyden, 1995, Heyden, 1998, Hartley, 1998b], and [Shashua and Wolf, 2000].

### *Rank conditions*

Previous derivations of the multilinear constraints were mostly based on the matrix $N_p$ or the matrix $W_p$. For instance, the derivation given in [Triggs, 1995, Heyden and Åström, 1997] was based on the rank constraint $\text{rank}(N_p) \leq m + 3$, from which multilinear constraints correspond to $(m + 4) \times (m + 4)$ minors of $N_p$. Algebraic geometric tools must be applied to eliminate redundant constraints and establish the relationship among them using algebraic varieties and ideals. The derivation in [Faugeras and Mourrain, 1995] was based on the rank condition $\text{rank}(W_p) \leq 3$, from which multilinear constraints are $4 \times 4$ minors of $W_p$. Grassman Cayley algebras [Faugeras and Papadopoulos, 1995, Faugeras and Luong, 2001] and the double algebra [Carlsson, 1994] were also used to establish algebraic relationships among the obtained constraints.

Matrix rank-based methods were mostly used in the study of approximated camera models such as orthographic [Tomasi and Kanade, 1992], affine [Quan and Kanade, 1996, Quan and Kanade, 1997, Kahl and Heyden, 1999], paraperspective [Poelman and Kanade, 1997, Basri, 1996], and weakly-perspective [Irani, 1999], even for nonrigid motions [Torresani et al., 2001]. For the perspective projection case, the rank-based approach presented in this chapter was based on [Ma et al., 2001a, Ma et al., 2002]. As we will show in the upcoming chapters, this approach easily generalizes to incidence relations among different types of features (Chapter 9) as well as to incorporation of scene knowledge (Chapter 10).

*Absolute quadric constraints and critical motions*

In the uncalibrated case, the rank condition leads to an effortless proof of the fact that we can obtain a consistent reconstruction up to a single transformation (captured by the $H$ matrix) from all the views. This makes the use of the absolute quadric natural. The absolute quadric constraint first showed up in the work [Heyden and Åström, 1996]. The related critical motion sequences were categorized by [Sturm, 1997]. Criticality and degeneracy in multiple-view reconstruction were also reported in the work of [Kahl, 1999, Torr et al., 1999]. A complete classification of multiple-view critical configurations for projective reconstruction can be found in [Hartley and Kahl, 2003].

*Reconstruction algorithms*

Besides the algorithm given in this chapter [Ma et al., 2002], there exist numerous multiple-view structure and motion reconstruction algorithms based on different technical and practical conditions, e.g., matching and reconstruction [Tsai, 1986b], relative 3-D reconstruction [Mohr et al., 1993], fundamental matrices [Faugeras and Laveau, 1994], parallax [Anandan et al., 1994], canonical representation [Luong and Vieville, 1994], projective-duality based methods [Carlsson and Weinshall, 1998], sequential updates [Beardsley et al., 1997], small baselines [Oliensis, 1999], varying focal length [Pollefeys et al., 1996], iterative [Chen and Medioni, 1999, Li and Brooks, 1999, Christy and Horaud, 1996, Ueshiba and Tomita, 1998], normalized epipolar constraint [Vidal et al., 2002a], and other rank-based factorization methods [Sturm and Triggs, 1996, Triggs, 1996, Morris and Kanade, 1998].

For algorithms that use both points and lines, see [Faugeras et al., 1987, Hartley, 1995]. There are also reconstruction algorithms designed for line features only [Weng et al., 1992b, Taylor and Kriegman, 1995]. Although there is no effective constraint for line features between two views, it was shown by [Zhang, 1995] that for line segments two views are sufficient to recover both camera motion and scene structure.

# Chapter 9
# Extension to General Incidence Relations

*Mathematics is the art of giving the same name to different things.*
– Henri Poincaré

This chapter[1] extends development in the previous chapter to the study of all *incidence relations* among different geometric primitives in 3-D space and in multiple images (e.g., intersection and coplanarity). We will demonstrate how incidence relations among multiple points, lines, and planes in space can be encoded in multiple images through the same matrix rank conditions. Such a generalization reveals additional instances that give rise to some nontrivial constraints among features in multiple views. This revelation will in turn lead to a more general class of techniques for structure and motion recovery that can use a multitude of geometric features simultaneously and exploit arbitrary incidence relations among them.

## 9.1 Incidence relations among points, lines, and planes

### *9.1.1 Incidence relations in 3-D space*

Examples of nontrivial incidence relations among geometric primitives – points, lines, and planes – in 3-D are illustrated in Figures 9.1, 9.2, and 9.3. By

[1]This chapter can be skipped at a first reading without loss of continuity.

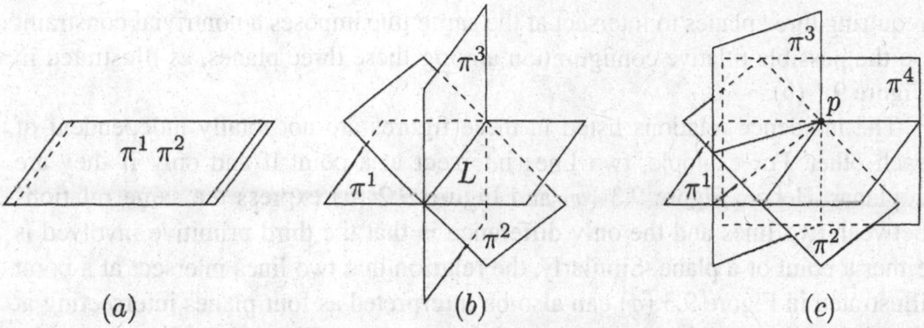


Figure 9.1. $(a)$ Two or more planes are coplanar; $(b)$ three or more planes intersect at the same line; $(c)$ four or more planes intersect at the same point.

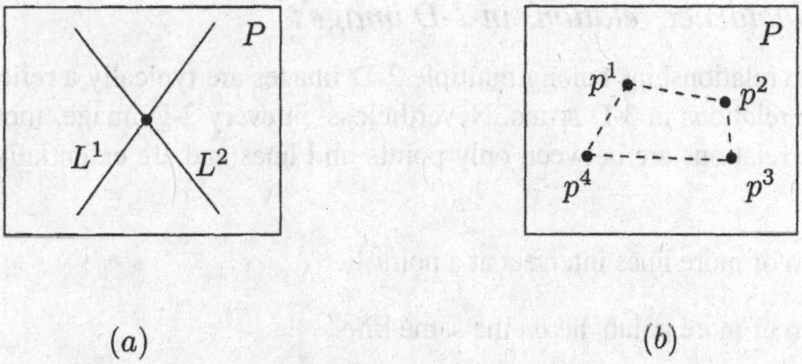


Figure 9.2. $(a)$ Two or more lines belong to the same plane (coplanar); $(b)$ four or more points belong to the same plane (coplanar).

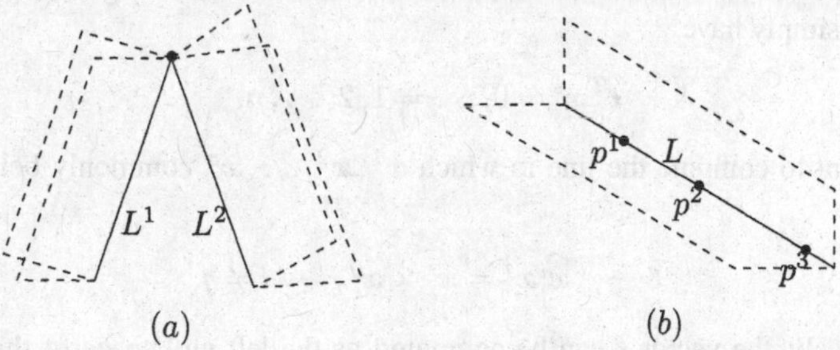


Figure 9.3. $(a)$ Two or more lines intersect at the same point; $(b)$ three or more points belong to, or are included by, the same line (collinear).

"nontrivial" we mean other than the usual hierarchy of inclusion:

$$\text{a point} \subset \text{a line} \subset \text{a plane}. \tag{9.1}$$

Nontrivial incidence relations usually require some minimal number of primitives in order to be effective. For instance, we know that two planes always intersect at a line,[2] but three planes in general do not intersect at the same line. Hence,

[2] We view two parallel planes as intersecting at a line at infinity.

requiring three planes to intersect at the same line imposes a nontrivial constraint on the possible relative configuration among these three planes, as illustrated in Figure 9.1 $(b)$.

The incidence relations listed in these figures are not totally independent of each other. For example, two lines intersect at a point if and only if they are coplanar. Hence, Figure 9.3 $(a)$ and Figure 9.2 $(a)$ express the same relations between two lines and the only difference is that the third primitive involved is either a point or a plane. Similarly, the relation that two lines intersect at a point illustrated in Figure 9.3 $(a)$ can also be interpreted as four planes intersecting at the same point, which identifies it with the relation in Figure 9.1 $(c)$.

### 9.1.2 *Incidence relations in 2-D images*

Geometric relationships among multiple 2-D images are typically a reflection of incidence relations in 3-D space. Nevertheless, in every 2-D image, meaningful incidence relations are between only points and lines and are essentially of two types:

1. Two or more lines intersect at a point;

2. Two or more points lie on the same line.

These relations can be easily described in terms of homogeneous representation of points and lines in the image. Let $\boldsymbol{x}^1, \boldsymbol{x}^2, \ldots, \boldsymbol{x}^n \in \mathbb{R}^3$ be a set of image points (in a single view). Suppose they all *belong to* the same (coimage) line $\boldsymbol{\ell} \in \mathbb{R}^3$. Then we simply have

$$\boldsymbol{\ell}^T \boldsymbol{x}^i = 0, \quad i = 1, 2, \ldots, n. \tag{9.2}$$

If we want to compute the line to which $\boldsymbol{x}^1, \boldsymbol{x}^2, \ldots, \boldsymbol{x}^n$ commonly belong, we have

$$\boldsymbol{\ell} \sim \widehat{\boldsymbol{x}^i} \boldsymbol{x}^j = \boldsymbol{x}^i \times \boldsymbol{x}^j, \quad \forall\, i \neq j. \tag{9.3}$$

Alternatively, the vector $\boldsymbol{\ell}$ can be computed as the left null space of the matrix $[\boldsymbol{x}^1, \boldsymbol{x}^2, \ldots, \boldsymbol{x}^n] \in \mathbb{R}^{3\times n}$. Similarly, let $\boldsymbol{\ell}^1, \boldsymbol{\ell}^2, \ldots, \boldsymbol{\ell}^n \in \mathbb{R}^3$ be a set of coimage lines. Suppose that they all *intersect* at the same image point $\boldsymbol{x} \in \mathbb{R}^3$. We simply have

$$\boldsymbol{x}^T \boldsymbol{\ell}^i = 0, \quad i = 1, 2, \ldots, n. \tag{9.4}$$

To compute the point that $\boldsymbol{\ell}^1, \boldsymbol{\ell}^2, \ldots, \boldsymbol{\ell}^n$ have in common, we can use

$$\boldsymbol{x} \sim \widehat{\boldsymbol{\ell}^i} \boldsymbol{\ell}^j = \boldsymbol{\ell}^i \times \boldsymbol{\ell}^j, \quad \forall\, i \neq j. \tag{9.5}$$

Alternatively, the vector $\boldsymbol{x}$ can be determined as the left null space of the matrix $[\boldsymbol{\ell}^1, \boldsymbol{\ell}^2, \ldots, \boldsymbol{\ell}^n] \in \mathbb{R}^{3\times n}$.

## 9.2 Rank conditions for incidence relations

As we have seen in the previous chapter, incidence relations among multiple images of a point or a line in space are governed by certain matrix rank conditions. For example, as we have mentioned in Examples 8.5 and 8.6 of Chapter 8, to test whether a family of $m$ planes satisfy one of the relations in Figure 9.1, we need to check the rank of the matrix

$$W = \begin{bmatrix} (\pi^1)^T \\ (\pi^2)^T \\ \vdots \\ (\pi^m)^T \end{bmatrix} \in \mathbb{R}^{m \times 4}, \tag{9.6}$$

where each row vector $\pi^i \in \mathbb{R}^4$ represents a hyperplane in space. Then the cases $(a)$, $(b)$, and $(c)$ correspond to $\text{rank}(W) = 1, 2$, and 3, respectively. Hence, the rank conditions on $W_p$ and $W_l$ (and correspondingly those on $M_p$ and $M_l$) that we have studied in the previous chapter are nothing but the incidence relations in Figure 9.1 at play. The interested reader may see Appendix 9.A for a complete match-up between rank conditions studied so far and the incidence relations given in the above figures.

In this section, we will show that remaining incidence relations, namely, *intersection* at a point (Figure 9.3 $(a)$) and *restriction* to a plane (Figure 9.2), can also be described by rank conditions (on extended multiple-view matrices). The situation with the collinear case (Figure 9.3 $(b)$) is more complicated but of less practical importance. We leave the discussion to Appendix 9.B and to the exercises (Exercises 9.6 and 9.7). Since these relationships can either be verified in each image or be given a priori in practice, such knowledge can be and should be exploited if a consistent 3-D reconstruction from multiple images is sought.

### 9.2.1 *Intersection of a family of lines*

We first consider the case that a family of lines, say $L^1, L^2, \ldots, L^m$, intersect at a single point, say $p$, in 3-D space, as shown in Figure 9.4. This type of incidence relation is very common in practice, for instance, three edges intersecting at the corner of a building. Now suppose $\boldsymbol{x}_1, \boldsymbol{x}_2, \ldots, \boldsymbol{x}_m$ are $m$ images of the point $p$ and $\boldsymbol{\ell}_1, \boldsymbol{\ell}_2, \ldots, \boldsymbol{\ell}_m$ are $m$ coimages of those lines. Here remember that these coimages do not have to correspond to the same line in 3-D, contrary to the

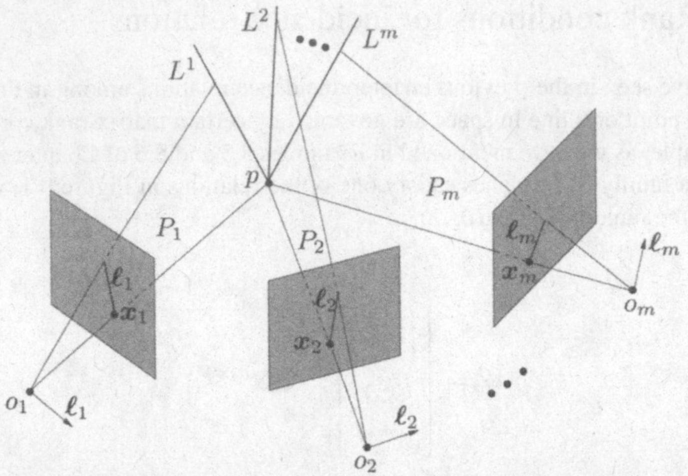


Figure 9.4. Images of a family of lines $L^1, L^2, \ldots, L^m$ intersect at a point $p$. (Preimage) planes $P_1, P_2, \ldots, P_m$ extended from the image lines might not intersect at the same line in 3-D. But all the planes must intersect at the same point.

situation described in Chapter 8. Note that since each row of the matrices[3]

$$W_p \doteq \begin{bmatrix} \widehat{x_1} & 0 \\ \widehat{x_2}R_2 & \widehat{x_2}T_2 \\ \vdots & \vdots \\ \widehat{x_m}R_m & \widehat{x_m}T_2 \end{bmatrix}, \quad W_l \doteq \begin{bmatrix} \ell_1^T & 0 \\ \ell_2^T R_2 & \ell_2^T T_2 \\ \vdots & \vdots \\ \ell_m^T R_m & \ell_m^T T_2 \end{bmatrix}, \quad \text{and} \quad W \doteq \begin{bmatrix} W_p \\ W_l \end{bmatrix}$$

represents a 3-D plane that passes through the same point $p$, naturally these matrices satisfy the rank condition

$$\operatorname{rank}(W_p) \leq 3, \quad \operatorname{rank}(W_l) \leq 3, \quad \operatorname{rank}(W) \leq 3. \tag{9.7}$$

The rank condition on the last matrix $W$ implies that any matrix with rows obtained from mixing those of either $W_p$ or $W_l$ would be bounded by the same rank condition.

From the previous chapter, we know that these rank conditions on $W$ lead to more concise relationships if we use the multiple-view matrix $M$ instead. Suppose we take only the images of the lines and construct a matrix that is similar to $M_l$ defined in the previous chapter:

$$M_l \doteq \begin{bmatrix} \ell_2^T R_2 \widehat{\ell_1} & \ell_2^T T_2 \\ \ell_3^T R_3 \widehat{\ell_1} & \ell_3^T T_3 \\ \vdots & \vdots \\ \ell_m^T R_m \widehat{\ell_1} & \ell_m^T T_m \end{bmatrix} \in \mathbb{R}^{(m-1)\times 4}. \tag{9.8}$$

[3]Here again we adopt the convention from the previous chapter and choose the first camera frame to be the reference.

Notice that each $\ell_i$ chosen in the $i$th view can be the coimage of *any* line that passes through the point $p$. Applying the same rank reduction technique used to prove Theorem 8.15, one can show still $\text{rank}(M_l) = \text{rank}(W_l) - 1$. This gives us the following statement.

**Lemma 9.1 (An intersecting family of lines).** *Given $m$ images of a family of lines intersecting at a 3-D point $p$, the matrix $M_l$ defined above satisfies*

$$\boxed{rank(M_l) = rank(W_l) - 1 \leq 2.} \tag{9.9}$$

*Furthermore, we have $rank(M_l) \leq 1$ if and only if all the lines in space coincide.*

The reader should be aware of the reason for the difference between this lemma and Theorem 8.15: Here $\ell_1, \ell_2, \ldots, \ell_m$ do not have to be the coimages of a single line in space, but instead each of them may correspond to any line in the intersecting family (Figure 9.4).

Since $\text{rank}(\widehat{\ell}) = 2$, the first three columns of the matrix $M_l$ have at most rank 2. So in general, the rank of the matrix $M_l$ does not exceed 3. With no surprise at all, each rank value of $M_l$ corresponds to a (qualitatively different) category of configurations for the $m$ image lines: $\text{rank}(M_l) = 3$ means that the $m$ images come from a general set of 3-D lines; $\text{rank}(M_l) = 2$ means that they come from a set of intersecting 3-D lines; $\text{rank}(M_l) = 1$ means that they come from a single 3-D line; $\text{rank}(M_l) = 0$ means the degenerate case in which the the line and all the camera centers are coplanar. We leave the detailed verification as an exercise to the reader (see Exercise 9.1).

Algebraically, the rank condition $\text{rank}(M_l) \leq 2$ is equivalent to every $3 \times 3$ submatrix of $M_l$ having determinant zero. Note that any such determinant must involve three rows of $M_l$ and image quantities from at least four views. This type of equation is beyond the previously studied multilinear relationships between two or three views. In some literature, this is referred to as the *quadrilinear constraints*.

Out of pure mathematical curiosity, then, we can ask whether there exist irreducible relationships beyond four images. The answer is yes, largely due to the incidence relation in Figure 9.3 $(b)$. But the resulting relationships are difficult to harness for practical purposes, we leave the discussion to Appendix 9.B.

Before we leave this section, we study two useful examples that are also related to a family of intersecting lines shown in Figure 9.4. The first example is a direct implication of the rank of the $M_l$ matrix, and the second example is of some practical importance.

**Example 9.2 (A line-point-point configuration).** Consider the situation of Figure 9.4. Suppose that in the reference view we do not observe the point $p$ but only an image $\ell_1$ of a line that passes through the point, and in the remaining views we do have images

$\boldsymbol{x}_2, \boldsymbol{x}_3, \ldots, \boldsymbol{x}_m$ of the point. We can also define the matrix

$$M_{lp} \doteq \begin{bmatrix} \widehat{\boldsymbol{x}_2} R_2 \widehat{\boldsymbol{\ell}_1} & \widehat{\boldsymbol{x}_2} T_2 \\ \widehat{\boldsymbol{x}_3} R_3 \widehat{\boldsymbol{\ell}_1} & \widehat{\boldsymbol{x}_3} T_3 \\ \vdots & \vdots \\ \widehat{\boldsymbol{x}_m} R_m \widehat{\boldsymbol{\ell}_1} & \widehat{\boldsymbol{x}_m} T_m \end{bmatrix} \in \mathbb{R}^{[3(m-1)]\times 4}, \tag{9.10}$$

called the *multiple-view matrix* for a line-point-point configuration. Notice that each row of $\widehat{\boldsymbol{x}}$ can be interpreted as a coimage $\boldsymbol{\ell}$ of some line that passes through the point, i.e. $\boldsymbol{\ell}^T \boldsymbol{x} = 0$. Each row of $M_{lp}$ is then of the form $[\boldsymbol{\ell}^T R \widehat{\boldsymbol{\ell}_1}, \boldsymbol{\ell}^T T]$, which conforms to the rows in the matrix $M_l$ defined in (9.8). Therefore, $\text{rank}(M_{lp}) \leq 2$. Furthermore, due to Sylvester's inequality (Appendix A), since $\text{rank}(\widehat{\boldsymbol{x}} R \widehat{\boldsymbol{\ell}}) \geq 1$, we have

$$\boxed{1 \leq \text{rank}(M_{lp}) \leq 2.} \tag{9.11}$$

The case $\text{rank}(M_{lp}) = 1$ occurs if and only if all the camera centers are on the same plane as the preimage of $\boldsymbol{\ell}_1$, and the preimages of $\boldsymbol{x}_2, \boldsymbol{x}_3, \ldots, \boldsymbol{x}_m$ are parallel (see Appendix 9.C.2). ■

**Example 9.3 (A point-line-line configuration).** Consider the situation of Figure 9.4. Suppose that you have the image $\boldsymbol{x}_1$ of a point $p$ in the reference view, but in the remaining views it is occluded, and you observe only the image lines $\boldsymbol{\ell}_2, \boldsymbol{\ell}_3, \ldots, \boldsymbol{\ell}_m$ that pass through the point. We can define the matrix

$$M_{pl} \doteq \begin{bmatrix} \boldsymbol{\ell}_2^T R_2 \boldsymbol{x}_1 & \boldsymbol{\ell}_2^T T_2 \\ \boldsymbol{\ell}_3^T R_3 \boldsymbol{x}_1 & \boldsymbol{\ell}_3^T T_3 \\ \vdots & \vdots \\ \boldsymbol{\ell}_m^T R_m \boldsymbol{x}_1 & \boldsymbol{\ell}_m^T T_m \end{bmatrix} \in \mathbb{R}^{(m-1)\times 2}, \tag{9.12}$$

called the *multiple-view matrix* for a point-line-line configuration. It is easy to see that rows of this matrix are simply linear combinations of the rows of the multiple-view matrix $M_p$ associated with the point $p$. Since $M_p$ has rank less than 1, we have

$$\boxed{0 \leq \text{rank}(M_{pl}) \leq 1.} \tag{9.13}$$

It is worth noting that for the rank condition to be true, the following constraints must hold among arbitrary triplets of given images:

$$(\boldsymbol{\ell}_i^T R_i \boldsymbol{x}_1)(\boldsymbol{\ell}_j^T T_j) - (\boldsymbol{\ell}_i^T T_i)(\boldsymbol{\ell}_j^T R_j \boldsymbol{x}_1) = 0, \quad i, j = 2, 3, \ldots, m. \tag{9.14}$$

These are the so-called *point-line-line* relationships among three views. Note, however, that here $\boldsymbol{\ell}_i$ and $\boldsymbol{\ell}_j$ do not have to be coimages of the same line in 3-D. This relaxes the notion of "corresponding" line features when we use this type of constraint in practice. ■

### *9.2.2 Restriction to a plane*

Another incidence relation commonly encountered in practice involves feature points and lines lying on a common plane, say $P$, in 3-D (see Figure 9.2). We have studied such a coplanarity constraint in the two-view setting in Section 5.3 through the notion of homography. We here extend the study to the multiple-view setting. In particular, a fundamental connection between the homography and the multiple-view rank condition will be revealed.

In general, a 3-D plane can be described by a vector $\pi \doteq [a, b, c, d]^T \in \mathbb{R}^4$ such that the homogeneous coordinates $\boldsymbol{X}$ of any point $p$ on this plane satisfy the equation

$$\pi^T \boldsymbol{X} = 0. \tag{9.15}$$

Although we assume such a constraint on the coordinates $\boldsymbol{X}$ of $p$, we do not assume that we know $\pi$. Similarly, consider a line $L = \{\boldsymbol{X} \mid \boldsymbol{X} = \boldsymbol{X}_o + \mu \boldsymbol{V}, \mu \in \mathbb{R}\}$. This line is on the plane $P$ if and only if

$$\pi^T \boldsymbol{X}_o = \pi^T \boldsymbol{V} = 0. \tag{9.16}$$

For convenience, given a plane $\pi = [a, b, c, d]^T$, we define $\pi_R = [a, b, c]^T \in \mathbb{R}^3$ and $\pi_T = d \in \mathbb{R}$.[4] It turns out that in order to take into account the planar restriction, we need only to slightly modify the definition of each multiple-view matrix, and all the rank conditions remain exactly the same. This is because in order to apply the rank conditions (8.21) and (8.26) with the planar constraints (9.15) and (9.16), we need only to change the definition of matrices $W_p$ and $W_l$ to

$$W_p \doteq \begin{bmatrix} \widehat{\boldsymbol{x}_1}\Pi_1 \\ \widehat{\boldsymbol{x}_2}\Pi_2 \\ \vdots \\ \widehat{\boldsymbol{x}_m}\Pi_m \\ \pi^T \end{bmatrix} \in \mathbb{R}^{(3m+1)\times 4} \quad \text{and} \quad W_l \doteq \begin{bmatrix} \boldsymbol{\ell}_1^T\Pi_1 \\ \boldsymbol{\ell}_2^T\Pi_2 \\ \vdots \\ \boldsymbol{\ell}_m^T\Pi_m \\ \pi^T \end{bmatrix} \in \mathbb{R}^{(m+1)\times 4}.$$

Such modifications do not change the ranks of $W_p$ and $W_l$ at all: we still have $\text{rank}(W_p) \leq 3$ and $\text{rank}(W_l) \leq 2$, since their null spaces will be the point and the line, respectively. Then one can easily follow previous proofs for all the rank conditions by carrying this extra row of (planar) constraint with the matrices, and the rank conditions on the resulting multiple-view matrices $M_p$ and $M_l$ remain the same as before. We leave this as an exercise for the reader (see Exercise 9.2). We summarize the results as the following statement.

**Corollary 9.4 (Rank conditions for coplanar features).** *Given a point $p$ and a line $L$ lying on a plane $P$ specified by the vector $\pi \in \mathbb{R}^4$, append the row $[\pi_R^T \boldsymbol{x}_1 \;\; \pi_T]$ to the matrices $M_p$; alternatively, append the row $[\pi_R^T \widehat{\boldsymbol{\ell}_1} \;\; \pi_T]$ to the matrices $M_l$. Then the rank conditions on the new matrices $M_p$ and $M_l$ remain the same as in Theorems 8.8, 8.15, or Lemma 9.1 for a family of coplanar lines intersecting at a point.*

---

[4]It will become clear later that the subscript $R$ indicates the part that plays a similar role as that of $R$ in $\Pi$, similarly for the subscript $T$. But $\pi_T$ is not to be confused with $\pi^T$, which is the transpose of $\pi$.

*Homography between pairs of views of coplanar features*

As examples for the above corollary, the multiple-view matrices $M_p$ and $M_l$ for coplanar point or line features are

$$M_p \doteq \begin{bmatrix} \widehat{\boldsymbol{x}_2} R_2 \boldsymbol{x}_1 & \widehat{\boldsymbol{x}_2} T_2 \\ \widehat{\boldsymbol{x}_3} R_3 \boldsymbol{x}_1 & \widehat{\boldsymbol{x}_3} T_3 \\ \vdots & \vdots \\ \widehat{\boldsymbol{x}_m} R_m \boldsymbol{x}_1 & \widehat{\boldsymbol{x}_m} T_m \\ \pi_R^T \boldsymbol{x}_1 & \pi_T \end{bmatrix}, \quad M_l \doteq \begin{bmatrix} \boldsymbol{\ell}_2^T R_2 \widehat{\boldsymbol{\ell}_1} & \boldsymbol{\ell}_2^T T_2 \\ \boldsymbol{\ell}_3^T R_3 \widehat{\boldsymbol{\ell}_1} & \boldsymbol{\ell}_3^T T_3 \\ \vdots & \vdots \\ \boldsymbol{\ell}_m^T R_m \widehat{\boldsymbol{\ell}_1} & \boldsymbol{\ell}_m^T T_m \\ \pi_R^T \widehat{\boldsymbol{\ell}_1} & \pi_T \end{bmatrix}. \tag{9.17}$$

The rank condition $\text{rank}(M_p) \leq 1$ implies not only the multilinear constraints as before, but also the following constraints (by considering the submatrix consisting of the $i$th group of three rows of $M_p$ and its last row):

$$\widehat{\boldsymbol{x}}_i T_i \pi_R^T \boldsymbol{x}_1 - \widehat{\boldsymbol{x}}_i R_i \boldsymbol{x}_1 \pi_T = 0, \quad i = 2, 3, \ldots, m. \tag{9.18}$$

When the plane $P$ does not pass through the camera center $o_1$, i.e. $\pi_T \neq 0$, these equations give exactly the homography that we have studied in Section 5.3.1 for two views of a planar scene

$$\widehat{\boldsymbol{x}}_i \left( R_i - \frac{1}{\pi_T} T_i \pi_R^T \right) \boldsymbol{x}_1 = 0, \tag{9.19}$$

here between the first and the $i$th views. The matrix

$$H_i = \left( R_i - \frac{1}{\pi_T} T_i \pi_R^T \right) \quad \in \mathbb{R}^{3\times 3} \tag{9.20}$$

in the above equation represents the *homography* between the two views. From equation (9.19), the vector $H_i \boldsymbol{x}_1$ is obviously in the null space of the matrix $\widehat{\boldsymbol{x}}_i$, hence it is proportional to $\boldsymbol{x}_i$. In the homogeneous representation, they both represent the same point in the image plane. Therefore, the relation can be rewritten as

$$\boldsymbol{x}_i \sim H_i \boldsymbol{x}_1. \tag{9.21}$$

From the rank condition on $M_l$, we can alternatively obtain the homography in terms of line features

$$\boldsymbol{\ell}_i^T \left( R_i - \frac{1}{\pi_T} T_i \pi_R^T \right) \widehat{\boldsymbol{\ell}}_1 = 0 \tag{9.22}$$

between the first and the $i$th views. Or equivalently,

$$\boldsymbol{\ell}_i^T H_i \sim \boldsymbol{\ell}_1. \tag{9.23}$$

*Rank duality between coplanar point and line features*

We know that on a plane $P$, any two points determine a line and any two lines determine a point. This dual relationship is inherited in the following relationship between the rank conditions on $M_p$ and $M_l$ defined above.

**Corollary 9.5 (Duality between coplanar points and lines).** *If the $M_p$ matrices of two distinct points on a plane are of rank less than or equal to 1, then the $M_l$ matrix associated with the line determined by the two points is of rank less than or equal to 1. On the other hand, if the $M_l$ matrices of two distinct lines on a plane are of rank less than or equal to 1, then the $M_p$ matrix associated with the intersection of the two lines is of rank less than or equal to 1.*

The proof is left as Exercise 9.3. An immediate implication of this corollary is that given a set of feature points sharing the same 3-D plane (see Figure 9.5); it really does not matter too much whether one uses the matrix $M_p$ of the points or the matrix $M_l$ of the lines from pairs of points. They essentially give the same set of constraints.

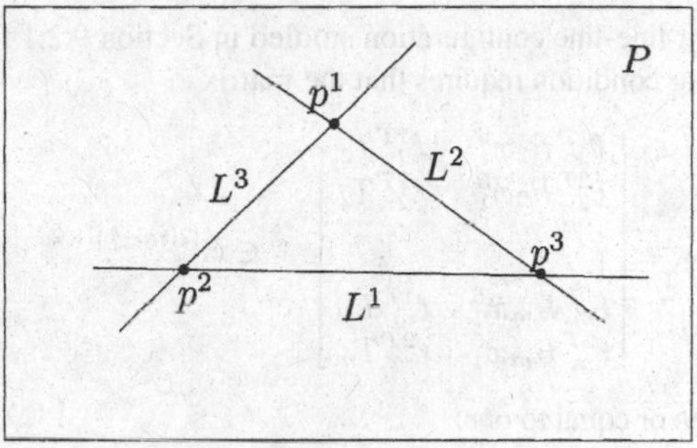


Figure 9.5. Duality between a set of three points and three lines on a plane $P$: the rank conditions associated with $p^1, p^2, p^3$ are exactly equivalent to those associated with $L^1, L^2, L^3$.

**Example 9.6 (Intrinsic rank condition for coplanar features).** The above approach for expressing a planar restriction relies explicitly on the parameters $\pi$ of the underlying plane $P$, which leads to the homography. There is, however, another intrinsic (but equivalent) way to express the planar restriction, by using combinations of the rank conditions that we have so far for point and line features. Since three points are always coplanar, at least four points are needed to make any planar restriction nontrivial (as shown in Figure 9.2 $(b)$). Suppose four feature points $p^1, p^2, p^3, p^4$ are known to be coplanar as shown in Figure 9.6, and their images are denoted by $\boldsymbol{x}^1, \boldsymbol{x}^2, \boldsymbol{x}^3, \boldsymbol{x}^4$. The (virtual) coimages $\boldsymbol{\ell}^1, \boldsymbol{\ell}^2$ of the two lines $L^1, L^2$ and the (virtual) image $\boldsymbol{x}^5$ of their intersection $p^5$ can be uniquely determined by $\boldsymbol{x}^1, \boldsymbol{x}^2, \boldsymbol{x}^3, \boldsymbol{x}^4$ as

$$\boldsymbol{\ell}^1 \sim \widehat{\boldsymbol{x}^1}\boldsymbol{x}^2, \quad \boldsymbol{\ell}^2 \sim \widehat{\boldsymbol{x}^3}\boldsymbol{x}^4, \quad \boldsymbol{x}^5 \sim \widehat{\boldsymbol{\ell}^1}\boldsymbol{\ell}^2. \tag{9.24}$$

Then, the coplanar constraint for $p^1, p^2, p^3, p^4$ can be expressed in terms of the intersection relation between $L^1, L^2$, and $p^5$ if we use $\boldsymbol{\ell}_i^j$ to denote the $i$th image of the line $j$, for $j = 1, 2$, and $i = 1, 2, \ldots, m$, and $\boldsymbol{x}_i^j$ is defined similarly.

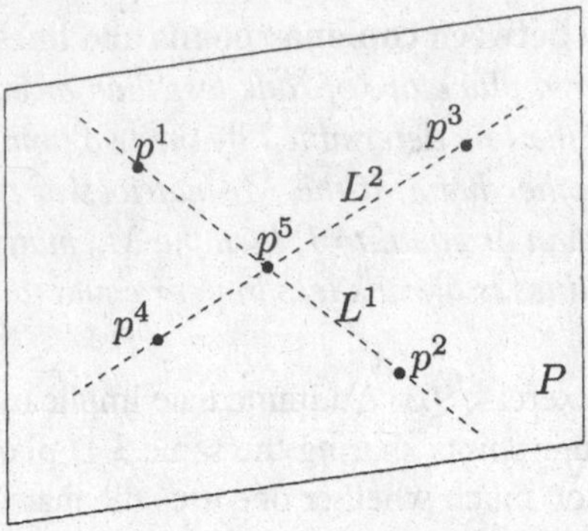


Figure 9.6. $p^1, p^2, p^3, p^4$ are four points on the same plane $P$ if and only if the two associated (virtual) lines $L^1$ and $L^2$ intersect at a (virtual) point $p^5$.

Applying the point-line-line configuration studied in Section 9.2.1 for the virtual point and lines, the coplanar condition requires that the matrix

$$M_{pl} = \begin{bmatrix} \boldsymbol{\ell}_2^{1T} R_2 \boldsymbol{x}_1^5 & \boldsymbol{\ell}_2^{1T} T_2 \\ \boldsymbol{\ell}_2^{2T} R_2 \boldsymbol{x}_1^5 & \boldsymbol{\ell}_2^{2T} T_2 \\ \vdots & \vdots \\ \boldsymbol{\ell}_m^{1T} R_m \boldsymbol{x}_1^5 & \boldsymbol{\ell}_m^{1T} T_m \\ \boldsymbol{\ell}_m^{2T} R_m \boldsymbol{x}_1^5 & \boldsymbol{\ell}_m^{2T} T_m \end{bmatrix} \in \mathbb{R}^{[2(m-1)]\times 2} \tag{9.25}$$

to be of rank less than or equal to one:

$$\operatorname{rank}(M_{pl}) \leq 1.$$

Any other coplanar relations among points and lines can be expressed in a similar fashion that does not explicitly depend on the plane parameter $\pi$ or on the homography $H$. For instance, two points and a line, or two lines, are coplanar. We leave these cases to the reader as an exercise (see Exercise 9.4). ■

Finally, the same rank technique applies to the case in which multiple points are restricted to a line in 3-D (Figure 9.3 ($b$)). Since that case is of less practical merit, we leave it to Appendix 9.B and Exercise 9.6.

## 9.3 Universal rank conditions on the multiple-view matrix

In this section, we summarize all rank conditions studied so far as special instances of a unified rank condition on a universal multiple-view matrix.

For $m$ images $\boldsymbol{x}_1, \boldsymbol{x}_2, \ldots, \boldsymbol{x}_m$ of a point $p$ on a line $L$ with its $m$ coimages $\boldsymbol{\ell}_1, \boldsymbol{\ell}_2, \ldots, \boldsymbol{\ell}_m$, we define the following two symbols:

$$\begin{array}{llllll} \text{Image:} & D_i & \doteq & \boldsymbol{x}_i \in \mathbb{R}^3 & \text{or} & \widehat{\boldsymbol{\ell}}_i \in \mathbb{R}^{3\times 3}, \\ \text{Coimage:} & D_i^{\perp} & \doteq & \widehat{\boldsymbol{x}}_i^T \in \mathbb{R}^{3\times 3} & \text{or} & \boldsymbol{\ell}_i \in \mathbb{R}^3. \end{array}$$

Then, depending on whether the available (or chosen) measurement from the $i$th image is a point feature $\boldsymbol{x}_i$ or a line feature $\boldsymbol{\ell}_i$, $D_i$ (or $D_i^\perp$) is assigned the corresponding value. That choice is completely independent of the other $D_j$ (or $D_j^\perp$) for $j \neq i$. $D_i^\perp$ can be viewed as the *orthogonal complement* to $D_i$ and it always represents a coimage of a point or a line.[5] Using the above definition of $D_i$ and $D_i^\perp$, we now formally define a *universal multiple-view matrix*

$$M \doteq \begin{bmatrix} (D_2^\perp)^T R_2 D_1 & (D_2^\perp)^T T_2 \\ (D_3^\perp)^T R_3 D_1 & (D_3^\perp)^T T_3 \\ \vdots & \vdots \\ (D_m^\perp)^T R_m D_1 & (D_m^\perp)^T T_m \end{bmatrix}. \tag{9.26}$$

Depending on the particular choice for each $D_i^\perp$ or $D_1$, the dimension of the matrix $M$ may vary. But no matter what the choice for each individual $D_i^\perp$ or $D_1$ is, $M$ will always be a valid multiple-view matrix of a certain dimension.

**Theorem 9.7 (Multiple-view rank conditions).** *Consider a point $p$ lying on a line $L$. Given the images $\boldsymbol{x}_1, \boldsymbol{x}_2, \ldots, \boldsymbol{x}_m \in \mathbb{R}^3$ of the point and coimages $\boldsymbol{\ell}_1, \boldsymbol{\ell}_2, \ldots, \boldsymbol{\ell}_m \in \mathbb{R}^3$ of the line relative to $m$ camera frames specified by $(R_i, T_i)$ for $i = 2, 3, \ldots, m$, then, for any choice of $D_i^\perp$ and $D_1$ in the definition of the general multiple-view matrix $M$, the rank of the resulting $M$ belongs to one of the following two cases:*

1. *If $D_1 = \widehat{\boldsymbol{\ell}_1}$ and $D_i^\perp = \widehat{\boldsymbol{x}_i}^T$ for some $i \geq 2$, then*

$$\boxed{1 \leq \text{rank}(M) \leq 2.} \tag{9.27}$$

2. *Otherwise,*

$$\boxed{0 \leq \text{rank}(M) \leq 1.} \tag{9.28}$$

*The matrix $M$ takes the lower-bound rank values if and only if degenerate configurations occur.*

A complete proof of this theorem is a straightforward combination and extension of Theorems 8.8, 8.15, and Lemma 9.1. Essentially, the above theorem gives a universal description of the incidence relation between a point and a line in terms of their $m$ images seen from $m$ vantage points.

In the following examples, we demonstrate how to obtain many more types of multiple-view matrices by instantiating $M$.

**Example 9.8 (Point-point-line constraints).** Let us choose $D_1 = \boldsymbol{x}_1, D_2^\perp = \widehat{\boldsymbol{x}_2}^T, D_3^\perp = \boldsymbol{\ell}_3$. We get a multiple-view matrix

$$M = \begin{bmatrix} \widehat{\boldsymbol{x}_2} R_2 \boldsymbol{x}_1 & \widehat{\boldsymbol{x}_2} T_2 \\ \boldsymbol{\ell}_3^T R_3 \boldsymbol{x}_1 & \boldsymbol{\ell}_3^T T_3 \end{bmatrix} \in \mathbb{R}^{4\times 2}. \tag{9.29}$$

[5]In fact, there are many equivalent matrix representations for $D_i$ and $D_i^\perp$. We choose $\widehat{\boldsymbol{x}_i}$ and $\widehat{\boldsymbol{\ell}_i}$ here because they are the simplest forms representing the orthogonal subspaces of $\boldsymbol{x}_i$ and $\boldsymbol{\ell}_i$ and also they are linear in $\boldsymbol{x}_i$ and $\boldsymbol{\ell}_i$, respectively.

Then, $\text{rank}(M) \leq 1$ gives

$$(\widehat{\boldsymbol{x}_2} R_2 \boldsymbol{x}_1)(\boldsymbol{\ell}_3^T T_3)^T - (\boldsymbol{\ell}_3^T R_3 \boldsymbol{x}_1)(\widehat{\boldsymbol{x}_2} T_2)^T = 0 \quad \in \mathbb{R}^3. \tag{9.30}$$

■

**Example 9.9 (Line-point-line constraints).** Let us choose $D_1 = \widehat{\boldsymbol{\ell}_1}, D_2^\perp = \widehat{\boldsymbol{x}_2}^T, D_3^\perp = \boldsymbol{\ell}_3$. We get a multiple-view matrix

$$M = \begin{bmatrix} \widehat{\boldsymbol{x}_2} R_2 \widehat{\boldsymbol{\ell}_1} & \widehat{\boldsymbol{x}_2} T_2 \\ \boldsymbol{\ell}_3^T R_3 \widehat{\boldsymbol{\ell}_1} & \boldsymbol{\ell}_3^T T_3 \end{bmatrix} \in \mathbb{R}^{4 \times 4}. \tag{9.31}$$

Then, $\text{rank}(M) \leq 1$ (which is a degenerate rank value for the so-defined $M$) gives

$$\left(\widehat{\boldsymbol{x}_2} R_2 \widehat{\boldsymbol{\ell}_1}\right)(\boldsymbol{\ell}_3^T T_3) - (\widehat{\boldsymbol{x}_2} T_2)\left(\boldsymbol{\ell}_3^T R_3 \widehat{\boldsymbol{\ell}_1}\right) = 0 \quad \in \mathbb{R}^{3 \times 3}. \tag{9.32}$$

■

**Example 9.10 (Point-line-line constraints).** Let us choose $D_1 = \boldsymbol{x}_1, D_2^\perp = \boldsymbol{\ell}_2, D_3^\perp = \boldsymbol{\ell}_3$. We get a multiple-view matrix

$$M = \begin{bmatrix} \boldsymbol{\ell}_2^T R_2 \boldsymbol{x}_1 & \boldsymbol{\ell}_2^T T_2 \\ \boldsymbol{\ell}_3^T R_3 \boldsymbol{x}_1 & \boldsymbol{\ell}_3^T T_3 \end{bmatrix} \in \mathbb{R}^{2 \times 2}. \tag{9.33}$$

Then $\text{rank}(M) \leq 1$ gives

$$(\boldsymbol{\ell}_2^T R_2 \boldsymbol{x}_1)(\boldsymbol{\ell}_3^T T_3) - (\boldsymbol{\ell}_3^T R_3 \boldsymbol{x}_1)(\boldsymbol{\ell}_2^T T_2) = 0 \quad \in \mathbb{R}. \tag{9.34}$$

■

**Example 9.11 (Line-point-point constraints).** Let us choose $D_1 = \widehat{\boldsymbol{\ell}_1}, D_2^\perp = \widehat{\boldsymbol{x}_2}^T, D_3^\perp = \widehat{\boldsymbol{x}_3}^T$. We get a multiple-view matrix

$$M = \begin{bmatrix} \widehat{\boldsymbol{x}_2} R_2 \widehat{\boldsymbol{\ell}_1} & \widehat{\boldsymbol{x}_2} T_2 \\ \widehat{\boldsymbol{x}_3} R_3 \widehat{\boldsymbol{\ell}_1} & \widehat{\boldsymbol{x}_3} T_3 \end{bmatrix} \in \mathbb{R}^{6 \times 4}. \tag{9.35}$$

Then, the condition $\text{rank}(M) \leq 2$ implies that all $3 \times 3$ submatrices of $M$ have determinant equal to zero. ■

Since there are practically infinitely many possible instances for the multiple-view matrix, it is impossible to provide a geometric description for each one of them. In Appendix 9.C of this chapter we study a few representative cases that will give the reader a clear idea about how the rank condition of the multiple-view matrix $M$ works geometrically. Understanding these representative cases will be sufficient for the reader to carry out a similar analysis to any other instances (e.g., see Exercise 9.10).

As we have demonstrated in the previous section, other incidence relations such as all features belonging to a plane $P$ can also be expressed in terms of the same type of rank conditions.

**Corollary 9.12 (Planar features and homography).** *Suppose that all features are in a plane and the coordinates $\boldsymbol{X}$ of any point on it satisfy the equation*

*$\pi^T X = 0$ for some vector $\pi \in \mathbb{R}^4$. Let $\pi = [\pi_R^T, \pi_T]^T$, with $\pi_R \in \mathbb{R}^3, \pi_T \in \mathbb{R}$. Then simply append the matrix*

$$\begin{bmatrix}\pi_R^T D_1 & \pi_T\end{bmatrix} \tag{9.36}$$

*to the matrix $M$ in its formal definition (9.26). The rank condition on the new planar multiple-view matrix $M$ remains exactly the same as in Theorem 9.7.*

The rank condition on the new (planar) multiple-view matrix $M$ then implies *all* constraints among multiple images of these coplanar features, including the homography. Of course, the above representation is not intrinsic; it depends on parameters $\pi$ that describe the 3-D location of the plane. Following the procedure given in Section 9.2.2, the above corollary can be reduced to rank conditions on matrices of the type in (9.25).

**Example 9.13 (Point-line-plane constraints).** Let us choose $D_1 = \boldsymbol{x}_1, D_2^{\perp} = \boldsymbol{\ell}_2$. We get a planar multiple-view matrix

$$M = \begin{bmatrix}\boldsymbol{\ell}_2^T R_2 \boldsymbol{x}_1 & \boldsymbol{\ell}_2^T T_2 \\ \pi_R^T \boldsymbol{x}_1 & \pi_T\end{bmatrix} \in \mathbb{R}^{2\times 2}. \tag{9.37}$$

Then $\text{rank}(M) \leq 1$ gives

$$(\boldsymbol{\ell}_2^T R_2 \boldsymbol{x}_1)\pi_T - (\boldsymbol{\ell}_2^T T_2)(\pi_R^T \boldsymbol{x}_1) = \boldsymbol{\ell}_2^T(\pi_T R_2 - T_2 \pi_R^T)\boldsymbol{x}_1 = 0 \quad \in \mathbb{R}. \tag{9.38}$$

■

**Example 9.14 (Features at infinity).** In Theorem 9.7, if the point $p$ and the line $L$ are on the plane at infinity $P_\infty = \mathbb{P}^3 \setminus \mathbb{R}^3$ (see Appendix 3.B), the rank condition on the associated multiple-view matrix $M$ remains *the same*. Therefore, the rank condition extends to features in the entire projective space $\mathbb{P}^3$, and it does not discriminate between Euclidean, affine, or projective spaces. We leave the details as an exercise (see Exercise 9.9). ■

**Example 9.15 (Incidence relations for a box).** Figure 9.7 shows a box. Let the corner $p$

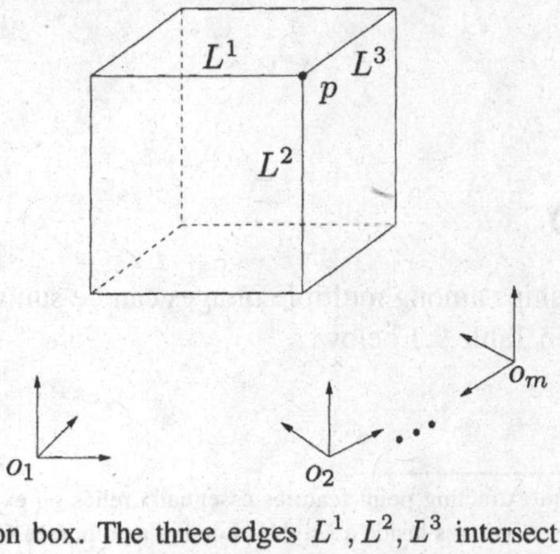


Figure 9.7. A common box. The three edges $L^1, L^2, L^3$ intersect at the corner $p$. The coordinate frames indicate that $m$ images are taken at these vantage points.

be the intersection of the three edges $L^1$, $L^2$, and $L^3$. From $m$ images of the cube, we have the multiple-view matrix $M$ associated with the point $p$:

$$M = \begin{bmatrix} \widehat{\boldsymbol{x}_2} R_2 \boldsymbol{x}_1 & \widehat{\boldsymbol{x}_2} T_2 \\ (\boldsymbol{\ell}_2^1)^T R_2 \boldsymbol{x}_1 & (\boldsymbol{\ell}_2^1)^T T_2 \\ (\boldsymbol{\ell}_2^2)^T R_2 \boldsymbol{x}_1 & (\boldsymbol{\ell}_2^2)^T T_2 \\ (\boldsymbol{\ell}_2^3)^T R_2 \boldsymbol{x}_1 & (\boldsymbol{\ell}_2^3)^T T_2 \\ \vdots & \vdots \\ \widehat{\boldsymbol{x}_m} R_m \boldsymbol{x}_1 & \widehat{\boldsymbol{x}_m} T_m \\ (\boldsymbol{\ell}_m^1)^T R_m \boldsymbol{x}_1 & (\boldsymbol{\ell}_m^1)^T T_m \\ (\boldsymbol{\ell}_m^2)^T R_m \boldsymbol{x}_1 & (\boldsymbol{\ell}_m^2)^T T_m \\ (\boldsymbol{\ell}_m^3)^T R_m \boldsymbol{x}_1 & (\boldsymbol{\ell}_m^3)^T T_m \end{bmatrix} \in \mathbb{R}^{[6(m-1)] \times 2}, \tag{9.39}$$

where $\boldsymbol{x}_i \in \mathbb{R}^3$ is the image of the corner in the $i$th view and $\boldsymbol{\ell}_i^j \in \mathbb{R}^3$ is the coimage of the $j$th edge in the $i$th view. Theorem 9.7 says that $\operatorname{rank}(M) = 1$. One can verify that $\alpha = [\lambda_1, 1]^T \in \mathbb{R}^2$ is in the null space of $M$. In addition to the multiple images $\boldsymbol{x}_1, \boldsymbol{x}_2, \ldots, \boldsymbol{x}_m$ of the corner $p$ itself, the extra rows associated with the line features $\boldsymbol{\ell}_i^j, j = 1, 2, 3, i = 1, 2, \ldots, m$, also help to determine the depth scale $\lambda_1$. ■

From the above example, we can already see one advantage of the rank condition: It can simultaneously handle multiple incidence conditions associated with the same feature.[6] Since such incidence relations among points, lines and planes occur frequently in practice, the use of the multiple-view matrix for mixed features is going to improve the quality of the overall reconstruction by explicitly and simultaneously taking into account all incidence relations among all features in all images. Furthermore, the multiple-view factorization algorithm given in the previous chapter can be easily generalized to perform this task. An example of reconstruction from mixed points and lines is shown in Figures 9.8 and 9.9.

Besides incidence relations, extra scene knowledge such as symmetry (including parallelism and orthogonality) can also be naturally accounted for using the multiple-view-matrix-based techniques. In the next chapter, we will show why this is the case.

## 9.4 Summary

Most of the relationships among multiple images can be summarized in terms of the rank conditions in Table 9.1 below.

[6] In fact, any algorithm extracting point features essentially relies on exploiting local incidence conditions on multiple edge features inside a neighborhood of a point (see Chapter 4). The structure of the $M$ matrix simply reveals a similar fact within a larger scale.

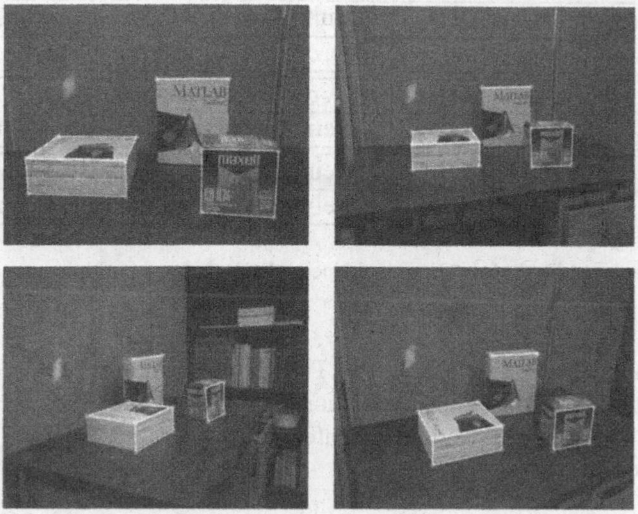

Figure 9.8. Four views from an eight-frame sequence of a desk with corresponding line and point features.

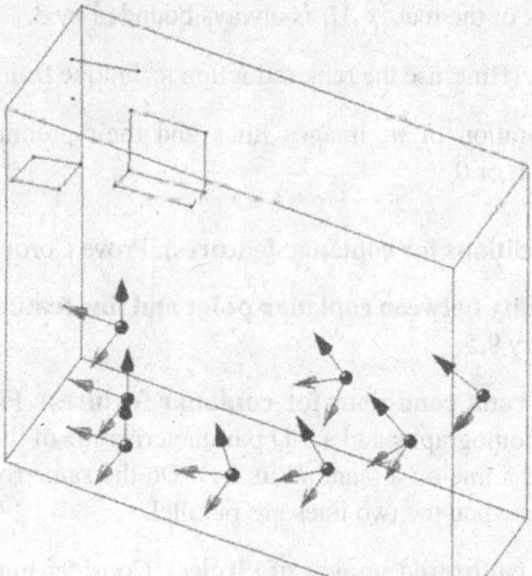

Figure 9.9. Recovered 3-D structure of the scene and motion of the camera, i.e. the eight camera coordinate frames. The direction of the optical axis is indicated by the black arrows.

All rank conditions can be expressed uniformly in terms of a formal multiple-view matrix

$$M \doteq \begin{bmatrix} (D_2^{\perp})^T R_2 D_1 & (D_2^{\perp})^T T_2 \\ (D_3^{\perp})^T R_3 D_1 & (D_3^{\perp})^T T_3 \\ \vdots & \vdots \\ (D_m^{\perp})^T R_m D_1 & (D_m^{\perp})^T T_m \end{bmatrix}$$

| Matrix $W$ | Matrix $M$ | Multiple-view incidence relations |
|---|---|---|
| $\text{rank}(W_l) = 1$ | $\text{rank}(M_l) = 0$ | $o_1, o_2, \ldots, o_m$ and $L$ coplanar |
| $\text{rank}(W_p) = 2$ | $\text{rank}(M_p) = 0$ | $o_1, o_2, \ldots, o_m$ and $p$ collinear |
| $\text{rank}(W_l) = 2$ | $\text{rank}(M_l) = 1$ | preimages of lines intersect at a line $L$ |
| $\text{rank}(W_p) = 3$ | $\text{rank}(M_p) = 1$ | preimages of points intersect at a point $p$ |
| $\text{rank}(W_l) = 3$ | $\text{rank}(M_l) = 2$ | preimages of lines intersect at a point |

Table 9.1. Rank conditions for various incidence relations among multiple images of points and lines.

such that $\text{rank}(M)$ is either 3, 2, 1, or 0. These rank conditions can either be used to test whether particular incidence relations in 3-D space are valid, or be used for reconstruction once such relations are enforced.

## 9.5 Exercises

**Exercise 9.1 (A family of lines).**

1. Show that the rank of the matrix $M_l$ is always bounded by 3.
2. Prove Lemma 9.1. (Hint: use the rank reduction technique from previous chapters.)
3. Draw the configuration of $m$ images lines and their preimages for each case: $\text{rank}(M_l) = 3, 2, 1,$ or $0$.

**Exercise 9.2 (Rank conditions for coplanar features).** Prove Corollary 9.4.

**Exercise 9.3 (Rank duality between coplanar point and line features).** Using results in Chapter 8, prove Corollary 9.5.

**Exercise 9.4 (Intrinsic rank conditions for coplanar features).** Find an *intrinsic* rank condition (i.e. no use of homography and a 3-D parameterization of the plane) for multiple images of two points and a line on a plane $P$ in 3-D. Do the same for two coplanar lines and discuss what happens when the two lines are parallel.

**Exercise 9.5 (Multiple calibrated images of circles).** Consider multiple calibrated images of a circle in 3-D, as shown in Figure 9.10. Suppose $\boldsymbol{\ell}_i$ corresponds to the shortest chord of the $i$th image of the circle. What type of rank conditions do they satisfy? Reason why we no longer have this if the camera is not calibrated.

**Exercise 9.6 (Rank conditions for a point moving on a straight line).** In this exercise we explore how to derive a rank condition among images of a point moving freely along a straight line $L$ in space, as shown in Figure 9.13. Let $\boldsymbol{X}$ be a base point on the line and let $\boldsymbol{V}$ be the direction vector of the line. Suppose $m$ images of a point moving on the line are taken. We have

$$\boldsymbol{x}_i \sim \Pi_i[\boldsymbol{X} + \lambda_i \boldsymbol{V}], \quad i = 1, 2, \ldots, m, \tag{9.40}$$

where $\Pi_i = [R_i, T_i]$ is the camera pose.

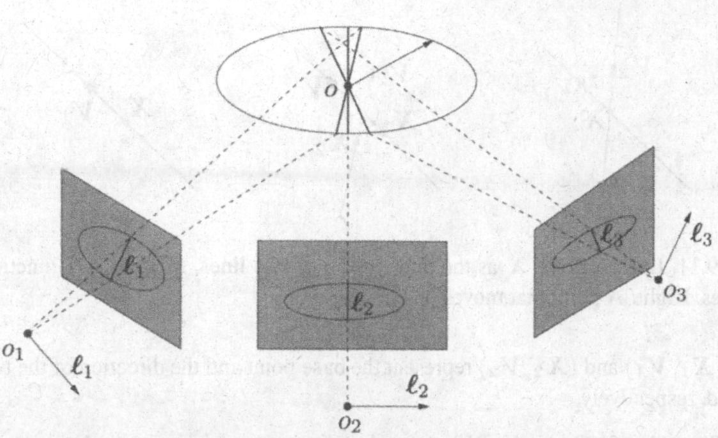


Figure 9.10. Multiple views of a circle.

1. Show that the matrix

$$W_L \doteq \begin{bmatrix} \boldsymbol{x}_1^T R_1 & \boldsymbol{x}_1^T \widehat{T}_1 R_1 \\ \boldsymbol{x}_2^T R_2 & \boldsymbol{x}_2^T \widehat{T}_2 R_2 \\ \vdots & \vdots \\ \boldsymbol{x}_m^T R_m & \boldsymbol{x}_m^T \widehat{T}_m R_m \end{bmatrix} \in \mathbb{R}^{m \times 6} \tag{9.41}$$

satisfies $\text{rank}(W_L) \leq 5$.

2. If we choose the first camera frame to be the reference frame; i.e. $[R_1, T_1] = [I, 0]$, show that the matrix

$$M_L \doteq \begin{bmatrix} \boldsymbol{x}_2^T R_2 \widehat{\boldsymbol{x}}_1 & \boldsymbol{x}_2^T \widehat{T}_2 R_2 \\ \boldsymbol{x}_3^T R_3 \widehat{\boldsymbol{x}}_1 & \boldsymbol{x}_3^T \widehat{T}_3 R_3 \\ \vdots & \vdots \\ \boldsymbol{x}_m^T R_m \widehat{\boldsymbol{x}}_1 & \boldsymbol{x}_m^T \widehat{T}_m R_m \end{bmatrix} \in \mathbb{R}^{(m-1) \times 6} \tag{9.42}$$

satisfies $\text{rank}(M_L) \leq 4$.

3. Show that

$$\boldsymbol{U}_0 = \begin{bmatrix} \boldsymbol{x}_1 \\ 0 \end{bmatrix}, \quad \boldsymbol{U}_1 = \begin{bmatrix} \lambda_1 \boldsymbol{V} \\ \boldsymbol{V} \end{bmatrix} \quad \in \mathbb{R}^6, \tag{9.43}$$

for some $\lambda_1 \in \mathbb{R}$, are two linearly independent vectors in the two-dimensional null space of $M_L$.

4. Conclude that any nontrivial constraint among these $m$ images involves at least five views.

**Exercise 9.7 (Rank conditions for a T-junction).** A T-junction, generated by one edge occluding another, can be viewed as the intermediate case between a fixed point and a point which can move freely along a straight line, as shown in Figure 9.11. Suppose that $m$ images of such a T-junction are taken. We have

$$\boldsymbol{x}_i \sim \Pi_i[\boldsymbol{X}_1 + \lambda_i \boldsymbol{V}_1] \sim \Pi_i[\boldsymbol{X}_2 + \gamma_i \boldsymbol{V}_2], \quad i = 1, 2, \ldots, m, \tag{9.44}$$

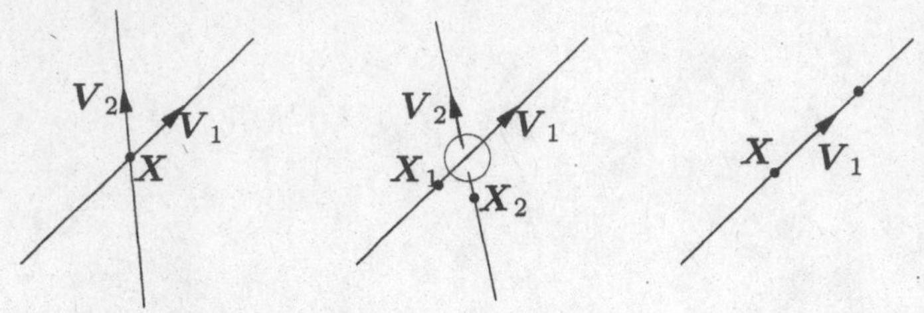


Figure 9.11. Left: A point $\boldsymbol{X}$ as the intersection of two lines. Middle: A T-junction from two lines. Right: A point that moves on one line.

where $(\boldsymbol{X}_1, \boldsymbol{V}_1)$ and $(\boldsymbol{X}_2, \boldsymbol{V}_2)$ represent the base point and the direction of the two lines involved, respectively.

1. Show that for the matrix $W_L$ defined in the previous exercise, we have
$$\text{rank}(W_L) \leq 4.$$
2. If we choose the first camera frame to be the reference, show that for the matrix $M_L$ defined in the previous exercise, we have
$$\text{rank}(M_L) \leq 3.$$
3. Show that the following three vectors
$$\boldsymbol{U}_0 = \begin{bmatrix} \boldsymbol{x}_1 \\ 0 \end{bmatrix}, \quad \boldsymbol{U}_1 = \begin{bmatrix} \lambda_1 \boldsymbol{V}_1 \\ \boldsymbol{V}_1 \end{bmatrix}, \quad \boldsymbol{U}_2 = \begin{bmatrix} \lambda_2 \boldsymbol{V}_2 \\ \boldsymbol{V}_2 \end{bmatrix} \quad \in \mathbb{R}^6, \tag{9.45}$$
for some scalars $\lambda_1, \lambda_2 \in \mathbb{R}$, are linearly independent and span the three-dimensional null space of $M_L$.
4. What is the null space of $M_L$ when the two lines actually intersect at the same point, as in the first case shown in Figure 9.11? Show that in this case:
   (a) We have $\text{rank}(M_L) \leq 2$.
   (b) Furthermore, this rank condition is equivalent to the multiple-view rank condition for the intersection point; i.e. $\text{rank}(M_p) \leq 1$.

**Exercise 9.8 (Irreducible constraints among $m$ images from spirals).** Consider a class of (spiral) curves specified by the coordinates
$$[x(t), y(t), z(t)] \doteq \left[r\sin(t), r\cos(t), \sum_{i=0}^{n} \alpha_i t^i\right], \quad t \in [0, \infty).$$
You may assume that $r \in \mathbb{R}_+, n \in \mathbb{Z}_+$ are given and fixed, but for the coefficients $\alpha_i \in \mathbb{R}$ could vary among the class. Consider multiple images $\boldsymbol{x}_1, \boldsymbol{x}_2, \ldots, \boldsymbol{x}_m$ whose rays are tangent to the supporting cylinder $[r\sin(t), r\cos(t), z]$ and whose intersection with such a spiral are given at different views, as shown in Figure 9.12. You should convince yourself that such points are easy to identify in images of the spiral. You can even assume that the images are ordered in a way that their indices grow monotonically in $t$ and the increase in $t$ between consecutive indices is less than $2\pi$. Show that:

1. Any $n$ such images (rays) tangent to the cylinder $[r\sin(t), r\cos(t), z]$ belong to some (in fact, infinitely many) spiral(s).

2. It takes at least $n+1$ such images (rays) in general position to determine all the coefficients $\alpha_i$ uniquely. (Hint: You may need to know that the Vandermonde matrix
$$\begin{bmatrix} 1 & t_0 & t_0^2 & \cdots & t_0^n \\ 1 & t_1 & t_1^2 & \cdots & t_1^n \\ \vdots & \vdots & \vdots & \ddots & \vdots \\ 1 & t_n & t_n^2 & \cdots & t_n^n \end{bmatrix} \in \mathbb{R}^{(n+1)\times(n+1)}$$
is nonsingular if $t_0, t_1, \ldots, t_n$ are distinct.)

3. Conclude that irreducible constraints among such images are up to $m = n+2$ views at least. Increasing the degree of the polynomial $\sum_{i=0}^{n} \alpha_i t^i$, such constraints can be among any number of views.

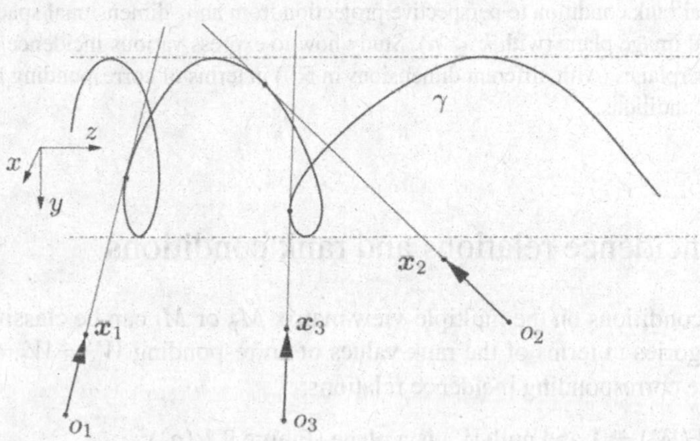


Figure 9.12. Images of points on a spiral curve $\gamma$ are taken at different vantage points. Also, we suppose that the image rays are tangent to the supporting cylinder of the curve.

**Exercise 9.9 (Features on the plane at infinity).** Show that the same rank conditions in Theorem 9.7 hold for multiple images of a feature (point or line) on the plane at infinity. Note the special structure of the multiple-view matrices associated with such features. (Note: Homogeneous coordinates of a point on the plane at infinity have the form $\boldsymbol{X} = [X, Y, Z, 0]^T \in \mathbb{R}^4$. Its image $\boldsymbol{x} \in \mathbb{R}^3$ is given by the same equation $\lambda \boldsymbol{x} = \Pi \boldsymbol{X}$ as for regular points.)

**Exercise 9.10 (Geometry of the multiple-view matrix).** Identify the configuration of features (relative to the camera frames) involved in each of the multiple-view matrices below and draw an illustrative picture:
$$\text{rank}\begin{bmatrix} \widehat{\boldsymbol{x}_2} R_2 \widehat{\boldsymbol{\ell}_1} & \widehat{\boldsymbol{x}_2} T_2 \\ \boldsymbol{\ell}_3^T R_3 \widehat{\boldsymbol{\ell}_1} & \widehat{\boldsymbol{x}_3} T_3 \\ \widehat{\boldsymbol{x}_4} R_4 \widehat{\boldsymbol{\ell}_1} & \widehat{\boldsymbol{x}_4} T_4 \\ \boldsymbol{\ell}_5^T R_5 \widehat{\boldsymbol{\ell}_1} & \boldsymbol{\ell}_5^T T_5 \end{bmatrix} = 1, \quad \text{rank}\begin{bmatrix} \boldsymbol{\ell}_2^{1T} R_2 \boldsymbol{x}_1 & \boldsymbol{\ell}_2^{1T} T_2 \\ \boldsymbol{\ell}_2^{2T} R_2 \boldsymbol{x}_1 & \boldsymbol{\ell}_2^{2T} T_2 \\ \widehat{\boldsymbol{x}_3} R_3 \boldsymbol{x}_1 & \widehat{\boldsymbol{x}_3} T_3 \\ \boldsymbol{\ell}_4^{1T} R_4 \boldsymbol{x}_1 & \boldsymbol{\ell}_4^{1T} T_4 \\ \boldsymbol{\ell}_4^{2T} R_4 \boldsymbol{x}_1 & \boldsymbol{\ell}_4^{2T} T_4 \end{bmatrix} = 1.$$
(Hint: Appendix 9.C.)

**Exercise 9.11 (Multiple-view factorization for coplanar points).** Follow a similar development in Section 8.3.3 and, using the rank condition on the coplanar multiple-view matrix $M_p$ given in Section 9.2.2, construct a multiple-view factorization algorithm for coplanar point features. In particular, answer the following questions:

1. How many points (in general position) are needed?
2. How does one initialize the algorithm?
3. How does ones update the structural information, i.e. the depth of the points and the parameters of the plane in each iteration?

The overall structure of the resulting algorithm should resemble the algorithm for the point case given in the previous chapter.

**Exercise 9.12 (Multiple-view rank conditions in high-dimensional spaces).** Generalize the universal rank condition to perspective projection from an $n$-dimensional space to a $k$-dimensional image plane (with $k < n$). Study how to express various incidence relations among hyperplanes (with different dimensions in $\mathbb{R}^n$) in terms of corresponding multiple-view rank conditions.

## 9.A Incidence relations and rank conditions

The rank conditions on the multiple-view matrix $M_p$ or $M_l$ can be classified into three categories in terms of the rank values of corresponding $W = W_p$ or $W = W_l$ and the corresponding incidence relations:

1. $\text{rank}(W) = 1$ and $\text{null}(W)$ is a plane (Figure 9.1 ($a$)):
$$\text{rank}(W_l) = 1 \quad \Leftrightarrow \quad \text{rank}(M_l) = 0.$$
This is the degenerate case in which all camera centers and the 3-D line are coplanar; the line can be anywhere on this plane. Since $\text{rank}(W_p) \geq 2$ always, this case does not apply to a point feature.

2. $\text{rank}(W) = 2$ and $\text{null}(W)$ is a line (Figure 9.1 ($b$)):
$$\begin{aligned}\text{rank}(W_l) = 2 \quad &\Leftrightarrow \quad \text{rank}(M_l) = 1,\\ \text{rank}(W_p) = 2 \quad &\Leftrightarrow \quad \text{rank}(M_p) = 0.\end{aligned}$$
For a line feature, this is the generic case, since the preimage (as the null space) is supposed to be a line; for a point feature, this is the degenerate case, since all camera centers and the point are collinear; the point can be anywhere on this line.

3. $\text{rank}(W) = 3$ and $\text{null}(W)$ is a point (Figure 9.1 ($c$)):
$$\begin{aligned}\text{rank}(W_l) = 3 \quad &\Leftrightarrow \quad \text{rank}(M_l) = 2,\\ \text{rank}(W_p) = 3 \quad &\Leftrightarrow \quad \text{rank}(M_p) = 1.\end{aligned}$$

For a line feature, this is the case of a family of lines intersecting at a point (Figure 9.4); for a point feature, this is the generic case, since the preimage (as the null space) is supposed to be a point.

We have also seen in Section 9.2.2 the incidence relations in Figure 9.2 in play. Then what about the remaining case in Figure 9.3 $(b)$? Does this incidence relation give useful 3-D information in a multiple-view setting, and how? Can it also be expressed in terms of some kind of rank conditions? The next appendix attempts to give an answer to these questions.

## 9.B Beyond constraints among four views

From the previous appendix, we know that all values of $\text{rank}(W)$ less than 3 correspond to 3-D incidence relations given in the beginning of this chapter. But it leaves us one unattended case: the collinear incidence relation shown in Figure 9.3 $(b)$. But this case could not correspond to any rank-deficient matrix $W$, since the null space of $W$ can never be a set of collinear points. This leaves us to the last possible rank value for $W$:

$$\text{rank}(W) = 4. \tag{9.46}$$

At first sight, this category seems to be meaningless. Weather $\text{rank}(W_p) = 4$ or $\text{rank}(W_l) = 4$ makes little sense: Nothing can be in their null space, hence features involved in the rows of each matrix do not even correspond to one another. Nevertheless, it merely implies that, in this scenario, the use of the rank of $W \in \mathbb{R}^{n \times 4}$ is no longer an effective way to impose constraints among multiple images. It by no means suggests that meaningful and useful (intrinsic) constraints cannot be imposed on the matrix $W$ and consequently on the multiple images, since effective constraints can still be imposed through rank conditions on its submatrices. Here we show how to use this type of submatrix rank condition to take into account the collinear incidence relation. As we will see, this will, in fact, lead to nontrivial intrinsic constraints across at least five images.

Consider the scenario where a point is moving on a straight line and multiple images of this point are taken from different vantage points. See Figure 9.13. Suppose the line is described by the matrix $\begin{bmatrix} (\pi^1)^T \\ (\pi^2)^T \end{bmatrix} \in \mathbb{R}^{2 \times 4}$; its null space contains the points on the line. It is easy to see that the matrix

$$W \doteq \begin{bmatrix} W_p \\ (\pi^1)^T \\ (\pi^2)^T \end{bmatrix} \in \mathbb{R}^{(3m+2) \times 4}$$

in general has

$$\text{rank}(W) = 4.$$

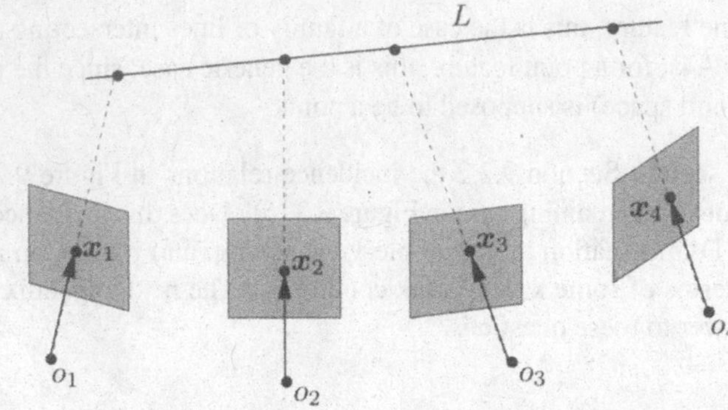


Figure 9.13. Images of points on a straight line $L$ are taken from different vantage points. In fact, there is always at least one line across any given four lines. Hence a fifth image is needed to impose nontrivial constraints.

Nevertheless, its submatrices satisfy

$$\operatorname{rank}\begin{bmatrix} \widehat{x}_i\Pi_i \\ (\pi^1)^T \\ (\pi^2)^T \end{bmatrix} \leq 3, \quad i = 1, 2, \ldots, m, \tag{9.47}$$

since the ray from the image point and the line intersect in space. This condition in fact imposes nontrivial constraints among the multiple images $x_1, x_2, \ldots, x_m$ and the camera configuration $\Pi_1, \Pi_2, \ldots, \Pi_m$. After the extrinsic parameters of the line are eliminated, we in fact will get another rank-deficient matrix of higher dimension. It will then be easy to see that nontrivial constraints will be among five views instead of two, three, or four. We leave the details as an exercise to the reader (see Exercise 9.6). This new type of rank conditions help characterize the multiple-view geometry of T-junctions (Exercise 9.7), which is of great practical importance.

Notice that the above rank-3 condition on the submatrices of $W$ is just one of many ways in which new relationships among multiple views can be introduced. In principle, there are practically infinitely many ways in which nontrivial (and irreducible) algebraic and geometric constraints can be imposed among *any number of views*.[7] It all depends on what type of geometric objects or primitives we consider and how much knowledge we (assume to) know about them. For instance, if we consider the primitives to be certain classes of 3-D curves, intrinsic relationships may exist among arbitrarily many views in the same spirit as the ones we have seen so far (e.g., see Exercise 9.8), although their practicality is debatable.

---

[7]It is standard belief that intrinsic constraints only exist among up to four images. This is true only for pure point and line features, without considering situations like Figure 9.13.

# 9.C Examples of geometric interpretation of the rank conditions

In this appendix, we illustrate in more detail the geometric interpretation of a few instances of the multiple-view matrix $M$ in Theorem 9.7.

## 9.C.1 Case 2: $0 \leq rank(M) \leq 1$

Let us first consider the more general case in which $0 \leq \text{rank}(M) \leq 1$, i.e. case 2 in Theorem 9.7. There are only two subcases, depending on the value of the rank of $M$:

$$(a)\ \text{rank}(M) = 1, \quad \text{and} \quad (b)\ \text{rank}(M) = 0. \tag{9.48}$$

*Subcase* $(a)$. When the rank of $M$ is 1, it corresponds to the generic case: all image points (if at least two are present in $M$) come from a unique point $p$ in space; all image lines (if at least three are present in $M$) come from a unique line $L$ in space; and if both point and line features are present in $M$, the point $p$ then must lie on the line $L$ in space. This is illustrated in Figure 9.14.

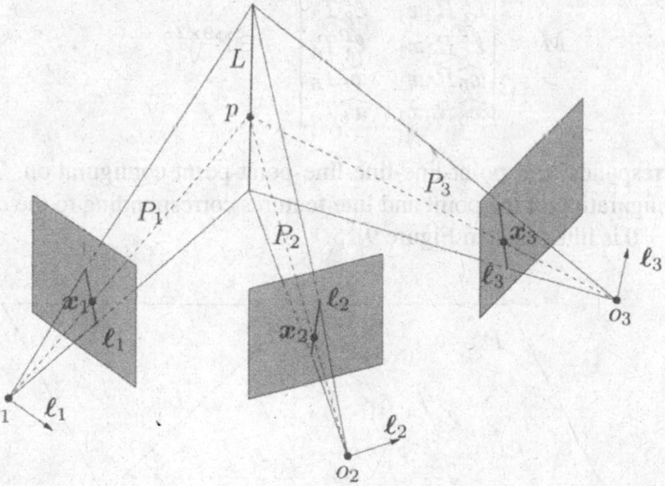


Figure 9.14. Generic configuration for the case rank$(M) = 1$. preimage planes extended from the (co)images $\boldsymbol{\ell}_1, \boldsymbol{\ell}_2, \boldsymbol{\ell}_3$ intersect at one line $L$ in space. preimage lines extended from the images $\boldsymbol{x}_1, \boldsymbol{x}_2, \boldsymbol{x}_3$ intersect at one point $p$ in space which must lie on $L$.

What happens if there are not enough image points or lines present in $M$? For example, there is only one (reference) image point $\boldsymbol{x}_1$ present in $M_{pl}$ (see Example 9.3). Then the rank of $M_{pl}$ being 1 implies that the line $L$ is uniquely determined by preimages of $\boldsymbol{\ell}_2, \boldsymbol{\ell}_3, \ldots, \boldsymbol{\ell}_m$. Hence, the point $p$ is determined by both $L$ and $\boldsymbol{x}_1$. On the other hand, if there is only one image line present in some

$M$ of this type, $L$ can be a family of lines lying on the preimage plane of the line and all passing through the point $p$ determined by the preimages of points in $M$. *Subcase* (*b*). When the rank of $M$ is 0, it implies that all the entries of $M$ are zero. It is easy to verify that this corresponds to a set of degenerate cases in which the 3-D location of the point or the line cannot be uniquely determined from their multiple (pre)images (no matter how many). In these cases, the best one can do is:

- When there are more than two image points present in $M$, preimages of these points are collinear;
- When there are more than three image lines present in $M$, preimages of these lines are coplanar;
- When both points and lines are present in $M$, the preimages of the points are collinear and coplanar with the preimages of the lines.

Let us demonstrate this last case with a concrete example. Suppose the number of views is $m = 6$ and we choose the matrix $M$ to be

$$M = \begin{bmatrix} \boldsymbol{\ell}_2^T R_2 \boldsymbol{x}_1 & \boldsymbol{\ell}_2^T T_2 \\ \boldsymbol{\ell}_3^T R_3 \boldsymbol{x}_1 & \boldsymbol{\ell}_3^T T_3 \\ \boldsymbol{\ell}_4^T R_4 \boldsymbol{x}_1 & \boldsymbol{\ell}_4^T T_4 \\ \widehat{\boldsymbol{x}_5} R_5 \boldsymbol{x}_1 & \widehat{\boldsymbol{x}_5} T_5 \\ \widehat{\boldsymbol{x}_6} R_6 \boldsymbol{x}_1 & \widehat{\boldsymbol{x}_6} T_6 \end{bmatrix} \in \mathbb{R}^{9\times 2}, \tag{9.49}$$

which corresponds to a point-line-line-line-point-point configuration. The geometric configuration of the point and line features corresponding to the condition $\text{rank}(M) = 0$ is illustrated in Figure 9.15.

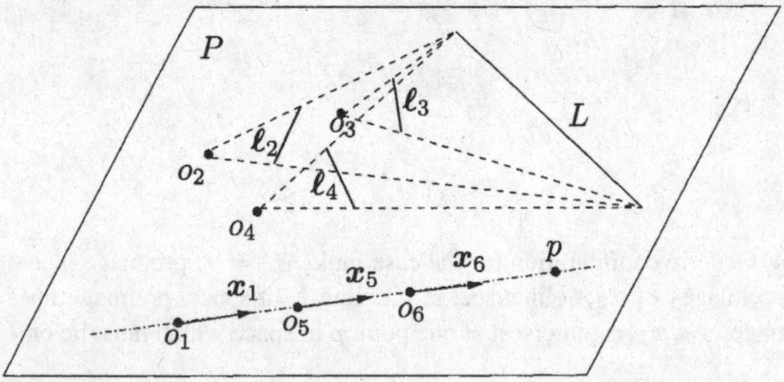


Figure 9.15. A degenerate configuration for the case $\text{rank}(M) = 0$. From the given rank condition, the line $L$ could be anywhere on the plane spanned by all the camera centers; the point $p$ could be anywhere on the line through $o_1, o_5, o_6$.

## 9.C.2 *Case 1:* $1 \leq rank(M) \leq 2$

We now discuss case 1 in Theorem 9.7, i.e. $1 \leq \text{rank}(M) \leq 2$. In this case, the matrix $M$ must contain at least one submatrix of the type

$$\begin{bmatrix} \widehat{\boldsymbol{x}}_i R_i \widehat{\boldsymbol{\ell}}_1 & \widehat{\boldsymbol{x}}_i T_i \end{bmatrix} \in \mathbb{R}^{3\times 4}, \tag{9.50}$$

for some $i \geq 2$. It is easy to verify that such a submatrix can never be zero; hence the only possible values for the rank of $M$ are

$$(a)\ \text{rank}(M) = 2, \quad \text{and} \quad (b)\ \text{rank}(M) = 1. \tag{9.51}$$

*Subcase* $(a)$. When the rank of $M$ is 2, it corresponds to the generic case. A representative example here is the matrix $M_{lp}$ given in (9.10). If $\text{rank}(M_{lp}) = 2$, it can be shown that the point $p$ must lie on the preimage plane of $\boldsymbol{\ell}_1$, and it is also the preimage of all the image points $\boldsymbol{x}_2, \boldsymbol{x}_3, \ldots, \boldsymbol{x}_m$. The line $L$, however, is determined only up to this plane, and the point $p$ does not have to be on this line. This is illustrated in Figure 9.16.

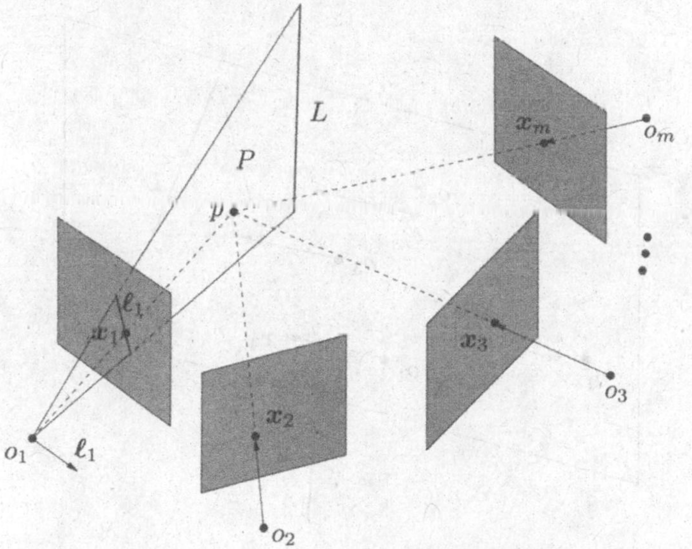


Figure 9.16. Generic configuration for the case $\text{rank}(M_{lp}) = 2$.

Beyond $M_{lp}$, if there are more than two image lines present in some $M$ of this type, the preimage (plane) of each line must pass through the point $p$. Hence $p$ must be on the intersection of these planes. Notice that in this case, adding more rows of line features to $M$ will not be useful to determine $L$ in space. This is because the incidence condition for multiple line features requires that the rank of the associated matrix $M_l$ be 1 not 2. If we only require rank 2 for the overall matrix $M$, we can at best determine the lines up to a family of lines intersecting at $p$.

*Subcase* (*b*). When the rank of $M$ is 1, it corresponds to a set of degenerate cases. For example, it is straightforward to show that $M_{lp}$ is of rank 1 if and only if all the vectors $R_i^{-1}\boldsymbol{x}_i$ are parallel to each other and they are all orthogonal to $\boldsymbol{\ell}_1$, and all $R_i^{-1}T_i$ are orthogonal to $\boldsymbol{\ell}_1$, $i = 2, 3, \ldots, m$. That means that all the camera centers lie on the same plane specified by $o_1$ and $\boldsymbol{\ell}_1$ and all the images $\boldsymbol{x}_2, \boldsymbol{x}_3, \ldots, \boldsymbol{x}_m$ (transformed to the reference camera frame) lie on the same plane and are parallel to each other. For example, suppose that $m = 5$ and choose $M$ to be

$$M = \begin{bmatrix} \widehat{\boldsymbol{x}_2} R_2 \widehat{\boldsymbol{\ell}_1} & \widehat{\boldsymbol{x}_2} T_2 \\ \widehat{\boldsymbol{x}_3} R_3 \widehat{\boldsymbol{\ell}_1} & \widehat{\boldsymbol{x}_3} T_3 \\ \widehat{\boldsymbol{x}_4} R_4 \widehat{\boldsymbol{\ell}_1} & \widehat{\boldsymbol{x}_4} T_4 \\ \boldsymbol{\ell}_5^T R_5 \widehat{\boldsymbol{\ell}_1} & \boldsymbol{\ell}_5^T T_5 \end{bmatrix} \in \mathbb{R}^{10 \times 4}. \tag{9.52}$$

The geometric configuration of the point and line features corresponding to the condition $\text{rank}(M) = 1$ is illustrated in Figure 9.17.

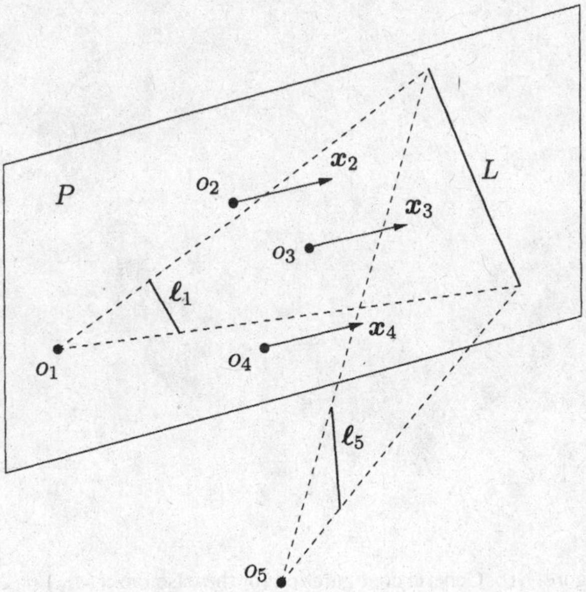


Figure 9.17. A degenerate configuration for the case $\text{rank}(M) = 1$: a line-point-point-point-line scenario.

Notice that since the rank condition $\text{rank}(M) = 1$ is the generic case for line features, preimages of lines intersect at a unique line $L$. But preimages of the points are parallel, and one can view them as if they intersected at a point $p$ at infinity. In general, the point $p$ does not have to lie on the line $L$, unless both the point $p$ and line $L$ are on the plane at infinity in the first place.

# Further readings

*Points, lines, and planes*

The constraints among multiple images of point and line features have been to a large extent studied separately, except for the work of [Faugeras et al., 1987, Spetsakis and Aloimonos, 1990b, Hartley, 1994a, Vieville et al., 1996], and more recently [Morris and Kanade, 1998]. Planes were studied almost exclusively using two-view homography but usually treated differently from points and lines; see [Hartley and Zisserman, 2000, Faugeras and Luong, 2001]. The first unified study of the relationships between the rank conditions and the incidence relations among points, lines, and planes was given by [Ma et al., 2001a, Ma et al., 2003], on which this chapter is based.

*Curves and surfaces*

Study on incidence relations with linear and conic trajectories can be found in [Avidan and Shashua, 1999, Avidan and Shashua, 2000], and their relations to the rank-based approach, as shown in this chapter, was given in [Huang et al., 2002]. Matching curves was studied by [Schmidt and Zisserman, 2000]. Effective reconstruction schemes for 3-D curves were also developed by [Berthilsson et al., 1999]. Generalization of multiple-view geometry (especially the two-view geometry) to curved surfaces can be found in [Cipolla et al., 1995, Åström et al., 1999].

*High-dimensional and non-Euclidean spaces*

[Wolf and Shashua, 2001a] showed that many interesting dynamic scenes can be embedded as a multiple-view geometric problem in higher-dimensional spaces. A systematic generalization of the rank-condition approach to dynamic scenes and higher-dimensional spaces was given by [Huang et al., 2002]. A further extension of multiple-view geometry to spaces of constant curvature (Euclidean, hyperbolic, and spherical) can be found in [Ma, 2003].

# Chapter 10

# Geometry and Reconstruction from Symmetry

*So their (the five platonic solids) combinations with themselves and with each other give rise to endless complexities, which anyone who is to give a likely account of reality must survey.*

– Plato, *The Timaeus, fourth century B.C.*

In Chapter 6 we have illustrated how prior assumptions on the scene can be exploited to simplify, or in some case enable, the reconstruction of camera pose and calibration. For instance, the presence of parallel lines and right angles in the scene allows one to upgrade the projective reconstruction to affine and even Euclidean. In this chapter, we generalize these concepts to the case where the scene contains objects that are *symmetric*. While we will make this notion precise shortly, the intuitive terms of "regular structures," (deterministic) "patterns," "tiles," etc. can all be understood in terms of symmetries (Figure 10.1).

The reason for introducing this chapter at this late stage (as opposed to presenting this material in Chapter 6) is because, as we will see, symmetry constraints on perspective images can be enforced very naturally within the context of rank conditions for multiple-view geometry, which we have studied in the previous chapters.

## 10.1 Symmetry and multiple-view geometry

Before we proceed formally, first let us pause and examine the images given in Figure 10.1 below. It is not hard to convince ourselves that even from a *single*

Figure 10.1. Man-made and natural symmetry. (Polyhedron image courtesy of Vladimir Bulatov, Physics Department, Oregon State University.)

image, we can perceive the 3-D structure and pose (orientation and location) of the objects. The goal of this chapter is to provide a computational framework for exploiting symmetries in the scene. The reader should be aware, however, that we are making a strong assumption on the scene. If this assumption is violated, as for instance the case shown in Figure 10.2, or Figure 6.1 that we have seen earlier, our reconstruction will necessarily be incorrect and lead to illusions.

Figure 10.2. Making assumptions on the scene results in gross perceptual errors when such assumptions are violated. The photograph above could come from an ordinary room inhabited by people of improbable stature, or from an improbable room with nonrectangular floors and windows that, when viewed from a particular vantage point, give the impression of symmetry. The Ames room is named after its designer Adelbert Ames. (Courtesy of (c) Exploratorium, www.exploratorium.edu.)

### 10.1.1 Equivalent views of symmetric structures

Symmetric structures are characterized by the fact that there exist several vantage points from which they appear identical. This notion is captured by the concept of *equivalent views*. Before we formulate this concept more formally later in this section, we illustrate the basic idea with a few examples.

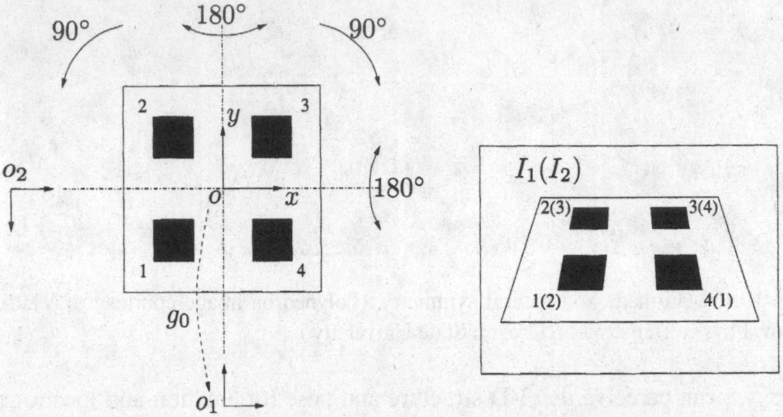


Figure 10.3. Left: a checkerboard exhibits different types of symmetries, including rotations about $o$ by 90° and reflections along the $x$- and $y$-axes. $g_0$ is the relative pose between the board and our vantage point. Right: an image $I_1$ taken at location $o_1$. Notice that the image would be identical if it was taken at $o_2$ instead ($I_2$).

**Example 10.1 (Rotational symmetry).** Consider a checkerboard as shown in Figure 10.3. There are at least four equivalent vantage points from which we would see exactly the same image. These are obtained by rotating the board about its normal. The only difference in an image seen from an equivalent vantage point is which features in the image correspond to those on the board. In this sense, such an image is in fact *different* from the original one. For simplicity, we call all such images *equivalent views*. For instance, in Figure 10.3, we label the corresponding corners for one of the equivalent views, to compare it with the original image. ■

**Example 10.2 (Reflective symmetry).** In addition to rotational symmetry, Figure 10.3 also exhibits (bilateral) *reflective symmetry*. This gives rise to additional equivalent views shown in Figure 10.4. Notice that in the figure, the two equivalent views with the four corners labeled by numbers in parentheses cannot be an image of the same board from any (physically viable) vantage point! One may argue that they are images taken from behind the board. This is true if the board is "transparent." If the symmetric object is a 3-D object rather than a 2-D plane, such an argument will fall apart. Nevertheless, as we will see below, just like the rotational symmetry, this type of equivalent views also encodes rich 3-D geometric information about the object. ■

**Example 10.3 (Translational symmetry).** Another type of symmetry, as shown in Figure 10.5, is due to a basic element (in this case a square) repeating itself indefinitely along one or more directions, called *infinite rapport*. Although equivalent images all appear identical, features (e.g., points, lines) in these images correspond to different physical features in the world. Therefore, an image like the third one in Figure 10.1 gives rise to many equivalent views. ■

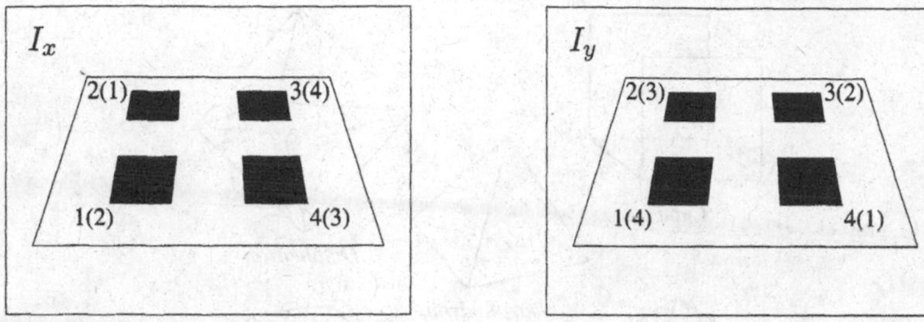


Figure 10.4. $I_x$: Correspondence between the original image of the board and an equivalent image with the board reflected about the $x$-axis by 180°; $I_y$: Correspondence between the original image of the board and an equivalent image with the board reflected about the $y$-axis by 180°.

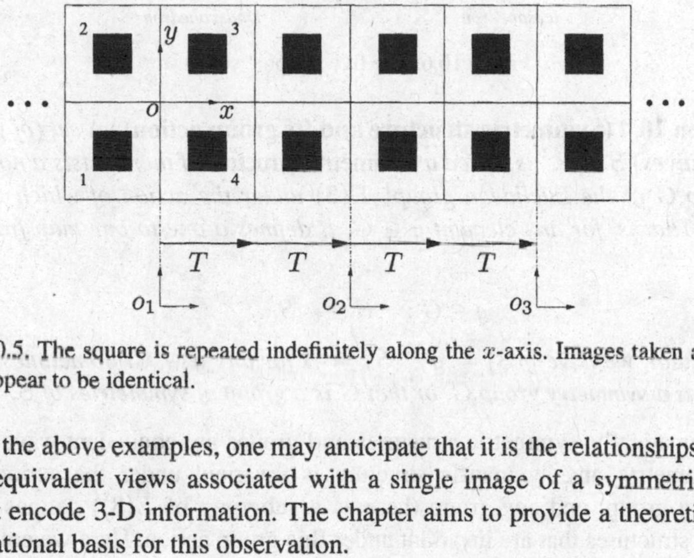


Figure 10.5. The square is repeated indefinitely along the $x$-axis. Images taken at $o_1, o_2$, and $o_3$ appear to be identical.

From the above examples, one may anticipate that it is the relationships among all the equivalent views associated with a single image of a symmetric structure that encode 3-D information. The chapter aims to provide a theoretical and computational basis for this observation.

### 10.1.2 Symmetric structure and symmetry group

Discussion on how many different types of 2-D or 3-D symmetric structures may exist has been documented in the antiquity. For instance, Egyptians certainly knew about all 17 possible ways of tiling the plane (we will see a few of them in Exercise 10.10); Pythagoras knew about the five platonic solids, shown in Figure 10.6. These solids are the ones that admit the only three fundamental types of (discrete) 3-D rotational symmetry (see Exercise 10.3), in addition to the planar ones (see Example 10.5). These observations, however, were not formalized until the nineteenth century [Fedorov, 1885, Fedorov, 1971]. In our applications, however, it suffices to know that rotational, reflective, and translational symmetries are the only isometric symmetries in Euclidean space; any other symmetry is just a combination of these three [Weyl, 1952]. More formally, we have the following definition:

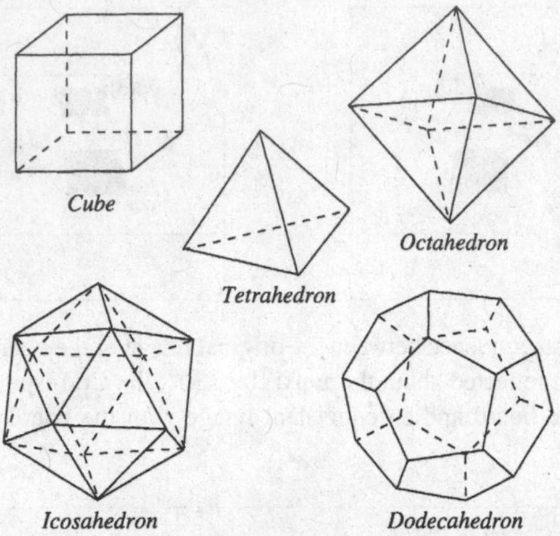


Figure 10.6. The five platonic solids.

**Definition 10.4 (Symmetric structure and its group action).** *A set (of geometric primitives) $S \subset \mathbb{R}^3$ is called a* symmetric structure *if there exists a nontrivial subgroup $G$ of the Euclidean group*[1] *$E(3)$ under the action of which $S$ is invariant. That is, for any element $g \in G$, it defines a one-to-one map from $S$ to itself:*

$$g \in G: \quad S \rightarrow S.$$

*In particular, we have $g(S) = g^{-1}(S) = S$ for any $g \in G$. Sometimes we say that $S$ has a symmetry group $G$, or that $G$ is a group of symmetries of $S$.*

Mathematically, symmetric structures and groups are equivalent ways to capture symmetry: any symmetric structure is invariant under the action of its symmetry group; and any group (here as a subgroup of $E(3)$) defines a class of (3-D) structures that are invariant under this group action. Here we emphasize that $G$ is in general a subgroup of the Euclidean group $E(3)$, and therefore, we consider only isometric symmetry groups. The reader should be aware that $G$ is in general not a subgroup of $SE(3)$. This is because many symmetric structures that we are going to consider are invariant under reflection, which is an element in $O(3)$ but not $SO(3)$.[2] For simplicity, in this chapter, we will consider primarily discrete symmetry groups.

**Example 10.5 (Symmetry groups of $n$-gons).** An $n$-gon is an equilateral and equal-angle $n$-sided (planar) polygon (e.g., a square is a 4-gon). Figure 10.7 shows a few common $n$-gons for $n = 3, 4, 5, 6, 8$. Each $n$-gon allows a finite rotation group that contains a primitive rotation $R$ around its center by $\theta = \frac{2\pi}{n}$ and its iteration $R^1, R^2, \ldots, R^n =$

[1]The Euclidean group $E(3)$ includes rigid-body motions and reflections, which do not preserve the orientation.

[2]Here $O(3)$ denotes the group of $3 \times 3$ orthogonal matrices including both rotations ($SO(3)$) and reflections, with determinant $-1$.

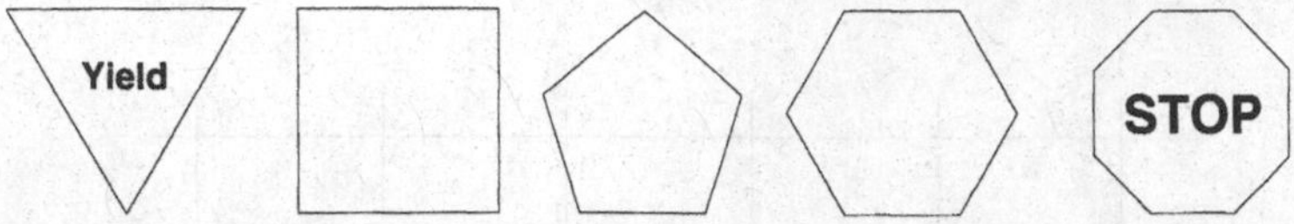


Figure 10.7. Equilateral triangle, square, pentagon, hexagon, and octagon.

identity. This rotational subgroup is called the *cyclic group* of order $n$, denoted by $C_n$. The $n$-gon also admits a total of $n$ reflections in $n$ lines forming angles of $\frac{1}{2}\theta$. Thus, there are a total of $2n$ elements in the symmetry group that an $n$-gon admits. This is the so-called *dihedral group* of order $n$, denoted by $D_n$. ■

Using the homogeneous representation of $E(3)$, any element $g$ in $G$ can be represented as a $4 \times 4$ matrix of the form

$$g = \begin{bmatrix} R & T \\ 0 & 1 \end{bmatrix} \in \mathbb{R}^{4\times 4}, \tag{10.1}$$

where $R \in \mathbb{R}^{3\times 3}$ is an orthogonal matrix ("$R$" stands for both rotation and reflection) and $T \in \mathbb{R}^3$ is a vector ("$T$" for translation). Note that in order to represent $G$, a world coordinate frame must be chosen. Interestingly, for a symmetric structure, there is typically a natural choice for this frame, called a *canonical frame*. For instance, if the symmetry is rotational, it is natural to choose the origin of the world frame to be the center of rotation and one of the coordinate axes, say the $z$-axis, to be the rotational axis. Often, such a choice results in a simple representation of the symmetry.

**Example 10.6 (Homogeneous representation for the symmetry group of a rectangle).** Rectangles are arguably the most ubiquitous symmetric objects in man-made environments. Choose an object coordinate frame in a rectangle in the following way: the $x$- and $y$-axes correspond to the two axes of reflection and the $z$-axis is the normal to the rectangle. With respect to this coordinate frame, the four elements in the symmetry group $G$ admitted by the rectangle can be expressed in homogeneous representation as

$$\begin{bmatrix} 1&0&0&0\\0&1&0&0\\0&0&1&0\\0&0&0&1 \end{bmatrix}, \begin{bmatrix} -1&0&0&0\\0&1&0&0\\0&0&1&0\\0&0&0&1 \end{bmatrix}, \begin{bmatrix} 1&0&0&0\\0&-1&0&0\\0&0&1&0\\0&0&0&1 \end{bmatrix}, \begin{bmatrix} -1&0&0&0\\0&-1&0&0\\0&0&1&0\\0&0&0&1 \end{bmatrix}.$$

From left to right, these four matrices correspond to the identity transformation, denoted by $g_e$, reflection along the $x$-axis, denoted by $g_x$, reflection along the $y$-axis, denoted by $g_y$, and rotation by 180° around the $z$-axis, denoted by $g_z$. Obviously, elements in the group $G = \{g_e, g_x, g_y, g_z\}$ satisfy the relations:

$$g_x^2 = g_y^2 = g_z^2 = g_e, \quad g_x g_y = g_z, \quad g_x g_z = g_z g_x = g_y, \quad g_y g_z = g_z g_y = g_x.$$

Note that the symmetry of a rectangle has no translational part (with respect to the chosen frame); i.e. $T = 0$ in all $g \in G$. ■

**Example 10.7 (Symmetry of a tiled plane).** For a 2-D plane tiled with congruent rectangles, as shown in Figure 10.8, its symmetry group $G$ consists of not only all the symmetries

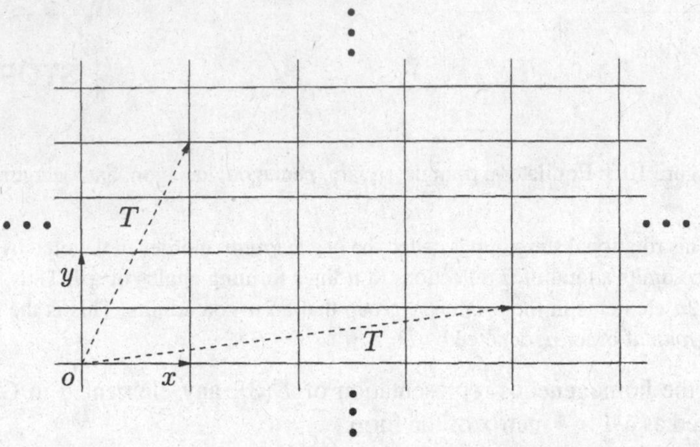


Figure 10.8. A 2-D plane tiled by congruent rectangles.

of the rectangle given in the above example but also the translational symmetries:

$$g_T = \begin{bmatrix} 1 & 0 & 0 & T_x \\ 0 & 1 & 0 & T_y \\ 0 & 0 & 1 & 0 \\ 0 & 0 & 0 & 1 \end{bmatrix} \in SE(3),$$

where $(T_x, T_y)$ can be the coordinates of any lattice point in the grid. Clearly, the set of all such translations forms a group. ■

### 10.1.3 *Symmetric multiple-view matrix and rank condition*

Now suppose that an image of a symmetric structure $S$ is taken at the vantage point $g_0 = (R_0, T_0) \in SE(3)$; we call it the *canonical pose* of the structure relative to the viewer or camera. Here $g_0$, though unknown, is assumed to be represented with respect to the same canonical coordinate frame. As we will soon see, this relative pose $g_0$ from the structure to the viewer can be uniquely determined from a single image as long as the symmetry admitted by the structure (or the object) is "rich" enough.

A (calibrated) perspective image of $S$ is simply a set of image points $I \subset \mathbb{R}^2$, and, in homogeneous coordinates, each image point $\boldsymbol{x} \in I$ satisfies

$$\lambda \boldsymbol{x} = \Pi_0 g_0 \boldsymbol{X}, \tag{10.2}$$

where $\boldsymbol{X} \in \mathbb{R}^4$ represents the homogeneous coordinates of a point $p \in S$. Now since $g(S) = S$ for all $g \in G$, we have $g(I) = I$. For example, in Figure 10.3, if $g$ is a rotation of $90°$ around the center of the board, we have $g(I_1) = I_2$. Although, after applying the transformation $g$ to $S$ in space, the image $\boldsymbol{x}$ of a point $p \in S$ will become a different point, say $\boldsymbol{x}'$, on the image plane, $\boldsymbol{x}'$ must coincide with one of the image points in $I$ (taken from the original vantage point $g_0$). That is, $\boldsymbol{x}' \in I$, and we denote $\boldsymbol{x}' = g(\boldsymbol{x})$. Thus, the group $G$ does nothing but permute image points in $I$, which is an action induced from its action on $S$ in space.

We can also interpret the above observation in a slightly different way: After applying the symmetry $g \in G$, equation (10.2) yields

$$\lambda' g(\boldsymbol{x}) = \Pi_0 g_0 g \boldsymbol{X} = \Pi_0 (g_0 g g_0^{-1})(g_0 \boldsymbol{X}), \quad \forall g \in G. \tag{10.3}$$

The second equality expresses the fact that the image of the structure $S$ remains the same if taken from a vantage point different from the original one by $g_0 g g_0^{-1}$. The transformation $g_0 g g_0^{-1}$ is exactly the action of $g$ on $S$ expressed with respect to the camera coordinate frame. This relationship can be concisely described by the following commutative diagram:

$$\begin{array}{ccccc} S & \xrightarrow{g_0} & g_0(S) & \xrightarrow{\Pi_0} & I \\ g \big\downarrow & & g_0 g g_0^{-1} \big\downarrow & & \big\downarrow g \\ S & \xrightarrow{g_0} & g_0(S) & \xrightarrow{\Pi_0} & I \end{array} \tag{10.4}$$

Therefore, given only one image $I$, if we know the type of symmetry $G$ in advance and how its elements act on points in $I$, the set $\{g(\boldsymbol{x}) : g \in G\}$ can be interpreted as the set of different images of the same point $\boldsymbol{X}$ seen at different vantage points. In this way, we effectively obtain as many as $|G|$ *equivalent views* of the same 3-D structure $S$.[3]

We remind the reader that here $g_0$ is not the relative motion between different vantage points but the relative pose from the structure to the viewer. As we will soon see, symmetry in general encodes strong 3-D information and often allows us to determine $g_0$. Of course, also because of symmetry, there is no unique solution to the initial pose $g_0$, since $\boldsymbol{x} \sim \Pi_0 g_0 g(g^{-1}\boldsymbol{X})$ for all $g \in G$.[4] That is, the image point $\boldsymbol{x}$ might as well be the image of the point $g^{-1}\boldsymbol{X}$ seen from the vantage point $g_0 g$. Hence, in principle, $g_0$ is recoverable only up to the form $g_0 g$ for an arbitrary $g \in G$.[5] Since in most cases we will be dealing with a finite group $G$, determining $g_0$ up to such a form will give us all the information we need about the relative pose of the symmetric structure.

Let $\{g_i = (R_i, T_i)\}_{i=1}^m$ be $m$ different elements in $G$. Then, one image $\boldsymbol{x} \sim \Pi_0(g_0\boldsymbol{X})$ of a symmetric structure with the symmetry $G$ is equivalent to at least $m$ equivalent views that satisfy the following equations:

$$\begin{aligned} g_1(\boldsymbol{x}) &\sim \Pi_0 g_0 g_1 g_0^{-1}(g_0\boldsymbol{X}), \\ g_2(\boldsymbol{x}) &\sim \Pi_0 g_0 g_2 g_0^{-1}(g_0\boldsymbol{X}), \\ &\vdots \\ g_m(\boldsymbol{x}) &\sim \Pi_0 g_0 g_m g_0^{-1}(g_0\boldsymbol{X}). \end{aligned}$$

[3] Here we use $|G|$ to denote the cardinality of $G$. In particular, when $G$ is finite, $|G|$ is the number of elements in $G$.

[4] Recall that we use the symbol $\sim$ to denote equality up to a scalar factor.

[5] $\{g_0 g : g \in G\}$ is called a *left coset* of $G$.

Notice that $g_i' \doteq g_0 g_i g_0^{-1}$ plays the role of the relative transformation between the original image and the $i$th equivalent view. From previous chapters, these equivalent views must be related by the multiple-view rank condition. That is, the symmetric multiple-view matrix

$$M_s(\boldsymbol{x}) \doteq \begin{bmatrix} \widehat{g_1(\boldsymbol{x})} R_1' \boldsymbol{x} & \widehat{g_1(\boldsymbol{x})} T_1' \\ \widehat{g_2(\boldsymbol{x})} R_2' \boldsymbol{x} & \widehat{g_2(\boldsymbol{x})} T_2' \\ \vdots & \vdots \\ \widehat{g_m(\boldsymbol{x})} R_m' \boldsymbol{x} & \widehat{g_m(\boldsymbol{x})} T_m' \end{bmatrix} \in \mathbb{R}^{3m \times 2} \tag{10.5}$$

with $g_i' = (R_i', T_i')$ and

$$\begin{cases} R_i' & \doteq & R_0 R_i R_0^T \quad \in O(3), \\ T_i' & \doteq & (I - R_0 R_i R_0^T) T_0 + R_0 T_i \quad \in \mathbb{R}^3, \end{cases} \qquad i = 1, 2, \ldots, m, \tag{10.6}$$

satisfies the rank condition

$$\boxed{\operatorname{rank}(M_s(\boldsymbol{x})) \leq 1, \quad \forall \boldsymbol{x} \in I.} \tag{10.7}$$

Note that this rank condition is independent of any particular order of the group elements $g_1, g_2, \ldots, g_m$, and it captures the *only* fundamental invariant that a perspective image of a symmetric structure admits.[6] We call it the *symmetric multiple-view rank condition*. Note that if $G \subseteq O(3)$ (i.e. $T_i = 0$ for all $i$), the expression for $T_i'$ is simplified to

$$T_i' = (I - R_0 R_i R_0^T) T_0, \quad i = 1, 2, \ldots, m. \tag{10.8}$$

To summarize, one image of a symmetric structure $S$ with its symmetry group $G$ is equivalent to $m = |G|$ images of $n = |S|$ feature points.[7] The reconstruction of $g_i' = (R_i', T_i')$ and the 3-D structure of $S$ can be easily solved by the multiple-view factorization algorithm given in the previous chapters. Nevertheless, in order to solve for the initial canonical pose $g_0 = (R_0, T_0)$, we need to further solve a system of Lyapunov type equations

$$\boxed{g_i' g_0 - g_0 g_i = 0, \quad g_i \in G,} \tag{10.9}$$

with $g_i$ and $g_i' = g_0 g_i g_0^{-1}$ known. The uniqueness of the solution for $g_0$ depends on the group $G$, the conditions for which will become clear later.

### 10.1.4 *Homography group for a planar symmetric structure*

According to the previous chapter, if the symmetric structure $S$ is planar, the multiple-view rank conditions associated with its equivalent views reduce to the

---

[6]Here "only" is in the sense of sufficiency: If a set of features satisfies the rank condition, it can always be interpreted as a valid image of an object with the symmetry $G$.

[7]It is possible that both $|G|$ and $|S|$ are infinite. In practice, one can conveniently choose a finite subset that is limited to the field of view. In this case, $G$ in fact becomes a "groupoid" instead of a "group"; see [Weinstein, 1996].

so-called homography constraint. Here we show how the homography helps simplifying the study of planar symmetry. Suppose that the supporting plane $P$ of $S$ is defined by the equation $N^T X = d$ for any point $X \in \mathbb{R}^3$ on $P$ in the camera frame. Geometrically, $N \in \mathbb{R}^3$ is the unit normal vector of the plane, and $d \in \mathbb{R}_+$ is its distance to the camera center. As we know from Chapter 5, the homography matrix $H_0 \doteq [R_0(1), R_0(2), T_0] \in \mathbb{R}^{3\times 3}$, where $R_0(1), R_0(2)$ are the first two columns of $R_0$, directly maps the plane $P \supseteq S$ to the image plane $I$.

Since $S$ is now planar, its symmetry group $G$ can be represented as a subgroup of the planar Euclidean group $E(2)$, and each element in $G$ can be represented as a $3 \times 3$ matrix.[8] Due to the symmetry of $S$, we have $g(S) = S$ and therefore $H_0(g(S)) = H_0(S)$. For a particular point $X \in S$, we have

$$H_0(g(X)) = H_0 g H_0^{-1}(H_0(X)). \tag{10.10}$$

Therefore, in the planar case, the diagram (10.4) simplifies to

$$\begin{array}{ccc} S & \xrightarrow{H_0} & I \\ g \downarrow & & \downarrow H_0 g H_0^{-1} \\ S & \xrightarrow{H_0} & I \end{array} \tag{10.11}$$

The group action of $G$ on the plane $P$ in space is then naturally represented by its conjugate group $G' \doteq H_0 G H_0^{-1}$ acting on the image plane. Or, equivalently, any element $H' = H_0 g H_0^{-1} \in G'$ represents the homography transformation between two equivalent views. Depending on the type of symmetry, the matrix $H'$ may represent either a *rotational*, *reflective*, or *translational* homography. Figure 10.9 shows a reflective homography induced from the symmetry of a rectangle. We call the group $G' = H_0 G H_0^{-1}$ the *homography group*.

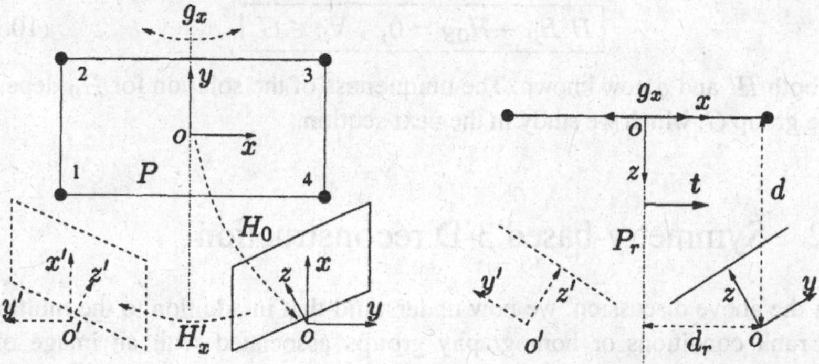


Figure 10.9. Homography between equivalent images of a rectangle, before and after a reflection $g_x$. Left: frontal view; right: top view. $P_r$ is the plane of reflection and $t$ is its (unit) normal vector.

[8] Since all the symmetry transformations are restricted to the $xy$-plane, the $z$-coordinate can be simply dropped from the representation.

**Example 10.8 (The homography group for an image of a rectangle).** For an image of a rectangle, whose symmetry was studied in Example 10.6, the homography group $G' = H_0 G H_0^{-1}$ is given by

$$\{I, H_0 g_x H_0^{-1}, H_0 g_y H_0^{-1}, H_0 g_z H_0^{-1}\} \doteq \{I, H'_x, H'_y, H'_z\},$$

and its elements satisfy the same set of relations as $G$ in Example 10.6:

$$(H'_x)^2 = (H'_y)^2 = (H'_z)^2 = I, \quad H'_x H'_y = H'_z,$$
$$H'_x H'_z = H'_z H'_x = H'_y, \quad H'_y H'_z = H'_z H'_y = H'_x.$$

One of the elements $H'_x$ of $G'$ is illustrated in Figure 10.9. ■

To compute the homography $H' \in G'$ from the image, let $\boldsymbol{x} = [x, y, z]^T \in \mathbb{R}^3$ be the (homogeneous) image of the point $\boldsymbol{X}$; i.e. $\boldsymbol{x} \sim H_0 \boldsymbol{X} \in \mathbb{R}^3$. Let $\boldsymbol{x}'$ be the image of its symmetric point $g(\boldsymbol{X})$. From (10.10) we get

$$\boldsymbol{x}' \sim H' \boldsymbol{x} \quad \Leftrightarrow \quad \boldsymbol{x}' \times (H' \boldsymbol{x}) = 0. \tag{10.12}$$

With four corresponding points between any pair of equivalent views, such as the four corner features of the rectangle in Figure 10.9, the homography matrix $H'$ can be linearly recovered from equation (10.12), using the techniques given in Chapter 5. We can decompose the recovered $H'$ into

$$H' \to \left\{R', \frac{1}{d}T', N\right\}$$

to obtain the relative pose $(R', T')$ between the two equivalent views (Exercises 10.6 and 10.7). The 3-D structure of $S$ can then be determined by a triangulation between the two views. Furthermore, we can use $H' = H_0 g H_0^{-1}$ to recover information about the homography matrix $H_0 = [R_0(1), R_0(2), T_0]$. The matrix $H_0$ obviously satisfies the following set of Lyapunov type linear equations

$$\boxed{H' H_0 - H_0 g = 0, \quad \forall g \in G,} \tag{10.13}$$

with both $H'$ and $g$ now known. The uniqueness of the solution for $H_0$ depends on the group $G$, which we study in the next section.

## 10.2 Symmetry-based 3-D reconstruction

From the above discussion, we now understand that in addition to the multiple-view rank conditions or homography groups associated with an image of a symmetric structure, we also have the Lyapunov equations (10.9) and (10.13), which give us some extra information about the camera initial pose $g_0$ relative to the canonical frame (centered at the object). In Section 10.2.1 we give necessary and sufficient conditions under which the pose $g_0$ is uniquely recoverable; then we study, in Section 10.2.2, what can be recovered if such conditions are not satisfied, and finally, we show in Section 10.2.3 how symmetry can facilitate 3-D reconstruction from multiple images.

### 10.2.1 Canonical pose recovery for symmetric structure

**Proposition 10.9 (Rotational and reflective symmetry).** *Given a (discrete) subgroup $G$ of $O(3)$, a rotation $R_0$ is uniquely determined from the pair of sets $(R_0GR_0^T, G)$ if and only if the only fixed point of $G$ acting on $\mathbb{R}^3$ is the origin.*

*Proof.* If $(R_0GR_0^T, G)$ are not sufficient to determine $R_0$, then there exists at least one other $R_1 \in SO(3)$ such that $R_0RR_0^T = R_1RR_1^T$ for all $R \in G$. Let $R_2 = R_1^TR_0$. Then, $R_2R = RR_2$ for all $R \in G$. Hence $R_2$ commutes with all elements in $G$. If $R_2$ is a rotation, all $R$ in $G$ must have the same rotation axis as $R_2$; if $R_2$ is a reflection, $R$ must have its axis normal to the plane that $R_2$ fixes. This is impossible for a group $G$ that fixes only the origin. On the other hand, if $(R_0GR_0^T, G)$ is sufficient to determine $R_0$, then the group $G$ cannot fix any axis (or a plane). Otherwise, simply choose $R_2$ to be a rotation with the same axis (or an axis normal to the plane). Then it commutes with $G$. The solution for $R_0$ cannot be unique. □

Once $R_0$ is determined, it is not difficult to show that with respect to the same group $G$, $T_0$ can be uniquely determined from the second equation in (10.6). Thus, as a consequence of the above proposition, we have the following theorem.

**Theorem 10.10 (Unique canonical pose from a symmetry group).** *Suppose that a symmetric structure $S$ admits a symmetry group $G$ that contains a rotational or reflective subgroup that fixes only the origin of $\mathbb{R}^3$. Then the canonical pose $g_0$ can always be uniquely determined from one image of $S$.*

Note that the group $G$ does not have to be the only symmetry that $S$ admits; as long as such a $G$ exists as a subgroup of the total symmetry group of $S$, one may claim uniqueness for the recovery of $g_0$.

The above statement, however, applies only to 3-D structures. The symmetry group of any 2-D (planar) symmetric structure, as a special 3-D symmetric group, does not satisfy the condition of the theorem. Since any planar structure $S$ is "symmetric" with respect to the reflection in its own supporting plane, we wonder whether this reflection can be added to the overall symmetry group $G$. The problem is that, even if we could add this reflection, say $R$, into the symmetry group $G$, it is not possible to recover its corresponding element $R_0RR_0^T$ in $R_0GR_0^T$, since features on the plane correspond to themselves under this reflection, and no other feature point outside the plane is available (by our own planar assumption). Thus, only elements in the planar symmetry group can be recovered from homographies between the equivalent views of the planar structure.

In order to give a correct statement for the planar case, for a reflection $R$ with respect to a plane, we call the normal vector to this plane of reflection the *axis of reflection*.[9] Using this notion and by restricting the argument of the proposition to the planar orthogonal group $O(2)$, one can reach the following conclusion.

---

[9]The role of the axis of a reflection is very similar to that of the axis of a rotation once we notice that for any reflection $R$, $-R$ is a rotation of angle $\theta = \pi$ about the same axis.

**Corollary 10.11 (Canonical pose from a planar symmetry group).** *If a planar symmetric structure $S$ allows a rotational or reflective symmetry subgroup $G$ (without the reflection with respect to the supporting plane of $S$ itself) with two independent rotation or reflection axes, the canonical pose $g_0$ can always be uniquely determined from one image of $S$ (with the canonical frame origin $o$ restricted in the plane and the $z$-axis chosen as the plane normal).*

As a consequence, to have a unique solution for $g_0$, a planar symmetric structure $S$ must admit at least two reflections with independent axes, or one reflection and one rotation (automatically with independent axes for a planar structure).

## 10.2.2 *Pose ambiguity from three types of symmetry*

In this section, we study ambiguities in recovering $g_0$ from a single reflective, rotational, or translational symmetry. The results will give the reader a clear understanding of the extent to which $g_0$ can be recovered if the conditions of the above theorem and corollary are not met.

*Reflective symmetry*

Many man-made objects, for example a building or a car, are symmetric with respect to a central plane (the plane or mirror of reflection). That is, the overall structure is invariant under a reflection with respect to this plane. Without loss of generality, suppose this plane is the $yz$-plane of a selected canonical coordinate frame. For instance, in Figure 10.3, the board is obviously symmetric with respect to the $yz$-plane if the $z$-axis is the normal to the board. Then a reflection in this plane can be described by a motion $g = (R, 0)$, where

$$R = \begin{bmatrix} -1 & 0 & 0 \\ 0 & 1 & 0 \\ 0 & 0 & 1 \end{bmatrix} \in O(3) \subset \mathbb{R}^{3\times 3} \tag{10.14}$$

is an element in $O(3)$ and it has $\det(R) = -1$. Notice that a reflection always fixes the plane of reflection. If one reflection is the only symmetry that a structure admits, then its symmetry group $G$ consists of only two elements $\{e, g\}$, where $e = g^2$ is the identity map.[10]

If one image of such a symmetric object is taken at $g_0 = (R_0, T_0)$, then we have the following two equations for each image point on this structure:

$$\lambda \boldsymbol{x} = \Pi_0 g_0 \boldsymbol{X}, \quad \lambda' \boldsymbol{x}' = \Pi_0 g_0 g \boldsymbol{X}, \tag{10.15}$$

where $\boldsymbol{x}' = g(\boldsymbol{x})$. To simplify the notation, define $R' \doteq R_0 R R_0^T$ and $T' \doteq (I - R_0 R R_0^T) T_0$. Then the symmetric multiple-view rank condition, in the two-view case, reduces to the well-known epipolar constraint $(\boldsymbol{x}')^T \widehat{T'} R' \boldsymbol{x} = 0$. In fact, if we normalize the length of $T'$ to be 1, one can show that $R' = I - 2T'(T')^T$ and

[10]In other words, $G$ is isomorphic to the group $\mathbb{Z}_2$.

therefore $\widehat{T'}R' = \widehat{T'}$. We leave the proof as an exercise to the reader (see Exercise 10.7). The epipolar constraint induced from a reflective symmetry becomes

$$(\boldsymbol{x}')^T \widehat{T'} \boldsymbol{x} = 0, \tag{10.16}$$

so that $T'$ can be recovered from two pairs of symmetric points, as opposed to eight points in the general case (in Chapter 5).

**Example 10.12 (Triangulation with the reflective symmetry).** For each pair of (reflectively) symmetric image points $(\boldsymbol{x}, \boldsymbol{x}')$, their 3-D depths can be uniquely determined from the equation

$$\begin{bmatrix} \widehat{\boldsymbol{t}}\boldsymbol{x} & -\widehat{\boldsymbol{t}}\boldsymbol{x}' \\ \boldsymbol{t}^T\boldsymbol{x} & \boldsymbol{t}^T\boldsymbol{x}' \end{bmatrix} \begin{bmatrix} \lambda \\ \lambda' \end{bmatrix} = \begin{bmatrix} 0_{3\times 1} \\ 2d_r \end{bmatrix}, \tag{10.17}$$

where $\boldsymbol{t} = T'$ is the unit normal vector of the plane of reflection $P_r$, and $d_r$ is the distance from the camera center $o$ to $P_r$ (see Figure 10.9).[11] ■

Once $R' = R_0 R R_0^T$ is obtained, we need to use $R'$ and $R$ to solve for $R_0$. The associated Lyapunov equation can be rewritten as

$$R' R_0 - R_0 R = 0 \tag{10.18}$$

with $R'$ and $R$ known.

**Lemma 10.13 (Reflective Lyapunov equation).** *Let $L : \mathbb{R}^{3\times 3} \to \mathbb{R}^{3\times 3};\ R_0 \to R'R_0 - R_0 R$ be the Lyapunov map associated with the above equation with $R$ a reflection and $R' = R_0 R R_0^T$ both known. The kernel $ker(L)$ of $L$, which is defined as the set $\{R \mid L(R) = 0\}$, is in general five-dimensional. Nevertheless, for orthogonal solutions of $R_0$, the intersection $ker(L) \cap SO(3)$ is only a one parameter family that corresponds to an arbitrary rotation in the plane of reflection.*

Using the property of Lyapunov maps, the proof is not difficult. We leave it to the reader as an exercise (see Exercise 10.4). But here we give out explicitly this one-parameter family solutions of $R_0$:

$$R_0 = [\pm v_1,\ v_2\cos(\alpha) + v_3\sin(\alpha),\ -v_2\sin(\alpha) + v_3\cos(\alpha)] \in SO(3),$$

where $v_1, v_2, v_3 \in \mathbb{R}^3$ are three (real) eigenvectors of $R'$ that correspond to the eigenvalues $-1$, $+1$, and $+1$, respectively, and $\alpha \in \mathbb{R}$ is an arbitrary angle.[12] Geometrically, the three columns of $R'$ can be interpreted as the three axes of the canonical coordinate frame that we attach to the structure. The ambiguity in $R_0$ then corresponds to an arbitrary rotation of the $yz$-plane around the $x$-axis (the reflection axis). If the structure further admits a reflection with respect to another plane, say the $xz$-plane as in the case of the checkerboard (Figure 10.3), this one-parameter family ambiguity can be eliminated.

[11] As long as the camera center $o$ is not on the plane of reflection $P_r$, we can normalize $d_r = 1$.

[12] Here, the "$\pm$" sign in front of $v_1$ is due to the fact that a particular choice of signs and orders for $v_1, v_2, v_3$ may not result in a rotation matrix $R_0$ with $\det(R_0) = +1$. But one of the choices is the correct one. The same convention will be adopted in the rest of this chapter.

After $R_0$ is recovered, $T_0$ is recovered up to the following form:

$$T_0 \in \left(I - R_0 R R_0^T\right)^\dagger T' + \text{null}(I - R_0 R R_0^T), \tag{10.19}$$

where $(I - R_0 R R_0^T)^\dagger$ is the pseudo-inverse[13] of $I - R_0 R R_0^T$ and $\text{null}(I - R_0 R R_0^T) = \text{span}\{v_2, v_3\}$, since both $v_2$ and $v_3$ are in the null space of the matrix $I - R_0 R R_0^T$. Such ambiguities in the recovered $g_0 = (R_0, T_0)$ are exactly what we should have expected: With a reflection with respect to the $yz$-plane, we can determine the $y$-axis and $z$-axis (including the origin) of the canonical coordinate frame only up to any orthonormal frame within the $yz$-plane, which obviously has three degrees of freedom, parameterized by $(\alpha, \beta, \gamma)$ (where $\alpha$ is the angle in the one-parameter family solutions of $R_0$, and $(\beta, \gamma)$ is the position of the frame origin in the $yz$-plane).

**Example 10.14 (Planar case: reflective homography).** In the case in which the structure $S$ is planar, by choosing a particular canonical coordinate frame, we can bypass the computation of the relative pose from the kernel of the Lyapunov map and determine the pose $(R_0, T_0)$ directly. By decomposing the associated reflective homography matrix $H' = R' + \frac{1}{d}T'N^T \in \mathbb{R}^{3\times3}$ (see Exercise 10.7), we obtain

$$H' \mapsto \left\{R', \frac{1}{d}T', N\right\}. \tag{10.20}$$

If we set the $x$-axis to be the axis of reflection and the $z$-axis to be the plane normal $N$, we get a solution for $R_0$:

$$R_0 = \left[\pm v_1,\ \pm\widehat{N} v_1,\ N\right] \quad \in SO(3),$$

where $v_1, v_2, v_3$ are eigenvectors of $R'$ as before. We may further choose the origin of the object frame to be in the plane. Thus, we can reduce the overall ambiguity in $g_0$ to a one-parameter family: only the origin $o$ may now translate freely along the $y$-axis, the intersection of the plane $P$ where $S$ resides, and the plane of reflection $P_r$. ■

To conclude our discussions on the reflective symmetry, we have the following result:

**Proposition 10.15 (Canonical pose from a reflective symmetry).** *Given an image of a structure $S$ with a reflective symmetry with respect to a plane in 3-D, the canonical pose $g_0$ can be determined up to an arbitrary choice of an orthonormal frame in this plane, which is a three-parameter family of ambiguities (i.e. $SE(2)$). However, if $S$ itself is in a (different) plane, $g_0$ is determined up to an arbitrary translation of the frame along the intersection line of the two planes (i.e. $\mathbb{R}$).*

**Example 10.16** Figure 10.10 demonstrates an experiment with the reflective symmetry. The checkerboard is a planar structure that is symmetric with respect to a central axis. ■

**Example 10.17 (Reflective stereo for the human face).** The human face, at least at a first approximation, is reflectively symmetric with respect its central plane. Hence, the above

[13] See Appendix A.

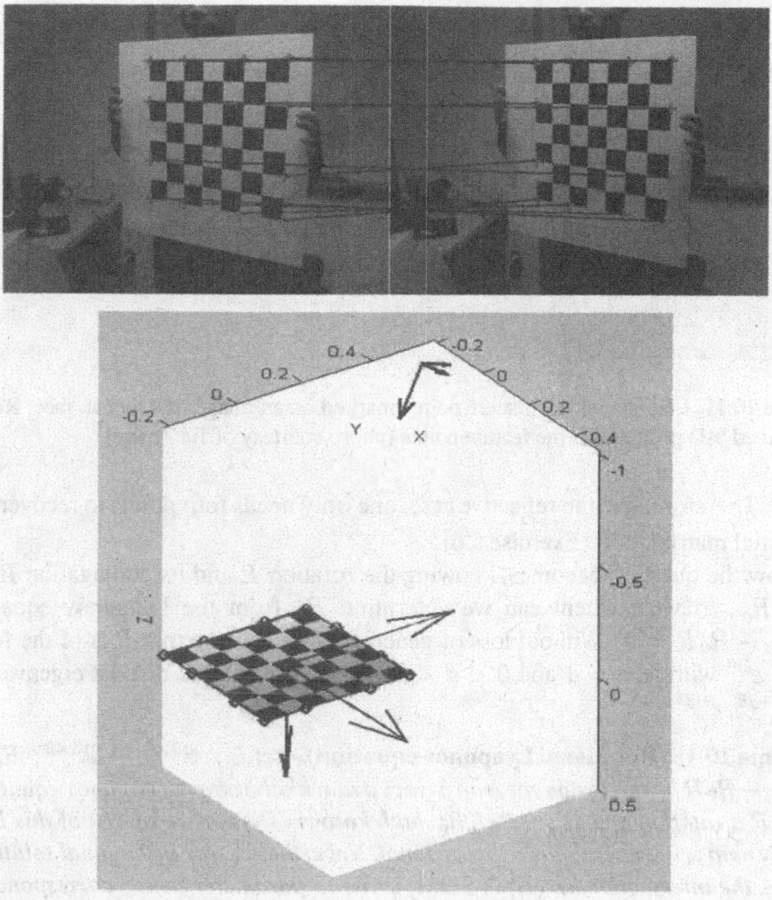


Figure 10.10. Top: An image of a reflectively symmetric checkerboard. We draw two identical images here to illustrate the correspondence more clearly. Points in the left image correspond to points in the right image by a reflective symmetry. Bottom: The reconstruction result from the reflective symmetry. The recovered structure is represented in the canonical world coordinate frame. From our discussion above, the origin $o$ of the world coordinate frame may translate freely along the $y$-axis. The smaller coordinate frame is the camera coordinate frame. The longest axis is the $z$-axis of the camera frame, which represents the optical axis of the camera.

technique allows us to recover a 3-D model of human face from a single view (see Figure 10.17). ■

*Rotational symmetry*

Now suppose that we replace the reflection $R$ above by a proper rotation. For instance, in Figure 10.3, the pattern is symmetric with respect to any rotation by a multiple of 90° around $o$ in the $xy$-plane. One can show that for a rotational symmetry, the motion between equivalent views is always an orbital motion (Exercise

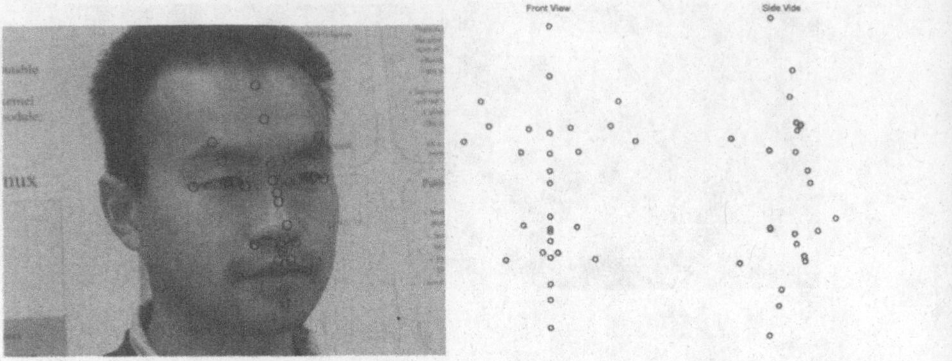


Figure 10.11. Left: symmetric feature points marked on an image of a human face; Right: recovered 3-D positions of the feature points (photo courtesy of Jie Zhang).

10.8). Therefore, like the reflective case, one only needs four points to recover the essential matrix $\widehat{T'}R'$ (Exercise 5.6).

Now the question becomes, knowing the rotation $R$ and its conjugation $R' = R_0 R R_0^T$, to what extent can we determine $R_0$ from the Lyapunov equation $R'R_0 - R_0 R = 0$? Without loss of generality, we assume that $R$ is of the form $R = e^{\widehat{\omega}\theta}$ with $\|\omega\| = 1$ and $0 < \theta < \pi$; hence it has three distinct eigenvalues $\{1, e^{+j\theta}, e^{-j\theta}\}$.

**Lemma 10.18 (Rotational Lyapunov equation).** *Let $L: \mathbb{R}^{3\times 3} \to \mathbb{R}^{3\times 3}$; $R_0 \to R'R_0 - R_0 R$ be the Lyapunov map associated with the above Lyapunov equation, with $R$ a rotation and $R' = R_0 R R_0^T$ both known. The kernel $ker(L)$ of this Lyapunov map is in general three-dimensional. Nevertheless, for orthogonal solutions of $R_0$, the intersection $ker(L) \cap SO(3)$ is a one-parameter family corresponding to an arbitrary rotation (of $\alpha$ radians) about the rotation axis $\omega$ of $R$.*

We also leave the proof to the reader as an exercise (see Exercise 10.5). We give only the solutions to $R_0$ here:

$$[-\mathrm{Im}(v_2)\cos(\alpha) - \mathrm{Re}(v_2)\sin(\alpha),\ \mathrm{Re}(v_2)\cos(\alpha) - \mathrm{Im}(v_2)\sin(\alpha),\ \pm v_1],$$

where $v_1, v_2, v_3 \in \mathbb{R}^3$ are three eigenvectors of $R'$ that correspond to the eigenvalues $+1$, $e^{+j\theta}$, and $e^{-j\theta}$, respectively, and $\alpha \in \mathbb{R}$ is an arbitrary angle. The ambiguity in $R_0$ then corresponds to an arbitrary rotation of the $xy$-plane around the $z$-axis (the rotation axis $\omega$).

This lemma assumes that $0 < \theta < \pi$. If $\theta = \pi$, $R$ has two repeated $-1$ eigenvalues, and the lemma and above formula for $R_0$ no longer apply. Nevertheless, we notice that $-R$ is exactly a reflection with two $+1$ eigenvalues, with a reflection plane orthogonal to the rotation axis of $R$. Thus, this case is essentially the same as the reflective case stated in Lemma 10.13. Although the associated Lyapunov map now has a five-dimensional kernel, its intersection with $SO(3)$ is the same as any other rotation: a one-parameter family that corresponds to an arbitrary rotation around the $z$-axis.

It can be verified directly that the null space of the matrix $I - R_0 R R_0^T$ is always one-dimensional (for $0 < \theta \leq \pi$) and $(I - R_0 R R_0^T)v_1 = 0$. Thus, the translation $T_0$ is recovered up to the form

$$T_0 \in \left(I - R_0 R R_0^T\right)^{\dagger} T' + \text{null}(I - R_0 R R_0^T), \tag{10.21}$$

where $\text{null}(I - R_0 R R_0^T) = \text{span}\{v_1\}$. Together with the ambiguity in $R_0$, $g_0$ is determined up to a so-called *screw motion* (see Chapter 2) about the rotation axis $\omega$.

The planar case can be dealt with in a similar way to the reflective homography in Example 10.14, except that only the $z$-axis can be fixed to be the rotation axis and the $x$- and $y$-axes are still free.

**Proposition 10.19 (Canonical pose from rotational symmetry).** *Given an image of a structure $S$ with a rotational symmetry with respect to an axis $\omega \in \mathbb{R}^3$, the canonical pose $g_0$ is determined up to an arbitrary choice of a screw motion along this axis, which is a two-parameter family of ambiguity (i.e. $SO(2) \times \mathbb{R}$). However, if $S$ itself is in a plane, $g_0$ is determined up to an arbitrary rotation around the axis (i.e. $SO(2)$).*

**Example 10.20** Figure 10.12 demonstrates an experiment with the rotational symmetry. Each face of the cube is a planar structure that is identical to another face by a rotation about the longest diagonal of the cube by 120°. ■

*Translational symmetry*

In the case of translational symmetry, since $R = I$ and $T \neq 0$, equation (10.6) is reduced to the following equations:

$$R' = R_0 I R_0^T = I, \quad T' = R_0 T. \tag{10.22}$$

Obviously, the first equation does not give any information on $R_0$ (since the associated Lyapunov map is trivial), nor on $T_0$. From the second equation, however, since both $T$ and $T'$ are known (up to a scalar factor), $R_0$ can be determined up to a one-parameter family of rotations.[14] Thus, the choice of the canonical frame (including $T_0$) is up to a four-parameter family.

Furthermore, if $S$ is planar, which is often the case for translational symmetry, the origin $o$ of the world frame can be chosen in the supporting plane, the plane normal as the $z$-axis, and $T'$ as the $x$-axis. Thus

$$R_0 = \left[T', \widehat{N}T', N\right] \in SO(3),$$

where both $T'$ and $N$ can be recovered by decomposing the translational homography $H = I + \frac{1}{d}T'N^T$ (Exercise 10.6). We end up with a two-parameter family of ambiguities in determining $g_0$: translating $o$ arbitrarily inside the plane (i.e.

[14] It consists of all rotations $\{R \in SO(3)\}$ such that $RT \sim T'$.

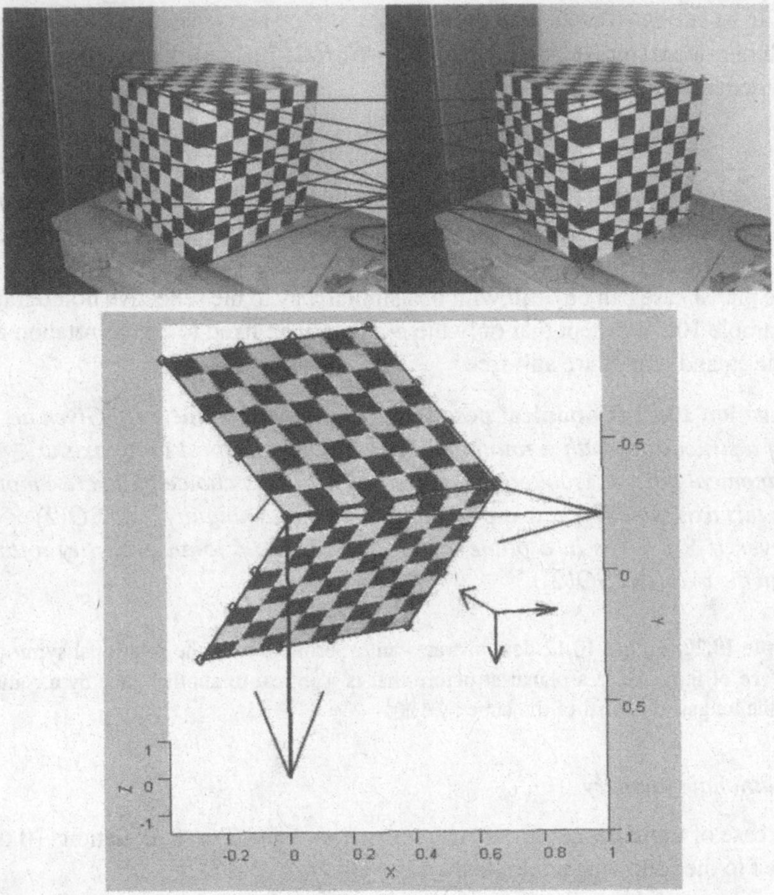


Figure 10.12. Top: An image of a cube that is rotationally symmetric about its longest diagonal axis. The symmetry is represented by some corresponding points. Points in the left images correspond to points in the right image by a rotational symmetry. Bottom: Reconstruction result from the rotational symmetry. The recovered structure is represented in the canonical world coordinate frame. From our discussion above, the origin $o$ of the world coordinate may translate freely along the $z$-axis, and the $xy$-axis can be rotated within the $xy$-plane freely. The smaller coordinate frame is the camera coordinate frame. The longest axis is the $z$-axis of the camera frame, which represents the optical axis of the camera.

$\mathbb{R}^2$). An extra translational symmetry along a different direction does not help reduce the ambiguities.

**Example 10.21** Figure 10.13 demonstrates an experiment with translational symmetry. A mosaic floor is a planar structure that is invariant with respect to the translation along proper directions. ■

We summarize in Table 10.1 the ambiguities in determining the pose $g_0$ from each of the three types of symmetry, for both generic and planar scenes.

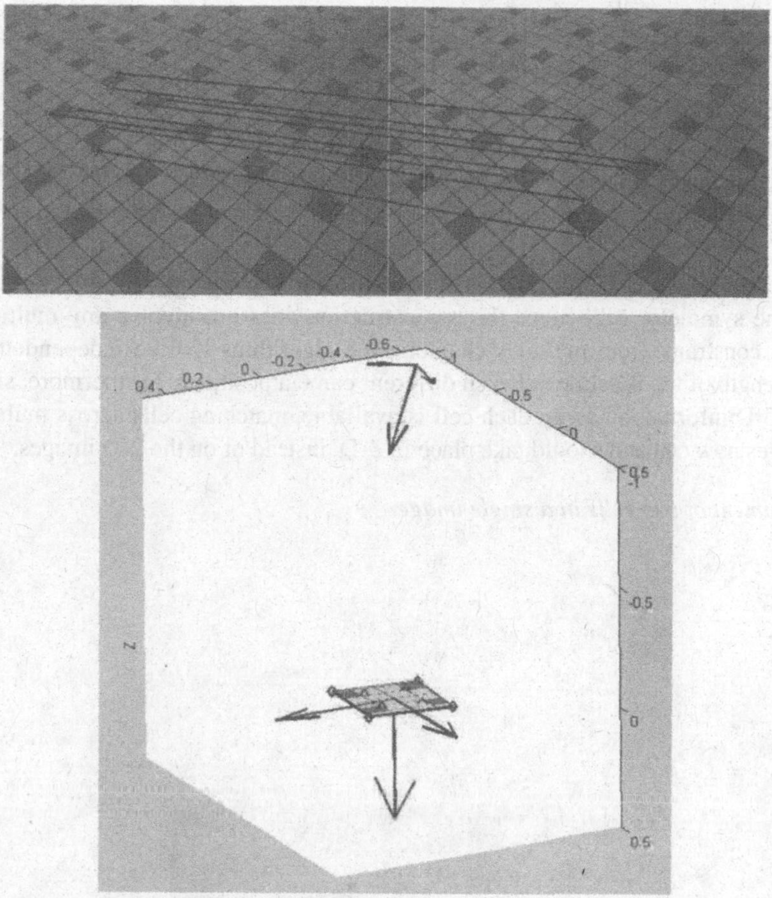

Figure 10.13. Top: An image of a mosaic floor that admits translational symmetry. The symmetry is represented by some corresponding points. We draw two identical images here to represent the correspondence more clearly: Points shown in the left images correspond to points shown in the right image by a translational symmetry. Bottom: Reconstruction result for the translational symmetry. The structure is represented in the canonical world coordinate frame. From our discussion above, the origin $o$ of the world coordinate may translate freely within the $xy$-plane. The smaller coordinate frame is the camera coordinate frame. The longest axis is the $z$-axis of the camera frame, which represents the optical axis of the camera.

### *10.2.3 Structure reconstruction based on symmetry*

In this section we give examples on how to exploit symmetric structures for reconstruction from multiple images. In particular, we use rectangles as an example, and we refer to a fundamental rectangular region as a "cell." Cells will now be the basic unit in the reconstruction of a 3-D model for symmetric objects.

From every image of every rectangular cell in the scene, we can determine up to scale its 3-D structure and 3-D pose. Knowing the pose $g_0$ allows us to

| Ambiguity | ker($L$) | $g_0$ (general scene) | $g_0$ (planar scene) |
|---|---|---|---|
| Reflective | 5-dimensional | $SE(2)$ | $\mathbb{R}$ |
| Rotational | 3-dimensional | $SO(2) \times \mathbb{R}$ | $SO(2)$ |
| Translational | 9-dimensional | $SO(2) \times \mathbb{R}^3$ | $\mathbb{R}^2$ |

Table 10.1. Ambiguity in determining canonical pose $g_0$ from one symmetry of each type.

represent all the 3-D information with respect to the canonical frame admitted by the symmetry cell. Since the reconstruction does not involve any multiple-view constraint from previous chapters, the algorithms will be independent of the length of the baseline between different camera positions. Furthermore, since full 3-D information about each cell is available, matching cells across multiple images now can and should take place in 3-D, instead of on the 2-D images.

*Alignment of two cells in a single image*

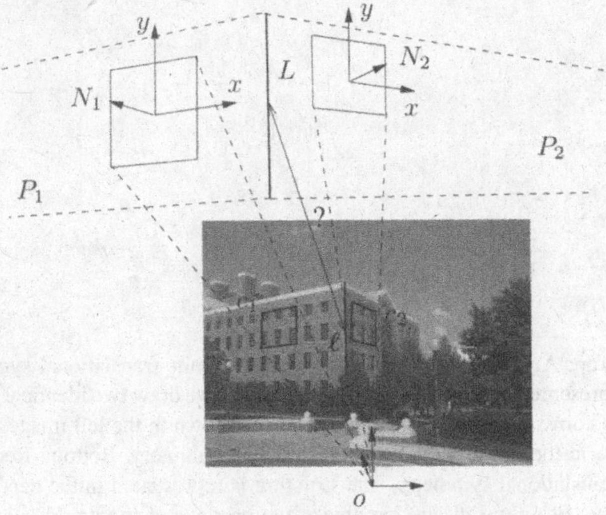


Figure 10.14. Two canonical coordinate frames determined using reflective symmetry of rectangles for window cells $c_1$ and $c_2$ do not necessarily conform to their true relative depth.

As shown in Figure 10.14, for an image with multiple rectangular cells in sight, the camera pose $(R_i, T_i)$ with respect to the $i$th cell frame calculated from the symmetry-based algorithm might have adopted a different scale for each $T_i$ (if the distance $d_i$ to every $P_i$ was normalized to 1 in the algorithm). Because of that, as shown in Figure 10.14, the intersection of the two planes in 3-D, the line $L$, does not necessarily correspond to the image line $\ell$. Thus, we must determine the correct ratio between distances of the planes to the optical center. A simple idea is to use the points on the intersection of the two planes. For every point $\boldsymbol{x}$ on the line $\boldsymbol{\ell}$, its depth with respect to the camera frame can be calculated using $\lambda_i = \frac{d_i}{N_i^T \boldsymbol{x}}$. By setting $\lambda_1 = \lambda_2$, we can obtain the ratio $\alpha$ between the true distances:

$$\boxed{\alpha = \frac{d_2}{d_1} = \frac{N_2^T \boldsymbol{x}}{N_1^T \boldsymbol{x}}.} \tag{10.23}$$

If we keep distance $d_1$ to $P_1$ to be 1, then $d_2 = \alpha$, and $T_2$ should be scaled accordingly: $T_2 \leftarrow \alpha T_2$.

*Alignment of two images through the same symmetry cell*

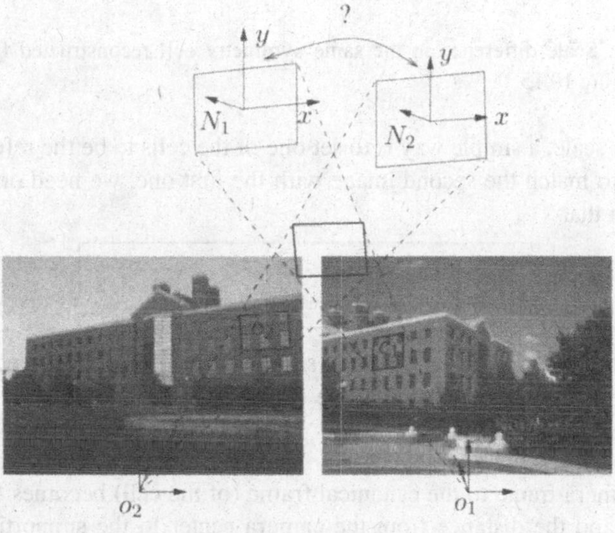


Figure 10.15. The canonical coordinate frames determined using reflective symmetry of the same window complex in two views $c_1$ and $c_2$ do not necessarily use the same scale.

We are now ready to show how symmetry can help us align two (or multiple) images of the same symmetry cell from different viewpoints. In Figure 10.15, by picking the same rectangular cell, its pose $(R_i, T_i)$ with respect to the $i$th camera $(i = 1, 2)$ can be determined separately. However, as before, the reconstructions might have used different scales. We need to determine these scales in order to properly align the relative pose between the two views.

For instance, Figure 10.16 shows the two 3-D structures of the same rectangular cell (windows) reconstructed from the previous two images. Orientation of the recovered cells is already aligned (which is easy for rectangles). Suppose the four corners of the cell recovered from the first image have coordinates $[x_i, y_i]^T$ and those from the second image have coordinates $[u_i, v_i]^T$, $i = 1, 2, 3, 4$. Due to the difference in scales used, the recovered coordinates of corresponding points differ by the same ratio. Since the corresponding points are essentially the same points, we need to find out the ratio of scales.

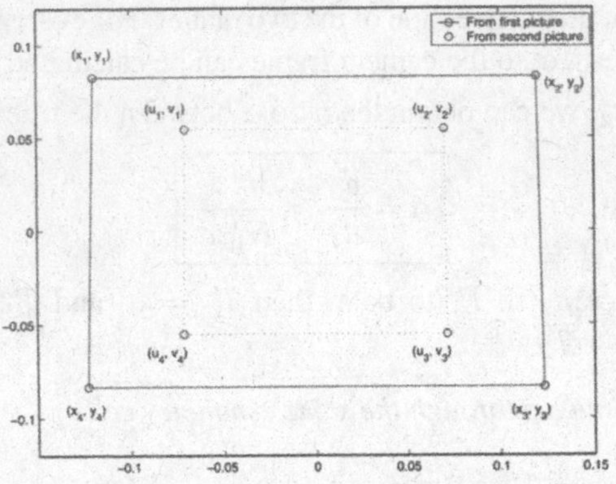

Figure 10.16. Scale difference in the same symmetry cell reconstructed from the two images in Figure 10.15.

To fix the scale, a simple way is to set one of the cells to be the reference. Say, if we want to match the second image with the first one, we need only to find a scale $\alpha$ such that

$$\begin{bmatrix} x_i - \bar{x} \\ y_i - \bar{y} \end{bmatrix} = \alpha \begin{bmatrix} u_i - \bar{u} \\ v_i - \bar{v} \end{bmatrix}, \quad i = 1, 2, 3, 4, \tag{10.24}$$

where $\bar{x}$ is the mean $\bar{x} \doteq \frac{1}{4}\sum_i x_i$ and similarly for $\bar{y}, \bar{u}, \bar{v}$. These equations are linear in the unknown $\alpha$, which can be easily computed using least squares, as described in Appendix A.

Thus, to set a correct relative scale for the second image, the relative pose from the camera frame to the canonical frame (of the cell) becomes $(R_2, T_2) \leftarrow (R_2, \alpha T_2)$, and the distance from the camera center to the supporting plane is $d_2 \leftarrow \alpha$.

*Alignment of coordinate frames from multiple images*

The reader should be aware that we deliberately avoid using the homography between the cells $c_1$ and $c_2$ in the two images (Figure 10.15) because that homography, unlike the reflective homography actually used, depends on the baseline between the two views. Symmetry allows us to find out the relative poses of the camera with respect to different symmetry cells (hence planes) in each image or the same symmetry cell (hence plane) in different images, without using multiple-view geometric constraints between different images at all.[15] The method, as mentioned before, is indeed independent of the baseline between different views.

Since our final goal is to build a global geometric model of the object, knowing the canonical frames and the pose of the camera relative to these frames is sufficient for us to arbitrarily transform coordinates among different views. Figure 10.17 shows a possible scheme for reconstructing three sides $P_1, P_2, P_3$ of any

[15] We used "multiple-view" geometric constraints only between the equivalent images.

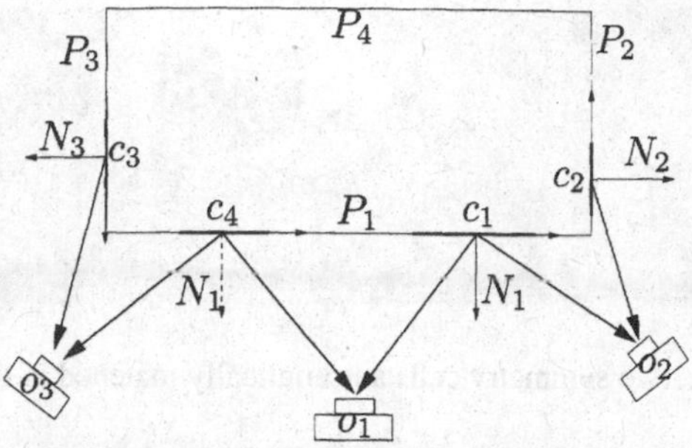


Figure 10.17. Alignment of multiple images based on symmetry cells.

box-like structure from as few as three images. Using properly chosen rectangular cells, we can easily recover the relative poses of the cameras with respect to any plane in any view. Notice that cells such as $c_1$ and $c_4$ on the same plane will always share the same canonical reference frame (only the origin of the frame can be translated on the plane, according to the center of each cell).

**Example 10.22 (Symmetry cell extraction, matching, and reconstruction).** To automatically extract symmetry cells, one can apply conventional image segmentation techniques and verify whether each polygonal region passes the test as the image of a symmetric structure in space (Exercise 10.12). Figure 10.18 shows a possible pipeline for symmetry cell extraction.

Figure 10.18. Left: original image. Middle: image segmentation and polygon fitting. Right: cells that pass the symmetry verification. An object frame is attached to each detected symmetry cell.

Once symmetry cells are extracted from each image, we can match them across multiple images in terms of their shape and color. Figure 10.19 shows that two symmetry cells are extracted and matched across three images of an indoor scene. With respect to this pair of matched cells, Figure 10.20 shows the recovered camera poses. Due to a lot of symmetry in the scene, the point-based matching methods introduced in Chapter 4 would have difficulty with these images because similar corners lead to many mis-matches and outliers. Notice that there is only a very large rotation between the first and second views but a very large translation between the third view and the first two views. Feature tracking is impossible for such large motions and robust matching techniques (to be introduced in Chapter 11) based on epipolar geometry would also fail since the near zero translation between the first two views makes the estimation of the fundamental matrix ill-conditioned. In fact, the translation estimated from different sets of point correspondences in the first two images could differ by up to 35°.

Figure 10.19. Two symmetry cells automatically matched in three images.

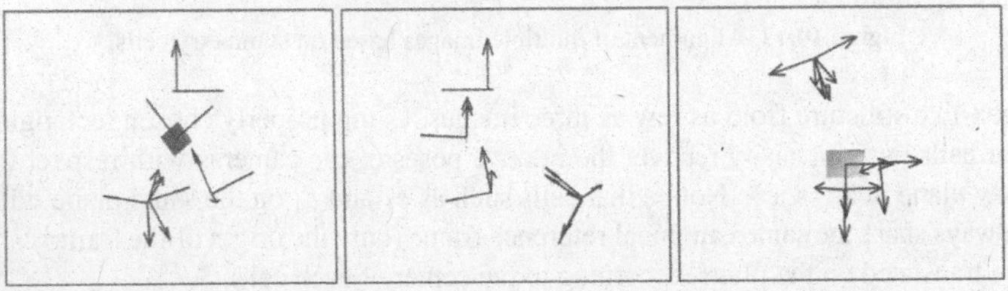

Figure 10.20. Camera poses and cell structure automatically recovered from the matched cells. From left to right: top, side, and frontal views of the cells and camera poses.

As we see, by using symmetry cells rather than points, the above difficulties can be effectively resolved. The ground truth for the length ratios of the white board and table are 1.51 and 1.00, and the recovered length ratios are 1.506 and 1.003, respectively. Error in all the right angles is less than 1.5°. ■

**Example 10.23 (Step-by-step 3-D reconstruction).** In this example, we show how to semiautomatically build a 3-D model from the five photos given in Figure 10.21 based on the techniques introduced in this chapter.

Figure 10.21. Images used for reconstruction of the building (Coordinated Science Laboratory, UIUC) from multiple views. The fifth image is used only to get the correct geometry of the roof (from its reflective symmetry) but not used to render the final 3-D model.

*1. Recovery of 3-D canonical frames.* Figure 10.22 shows the results for recovery of the canonical frames of two different rectangular cells on two planes (in the second image in Figure 10.21). The "images" of the canonical frames are drawn on the image according to the correct perspective based on recovered poses. To recover the correct relative depth ratio of the supporting planes of the two cells, we use (10.23) to obtain the ratio $\alpha = \frac{d_2}{d_1} = 0.7322$ and rescale all scale-related entities associated with the second cell so that everything is with respect to a single canonical frame (of the first cell). We point out that the angle between recovered normal vectors (the $z$-axes of both frames) of the two supposedly orthogonal planes is $90.36°$, within $1°$ error to $90°$.

Figure 10.22. Recovery of the canonical frame axes associated with symmetry cells on two planes. Solid arrows represent normal vectors of the planes.

*2. Alignment of different images.* Figure 10.23 shows the alignment of the first two images in Figure 10.21 through a common symmetry cell (Figure 10.15). The correct cross-view scale ratio computed from (10.24) is $\alpha = 0.7443$. Figure 10.15 shows how the front side of the building in each view is warped by the recovered 3-D transformation onto the other one. As we see, the entire front side of the building is re-generated in this way. Parts that are not on the front side are not correctly mapped (however, they can be corrected by stereo matching once the camera pose is known).

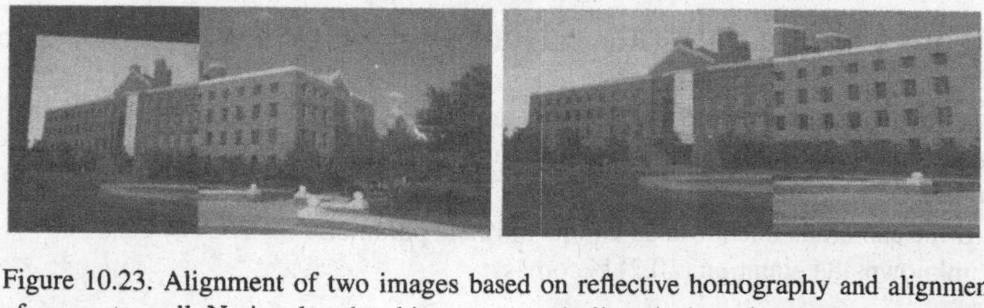

Figure 10.23. Alignment of two images based on reflective homography and alignment of symmetry cell. Notice that the chimneys are misaligned, since they are not part of the symmetric structure of the scene.

*3. Reconstruction of geometric models of the building.* Equipped with the above techniques, one can now easily obtain a full 3-D model of the building from the first four images in Figure 10.21 taken around the building. The user only needs to point out which cells

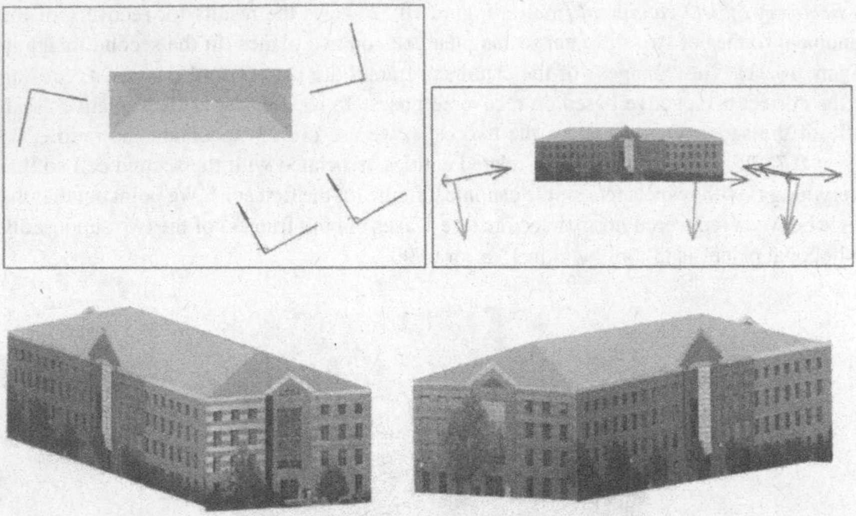

Figure 10.24. Top: The four coordinate frames show the recovered camera poses from the four images. Arrows that point towards the building are the camera optical axes. Bottom: Reconstructed 3-D model rendered with the original set of four images.

correspond to which in the images, and a consistent set of camera poses from the matched cells can be obtained, as Figure 10.24 top shows. Based on the recovered camera poses, the subsequent 3-D structure reconstruction is shown in Figure 10.24 bottom. The 3-D model is rendered as piecewise planar with parts on the building manually specified in the images. The original color of the images is kept to show from which image each patch on the model is rendered. The fifth image in Figure 10.21 is used only to get the geometry of the roof from its reflective symmetry and is not used for rendering. The chimney-like structure on the roof is missing, since we do not have enough information about its geometry from the photos. ■

## 10.3 Camera calibration from symmetry

Results given in the preceding sections are all based on the assumption that the camera has been calibrated. In this section, we study the effect of symmetry on camera calibration.

If the camera is uncalibrated and its intrinsic parameter matrix, say $K \in \mathbb{R}^{3\times 3}$, is unknown, the equation (10.2) becomes

$$\lambda \boldsymbol{x} = K\Pi_0 g_0 \boldsymbol{X}.$$

From the epipolar constraints between pairs of equivalent views, instead of the essential matrix $E = \widehat{T'}R'$, we can recover only the fundamental matrix

$$F \sim K^{-T}\widehat{T'}R'K^{-1}, \tag{10.25}$$

where, as before, $R' = R_0 R R_0^T \in O(3)$ and $T' = (I - R_0 R R_0^T)T_0 + R_0 T \in \mathbb{R}^3$. Notice that here $K$ is automatically the same for the original image and all the equivalent ones.

## 10.3.1 Calibration from translational symmetry

In the translational symmetry case, we have $R' = I$ and $T' = R_0 T$. Given three mutually orthogonal translations $T_1, T_2, T_3 \in \mathbb{R}^3$ under which the structure is invariant, from the fundamental matrix $F \sim K^{-T}\widehat{T'}K^{-1} = \widehat{KT'} = \widehat{KR_0T}$ we get vectors

$$v_i \sim K R_0 T_i, \quad i = 1, 2, 3. \tag{10.26}$$

That is, $v_i$ is equal to $KR_0T_i$ up to an (unknown) scale. Since $T_1, T_2, T_3$ are assumed to be mutually orthogonal, we have

$$v_i^T K^{-T} K^{-1} v_j = 0, \quad \forall i \neq j. \tag{10.27}$$

We get three linear equations on entries of the matrix $K^{-T}K^{-1}$. If there are fewer than three unknown parameters in $K$,[16] the calibration $K$ can be determined from these three linear equations (all from a single image). The reader should be aware that these three orthogonal translations correspond precisely to the notion of vanishing point, exploited for calibration in Chapter 6.

**Example 10.24 ("Vanishing point" from translational homography).** Notice that if the structure is planar, instead of the fundamental matrix, we get the (uncalibrated) translational homography matrix

$$\tilde{H} = KHK^{-1} = K\left(I + \frac{1}{d}T'N^T\right)K^{-1} = I + \frac{1}{d}(KT')(K^{-T}N)^T. \tag{10.28}$$

Again, the vector $v \sim KT'$ can be recovered up to scale from decomposing the homography matrix $\tilde{H}$. ■

## 10.3.2 Calibration from reflective symmetry

In the reflective symmetry case, if $R$ is a reflection, we have $R^2 = (R')^2 = I$ and $R'T' = -T'$. Thus, $\widehat{T'}R' = \widehat{T'}$. Then $F \sim \widehat{KT'}$ is of the same form as a fundamental matrix for the translational case. Thus, if we have reflective symmetry along mutually orthogonal directions, the camera calibration $K$ can be recovered similarly to the translational case.

**Example 10.25 ("Vanishing point" from the reflective homography).** If the structure is planar, in the uncalibrated camera case, we get the uncalibrated version $\tilde{H}$ of the reflective homography $H$ in Example 10.14:

$$\tilde{H} \doteq KHK^{-1} = KR'K^{-1} + \frac{1}{d}KT'(K^{-T}N)^T \in \mathbb{R}^{3\times3}.$$

[16] For example, pixels are square.

Since $R'T' = -T'$ and $N^T T' = 0$, it is straightforward to check that $HT' = -T'$, or equivalently, $\tilde{H}v = -v$ for $v \sim KT'$. Furthermore, $v$ is the only eigenvector that corresponds to the eigenvalue $-1$ of $\tilde{H}$.[17]

If the object is a rectangle, admitting two reflections, we then get two vectors $v_i \sim KT'_i, i = 1, 2$, for $T'_1 \perp T'_2$. They obviously satisfy the equation $v_2 K^{-T} K^{-1} v_1 = 0$. ■

### 10.3.3 Calibration from rotational symmetry

A less trivial case where symmetry may help with self-calibration is that of rotational symmetry. In this case, it is easy to show that the axis of the rotation $R'$ is always perpendicular to the translation $T'$ (see Exercise 10.8). According to Chapter 6 (Appendix 6.B), the fundamental matrix $F$ must be of the form

$$F = \lambda \hat{e} K R' K^{-1}, \tag{10.29}$$

where $e \in \mathbb{R}^3$ of unit length is the (left) epipole of $F$, and the scalar $\lambda$ is one of the two nonzero eigenvalues of the matrix $F^T \hat{e}$ (see Theorem 6.15). Then the calibration matrix $K$ satisfies the so-called normalized Kruppa's equation

$$F K K^T F^T = \lambda^2 \hat{e} K K^T \hat{e}^T, \tag{10.30}$$

with $F, e, \lambda$ known and only $KK^T$ unknown. This equation, as shown in Chapter 6 (Theorem 6.16), gives two linearly independent constraints on $KK^T$. For instance, if only the camera focal length $f$ is unknown, we may rewrite the above equation as

$$F \begin{bmatrix} f^2 & 0 & 0 \\ 0 & f^2 & 0 \\ 0 & 0 & 1 \end{bmatrix} F^T = \widehat{\lambda e} \begin{bmatrix} f^2 & 0 & 0 \\ 0 & f^2 & 0 \\ 0 & 0 & 1 \end{bmatrix} \widehat{\lambda e}^T, \tag{10.31}$$

which is a linear equation in $f^2$. It is therefore possible to recover the focal length from a single image (of some object with rotational symmetry).

**Example 10.26 (Focal length from rotational symmetry: a numerical example).** For a rotational symmetry, let

$$K = \begin{bmatrix} 2 & 0 & 0 \\ 0 & 2 & 0 \\ 0 & 0 & 1 \end{bmatrix}, \quad R = \begin{bmatrix} \cos(2\pi/3) & 0 & -\sin(2\pi/3) \\ 0 & 1 & 0 \\ \sin(2\pi/3) & 0 & \cos(2\pi/3) \end{bmatrix},$$

$$R_0 = \begin{bmatrix} \cos(\pi/6) & \sin(\pi/6) & 0 \\ -\sin(\pi/6) & \cos(\pi/6) & 0 \\ 0 & 0 & 1 \end{bmatrix}, \quad T_0 = \begin{bmatrix} 2 \\ 10 \\ 1 \end{bmatrix}.$$

[17] The other two eigenvalues are $+1$. Also notice that if $\tilde{H}$ is recovered up to an arbitrary scale from the homography (10.12), it can be normalized using the fact $\tilde{H}^2 = H^2 = I$.

Note that the rotation $R$ by $2\pi/3$ corresponds to the rotational symmetry that a cube admits. Then we have

$$F = \begin{bmatrix} -0.3248 & -0.8950 & -1.4420 \\ 0.5200 & 0.3248 & -2.4976 \\ -1.0090 & -1.7476 & -0.0000 \end{bmatrix}, \ \widehat{e} = \begin{bmatrix} 0 & -0.4727 & 0.4406 \\ 0.4727 & 0 & 0.7631 \\ -0.4406 & -0.7631 & 0 \end{bmatrix},$$

and $\lambda = 2.2900$ (with the other eigenvalue $-2.3842$ of $F^T\widehat{e}$ rejected). Then the equation (10.31) gives a linear equation in $f^2$:

$$-0.2653f^2 + 1.0613 = 0.$$

This gives $f = 2.0001$, which is, within the numerical accuracy, the focal length given by the matrix $K$ in the first place. ■

In the planar case, we knew from Example 10.5 that any rotationally symmetric structure in fact admits a dihedral group $D_n$ that includes the rotations $C_n$ as a subgroup.[18] Therefore, in principle, any information about calibration that one can extract from the rotational homography, can also be extracted from the reflective ones.

## 10.4 Summary

Any single image of a symmetric object is equivalent to multiple equivalent images that are subject to multiple-view rank conditions or homography constraints. In general, we only need four points to recover the essential matrix or homography matrix between any two equivalent views. This allows us to recover the 3-D structure and canonical pose of the object from a single image. There are three fundamental types of isometric 3-D symmetry: *reflective*, *rotational*, and *translational*, which any isometric symmetry is composed of. A comparison of ambiguities associated with these three types of symmetry in recovering the canonical pose between the object and the camera was given in Table 10.1. A comparison between 3-D and 2-D symmetric structures is summarized in Table 10.2.

| | 3-D symmetry | 2-D symmetry |
|---|---|---|
| Symmetry group | $G = \{g\} \subset E(3)$ | $G = \{g\} \subset E(2)$ |
| Initial pose | $g_0 : (R_0, T_0) \in SE(3)$ | $H_0 : [R_0(1), R_0(2), T_0]$ |
| Equivalent views | $g' = g_0 g g_0^{-1}$ | $H' = H_0 g H_0^{-1}$ |
| Lyapunov equation | $g' g_0 - g g_0 = 0$ | $H' H_0 - H_0 g = 0$ |
| Unique $g_0$ recovery | two independent rotations or reflections in $G$ | |

Table 10.2. Comparison between 3-D and 2-D symmetry under perspective imaging.

[18] In fact, in $\mathbb{R}^3$ each rotation can be decomposed into two reflections; see Exercise 10.9.

If multiple images are given, the presence of symmetric structures may significantly simplify and improve the 2-D matching and 3-D reconstruction. In case the camera is not calibrated, symmetry also facilitates automatic retrieval of such information directly from the images.

## 10.5 Exercises

**Exercise 10.1** What type(s) of symmetries do you identify from each image in Figure 10.1?

**Exercise 10.2** Exactly how many different equivalent views can one get from

1. the image of the checkerboard shown in Figure 10.3?
2. a generic image of the cube shown in Figure 10.25, assuming that the cube is $a)$ opaque and $b)$ transparent?

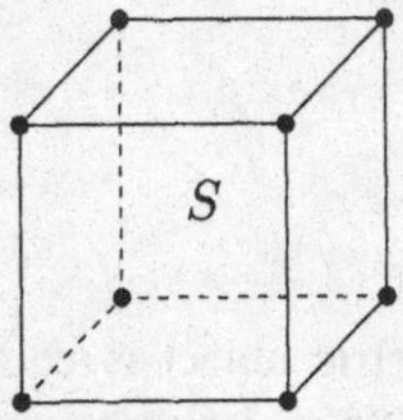


Figure 10.25. A cube consists of 8 vertices, 12 edges, and 6 faces.

**Exercise 10.3 (Nonplanar 3-D rotational symmetry).** Show that among the five platonic solids in Figure 10.6, the cube and the octahedron share the same rotational symmetry group, and the icosahedron and the dodecahedron share the same rotational symmetry group.

**Exercise 10.4** Prove Lemma 10.13.

**Exercise 10.5** Prove Lemma 10.18.

**Exercise 10.6 (Translational homography).** Given a homography matrix $H$ of the form $H \sim I + T'N^T$, the type of homography matrix that one expects to get from a translational homography, without using the decomposition algorithm given in Chapter 5, find a more efficient way to retrieve $T'$ and $N$ from $H$.

**Exercise 10.7 (Relative camera pose from reflective symmetry).** Given a reflection $R$ and camera pose $(R_0, T_0)$, the relative pose between a given image and an equivalent one is $R' = R_0 R R_0^T$ and $T' = (I - R_0 R R_0^T)T_0$.

1. Show that $R'T' = -T'$ and $R'\widehat{T'} = \widehat{T'}R' = \widehat{T'}$.
2. Let $n = T'/\|T'\|$. Show that $R' = I - 2nn^T$.
3. Explain how this may simplify the decomposition of a reflective homography matrix, defined in Example 10.14.

**Exercise 10.8 (Relative camera pose from rotational symmetry).** Given a rotation $R$ (from a rotational symmetry) and camera pose $(R_0, T_0)$, the relative pose between a given image and an equivalent one is $R' = R_0 R R_0^T$ and $T' = (I - R_0 R R_0^T) T_0$.

1. Show that if $\omega \in \mathbb{R}^3$ is the rotational axis of $R'$, i.e. $R' = e^{\widehat{\omega}}$, we have $\omega \perp T'$.
2. Therefore, $(R', T')$ is an orbital motion. What is the form for the essential matrix $E = \widehat{T'} R'$?
3. Explain geometrically what the relationship between $R'$ and $T'$ is. (Hint: proofs for 1 and 2 become simple if you can visualize the relationship.)

**Exercise 10.9 (Rotational homography).** Consider a homography matrix $H$ of the form $H = R' + T' N^T$, the type of homography matrix one gets from a rotational homography.

1. What are the relationships between $R'$, $T'$, and $N$?
2. Show that $H^m = I$ for some $n$, which is a factor of the order $n$ of the cyclic group associated with the rotational symmetry.
3. If an uncalibrated version $\tilde{H} \sim KHK^{-1}$ is recovered up to scale, show that the above fact helps normalize $\tilde{H}$ to be equal to $KHK^{-1}$.
4. Show that $H$ can always be decomposed into two matrices $H = H_1 H_2$, where $H_1$ and $H_2$ each correspond to a reflective homography of the same (planar) symmetric object.

**Exercise 10.10 (2-D symmetric patterns).** For the 2-D patterns shown in Figure 10.26, what is the symmetry group associated with each pattern? Give a matrix representation of the group. Discuss what the most pertinent symmetry-based algorithm would be for recovering the structure and pose of a plane textured with such a pattern. What is the minimum number of features (points or lines) you need? (Note: It has been shown that there are essentially only 17 different 2-D symmetric patterns that can tile the entire plane with no overlaps and gaps. The reason why they are different is exactly because they admit different symmetry groups; see [Weyl, 1952]).

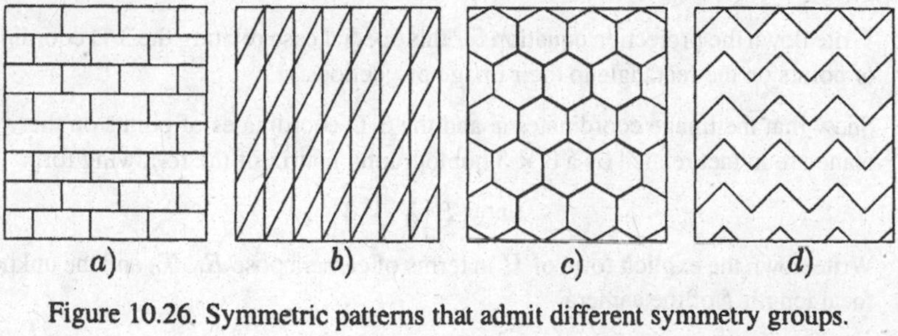

Figure 10.26. Symmetric patterns that admit different symmetry groups.

**Exercise 10.11 (Degenerate cases for reflective and rotational homography)** There are some degenerate cases associated with the homography induced from the reflective symmetry or the rotational symmetry. To clarify this, answer the following questions:

1. What is the form of a reflective homography if the camera center is in the plane of reflection? Explain how the reconstruction process should be adjusted in this case.

2. What is the form of a rotational homography if the camera center is on the axis of rotation? Explain how the reconstruction process should be adjusted in this case.

**Exercise 10.12 (Symmetry testing).** To test whether a set of point or line features can be interpreted as a calibrated image of that of a symmetric object, show the following facts:

1. Any four-side polygon can be the image of a parallelogram in space.
2. Let $\boldsymbol{v}_1$ and $\boldsymbol{v}_2$ be the intersections of the two pairs of opposite edges of a four-sided polygon in the image plane, respectively. If $\boldsymbol{v}_1 \perp \boldsymbol{v}_2$, then the polygon can be the image of a rectangle in space. Furthermore, $\boldsymbol{v}_1, \boldsymbol{v}_2$ are the two vanishing points associated with the rectangle.
3. Argue how to test if an $n$-side polygon can be the image of a (standard) $n$-gon.

**Exercise 10.13 (A calibrated image of a circle).** Given one calibrated perspective image of a circle, argue that you are able to completely recover its 3-D pose and structure, relative to your viewpoint. Design a step-by-step procedure to accomplish this task based on our knowledge on the reflective symmetry; you need to clarify to what "features" from the circle you apply the reflective symmetry. Argue why this no longer works if the image is uncalibrated. [Note: the circle admits a continuous symmetry group $O(2)$. However, by using only one element from this group, a reflection, we are able to recover its full 3-D information from a single image.]

**Exercise 10.14 (Pose recovery with partial knowledge in the structure and $K$).** Consider a camera with calibration matrix

$$K = \begin{bmatrix} f & 0 & 0 \\ 0 & f & 0 \\ 0 & 0 & 1 \end{bmatrix},$$

where the only unknown parameter is the focal length $f$. Suppose that you have a single view of a rectangular structure, whose four corners in the world coordinate frame have coordinates $\boldsymbol{X}_1 = [0,0,0,1]^T$, $\boldsymbol{X}_2 = [\alpha b, 0, 0, 1]^T$, $\boldsymbol{X}_2 = [0, b, 0, 1]^T$, $\boldsymbol{X}_4 = [\alpha b, b, 0, 1]^T$, where one of the dimensions $b$ and the ratio $\alpha$ between the two dimensions of the rectangle are unknown.

1. Write down the projection equation for this special case relating the 3-D coordinates of points on the rectangle to their image projections.
2. Show that the image coordinates $\boldsymbol{x}$ and the 3-D coordinates of points on the world plane are in fact related by a $3 \times 3$ homography matrix of the following form

   $$\lambda \boldsymbol{x} = H[X, Y, 1]^T.$$

   Write down the explicit form of $H$ in terms of camera pose $R_0, T_0$ and the unknown focal length $f$ of the camera.
3. Once $H$ is recovered up to a scale factor, say $\tilde{H} \sim H$, describe the steps that would enable you to decompose it and recover the unknown focal length $f$ and rotation $R$. Also recover the translation $T$ and ratio $\alpha$ up to a universal scale factor.

**Exercise 10.15 (Programming exercise).** Implement (e.g., in Matlab) the reconstruction scheme described in Section 10.2.3. Further integrate into your system the following extra features:

1. Translational symmetry.
2. Rotational symmetry.
3. A combination of the three fundamental types of symmetry, like the relation between cell $c_1$ and cell $c_2$ in Figure 10.14.
4. Take a few photos of your favorite building or house and reconstruct a full 3-D model of it.
5. Calibrate your camera using symmetry in the scene, and compare the results to the ground truth.

# Further readings

*Symmetry groups*

The mathematical study of possible symmetric patterns in an arbitrary $n$-dimensional space is known as Hilbert's 18th problem. Answers to the special 2-D and 3-D cases were given by Fedorov in the nineteenth century [Fedorov, 1971] and proven independently by George Pólya in 1924. A complete answer to Hilbert's problem for high-dimensional spaces was given by [Bieberbach, 1910]. A good general introduction to this subject is [Weyl, 1952] or, for a stronger mathematical flavor, [Grünbaum and Shephard, 1987, Goodman, 2003]. A good survey on a generalization to the notion of symmetry groups, i.e. the so-called groupoids, is [Weinstein, 1996].

*Symmetry and vision*

Symmetry, as a useful visual cue for extracting 3-D information from images, has been extensively discussed in psychology [Marr, 1982, Plamer, 1999]. Its advantages in computational vision were first explored in the statistical context, such as the study of isotropic texture (e.g., for the third image of Figure 10.1 top) [Gibson, 1950, Witkin, 1988, Malik and Rosenholtz, 1997]. It was the work of [Gärding, 1992, Gärding, 1993, Malik and Rosenholtz, 1997] that provided a wide range of efficient algorithms for recovering the shape (i.e. the slant and tilt) of a textured plane based on the assumption of isotropy or weak isotropy.

On the geometric side, [Mitsumoto et al., 1992] were among the first to study how to reconstruct a 3-D object using reflective symmetry induced by a mirror. [Zabrodsky and Weinshall, 1997] used (bilateral) reflective symmetry to improve 3-D reconstruction from image sequences. Shape from symmetry for affine images was studied by [Mukherjee et al., 1995]. [Zabrodsky et al., 1995] provided a good survey on studies of reflective symmetry and rotational symmetry in computer vision at the time. For object and pose recognition, [Rothwell et al., 1993] pointed out that the assumption of reflective symmetry could also be useful in the construction of projective invariants. Certain invariants can also be formulated in tensorial terms using the double algebra, as pointed out by [Carlsson, 1998].

More recently, [Huynh, 1999] showed how to obtain an affine reconstruction from a single view in the presence of symmetry. Based on reflective epipolar geometry, [François et al., 2002] demonstrated how to recover 3-D symmetric objects (e.g., a face) from a single view. Bilateral reflective symmetry has also been exploited in the context of photometric stereo or shape-from-shading by [Shimoshoni et al., 2000, Zhao and Chellappa, 2001].

The unification between symmetry and multiple-view geometry was done in [Hong et al., 2002], and its implication in multiple-view reconstruction was soon pointed out by [Huang et al., 2003], upon which this chapter is based. As an effort to automatically extract and match symmetric objects across multiple images, [Yang et al., 2003, Huang et al., 2003] have shown that extracting and matching a group of features with symmetry is possible and often is much better conditioned than matching individual point and line features using epipolar or other multilinear constraints.

*Other scene knowledge*

It has long been known in computer vision that with extra scene knowledge, it is possible to recover 3-D structure from a single view [Kanade, 1981, Sparr, 1992]. As we have seen in this chapter, symmetry clearly reveals the reason why vanishing points (often caused by reflective and translational symmetries) are important for both camera calibration and scene reconstruction. Like symmetry, other scene knowledge such as orthogonality [Svedberg and Carlsson, 1999] can be very useful in recovering calibration, pose, and structure from a single view. [Liebowitz and Zisserman, 1998, Criminisi et al., 1999, Criminisi, 2000] have shown that scene knowledge such as length ratio and vanishing line also allows accurate reconstruction of 3-D structural metric and camera pose from single or multiple views. [Jelinek and Taylor, 1999] showed that it is possible to reconstruct a class of linearly parameterized models from a single view despite an unknown focal length.

# Part IV

# Applications

# Chapter 11
# Step-by-Step Building of a 3-D Model from Images

> *In respect of military method, we have, firstly, Measurement; secondly, Estimation of quantity; thirdly, Calculation; fourthly, Balancing of chances; fifthly, Victory.*
>
> – Sun Tzu, *Art of War, fourth century* B.C.

This chapter serves a dual purpose. For those who have been following the book up to this point, it provides hands-on experience by guiding them through the application of various algorithms on concrete examples.

For practitioners who are interested in building a system to reconstruct models from images, without necessarily wanting to delve into the material of this book, this chapter provides a step-by-step description of several algorithms. Together with the software distribution at `http://vision.ucla.edu/MASKS`, and reference to specific material in previous chapters, this serves as a "recipe book" to build a complete if rudimentary software system.

Naturally, the problem of building geometric models from images is multifaceted: the algorithms to be implemented depend on whether one has access to the camera and a calibration rig, whether the scene is Lambertian, whether there are multiple moving objects, whether one can afford to process data in a batch, whether the interframe motion is small, etc. One should not expect *one* set of algorithms or system to work satisfactorily in all possible practical conditions or scenarios, at least not yet. The particular system pipeline presented in this chapter is only one of many possible implementations implied by the general theory developed in earlier chapters. Therefore, we also provide a discussion of the domain of applicability of each algorithm, as well as reference to previous chapters of the book or the literature where more detailed explanations and derivations can be found.

*Agenda for this chapter*

Suppose that we are given a sequence of images of the type shown in Figure 11.1, and we are asked to reconstruct a geometric model of the scene, together with a texture map that can be used to render the scene from novel viewpoints.

Figure 11.1. Sample images from video sequences used in this chapter to illustrate the reconstruction of 3-D structure and motion. We will alternate among these two sequences to highlight features and difficulties of each algorithm.

Figure 11.2 outlines a possible system pipeline for achieving such a task.[1] For each component in the pipeline, we give a brief description below:

**Feature selection:** Automatically detect geometric features in each individual image (Section 11.1).

**Feature correspondence:** Establish the correspondence of features across different images (Section 11.2). The algorithm to be employed depends on whether the interframe motion ("baseline") is small (Section 11.2.1) or large (Section 11.2.2).

**Projective reconstruction:** Retrieve the camera motion and 3-D position of the matched feature points up to a projective transformation (Section 11.3). Lacking additional information on the camera or the scene, this is the best one can do.

**Euclidean reconstruction with partial camera knowledge:** Assume that all images are taken by the same camera or some of the camera parameters are

[1]"Possible" in the sense that, when practical conditions change, some of the components in the pipeline could be modified, simplified, or even bypassed.

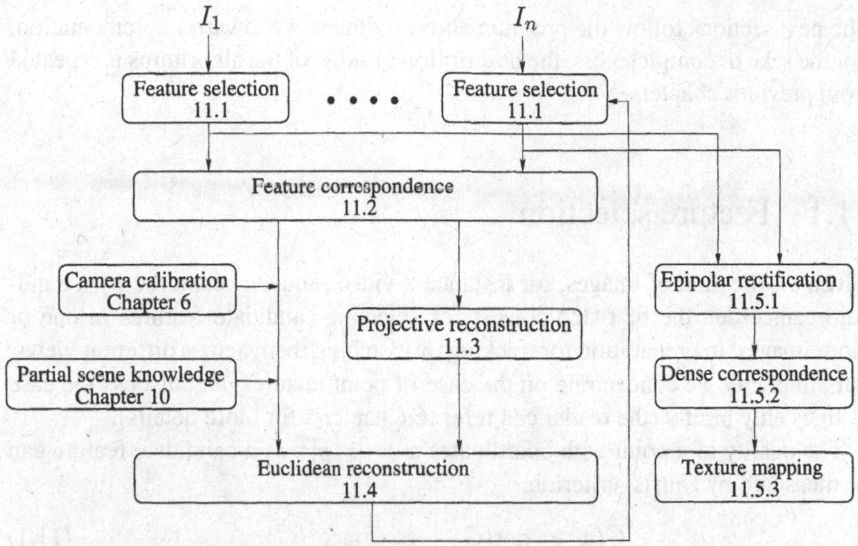


Figure 11.2. Organization of the reconstruction procedure and corresponding sections where each component is described.

known, and recover the Euclidean structure of the scene (Section 11.4). As a side benefit, also retrieve the calibration of the camera.

**Camera calibration:** Assume that one has full access to the camera and a calibration rig. Since several calibration techniques were discussed in Chapter 6, we refer the reader to that chapter for details and references to existing software packages (e.g., Section 6.5.2 and Section 6.5.3).

**Reconstruction with partial scene knowledge:** Assume that certain knowledge about the scene, such as symmetry, is available a priori. Then, feature correspondence, camera calibration, and scene structure recovery can be significantly simplified. Since this topic was addressed extensively in Chapter 10, we refer the reader to that chapter for examples (e.g., those in Section 10.2.2, Section 10.2.3, and Section 10.3).

**Visualization:** Once the correspondence of a handful of features is available, rectify the epipolar geometry so that epipolar lines correspond to image scan lines (Section 11.5.1). This facilitates establishing correspondence for many if not most pixels in each image for which the 3-D position can be recovered (Section 11.5.2). This yields a dense, highly irregular polygonal mesh model of the scene. Standard mesh-manipulation tools from computer graphics can be used to simplify and smooth the mesh, yielding a surface model of the scene. Given the surface model, texture-map the images onto it so as to generate a view of the scene from novel vantage points (Section 11.5.3).

The next sections follow the program above, with an eye toward implementation. For the sake of completeness, the description of some of the algorithms is repeated from previous chapters.

## 11.1 Feature selection

Given a collection of images, for instance a video sequence captured by a hand-held camcorder, the first step consists of selecting candidate features in one or more images, in preparation for tracking or matching them across different views. For simplicity we concentrate on the case of point features, and discuss the case of lines only briefly (the reader can refer to Chapter 4 for more details).

The quality of a point with coordinates $\boldsymbol{x} = [x, y]^T$ as a candidate feature can be measured by Harris' criterion,

$$C(\boldsymbol{x}) = \det(G) + k \times \text{trace}^2(G), \tag{11.1}$$

defined in Chapter 4, equation (4.31), computed on a region $W(\boldsymbol{x})$, for instance a rectangular window centered at $\boldsymbol{x}$, of size between $3 \times 3$ and $11 \times 11$ pixels; we use $7 \times 7$ in this example. In the above expression, $k$ is a constant to be chosen by the designer,[2] and $G$ is a $2 \times 2$ matrix that depends on $\boldsymbol{x}$, given by

$$G = \begin{bmatrix} \sum_{W(\boldsymbol{x})} I_x^2 & \sum_{W(\boldsymbol{x})} I_x I_y \\ \sum_{W(\boldsymbol{x})} I_x I_y & \sum_{W(\boldsymbol{x})} I_y^2 \end{bmatrix} \in \mathbb{R}^{2\times 2},$$

where $I_x, I_y$ are the gradients obtained by convolving the image $I$ with the derivatives of a pair of Gaussian filters (Section 4.A). A point feature $\boldsymbol{x}$ is selected if $C(\boldsymbol{x})$ exceeds a certain *threshold* $\tau$. Selection based on a single global threshold, however, is not a good idea, because one region of the image may contain objects with strong texture, whereas another region may appear more homogeneous and therefore it may not trigger any selection (see Figure 11.3). Therefore, we recommend partitioning the image into *tiles* (e.g., $10 \times 10$ regions of $64 \times 48$ pixels each, for a $640 \times 480$ image), sorting the features according to their quality $C(\boldsymbol{x})$ in each region, and then selecting as many features as desired, provided that they exceed a minimum threshold (to avoid forcibly selecting features where there are none). For instance, we typically start with selecting about 200 to 500 point features.

In addition, to avoid associating multiple features with the same point, we also impose a *minimum separation* between features. Consider, for instance, a white patch with a black dot in the middle. Every window of size, say, $11 \times 11$, centered at a point within 5 pixels of the black dot, will satisfy the requirements above and pass the threshold. However, we want to select only *one* point feature for this dot. Therefore, once the best point feature is selected, according to the criterion $C(\boldsymbol{x})$, we need to "suppress" feature selection in its neighborhood.

[2] A value that is often used is $k = 0.03$, which is empirically verified to yield good results.

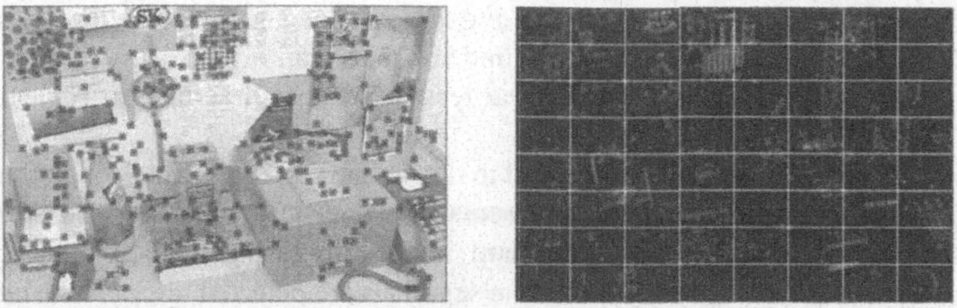

Figure 11.3. Examples of detected features (left) and quality criterion $C(\boldsymbol{x})$ (right). As it can be seen, certain regions of the image "attract" more features than others. To promote uniform selection, we divide the image into tiles and select a number of point features in each tile, as long as they exceed a minimum threshold.

The choice of quality criterion, window size, threshold, tile size, and minimum separation are all part of the design process. There is no right or wrong choice at this stage, and one should experiment with various choices to obtain the best results on the data at hand. The overall feature selection process is summarized in Algorithm 11.1.

**Algorithm 11.1 (Point feature detection).**

1. Compute image gradient $\nabla I = [I_x, I_y]^T$ as in Section 4.A.
2. Choose a size of the window $W(\boldsymbol{x})$ (e.g., $7 \times 7$). Compute the quality of each pixel location $\boldsymbol{x}$, using the quality measure $C(\boldsymbol{x})$ defined in equation (11.1).
3. Choose a threshold $\tau$; sort all locations $\boldsymbol{x}$ that exceed the threshold $C(\boldsymbol{x}) > \tau$, in decreasing order of $C(\boldsymbol{x})$.
4. Choose a tile size (e.g., $64 \times 48$). Partition the image into tiles. Within each tile, choose a minimum separation space (e.g., 10 pixels) and the maximum number of features to be selected within each tile (e.g., 5). Select the highest-scoring feature and store its location. Go through the list of features in decreasing order of quality; if the feature does not fall within the minimum separation space of any previously selected features, then select it. Otherwise, discard it.
5. Stop when the number of selected features has exceeded the maximum, or when all the features exceeding the threshold have been discarded.

*Further issues*

In order to further improve feature localization one can interpolate the function $C(\boldsymbol{x})$ *between pixels*, for instance using quadratic polynomial functions, and choose the point $\boldsymbol{x}$ that maximizes it. In general, the location of the maximum will not be exactly on the pixel grid, thus yielding *subpixel accuracy* in feature localization, at the expense of additional computation. In particular, the approximation of $C$ via a quadratic polynomial can be written as $C(\tilde{\boldsymbol{x}}) =$

$a\tilde{x}^2 + b\tilde{y}^2 + c\tilde{x}\tilde{y} + d\tilde{x} + e\tilde{y} + f$ for all $\tilde{\boldsymbol{x}} \in W(\boldsymbol{x})$. As long as the window chosen is of size greater than $3 \times 3$, we can find the minimum of $C(\tilde{\boldsymbol{x}})$ in $W(\boldsymbol{x})$ with respect to $a, b, c, d, e$ and $f$ using linear least-squares as described in Appendix A.

A similar procedure can be followed to detect line segments. The Hough transform is often used to map line segments to points in the transformed space. More details can be found in standard image processing textbooks such as [Gonzalez and Woods, 1992]. Once line segments are detected, they can be clustered into longer line segments that can then be used for matching. Since matching line segments is computationally more involved, we do not emphasize it here and refer the reader to [Schmidt and Zisserman, 2000] instead.

## 11.2 Feature correspondence

Once candidate point features are selected, the goal is to track or match them across different images. We first address the case of small baseline (e.g., when the sequence is taken from a moving video camera), and then the case of moderate baseline (e.g., when the sequence is a collection of snapshots taken from disparate vantage points).

### *11.2.1 Feature tracking*

We first describe the simplest feature-tracking algorithm for small interframe motion based on a purely translational model. We then describe a more elaborate but effective tracker that also compensates for contrast and brightness changes.

#### *Translational motion model: the basic tracker*

The displacement $\boldsymbol{d} \in \mathbb{R}^2$ of a feature point of coordinates $\boldsymbol{x} \in \mathbb{R}^2$ between consecutive frames can be computed by minimizing the sum of squared differences (SSD) between the two images $I_i(\boldsymbol{x})$ and $I_{i+1}(\boldsymbol{x} + \boldsymbol{d})$ in a small window $W(\boldsymbol{x})$ around the feature point $\boldsymbol{x}$. For the case of two views, $i = 1$ and $i + 1 = 2$, we can look for the displacement $\boldsymbol{d}$ that solves the following minimization problem (for multiple views taken by a moving camera, we will consider correspondence of two views at a time)

$$\min_{\boldsymbol{d}} E(\boldsymbol{d}) \doteq \sum_{\tilde{\boldsymbol{x}} \in W(\boldsymbol{x})} \left[ I_2(\tilde{\boldsymbol{x}} + \boldsymbol{d}) - I_1(\tilde{\boldsymbol{x}}) \right]^2. \tag{11.2}$$

As we have seen in Chapter 4, the closed-form solution to this problem is given by

$$\boldsymbol{d} = -G^{-1}\boldsymbol{b}, \tag{11.3}$$

where

$$G \doteq \begin{bmatrix} \sum_{W(\boldsymbol{x})} I_x^2 & \sum_{W(\boldsymbol{x})} I_x I_y \\ \sum_{W(\boldsymbol{x})} I_x I_y & \sum_{W(\boldsymbol{x})} I_y^2 \end{bmatrix}, \quad \boldsymbol{b} \doteq \begin{bmatrix} \sum_{W(\boldsymbol{x})} I_x I_t \\ \sum_{W(\boldsymbol{x})} I_y I_t \end{bmatrix},$$

and $I_t \doteq I_2 - I_1$ is an approximation of the temporal derivative,[3] computed as a first-order difference between two views. Notice that $G$ is the same matrix we have used to compute the quality index of a feature in the previous section, and therefore it is guaranteed to be invertible, although for tracking we may want to select a different window size and a different threshold. In order to obtain satisfactory results, we need to refine this primitive scheme in a number of ways.

First, when the displacement of features between views exceeds 2 to 3 pixels, first-order differences of pixel values cannot be used to compute temporal derivatives, as we have suggested doing for $I_t$ just now. The proposed tracking scheme needs to be implemented in a *multiscale* fashion. This can be done by constructing a "pyramid of images," by smoothing and downsampling the original image, yielding, say, $I^1, I^2, I^3$, and $I^4$, of size $640\times480$, $320\times240$, $160\times120$, $80\times60$, respectively.[4] Then, the basic scheme just described is first applied at the coarsest level to the pair of images $(I_1^4, I_2^4)$, resulting in an estimate of the displacement $\boldsymbol{d}^4 = -G^{-1}\boldsymbol{b}$. This displacement is scaled up (by a factor of two) and the window $W(\boldsymbol{x})$ is moved to $W(\boldsymbol{x}+2\boldsymbol{d}^4)$ at the next level ($I^3$) via a warping of the image[5] $\tilde{I}_2^3(\boldsymbol{x}) \doteq I_2^3(\boldsymbol{x}+2\boldsymbol{d}^4)$. We then apply the same scheme to the pair $(I_1^3, \tilde{I}_2^3)$ in order to estimate the displacement $\boldsymbol{d}^3$. The algorithm is then repeated until the finest level, where it is applied to the pair $I_1^1(\boldsymbol{x})$ and $\tilde{I}_2^1(\boldsymbol{x}) \doteq I_2^1(\boldsymbol{x}+2\boldsymbol{d}^2)$. Once the displacement $\boldsymbol{d}^1$ is computed, the total estimated displacement is given by $\boldsymbol{d} \doteq \boldsymbol{d}^1 + 2\boldsymbol{d}^2 + 4\boldsymbol{d}^3 + 8\boldsymbol{d}^4$. In most sequences captured with a video camera, two to four levels of the pyramid are typically sufficient.

Second, in the same fashion in which we have performed the iteration across different scales (i.e. by warping the image using the estimated displacement and reiterating the algorithm), we can perform the iteration repeatedly at the finest scale: In this iteration, $\boldsymbol{d}^{i+1}$ is computed between $I_1(\boldsymbol{x})$ and the warped interpolated[6] image $\tilde{I}_2^i(\boldsymbol{x}) = I_2(\boldsymbol{x}+\boldsymbol{d}^1+\cdots+\boldsymbol{d}^i)$. Typically, 5 to 6 iterations of this kind[7] are sufficient to yield a localization error of a tenth of a pixel with a window

[3] A better approximation of the temporal derivative can be computed using the derivative filter, which involves three images, following the guidelines of Appendix 4.A.

[4] A more detailed description of multiscale image representation can be found in [Simoncelli and Freeman, 1995] and references therein.

[5] Notice that interpolation of the brightness values of the original image is necessary: Even if we assume that point features were detected at the pixel level (and therefore $\boldsymbol{x}$ belongs to the pixel grid), there is no reason why $\boldsymbol{x}+\boldsymbol{d}$ should belong to the pixel grid. In general, $\boldsymbol{d}$ is not an integer, and therefore, at the next level, all the intensities in the computation of the gradients in $G$ and $\boldsymbol{b}$ must be interpolated outside the pixel grid. For our purposes here, it suffices to use standard linear or bilinear interpolation schemes.

[6] As usual, the image must be interpolated outside the pixel grid in order to allow computation of $G$ and $\boldsymbol{b}$.

[7] This type of iteration is similar in spirit to Newton-Raphson methods.

size of $7 \times 7$. This feature-tracking algorithm is summarized as Algorithm 11.2. An example of tracked features using the purely translational model is given in Figure 11.4.

**Algorithm 11.2 (Multiscale iterative feature tracking).**

1. Detect a set of candidate features in the first frame using Algorithm 11.1. Choose a size for the tracking window $W$ (this can in principle be different from the window used for selection).
2. Select a maximum number of levels (e.g., $k = 3$), and build a pyramid of images by successively smoothing (Appendix 4.A) and downsampling the images.
3. Starting from the coarsest level $k$ of the pyramid (smaller image) iterate the following steps until the finest level is reached.
   - Compute $\boldsymbol{d}^k = -G^{-1}\boldsymbol{b}$ defined by equation (11.3) for the image pair $(I_1^k, \tilde{I}_2^k)$.
   - Move the window $W(\boldsymbol{x})$ by $2\boldsymbol{d}^k$ through warping the second image $\tilde{I}_2^{k-1}(\boldsymbol{x}) = I_2^{k-1}(\boldsymbol{x} + 2\boldsymbol{d}^k)$.
   - Update the displacement $\boldsymbol{d} \leftarrow \boldsymbol{d} + 2\boldsymbol{d}^k$ and the index $k \leftarrow k - 1$. Repeat the steps above until $k = 0$.
   - Now let $\boldsymbol{d} = \boldsymbol{d}^1$ and update repeatedly $\boldsymbol{d} \leftarrow \boldsymbol{d} + \boldsymbol{d}^{i+1}$ with $\boldsymbol{d}^{i+1} = -G^{-1}\boldsymbol{b}$ computed from the pair $(I_1, \hat{I}_2^i)$ until the incremental displacement $\boldsymbol{d}^i$ is "small" (i.e. its norm is below a chosen threshold), or choose a fixed number of iterations, for example, $i = 5$.
4. Evaluate the quality of the features via Harris' criterion as per Section 11.1, and verify that each tracked feature exceeds the chosen threshold. Update the set of successfully tracked features, acquire the next frame, and go to step 3.

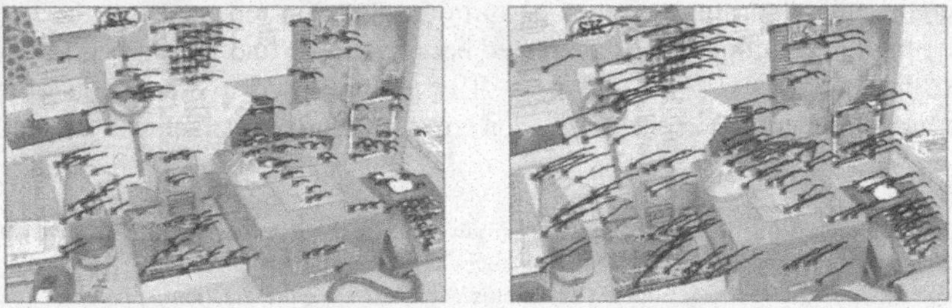

Figure 11.4. Feature points selected in the first frame are tracked in consecutive images: the figure shows the trace of features successfully tracked until the last frame of the video sequence.

*Affine tracker with contrast compensation*

In general, there is a fundamental tradeoff in the choice of the window size $W$: one would like it as large as possible, to counteract the effects of noise in the measurements, but also as small as possible, so that the image deformation between frames can be approximated to a reasonable extent by a simple translation. This tradeoff can in part be eased by choosing a richer deformation model. For instance, instead of the purely translational model of Algorithm 11.2, one can consider an *affine* deformation model, that represents a good compromise between complexity of the model and simplicity of the computation, as we have discussed in Section 4.2.1. For instance, it allows modeling in-plane rotation of feature windows $W(\boldsymbol{x})$ as well as skew transformations. In this case, we assume $\boldsymbol{x}$ to be transformed by $A\boldsymbol{x} + \boldsymbol{d}$ for some $A \in \mathbb{R}^{2\times 2}$, rather than simply by $\boldsymbol{x} + \boldsymbol{d}$. In addition, as objects move relative to the light, their appearance may change in a very complicated way. However, some of the macroscopic changes can be captured by an offset of the intensity $\delta_E$ in the window, and a change in the contrast factor $\lambda_E$. The effects of changes in intensity and contrast during tracking are visible in Figure 11.5. A more robust model to infer the displacement $\boldsymbol{d}$ is to estimate it along with $A$, $\lambda_E$, and $\delta_E$ by minimizing a weighted intensity residual

$$E(A, \boldsymbol{d}, \lambda_E, \delta_E) = \sum_{\tilde{\boldsymbol{x}} \in W(\boldsymbol{x})} w(\tilde{\boldsymbol{x}}) \left[ I(\tilde{\boldsymbol{x}}, 0) - (\lambda_E I(A\tilde{\boldsymbol{x}} + \boldsymbol{d}, t) + \delta_E) \right]^2, \quad (11.4)$$

where for clarity we indicate with $I(\tilde{\boldsymbol{x}}, t)$ the image intensity at the pixel $\tilde{\boldsymbol{x}}$ at time $t$, and $w(\cdot)$ is a weight function of the designer's choice. In the simplest case, $w \equiv 1$; a more common choice is a Gaussian function in $\tilde{\boldsymbol{x}}$. Notice that the displacement in (11.4) is computed between the camera at time 0 and time $t$. As we have discussed, this works only for small overall displacements; for larger displacements one can compute the inter-frame motion by substituting 0 with $t$, and $t$ with $t+1$. More in general, one can consider time $t$ and $t + \Delta t$ for a chosen interval $\Delta t$.

Although for simplicity the results in this chapter are shown for the basic translational model, we report the equations of the affine model, and readers can find its implementation at the website `http://vision.ucla.edu/MASKS`. Let $\lambda = \lambda_E - 1 \in \mathbb{R}$ and $C = A - I \in \mathbb{R}^{2\times 2}$. Then the first-order Taylor expansion for the second term of the objective function $E$ with respect to all the unknowns is

$$(\lambda_E I(A\tilde{\boldsymbol{x}} + \boldsymbol{d}, t) + \delta_E) \approx I(\tilde{\boldsymbol{x}}, t) + f(\tilde{\boldsymbol{x}}, t)^T \boldsymbol{z}, \quad (11.5)$$

where

$$f(\boldsymbol{x}, t) \doteq [xI_x \; yI_x \; xI_y \; yI_y \; I_x \; I_y \; I \; 1]^T, \text{ and } \boldsymbol{z} \doteq [a_{11} \; a_{12} \; a_{21} \; a_{22} \; d_x \; d_y \; \lambda \; \delta_E]^T \quad (11.6)$$

collect all the parameters of interest. Then the first-order approximation of the objective function $E$ becomes

$$E(\boldsymbol{z}) = \sum_{\tilde{\boldsymbol{x}} \in W(\boldsymbol{x})} w(\tilde{\boldsymbol{x}}) \left[ I(\tilde{\boldsymbol{x}}, 0) - I(\tilde{\boldsymbol{x}}, t) - f(\tilde{\boldsymbol{x}}, t)^T \boldsymbol{z} \right]^2. \quad (11.7)$$

For $z$ to minimize $E(z)$, it is necessary that $\frac{\partial E(z)}{\partial z} = 0$. This gives a system of eight linear equations

$$Sz = c, \tag{11.8}$$

where

$$c \doteq \sum_{\tilde{x} \in W(x)} w(\tilde{x})\big(I(\tilde{x},0) - I(\tilde{x},t)\big)f(\tilde{x},t) \quad \in \mathbb{R}^8, \tag{11.9}$$

and

$$S \doteq \sum_{\tilde{x} \in W(x)} w(\tilde{x})f(\tilde{x},t)f(\tilde{x},t)^T \quad \in \mathbb{R}^{8\times 8}. \tag{11.10}$$

In particular, the $8 \times 8$ matrix $f(x,t)f(x,t)^T$ is defined by

$$\begin{bmatrix} x^2I_x^2 & xyI_x^2 & x^2I_xI_y & xyI_xI_y & xI_x^2 & xI_xI_y & xI_xI & xI_x \\ xyI_x^2 & y^2I_x^2 & xyI_xI_y & y^2I_xI_y & yI_x^2 & yI_xI_y & yI_xI & yI_x \\ x^2I_xI_y & xyI_xI_y & x^2I_y^2 & xyI_y^2 & xI_xI_y & xI_y^2 & xI_yI & xI_y \\ xyI_xI_y & y^2I_xI_y & xyI_y^2 & y^2I_y^2 & yI_xI_y & yI_y^2 & yI_yI & yI_y \\ xI_x^2 & yI_xI_y & xI_xI_y & yI_xI_y & I_x^2 & I_xI_y & I_xI & I_x \\ xI_xI_y & yI_y^2 & xI_y^2 & yI_y^2 & I_xI_y & I_y^2 & I_yI & I_y \\ xI_xI & yI_xI & xI_yI & yI_yI & I_xI & I_yI & I^2 & I \\ xI_x & yI_x & xI_y & yI_y & I_x & I_y & I & 1 \end{bmatrix}.$$

The solution to $z$ is simply $z = S^{-1}c$, from which we can recover $d$ assuming that $S$ is invertible. If it is not, just discard that feature.[8] This procedure can substitute the simple computation of $d = -G^{-1}b$ in step 3 of Algorithm 11.2. Although the affine model typically gives substantially better tracking results, it still requires that the motion of the camera be "small." Figure 11.5 shows an example of tracking results using the affine model.

A radically different way to proceed consists in foregoing the computation of temporal derivatives altogether, and formulating the tracking problem as matching the local appearance of features. We discuss this in the next subsection.

### *Caveats*

When tracking features over a long sequence by determining correspondence among adjacent views, one can notice a *drift* of the position of the features as the sequence progresses. If $W(x(t))$ is the window centered at the location of a feature at time $t$, then in general at time $t+1$ the window $W(x(t+1))$ will be different from $W(x(t))$, so what is being tracked actually changes at each instant of time. One could think of tracking features from the initial time, rather than among adjacent frames, that is substitute the pair $I(x,t), I(x,t+1)$ with $I(x,1), I(x,t)$. Unfortunately, for most sequences, this will not work because

[8] Note that the affine tracker needs significantly larger windows than the translational tracker, in order to guarantee that $S$ is invertible.

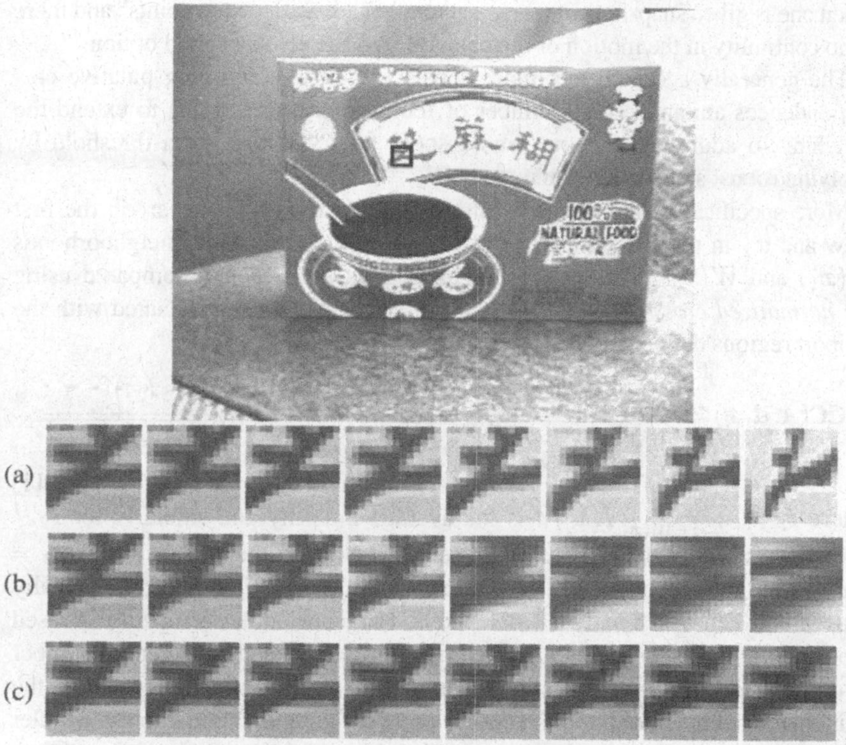


Figure 11.5. Some snapshots of the region inside the black square in the *sesame* sequence (top): (a) is the original sequence as it evolves in time (tracked by hand); (b) tracked window, warped back to the original position, using the affine deformation model $(A, \boldsymbol{d})$ but no illumination parameters; (c) tracked window warped back to the original position using an affine model with contrast and brightness parameters $\lambda_E, \delta_E$.

the deformation from the first to the current image will be so large as to not allow using the differential methods we have described so far.

If one can afford to post-process the sequence, we recommend reestimating the correspondence by applying the matching procedure described in the next subsection, i.e. an exhaustive search based on the correlation matching score, using the results of the feature tracking algorithm as an initial estimate.

### 11.2.2 *Robust matching across wide baselines*

An alternative to tracking features from one frame to another is to detect features *independently* in each image, and then search for corresponding features on the basis of a "matching score" that measures how likely it is for two features to correspond to the same point in space. This is a computationally intensive, error-prone, uncertainty-laden problem for which there is no easy solution. However, when one is given snapshots of a scene taken from disparate viewpoints[9] and there is no continuity in the motion of the camera, one has no other good option.[10]

The generally accepted technique consists in first establishing putative correspondences among a small number of features, and then trying to extend the matching to additional features whose score falls within a given threshold by applying robust statistical techniques.

More specifically, the selected features with their coordinates $\boldsymbol{x}_1$ in the first view and $\boldsymbol{x}_2$ in the second are characterized by their associated neighborhoods $W(\boldsymbol{x}_1)$ and $W(\boldsymbol{x}_2)$. The feature neighborhoods are typically compared using the *normalized cross-correlation* (NCC) of two 2-D signals associated with the support regions defined in Chapter 4 as (see Figure 11.6)

$$\mathrm{NCC}(A, \boldsymbol{d}, \boldsymbol{x}) = \frac{\sum_{\tilde{\boldsymbol{x}} \in W(\boldsymbol{x})} \left(I_1(\tilde{\boldsymbol{x}}) - \bar{I}_1\right)\left(I_2(A\tilde{\boldsymbol{x}} + \boldsymbol{d}) - \bar{I}_2)\right)}{\sqrt{\sum_{\tilde{\boldsymbol{x}} \in W(\boldsymbol{x})} \left(I_1(\tilde{\boldsymbol{x}}) - \bar{I}_1\right)^2 \sum_{\tilde{\boldsymbol{x}} \in W(\boldsymbol{x})} \left(I_2(A\tilde{\boldsymbol{x}} + \boldsymbol{d}) - \bar{I}_2\right)^2}}, \tag{11.11}$$

where $\bar{I}_1 \doteq \sum_{\tilde{\boldsymbol{x}} \in W(\boldsymbol{x})} I(\tilde{\boldsymbol{x}})/n$ is the average intensity and similarly for $\bar{I}_2$.[11] The windows $W(\boldsymbol{x}_1)$ and $W(\boldsymbol{x}_2)$ are centered at $\boldsymbol{x}_1$ in the first view and $\boldsymbol{x}_2 = A\boldsymbol{x}_1 + \boldsymbol{d}$ in the second view. The window size of the neighborhoods is typically chosen between $5 \times 5$ and $21 \times 21$ pixels. Larger windows result in increased robustness at the expense of a higher computational cost and an increased number of outliers. Due to the computational complexity of the matching procedure, the affine parameters $A$ are often fixed to $A = I$, reducing the model to the simpler translational one $\boldsymbol{x} \to \boldsymbol{x} + \boldsymbol{d}$; we summarize this procedure in Algorithm 11.3.

**Algorithm 11.3 (Exhaustive feature matching).**

1. Select features in two views $I_1$ and $I_2$ using Algorithm 11.1. Choose a threshold $\tau_e$.
2. For each feature in the first view with coordinates $\boldsymbol{x}_1$ find the feature $\boldsymbol{x}_2 = \boldsymbol{x}_1 + \boldsymbol{d}$ in the second view that maximizes the $\mathrm{NCC}(I, \boldsymbol{d}, \boldsymbol{x}_1)$ similarity between $W(\boldsymbol{x}_1)$ and $W(\boldsymbol{x}_2)$, defined in equation (11.11). If no feature in the second image has a score that exceeds the threshold $\tau_e$, the selected feature in the first image is declared an "orphan."

Note that Algorithm 11.3 is realistic only for computing correspondence between a few distinctive features across the views. This set of correspondences is usually sufficient for the initial estimation of the camera displacement.

[9]Even when correspondence is available through tracking, as we have seen in the previous subsection, one may want to correct for the drift due to tracking only inter-frame motion.

[10]This difficulty arises mostly because we have decided to associate point features with a very simple local signature, that is the intensity value of neighboring pixels, those in the window $W(\boldsymbol{x})$. If one uses more sophisticated signatures, such as affine invariants [Schaffalitzky and Zisserman, 2001], symmetry [Yang et al., 2003, Huang et al., 2003], feature topology [Tell and Carlsson, 2002] or richer sensing data, for instance from hyperspectral images, this problem can be considerably simplified.

[11]Here, $n$ is the number of pixels in the window $W$.

Figure 11.6. Two widely separated views (top row) and the normalized cross-correlations similarity measure (bottom row) for two selected feature points and their associated neighborhoods (middle row). The brighter values correspond to higher normalized cross-correlation scores. Note that for the patch on the left there is a distinct peak at the corresponding point, but there are several local maxima,which makes for difficult matching. The patch on the right results in an even flatter NCC profile, making it difficult to find an unambiguous match.

Figure 11.7. Feature points successfully tracked between first and last frame (left) and mismatches, or outliers, in the last frame (right).

Once the displacement is obtained, it can be used for establishing additional matches. The process of establishing correspondences and simultaneously estimating the camera displacement is best accomplished in the context of a robust matching framework. The two most commonly used approaches are RANSAC and LMedS. More details on LMedS can be found in [Zhang et al., 1995].

Here we focus on random sample consensus (RANSAC), as proposed by [Fischler and Bolles, 1981].

*RANSAC*

The main idea behind the technique is to use a minimal number of data points needed to estimate the model, and then count how many of the remaining data points are compatible with the estimated model, in the sense of falling within a chosen threshold. The underlying model here is that two corresponding image points have to satisfy the epipolar constraint

$$\boldsymbol{x}_2^{jT} F \boldsymbol{x}_1^j = 0, \quad \text{for} \quad j = 1, 2, \dots, n,$$

where $F$ is the fundamental matrix introduced in Chapter 6, equation (6.10). Given at least $n = 8$ correspondences,[12] *$F$ can be estimated using the linear eight-point algorithm,* Algorithm 6.1 (or an refined nonlinear version, Algorithm 11.5). RANSAC consists of randomly selecting eight corresponding pairs of points (points that maximize their NCC matching score) and computing the associated $F$. The quality of the putative matches is assessed by counting the number of inliers, that is, the number of pairs for which some residual error $d$ is less than a chosen threshold, $\tau$. The choice of residual error depends on the objective function used for computing $F$. Ideally we would like to choose $d$ to be the reprojection error defined in Chapter 6, equation (6.81). However, as we discussed in Appendix 6.A, the reprojection error can be approximated by the Sampson distance [Sampson, 1982], which is easier to compute[13]

$$d^j \doteq \frac{(\boldsymbol{x}_2^{jT} F \boldsymbol{x}_1^j)^2}{\|\widehat{e}_3 F \boldsymbol{x}_1^j\|^2 + \|\boldsymbol{x}_2^{jT} F \widehat{e}_3\|^2}. \tag{11.12}$$

The algorithm for feature matching and simultaneous robust estimation of $F$ is described in Algorithm 11.4.

Notice that in Algorithm 11.4, even though $F_k$ is computed using a *linear* algorithm, rejection is performed based on an approximation of the reprojection error, which is nonlinear.

The number of iterations depends on the percentage of outliers in the set of corresponding points. It must be chosen sufficiently large to make sure that at least one of the samples has a very high probability of being free of outliers. Figure 11.8 shows the percentage of correct and incorrect matches as a function of the threshold $\tau$ and the baseline.[14] As the threshold $\tau$ increases, the percentage

---

[12] Seven correspondences are already sufficient if one imposes the constraint $\det(F) = 0$ when estimating $F$, as discussed in Chapter 6. For a calibrated camera, five points are sufficient to determine the essential matrix (up to 10 solutions). An efficient combination of RANSAC and a five-point algorithm can be found in [Nistér, 2003].

[13] The computation of the reprojection error would require explicit computation of 3-D structure. Sampson's distance bypasses this computation, although it is only accurate to first order.

[14] Points were selected independently in different images and then matched combinatorially, and the procedure repeated for images that were separated by a larger and larger baseline.

**Algorithm 11.4 (RANSAC feature matching).**

- Set $k = 0$ and select the initial set of potential matches $S_0$ as follows:
  - Select points individually in each image, for instance via Algorithm 11.1. Select a threshold $\tau_e$.
  - For all points selected in one image, compute the NCC score (11.11) with respect to all other selected points in the other image(s).
  - If at least one point has an NCC score that exceeds the threshold $\tau_e$, choose as correspondent the point that maximizes the NCC score. Otherwise, the selected point in the first image has no correspondent.
  - Choose $S_0$ to be a subset of eight successful matches, selected at random (e.g., with a uniform distribution on the index of the points).
- Choose a threshold $\tau$.
- Repeat $N$ times the following steps:
  1. Estimate $F_k$ given the subset $S_k$ using the eight-point algorithm, Algorithm 6.1 or 11.5.
  2. Given $F_k$ determine a subset of correspondences $j = 1, 2, \ldots, n$ for which the residual $d^j$ defined by equation (11.12) satisfies $d^j < \tau$ pixels. This is called the *consensus set* of $S_k$.
  3. Count the number of points in the consensus set.
  4. Randomly select eight matches as the new set $S_{k+1}$.
- Choose $F_k$ with the largest consensus set.
- Reestimate $F$ using the all the inliers, i.e. all pairs of corresponding points in the consensus set.

of correct matches increases, but the absolute number of pairs whose NCC score passes the threshold decreases. The effect becomes more dramatic as the baseline increases.

*Caveats*

Feature matching is the most difficult and often most fragile component of the pipeline described in this chapter. Real scenes often present occlusions, specular reflections, cast shadows, changing illumination and other factors that make feature matching across wide baselines difficult. Such difficulties, however, can be eased if richer descriptors are used than simply the intensity of each pixels in a neighborhood: For instance, one can use affine invariant descriptors, or signatures based on hyperspectral images, or known landmarks in the scene.

When no intervention on the scene is possible, the algorithms described in this book can provide reasonable results if used properly. Improved robust matching can be achieved if the basic least-squares solution described in Appendix A is substituted with total least-squares, as described in [Golub and Loan, 1989]. The effect of outliers on the final estimate can also be reduced by replacing the least-squares criterion with a robust metric. A popular choice leads to *M-estimators*,

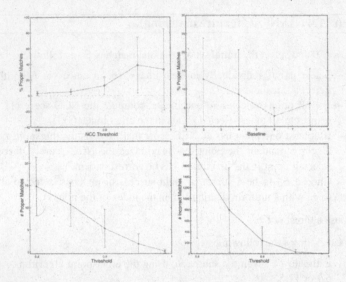

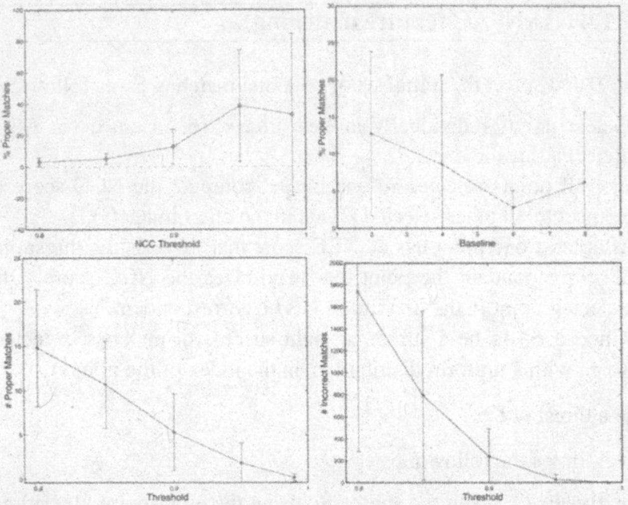


Figure 11.8. Effects of (*top left*) normalized cross correlation threshold $\tau$, (*top right*) baseline (meters) of viewpoint change on the percentage of proper matches, (*bottom left*) the number of physically correct and (*bottom right*) incorrect comparisons that passed the NCC threshold for various thresholds and a small baseline. In general, the workspace depths were in the range of 5 to 20 meters. Error bars indicate 1 standard deviation. A total of eight sequences of five images each were used to generate these charts. All results are shown for image patches of $21 \times 21$ pixels. It should be noted that the number of incorrect matches (or similarly, the total number of matches) decreases exponentially with the threshold, whereas the number of physically correct matches decreases linearly. This produces the observed peak in (*top left*) at a threshold of 0.95, as the small number of total matches at high threshold gives rise to the large standard deviations.

and another to the *least-median-of-squares* (LMedS) method, both of which have been explored in the context of this problem by [Zhang et al., 1995].

Additional issues arise when the feature matching is extended beyond two views. One option is to integrate the feature matching with the motion estimation and reconstruction process. In such a case, the partial reconstruction obtained from two views and the current motion estimate can be used to predict and/or validate the detected features in subsequent views. Alternatively, the multiple-view matching test based on the rank condition described in Chapter 8 (more details can be found in [Kosecká and Ma, 2002]) is a good indicator of the quality of feature correspondence in multiple views. After features disappear due to occlusions, one can maintain a database of features, and if the viewpoint is close enough in later views, one can attempt to rematch features in later stages, effectively reintroducing previous features. This approach has been suggested by [Pollefeys, 2000] and applied successfully to progressive scans of scenes.

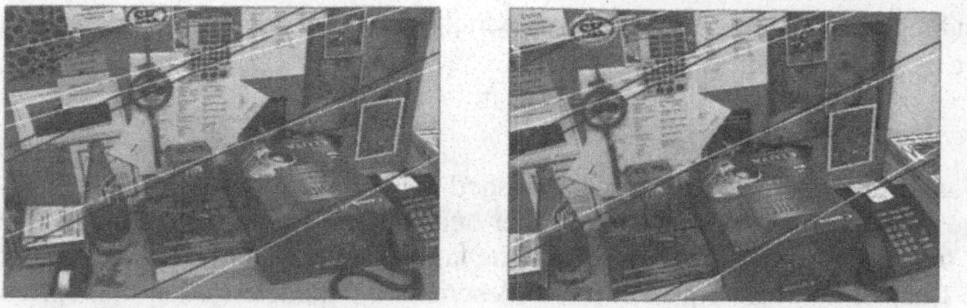

Figure 11.9. Epipolar lines estimated between the first and the 50th frame of the video sequence. The black lines correspond to the estimates obtained with the linear algorithm. The white lines are obtained from the estimate of $F$ obtained by robust matching and nonlinear refinement.

## 11.3 Projective reconstruction

Once the correspondence between features in different images has been established, we can directly recover the 3-D structure of the scene up to a projective transformation. In the absence of any additional information, this is the best one can do. We will start with the case of two views, and use the results to initialize a multiple-view algorithm.

### *11.3.1 Two-view initialization*

Recall the notation introduced in Chapter 6: The generic point $p \in \mathbb{E}^3$ has coordinates $\boldsymbol{X} = [X, Y, Z, 1]^T$ relative to a fixed ("world") coordinate frame. Given two views of the scene related by a rigid-body motion $g = (R, T)$, the 3-D coordinate $\boldsymbol{X}$ and image measurements $\boldsymbol{x}_1'$ and $\boldsymbol{x}_2'$ are related by the camera projection matrices $\Pi_1, \Pi_2 \in \mathbb{R}^{3\times 4}$ in the following way:

$$\lambda_1 \boldsymbol{x}_1' = \Pi_1 \boldsymbol{X}, \quad \lambda_2 \boldsymbol{x}_2' = \Pi_2 \boldsymbol{X}, \quad \Pi_1 = [K, 0], \quad \Pi_2 = [KR, KT],$$

where $\boldsymbol{x}' = [x', y', 1]^T$ is measured (in pixels) and $\lambda$ is an unknown scalar (the "projective depth" of the point). The calibration matrix $K$ is unknown and has the general form of equation (3.14) in Chapter 3. In case the camera intrinsic parameters are known, i.e. $K = I$, the fundamental matrix $F$ becomes the essential matrix $E = \widehat{T}R$, and the motion of the camera $g = (R, T)$ (with translation $T$ rescaled to unit norm) can be obtained directly, together with $\boldsymbol{X}$, using Algorithm 5.1 in Chapter 5.[15] In general, however, the intrinsic parameter matrix $K$ is not known. From image correspondence, one can still compute the fundamental matrix $F$ using the eight-point algorithm introduced in the appendix of Chapter 6,

[15] In such a case, the projection matrices and image coordinates are related by

$$\Pi_1 = [I, 0], \quad \Pi_2 = [R, T], \quad \lambda_1 \boldsymbol{x}_1 = \boldsymbol{X}, \quad \lambda_2 \boldsymbol{x}_2 = R\boldsymbol{X} + T.$$

and reported here as Algorithm 11.5, which includes a nonlinear refinement that we describe in the next paragraph.

*Nonlinear refinement of F*

Nonlinear refinement can be accomplished by a gradient-based minimization, of the kind described in Appendix C, after one chooses a suitable *cost function* and a *"minimal" parameterization* of $F$. The latter issue is described in the appendix to Chapter 6, where equation (6.80) describes a parameterization that enforces the constraint $\det(F) = 0$, at the price of introducing singularities. Here we present an alternative parameterization of the fundamental matrix in terms of both epipoles, written in homogeneous coordinates. Since we are free to choose such coordinates up to a scale, for convenience we will choose the scale so that the third component is $-1$ (this simplifies the derivation); then we write the coordinates of the epipoles as $\boldsymbol{e}_1 = [\alpha_1, \beta_1, -1]^T$ and $\boldsymbol{e}_2 = [\alpha_2, \beta_2, -1]^T$. With this choice of notation, the fundamental matrix can be parameterized in the following way

$$F = \begin{bmatrix} f_1 & f_4 & \alpha_1 f_1 + \beta_1 f_4 \\ f_2 & f_5 & \alpha_1 f_2 + \beta_1 f_5 \\ \alpha_2 f_1 + \beta_2 f_2 & \alpha_2 f_4 + \beta_2 f_5 & \alpha_1\alpha_2 f_1 + \alpha_1\beta_2 f_2 + \beta_1\alpha_2 f_4 + \beta_1\beta_2 f_5 \end{bmatrix}. \tag{11.13}$$

This parameterization, like the one proposed in equation (6.80) in Chapter 6, has singularities when the epipoles are at infinity. In such a case, one can simply apply a transformation $H$ to the homogeneous coordinates of each point, in such a way as to relocate the epipole away from infinity. For instance, if one of the two epipoles, say $\boldsymbol{e}$, is at infinity, so that $\boldsymbol{e} = [\alpha,\ \beta, 0]^T$, the following choice of $H$, which represents a rotation about the axis $[\beta,\ -\alpha,\ 0]^T$ by $\pi/2$ radians, moves $\boldsymbol{e}$ to $H\boldsymbol{e} = [0, 0, 1]^T$:[16]

$$H = \frac{1}{2}\begin{bmatrix} 1 + \beta^2 - \alpha^2 & -2\alpha\beta & -2\alpha \\ -2\alpha\beta & 1 - \beta^2 + \alpha^2 & -2\beta \\ 2\alpha & 2\beta & 1 - \beta^2 - \alpha^2 \end{bmatrix}. \tag{11.14}$$

As for the cost function, ideally one would want to use the reprojection error; however, this would render the computation of the gradient expensive, so we settle for the first-order approximation of the reprojection error, which we have introduced under the name of "Sampson distance" in equation (11.12). With these choices, one can easily set up an iterative minimization using standard software packages, for instance Matlab's `fmin`. We summarize this discussion in Algorithm 11.5.

*Recovering projection matrices and projective structure*

Given the fundamental matrix $F$ estimated via Algorithm 11.5, there are several ways to decompose it in order obtain projection matrices and 3-D structure from the two views. In fact, as we have seen in Chapter 6, since $F = \widehat{T'}KRK^{-1}$,

[16]Without loss of generality, here we assume that $\boldsymbol{e}$ is normalized, i.e. $\alpha^2 + \beta^2 = 1$.

**Algorithm 11.5 (Eight-point algorithm with refinement)**

Given a set of initial point feature correspondences expressed in pixel coordinates $(\boldsymbol{x}_1'^j, \boldsymbol{x}_2'^j)$ for $j = 1, 2, \ldots, n$:

- **(Optional)** normalize the image coordinates by $\tilde{\boldsymbol{x}}_1 = H_1\boldsymbol{x}_1'$ and $\tilde{\boldsymbol{x}}_2 = H_2\boldsymbol{x}_2'$, where $H_1$ and $H_2$ are normalizing transformations derived in Section 6.A.
- **A first approximation of the fundamental matrix:** Construct the matrix $\chi \in \mathbb{R}^{n\times 9}$ from the transformed correspondences $\tilde{\boldsymbol{x}}_1^j \doteq [\tilde{x}_1^j,\ \tilde{y}_1^j,\ 1]^T$ and $\tilde{\boldsymbol{x}}_2^j \doteq [\tilde{x}_2^j,\ \tilde{y}_2^j,\ 1]^T$ as in equation (6.76), where the $j$th row of $\chi$ is given by
$$[\tilde{x}_1^j\tilde{x}_2^j, \tilde{x}_1^j\tilde{y}_2^j, \tilde{x}_1^j, \tilde{y}_1^j\tilde{x}_2^j, \tilde{y}_1^j\tilde{y}_2^j, \tilde{y}_1^j, \tilde{x}_2^j, \tilde{y}_2^j, 1]^T \quad \in \mathbb{R}^9.$$
Find the vector $F^s \in \mathbb{R}^9$ of unit length such that $\|\chi F^s\|$ is minimized as follows: Compute the singular value decomposition (SVD, Appendix A) of $\chi = U\Sigma V^T$ and define $F^s$ to be the ninth column of $V$. Unstack the nine elements of $F^s$ into a square $3\times 3$ matrix $\tilde{F}$. Apply the inverse normalizing transformations, if applicable, to obtain $F = H_2^T\tilde{F}H_1$. Note that this matrix will in general *not* be a fundamental matrix.
- **Imposing the rank-2 constraint:** Compute the SVD of the matrix $F$ recovered from data to be $F = U_F\text{diag}\{\sigma_1, \sigma_2, \sigma_3\}V_F^T$. Impose the rank-2 constraint by letting $\sigma_3 = 0$ and reset the fundamental matrix to be
$$F = U_F\text{diag}\{\sigma_1, \sigma_2, 0\}V_F^T.$$
- **Nonlinear refinement:** Iteratively minimize the Sampson distance (11.12) with respect to the parameters of the fundamental matrix $f_1, f_2, f_4, f_5, \alpha_1, \beta_1, \alpha_2, \beta_2$, using a gradient descent algorithm as described in Appendix C. Reconstruct the fundamental matrix using equation (11.13), and – if the epipoles are at infinity – the transformation defined in equation (11.14).

---

all projection matrices $\Pi_p = [KRK^{-1} + T'v^T, v_4T']$ yield the same fundamental matrix for any value of $v = [v_1, v_2, v_3]^T$ and $v_4$, and hence there is a four-parameter family of possible choices. One common choice, known as the canonical decomposition, has been described in Section 6.4.2 in Chapter 6, and has the following form

$$\Pi_{1p} = [I, 0], \quad \Pi_{2p} = [(\widehat{T'})^T F, T'], \quad \lambda_1\boldsymbol{x}_1' = \boldsymbol{X}_p, \quad \lambda_2\boldsymbol{x}_2' = (\widehat{T'})^T F\boldsymbol{X}_p + T'. \tag{11.15}$$

Now, different choices of $v$ and $v_4$ result in different projection matrices $\Pi_p$, which in turn result in different projective coordinates $\boldsymbol{X}_p$, and hence different reconstructions. Some of these reconstructions may be more "distorted" than others, in the sense of being farther from the "true" Euclidean reconstruction. In order to minimize the amount of projective distortion and obtain the initial reconstruction as close as possible to the Euclidean one, we can play with the choice of $v$ and $v_4$, as suggested in [Beardsley et al., 1997]. In practice, it is common to assume that the optical center is at the center of the image, that the focal length is roughly known (for instance from previous calibrations of the camera), and that the pixels are square with no skew. Therefore, one can start with a rough approximation of the intrinsic parameter matrix $K$, call it $\tilde{K}$. This initial guess $\tilde{K}$ can be used instead of the normalizing transformation $H$ in Algorithm 11.5. After doing so, we can choose $v \in \mathbb{R}^3$ and $v_4$ by requiring that the first block of the projection matrix be as close as possible to the rotation matrix between two views, $R \approx v_4(\widehat{T'})^T F + T'v^T$. In case the actual rotation $R$ between the views is small, we can start by choosing $\tilde{R} \approx I$, and solve linearly for $v$ and $v_4$. In case of general rotation, one can still solve the equation $R = v_4(\widehat{T'})^T F + T'v^T$ for $v$, provided a guess for the rotation $\tilde{R}$ is available. If this is not the case, we have described in Exercise 6.10 an alternative method for improved canonical decomposition (although note that the solution suggested there will in general not guarantee that the estimated rotation is close to the actual rotation).

Once we have a choice of $v$ and $v_4$, and hence of projection matrices, the 3-D structure can be recovered. Ideally, if our guess $\tilde{K}$ was accurate, all points should be visible; i.e. all estimated scales should be positive. If this is not the case, different values for the focal length can be tested until the majority of points have positive depth. This procedure follows from considerations on Cheirality constraints and quasi-affine reconstruction that we have discussed in Appendix 6.B.2. This procedure is summarized as Algorithm 11.6, and an example of the reconstruction is shown in Figure 11.10.

Figure 11.10. Projective reconstruction of a simple scene, which allows easy visualization of the projective distortion. The positions of the camera are indicated by two coordinate frames.

## 11.3.2 Multiple-view reconstruction

When more than two views are available, they can be added one at a time or simultaneously using the multiple-view algorithms described in Chapter 8. Once the 3-D structure has been initialized, the calibrated and uncalibrated cases differ only in a single step; hence we will treat them simultaneously here. For the pur-

**Algorithm 11.6 (Projective reconstruction – two views)**

Given a set of initial point feature correspondences expressed in pixel coordinates $(\boldsymbol{x}_1'^j, \boldsymbol{x}_2'^j)$ for $j = 1, 2, \ldots, n$,

1. Guess a calibration matrix $\tilde{K}$ by choosing the optical center at the center of the image, assuming the pixels to be square, and guessing the focal length $\tilde{f}$. For example, for an image plane of size $(D_x \times D_y)$ pixels, a typical guess is
$$\tilde{K} = \begin{bmatrix} \tilde{f} & 0 & D_x/2 \\ 0 & \tilde{f} & D_y/2 \\ 0 & 0 & 1 \end{bmatrix}$$
with $\tilde{f} = k \times D_x$, where $k$ is typically chosen in the interval [0.5, 2].
2. Estimate the fundamental matrix $F$ using the eight-point algorithm (Algorithm 11.5) with the matrix $\tilde{K}^{-1}$ in place of the normalizing transformation $H$. The normalized coordinates are $\tilde{\boldsymbol{x}}_1 = \tilde{K}^{-1}\boldsymbol{x}_1'$ and $\tilde{\boldsymbol{x}}_2 = \tilde{K}^{-1}\boldsymbol{x}_2'$.
3. Compute the epipole $T'$ as the null space of $F^T$: From the SVD of $F^T = USV^T$, set $T'$ to be the last (third) column of the matrix $V$.
4. Choose $v \in \mathbb{R}^3$ and $v_4 \in \mathbb{R}$ so that the rotational part of the fundamental matrix $v_4(\widehat{T'})^T F + T'v^T$ is as close as possible to a (small) rotation:
   - Assume that $\tilde{R} \approx I$.
   - Solve the equation $\tilde{R} = v_4(\widehat{T'})^T F + T'v^T$ for $v$ and $v_4$ in the least-squares sense, using the SVD.
5. Setting the first frame as a reference, the projection matrices are given by
$$\Pi_{1p} = [I, 0], \quad \Pi_{2p} = \left[v_4(\widehat{T'})^T F + T'v^T, T'\right] = [R, T'].$$
6. The 3-D projective structure $\boldsymbol{X}_p$ for each $j = 1, 2, \ldots, n$ can now be estimated as follows:
   - Denote the projection matrices by $\Pi_{1p} = [\pi_1^{1T}, \pi_1^{2T}, \pi_1^{3T}]^T$ and $\Pi_{2p} = [\pi_2^{1T}, \pi_2^{2T}, \pi_2^{3T}]^T$ written in terms of their three row vectors; Let $\tilde{\boldsymbol{x}}_1 = [\tilde{x}_1, \tilde{y}_1, 1]^T$ and $\tilde{\boldsymbol{x}}_2 = [\tilde{x}_2, \tilde{y}_2, 1]^T$ be corresponding points in two views. The unknown structure satisfies the following constraints (see the paragraph following equation (6.33) in Chapter 6):
$$(\tilde{x}_1\pi_1^{3T} - \pi_1^{1T})\boldsymbol{X}_p = 0, \quad (\tilde{y}_1\pi_1^{3T} - \pi_1^{2T})\boldsymbol{X}_p = 0,$$
$$(\tilde{x}_2\pi_2^{3T} - \pi_2^{1T})\boldsymbol{X}_p = 0, \quad (\tilde{y}_2\pi_2^{3T} - \pi_2^{2T})\boldsymbol{X}_p = 0.$$
   - The projective structure can then be recovered as the least-squares solution of a linear system of equations $M\boldsymbol{X}_p = 0$. The solution for each point is given by the eigenvector of $M^T M$ that corresponds to its smallest eigenvalue, and computed again using the SVD.
   - The unknown scales $\lambda_1^j$ are simply the third coordinate of the homogeneous representation of $\boldsymbol{X}_p^j$ (with the fourth coordinate of $\boldsymbol{X}_p$ normalized to 1), such that $\boldsymbol{X}_p^j = \lambda_1^j \tilde{\boldsymbol{x}}_1^j$ for all points $j = 1, 2, \ldots, n$.

pose of clarity we will drop the superscript $'$ from $\boldsymbol{x}'$ and simply write $\boldsymbol{x}$ for both the calibrated and uncalibrated entities.

For the multiple-view setting (using the notation introduced in Chapter 8) we have

$$\lambda_i^j \boldsymbol{x}_i^j = \Pi_i \boldsymbol{X}^j, \quad i = 1, 2, \ldots, m, \; j = 1, 2, \ldots, n. \tag{11.16}$$

The matrix $\Pi_i = K_i \Pi_0 g_i$ is a $3 \times 4$ camera projection matrix that relates the $i$th (measured) image of the point $p$ to its (unknown) 3-D coordinates $\boldsymbol{X}^j$ with respect to the world reference frame. The intrinsic parameter matrix $K_i$ is upper triangular;[17] $\Pi_0 = [I, 0] \in \mathbb{R}^{3\times 4}$ is the standard projection matrix; and $g_i \in SE(3)$ is the rigid-body displacement of the camera with respect to the world reference frame. The goal of this step is to recover all the camera poses for the $m$ views and the 3-D structure of any point that appears in at least two views. For convenience, we will use the same notation $\Pi_i = [R_i, T_i]$ with $R_i \in \mathbb{R}^{3\times 3}$ and $T_i \in \mathbb{R}^3$ for both the calibrated and uncalibrated case.[18]

The core of the multiple-view algorithm consists of exploiting the following equation, derived from the multiple-view rank conditions studied in Chapter 8:

$$P_i \begin{bmatrix} R_i^s \\ T_i \end{bmatrix} \doteq \begin{bmatrix} {\boldsymbol{x}_1^1}^T \otimes \widehat{\boldsymbol{x}_i^1} & \alpha^1 \widehat{\boldsymbol{x}_i^1} \\ {\boldsymbol{x}_1^2}^T \otimes \widehat{\boldsymbol{x}_i^2} & \alpha^2 \widehat{\boldsymbol{x}_i^2} \\ \vdots & \vdots \\ {\boldsymbol{x}_1^n}^T \otimes \widehat{\boldsymbol{x}_i^n} & \alpha^n \widehat{\boldsymbol{x}_i^n} \end{bmatrix} \begin{bmatrix} R_i^s \\ T_i \end{bmatrix} = 0 \in \mathbb{R}^{3n}, \tag{11.17}$$

where $\otimes$ is the *Kronecker product* (Appendix A, equation (A.14)). Since $\alpha^j = 1/\lambda_1^j$, the inverse depth of $\boldsymbol{X}_1^j$ with respect to the first view, is known from the initialization stage from two views (Algorithm 11.6), the matrix $P_i \in \mathbb{R}^{3n\times 12}$ is of rank 11 if more than $n \geq 6$ points in general position are given, and the unknown motion parameters lie in the null space of $P_i$. This leads to Algorithm 11.7, which alternates between estimation of camera motion and 3-D structure, exploiting multiple-view constraints available in all views. After the algorithm has converged, the camera motion is given by $[R_i, T_i], i = 2, 3, \ldots, m$, and the depth of the points (with respect to the first camera frame) is given by $\lambda_1^j = 1/\alpha^j, j = 1, 2, \ldots, n$. Some bookkeeping is necessary for features that appear and disappear during the duration of the sequence. The resulting projection matrices and the 3-D structure obtained by the above iterative procedure can be then refined using a nonlinear optimization algorithm, as we describe next.

---

[17] In case the camera is calibrated, we have $K_i = I$.

[18] In the calibrated case, $R_i \in SO(3)$ is a rotation matrix. It describes the rotation from the $i$th view to the first.

**Algorithm 11.7 (Multiple-view structure and motion estimation algorithm).**

Given $m$ images $\boldsymbol{x}_1^j, \boldsymbol{x}_2^j, \ldots, \boldsymbol{x}_m^j$ of $n$ points, $j = 1, 2, \ldots, n$, estimate the projection matrix $\Pi_i = [R_i, T_i]$, $i = 2, 3, \ldots, m$ as follows:

1. Initialization: $k = 0$; Let $\alpha_0^j = 1/\lambda_1^j$ be the scales recovered from the two-view initialization algorithm, Algorithm 11.6.
2. For any given $k$, set $\alpha^j = \alpha_k^j / \alpha_k^1$ for $j = 1, 2, \ldots, n$.
3. Assemble the matrix $P_i$, using the scales $\alpha^j$, as per equation (11.17) and compute the singular vector $v_{12}$ associated with the smallest singular value of $P_i$, $i = 2, 3, \ldots, m$. Unstack the first nine entries of $v_{12}$ to obtain $\tilde{R}_i$; the last three entries of $v_{12}$ are $\tilde{T}_i$.
4. (a) In case of *calibrated cameras*, compute the SVD of the matrix $\tilde{R}_i = U_i S_i V_i^T$ and set the current rotation and translation estimates to be
$$R_i = \operatorname{sign}(\det(U_i V_i^T)) U_i V_i^T \quad \in SO(3),$$
$$T_i = \frac{\operatorname{sign}(\det(U_i V_i^T))}{\sqrt[3]{\det(S_i)}} \tilde{T}_i \quad \in \mathbb{R}^3.$$
(b) In case of *uncalibrated cameras*, set the current estimates of $(R_i, T_i)$ simply to $(\tilde{R}_i, \tilde{T}_i)$, for $i = 2, 3, \ldots, m$.
5. Let $\Pi_{ik+1} = [R_i, T_i]$.
6. Given all the motions, recompute the scales $\alpha_{k+1}^j$ via
$$\alpha_{k+1}^j = -\frac{\sum_{i=2}^m \left(\widehat{\boldsymbol{x}_i^j} T_i\right)^T \widehat{\boldsymbol{x}_i^j} R_i \boldsymbol{x}_1^j}{\sum_{i=2}^m \left\|\widehat{\boldsymbol{x}_i^j} T_i\right\|^2}, \quad j = 1, 2, \ldots, n,$$
and hence recompute the 3-D coordinates of each point $\boldsymbol{X}_{k+1}^j = \lambda_{1k+1}^j \boldsymbol{x}_1^j$.
7. Compute the reprojection error
$$e_r = \frac{1}{mn} \sum_{i=1}^m \sum_{j=1}^n \left\|\boldsymbol{x}_i^j - \pi(\Pi_{ik+1} \boldsymbol{X}_{k+1}^j)\right\|^2.$$
If $e_r > \epsilon$, for a specified $\epsilon > 0$, then $k \leftarrow k + 1$ and go to step 2, else stop.

### 11.3.3 *Gradient descent nonlinear refinement ("bundle adjustment")*

The quality of the estimates obtained by Algorithm 11.7 is measured by the reprojection error[19]

$$e_r = \frac{1}{mn} \sum_{i=1}^m \sum_{j=1}^n \left\|\boldsymbol{x}_i^j - \pi(\Pi_i \boldsymbol{X}^j)\right\|^2. \tag{11.18}$$

[19]We use $\pi$ to denote the standard planar projection $\pi : [X, Y, Z]^T \mapsto [X/Z, Y/Z, 1]^T$, introduced in Chapter 3.

If the reprojection error is still large (the typical average reprojection error is between 0.01 to 0.3 pixels, depending on the quality of the measurements) after following the procedure outlined so far, the estimates can be refined using a nonlinear optimization procedure that simultaneously updates the estimates of both motion and structure parameters $\xi \doteq \{\Pi_i, \boldsymbol{X}^j\}$. The total dimension of the parameter space is $11(m-1)+3n$ for $m$ views and $n$ points.[20] The iterative optimization scheme updates the current estimate $\xi^k$ via

$$\xi^{k+1} = \xi^k - \alpha_k D_k \nabla e_r(\xi^k).$$

Various choices of the step size $\alpha_k$ and weight matrix $D_k$ are discussed in Appendix C. A popular algorithm is that of Levenberg and Marquardt, where $D_k$ is of the form described in equation (C.17). Appendix C gives details on gradient-based optimization, including how to adaptively update the stepsize. However, the nature of our objective function $e_r$ is special. The weight matrix has a very sparse structure due to the quasi block-diagonal nature of the Jacobian. Therefore, the resulting algorithm can be considerably speeded up, despite the large parameter space [Triggs and Fitzgibbon, 2000]. In the case of a calibrated camera, the recovered structure is the true Euclidean one and we can proceed directly to Section 11.5.1. In the uncalibrated case, the projective structure thus obtained has to be upgraded to Euclidean, as we describe in the next section.

*Caveats*

Due to occlusions, features tend to disappear during a sequence. As more and more features do so, there is an increased risk that the few remaining features will align in a singular configuration, most notably a plane. In order to avoid such a situation, one may periodically run a feature selection algorithm in order to generate new correspondences, or even check for occluded features to reappear as suggested by [Pollefeys, 2000]. Another issue that should be further addressed in a practical implementation is the choice of reference frame: So far we have associated it to the first camera, but a more proper assignment would entail averaging over all views.

## 11.4 Upgrade from projective to Euclidean reconstruction

The projective reconstruction $\boldsymbol{X}_p$ obtained with Algorithm 11.7, in the absence of calibration information, is related to the Euclidean structure $\boldsymbol{X}_e$ by a linear

[20] There is some redundancy in this parameterization, as a minimal parameterization has less degrees of freedom (Proposition 8.21, Chapter 8). This, however, is not a problem provided that one uses gradient descent algorithms that do not require the Jacobian matrix to be nonsingular, as for instance the Levenberg-Marquardt algorithm. See Appendix C.

transformation $H \in \mathbb{R}^{4\times4}$, as discussed in detail in Section 6.4

$$\Pi_{ip} \sim \Pi_{ie} H^{-1}, \quad \boldsymbol{X}_p \sim H \boldsymbol{X}_e, \quad i = 1, 2, \ldots, m, \tag{11.19}$$

where $\sim$ indicates equality up to a scale factor, $\Pi_{1p} = [I, \ 0]$ and $H$ has the form

$$H = \begin{bmatrix} K_1 & 0 \\ -v^T K_1 & 1 \end{bmatrix} \in \mathbb{R}^{4\times4}. \tag{11.20}$$

### 11.4.1 Stratification with the absolute quadric constraint

Examine the equation

$$\Pi_{ip} H \sim \Pi_{ie} = [K_i R_i, K_i T_i],$$

where we now use $[R_i, T_i]$ to denote the Euclidean motion[21] between the $i$th and the first camera frame. Since the last column gives three equations, but adds three unknowns, it is useless as far as providing constraints on $H$. Therefore, we can restrict our attention to the leftmost $3 \times 3$ block

$$\Pi_{ip} \begin{bmatrix} K_1 \\ -v^T K_1 \end{bmatrix} \sim K_i R_i. \tag{11.21}$$

One can then eliminate the unknown rotation matrix $R_i$ by multiplying both sides by their transpose:

$$\Pi_{ip} \begin{bmatrix} K_1 K_1^T & -K_1 K_1^T v \\ -v^T K_1 K_1^T & v^T K_1 K_1^T v \end{bmatrix} \Pi_{ip}^T \sim K_i K_i^T. \tag{11.22}$$

If we define $S_i^{-1} \doteq K_i K_i^T \in \mathbb{R}^{3\times3}$, and

$$Q \doteq \begin{bmatrix} K_1 K_1^T & -K_1 K_1^T v \\ -v^T K_1 K_1^T & v^T K_1 K_1^T v \end{bmatrix} \in \mathbb{R}^{4\times4}, \tag{11.23}$$

then we obtain the *absolute quadric constraint* introduced in Chapter 6:

$$\Pi_{ip} Q \Pi_{ip}^T \sim S_i^{-1}. \tag{11.24}$$

If we assume that $K$ is constant, so that $K_i = K$ for all $i$, then we can minimize the angle between the vectors composing the matrices on the left-hand side and those on the right-hand side[22] with respect to the unknowns, $K$ and $v$, using for instance a gradient descent procedure. Alternatively, we could first estimate $Q$ and $K_i$ from this equation by ignoring its internal structure; then, $H$ and $K$ can be extracted from $Q$, and subsequently the recovered structure and motion can be upgraded to Euclidean.

As we have discussed in great detail in Chapters 6 and 8, in order to have a unique solution, it is necessary for the scene to be generic and for the camera

[21] In particular, $R_i R_i^T = R_i^T R_i = I$.

[22] Notice that the above equality is up to an unknown scalar factor, so we cannot simply take the norm of the difference of the two sides, but we can consider angles instead.

motion to be "rich enough," in particular it must include rotation about two independent axes.[23] Partial knowledge about the camera parameters helps simplifying the solution. In the rest of this section, we describe a useful case in which the intrinsic camera parameters are known, with the exception of the focal length, which is allowed to vary during the sequence, for instance as a result of zooming or focusing. The method presented in the next paragraph and originally introduced by [Pollefeys et al., 1998] is often used for initialization of more elaborate nonlinear estimation schemes.

*The case of changing focal length*

When the calibration parameters are known, for instance because the camera has been calibrated using a calibration rig, but the lens is moved to zoom or focus during the sequence, one can use a simple algorithm outlined below. The reader should be advised, however, that zooming or focusing often causes the principal point, i.e. the intersection of the optical axis with the image plane, to move as well, often by several pixels, following a spiral trajectory. It is very rare for a commercial camera to have the principal point close to the center of the image and have the optical axis perfectly orthogonal to the image plane. Nevertheless, it is common to assume, as an initial approximation, that the optical axis is orthogonal to the image plane and intersects it at its center, and that the pixels are square. Under these assumptions one can obtain reasonable estimates of the intrinsic parameters, which can be further refined using nonlinear optimization schemes. If we accept these assumptions, then the absolute quadric constraint (11.24) takes a particularly simple form

$$\Pi_{ip} \begin{bmatrix} a_1 & 0 & 0 & a_2 \\ 0 & a_1 & 0 & a_3 \\ 0 & 0 & 1 & a_4 \\ a_2 & a_3 & a_4 & a_5 \end{bmatrix} \Pi_{ip}^T \sim \begin{bmatrix} f_i^2 & 0 & 0 \\ 0 & f_i^2 & 0 \\ 0 & 0 & 1 \end{bmatrix}. \tag{11.25}$$

Note that the entries of $S_i^{-1}$ satisfy the following relationships:

$$s_{11} = s_{22}, \quad s_{12} = s_{13} = s_{23} = 0, \quad s_{21} = s_{31} = s_{32} = 0.$$

These can be directly translated into constraints on the matrix $Q$:

$$\begin{cases} \pi_i^{1T} Q \pi_i^1 &= \pi_i^{2T} Q \pi_i^2, \\ \pi_i^{1T} Q \pi_i^2 &= 0, \\ \pi_i^{1T} Q \pi_i^3 &= 0, \\ \pi_i^{2T} Q \pi_i^3 &= 0, \end{cases} \tag{11.26}$$

where $\Pi_i = [\pi_i^{1T}, \pi_i^{2T}, \pi_i^{3T}]^T$ is the projection matrix written in terms of its rows. If $Q$ is parameterized as in equation (11.25), the five unknowns can be recovered linearly. Since each pair of views gives four constraints, at least three views are

[23] When these hypotheses are not satisfied, the camera is undergoing critical motions, in which case refer to Section 8.5.2.

necessary for a unique solution. This is summarized as Algorithm 11.8. A similar linear algorithm can be devised for the case of unknown aspect ratio (the pixels are rectangular, but not square). Finally, once $Q$ is recovered, $K$ and $v$ can be extracted from equation (11.23), and hence $H$, the projective upgrade, recovered from equation (11.20). The reconstruction result for some features in the house scene from Figure 11.1 is shown in Figure 11.11.

**Algorithm 11.8 (Recovery of the absolute quadric and Euclidean upgrade).**

1. Given $m$ projection matrices $\Pi_i, i = 1, 2, \ldots, m$ recovered by Algorithm 11.7, for each projection matrix set up the linear constraints in $Q \doteq \begin{bmatrix} a_1 & 0 & 0 & a_2 \\ 0 & a_1 & 0 & a_3 \\ 0 & 0 & 1 & a_4 \\ a_2 & a_3 & a_4 & a_5 \end{bmatrix}$. Let $Q^s \doteq [a_1,\ a_2,\ a_3,\ a_4,\ a_5]^T \in \mathbb{R}^5$ be the stacked version of $Q$.
2. Form a matrix $\chi \in \mathbb{R}^{4m \times 5}$ by stacking together $m$ of the following $4 \times 5$ block of rows, one for each $i = 1, 2, \ldots, m$, where each row of the block corresponds to one of the constraints from equation (11.26)
$$\begin{bmatrix} u_1^2 + u_2^2 - v_1^2 - v_2^2 & 2u_4u_1 - 2v_1v_4 & 2u_4u_2 - 2v_2v_4 & 2u_4u_3 - 2v_3v_4 & u_4^2 - v_4^2 \\ u_1v_1 + u_2v_2 & u_4v_1 + u_1v_4 & u_4v_2 + u_2v_4 & u_4v_3 + u_3v_4 & u_4v_4 \\ u_1w_1 + u_2w_2 & u_4w_1 + u_1w_4 & u_4w_2 + u_2w_4 & u_4w_3 + u_3w_4 & u_4w_4 \\ v_1w_1 + v_2w_2 & v_4w_1 + v_1w_4 & v_4w_2 + v_2w_4 & v_4w_3 + v_3w_4 & v_4w_4 \end{bmatrix},$$
where $u = [u_1,\ u_2,\ u_3, u_4] \doteq \pi_i^1,\ v \doteq \pi_i^2,\ w \doteq \pi_i^3$ are the three rows of the projection matrix $\Pi_i = [\pi_i^1; \pi_i^2; \pi_i^3]$, respectively.
3. Similarly form a vector $b \in \mathbb{R}^{4m}$ by stacking together $m$ of the following four-dimensional blocks
$$\begin{bmatrix} -u_3^2 + v_3^2 & -u_3v_3 & -u_3w_3 & -v_3w_3 \end{bmatrix}^T.$$
4. Solve for $Q^s$ in the least-squares sense: $\hat{Q}^s \doteq \chi^\dagger b$, where $\dagger$ denotes the pseudo-inverse (Appendix A).
5. Unstack $Q^s$ to a matrix $\tilde{Q}$ according to the definition in step 1.
6. Enforce the rank-3 constraint on $\tilde{Q}$, by computing its SVD $\tilde{Q} = U_Q \text{diag}\{\sigma_1, \sigma_2, \sigma_3, \sigma_4\} V_Q^T$. Obtain $Q$ by setting the smallest singular value of $\tilde{Q}$ to zero
$$Q = U_Q \text{diag}\{\sigma_1, \sigma_2, \sigma_3, 0\} V_Q.$$
7. Once $Q$ has been recovered using the above algorithm, the focal lengths $f_i$ of the individual cameras can be obtained by substitution into equation (11.25).
8. Perform the Euclidean upgrade using $H$ in equation (11.20), with $K_1$ and $v$ computed from the parameters of the absolute quadric $Q$ via
$$K_1 = \begin{bmatrix} \sqrt{a_1} & 0 & 0 \\ 0 & \sqrt{a_1} & 0 \\ 0 & 0 & 1 \end{bmatrix} \quad \text{and} \quad v = -[a_2/a_1, a_3/a_1, a_4]^T.$$

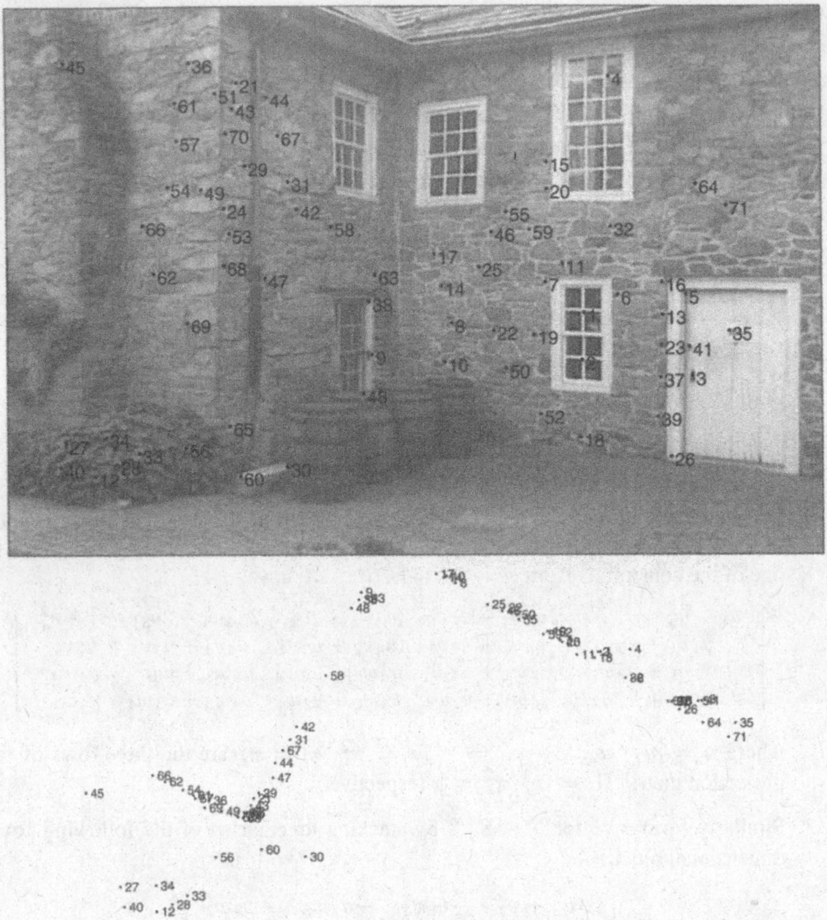

Figure 11.11. Euclidean reconstruction from multiple views of the house scene of Figure 11.1: Labeled features (top), top view of the reconstruction with labeled points (bottom). The unlabeled reconstruction is shown in Figure 11.12 (left).

After running Algorithm 11.8, one can perform additional steps of optimization as we discuss next.

### *11.4.2 Gradient descent nonlinear refinement ("Euclidean bundle adjustment")*

In order to refine the estimate obtained so far, we can set up an iterative optimization, known as Euclidean bundle adjustment, by minimizing the reprojection error as in Section 11.4.2, but this time with respect to *all and only* the unknown Euclidean parameters: structure, motion, and calibration.

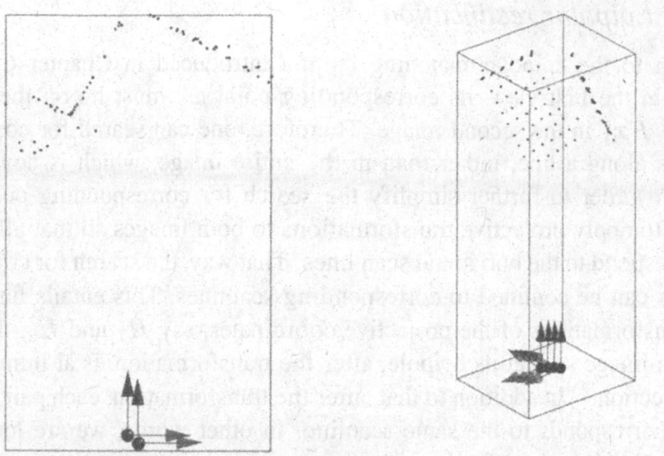

Figure 11.12. Euclidean reconstruction from multiple views of the house scene (left) of Figure 11.1 and the calibration scene (right) of Figure 11.10. The camera position and orientation through the sequence is indicated by its moving reference frame. Compare the calibration scene (right) with the projective reconstruction displayed in Figure 11.10.

The reprojection error is still given by[24]

$$e_r = \frac{1}{mn}\sum_{i=1}^{m}\sum_{j=1}^{n}\left\|\boldsymbol{x}_i^j - \pi\big(K_i(R_i\boldsymbol{X}^j + T_i)\big)\right\|^2. \qquad (11.27)$$

However, this time the parameters are given by $\xi \doteq \{K_i, \omega_i, T_i, \boldsymbol{X}^j\}$, where $\omega_i$ are the exponential coordinates of rotation $R_i = e^{\widehat{\omega}_i}$ and are computed via Rodrigues' formula (2.16). The total dimension of the parameter space is $5 + 6(m-1) + 3n$ for $m$ views and $n$ points. The iterative optimization scheme updates the current estimate $\xi^i$ via the same iteration described in Section 11.3.3, and the same considerations apply here.

## 11.5 Visualization

In order to render the scene from novel viewpoints, we need a model of its surfaces, so that we can texture-map images onto it. Clearly, the handful of point features we have reconstructed so far is not sufficient. We will follow a procedure to first simplify the correspondence process and obtain many more matches, then interpolate these matches with planar patches or smooth surfaces, and finally texture-map images onto them.

[24] Again, here $\pi$ denotes the standard planar projection $\pi : [X, Y, Z]^T \mapsto [X/Z, Y/Z, 1]^T$.

### *11.5.1 Epipolar rectification*

According to the epipolar matching lemma introduced in Chapter 6, given a point $\boldsymbol{x}_1'$ in the first view, its corresponding point $\boldsymbol{x}_2'$ must lie on the epipolar line $\boldsymbol{\ell}_2' \sim F\boldsymbol{x}_1'$ in the second image. Therefore, one can search for corresponding points along a line, rather than in the entire image, which is considerably simpler. In order to further simplify the search for corresponding points, it is desirable to apply projective transformations to both images so that all epipolar lines correspond to the horizontal scan lines. That way, the search for corresponding points can be confined to corresponding scanlines. This entails finding two linear transformations of the projective coordinates, say $H_1$ and $H_2$, that transform each image so that its epipole, after the transformation, is at infinity in the $x$-axis direction.[25] In addition to that, after the transformation, each pair of epipolar lines corresponds to the same scanline. In other words, we are looking for $H_1, H_2 \in \mathbb{R}^{3\times3}$ that satisfy

$$H_1\boldsymbol{e}_1 \sim [1,0,0]^T, \quad H_2\boldsymbol{e}_2 \sim [1,0,0]^T, \tag{11.28}$$

where $\boldsymbol{e}_1$ and $\boldsymbol{e}_2$ are the right and left null-spaces of $F$, i.e. the left and right epipole respectively, and for any pair of matched feature points, say $(\boldsymbol{x}_1', \boldsymbol{x}_2')$, after the transformation, they result in the same $y$-coordinate on the image plane. This process is called *epipolar rectification*, and its effects are shown in Figure 11.13.

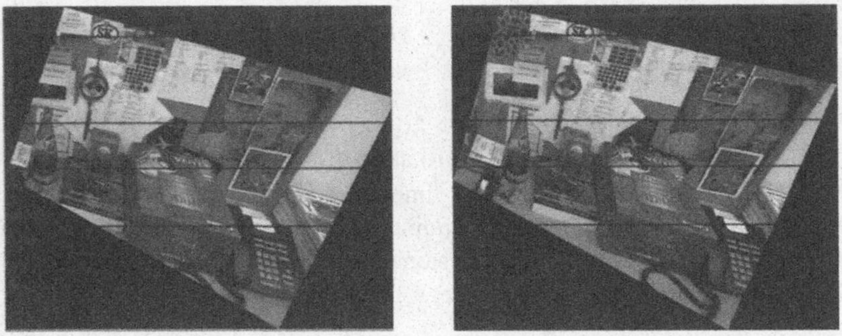

Figure 11.13. Epipolar rectification of a pair of images: Epipolar lines correspond to scanlines, which reduces the correspondence process to a one-dimensional search.

In general, there are many transformations that satisfy the requirements just described, and therefore we have to make a choice. Here we first find a transformation $H_2$ that maps the second epipole $\boldsymbol{e}_2$ to infinity and aligns the epipolar lines with the scanlines; once the transformation $H_2$ has been found, a corresponding transformation $H_1$ for the first view, called the *matching homography*, is obtained via the fundamental matrix $F$.

[25]The reader should be aware that we here are looking for an $H$ that has the opposite effect than the one used in Section 11.3.1 for nonlinear refinement of the fundamental matrix $F$.

*Mapping the epipole to infinity*

To map the epipole $e_2$ to infinity $[1,0,0]^T$, the matrix $H_2$ only needs to satisfy the constraint

$$H_2 e_2 \sim [1,0,0]^T,$$

which still leaves at least six degrees of freedom in $H_2$. To minimize distortion, one can choose the remaining degrees of freedom so as to have $H_2$ as close as possible to a rigid-body transformation. One such choice for $H_2$, as suggested by [Hartley, 1997], can be computed as follows:

- $G_T \in \mathbb{R}^{3\times3}$, defined as
$$G_T \doteq \begin{bmatrix} 1 & 0 & -o_x \\ 0 & 1 & -o_y \\ 0 & 0 & 1 \end{bmatrix},$$
translates the image center $[o_x, o_y, 1]^T$ to the origin $[0,0,1]^T$;
- $G_R \in SO(3)$ is a rotation around the $z$-axis that rotates the translated epipole onto the $x$-axis; i.e. $G_R G_T e_2 = [x_e, 0, 1]^T$;
- $G \in \mathbb{R}^{3\times3}$, defined as
$$G \doteq \begin{bmatrix} 1 & 0 & 0 \\ 0 & 1 & 0 \\ -1/x_e & 0 & 1 \end{bmatrix},$$
transforms the epipole from the $x$-axis in the image plane to infinity $[1,0,0]^T$; i.e. $G[x_e, 0, 1]^T \sim [1,0,0]^T$.

The resulting rectifying homography $H_2$ for the second view then is

$$H_2 \doteq G G_R G_T \quad \in \mathbb{R}^{3\times3}. \tag{11.29}$$

*Matching homography*

The matching rectifying homography for the first view $H_1$ can then be obtained as $H_1 = H_2 H$, where $H$ can be any homography which is compatible with the fundamental matrix $F$, i.e. $\widehat{T'}H \sim F$ (with $T' \sim e_2$). Given the two conditions

$$H_2 e_2 \sim [1,0,0]^T \quad \text{and} \quad H_1 = H_2 H,$$

it is quite easy to show that this choice of $H_1$ and $H_2$ indeed rectifies the images as we had requested.[26] Due to the nature of the decomposition of $F$, the choice in $H$

[26] Recall from Chapter 6 that, given a pair of views, a decomposition of $F = \widehat{T'}H$ yields the following relationship between the pixel coordinates of corresponding points in the two images: $\lambda_2 \boldsymbol{x}_2' = \lambda_1 H \boldsymbol{x}_1' + \gamma e_2$. Multiplying both sides of the above equation by $H_2$ and denoting with $\bar{\boldsymbol{x}}_2 = H_2 \boldsymbol{x}_2'$ and $\bar{\boldsymbol{x}}_1 = H_2 H \boldsymbol{x}_1'$, we obtain $\lambda_2 \bar{\boldsymbol{x}}_2 = \lambda_1 \bar{\boldsymbol{x}}_1 + \bar{\gamma}[1,0,0]^T$. Therefore, in the rectified coordinates $\bar{\boldsymbol{x}}$, if we normalize the $z$-coordinate to be 1, the $y$-coordinates of corresponding points become the same, and the discrepancy is solely along the $x$-axis.

is not unique. As we know from earlier sections, as well as Chapter 6, there is an entire three-parameter family of homographies $H = (\widehat{T'})^T F + T'v^T$ compatible with the fundamental matrix $F$, since $v \in \mathbb{R}^3$ can be arbitrary. While, in principle, any $H$ compatible with $F$ could do, in order to minimize the distortion induced by the rectifying transformations, $H$ has to be chosen with care. A common choice is to set the free parameters $v \in \mathbb{R}^3$ in such a way that the distance between $\boldsymbol{x}_2'$ and $H\boldsymbol{x}_1'$ for previously matched feature points is minimized. This criterion is captured by the algebraic error associated with the homography transfer equation

$$\widehat{\boldsymbol{x}_2'} H \boldsymbol{x}_1' = \widehat{\boldsymbol{x}_2'}\big((\widehat{T'})^T F + T'v^T\big)\boldsymbol{x}_1' \approx 0. \tag{11.30}$$

The unknown parameter $v$ can then be solved by a simple linear least-squares minimization problem of the following objective function

$$\min_v \sum_{j=1}^{n} \|\widehat{\boldsymbol{x}_2'^j}\big((\widehat{T'})^T F + T'v^T\big)\boldsymbol{x}_1'^j\|^2. \tag{11.31}$$

The overall rectification algorithm is summarized in Algorithm 11.9, and an example of the rectification results is shown in Figure 11.13.

---

**Algorithm 11.9 (Epipolar rectification).**

1. Compute the fundamental matrix $F$ for the two views and the epipole $\boldsymbol{e}_2$.
2. Compute the rectifying transform $H_2$ for the second view from equation (11.29).
3. Choose $H$ according to the criterion described before equation (11.30).
4. Compute the least-squares solution $v$ from equation (11.30) using the set of matched feature points, and determine the matching homography $H_1 = H_2 H$.
5. Transform image coordinates $\bar{\boldsymbol{x}}_1 = H_1\boldsymbol{x}_1'$, $\bar{\boldsymbol{x}}_2 = H_2\boldsymbol{x}_2'$, normalize the $z$-coordinate to be 1, which rectifies the images as desired. The transformations $H_1$ and $H_2$ are then applied to the entire image. Since the transformed coordinates will be outside the pixel grid, the intensity values of the image must be interpolated, for instance standard (linear or bilinear) interpolation.

---

The above approach works well in practice as long as the epipoles are outside the images. Otherwise, by pushing the epipoles to infinity, the neighborhood around the epipoles goes to infinity, if we insist in using a planar image representation.[27] In this case, an alternative nonlinear rectification procedure using a polar parameterization around the epipoles can be adopted, as suggested by [Pollefeys, 2000].[28] After the images are rectified, all epipolar lines are paral-

[27]This problem does not apply for spherical imaging models.

[28]The advantage of such a nonlinear rectification is that it is applicable to general viewing configurations, while minimizing the amount of distortion. This, however, is at the expense of a linear solution. Indeed, it is common to specify the nonlinear rectifying transformation using a lookup table.

lel, and corresponding points have the same $y$-coordinate, which makes the dense matching procedure that we describe below significantly simpler.

Note that the rectification procedure described above does not require knowledge of the camera pose and intrinsic parameters and is based solely on the epipolar geometry captured by the fundamental matrix $F$. However due to the arbitrary choice of $v$ obtained by minimizing the objective function (11.31), for scenes with large depth variation the above procedure can introduce distortions which can in some difficult cases complicate the search for correspondences. Since at this stage the camera pose and intrinsic parameters are already available, one can alternatively apply Euclidean rectification, where the rectifying homography $H_2$ and the matching homography $H_1$ can be chosen easily as long as the epipoles are not on the image. One such approach is described in [Fusiello et al., 1997].

### *11.5.2 Dense matching*

Given a rectified pair of images, the standard matching techniques described in Section 11.2 can be used to compute the correspondence for almost all pixels. The search is now restricted to each one-dimensional horizontal scanline. Several additional constraints can be used to speed up the search:

- Ordering constraint: The points along epipolar lines appear in the same order in both views, assuming that the objects in the scene are opaque.
- Disparity constraint: The disparity varies smoothly away from occluding boundaries, and typically a limit on the disparity can be imposed to reduce the search.
- Uniqueness constraint: Each point has a unique match in the second view.

When points become occluded from one image to another, there is a discontinuity in the disparity. One can construct a map of the NCC score along corresponding scan-lines in two images, and then globally estimate the path of maxima, which determines the correspondence. Horizontal or vertical segments of this path correspond to occlusions: horizontal for points seen in the first image but not the second, vertical for points seen in the second but not the first, as illustrated in Figure 11.14. Dynamic programming [Bellman, 1957] can be used for this purpose.

Extensions of dense matching to multiple views are conceptually straightforward, although some important technical issues are discussed in [Pollefeys, 2000].

Once many corresponding points are available, one can just apply the usual eight-point algorithm and bundle adjustment to compute their position, and finally triangulate a mesh to obtain a surface model. In general, due to errors and outliers, this mesh will be highly irregular. Standard techniques and software packages for mesh simplification are available. For instance, the reader can refer to [Hoppe, 1996]. The results are shown in Figure 11.16 for the scene in Figure 11.1.

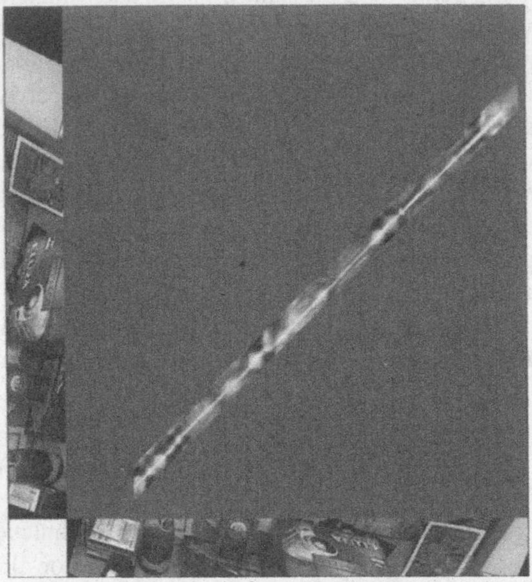

Figure 11.14. Correlation score for pixels along corresponding scan-lines. High correlation score are indicated as bright values; regions where the bright region is "thick" indicates lack of texture. In this case it is impossible to establish correspondence. Horizontal and vertical segments of the path of maxima indicate that a pixel on the first image does not have a corresponding one on the second (vertical), or vice versa (horizontal).

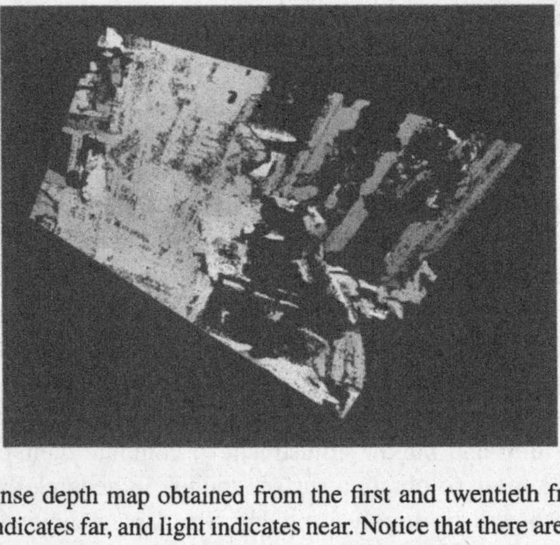

Figure 11.15. Dense depth map obtained from the first and twentieth frames of the desk sequence. Dark indicates far, and light indicates near. Notice that there are several gaps, due to lack of reliable features in regions of constant intensity. Further processing is necessary in order to fill the gaps and arrive at a dense reconstruction.

Figure 11.16. Dense reconstruction of the house sequence in Figure 11.1, obtained by interpolating sparse points into a triangulated mesh (courtesy of Hailin Jin).

### *11.5.3 Texture mapping*

Once a surface model is available, we have a number of points and a plane between any three of them. The goal of texture mapping is to attach an image patch to each of these planes. There are essentially two ways of doing so: one is a *view-independent* texture map; the other is a *view-dependent* texture map. Both techniques are standard in computer graphics, and we therefore concentrate only on view-independent ones for simplicity.

Given three vectors, their centroid can be easily computed, as well as the normal vector to the plane they identify. Since the relative pose of the cameras and the position of the points are now known, it is straightforward to compute where that point projects in each image and, moreover, where the triangular patch projects. We can then simply take the median[29] of the pixel values of all the views. While this algorithm is conceptually straightforward, several improvements can be applied in order to speed up the construction. We refer the reader to standard textbooks in computer graphics. Here we limit ourselves to showing, in Figure 11.17, the results of applying a simple algorithm to the images in Figure 11.1. Many of these functions are taken into account automatically when the final model is represented in VRML, a modeling language commonly used for visualization [VRML Consortium, 1997].

## 11.6 Additional techniques for image-based modeling

If the final goal is to generate novel views of a scene, a model of its geometry may not be strictly necessary; instead, image data can be manipulated directly to generate the appearance of a scene from novel viewpoints. Most instances of this approach relate to the notion of *plenoptic sampling* [Levoy and Hanrahan, 1996, Gortler et al., 1996] which is inspired by the plenoptic function introduced by

[29]The median, as opposed to the mean, helps to mitigate the effects of outliers, for instance, due to specularities.

Figure 11.17. Texture-mapped reconstruction of the model in Figure 11.16. The top of the barrels is not accurately reconstructed and texture-mapped, as expected from the lack of information from the original sequence in Figure 11.1 (courtesy of Hailin Jin).

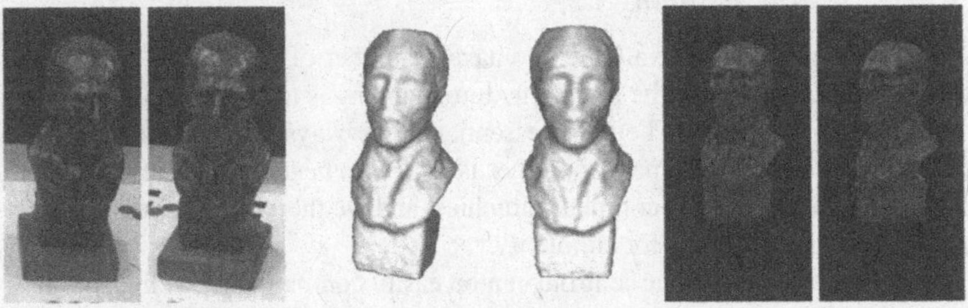

Figure 11.18. Once the camera motion and calibration have been recovered as we have described in this chapter, they can be fed, together with the original images, to any stereo algorithm. For instance, a few feature point correspondences from the images on the left can be exploited to estimate camera motion and calibration. That in turn can be used to initialize a dense stereo algorithm such as [Jin et al., 2003]. This allows handling scenes with complex photometric properties and yields an estimate of the shape (center) and reflection of the scene, which can finally be used to render it from any viewpoint (right).

[Adelson and Bergen, 1991]. Image mosaics can be viewed as a special case of plenoptic sampling. While the resulting representations can be very visually pleasant, the capability of this approach to extrapolating novel views is limited. Of course, when the goal is not just to generate novel images, but to have a model of the scene for the purpose of interaction or control, one has no choice but to infer a model of the structure of the scene, for instance by following the pipeline proposed in this chapter. The *quasi-dense approach*, which is different from the approach introduced in this chapter, provides another alternative to surface reconstruction from uncalibrated images [Lhuillier and Quan, 2002]. Encouraging results along the lines of modeling and estimating non-Lambertian reflection along with shape from moving images have recently been obtained by [Jin et al., 2003, Magda et al., 2001].

An alternative approach to obtaining models from images is to represent the scene as a collection of "voxels," and to reconstruct volumes rather than points. The representation of a scene in a convex hull of the cameras (the intersection of the viewing cones) is obtained by carving a 3-D volume in space. Space-carving techniques use reflectance information directly for matching [Kutulakos and Seitz, 1999] or contour fitting in multiple views [Cipolla et al., 1999, McMillan and Bishop, 1995, Yezzi and Soatto, 2003]. The majority of these approaches require accurate knowledge of the relative pose of the cameras, which leads us back to the reconstruction techniques discussed earlier in this chapter.

Additional techniques that have enjoyed significant success in limited domains of application entail a specialization of the techniques described in this chapter to the case of partially structured environments. Examples of such systems are Photomodeler (`http://www.photomodeler.com`) and Façade [Debevec et al., 1996]. Since the choice of model primitives largely depends on the application domain, these techniques have been successfully used in building 3-D models of architectural environments that are naturally parameterized by cubes, tetrahedra, prisms, arches, surfaces of revolutions, and their combinations.

# Chapter 12
# Visual Feedback

*I hope that posterity will judge me kindly, not only as to the things which I have explained, but also to those which I have intentionally omitted so as to leave to others the pleasure of discovery.*
– René Descartes, *La Géométrie, 1637*

In the introduction to this book we have emphasized the role of vision as a sensor for machines to interact with complex, unknown, dynamic environments, and we have given examples of successful application of vision techniques to autonomous driving and helicopter landing. Interaction with a dynamically changing environment requires action based on the current assessment of the situation, as inferred from sensory data. For instance, driving a car on the freeway requires inferring the position of neighboring vehicles as well as the *ego-motion* within the lane in order to adjust the position of the steering wheel and act on the throttle or the breaks. In order to be able to implement such a "sensing and action loop," sensory information must be processed *causally* and in real time. That is, the situation at time $t$ has to be assessed based on images *up to time* $t$. If we were to follow the guidelines and the algorithms described so far in designing an automated driving system, we would have first to collect a sufficient number of images, then organize them into multiple-view matrices, and finally iterate reconstruction algorithms. Each of these steps introduces a delay that compromises the sensing and action loop: when we drive a car, we cannot wait until we have collected enough images before processing them to decide that we needed to swerve or stop. Therefore, we need to adjust our focus and develop algorithms that are suitable for causal, real-time processing. Naturally, the study of the geometry of multiple views is fundamental, and we will exploit it to aid the design of causal algorithms.

Developing algorithms for real-time interaction is not just a matter of speeding up the computation: no matter how fast our implementation, an algorithm that requires the collection of a batch of data before processing it is not suitable for real-time implementation, since the data-gathering delay cannot be forgiven (nor reduced by the advent of faster computers). So, we could conceive of deriving *recursive* or "windowed" versions of the algorithms described in previous sections that update the estimate based on the last measurement gathered, after proper initialization. Simple implementations of these algorithms on real vision-based control systems will be discussed in the last two sections of this chapter.

However, recall that in all of our previous derivations, $t$ is just an index of the particular frame, and there is no constraint on the ordering of the data acquired. This means that we can take a sequence of images, scramble the order of the frames, and all the algorithms described so far would work just as well. In other words, the algorithms we have described so far do not exploit the fact that in a video sequence acquired for the purpose of control and interaction there is a natural ordering in the data, which is due to the fact that the motion of objects follows the basic rules of Newtonian mechanics, governed by forces, inertias, and other physical constraints. Since motion is obtained by integrating forces, and forces are necessarily limited, objects cannot "jump" from one place to another, and their motion is naturally subject to a certain degree of regularity. In this chapter we show how to develop algorithms that exploit the *natural ordering of the data,* by imposing some degree of smoothness in the motion estimates. This will be done by modeling motion as the integral of forces or accelerations.

If the forces or accelerations are known (or measured), we can use that information in the model. Otherwise, they are treated as uncertainty. In this chapter, we will discuss very simple statistical models of uncertainty, with the idea that interested readers, once grasped the simple models, can work to extend them to suit the needs of their application of interest. So, while the geometry of the problem is identical to that discussed in previous chapters, here for the first time we will need to talk about uncertainty in our model, since the forces are typically unknown, or too complex to account for explicitly. In this chapter, we concentrate on a simple model that represents uncertainty in the forces (or accelerations) as the realization of a white, zero-mean Gaussian random vector. Even for such a simple model, however, the optimal inference of 3-D structure and motion is elusive. We will therefore resort to a local approximation of the optimal filter to recursively estimate structure and motion implemented in real time.

### *Overview of this chapter*

In the next section we formulate the problem of 3-D structure and motion estimation in the context of a causal model. As we will see, by just listing the ingredients of the problem (rigid-body motion, perspective projection, *and* motion as the integral of acceleration), we end up with a *dynamical system*, whose "state" describes the unknown structure and motion, and whose "output" describes the measured images. Inferring structure and motion, therefore, can be formulated as the estimation of the state of a dynamical model given its output, in the presence of uncertainty. This problem is known as *filtering*. For the simplest case of linear models, this problem is solved optimally by the Kalman filter, which we describe in Appendix B. There, the reader who is unfamiliar with the Kalman filter can find the derivation, a discussion geared toward developing some intuition, as well as a "recipe" to implement it. Unfortunately, as we will see shortly, our model is *not* linear. Therefore, in the sections to follow, we discuss how to rephrase the problem so that it can be cast within the framework of Kalman filtering. In Appendix B, we also derive an extension of the Kalman filter to nonlinear models.

Finally, in the rest of this chapter, we demonstrate the application of these techniques, as well as techniques developed in earlier chapters, on real testbeds for automated virtual insertion, vehicle driving, and helicopter landing.

## 12.1 Structure and motion estimation as a filtering problem

In this section, we list the ingredients of the problem of causal structure and motion inference, and show how they naturally lead to a nonlinear dynamical system. The reader who is not familiar with the basic issues of Kalman filtering should consult Appendix B before reading this section.

Consider an $N$-tuple of points in three-dimensional Euclidean space and their coordinates represented as a matrix

$$\boldsymbol{X} \doteq [\ \boldsymbol{X}^1,\ \boldsymbol{X}^2,\ \ldots,\ \boldsymbol{X}^N\ ] \quad \in \mathbb{R}^{3\times N}, \tag{12.1}$$

and let them move under the action of a rigid-body motion between two adjacent time instants $g(t+1) = \exp\left(\widehat{\xi}(t)\right)g(t)$; $\widehat{\xi}(t) \in se(3)$. Here the "twist" $\widehat{\xi}$ plays the role of *velocity* (as we have described in Chapter 2, equation (2.20)), which in turn is the integral of *acceleration* $\alpha$, which we do not know. We assume that we can measure the (noisy) projection[1] of the points $\boldsymbol{X}^i$:

$$\boldsymbol{x}^i(t) = \pi\left(g(t)\boldsymbol{X}^i\right) + \boldsymbol{n}^i(t) \quad \in \mathbb{R}^2, \quad \forall\, i = 1, 2, \ldots, N. \tag{12.2}$$

By organizing the time-evolution of the configuration of points and their motion, we end up with a discrete-time, nonlinear dynamical system:

$$\begin{cases} \boldsymbol{X}(t+1) = \boldsymbol{X}(t), & \boldsymbol{X}(0) = \boldsymbol{X}_0 \in \mathbb{R}^{3\times N}, \\ g(t+1) = \exp\left(\widehat{\xi}(t)\right)g(t), & g(0) = g_0 \in SE(3), \\ \xi(t+1) = \xi(t) + \alpha(t), & \xi(0) = \xi_0 \in \mathbb{R}^6, \\ \boldsymbol{x}^i(t) = \pi\left(g(t)\boldsymbol{X}^i(t)\right) + \boldsymbol{n}^i(t), & \boldsymbol{n}^i(t) \sim \mathcal{N}(0, \Sigma_n). \end{cases} \tag{12.3}$$

[1]Here the projection, denoted by $\pi$, can be either the canonical perspective projection or a spherical projection, both described in Chapter 3.

Here, $\boldsymbol{X}(t)$ denotes the coordinates of the points in the world reference frame; since we assume that the scene is static, these coordinates do not change in time (first line). The motion of the scene relative to the camera, on the other hand, does change in time. The second two lines express the fact that the pose $g(t)$ is the integral of velocity $\widehat{\xi}(t)$, whereas velocity is the integral of acceleration (third line). Here $\sim \mathcal{N}(M, S)$ indicates a normal distribution with mean $M$ and covariance matrix $S$. Our assumption here is that $\alpha$, the relative acceleration between the viewer and the scene (or, equivalently, the force acting on the scene) is white, zero-mean Gaussian noise. This choice is opportunistic, because it will allow us to fit the problem of inferring structure and motion within the framework of Kalman filtering. However, the reader should be aware that if some prior modeling information is available (for instance, when the camera is mounted on a vehicle or on a robot arm), this is the place to use it. Otherwise, a statistical model can be employed. In particular, our formalization above encodes the fact that no information is available on acceleration, and therefore velocity is a Brownian motion process. We wish to emphasize that this choice is not crucial for the results of this chapter. Any other model would do, as long as certain modeling assumptions are satisfied, which we discuss in Section 12.1.1.

In principle one would like, at least for this simplified formalization of the problem, to find the "optimal solution," that is, the description of the *state* of the above system $\{\boldsymbol{X}, g, \xi\}$ given a sequence of *output* measurements (correspondences) $\boldsymbol{x}^i(t)$ over an interval of time. Since the measurements are noisy and the model is uncertain, the description of the state consists in its probability density conditioned on the measurements. We call an algorithm that delivers the conditional density of the state at time $t$ causally (i.e. based upon measurements up to time $t$) the *optimal filter*. Unfortunately, no finite-dimensional optimal filter is known for this problem. Therefore, in this chapter we will concentrate on an approximate filter that can guarantee a bounded estimation error. In order for any filter to work, the model has to satisfy certain conditions, which we describe in the next subsection.

### *12.1.1 Observability*

To what extent can the 3-D structure and motion of a scene be reconstructed *causally* from measurements of the motion of its projection onto the sensor? We already know that without the causality constraint, if the cameras are calibrated and a sufficient number of points in general position are given, we can recover the structure and motion of the scene up to an arbitrary choice of the Euclidean reference frame and a scale factor (see Chapter 5). In this section, we carry out a similar analysis, under the constraint of causal processing. We start by establishing some notation that will be used throughout the rest of the chapter.

*Similarity transformation*

As usual, let a rigid-body motion $g \in SE(3)$ be represented by a translation vector $T \in \mathbb{R}^3$ and a rotation matrix $R \in SO(3)$. Let the *similarity group* be the composition of a rigid-body motion and a scaling, denoted by $SE(3) \times \mathbb{R}_+$. Let $\beta \neq 0$ be a scalar; an element $g_\beta \in SE(3) \times \mathbb{R}_+$ acts on points $\boldsymbol{X}$ in $\mathbb{R}^3$ as follows:

$$g_\beta(\boldsymbol{X}) = \beta R \boldsymbol{X} + \beta T. \tag{12.4}$$

We also define an action of $g_\beta$ on $g' = (R', T') \in SE(3)$ as[2]

$$g_\beta(g') = (RR', \beta RT' + \beta T) \tag{12.5}$$

and an action on $se(3)$, whose element is represented by $\xi = (\omega, v)$, as

$$g_\beta(\xi) = (\omega, \beta v). \tag{12.6}$$

We say that two configurations of points $\boldsymbol{X}$ and $\boldsymbol{Y} \in \mathbb{R}^{3 \times N}$ are *equivalent* if there exists a similarity transformation $g_\beta$ that brings one onto the other: $\boldsymbol{Y} = g_\beta(\boldsymbol{X})$. Then the similarity group, acting on the $N$-tuple of points $\boldsymbol{X} \in \mathbb{R}^{3 \times N}$, generates an equivalence class

$$[\boldsymbol{X}] \doteq \{\boldsymbol{Y} \in \mathbb{R}^{3 \times N} \mid \exists\, g_\beta,\ \boldsymbol{Y} = g_\beta(\boldsymbol{X})\}. \tag{12.7}$$

Each equivalence class as in (12.7) can be represented by any of its elements. However, in order to compare the equivalence classes, we need a way to choose a representative element that is consistent across different classes. This corresponds to choosing a reference frame for the similarity group, which in our context entails choosing a "minimal" model, in a way that we will discuss shortly.

*Observability up to a group transformation*

Consider a discrete-time nonlinear dynamical system of the general form

$$\begin{cases} x(t+1) = f(x(t)), & x(t_0) = x_0, \\ y(t) = h(x(t)), \end{cases} \tag{12.8}$$

and let $y(t; t_0, x_0)$ indicate the output of the system at time $t$, starting from the initial condition $x_0$ at time $t_0$. We want to characterize to what extent the states $x$ can be reconstructed from the measurements $y$. Such a characterization depends on the structure of the system $f(\cdot)$ and $h(\cdot)$ but not on the measurement noise, which is therefore assumed to be absent for the purpose of the analysis in this section.

**Definition 12.1.** *Consider a system in the form (12.8) and a point in the state space $x_0$. We say that $x_0$ is* indistinguishable *from $x_0'$ if $y(t; t_0, x_0') =$*

---

[2]Notice that this is not the action induced from the natural group structure of the similarity group restricted to $SE(3)$.

$y(t; t_0, x_0), \forall\, t, t_0$. *We indicate with* $\mathcal{I}(x_0)$ *the set of initial conditions that are indistinguishable from* $x_0$.

**Definition 12.2.** *We say that the system (12.8) is* observable up to a group transformation $G$ *if*

$$\mathcal{I}(x_0) = [x_0] \doteq \{x_0' \mid \exists\, g \in G,\ x_0' = g(x_0)\}. \tag{12.9}$$

For a system that is observable up to a group transformation, the state-space can be represented as a collection of equivalence classes: each class corresponds to a set of states that are indistinguishable. Clearly, from measurements of the output $y(t)$ over any period of time, it is possible to recover at most the equivalence class where the initial condition belongs, that is, $\mathcal{I}(x_0)$, but not $x_0$ itself. The only case in which the true initial condition can be recovered is that in which the system is observable up to the *identity* transformation, i.e. $G = \{e\}$. In this case we have $\mathcal{I}(x_0) = \{x_0\}$, and we say that the system is *observable*.

*Observability of structure and motion*

It can be shown (see [Chiuso et al., 2002]) that the model (12.3) where the points $\boldsymbol{X}$ are in general position is observable up to a similarity transformation of $\boldsymbol{X}$, provided that $v_0 \neq 0$. More specifically, the set of initial conditions that are indistinguishable from $\{\boldsymbol{X}_0, g_0, \xi_0\}$, where $g_0 = (R_0, T_0)$ and $\xi_0 = (\omega_0, v_0)$, is given by $\{\beta\tilde{R}\boldsymbol{X}_0 + \beta\tilde{T}, \tilde{g}_0, \tilde{\xi}_0\}$, where $\tilde{g}_0 = (R_0\tilde{R}^T, \beta T_0 - \beta R_0\tilde{R}^T\tilde{T})$ and $\tilde{\xi}_0 = (\omega_0, \beta v_0)$.

Since the observability analysis refers to the model where velocity is constant, the relevance of the statement above for the practical estimation of structure and motion is that one can, in principle, solve the problem using the above model only when velocity varies slowly compared to the sampling frequency. If, however, some information on acceleration becomes available (as for instance if the camera is mounted on a support with some inertia), then the restriction on velocity can be lifted to a restriction on acceleration. This framework, however, will not hold if the data $y(t)$ are snapshots of a scene taken from sparse viewpoints, rather than a sequence of images taken at adjacent time instants while the camera is moving in a continuous trajectory.

The arbitrary choice of the reference frame is a nuisance in designing a filter. Therefore, one needs to fix the arbitrary degrees of freedom in order to arrive at an observable model (i.e. one that is observable up to the identity transformation). When we interpret the state-space as a collection of equivalence classes under the similarity group, fixing the direction of three points and one depth scale identifies a representative for each equivalence class: Without loss of generality (i.e. modulo a reordering of the states), we will assume the indices of such three points to be 1, 2, and 3. We consider a point $\boldsymbol{X}$ as parameterized by its direction $\boldsymbol{x}$ and depth $\lambda$, so that $\boldsymbol{X} = \lambda\boldsymbol{x}$, in the context of calibrated cameras. Then, given the direction of three non-collinear image points, written in homogeneous coordinates $\boldsymbol{x}^1, \boldsymbol{x}^2, \boldsymbol{x}^3 \in \mathbb{R}^3$ and the scale of one point $\lambda^1 > 0$, and given points in space $\{\boldsymbol{X}^i\}_{i=1}^N \subset \mathbb{R}^3$, there are four isolated solutions for the motion $g = (R, T) \in$

$SE(3)$ and the scales $\lambda^i \in \mathbb{R}$ that solve

$$\beta R \lambda^i \boldsymbol{x}^i + \beta T = \boldsymbol{X}^i, \quad \forall\, i = 1, 2, \ldots, N \geq 3. \tag{12.10}$$

This means that the remaining points $\boldsymbol{x}^i$, $i \geq 4$ and scales $\lambda^i$, $i \geq 2$ are not free to vary arbitrarily. The above fact is just another incarnation of the arbitrary choice of Euclidean reference frame and scale factor that we have discussed in Chapter 5, except that in our case we have to consider the *causality constraint*.

## *12.1.2 Realization*

Because of the uncertainty $\alpha$ in the model and the noise $\boldsymbol{n}$ in the measurements, the state of the model (12.3) is described as a random process. In order to describe such a process based on our observations of the output, we need to specify the conditional density of the state of (12.3) at time $t$, given measurements of the output up to $t$. Such a density is a function that, in general, lives in an infinite-dimensional space. Only in a few fortunate cases does this function retain its general form over time, so that one can describe its evolution using a finite number of parameters. For instance, in the Kalman filter such a conditional density is Gaussian, and therefore one can describe it by the evolution of its mean and its covariance matrix, as we discuss in Appendix B.

In general, however, one cannot describe the evolution of the conditional density using a finite number of parameters. Therefore, one has to resort to a finite-dimensional approximation to the optimal filter. The first step in designing such an approximation is to have a parameterized observable model. How to obtain it is the subject of this section.

### *Local coordinates*

Our first step consists in characterizing the local-coordinate representation of the model (12.3). To this end, we represent $SO(3)$ locally in its canonical exponential coordinates as described in Chapter 2: Let $\Omega$ be a three-dimensional real vector ($\Omega \in \mathbb{R}^3$); $\frac{\Omega}{\|\Omega\|}$ specifies the direction of rotation, and $\|\Omega\|$ specifies the angle of rotation in radians. Then a rotation matrix can be represented by its exponential coordinates $\widehat{\Omega} \in so(3)$ such that $R \doteq \exp(\widehat{\Omega}) \in SO(3)$. The three-dimensional coordinate $\boldsymbol{X}^i \in \mathbb{R}^3$ is represented by its projection onto the image plane $\boldsymbol{x}^i \in \mathbb{R}^2$ and its depth $\lambda^i \in \mathbb{R}$, so that

$$\boldsymbol{x}^i \doteq \pi(\boldsymbol{X}^i) \doteq \begin{bmatrix} \frac{X_1^i}{X_3^i} \\ \frac{X_2^i}{X_3^i} \end{bmatrix}, \quad \lambda^i = X_3^i. \tag{12.11}$$

Such a representation has the advantage of decomposing the uncertainty in the measured directions $\boldsymbol{x}$ (low) from the uncertainty in depth $\lambda$ (high). The model

(12.3) in local coordinates is therefore

$$\begin{cases} \boldsymbol{x}_0^i(t+1) = \boldsymbol{x}_0^i(t), \quad i = 1,2,\dots,N, & \boldsymbol{x}_0^i(0) = \boldsymbol{x}_0^i, \\ \lambda^i(t+1) = \lambda^i(t), \quad i = 1,2,\dots,N, & \lambda^i(0) = \lambda_0^i, \\ T(t+1) = \exp(\widehat{\omega}(t))T(t) + v(t), & T(0) = T_0, \\ \Omega(t+1) = \log_{SO(3)}\big(\exp(\widehat{\omega}(t))\exp(\widehat{\Omega}(t))\big), & \Omega(0) = \Omega_0, \\ v(t+1) = v(t) + \alpha_v(t), & v(0) = v_0, \\ \omega(t+1) = \omega(t) + \alpha_\omega(t), & \omega(0) = \omega_0, \\ \boldsymbol{x}^i(t) = \pi\left(\exp(\widehat{\Omega}(t))\boldsymbol{x}_0^i(t)\lambda^i(t) + T(t)\right) + \boldsymbol{n}^i(t), & i = 1,2,\dots,N. \end{cases} \tag{12.12}$$

The notation $\log_{SO(3)}(R)$ stands for $\Omega$ such that $R = e^{\widehat{\Omega}}$ and is computed by inverting Rodrigues' formula as described in Chapter 2, equations (2.15) and (2.16).

*Minimal realization*

In linear time-invariant systems one can decompose the state space into an observable subspace and its (unobservable) complement. In the case of our system, which is nonlinear and observable up to a group transformation, we can exploit the structure of the state-space to realize a similar decomposition: the representative of each equivalence class is observable, while individual elements in the class are not. Therefore, in order to restrict our attention to the observable component of the system, we need only to choose a representative for each class. As we have anticipated in the previous section, one way to render the model (12.12) observable is to eliminate some states, in order to fix the similarity transformation. In particular, the following model, which is obtained by eliminating $\boldsymbol{x}_0^1(t), \boldsymbol{x}_0^2(t), \boldsymbol{x}_0^3(t)$, and $\lambda^1(t)$ from the state of (12.12), is observable:

$$\begin{cases} \boldsymbol{x}_0^i(t+1) = \boldsymbol{x}_0^i(t), \quad i = \mathbf{4},5,\dots,N, & \boldsymbol{x}_0^i(0) = \boldsymbol{x}_0^i, \\ \lambda^i(t+1) = \lambda^i(t), \quad i = \mathbf{2},3,\dots,N, & \lambda^i(0) = \lambda_0^i, \\ T(t+1) = \exp(\widehat{\omega}(t))T(t) + v(t), & T(0) = T_0, \\ \Omega(t+1) = \log_{SO(3)}\big(\exp(\widehat{\omega}(t))\exp(\widehat{\Omega}(t))\big), & \Omega(0) = \Omega_0, \\ v(t+1) = v(t) + \alpha_v(t), & v(0) = v_0, \\ \omega(t+1) = \omega(t) + \alpha_\omega(t), & \omega(0) = \omega_0, \\ \boldsymbol{x}^i(t) = \pi\left(\exp(\widehat{\Omega}(t))\boldsymbol{x}_0^i(t)\lambda^i(t) + T(t)\right) + \boldsymbol{n}^i(t), & i = 1,2,\dots,N. \end{cases} \tag{12.13}$$

The problem of estimating the motion, velocity, and structure of the scene s then equivalent to the problem of estimating the state of the model (12.13).

To simplify the notation for further discussion, we collect all state and output variables in the model (12.13) into two new variables $x \in \mathbb{R}^{3N+5}$ and $y \in \mathbb{R}^{2N}$ respectively:[3]

$$\begin{aligned} x(t) &\doteq \left[\boldsymbol{x}_0^4(t)^T, \dots, \boldsymbol{x}_0^N(t)^T,\ \lambda^2(t), \dots, \lambda^N(t), T^T(t), \Omega^T(t), v^T(t), \omega^T(t)\right]^T, \\ y(t) &\doteq \left[\boldsymbol{x}^1(t)^T, \dots, \boldsymbol{x}^N(t)^T\right]^T. \end{aligned}$$

[3]Note that $\boldsymbol{x}_0^i$ is represented in $x$ and $y$ using Cartesian, as opposed to homogeneous, coordinates, so that $\boldsymbol{x}_0^i \in \mathbb{R}^2$.

Adopting the notation in Appendix B, we use $f(\cdot)$ and $h(\cdot)$ to denote the state and measurement models, respectively. Then the model (12.13) can be written concisely as

$$\begin{cases} x(t+1) = f(x(t)) + w(t), & w(t) \sim \mathcal{N}(0, \Sigma_w(t)), \\ y(t) = h(x(t)) + n(t), & n(t) \sim \mathcal{N}(0, \Sigma_n(t)), \end{cases} \tag{12.14}$$

where $w$ and $n$ collect the model error and measurement noise $\boldsymbol{n}^i$, respectively (see Appendix B). The covariance of the model error, $\Sigma_w(t)$, is a design parameter that is available for tuning. The covariance $\Sigma_n(t)$ is usually available from the analysis of the feature-tracking algorithm (see Chapter 11). We assume that the tracking error is independent for each point, and therefore $\Sigma_n(t)$ is block diagonal. We choose each block to be the covariance of the measurement $\boldsymbol{x}^i(t)$ (e.g., 1 pixel standard deviation).

Notice that the above model (12.14) is observable and already in the form used to derive the nonlinear extension of the Kalman filter in Appendix B. However, before we can implement a filter for the model (12.13), we have to address a number of issues, the most crucial being the fact that points appear and disappear due to occlusions, as we have seen in Chapter 11. In the next section, we address these implementation issues, and we summarize the complete algorithm in Section 12.1.4.

In the discussions that follow, we use the "hat" notation, introduced in Appendix B, to indicate estimated quantities. For instance, $\hat{x}(t|t)$ (or simply $\hat{x}(t)$) is the estimate of $x$ at time $t$, and $\hat{x}(t+1|t)$ is the one-step prediction of $x$ at time $t$. This "hat" is not to be confused with the "wide hat" operator that indicates a skew-symmetric matrix, for instance, $\widehat{\Omega}$. When writing the estimate of $\widehat{\Omega}$, however, in order to avoid a "double hat," we simply write $\widehat{\Omega}(t|t)$ or $\widehat{\Omega}(t+1|t)$ depending on whether we are interested in the current estimate or the prediction. The recursion to update the state estimate $\hat{x} \in \mathbb{R}^{3N+5}$ and the covariance $P \in \mathbb{R}^{(3N+5)\times(3N+5)}$ of the estimation error $\tilde{x} = x - \hat{x}$ is given by the extended Kalman filter introduced in Appendix B.

## *12.1.3 Implementation issues*

This section discusses issues that occur when trying to implement an extended Kalman filter for estimating structure and motion in practice. The reader who is not interested in implementation can skip this section without loss of continuity.

### *Occlusion and drift*

When a feature point, say $\boldsymbol{X}^i$, becomes occluded, the corresponding measurement $\boldsymbol{x}^i(t)$ becomes unavailable. In order to avoid useless computation and ill-conditioned inverses, we can simply eliminate the states $\boldsymbol{x}_0^i(t)$ and $\lambda^i(t)$ altogether, thereby reducing the dimension of the state-space. We can do so because of the diagonal structure of the model (12.13): The states $\boldsymbol{x}_0^i(t)$ and $\lambda^i(t)$ are decoupled from other states, and therefore it is sufficient to remove them and delete the corresponding rows from the gain matrix of the Kalman filter and from the covariance of the noise for all $t$ past the disappearance of the feature.

The only case in which losing a feature constitutes a problem occurs when the feature is used as reference to fix the observable component of the state-space (in our notation, $i = 1, 2, 3$).[4] The most obvious choice consists in associating the reference with any other visible point, saturating the corresponding state, and assigning as reference value the current best estimate. In particular, if feature $i$ is lost at time $\tau$, and we want to switch the reference index to feature $j$, we eliminate $(\boldsymbol{x}_0^i(t), \lambda^i(t))$ from the state, and set the corresponding diagonal block of the noise covariance and the initial covariance of the state $(\boldsymbol{x}_0^j(t), \lambda^j(t))$ to zero. One can easily verify that, as a consequence

$$\hat{\boldsymbol{x}}_0^j(\tau + t) = \hat{\boldsymbol{x}}_0^j(\tau), \quad \forall\, t > 0. \tag{12.15}$$

If $\hat{\boldsymbol{x}}_0^j(\tau)$ were equal to $\boldsymbol{x}_0^j(\tau) = \boldsymbol{x}_0^j$, switching the reference feature would have no effect on the other states, and the filter would evolve on the same observable component of the state-space determined by the reference feature $i$.

However, in general, the difference $\tilde{\boldsymbol{x}}_0^j(\tau) \doteq \boldsymbol{x}_0^j(\tau) - \hat{\boldsymbol{x}}_0^j(\tau)$ is a random variable with covariance $\Sigma_\tau$, which is available from the corresponding diagonal block of $P(\tau|\tau)$. Therefore, switching the reference to feature $j$ causes the observable component of the state-space to move by an amount proportional to $\tilde{\boldsymbol{x}}_0^j(\tau)$. When a number of switches have occurred, we can expect, on average, the state-space to move by an amount proportional to $\|\Sigma_\tau\|$ multiplied by the number of switches. This is unavoidable. What we can do is at most try to keep the bias to a minimum by switching the reference to the state that has the lowest covariance.[5]

Of course, should the original reference feature $i$ become available again, one can immediately switch the reference to it, and therefore recover the original base and annihilate the bias [Favaro et al., 2003, Rahimi et al., 2001].

*New features*

When a new feature point is selected, it is not possible to simply insert it into the state of the model, since the initial condition is unknown. Any initialization error will disturb the current estimate of the remaining states, since it is fed back into the update equation for the filter, and generates a spurious transient. One can address this problem by running a separate filter in parallel for each point using the current estimates of motion from the main filter, running with existing features, in order to reconstruct the initial condition. Such a "subfilter" is based

[4]When the scale factor is not directly associated with one feature, but is associated with a function of a number of features (for instance, the depth of the centroid, or the average inverse depth), then losing any of these features causes a drift.

[5]Just to give the reader an intuitive feeling of the numbers involved, we find that in practice the average lifetime of a tracked feature is around 10 to 30 frames. The covariance of the estimation error for $\boldsymbol{x}_0^i$ is of order $10^{-6}$ units, while the covariance of $\lambda^i$ is of order $10^{-4}$ units for noise levels commonly encountered with commercial cameras.

upon the following model, where we assume that $N_\tau$ features appear at time $\tau$:

$$\begin{cases} \boldsymbol{x}_\tau^i(t+1) = \boldsymbol{x}_\tau^i(t) + \eta_{x^i(t)}, & \boldsymbol{x}_\tau^i(0) \sim \mathcal{N}(\boldsymbol{x}^i(\tau), \Sigma_{n^i}), \quad t > \tau, \\ \lambda_\tau^i(t+1) = \lambda_\tau^i(t) + \eta_{\lambda^i(t)}, & \lambda^i(0) \sim \mathcal{N}(1, P_\lambda(0)), \\ \boldsymbol{x}^i(t) = \pi\big(\exp(\widehat{\Omega}(t))\exp(\widehat{\Omega}(\tau))^{-1}[\boldsymbol{x}_\tau^i(t)\lambda_\tau^i(t) - T(\tau)] + T(t)\big) + \boldsymbol{n}^i(t), \end{cases} \tag{12.16}$$

for $i = 1, 2, \ldots, N_\tau$, where $\Omega(t) = \Omega(t|t)$ and $T(t) = T(t|t)$ are the current best estimates of $\Omega$ and $T$, and similarly $\Omega(\tau)$ and $T(\tau)$ are the best estimates of $\Omega$ and $T$ at $t = \tau$, both available from the main filter. Note that the covariance of the model error for $\boldsymbol{x}_\tau^i$ is the same as for the measurement $\boldsymbol{x}^i$, i.e. that of the noise $\boldsymbol{n}$.

Several heuristics can be employed in order to decide when the state estimate from the subfilter is good enough for it to be inserted into the main Kalman filter. A simple criterion is the covariance of the estimation error of $\lambda_\tau^i$ in the subfilter being comparable to the covariance of $\lambda_0^j$ for $j \neq i$ in the main filter.

*Partial autocalibration*

The model (12.13) proposed above can be extended to account for changes in calibration. For instance, if we consider an imaging model with focal length $f \in \mathbb{R}$,[6]

$$\boldsymbol{x} = \pi_f(\boldsymbol{X}) = f \begin{bmatrix} \frac{X_1}{X_3} \\ \frac{X_2}{X_3} \end{bmatrix}, \tag{12.17}$$

where the focal length can change in time, but no prior knowledge on how it does so is available, one can model its evolution as a random walk

$$f(t+1) = f(t) + \alpha_f(t), \quad \alpha_f(t) \sim \mathcal{N}(0, \sigma_f^2), \tag{12.18}$$

and insert it into the state of the model (12.13). It can be shown that the overall system is still observable, and therefore all the conclusions reached in the previous sections will hold. In other words, the realization remains minimal if we add into the model the focal length parameter.

### 12.1.4 Complete algorithm

This section summarizes the implementation of a nonlinear filter for motion and structure estimation based on the model (12.13), or equivalently (12.14).

*Main Kalman filter*

We first properly initialize the state of the main Kalman filter according to Algorithm 12.1. During the first transient of the filter, we do not allow for new features to be acquired. When a feature is lost, its state is simply removed from the model. If the lost feature was among one of the chosen three, this causes a drift, and one can proceed as discussed in Section 12.1.3. The transient can be tested as either

[6] This $f$ is not to be confused with the function $f(\cdot)$ in the generic state equation (12.14).

**Algorithm 12.1 (Structure and motion filtering: Initialization).**

With a set of selected and tracked feature points $\{\boldsymbol{x}_0^i\}_{i=1}^N$, we choose the initial conditions for the extended Kalman filter to be

$$\begin{cases} \hat{x}(0|0) = [\boldsymbol{x}_0^{4^T}, \ldots, \boldsymbol{x}_0^{N^T}, 1, \ldots, 1, 0_{1\times3}, 0_{1\times3}, 0_{1\times3}, 0_{1\times3}]^T & \in \mathbb{R}^{3N+5}, \\ P(0|0) = P_0 \in \mathbb{R}^{(3N+5)\times(3N+5)}. \end{cases} \tag{12.19}$$

For the initial covariance $P_0$, we choose it to be block diagonal with positive-definite blocks $\Sigma_{n^i}(0) \in \mathbb{R}^{2\times2}$ corresponding to $\boldsymbol{x}_0^i$, a large positive number $M \in \mathbb{R}_+$ (typically 100 to 1000 units) corresponding to $\lambda_0^i$, zeros corresponding to $T_0$ and $\Omega_0$.[7] We also choose a diagonal matrix with a large positive number $W \in \mathbb{R}_+$ for the initial covariance blocks corresponding to $v_0$ and $\omega_0$.

---

a threshold on the innovation, a threshold on the covariance of the estimates, or a fixed time interval. We choose a combination with the time set to 30 frames, corresponding to one second of video.

Once the filter is initialized according to Algorithm 12.1, the recursive iteration proceeds as described in Algorithm 12.2.

**Algorithm 12.2 (Structure and motion filtering: Iteration).**

**Linearization:**

$$\begin{cases} F(t) \doteq \frac{\partial f}{\partial x}(\hat{x}(t|t)) & \in \mathbb{R}^{(3N+5)\times(3N+5)}, \\ H(t+1) \doteq \frac{\partial h}{\partial x}(\hat{x}(t+1|t)) & \in \mathbb{R}^{2N\times(3N+5)}, \end{cases} \tag{12.20}$$

where $F(t)$ is the linearization of the state function $f(\cdot)$ and is given in equation (12.24), and $H(t)$ is the linearization of the output function $h(\cdot)$ and is given in equation (12.25).

**Prediction:**

$$\begin{cases} \hat{x}(t+1|t) = f(\hat{x}(t|t)), \\ P(t+1|t) = F(t)P(t|t)F^T(t) + \Sigma_w(t). \end{cases} \tag{12.21}$$

**Update:**

$$\begin{cases} \hat{x}(t+1|t+1) = \hat{x}(t+1|t) + L(t+1)\big(y(t+1) - h(\hat{x}(t+1|t))\big), \\ P(t+1|t+1) = \Gamma(t+1)P(t+1|t)\Gamma^T(t+1) + L(t+1)\Sigma_n(t+1)L^T(t+1). \end{cases} \tag{12.22}$$

**Gain:**

$$\begin{cases} \Gamma(t+1) \doteq I - L(t+1)H(t+1), \\ L(t+1) \doteq P(t+1|t)H^T(t+1)\Lambda^{-1}(t+1), \\ \Lambda(t+1) \doteq H(t+1)P(t+1|t)H^T(t+1) + \Sigma_n(t+1). \end{cases} \tag{12.23}$$

---

To compute $F(t)$, the linearization of the state equation of the model (12.13), we need to compute the derivatives of the logarithm function in $SO(3)$, which is the inverse function of the exponential. The derivatives can be readily computed using the inverse function theorem. We shall use the following notation:

$$\frac{\partial \log_{SO(3)}(R)}{\partial R} \doteq \left[ \frac{\partial \log_{SO(3)}(R)}{\partial r_{11}} \quad \frac{\partial \log_{SO(3)}(R)}{\partial r_{21}} \quad \cdots \quad \frac{\partial \log_{SO(3)}(R)}{\partial r_{33}} \right] \in \mathbb{R}^{3\times 9},$$

where $r_{ij}$ is the $(i,j)$th entry of $R$. Let us define $R \doteq e^{\widehat{\omega}} e^{\widehat{\Omega}}$; the linearization of the state equation can be written in the following form:

$$F \doteq \begin{bmatrix} I_{2N-6} & 0 & 0 & 0 & 0 & 0 \\ 0 & I_{N-1} & 0 & 0 & 0 & 0 \\ 0 & 0 & e^{\widehat{\omega}} & 0 & I & \left[\frac{\partial e^{\widehat{\omega}}}{\partial \omega_1} T \quad \frac{\partial e^{\widehat{\omega}}}{\partial \omega_2} T \quad \frac{\partial e^{\widehat{\omega}}}{\partial \omega_3} T\right] \\ 0 & 0 & 0 & \frac{\partial \log_{SO(3)}(R)}{\partial R}\frac{\partial R}{\partial \Omega} & 0 & \frac{\partial \log_{SO(3)}(R)}{\partial R}\frac{\partial R}{\partial \omega} \\ 0 & 0 & 0 & 0 & I & 0 \\ 0 & 0 & 0 & 0 & 0 & I \end{bmatrix}, \tag{12.24}$$

where

$$\frac{\partial R}{\partial \Omega} \doteq \left[ \left(e^{\widehat{\omega}} \frac{\partial e^{\widehat{\Omega}}}{\partial \Omega_1}\right)^s \quad \left(e^{\widehat{\omega}} \frac{\partial e^{\widehat{\Omega}}}{\partial \Omega_2}\right)^s \quad \left(e^{\widehat{\omega}} \frac{\partial e^{\widehat{\Omega}}}{\partial \Omega_3}\right)^s \right] \in \mathbb{R}^{9\times 3}$$

and

$$\frac{\partial R}{\partial \omega} \doteq \left[ \left(\frac{\partial e^{\widehat{\omega}}}{\partial \omega_1} e^{\widehat{\Omega}}\right)^s \quad \left(\frac{\partial e^{\widehat{\omega}}}{\partial \omega_2} e^{\widehat{\Omega}}\right)^s \quad \left(\frac{\partial e^{\widehat{\omega}}}{\partial \omega_3} e^{\widehat{\Omega}}\right)^s \right] \in \mathbb{R}^{9\times 3}.$$

Recall that the notation $(\cdot)^s$ indicates that a $(3 \times 3)$ matrix has been rearranged by stacking the columns on top of each other (see Appendix A).

To compute $H(t)$, the linearization of the output equation of the model (12.13), we define $\boldsymbol{X}^i(t) \doteq e^{\widehat{\Omega}(t)} \boldsymbol{x}_0^i(t) \lambda^i(t) + T(t)$, $Z^i(t) \doteq [0,0,1]\boldsymbol{X}^i(t)$. The $i$th block-row $H_i(t) \in \mathbb{R}^{2\times(3N+5)}$ of the matrix $H(t)$ can be computed as

$$H_i = \frac{\partial \boldsymbol{x}^i}{\partial \boldsymbol{X}^i} \frac{\partial \boldsymbol{X}^i}{\partial x} \doteq \Pi_i \frac{\partial \boldsymbol{X}^i}{\partial x}, \tag{12.25}$$

where the time argument $t$ has been omitted for simplicity of notation. It is easy to check that $\Pi_i = \frac{1}{Z^i} \left[\, I_2 \quad -\pi(\boldsymbol{X}^i) \,\right] \in \mathbb{R}^{2\times 3}$ and

$$\frac{\partial \boldsymbol{X}^i}{\partial x} = \Bigg[ \underbrace{\left[0, \ldots, \frac{\partial \boldsymbol{X}^i}{\partial \boldsymbol{x}_0^i}, \ldots, 0\right]}_{3\times(2N-6)}, \underbrace{\left[0, \ldots, \frac{\partial \boldsymbol{X}^i}{\partial \lambda^i}, \ldots, 0\right]}_{3\times(N-1)}, \underbrace{\frac{\partial \boldsymbol{X}^i}{\partial T}}_{3\times 3}, \underbrace{\frac{\partial \boldsymbol{X}^i}{\partial \Omega}}_{3\times 3}, \underbrace{0}_{3\times 3}, \underbrace{0}_{3\times 3} \Bigg].$$

The partial derivatives in the previous expression are given by

$$\begin{cases} \frac{\partial \boldsymbol{X}^i}{\partial \boldsymbol{x}_0^i} = e^{\widehat{\Omega}} \begin{bmatrix} I_2 \\ 0 \end{bmatrix} \lambda^i, \quad \frac{\partial \boldsymbol{X}^i}{\partial \lambda^i} = e^{\widehat{\Omega}} \boldsymbol{x}_0^i, \quad \frac{\partial \boldsymbol{X}^i}{\partial T} = I, \\ \frac{\partial \boldsymbol{X}^i}{\partial \Omega} = \left[ \frac{\partial e^{\widehat{\Omega}}}{\partial \Omega_1} \boldsymbol{x}_0^i \lambda^i \quad \frac{\partial e^{\widehat{\Omega}}}{\partial \Omega_2} \boldsymbol{x}_0^i \lambda^i \quad \frac{\partial e^{\widehat{\Omega}}}{\partial \Omega_3} \boldsymbol{x}_0^i \lambda^i \right]. \end{cases}$$

*Subfilter*

Whenever a feature disappears, we simply remove it from the state as during the transient. After the transient, a feature-selection module works in parallel with the filter to select new features so as to maintain roughly a constant number (equal to the maximum that the hardware can handle in real time), and to maintain a distribution as uniform as possible across the image plane. We can implement this by randomly sampling points on the plane, searching then around that point for a feature with sufficient brightness variation (using, for instance, the corner detectors described in Chapter 4).

Based on the model (12.16), a subfilter for the newly acquired features is given in Algorithm 12.3. In practice, rather than initializing $\lambda$ to 1 in the subfilter,

**Algorithm 12.3 (Structure and motion filtering: Subfilter).**

**Initialization:**

$$\begin{cases} \hat{\boldsymbol{x}}^i_\tau(\tau|\tau) = \boldsymbol{x}^i(\tau), \\ \hat{\lambda}^i_\tau(\tau|\tau) = 1, \\ P^i_\tau(\tau|\tau) = \begin{bmatrix} \Sigma_{n^i}(\tau) & 0 \\ 0 & M \end{bmatrix}. \end{cases} \tag{12.26}$$

**Prediction:**

$$\begin{cases} \hat{\boldsymbol{x}}^i_\tau(t+1|t) = \hat{\boldsymbol{x}}^i_\tau(t|t), \\ \hat{\lambda}^i_\tau(t+1|t) = \hat{\lambda}^i_\tau(t|t), \\ P^i_\tau(t+1|t) = P^i_\tau(t|t) + \Sigma_w(t), \end{cases} \qquad t > \tau. \tag{12.27}$$

**Update:**

$$\begin{bmatrix} \hat{\boldsymbol{x}}^i_\tau(t+1|t+1) \\ \hat{\lambda}^i_\tau(t+1|t+1) \end{bmatrix} = \begin{bmatrix} \hat{\boldsymbol{x}}^i_\tau(t+1|t) \\ \hat{\lambda}^i_\tau(t+1|t) \end{bmatrix} + L^i_\tau(t+1) \times \Big( \boldsymbol{x}^i(t+1) - \\ \pi\big(\exp(\widehat{\Omega}(t+1))\exp(\widehat{\Omega}(\tau))^{-1}[\hat{\boldsymbol{x}}^i_\tau(t+1|t)\hat{\lambda}^i_\tau(t+1|t) - T(\tau)] + T(t+1)\big)\Big).$$

In the above, estimates of $\widehat{\Omega}(t)$ and $T(t)$ are obtained from the main filter, the error covariance $P^i_\tau(t+1|t+1)$ is updated according to a Riccati equation similar to (12.22), and the gain $L^i_\tau(t+1)$ is updated according to the usual equation (12.23).

one can compute a first approximation by triangulating on two adjacent views, and compute the covariance of the initialization error from the covariance of the current estimates of motion.

After a probation period, whose length is chosen according to the same criterion adopted for the transient of the main filter, the new feature $i$ is inserted back into the main filter state using

$$\hat{\boldsymbol{X}}^i_0 = \big[\exp(\widehat{\Omega}(\tau|\tau))\big]^{-1}\big[\hat{\boldsymbol{x}}^i_\tau(t|t)\hat{\lambda}^i_\tau(t|t) - T(\tau|\tau)\big]. \tag{12.28}$$

In the main filter, the initial covariance of the state $(\boldsymbol{x}^i_0(t), \lambda^i(t))$ associated with the new feature $i$ can be set to be the covariance of the estimation error of the subfilter.

*Tuning*

The model error covariance $\Sigma_w(t)$ of the model (12.14) is a design parameter in the Kalman filter. We choose it to be block diagonal, with the blocks corresponding to $T(t)$ and $\Omega(t)$ equal to zero (a deterministic integrator). We choose the remaining parameters using standard statistical tests, such as the cumulative periodogram [Bartlett, 1956]. The idea is that the parameters in $\Sigma_w(t)$ are changed until the innovation process $e(t) \doteq y(t) - h(\hat{x}(t))$ is as close as possible to being white. The periodogram is one of many ways to test the "whiteness" of a random process. In practice, we choose the blocks corresponding to $\boldsymbol{x}_0^i$ equal to the covariance of the measurements, and the elements corresponding to $\lambda^i$ all equal to $\sigma_\lambda$. We then choose the blocks corresponding to $v$ and $\omega$ to be diagonal with element $\sigma_v$, and then we change $\sigma_v$ relative to $\sigma_\lambda$ depending on whether we want to allow for more or less regular motions. We then change both, relative to the covariance of the measurement noise, depending on the level of desired smoothness in the estimates.

Tuning nonlinear filters is an art, and this is not the proper venue to discuss this issue. Suffice it to say that we have performed the procedure only once and for all. We then keep the same tuning parameters no matter what the motion, structure, and noise in the measurements.

## 12.2 Application to virtual insertion in live video

The extended Kalman filter algorithm described in the previous section can be implemented on a standard laptop PC and run in real time for a number of feature $N$ in the order of 40-50. For instance, we have implemented it on a 1GHz laptop PC, connected to a digital camera via firewire. This system can be mounted onboard a moving vehicle, for instance a mobile robot, for ego-motion estimation. In this section, however, we illustrate the use of this system for the purpose of real-time virtual insertion into live video. Figure 12.1 shows the interface of a system, that is used to superimpose a computer-generated object onto live footage of a static scene and make it appear as if it is moving with the scene.

Assuming that the camera is moving in front of a static scene (or that an object is moving relative to the camera), the algorithm just described estimates the position of a number of point features relative to the reference frame of the camera at the initial time instant, along with the position of the camera at each subsequent time. The estimate is computed in real time (the current implementation runs at 30 Hz on a laptop PC). A virtual object (i.e. a computer-generated object whose geometry is known) can then be positioned on the plane identified by three point features of choice (selected by the user by clicking a location on the image) and moved according to the estimated motion so that it appears to belong to the scene. Naturally, if the scene changes, or if the features associated with the scale factor disappear, the visualization needs to be reinitialized.

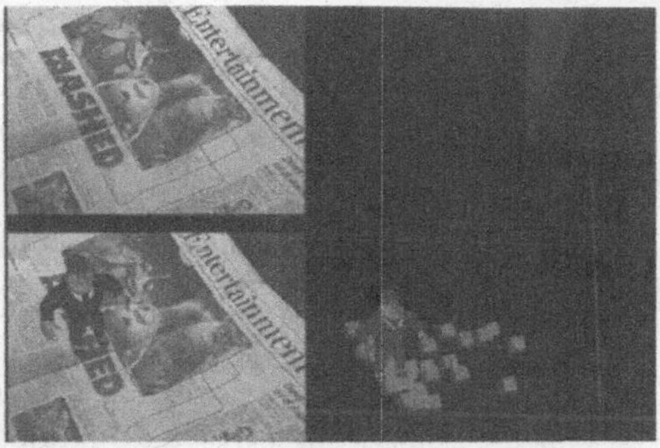


Figure 12.1. (Top left) Live video of a static scene can be used to infer camera motion and the position of a number of point features. This can in turn be used to insert a virtual object and move it according to the estimated motion (right), so that it appears to be part of the scene (bottom left).

Here, the "sensing and action loop" consists in the camera capturing images of the scene (sensing), transferring them to the laptop for inference of structure and motion (computation), and manipulating the virtual object for visualization (control). The system was first demonstrated by researchers at the University of California at Los Angeles and Washington University at the IEEE Conference on Computer Vision and Pattern Recognition in 2000. Code is available for distribution from `http://vision.ucla.edu`.

## 12.3 Visual feedback for autonomous car driving

Figure 12.2 shows a few images of a vision-based autonomous guidance system developed as part of the California PATH program. This system was used in an experimental demonstration of the National Automated Highway Systems Consortium (NAHSC), which took place in August 1997 in San Diego. The overall system was demonstrated both as a part of the main highway scenario and as part of a small public demonstration of the vision-based lateral control on a highly curved test track (with the car running typically at 50 to 75 miles per hour).

The "sensing and control loop" here consists of pair of cameras in rigid configuration (sensing) that provide synchronized image pairs that are used by the on-board computer to provide a depth map and the location of the road markers (computation); the resulting estimates are then used to actuate the steering wheel and the throttle/brakes (control). In this particular application, for the sake of robustness, various sensors were used in addition to vision. The next subsection describes the system in greater detail.

Figure 12.2. Vision-based highway vehicle control system. Left: An automated Honda Accord LX sedan developed for experiments and demonstrations. Right: View from inside the automated Honda Accord showing the mounting of the stereo cameras.

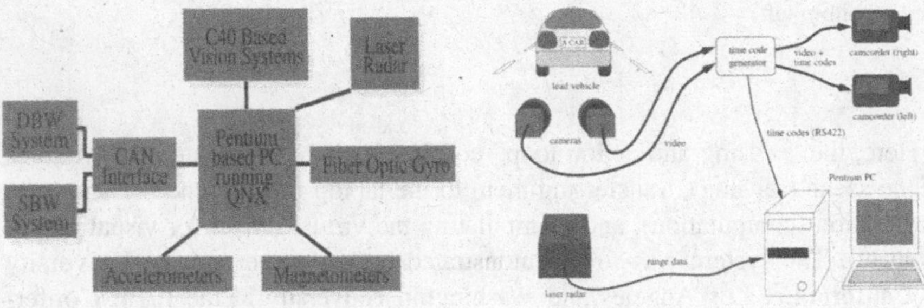


Figure 12.3. Left: System diagram. Right: Experimental setup.

### 12.3.1 System setup and implementation

Figure 12.3 (left) shows the major components of the autonomous vehicle control system, which is implemented on the Honda Accord LX shown in Figure 12.2. This system takes input from a range of sensors, which provide information about its own motion (speedometer, yaw rate sensor, and accelerometers), its position in the lane (vision system and magnetic nail sensors), and its position with respect to other vehicles in the roadway (vision system and laser range sensors). All the sensors are interfaced through an Intel-based industrial computer, which runs the QNX real-time operating system. All of the control algorithms and most of the sensor processing are performed by the host computer.

In particular, for the vision system, the real-time lane extraction operation is carried out on a network of TMS320C40 digital signal processors that is hosted on the bus of the main computer. The experimental setup for the vision-based tracking and range estimation for longitudinal control is illustrated in Figure 12.3 (right). The off-line version of the tracking algorithm is typically tested on approximately 20 minutes of synchronized video and laser radar data. We discuss below in more detail the design of components of the vision system.

## 12.3.2 *Vision system design*

*Lane extraction for lateral control*

The lane recognition and extraction module is responsible for recovering estimates for the position and orientation of the car within the lane from the image data acquired by the camera (Figure 12.2). The roadway can be modeled as a planar surface, which implies that there is a simple homography between the image plane, with coordinates $\boldsymbol{x} = [x, y, 1]^T$, and the ground plane, with coordinates $\boldsymbol{X} = [X, Y, 1]^T$:

$$\boldsymbol{x} \sim H\boldsymbol{X}. \tag{12.29}$$

The $3 \times 3$ homography matrix $H$ can be recovered through an offline calibration procedure. This model is adequate for our imaging configuration where a camera with a fairly wide field of view (approximately 30 degrees) monitors the area immediately in front of the vehicle (4 to 25 meters).

The first stage of the lane recognition process is responsible for detecting and localizing possible lane markers on each row of an input image. The lane markers are modeled as white bars of a particular width against a darker background. Regions in the image that satisfy this intensity profile can be identified through a template-matching procedure similar to the feature-detection algorithms described in Chapter 4. It is important to notice that the width of the lane markers in the image changes linearly as a function of the pixel row. This means that different templates or feature detectors must be used for different pixel rows.

Once a set of candidate lane markers has been extracted, a robust fitting procedure is used to find the best-fitting straight line through these points on the image plane. A robust fitting strategy is essential in this application because in real highway traffic scenes, the feature extraction process will almost always generate extraneous features that are not part of the lane structure. These extra features can come from a variety of sources such as other vehicles on the highway, shadows or cracks on the roadway, and other road markings. They can confuse naive estimation procedures that are based on simple least-squares techniques.[8]

The lane extraction system is able to process images from the video camera at a rate of 30 frames per second with a latency of 57 milliseconds. This latency refers to the interval between the instant when the shutter of the camera closes and the instant when a new estimate for the vehicle position computed from that image is available to the control system. This system has been used successfully even in the presence of difficult lane markings like the "Bott's dot" reflectors on a concrete surface (see Figure 12.4).

Once the lane markers are extracted from each image frame, the car's states (position, heading, and velocity relative to the lane) are recovered using a Kalman filter, as described in this chapter. To improve the estimation, the Kalman filter can be designed and implemented in such a way that the dynamical model

[8] In the current implementation, the Hough transform is used for line fitting.

Figure 12.4. Automatic lane extraction (right) on a typical input image (left). (Courtesy of C.J. Taylor.)

of the car is also accounted for. These, however, are application-specific details that are beyond the scope of this book, and interested readers may refer to [Kosecká et al., 1997]. The recovered state estimates are then used to design state feedback laws for lateral control.

*Stereo tracking and range estimation for longitudinal control*

Another modality of the automated vehicle is a longitudinal control system that combines both laser radar and vision sensors, enabling throttle and brake control to maintain a safe distance and speed relative to a car in front. To properly employ any longitudinal controllers, we need reliable estimates of the range of the car in front. To achieve this, the vision system needs to track features on the car in front over an extended time period and simultaneously estimate the range of the car from stereo disparity in the images of these features (see Figure 12.5).[9]

**1** **421...**

Figure 12.5. Example stereo pairs from a tracking sequence. (Courtesy of P. McLauchlan.)

Here we have a very special stereo problem at hand: only the rear of the leading vehicle is visible to the camera, typically with little change in orientation (see Figure 12.5). Therefore, the 3-D depth relief in all visible features is very small, and we can assume that from the rear of the vehicle to each image plane is an affine projection of a planar object (almost parallel to the image plane):

$$\boldsymbol{x} = A\boldsymbol{X} : \quad \begin{bmatrix} x \\ y \\ 1 \end{bmatrix} = \begin{bmatrix} a_{11} & a_{12} & t_x \\ a_{21} & a_{22} & t_y \\ 0 & 0 & 1 \end{bmatrix} \begin{bmatrix} X \\ Y \\ 1 \end{bmatrix}, \tag{12.30}$$

[9]Although we do not really need stereo for lateral control, stereo is used for longitudinal control.

where for the left and right views the matrix $A$ is different, say $A_l$ and $A_r$, respectively. The smallest range we consider in our experiments is about 10 m, so that a car of size 2 m will span an angle of at most 10°, further justifying the assumption of a (planar) affine projection from scene to image. Experimental results are used to validate these simplifying assumptions.

A major issue with all reconstruction techniques is their reliance on high-quality, essentially outlier-free input data. We therefore apply the RANSAC algorithm, described in Chapter 11, to compute in each view a large subset of feature matches consistent with a single set of the affine transformation parameters (in the above equation). The stereo matching between the two views also adopts a similar robust matching technique. Given a set of well-tracked or matched features, we can compute their center of mass $\boldsymbol{x}_f$ in each image, the so-called *fixation point*. The fixation point ideally should be very robust to the loss or gain of individual features. The range of the leading car can be robustly estimated from the stereo disparity of the two fixation points in the two views. Since feature-based robust matching is relatively time-consuming, this algorithm is currently running at 3 to 5 Hz depending on the actual size of the region (maximum $140 \times 100$ pixels in our implementation) to which corner detection is applied.

A frame-rate (30 Hz) performance of the overall tracker is achieved from coordinating the above robust tracking algorithm with a separate tracking algorithm: a frame-rate tracking and matching algorithm, the so-called *correlator*, based on normalized cross-correlation (NCC) only (see Chapter 4). The two algorithms run in parallel on separate processors, and are coordinated in such a way that the correlator is always using an image region centered at the latest fixation point. The two processes communicate whenever the fixation algorithm has a new output to pass on. In addition, the laser radar provides the fixation algorithm with the initial bounding boxes around the vehicle in front. Thus, the radar may be considered a third layer of the tracker, which provides the system with a great deal of robustness. If the correlator fails for any reason (usually due to not finding a matching with a high enough correlation score), it simply waits for the fixation algorithm to provide it with a new template/position pair. If the fixation algorithm fails, then it also must wait for the laser to provide it with a new bounding box pair.

### 12.3.3 System test results

Figure 12.5 shows some example images, with the tracking results superimposed. The corner features are shown as small crosses, white for those matched over time or in stereo, and black for unmatched features. Black and white circle indicates the position of the fixation point, which ideally should remain at the same point on the car throughout the sequence. White rectangle describes the latest estimate of the bounding box for the car. Images 1 and 2 show the first stereo pair in the sequence, where the vehicle is close (about 17 m) to the camera and range estimates from stereo disparity may be expected to be accurate. By contrast images 421 and 422 are taken when the car is about 60 m away from the camera (the largest distance during the given test sequence). We may expect that depth estimates from stereo will be unreliable, since the disparity relative to infinity is less than a few pixels and difficult to measure in the presence of noise. However, it will still be feasible to use the change in apparent size to obtain reasonable range estimates.

We also compute the range and bearing estimated from the laser radar range finder and plot them together with the corresponding data collected from the vision algorithms in Figures 12.6 and 12.7. Depth from stereo is computed by inverting the projection of the fixation point at each image pair and finding the closest point of intersection of the two resulting space rays.[10]

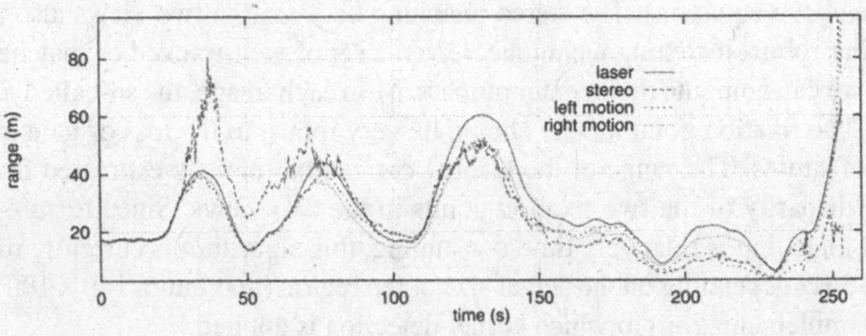


Figure 12.6. Comparison of range estimates from laser radar and vision.

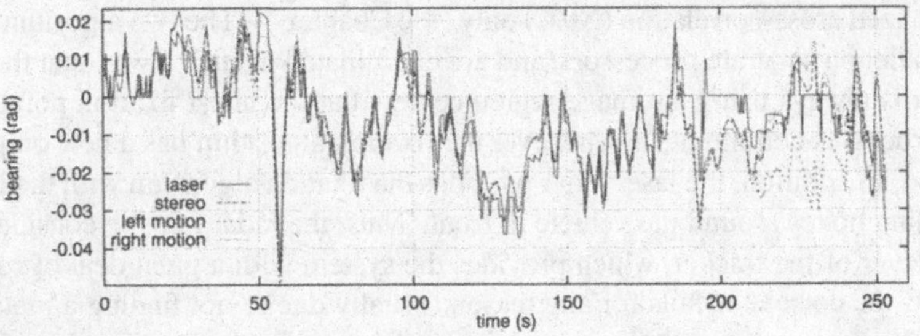


Figure 12.7. Comparison of bearing estimates from laser radar and vision (courtesy of P. McLauchlan).

## 12.4 Visual feedback for autonomous helicopter landing

In addition to its use in unmanned ground vehicles (UGVs), vision is also becoming a standard sensor in unmanned aerial vehicle (UAV) control and navigation, either replacing or augmenting conventional navigation sensors such as gyroscopes, sonar, electronic compass, and global positioning systems (GPS). One

[10]One can also use a better triangulation scheme discussed in Chapter 5, at a higher computational cost.

of the most important yet challenging tasks for controlling UAVs is autonomous landing or takeoff based on visual guidance only. An autonomous helicopter shown in Figure 12.8 developed by researchers at the University of California at Berkeley, the Berkeley Aerial Robot (BEAR) project, has accomplished such tasks. In this section we describe its basic components, system architecture, and performance. As in the previous section, the goal here is to provide the reader a basic picture about how to build an autonomous system that utilizes visual feedback in real time. However, the difference is that here we get to design the environment, i.e. the landing pad.

Figure 12.8. Berkeley unmanned aerial vehicle test-bed with on-board vision system: Yamaha R-50 helicopter (left) with Sony pan/tilt camera (top right) and LittleBoard computer (bottom right) hovering above a landing platform (left).

## 12.4.1 System setup and implementation

The helicopter is a Yamaha R-50 (see Figure 12.8) on which we have mounted:

- *Navigation Computer*: Pentium 233MHz Ampro LittleBoard running QNX real-time OS, responsible for low-level flight control;
- *Inertial Measurement Unit*: NovAtel MillenRT2 GPS system (2 cm accuracy) and Boeing DQI-NP INS/GPS integration system;
- *Vision Computer*: Pentium 233MHz Ampro LittleBoard running Linux, responsible for grabbing images, vision algorithms and camera control;
- *Camera*: Sony EVI-D30 Pan/Tilt/Zoom camera;
- *Frame-Grabber*: Imagenation PXC200 for capturing $320 \times 240$ resolution images at 30 Hz;
- *Wireless Ethernet*: WaveLAN IEEE 802.11b for communications between the helicopter and the monitoring base station.

The frame-grabber captures images at 30 Hz, which sets an upper bound on the rate of estimates from the vision system. The interrelationship of this hardware as it is mounted on the UAV is depicted in Figure 12.9 (left).

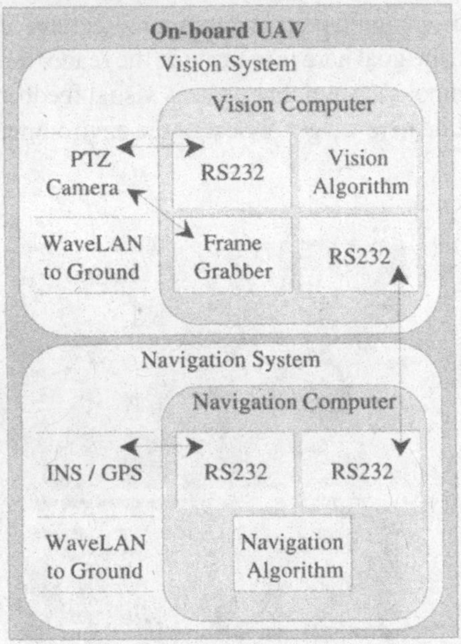


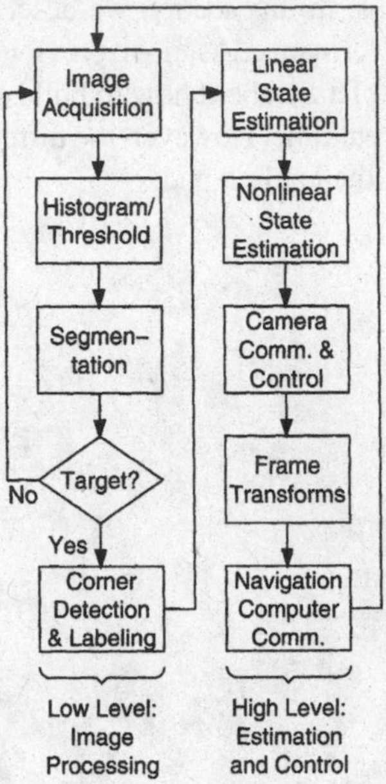


Figure 12.9. Left: Organization of hardware on the UAV. Right: Vision system software flow chart: image processing followed by pose estimation and subsequent control.

## 12.4.2 *Vision system design*

The vision system software consists of two main stages: image processing and pose estimation, each with a sequence of subroutines. Figure 12.9 (right) shows a flowchart of the algorithm. The helicopter pose estimation is a module that takes as input the features extracted from the low-level image processing and returns as output the helicopter pose relative to the landing pad. This is a problem that has been extensively discussed before in this book.[11] We discuss how a customized low-level image processing module is built in order to meet the real-time requirement.

The goal of low-level image processing is to locate the landing target and then extract and label its feature points. This process is done using standard thresholding techniques, since the shape and appearance of the landing pattern are known.

[11] For pose estimation from features of the landing pad, a Kalman filter can certainly be used, and, to obtain better estimates, one can also incorporate the coplanar and symmetry constraints among features on the landing pad.

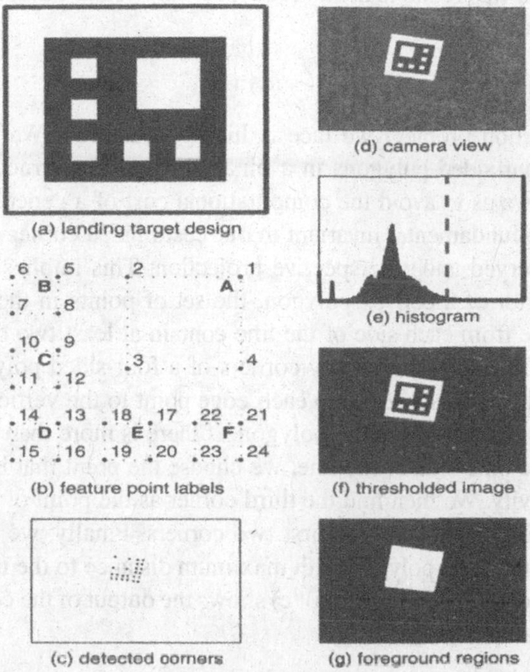


Figure 12.10. Landing target design and image processing. The target design (a) is made for simple feature matching (b), robust feature point extraction (c), and simplified image segmentation (d) to (g) (courtesy of Omid Shakernia).

In practice, the image is thresholded, so that only the light parts of the landing target are detected, and the resulting mask is compared with the stored model of the target. In order to simplify the image processing, the design of the landing target must make it easy to identify and segment from the background, provide distinctive feature points, simplify feature matching, and allow for algorithms that can operate in real time using off-the-shelf hardware. Figure 12.10 (a) shows the landing target design. Figure 12.10 (b) shows the feature point labeling on the corners of the interior white squares of the landing target. We choose corner detection (see Chapter 4) over other forms of feature-point-extraction because it is simple and robust, and it provides a high density of feature points per image pixel area. We choose squares over other $n$-sided polygons because they maximize the quality of the corners under perspective projection and pixel quantization.[12] Moreover, our particular organization of the squares in the target allows for a straightforward feature point matching, invariant of Euclidean motion and perspective projection, as we discuss below. Color tracking was also explored as a cue for feature

[12] Squares also have nice symmetry, which can be used to improve pose estimation using techniques studied in Chapter 10.

points; however, it was found to be not robust because of the variability of outdoor lighting conditions.

*Corner detection*

The corner detection problem we face is highly structured: We need to detect the corners of four-sided polygons in a binary image. The structured nature of the problem allows us to avoid the computational cost of a general-purpose corner detector. The fundamental invariant in our customized corner detector is that *convexity* is preserved under perspective projection. This implies that for a line through the interior of a convex polygon, the set of points in the polygon with maximal distance from each side of the line contain at least two distinct corners of the polygon. To find two arbitrary corners of a four-sided polygon, we compute the perpendicular distance from each edge point to the vertical line passing through the center of gravity of the polygon. If there is more than one point with maximal distance on a side of the line, we choose the point that is farthest from the center of gravity. We then find the third corner as the point of maximum distance from the line connecting the first two corners. Finally, we find the fourth corner as the point of the polygon with maximum distance to the triangle defined by the first three corners. Figure 12.10 (c) shows the output of the corner detection algorithm on a sample image.

*Feature matching*

To speed up the feature-matching process, we also exploit the design of the landing pad. The speedup is based on the fact that, like convexity, the *ordering of angles* (not the angles themselves) among image features from a plane (the landing pad) is always preserved in viewing from one side of the plane. To be more specific, let $q_1, q_2, q_3$ be three points on the plane, and consider the angle $\theta$ between vectors $(q_2 - q_1)$ and $(q_3 - q_1)$, where $\theta > 0$ when the view is from one side of the plane. Given any image of points $q_1, q_2, q_3$ taken from the same side of the plane, if $\theta'$ is the corresponding angle in the image of those points, then $\text{sign}(\theta') = \text{sign}(\theta)$. This property guarantees that correspondence matching based on counterclockwise ordering of feature points on the plane will work for any given image of those feature points. In particular, to identify the corners of a square, we calculate the vectors between its corners and the center of another particular square in the landing target. One such vector will always be first on counterclockwise ordering, and we identify the associated corner this way. We determine the labeling of the remaining corners by ordering them counterclockwise from the identified corner.

### 12.4.3 System performance and evaluation

We now show the performance of the overall vision-based landing system from a real flight test that took place in early 2002: the UAV hovered autonomously above a stationary landing pad with the vision system running in real time. The vision-based state estimates were used by the supervisory controller (at an update rate of 10 Hz) to command the UAV to hover above the landing target, making it a truly closed-loop vision-controlled flight experiment. State estimates from the INS/GPS navigation system were synchronously gathered only for comparison.

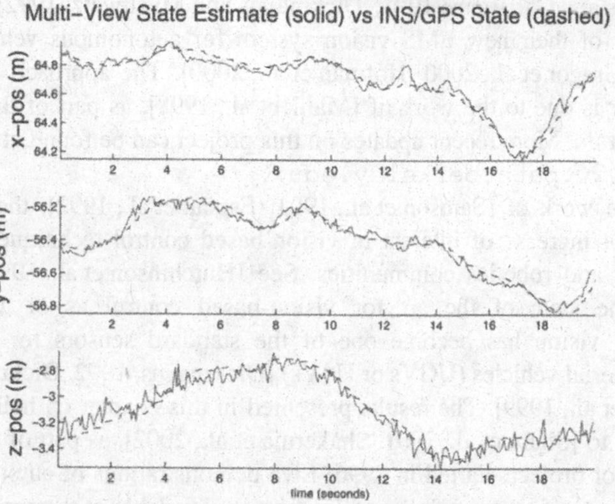


Figure 12.11. Comparing a vision-based motion estimation algorithm with inertial navigation system measurements in a real flight test (courtesy of Omid Shakernia and René Vidal).

Figure 12.11 shows the results from this flight test, comparing the output of the vision-based motion estimation algorithm with the measurements from the INS/GPS navigation system onboard (which are accurate to within 2 cm). Overall, the vision algorithm achieves an estimate of the helicopter's position with an error less than 5 cm and orientation less than 4 degrees, and its performance is still being improved by implementing the estimation algorithms within a filtering framework.

# Further readings

The Kalman-Bucy filter [Kalman, 1960, Bucy, 1965] provides an optimal state estimator for linear (stochastic) dynamical systems. Its extensions to nonlinear systems are often referred to as extended Kalman filters (EKF). Appendix B gives a brief review, and we recommend the book of [Jazwinski, 1970] for more details on this subject. There has been a long history of using Kalman filters for solving motion estimation problems in computer vision (see [Dickmanns and Graefe, 1988b, Broida and Chellappa, 1986a, Matthies et al., 1989, Broida et al., 1990], or more recently [Soatto et al., 1996] and references therein). The method presented in this chapter is due to the work of [Chiuso et al., 2002, Favaro et al., 2003], and the implementation of the system is also publicly available at the website: `http://vision.ucla.edu/`.

The use of vision-based systems for driving cars on highways dates back to the 1980s in the work of Dickmanns and his coworkers [Dickmanns and Graefe, 1988a, Dickmanns and Graefe, 1988b] and some of their later work in the 1990s [Dickmanns and Christians, 1991, Dickmanns and Mysliwetz, 1992]. An extensive survey of their new EMS vision system for autonomous vehicles can be found in [Gregor et al., 2000, Hofman et al., 2000]. The approach presented in this chapter is due to the work of [Malik et al., 1998], as part of the California PATH program. More recent updates on this project can be found at the website: `http://www.path.berkeley.edu/`.

Since the work of [Samson et al., 1991, Espiau et al., 1992], there has been a significant increase of interest in vision-based control techniques from both the control and robotics communities. See [Hutchinson et al., 1996] for a review of the state of the art for vision-based control as of 1996. In recent years, vision has become one of the standard sensors for autonomous ground or aerial vehicles (UGVs or UAVs) [Dickmanns, 1992, Dickmanns, 2002, Shakernia et al., 1999]. The results presented in this chapter on helicopter landing are due to [Sharp et al., 2001, Shakernia et al., 2002], as part of the Berkeley Aerial Robot project, started in 1996. Live demonstrations of autonomous helicopters taking off, cooperating, and landing are available at the project website: `http://robotics.eecs.berkeley.edu/bear/`.

# Part V

# Appendices

# Appendix A
# Basic Facts from Linear Algebra

*Algebra is very generous: she often gives more than is asked for.*
– Jean d'Alembert

We assume that the reader already has basic training in linear algebra.[1] This appendix provides a brief review of some of the important facts that are used in this book. For a more complete introduction, the reader should resort to a book such as [Strang, 1988].

In this book, we deal mostly with finite-dimensional (typically two-, three-, or four-dimensional) *linear spaces*, which are also often called linear vector spaces. For generality, we consider the linear space to be $n$-dimensional. A linear space is typically denoted by the letter $V$ (for "vector space"). Although most of the time we will deal with vectors of real numbers $\mathbb{R}$, occasionally, we will encounter vectors of complex numbers[2] $\mathbb{C}$. For simplicity, our review will be conducted for linear spaces over the field $\mathbb{R}$ of real numbers with the understanding that most definitions and results generalize to the complex case with little change.

**Definition A.1 (A linear space or a vector space).** *A set (of vectors) $V$ is considered* a linear space *over the field $\mathbb{R}$ if its elements, called* vectors, *are closed under two basic operations: scalar multiplication and vector summation "+." That is, given any two vectors $v_1, v_2 \in V$ and any two scalars $\alpha, \beta \in \mathbb{R}$, the linear combination $v = \alpha v_1 + \beta v_2$ is also a vector in $V$. Furthermore, the ad-*

[1] Some familiarity with the numerical software Matlab is also encouraged.

[2] For instance, the eigenvalues or eigenvectors of a real matrix could be complex.

*dition is commutative and associative, it has an identity* 0, *and each element has an inverse, "*$-v$*," such that* $v + (-v) = 0$. *The scalar multiplication respects the structure of* $\mathbb{R}$*; i.e.* $\alpha(\beta)v = (\alpha\beta)v$, $1v = v$ *and* $0v = 0$. *The addition and scalar multiplication are related by the distributive laws:* $(\alpha + \beta)v = \alpha v + \beta v$ *and* $\alpha(v + u) = \alpha v + \alpha u$.

For example, $\mathbb{R}^n$ is a linear space over the field of real numbers $\mathbb{R}$. To be consistent, we always use a column to represent a vector:

$$[x_1, x_2, \ldots, x_n]^T = \begin{bmatrix} x_1 \\ x_2 \\ \vdots \\ x_n \end{bmatrix} \in \mathbb{R}^n, \tag{A.1}$$

where $[x_1, x_2, \ldots, x_n]^T$ means "the (row) vector $[x_1, x_2, \ldots, x_n]$ transposed." Given two scalars $\alpha, \beta \in \mathbb{R}$ and two vectors $v_1 = [x_1, x_2, \ldots, x_n]^T \in \mathbb{R}^n$ and $v_2 = [y_1, y_2, \ldots, y_n]^T \in \mathbb{R}^n$, their linear combination is a componentwise summation weighted by $\alpha$ and $\beta$:

$$\begin{aligned} \alpha v_1 + \beta v_2 &= \alpha[x_1, x_2, \ldots, x_n]^T + \beta[y_1, y_2, \ldots, y_n]^T \\ &= [\alpha x_1 + \beta y_1, \alpha x_2 + \beta y_2, \ldots, \alpha x_n + \beta y_n]^T. \end{aligned}$$

## A.1 Basic notions associated with a linear space

We will now provide a brief review of basic notions and frequently used notation associated with a linear vector space $V$ (i.e. $\mathbb{R}^n$).

### *A.1.1 Linear independence and change of basis*

**Definition A.2 (Subspace).** *A subset* $W$ *of a linear space* $V$ *is called* a subspace *if the zero vector* 0 *is in* $W$ *and* $\alpha w_1 + \beta w_2 \in W$ *for all* $\alpha, \beta \in \mathbb{R}$ *and* $w_1, w_2 \in W$.

**Definition A.3 (Spanned subspace).** *Given a set of vectors* $S = \{v_i\}_{i=1}^m$, the subspace spanned by $S$ *is the set of all finite linear combinations* $\sum_{i=1}^m \alpha_i v_i$ *for all* $\alpha_1, \alpha_2, \ldots, \alpha_m \in \mathbb{R}$. *This subspace is usually denoted by span*$(S)$.

For example, the two vectors $v_1 = [1, 0, 0]^T$ and $v_2 = [1, 1, 0]^T$ span a subspace of $\mathbb{R}^3$ whose vectors are of the general form $v = [x, y, 0]^T$.

**Definition A.4 (Linear independence).** *A set of vectors* $S = \{v_i\}_{i=1}^m$ *is* linearly independent *if the equation*

$$\alpha_1 v_1 + \alpha_2 v_2 + \cdots + \alpha_m v_m = 0$$

*implies*

$$\alpha_1 = \alpha_2 = \cdots = \alpha_n = 0.$$

On the other hand, a set of vectors $\{v_i\}_{i=1}^m$ is said to be *linearly dependent* if there exist $\alpha_1, \alpha_2 \ldots, \alpha_m \in \mathbb{R}$ not all zero such that

$$\alpha_1 v_1 + \alpha_2 v_2 + \cdots + \alpha_m v_m = 0.$$

**Definition A.5 (Basis).** *A set of vectors $B = \{b_i\}_{i=1}^n$ of a linear space $V$ is said to be* a basis *if $B$ is a linearly independent set and $B$ spans the entire space $V$; i.e. $V = span(B)$.*

**Fact A.6 (Properties of a basis).** *Suppose $B$ and $B'$ are two bases for a linear space $V$. Then:*

1. *$B$ and $B'$ contain exactly the same number of linearly independent vectors. This number, say $n$, is the* dimension *of the linear space $V$.*
2. *Let $B = \{b_i\}_{i=1}^n$ and $B' = \{b'_i\}_{i=1}^n$. Then each basis vector of $B$ can be expressed as a linear combination of those in $B'$; i.e.*
$$b_j = a_{1j}b'_1 + a_{2j}b'_2 + \cdots + a_{nj}b'_n = \sum_{i=1}^n a_{ij}b'_i, \tag{A.2}$$
*for some $a_{ij} \in \mathbb{R}, i, j = 1, 2, \ldots, n$.*
3. *Any vector $v \in V$ can be written as a linear combination of vectors in either of the bases:*
$$v = x_1b_1 + x_2b_2 + \cdots + x_nb_n = x'_1b'_1 + x'_2b'_2 + \cdots + x'_nb'_n, \tag{A.3}$$
*where the coefficients $\{x_i \in \mathbb{R}\}_{i=1}^n$ and $\{x'_i \in \mathbb{R}\}_{i=1}^n$ are uniquely determined and are called the* coordinates *of $v$ with respect to each basis.*

In particular, if $B$ and $B'$ are two bases for the linear space $\mathbb{R}^n$, we may put the basis vectors as columns of two $n \times n$ matrices and also call them $B$ and $B'$, respectively:

$$B \doteq [b_1, b_2, \ldots, b_n], \quad B' \doteq [b'_1, b'_2, \ldots, b'_n] \quad \in \mathbb{R}^{n\times n}. \tag{A.4}$$

Then we can express the relationship between them in the matrix form $B = B'A$ as

$$[b_1, b_2, \ldots, b_n] = [b'_1, b'_2, \ldots, b'_n] \begin{bmatrix} a_{11} & a_{12} & \cdots & a_{1n} \\ a_{21} & a_{22} & \cdots & a_{2n} \\ \vdots & \vdots & \ddots & \vdots \\ a_{n1} & a_{n2} & \cdots & a_{nn} \end{bmatrix}. \tag{A.5}$$

The role of the $n \times n$ matrix $A$ is to transform one basis ($B'$) to the other ($B$). Since such a transformation can go the opposite way, the matrix $A$ must be invertible. So we can also write $B' = BA^{-1}$.

If $v$ is a vector in $V$, it can be expressed in terms of linear combinations of either basis as

$$v = x_1b_1 + x_2b_2 + \cdots + x_nb_n = x'_1b'_1 + x'_2b'_2 + \cdots + x'_nb'_n. \tag{A.6}$$

Thus, we have

$$v = [b_1, b_2, \ldots, b_n] \begin{bmatrix} x_1 \\ x_2 \\ \vdots \\ x_n \end{bmatrix} = [b_1', b_2', \ldots, b_n'] \begin{bmatrix} a_{11} & a_{12} & \cdots & a_{1n} \\ a_{21} & a_{22} & \cdots & a_{2n} \\ \vdots & \vdots & \ddots & \vdots \\ a_{n1} & a_{n2} & \cdots & a_{nn} \end{bmatrix} \begin{bmatrix} x_1 \\ x_2 \\ \vdots \\ x_n \end{bmatrix}.$$

Since the coordinates of $v$ with respect to $B'$ are unique, we obtain the following transformation of coordinates of a vector from one basis to the other:

$$\begin{bmatrix} x_1' \\ x_2' \\ \vdots \\ x_n' \end{bmatrix} = \begin{bmatrix} a_{11} & a_{12} & \cdots & a_{1n} \\ a_{21} & a_{22} & \cdots & a_{2n} \\ \vdots & \vdots & \ddots & \vdots \\ a_{n1} & a_{n2} & \cdots & a_{nn} \end{bmatrix} \begin{bmatrix} x_1 \\ x_2 \\ \vdots \\ x_n \end{bmatrix}. \tag{A.7}$$

Let $x = [x_1, x_2, \ldots, x_n]^T \in \mathbb{R}^n$ and $x' = [x_1', x_2', \ldots, x_n']^T \in \mathbb{R}^n$ denote the two coordinate vectors. We may summarize in matrix form the relationships between two bases and coordinates with respect to the bases as

$$\boxed{B' = BA^{-1}, \quad x' = Ax.} \tag{A.8}$$

*Be aware of the difference in transforming bases from transforming coordinates!*

## A.1.2 Inner product and orthogonality

**Definition A.7 (Inner product).** *A function*

$$\langle \cdot, \cdot \rangle : \mathbb{R}^n \times \mathbb{R}^n \to \mathbb{R}$$

*is an* inner product[3] *if*

1. $\langle u, \alpha v + \beta w \rangle = \alpha \langle u, v \rangle + \beta \langle u, w \rangle, \quad \forall\, \alpha, \beta \in \mathbb{R}$,
2. $\langle u, v \rangle = \langle v, u \rangle$,
3. $\langle v, v \rangle \geq 0$, *and* $\langle v, v \rangle = 0 \Leftrightarrow v = 0$.

*For each vector* $v$, $\sqrt{\langle v, v \rangle}$ *is called its* norm.

The inner product is also called a *metric*, since it can be used to measure length and angles.

For simplicity, a standard basis is often chosen for the vector space $\mathbb{R}^n$ as the set of vectors

$$e_1 = [1, 0, 0, \ldots, 0]^T, \quad e_2 = [0, 1, 0, \ldots, 0]^T, \quad e_n = [0, 0, \ldots, 0, 1]^T. \tag{A.9}$$

The matrix $I = [e_1, e_2, \ldots, e_n]$ with these vectors as columns is exactly the $n \times n$ identity matrix.

[3]In some literature, an inner product is also called a *dot product*, denoted by $u \cdot v$. However, in this book, we will *not* use that name.

**Definition A.8 (Canonical inner product on $\mathbb{R}^n$).** *Given any two vectors* $x = [x_1, x_2, \ldots, x_n]^T$ *and* $y = [y_1, y_2, \ldots, y_n]^T$ *in* $\mathbb{R}^n$*, we define* the canonical inner product *to be*

$$\langle x, y \rangle \doteq x^T y = x_1 y_1 + x_2 y_2 + \cdots + x_n y_n. \tag{A.10}$$

*This inner product induces the* standard 2-norm, *or* Euclidean norm, $\|\cdot\|_2$, *which measures the length of each vector as*

$$\|x\|_2 \doteq \sqrt{x^T x} = \sqrt{x_1^2 + x_2^2 + \cdots + x_n^2}. \tag{A.11}$$

Notice that if we choose another basis $B'$ related to the above standard basis $I$ as $I = B'A$, then the coordinates of the vectors $x, y$ related to the new basis are $x'$ and $y'$, respectively, and they relate to $x, y$ by $x' = Ax$ and $y' = Ay$. The inner product in terms of the new coordinates becomes

$$\langle x, y \rangle = x^T y = (A^{-1}x')^T (A^{-1}y') = (x')^T A^{-T} A^{-1} (y'). \tag{A.12}$$

We denote this expression of the inner product with respect to the new basis by

$$\langle x', y' \rangle_{A^{-T}A^{-1}} \doteq (x')^T A^{-T} A^{-1} (y'). \tag{A.13}$$

This is called an *induced inner product* from the matrix $A$. Knowing the matrix $A^{-T}A^{-1}$, we can compute the canonical inner product directly using coordinates with respect to the nonstandard basis $B'$.

**Definition A.9 (Orthogonality).** *Two vectors $x, y$ are said to be orthogonal if their inner product is zero: $\langle x, y \rangle = 0$. This is often indicated as $x \perp y$.*

### *A.1.3 Kronecker product and stack of matrices*

**Definition A.10 (Kronecker product of two matrices).** *Given two matrices $A \in \mathbb{R}^{m \times n}$ and $B \in \mathbb{R}^{k \times l}$, their Kronecker product, denoted by $A \otimes B$, is a new matrix*

$$A \otimes B \doteq \begin{bmatrix} a_{11}B & a_{12}B & \cdots & a_{1n}B \\ a_{21}B & a_{22}B & \cdots & a_{2n}B \\ \vdots & \vdots & \ddots & \vdots \\ a_{m1}B & a_{m2}B & \cdots & a_{mn}B \end{bmatrix} \in \mathbb{R}^{mk \times nl}. \tag{A.14}$$

*If $A$ and $B$ are two vectors, i.e. $n = l = 1$, the product $A \otimes B$ is also a vector but of dimension $mk$.*

In Matlab, one can easily compute the Kronecker product by using the command `C = kron(A,B)`.

**Definition A.11 (Stack of a matrix).** *Given an $m \times n$ matrix $A \in \mathbb{R}^{m \times n}$, the* stack *of the matrix $A$ is a vector, denoted by $A^s$, in $\mathbb{R}^{mn}$ obtained by stacking its*

*$n$ column vectors, say $a_1, a_2, \ldots, a_n \in \mathbb{R}^m$, in order:*

$$A^s \doteq \begin{bmatrix} a_1 \\ a_2 \\ \vdots \\ a_n \end{bmatrix} \in \mathbb{R}^{mn}. \tag{A.15}$$

*As mutually inverse operations, $A^s$ is called $A$ "stacked," and $A$ is called $A^s$ "unstacked."*

The Kronecker product and stack of matrices together allow us to rewrite algebraic equations that involve multiple vectors and matrices in many different but equivalent ways. For instance, the equation

$$u^T A v = 0 \tag{A.16}$$

for two vectors $u, v$ and a matrix $A$ of proper dimensions can be rewritten as

$$(v \otimes u)^T A^s = 0. \tag{A.17}$$

The second equation is particularly useful when $A$ is the only unknown in the equation.

## A.2 Linear transformations and matrix groups

Linear algebra studies the properties of the *linear transformations*, or *linear maps*, between different linear spaces. Since such transformations can be represented as matrices, linear algebra to a large extent studies the properties of matrices.

**Definition A.12 (Linear transformation).** *A* linear transformation *from a linear (vector) space $\mathbb{R}^n$ to $\mathbb{R}^m$ is defined as a map $L : \mathbb{R}^n \to \mathbb{R}^m$ such that*

- $L(x + y) = L(x) + L(y), \quad \forall x, y \in \mathbb{R}^n$;
- $L(\alpha x) = \alpha L(x), \quad \forall x \in \mathbb{R}^n, \alpha \in \mathbb{R}$.

With respect to the standard bases of $\mathbb{R}^n$ and $\mathbb{R}^m$, the map $L$ can be *represented* by a matrix $A \in \mathbb{R}^{m \times n}$ such that

$$L(x) = Ax, \quad \forall x \in \mathbb{R}^n. \tag{A.18}$$

The $i$th column of the matrix $A$ is then nothing but the image of the standard basis vector $e_i \in \mathbb{R}^n$ under the map $L$; i.e.

$$A = [L(e_1), L(e_2), \ldots, L(e_n)] \quad \in \mathbb{R}^{m \times n}.$$

The set of all (real) $m \times n$ matrices is denoted by $\mathcal{M}(m, n)$. When viewed as a linear space, $\mathcal{M}(m, n)$ can be identified as the space $\mathbb{R}^{mn}$. When there is little ambiguity, we refer to a linear map $L$ by its matrix representation $A$. If $n = m$, the set $\mathcal{M}(n, n) \doteq \mathcal{M}(n)$ forms an algebraic structure called a *ring* (over the field $\mathbb{R}$). That is, matrices in $\mathcal{M}(n)$ are closed under both matrix multiplication and summation: If $A, B$ are two $n \times n$ matrices, so are $C = AB$ and $D = A + B$.

Linear maps or matrices that we encounter in computer vision often have a special algebraic structure called a *group*.

**Definition A.13 (Group).** *A* group *is a set $G$ with an operation "$\circ$" on the elements of $G$ that:*

- is closed : *If $g_1, g_2 \in G$, then also $g_1 \circ g_2 \in G$;*
- is associative: *$(g_1 \circ g_2) \circ g_3 = g_1 \circ (g_2 \circ g_3)$, for all $g_1, g_2, g_3 \in G$;*
- has a unit element *e: $e \circ g = g \circ e = g$, for all $g \in G$;*
- is invertible: *For every element $g \in G$, there exists an element $g^{-1} \in G$ such that $g \circ g^{-1} = g^{-1} \circ g = e$.*

**Definition A.14 (The general linear group $GL(n)$).** *The set of all $n \times n$ nonsingular (real) matrices with matrix multiplication forms a group. Such a group of matrices is usually called the general linear group and denoted by $GL(n)$.*

**Definition A.15 (Matrix representation of a group).** *A group $G$ has a* matrix representation *or can be realized as a matrix group if there exists an injective map*[4]

$$\mathcal{R}: \; G \to GL(n); \quad g \mapsto \mathcal{R}(g),$$

*which preserves the group structure*[5] *of $G$. That is, the inverse and composition of elements in $G$ are preserved by the map in the following way:*

$$\mathcal{R}(e) = I_{n\times n}, \quad \mathcal{R}(g \circ h) = \mathcal{R}(g)\mathcal{R}(h), \quad \forall g, h \in G. \tag{A.19}$$

Below, we identify a few important subsets of $\mathcal{M}(n)$ that have special algebraic structures (as examples of matrix groups) and nice properties.

The group $GL(n)$ itself can be identified as the set of all invertible linear transformations from $\mathbb{R}^n$ to $\mathbb{R}^n$ in the sense that for every $A \in GL(n)$, we obtain a linear map

$$L: \mathbb{R}^n \to \mathbb{R}^n; \quad x \mapsto Ax. \tag{A.20}$$

Notice that if $A \in GL(n)$, then so is its inverse: $A^{-1} \in GL(n)$. We know that an $n \times n$ matrix $A$ is invertible if and only if its determinant is nonzero. Therefore, we have

$$\det(A) \neq 0, \quad \forall A \in GL(n). \tag{A.21}$$

The general linear group, when matrices are considered to be known only up to a scalar factor, $GL(n)/\mathbb{R}$, is referred to as the *projective transformation group*, whose elements are called *projective matrices* or *homographies*.

[4]A map $f(\cdot)$ is called *injective* if $f(x) \neq f(y)$ as long as $x \neq y$.

[5]Such a map is called a *group homomorphism* in algebra.

Matrices in $GL(n)$ of determinant $+1$ form a subgroup called the *special linear group*, denoted by $SL(n)$. That is, $\det(A) = +1$ for all $A \in SL(n)$. It is easy to verify that if $A \in SL(n)$, then so is $A^{-1}$, since $\det(A^{-1}) = \det(A)^{-1}$.

**Definition A.16 (The affine group $A(n)$).** *An affine transformation $L$ from $\mathbb{R}^n$ to $\mathbb{R}^n$ is defined jointly by a matrix $A \in GL(n)$ and a vector $b \in \mathbb{R}^n$ such that*

$$L : \mathbb{R}^n \to \mathbb{R}^n; \quad x \mapsto Ax + b. \tag{A.22}$$

*The set of all such affine transformations is called the affine group of dimension $n$ and is denoted by $A(n)$.*

Notice that the map $L$ so-defined is *not* a linear map from $\mathbb{R}^n$ to $\mathbb{R}^n$ unless $b = 0$. Nevertheless, we may "embed" this map into a space one dimension higher so that we can still represent it by a single matrix. If we identify an element $x \in \mathbb{R}^n$ with $\begin{bmatrix} x \\ 1 \end{bmatrix} \in \mathbb{R}^{n+1}$,[6] then $L$ becomes a map from $\mathbb{R}^{n+1}$ to $\mathbb{R}^{n+1}$ in the following sense:

$$L : \mathbb{R}^{n+1} \to \mathbb{R}^{n+1}; \quad \begin{bmatrix} x \\ 1 \end{bmatrix} \mapsto \begin{bmatrix} A & b \\ 0 & 1 \end{bmatrix} \begin{bmatrix} x \\ 1 \end{bmatrix}. \tag{A.23}$$

Thus, a matrix of the form

$$\begin{bmatrix} A & b \\ 0 & 1 \end{bmatrix} \in \mathbb{R}^{(n+1)\times(n+1)}, \quad A \in GL(n),\ b \in \mathbb{R}^n, \tag{A.24}$$

fully describes an affine map, and we call it an *affine matrix*. This matrix is an element in the general linear group $GL(n+1)$. In this way, $A(n)$ is identified as a subset (and in fact a subgroup) of $GL(n+1)$. The multiplication of two affine matrices in the set $A(n)$ is

$$\begin{bmatrix} A_1 & b_1 \\ 0 & 1 \end{bmatrix} \begin{bmatrix} A_2 & b_2 \\ 0 & 1 \end{bmatrix} = \begin{bmatrix} A_1 A_2 & A_1 b_2 + b_1 \\ 0 & 1 \end{bmatrix} \in \mathbb{R}^{(n+1)\times(n+1)}, \tag{A.25}$$

which is also an affine matrix in $A(n)$ and represents the composition of two affine transformations.

Given $\mathbb{R}^n$ and its standard inner product structure, $\langle x, y \rangle = x^T y, \forall x, y \in \mathbb{R}^n$, let us consider the set of linear transformations (or matrices) that preserve the inner product.

**Definition A.17 (The orthogonal group $O(n)$).** *An $n \times n$ matrix $A$ (representing a linear map from $\mathbb{R}^n$ to itself) is called* orthogonal *if it preserves the inner product, i.e.*

$$\langle Ax, Ay \rangle = \langle x, y \rangle, \quad \forall\, x, y \in \mathbb{R}^n. \tag{A.26}$$

*The set of all $n \times n$ orthogonal matrices forms the* orthogonal group *of dimension $n$, and it is denoted by $O(n)$.*

---

[6] This is the so-called homogeneous representation of $x$. Notice that this identification does not preserve the vector structure of $\mathbb{R}^n$.

Obviously, $O(n)$ is a subset (and in fact a subgroup) of $GL(n)$. If $R$ is an orthogonal matrix, we must have $R^T R = RR^T = I$. Therefore, the orthogonal group $O(n)$ can be characterized as

$$O(n) = \{R \in GL(n) \mid R^T R = I\}. \tag{A.27}$$

The determinant $\det(R)$ of an orthogonal matrix $R$ can be either $+1$ or $-1$. The subgroup of $O(n)$ with determinant $+1$ is called the *special orthogonal group* and is denoted by $SO(n)$. That is, for any $R \in SO(n)$, we have $\det(R) = +1$. Equivalently, one may define $SO(n)$ as the intersection $SO(n) = O(n) \cap SL(n)$. In the case $n = 3$, the special orthogonal matrices are exactly the $3 \times 3$ rotation matrices (studied in Chapter 2).

The affine version of the orthogonal group gives the *Euclidean (transformation) group*:

**Definition A.18 (The Euclidean group $E(n)$).** *A Euclidean transformation $L$ from $\mathbb{R}^n$ to $\mathbb{R}^n$ is defined jointly by a matrix $R \in O(n)$ and a vector $T \in \mathbb{R}^n$ such that*

$$L : \mathbb{R}^n \to \mathbb{R}^n; \quad x \mapsto Rx + T. \tag{A.28}$$

*The set of all such transformations is called the Euclidean group of dimension $n$ and is denoted by $E(n)$.*

Obviously, $E(n)$ is a subgroup of $A(n)$. Therefore, it can also be embedded into a space one-dimension higher and has a matrix representation

$$\begin{bmatrix} R & T \\ 0 & 1 \end{bmatrix} \in \mathbb{R}^{(n+1)\times(n+1)}, \quad R \in O(n),\ T \in \mathbb{R}^n. \tag{A.29}$$

If $R$ further belongs to $SO(n)$, such transformations form the *special Euclidean group*, which is traditionally denoted by $SE(n)$. When $n = 3$, $SE(3)$ represents the conventional rigid-body motion in $\mathbb{R}^3$, where $R$ is the rotation of a rigid body and $T$ is the translation (with respect to a chosen reference coordinate frame).

Since all the transformation groups introduced so far have natural matrix representations, they are *matrix groups*.[7] To summarize their relationships, we have

$$\boxed{SO(n) \subset O(n) \subset GL(n), \quad SE(n) \subset E(n) \subset A(n) \subset GL(n+1).} \tag{A.30}$$

## A.3 Gram-Schmidt and the QR decomposition

A matrix in $GL(n)$ has $n$ independent rows (or columns). A matrix in $O(n)$ has orthonormal rows (or columns). The Gram-Schmidt procedure can be viewed as a map from $GL(n)$ to $O(n)$, for it transforms a nonsingular matrix into an orthogonal one. Call $\mathcal{L}_+(n)$ the subset of $GL(n)$ consisting of lower triangular matrices

[7] Since these groups themselves admit a differential structure, they belong to the *Lie groups*.

with positive elements along the diagonal. Such matrices form a subgroup of $GL(n)$.

**Theorem A.19 (Gram-Schmidt procedure).** *For every $A \in GL(n)$, there exists a lower triangular matrix $L \in \mathbb{R}^{n\times n}$ and an orthogonal matrix $E \in O(n)$ such that*

$$A = LE. \tag{A.31}$$

*Proof.* Contrary to the convention of the book, for simplicity in this proof all vectors indicate row vectors. That is, if $v$ is an $n$-dimensional row vector, it is of the form: $v = [v_1, v_2, \ldots, v_n] \in \mathbb{R}^n$. Denote the $i$th row vector of the given matrix $A$ by $a_i$ for $i = 1, 2, \ldots, n$. The proof consists in constructing $L$ and $E$ iteratively from the row vectors $a_i$:

$$\begin{array}{lclcl} l_1 & \doteq & a_1 & \longrightarrow & e_1 \doteq l_1/\|l_1\|_2, \\ l_2 & \doteq & a_2 - \langle a_2, e_1\rangle e_1 & \longrightarrow & e_2 \doteq l_2/\|l_2\|_2, \\ \vdots & \vdots & \vdots & \vdots & \vdots \\ l_n & \doteq & a_n - \sum_{i=1}^{n-1}\langle a_{i+1}, e_i\rangle e_i & \longrightarrow & e_n \doteq l_n/\|l_n\|_2. \end{array}$$

Then $E = [e_1^T, \ldots, e_n^T]^T$, and the matrix $L$ is obtained as

$$L = \begin{bmatrix} \|l_1\|_2 & 0 & \cdots & 0 \\ \langle a_2, e_1\rangle & \|l_2\|_2 & \cdots & 0 \\ \vdots & \ddots & \ddots & \vdots \\ \langle a_2, e_1\rangle & \cdots & \langle a_n, e_{n-1}\rangle & \|l_n\|_2 \end{bmatrix}.$$

By construction $E$ is orthogonal; i.e. $EE^T = E^TE = I$. □

**Remark A.20.** *The Gram-Schmidt's procedure has the peculiarity of being* causal, *in the sense that the ith row of the transformed matrix E depends only upon rows with index $j \leq i$ of the original matrix A. The choice of the name E for the orthogonal matrix above is not accidental. In fact, we will view the Kalman filter (to be reviewed in the next appendix) as a way to perform a Gram-Schmidt orthonormalization in a special Hilbert space, and the outcome E of the procedure is traditionally called the* innovation.

There are a few useful variations to Gram-Schmidt procedure. By transposing $A = LE$, we get $A^T = E^TL^T \doteq QR$. Notice that $R = L^T$ is an upper triangular matrix. Thus, by applying Gram-Schmidt procedure to the transpose of a matrix, we can also decompose it into the form $QR$ where $Q$ is an orthogonal matrix and $R$ an upper triangular matrix. Such a decomposition is called the *QR decomposition*. In Matlab, this can be done by the command `[Q,R] = qr(A)`. Furthermore, by inverting $A^T = E^TL^T$, we get $A^{-T} = L^{-T}E \doteq KE$. Notice that $K = L^{-T}$ is still an upper triangular matrix. Thus, we can also decompose any matrix into the form of an upper triangular matrix followed by an orthogonal one. The latter one is the kind of "QR decomposition" we use in Chapter 6 for camera calibration.

## A.4 Range, null space (kernel), rank and eigenvectors of a matrix

Let $A$ be a general $m \times n$ matrix that also conveniently represents a linear map from the vector space $\mathbb{R}^n$ to $\mathbb{R}^m$.

**Definition A.21 (Range, span, null space, and kernel).** *Define the* range *or* span *of A, denoted by range(A) or span(A), to be the subspace of* $\mathbb{R}^m$ *such that* $y \in range(A)$ *if and only if* $y = Ax$ *for some* $x \in \mathbb{R}^n$. *Define the* null space *of A, denoted by null(A), to be the subspace of* $\mathbb{R}^n$ *such that* $x \in null(A)$ *if and only if* $Ax = 0$. *When A is viewed as an abstract linear map, null(A) is also referred to as the* kernel *of the map, denoted by ker(A).*

Notice that the range of a matrix $A$ is exactly the span of all its column vectors; the null space of a matrix $A$ is exactly the set of vectors which are orthogonal to all its row vectors (for a definition of orthogonal vectors see Definition A.9). The notion of range or null space is useful whenever the solution to a linear equation of the form $Ax = b$ is considered. In terms of range and null space, this equation will have a solution if $b \in \text{range}(A)$ and will have a unique solution only if $\text{null}(A) = \emptyset$ (the empty set).

In Matlab, the null space of a matrix $A$ can be computed using the command `Z = null(A)`.

**Definition A.22 (Rank of a matrix).** *The* rank *of a matrix is the dimension of its range:*

$$rank(A) \doteq \dim(range(A)). \tag{A.32}$$

**Fact A.23 (Properties of matrix rank).** *For an arbitrary $m \times n$ matrix A, its rank has the following properties:*

1. $rank(A) = n - dim(null(A))$.
2. $0 \leq rank(A) \leq \min\{m, n\}$.
3. *rank(A) is equal to the maximum number of linearly independent column (or row) vectors of A.*
4. *rank(A) is the highest order of a nonzero minor*[8] *of A.*
5. **Sylvester's inequality***: Let B be an $n \times k$ matrix. Then AB is an $m \times k$ matrix and*

$$rank(A) + rank(B) - n \leq rank(AB) \leq \min\{rank(A), rank(B)\}. \tag{A.33}$$

[8] A minor of order $k$ is the determinant of a $k \times k$ submatrix of $A$.

*6. For any nonsingular matrices $C \in \mathbb{R}^{m \times m}$ and $D \in \mathbb{R}^{n \times n}$, we have*

$$\text{rank}(A) = \text{rank}(CAD). \tag{A.34}$$

In Matlab, the rank of a matrix $A$ is just `rank(A)`.

**Definition A.24 (Orthogonal complement to a subspace).** *Given a subspace $S$ of $\mathbb{R}^n$, we define its* orthogonal complement *to be the subspace $S^\perp \subseteq \mathbb{R}^n$ such that $x \in S^\perp$ if and only if $x^T y = 0$ for all $y \in S$. We write $\mathbb{R}^n = S \oplus S^\perp$.*

The notion of orthogonal complement is used in this book to define the "coimage" of an image of a point or a line. Also, with respect to any linear map $A$ from $\mathbb{R}^n$ to $\mathbb{R}^m$, the space $\mathbb{R}^n$ can be decomposed as a direct sum of two subspaces,

$$\mathbb{R}^n = \text{null}(A) \oplus \text{null}(A)^\perp,$$

and $\mathbb{R}^m$ can be decomposed similarly as

$$\mathbb{R}^m = \text{range}(A) \oplus \text{range}(A)^\perp.$$

We also have the following not so obvious relationships:

**Theorem A.25.** *Let $A$ be a linear map from $\mathbb{R}^n$ to $\mathbb{R}^m$. Then:*

*(a) $\text{null}(A)^\perp = \text{range}(A^T)$,*

*(b) $\text{range}(A)^\perp = \text{null}(A^T)$,*

*(c) $\text{null}(A^T) = \text{null}(AA^T)$,*

*(d) $\text{range}(A) = \text{range}(AA^T)$.*

*Proof.* To prove part *c*: $\text{null}(AA^T) = \text{null}(A^T)$, we have

- $AA^T x = 0 \Rightarrow \langle x, AA^T x\rangle = \|A^T x\|^2 = 0 \Rightarrow A^T x = 0$, hence $\text{null}(AA^T) \subseteq \text{null}(A^T)$.
- $A^T x = 0 \Rightarrow AA^T x = 0$; hence $\text{null}(AA^T) \supseteq \text{null}(A^T)$.

To prove part *d*, $\text{range}(AA^T) = \text{range}(A)$, we first need to prove that $\mathbb{R}^n$ is a direct sum of $\text{range}(A^T)$ and $\text{null}(A)$, i.e. part *a* of the theorem. Part *b* can then be proved similarly. We prove this by showing that a vector $x$ is in $\text{null}(A)$ if and only if it is orthogonal to $\text{range}(A^T)$: $x \in \text{null}(A) \Leftrightarrow \langle Ax, y\rangle = 0, \forall y \Leftrightarrow \langle x, A^T y\rangle = 0, \forall y$. Hence $\text{null}(A)$ is exactly the subspace that is the orthogonal complement to $\text{range}(A^T)$ (denoted by $\text{range}(A^T)^\perp$). Therefore, $\mathbb{R}^n$ is a direct sum of $\text{range}(A^T)$ and $\text{null}(A)$. Now to complete our proof of part *d*, let $\text{Img}_A(S)$ denote the image of a subspace $S$ under the map $A$. Then we have $\text{range}(A) = \text{Img}_A(\mathbb{R}^n) = \text{Img}_A(\text{range}(A^T)) = \text{range}(AA^T)$ (in the second equality we used the fact that $\mathbb{R}^n$ is a direct sum of $\text{range}(A^T)$ and $\text{null}(A)$). These relations are depicted by Figure A.1. □

In fact, the same result holds even if the domain of the linear map $A$ is replaced by an infinite-dimensional linear space with an inner product (i.e. $\mathbb{R}^n$ is replaced by a Hilbert space). In that case, this theorem is also known as the *finite-rank*

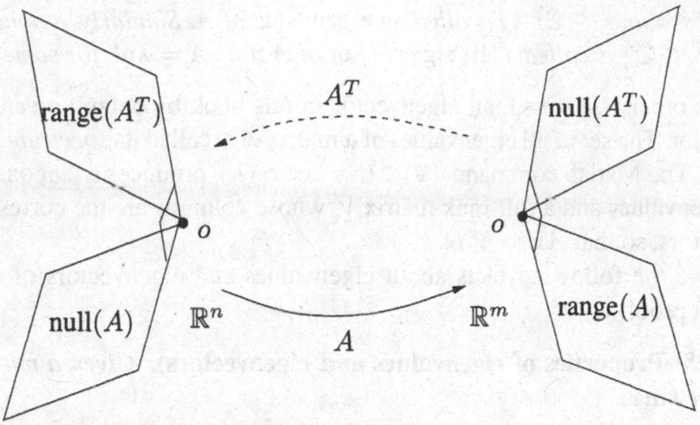


Figure A.1. The orthogonal decomposition of the domain and codomain of a linear map $A$.

*operator fundamental lemma* [Callier and Desoer, 1991]. We will later use this result to prove the singular value decomposition. But already it implies a result that is extremely useful in the study of multiple-view geometry:

**Lemma A.26 (Rank reduction lemma).** *Let $A \in \mathbb{R}^{n\times n}$ be a matrix and let $W$ be a matrix of the form*

$$W = \begin{bmatrix} M & 0 \\ AB & AA^T \end{bmatrix} \in \mathbb{R}^{(m+n)\times(k+n)} \tag{A.35}$$

*for some matrices $M \in \mathbb{R}^{m\times k}$ and $B \in \mathbb{R}^{n\times k}$. Then, regardless of what $B$ is, we always have*

$$\boxed{rank(M) = rank(W) - rank(A).} \tag{A.36}$$

The proof is easy using the fact range$(AB) \subseteq$ range$(A) =$ range$(AA^T)$ with the second identity from the previous theorem, and we leave the rest of the proof to the reader as an exercise.

A linear map from $\mathbb{R}^n$ to itself is represented by a square $n \times n$ matrix $A$. For such a map, we sometimes are interested in subspaces of $\mathbb{R}^n$ that are "invariant" under the map.[9] This notion turns out to be closely related to the *eigenvectors* of the matrix $A$.

**Definition A.27 (Eigenvalues and eigenvectors of a matrix).** *Let $A$ be an $n \times n$ complex matrix*[10] *in $\mathbb{C}^{n\times n}$. A nonzero vector $v \in \mathbb{C}^n$ is said to be its (right)* eigenvector *if*

$$Av = \lambda v \tag{A.37}$$

[9] More rigorously speaking, a subspace $S \subset \mathbb{R}^n$ is $A$-invariant if $A(S) \subseteq S$.

[10] Although $A$ will mostly be a real matrix in this book, to talk about its eigenvectors, it is more convenient to think of it as a complex matrix (with all entries that happen to be real).

*for some scalar* $\lambda \in \mathbb{C}$; $\lambda$ *is called* an eigenvalue *of* $A$. *Similarly, a nonzero row vector* $\eta^T \in \mathbb{C}^n$ *is called* a left eigenvector *of* $A$ *if* $\eta^T A = \lambda \eta^T$ *for some* $\lambda \in \mathbb{C}$.

Unless otherwise stated, an eigenvector in this book by default means a right eigenvector. The set of all eigenvalues of a matrix $A$ is called its *spectrum*, denoted by $\sigma(A)$. The Matlab command `[V,D] = eig(A)` produces a diagonal matrix $D$ of eigenvalues and a full-rank matrix $V$ whose columns are the corresponding eigenvectors, so that $AV = VD$.

We give the following facts about eigenvalues and eigenvectors of a matrix without a proof.

**Fact A.28 (Properties of eigenvalues and eigenvectors).** *Given a matrix* $A \in \mathbb{R}^{n \times n}$, *we have:*

1. *If* $Av = \lambda v$, *then for the same eigenvalue* $\lambda$, *there also exists a left eigenvector* $\eta^T$ *such that* $\eta^T A = \lambda \eta^T$ *and vice versa. Hence* $\sigma(A) = \sigma(A^T)$.

2. *The eigenvectors of* $A$ *associated with different eigenvalues are linearly independent.*

3. *All its eigenvalues* $\sigma(A)$ *are the roots of the (characteristic) polynomial equation* $\det(\lambda I - A) = 0$. *Hence* $\det(A)$ *is equal to the product of all eigenvalues of* $A$.

4. *If* $B = PAP^{-1}$ *for some nonsingular matrix* $P$, *then* $\sigma(B) = \sigma(A)$.

5. *If* $A$ *is a real matrix, then* $\lambda \in \mathbb{C}$ *is an eigenvalue implies that its conjugate* $\bar{\lambda}$ *is also an eigenvalue. Simply put,* $\sigma(A) = \bar{\sigma}(A)$ *for real matrices.*

## A.5 Symmetric matrices and skew-symmetric matrices

**Definition A.29 (Symmetric matrix).** *A matrix* $S \in \mathbb{R}^{n \times n}$ *is called* symmetric *if* $S^T = S$. *A symmetric matrix* $S$ *is called* positive (semi-)definite, *if* $x^T S x > 0$ *(or* $x^T S x \geq 0$*) for all* $x \in \mathbb{R}^n$, *denoted by* $S > 0$ *(or* $S \geq 0$*).*

**Fact A.30 (Properties of symmetric matrices).** *If* $S$ *is a real symmetric matrix, then:*

1. *All eigenvalues of* $S$ *must be real, i.e.* $\sigma(S) \subset \mathbb{R}$.

2. *Let* $(\lambda, v)$ *be an eigenvalue-eigenvector pair. If* $\lambda_i \neq \lambda_j$, *then* $v_i \perp v_j$; *i.e. eigenvectors corresponding to distinct eigenvalues are orthogonal.*

3. *There always exist* $n$ *orthonormal eigenvectors of* $S$, *which form a basis for* $\mathbb{R}^n$.

4. $S > 0$ $(S \geq 0)$ *if* $\lambda_i > 0$ $(\lambda_i \geq 0)$ $\forall i = 1, 2, \ldots, n$; *i.e.* $S$ *is positive (semi-)definite if all eigenvalues are positive (nonnegative).*

5. *If* $S \geq 0$ *and* $\lambda_1 \geq \lambda_2 \geq \cdots \geq \lambda_n$, *then* $\max_{\|x\|_2=1}\langle x, Sx\rangle = \lambda_1$ *and* $\min_{\|x\|_2=1}\langle x, Sx\rangle = \lambda_n$.

From point 3, we see that if $V = [v_1, v_2, \ldots, v_n] \in \mathbb{R}^{n\times n}$ is the matrix of all the eigenvectors, and $\Lambda = \text{diag}\{\lambda_1, \lambda_2, \ldots, \lambda_n\}$ is the diagonal matrix of the corresponding eigenvalues, then we can write

$$S = V\Lambda V^T,$$

where $V$ is an orthogonal matrix. In fact, $V$ can be further chosen to be in $SO(n)$ (i.e. of determinant $+1$) if $n$ is odd, since $V\Lambda V^T = (-V)\Lambda(-V)^T$ and $\det(-V) = (-1)^n \det(V)$.

**Definition A.31 (Induced 2-norm of a matrix).** *Let* $A \in \mathbb{R}^{m\times n}$. *We define* the induced 2-norm *of* $A$ *(as a linear map from* $\mathbb{R}^n$ *to* $\mathbb{R}^m$*) as*

$$\|A\|_2 \doteq \max_{\|x\|_2=1} \|Ax\|_2 = \max_{\|x\|_2=1} \sqrt{\langle x, A^T Ax\rangle}.$$

Similarly, other induced operator norms on $A$ can be defined starting from different norms on the domain and codomain spaces on which $A$ operates.

Let $A$ be as above. Then $A^T A \in \mathbb{R}^{n\times n}$ is clearly symmetric and positive semi-definite, so it can be diagonalized by a orthogonal matrix $V$. The eigenvalues, being nonnegative, can be written as $\sigma_i^2$. By ordering the columns of $V$ so that the eigenvalue matrix $\Lambda$ has decreasing eigenvalues on the diagonal, we see, from point 5 of the preceding fact, that $A^T A = V\text{diag}\{\sigma_1^2, \sigma_2^2, \ldots, \sigma_n^2\}V^T$ and

$$\|A\|_2 = \sigma_1.$$

The induced 2-norm of a matrix $A \in \mathbb{R}^{m\times n}$ is different from the "2-norm" of $A$ viewed as a vector in $\mathbb{R}^{mn}$. To distinguish them, the latter one is conventionally called the *Frobenius norm* of $A$, precisely defined as $\|A\|_f = \sqrt{\sum_{i,j} a_{ij}^2}$. Notice that $\sum_{i,j} a_{ij}^2$ is nothing but the trace of $A^T A$ (or $AA^T$). Thus, we have

$$\|A\|_f \doteq \sqrt{\text{trace}(A^T A)} = \sqrt{\sigma_1^2 + \sigma_2^2 + \cdots + \sigma_n^2}.$$

The inverse problem of retrieving $A$ from the symmetric matrix $S = A^T A$ is usually solved by *Cholesky facotorization*. For the given $S$, its eigenvalues must be nonnegative. Thus, we have $S = V\Lambda V^T = A^T A$ for $A = \Lambda^{(\frac{1}{2})}V^T$, where $\Lambda^{(\frac{1}{2})} = \text{diag}\{\sigma_1, \sigma_2, \ldots, \sigma_n\}$ is the "square root" of the diagonal matrix $\Lambda$. Since $R^T R = I$ for any orthogonal matrix, the solution for $A$ is not unique: $RA$ is also a solution. Cholesky factorization then restricts the solution to be an upper-triangular matrix (exactly what we need for camera calibration in Chapter 6). In Matlab, the Cholesky factorization is given by the command `A = chol(S)`.

**Definition A.32 (Skew-symmetric (or antisymmetric) matrix).** *A matrix* $A \in \mathbb{R}^{n\times n}$ *is called* skew-symmetric *(or* antisymmetric*) if* $A^T = -A$.

**Fact A.33 (Properties of a skew-symmetric matrix).** *If* $A$ *is a real skew-symmetric matrix, then:*

1. *All eigenvalues of $A$ are either zero or purely imaginary, i.e. of the form $i\omega$ for $i = \sqrt{-1}$ and some $\omega \in \mathbb{R}$.*

2. *There exists an orthogonal matrix $V$ such that*

$$A = V\Lambda V^T, \tag{A.38}$$

*where $\Lambda$ is a block-diagonal matrix $\Lambda = diag\{A_1, \ldots, A_m, 0, \ldots, 0\}$, where each $A_i$ is a $2 \times 2$ real skew-symmetric matrix of the form*

$$A_i = \begin{bmatrix} 0 & a_i \\ -a_i & 0 \end{bmatrix} \in \mathbb{R}^{2\times 2}, \quad i = 1, 2, \ldots, m. \tag{A.39}$$

From point 2, we conclude that the rank of any skew-symmetric matrix must be even. A commonly used skew-symmetric matrix in computer vision is associated with a vector $u \in \mathbb{R}^3$, denoted by

$$\widehat{u} = \begin{bmatrix} 0 & -u_3 & u_2 \\ u_3 & 0 & -u_1 \\ -u_2 & u_1 & 0 \end{bmatrix} \in \mathbb{R}^{3\times 3}. \tag{A.40}$$

The reason for such a definition is that $\widehat{u}v$ is equal to the conventional cross product $u \times v$ of two vectors in $\mathbb{R}^3$. Then we have $\text{rank}(\widehat{u}) = 2$ if $u \neq 0$ and the (left and right) null space of $\widehat{u}$ is exactly spanned by the vector $u$ itself. That is, $\widehat{u}u = 0$ and $u^T\widehat{u} = 0$. In other words, columns and rows of the matrix $\widehat{u}$ are always orthogonal to $u$.

Obviously, $A^T\widehat{u}A$ is also a skew-symmetric matrix. Then $A^T\widehat{u}A = \widehat{v}$ for some $v \in \mathbb{R}^3$. We want to know what the relationship between $v$ and $A, u$ is.

**Fact A.34 (Hat operator).** *If $A$ is a $3 \times 3$ matrix of determinant 1, then we have*

$$A^T\widehat{u}A = \widehat{A^{-1}u}. \tag{A.41}$$

This is an extremely useful fact, which will be extensively used in our book. For example, this property allows us to "push" a matrix through a skew-symmetric matrix in the following way: $\widehat{u}A = A^{-T}A^T\widehat{u}A = A^{-T}\widehat{A^{-1}u}$. We leave to the reader as an exercise to think about how this result needs to be modified when the determinant of $A$ is not 1, or when $A$ is not even invertible.

## A.6 Lyapunov map and Lyapunov equation

An important type of linear equation that we will encounter in our book is of Lyapunov type:[11] Find a matrix $X \in \mathbb{C}^{n\times n}$ that satisfies the equation

$$AX + XB = 0 \tag{A.42}$$

[11] It is also called Sylvester equation in some literature.

for a given pair of matrices $A, B \in \mathbb{C}^{n\times n}$. Although solutions to this type of equation can be difficult in general, simple solutions exist when both $A$ and $B$ have $n$ independent eigenvectors. Suppose $\{u_i \in \mathbb{C}^n\}_{i=1}^n$ are the $n$ right eigenvectors of $A$, and $\{v_j \in \mathbb{C}^n\}_{j=1}^n$ are the $n$ left eigenvectors of $B$; i.e.

$$Au_i = \lambda_i u_i; \quad v_j^* B = \eta_j v_j^* \tag{A.43}$$

for eigenvalues $\lambda_i, \eta_j$ for each $i, j$. Here $v^*$ means the complex-conjugate and transpose of $v$, since $v$ can be complex.

**Fact A.35 (Lyapunov map).** *For the above matrix A and B, the $n^2$ eigenvectors of the Lyapunov map*

$$L: X \mapsto AX + XB \tag{A.44}$$

*are exactly $X_{ij} = u_i v_j^* \in \mathbb{C}^{n\times n}$, and the corresponding eigenvalues are $\lambda_i + \eta_j \in \mathbb{C}$, $i, j = 1, 2, \ldots, n$.*

*Proof.* The $n^2$ matrices $\{X_{ij}\}_{i,j=1}^n$ are linearly independent, and they must be all the eigenvectors of $L$. □

Due to this fact, any matrix $X$ that satisfies the Lyapunov equation $AX+XB = 0$ must be in the subspace spanned by eigenvectors $X_{ij}$ that have zero eigenvalues: $\lambda_i + \eta_j = 0$. In Matlab, the command `X = lyap(A,B,C)` solves the more general Lyapunov equation $AX + XB = -C$.

In this book, we often look for solutions $X$ with extra requirements on its structure. For instance, $X$ needs to be real and symmetric (Chapter 6), or $X$ has to be a rotation matrix (Chapter 10). If so, we have only to take the intersection of the space of solutions to the Lyapunov equation with the space of symmetric matrices or rotation matrices.

## A.7 The singular value decomposition (SVD)

The singular value decomposition (SVD) is a useful tool to capture essential features of a matrix (that represents a linear map), such as its rank, range space, null space, and induced norm, as well as to "generalize" the concept of "eigenvalue– eigenvector" pair to non-square matrices. The computation of the SVD is numerically well conditioned, making it extremely useful for solving many linear-algebraic problems such as matrix inversion, calculation of the rank, linear least-squares estimate, projections, and fixed-rank approximations. Since this book will use the SVD quite extensively, here we give a detailed proof of its properties.

### A.7.1 *Algebraic derivation*

Given a matrix $A \in \mathbb{R}^{m\times n}$, we have the following theorem.

**Theorem A.36 (Singular value decomposition of a matrix).** *Let $A \in \mathbb{R}^{m \times n}$ have rank p. Furthermore, suppose, without loss of generality, that $m \geq n$. Then*

- *$\exists U \in \mathbb{R}^{m \times p}$ whose columns are orthonormal,*
- *$\exists V \in \mathbb{R}^{n \times p}$ whose columns are orthonormal, and*
- *$\exists \Sigma \in \mathbb{R}^{p \times p}, \Sigma = diag\{\sigma_1, \sigma_2, \cdots, \sigma_p\}$ diagonal with $\sigma_1 \geq \sigma_2 \geq \cdots \geq \sigma_p$*

*such that $A = U\Sigma V^T$.*

*Proof.* We prove the claim by construction.

- Compute $A^T A$: it is symmetric and positive semi-definite of dimension $n \times n$. Then order its eigenvalues in decreasing order and call them $\sigma_1^2 \geq \cdots \geq \sigma_p^2 \geq \cdots \sigma_n^2 \geq 0$. Call the $\sigma_i$'s *singular values*.
- From an orthonormal set of eigenvectors of $A^T A$ create an orthonormal basis for $\mathbb{R}^n$ such that

  $$\text{span}\{v_1, v_2, \dots, v_p\} = \text{range}(A^T), \quad \text{span}\{v_{p+1}, \dots, v_n\} = \text{null}(A).$$

  Note that the latter eigenvectors correspond to the zero singular values, since $\text{null}(A^T A) = \text{null}(A)$ (according to Theorem A.25).
- Define $u_i$ such that $Av_i = \sigma_i u_i, \forall i = 1, 2, \dots, p$, and see that the set $\{u_i\}_{i=1}^p$ is orthonormal (proof left as exercise).
- Complete the basis $\{u_i\}_{i=1}^p$, which spans $\text{range}(A)$ (by construction), to all $\mathbb{R}^m$.
- Then,

  $$A[v_1, v_2, \dots, v_n] = [u_1, u_2, \dots, u_m] \begin{bmatrix} \sigma_1 & 0 & \cdots & \cdots & \cdots & 0 \\ 0 & \sigma_2 & \cdots & \cdots & \cdots & 0 \\ \vdots & \vdots & \ddots & \vdots & \vdots & \vdots \\ \vdots & \vdots & \vdots & \sigma_p & \vdots & \vdots \\ \vdots & \vdots & \vdots & \vdots & \ddots & \vdots \\ 0 & 0 & 0 & \vdots & \vdots & 0_n \\ \vdots & \vdots & \vdots & \vdots & \vdots & \vdots \\ 0 & 0 & \cdots & \cdots & \cdots & 0_m \end{bmatrix},$$

  which we name $A\tilde{V} = \tilde{U}\tilde{\Sigma}$.
- Hence, $A = \tilde{U}\tilde{\Sigma}\tilde{V}^T$.

Then the claim follows by deleting the columns of $\tilde{U}$ and the rows of $\tilde{V}^T$ that multiply the zero singular values. □

In Matlab, to compute the SVD of a given $m \times n$ matrix $A$, simply use the command `[U,S,V] = svd(A)`, which returns matrices $U, S, V$ satisfying $A = USV^T$ (where $S$ replaces $\Sigma$). Notice that in the standard SVD routine, the orthogonal matrices $U$ and $V$ do not necessarily have determinant $+1$. So the reader should exercise extra caution when using the SVD to characterize the essential matrix in epipolar geometry in this book.

## A.7.2 Geometric interpretation

Notice that in the SVD of a square matrix $A = U\Sigma V^T \in \mathbb{R}^{n\times n}$, columns of $U = [u_1, u_2, \ldots, u_n]$ and columns of $V = [v_1, v_2, \ldots, v_n]$ form orthonormal bases for $\mathbb{R}^n$. The SVD essentially states that if $A$ (as a linear map) maps a point $x$ to $y$, then coordinates of $y$ with respect to the basis $U$ are related to coordinates of $x$ with respect to the basis $V$ by the diagonal matrix $\Sigma$ that scales each coordinate by the corresponding singular value.

**Theorem A.37.** *Let $A \in \mathbb{R}^{n\times n} = U\Sigma V^T$ be a square matrix. Then $A$ maps the unit sphere $\mathbb{S}^{n-1} \doteq \{x \in \mathbb{R}^n : \|x\|_2 = 1\}$ to an ellipsoid with semi-axes $\sigma_i u_i$, where $u_i$ is the $i$th column of $U$.*

*Proof.* let $x, y$ be such that $Ax = y$. The set $\{v_i\}_{i=1}^n$ is an orthonormal basis for $\mathbb{R}^n$. With respect to such a basis $x$ has coordinates

$$[\alpha_1, \alpha_2, \ldots, \alpha_n]^T = [\langle v_1, x\rangle, \langle v_2, x\rangle, \ldots, \langle v_n, x\rangle]^T.$$

That is, $x = \sum_{i=1}^n \alpha_i v_i$. With respect to the basis $\{u_i\}_{i=1}^n$, $y$ has coordinates

$$[\beta_1, \beta_2, \ldots, \beta_n]^T = [\langle u_1, y\rangle, \langle u_2, y\rangle, \ldots, \langle u_n, y\rangle]^T.$$

We also have $y = \sum_{i=1}^n \beta_i u_i = Ax = \sum_{i=1}^n \sigma_i u_i v_i^T x = \sum_{i=1}^n \sigma_i \langle v_i, x\rangle u_i$. Hence $\sigma_i \alpha_i = \beta_i$. Now $\|x\|_2^2 = \sum_{i=1} \alpha_i^2 = 1$, $\forall x \in \mathbb{S}^{n-1}$, and so we have $\sum_{i=1}^n \beta_i^2/\sigma_i^2 = 1$, which implies that the point $y$ satisfies the equation of an ellipsoid with semi-axes of length $\sigma_i$. This is illustrated in Figure A.2 for the case $n = 2$. □

## A.7.3 Some properties of the SVD

Problems involving orthogonal projections onto invariant subspaces of $A$, such as the linear least-squares (LLS) problem, can be easily solved using the SVD.

**Definition A.38 (Generalized (Moore Penrose) inverse).** *Given a matrix $A \in \mathbb{R}^{m\times n}$ of rank $r$ with its SVD $A = U\Sigma V^T$, we then define* the generalized inverse *of $A$ to be*

$$A^\dagger = V\Sigma^\dagger U^T, \quad \textit{where} \quad \Sigma^\dagger = \begin{bmatrix} \Sigma_1^{-1} & 0 \\ 0 & 0 \end{bmatrix}_{n\times m}.$$

*The generalized inverse is sometimes also called the* pseudo-inverse.

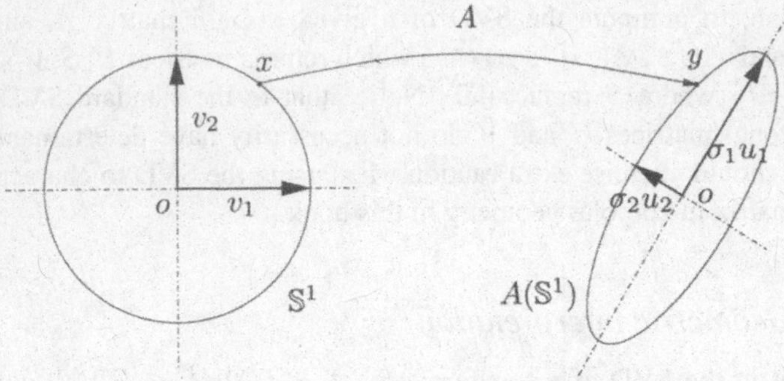


Figure A.2. The image of a unit sphere on the left under a nonsingular map $A \in \mathbb{R}^{2\times 2}$ is an ellipsoid on the right.

In Matlab, the pseudo-inverse of a matrix is computed by the command `X = pinv(A)`.

**Fact A.39 (Properties of generalized inverse).**

- $AA^\dagger A = A, \quad A^\dagger A A^\dagger = A^\dagger.$

The generalized inverse can then be used to solve linear equations in general.

**Proposition A.40 (Least-squares solution of a linear systems).** *Consider the problem $Ax = b$ with $A \in \mathbb{R}^{m\times n}$ of rank $r \leq \min(m, n)$. The solution $x^*$ that minimizes $\|Ax - b\|_2$ is given by $x^* = A^\dagger b$.*

The following two results have something to do with the sensitivity of solving linear equations of the form $Ax = b$.

**Proposition A.41 (Perturbations).** *Consider a nonsingular matrix $A \in \mathbb{R}^{n\times n}$. Let $\delta A$ be a full-rank perturbation. Then*

- $|\sigma_k(A + \delta A) - \sigma_k(A)| \leq \sigma_1(\delta A), \quad \forall k = 1, 2, \ldots, n,$
- $\sigma_n(A\delta A) \geq \sigma_n(A)\sigma_n(\delta A),$
- $\sigma_1(A^{-1}) = \frac{1}{\sigma_n(A)},$

*where $\sigma_i$ denotes the ith singular value.*

**Proposition A.42 (Condition number).** *Consider the problem $Ax = b$, and consider a "perturbed" full-rank problem $(A+\delta A)(x+\delta x) = b$. Since $Ax = b$, then to first-order approximation, $\delta x = -A^\dagger \delta A x$. Hence $\|\delta x\|_2 \leq \|A^\dagger\|_2 \|\delta A\|_2 \|x\|_2$, from which*

$$\frac{\|\delta x\|_2}{\|x\|_2} \leq \|A^\dagger\|_2 \|A\|_2 \frac{\|\delta A\|_2}{\|A\|_2} \doteq k(A) \frac{\|\delta A\|_2}{\|A\|_2},$$

*where $k(A) = \|A^\dagger\|_2 \|A\|_2$ is called the* condition number *of A. It easy to see that $k(A) = \sigma_1/\sigma_n$ if A is invertible.*

Last but not the least, one of the most important properties of the SVD is related to a fixed-rank approximation of a given matrix. Given a matrix $A$ of rank $r$, we want to find a matrix $B$ such that it has fixed rank $p < r$ and the Frobenius norm of the difference $\|A - B\|_f$ is minimal. The solution to this problem is given simply by setting all but the first $p$ singular values to zero

$$B \doteq U\Sigma_{(p)}V^T,$$

where $\Sigma_{(p)}$ denotes the matrix obtained from $\Sigma$ by setting to zero its elements on the diagonal after the $p$th entry. The matrix $B$ has exactly the same induced 2-norm of $A$, i.e. $\sigma_1(A) = \sigma_1(B)$, and satisfies the requirement on the rank.

**Proposition A.43 (Fixed rank approximation).** *Let $A, B$ be defined as above. Then $\|A - B\|_f^2 = \sigma_{p+1}^2 + \cdots + \sigma_r^2$. Furthermore, such a norm is the minimum achievable.*

The proof is an easy exercise that follows directly from the properties of orthogonal projection and the properties of the SVD given above.

# Appendix B
## Least-Variance Estimation and Filtering

*The theory of probability as a mathematical discipline can and should be developed from axioms in exactly the same way as geometry and algebra.*
– Andrey Kolmogorov, *Foundations of the Theory of Probability*

In Chapter 12 we formulated the problem of estimating 3-D structure and motion in a causal fashion as a filtering problem. Since the measurements are corrupted by noise, and there is uncertainty in the model, the task is naturally formulated in a probabilistic framework. Statistical estimation theory is a vast subject, which we have no hope of covering to any reasonable degree here. Instead, this appendix reviews some of the basic concepts of estimation theory as they pertain to Chapter 12, but in line with the spirit of the book, we formulate them in geometric terms, as orthogonal projections onto certain linear spaces.

We assume that the reader is familiar with the basic notions of random variables, random vectors, random processes, and (conditional) expectation. However, since we are going to rephrase most results in geometric terms, a summary knowledge suffices. The reader interested in a more thorough introduction to these topics can consult, for instance, [van Trees, 1992, Stark and Woods, 2002].

Throughout this appendix, $x$ typically indicates the unknown quantity of interest, whereas $y$ indicates the measured quantity. An estimator, in general, seeks to infer properties of $x$ from measurements of $y$. For the sake of clarity, we indicate vector quantities in boldface, for instance, $\boldsymbol{x}$, whereas scalars are in normal font; for instance, the components of $\boldsymbol{y}$ are $y_i$. A "hat" usually denotes an estimated quantity;[1] for instance, $\hat{x}$ indicates the estimate of $x$. Let us start with the simplest case of least-variance estimation.

# B.1 Least-variance estimators of random vectors

Let $x \mapsto y$ be a map between two spaces of random vectors with samples in $\mathbb{R}^m$ and $\mathbb{R}^n$ (the "model generating the data"). We are interested in building an estimator for the random vector $x$, given measurements of samples of the random vector $y$. An estimator is a function[2] $T : \mathbb{R}^m \to \mathbb{R}^n;\ y \mapsto \hat{x} = T(y)$, that solves an optimization problem of the form

$$\hat{T} \doteq \arg\min_{T\in\mathcal{T}} \mathcal{C}(x - T(y)), \tag{B.1}$$

where $\mathcal{T}$ is a suitably chosen class of functions and $\mathcal{C}(\cdot)$ some cost in the $x$-space $\mathbb{R}^n$.

We concentrate on one of the simplest possible choices, which corresponds to *affine least-variance* estimators,

$$\mathcal{T} \doteq \{A \in \mathbb{R}^{n\times m}; b \in \mathbb{R}^n \mid T(y) = Ay + b\}, \tag{B.2}$$

$$\mathcal{C}(\cdot) \doteq E\left[\|\cdot\|^2\right], \tag{B.3}$$

where the latter operator takes the expectation of the squared Euclidean norm of a random vector. Therefore, we seek

$$\hat{A}, \hat{b} \doteq \arg\min_{A,b} E\left[\|x - (Ay + b)\|^2\right]. \tag{B.4}$$

We set $\mu_x \doteq E[x] \in \mathbb{R}^n$ and $\Sigma_x \doteq E[xx^T] \in \mathbb{R}^{n\times n}$, and similarly for $y$. First notice that if $\mu_x = \mu_y = 0$, then $\hat{b} = 0$, and the affine least-variance estimator reduces to a *linear* one. Now observe that, if we call the centered vectors $\bar{x} \doteq x - \mu_x$ and $\bar{y} \doteq y - \mu_y$, we have

$$\begin{aligned} E\left[\|x - Ay - b\|^2\right] &= E\left[\|A\bar{y} - \bar{x} + (A\mu_y + b - \mu_x)\|^2\right] \\ &= E\left[\|\bar{x} - A\bar{y}\|^2\right] + \|A\mu_y + b - \mu_x\|^2. \end{aligned} \tag{B.5}$$

Hence, if we assume for a moment that we have found $\hat{A}$ that solves the problem (B.4), then trivially

$$\hat{b} = \mu_x - \hat{A}\mu_y \tag{B.6}$$

annihilates the second term of equation (B.5). Therefore, without loss of generality, we can restrict our attention to finding $A$ that minimizes the first term of equation (B.5):

$$\hat{A} \doteq \arg\min_{A} E\left[\|\bar{x} - A\bar{y}\|^2\right]. \tag{B.7}$$

[1] This is not to be confused with the symbol "wide had" used to indicate a skew-symmetric matrix, for example $\widehat{\omega}$.

[2] This "$T$" is not to be confused with a translation vector.

Therefore, we will concentrate on the case $\mu_x = \mu_y = 0$ without loss of generality. In other words, if we know how to solve for the *linear* least-variance estimator, we also know how to solve for the *affine* one.

### B.1.1 Projections onto the range of a random vector

The set of all random variables $z_i$ defined on the same probability space, with *zero mean* $E[z_i] = 0$ and *finite variance* $\Sigma_{z_i} < \infty$, is a *linear space*, and in particular, it is a *Hilbert space* with the inner product given by

$$\langle z_i, z_j \rangle_{\mathcal{H}} \doteq \Sigma_{z_i z_j} = E[z_i z_j]. \tag{B.8}$$

Therefore, we can exploit everything we know about vectors, distances, angles, and orthogonality, as we learned in Chapter 2 and Appendix A. In particular, in this space the notion of orthogonality corresponds to the notion of *uncorrelatedness*. The components of a random vector $\boldsymbol{y} = [y_1, y_2, \ldots, y_m]^T$ define a subspace of such a Hilbert space:

$$\mathcal{H}(\boldsymbol{y}) = \text{span}\,[y_1, y_2, \ldots, y_m], \tag{B.9}$$

where the span is intended over the set of real numbers.[3] We say that the subspace $\mathcal{H}(\boldsymbol{y})$ has *full rank* if $\Sigma_{\boldsymbol{y}} = E[\boldsymbol{y}\boldsymbol{y}^T] > 0$.

The structure of a Hilbert space allows us to make use of the concept of *orthogonal projection* of a random variable, say $x \in \mathbb{R}$, onto the span of a random vector, say $\boldsymbol{y} \in \mathbb{R}^m$. It is the random variable $\hat{x} \in \mathbb{R}$ that satisfies the following canonical equations:

$$\begin{aligned} \hat{x} = \text{pr}_{\langle \mathcal{H}(\boldsymbol{y}) \rangle}(x) \quad &\Leftrightarrow \quad \langle x - \hat{x}, z \rangle_{\mathcal{H}} = 0, \quad \forall\, z \in \mathcal{H}(\boldsymbol{y}); \\ &\Leftrightarrow \quad \langle x - \hat{x}, y_i \rangle_{\mathcal{H}} = 0, \quad \forall\, i = 1, 2, \ldots, m; \qquad \text{(B.10)} \\ &\doteq \quad \hat{E}[x|\boldsymbol{y}]. \qquad \text{(B.11)} \end{aligned}$$

The notation $\hat{x}(\boldsymbol{y}) = \hat{E}[x|\boldsymbol{y}]$ is often used for the projection of $x$ over the span of $\boldsymbol{y}$.[4]

### B.1.2 Solution for the linear (scalar) estimator

Let $\hat{x} = A\boldsymbol{y} \in \mathbb{R}$ be a linear estimator for the random *variable* $x \in \mathbb{R}$; $A \in \mathbb{R}^m$ is a row vector, and $\boldsymbol{y} \in \mathbb{R}^m$ an $m$-dimensional column vector. The least-squares[5] estimate $\hat{x}$ is given by the choice of $A$ that solves the following problem:

$$\hat{A} = \arg\min_{A} \|A\boldsymbol{y} - x\|^2_{\mathcal{H}}, \tag{B.12}$$

---

[3] So $\mathcal{H}(\boldsymbol{y})$ is the space of random variables that are linear combinations of $y_i$, $i = 1, 2, \ldots, m$.

[4] The resemblance of this notation to the symbol that denotes conditional expectation is due to the fact that for Gaussian random vectors, such a projection is indeed the conditional expectation.

[5] Note that now that we have the structure of a Hilbert space, we can talk about least-variance estimators as least-squares estimators, since they seek to minimize the squared norm induced by the new inner product.

where $\|\cdot\|_{\mathcal{H}}^2 = E\left[\|\cdot\|^2\right]$ is the norm induced by the inner product $\langle\cdot,\cdot\rangle_{\mathcal{H}}$. That is, $\|z\|_{\mathcal{H}} \doteq \sqrt{\langle z, z\rangle_{\mathcal{H}}}$.

**Proposition B.1.** *The solution $\hat{x} = \hat{A}\boldsymbol{y}$ to the problem (B.12) exists, is unique, and corresponds to the orthogonal projection of $x$ onto the span of $\boldsymbol{y}$:*

$$\hat{x} = \mathrm{pr}_{\langle\mathcal{H}(\boldsymbol{y})\rangle}(x). \tag{B.13}$$

The proof is an easy exercise. Here we report an explicit construction of the best estimator $\hat{A}$. From substituting the expression of the estimator into the definition of orthogonal projection (B.11), we get

$$0 = \langle x - \hat{A}\boldsymbol{y}, y_i\rangle_{\mathcal{H}} = E[(x - \hat{A}\boldsymbol{y})y_i], \tag{B.14}$$

which holds if and only if $E[xy_i] = \hat{A}E[\boldsymbol{y}y_i]$, $\forall\, i = 1, 2, \ldots, m$. In row-vector notation we write

$$\begin{aligned} E[x\boldsymbol{y}^T] &= \hat{A}E[\boldsymbol{y}\boldsymbol{y}^T], \\ \Sigma_{x\boldsymbol{y}} &= \hat{A}\Sigma_{\boldsymbol{y}}, \end{aligned} \tag{B.15}$$

which, provided that $\mathcal{H}(\boldsymbol{y})$ is of full rank, gives $\hat{A} = \Sigma_{x\boldsymbol{y}}\Sigma_{\boldsymbol{y}}^{-1}$.

## *B.1.3 Affine least-variance estimator*

Suppose we want to compute the best estimator of a zero-mean *random vector* $\boldsymbol{x} \in \mathbb{R}^n$ as a *linear* map of the zero-mean random vector $\boldsymbol{y}$. We just have to repeat the construction reported in the previous section for each component $x_i$ of $\boldsymbol{x}$, so that the rows $\hat{A}_{i.} \in \mathbb{R}^{1\times m}$ of the now *matrix* $\hat{A} \in \mathbb{R}^{n\times m}$ are given by

$$\begin{aligned} \hat{A}_{1.} &= \Sigma_{x_1\boldsymbol{y}}\Sigma_{\boldsymbol{y}}^{-1}, \\ \vdots &= \vdots \\ \hat{A}_{n.} &= \Sigma_{x_n\boldsymbol{y}}\Sigma_{\boldsymbol{y}}^{-1}, \end{aligned} \tag{B.16}$$

which eventually gives us

$$\hat{A} = \Sigma_{\boldsymbol{x}\boldsymbol{y}}\Sigma_{\boldsymbol{y}}^{-1}. \tag{B.17}$$

If now the vectors $\boldsymbol{x}$ and $\boldsymbol{y}$ are not of zero mean, $\mu_{\boldsymbol{x}} \neq 0$ , $\mu_{\boldsymbol{y}} \neq 0$, we first transform the problem by defining $\bar{\boldsymbol{y}} \doteq \boldsymbol{y} - \mu_{\boldsymbol{y}}$, $\bar{\boldsymbol{x}} \doteq \boldsymbol{x} - \mu_{\boldsymbol{x}}$, then solve for the linear least-variance estimator $\hat{A} = \Sigma_{\bar{\boldsymbol{x}}\bar{\boldsymbol{y}}}\Sigma_{\bar{\boldsymbol{y}}}^{-1} \doteq \Sigma_{\boldsymbol{x}\boldsymbol{y}}\Sigma_{\boldsymbol{y}}^{-1}$, and then substitute to get

$$\hat{\boldsymbol{x}} = \mu_{\boldsymbol{x}} + \Sigma_{\boldsymbol{x}\boldsymbol{y}}\Sigma_{\boldsymbol{y}}^{-1}(\boldsymbol{y} - \mu_{\boldsymbol{y}}), \tag{B.18}$$

which is the least-variance affine estimator

$$\hat{\boldsymbol{x}} \doteq \hat{E}[\boldsymbol{x}|\boldsymbol{y}] = \hat{A}\boldsymbol{y} + \hat{b}, \tag{B.19}$$

where

$$\hat{A} = \Sigma_{xy}\Sigma_y^{-1}, \tag{B.20}$$
$$\hat{b} = \mu_x - \Sigma_{xy}\Sigma_y^{-1}\mu_y. \tag{B.21}$$

It is an easy exercise to compute the covariance of the estimation error $\tilde{x} \doteq x - \hat{x}$:

$$\Sigma_{\tilde{x}} = \Sigma_x - \Sigma_{xy}\Sigma_y^{-1}\Sigma_{yx}. \tag{B.22}$$

If we interpret the covariance of $x$ as the "prior uncertainty," and the covariance of $\tilde{x}$ as the "posterior uncertainty," we may interpret the second term (which is positive semi-definite) of the above equation as a "decrease" in the uncertainty.

### B.1.4 *Properties and interpretations of the least-variance estimator*

The covariance of the estimation error in equation (B.22) is by construction the smallest that can be achieved *with an affine estimator.* Of course, if we consider a broader class $\mathcal{T}$ of estimators, the estimation error can be further decreased, unless the model that generates the data $T$ is itself affine:

$$y = T(x) = Fx + w. \tag{B.23}$$

In such a case, it is easy to compute the expression of the optimal (affine) estimator that depends only upon $\Sigma_x$, $\Sigma_w$, and $F$,

$$\hat{x} = \Sigma_x F^T(F\Sigma_x F^T + \Sigma_w)^{-1}y, \tag{B.24}$$

which achieves a covariance of the estimation error equal to[6]

$$\Sigma_{\tilde{x}} = \Sigma_x - \Sigma_x F^T(F\Sigma_x F^T + \Sigma_w)^{-1}F\Sigma_x. \tag{B.25}$$

*Projection onto an orthogonal sum of subspaces*

Let $y = \begin{bmatrix} y_1 \\ y_2 \end{bmatrix} \in \mathbb{R}^{m_1+m_2}$, where $m_1 + m_2 = m$, be such that

$$\mathcal{H}(y) = \mathcal{H}(y_1) \oplus \mathcal{H}(y_2), \tag{B.26}$$

where "$\oplus$" indicates the direct sum of subspaces. It is important, for later developments, to understand under what conditions we can decompose the projection onto the span of $y$ as the sum of the projection onto its components $y_1$ and $y_2$:

$$\hat{E}[x|y] = \hat{E}[x|y_1] + \hat{E}[x|y_2]. \tag{B.27}$$

After an easy calculation one can see that the above is true if and only if $E[y_1 y_2^T] = 0$, that is, if and only if

$$\mathcal{H}(y_1) \perp \mathcal{H}(y_2). \tag{B.28}$$

[6]This expression can be manipulated using the *matrix inversion lemma*, which states that if $A, B, C, D$ are real matrices of the appropriate dimensions with $A$ and $C$ invertible, then $(A + BCD)^{-1} = A^{-1} - A^{-1}B(C^{-1} + DA^{-1}B)^{-1}DA$.

*Change of basis*

Suppose that instead of measuring the samples of a random vector $\boldsymbol{y} \in \mathbb{R}^m$, we measure another random vector $\boldsymbol{z} \in \mathbb{R}^m$ that is related to $\boldsymbol{y}$ via a change of basis: $\boldsymbol{z} = T\boldsymbol{y} \mid T \in GL(m)$. If we write $\hat{E}[\boldsymbol{x}|\boldsymbol{y}] = \hat{A}\boldsymbol{y}$, then it is immediate to see that

$$\begin{aligned}\hat{E}[\boldsymbol{x}|\boldsymbol{z}] &= \Sigma_{\boldsymbol{x}\boldsymbol{z}}\Sigma_{\boldsymbol{z}}^{-1}\boldsymbol{z} \\ &= \Sigma_{\boldsymbol{x}\boldsymbol{y}}T^T(T^{-T}\Sigma_{\boldsymbol{y}}T^{-1})\boldsymbol{z} \\ &= \Sigma_{\boldsymbol{x}\boldsymbol{y}}\Sigma_{\boldsymbol{y}}^{-1}T^{-1}\boldsymbol{z}.\end{aligned} \tag{B.29}$$

*Innovation*

The linear least-variance estimator involves the computation of the inverse of the output covariance matrix $\Sigma_{\boldsymbol{y}}$. It may be interesting to look for changes of bases $T$ that transform the output $\boldsymbol{y}$ into $\boldsymbol{z} = T\boldsymbol{y}$ such that $\Sigma_{\boldsymbol{z}} = I$. In such a case the optimal estimator is simply

$$\hat{E}[\boldsymbol{x}|\boldsymbol{z}] = \Sigma_{\boldsymbol{x}\boldsymbol{z}}\boldsymbol{z}. \tag{B.30}$$

Let us pretend for a moment that the components of the vector $\boldsymbol{y}$ are instances of a random process taken over time, $y_i = y(i)$, and call $\boldsymbol{y}^t = [y_1, y_2, \ldots, y_t]^T$ the history of the process *up to time* $t$. When we want to emphasize the (Hilbert) subspace spanned by the components, we also write $\boldsymbol{y}^t$ as $\mathcal{H}(\boldsymbol{y}^t)$, or equivalently $\mathcal{H}_t(\boldsymbol{y})$. Each component (sample) is an element of the Hilbert space $\mathcal{H}$, which has a well-defined notion of orthogonality, and where we can apply Gram-Schmidt procedure (Appendix A, Theorem A.19) in order to make the "vectors"[7] $y(i)$ orthogonal (i.e. uncorrelated; we neglect the subscript $\mathcal{H}$ from the norm and the inner product for simplicity):

$$\begin{array}{lcllcl} v_1 & \doteq & y(1) & \longrightarrow & e_1 \doteq v_1/\|v_1\|, \\ v_2 & \doteq & y(2) - \langle y(2), e_1\rangle e_1 & \longrightarrow & e_2 \doteq v_2/\|v_2\|, \\ \vdots & \doteq & \vdots & \longrightarrow & \vdots \\ v_t & \doteq & y(t) - \sum_{i=1}^{t-1}\langle y(i), e_i\rangle e_i & \longrightarrow & e_t \doteq v_t/\|v_t\|. \end{array}$$

The process $\{\boldsymbol{e}\}$, whose instances up to time $t$ are collected into the vector $\boldsymbol{e}^t = [e_1, e_2, \ldots, e_t]^T$, has a number of important properties:

1. The components of $\boldsymbol{e}^t$ are *orthonormal* in $\mathcal{H}$ (or equivalently, $\{\boldsymbol{e}\}$ is an uncorrelated (random) process). This holds by construction.
2. The transformation from $\boldsymbol{y}$ to $\boldsymbol{e}$ is *causal*, in the sense that if we represent it as a matrix $L_t$ such that

$$\boldsymbol{y}^t = L_t\boldsymbol{e}^t, \tag{B.31}$$

then $L_t$ is lower triangular with positive diagonal. This follows from the Gram-Schmidt procedure.

[7] Vectors here is intended as elements of a vector space, i.e. the Hilbert space of random variables with zero mean and finite variance. The realization $y(i) \in \mathbb{R}$, however, is a scalar.

3. The process $\{e\}$ is *equivalent* to $\{y\}$ in the sense that they span the same subspace
$$\mathcal{H}(y^t) = \mathcal{H}(e^t). \tag{B.32}$$
This property follows from the fact that $L_t$ is non singular.

4. If we write $y^t = L_t e^t$ in matrix form as $y = Le$, then $\Sigma_y = LL^T$.

The meaning of the components $v_t$, and the name *innovation*, comes from the fact that we can interpret
$$v_t \doteq y(t) - \hat{E}[y(t)|y^{t-1}] \tag{B.33}$$
as a *one-step prediction error*. The process $e$ is a scaled version of $v$ such that its covariance is the identity.

We may now wonder whether each process $\{y\}$ has an innovation and, if so, whether it is unique. The answer is positive if the covariance matrix $\Sigma_y$ can be written in the form $\Sigma_y = LL^T$, where $L$ that satisfies the conditions above is called the *Cholesky factor* (see Appendix A, Section A.5). The Cholesky factor can be interpreted as a "whitening filter," in the sense that it acts on the components of the vector $y$ in a causal fashion to make them uncorrelated. We may therefore consider a two-step solution to the problem of finding the least-squares filter: a "whitening step"
$$e = L^{-1}y, \tag{B.34}$$
where $\Sigma_e = I$, and a projection onto $\mathcal{H}(e)$:
$$\hat{x}(y) = \Sigma_{xe}L^{-1}y. \tag{B.35}$$
This procedure will be useful in the calculation of the Kalman gain in the next section.

## B.2 The Kalman-Bucy filter

In this section we extend the ideas of least-variance estimation to random processes, with an explicit dependence on time. We start by restricting our attention to a special class of processes, for which we can easily derive the estimator. Once we have done that, before delving into the derivation of the Kalman-Bucy filter [Kalman, 1960, Bucy, 1965], we give some intuition on the structure behind it.

### *B.2.1 Linear Gaussian dynamical models*

A linear finite-dimensional stochastic process is defined as the output of a linear dynamical system driven by white Gaussian noise. Let $A(t)$, $B(t)$, $C(t)$, $D(t)$

be time-varying matrices of suitable dimensions,[8] $\{n(t)\} \sim \mathcal{N}(0, I)$ such that $E[n(t)n^T(s)] = I\delta(t-s)$ a white, zero-mean Gaussian noise, and $x_0 \in \mathbb{R}^n$ a random vector that is uncorrelated from $\{n\}$: $E[x_0 n^T(t)] = 0$, $\forall\, t$. Then $\{y(t)\}$ is a linear Gaussian model if there exists $\{x(t)\}$ such that

$$\begin{cases} x(t+1) = A(t)x(t) + B(t)n(t), & x(t_0) = x_0, \\ y(t) = C(t)x(t) + D(t)n(t). \end{cases} \tag{B.36}$$

We call $\{x\}$ the *state process*, $\{y\}$ the *output (or measurement) process*, and $\{n\}$ the *input (or driving) noise*. The time evolution of the state process can be written as the orthogonal sum of the past history (prior to the initial condition) and the present history (from the initial condition until the present time),

$$\begin{aligned} x(t) &= \Phi(t, t_0)x_0 + \sum_{k=t_0}^{t-1} \Phi(t, k+1)B(t)n(t) \\ &= \hat{E}[x(t)|\mathcal{H}(x^{t_0})] + \hat{E}[x(t)|x(t_0), \dots, x(t-1)], \end{aligned}$$

where $\Phi$, the *state transition matrix*, denotes a fundamental set of solutions of the differential equation

$$\begin{cases} \Phi(t+1, s) = A(t)\Phi(t, s), \\ \Phi(t, t) = I. \end{cases} \tag{B.37}$$

In the case of a time-invariant system $A(t) = A$, $\forall t$, then $\Phi(t, s)$ is given by $\Phi(t, s) = A^{(t-s)}$.

**Remark B.2.** *As a consequence of the definitions, the orthogonality*[9] *between the state and the input noise propagates up to the current time:*

$$n(t) \perp_{\mathcal{H}} x(s), \quad \forall s \leq t. \tag{B.38}$$

*Moreover, the past history up to time s is always summarized by the value of the state at that time (Markov property):*

$$\hat{E}[x(t)|\mathcal{H}(x^s)] = \hat{E}[x(t)|x(s)] = \Phi(t, s)x(s), \quad \forall t \geq s. \tag{B.39}$$

These properties will turn out to be very useful in the derivation of the equations of the Kalman filter.

### B.2.2 A little intuition

The linear Gaussian model (B.36) can be described by the flow chart of Figure B.1. The input noise $n(t)$, passed through $B(t)$, is added to the state $x(t)$, passed

[8] From now on, for simplicity, we drop the boldface notation; all lowercase quantities are vectors, whose dimensions can be deduced from the context.

[9] Orthogonality is intended in the sense described in the previous section, i.e. uncorrelatedness.

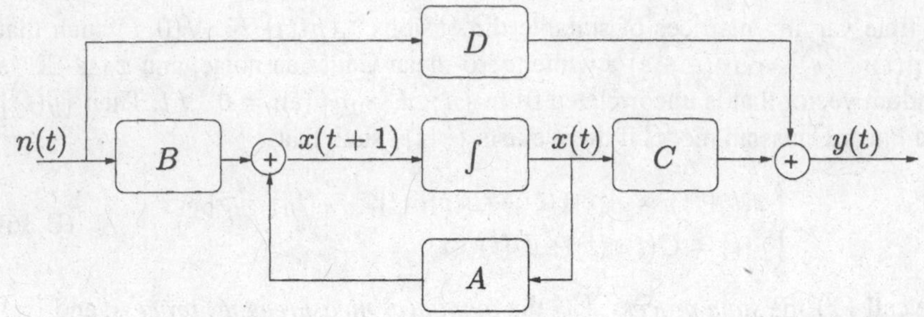


Figure B.1. Block diagram of a linear finite-dimensional stochastic process.

through $A(t)$, to make the next value of the state $x(t+1)$. From this, the measurement is obtained through $C(t)$, after adding output noise $D(t)n(t)$. Note that the input and output noises do not need to be correlated. For instance, we could have $B = [I,\ 0]$, $D = [0,\ I]$ and $n(t) = [n_1(t),\ n_2(t)]^T$ with $n_1 \perp_{\mathcal{H}} n_2$ (see Section B.2.4 for more details on this issue). Now, this is a pictorial description of an unknown process $x$, for instance the 3-D structure and motion of a scene, and its relationship to measurable quantities $y$, for instance images. In general, $x$ is not available for us to measure, so we have to infer it from $y$. *If the model were perfect*, and if noise were absent $n(t) = 0$, we could just duplicate the model

$$\begin{cases} \hat{x}(t+1) = A(t)\hat{x}(t), & \hat{x}(t_0) = x_0, \\ \hat{y}(t) = C(t)\hat{x}(t), \end{cases} \tag{B.40}$$

on a computer, as shown in Figure B.2. Now, while $x$ is some physical quantity that we do not have access to, $\hat{x}$ is a value in a register of our computer, which we can read off. But is it fair to take $\hat{x}$ as an approximation of $x$? Even if noise were absent (so that $n(t) = 0$ and the block diagram reduces to the upper part of Figure B.2), we would have to initialize our computer program with the "true" initial condition $\hat{x}(t_0) = x_0$, which we do not know. Furthermore, the model (B.36) is often an idealization of the true physical process,[10] and therefore there is no reason to believe that even if we started from the right initial condition, we would stay close to the true state as time goes by.

However, our computer program produces, in addition to the "estimated state" $\hat{x}(t)$, an "estimated output" $\hat{y}(t)$. While we cannot compare $\hat{x}(t)$ directly with $x(t)$, we can indeed compare $\hat{y}(t)$ with $y(t)$. Let us call the difference between the measured output and the estimated output $e(t) \doteq y(t) - \hat{y}(t)$. The least we can ask of our filter is that it make $e(t)$ "small" in some sense, or "uninformative" (e.g. white). If this is not so, one could think of "feeding back" $e(t)$ to the filter through a gain $K(t)$ that can be designed to drive $e(t)$ toward the goal (Figure B.2, dashed lines). In formulas, we have

$$\hat{x}(t+1) = A(t)\hat{x}(t) + K(t)(y(t) - C\hat{x}(t)). \tag{B.41}$$

[10]For instance, $x(t+1) = Ax(t)$ could come from the linearization of a model of the form $x(t+1) = f(x(t))$, with $A \doteq \frac{\partial f}{\partial x}$.

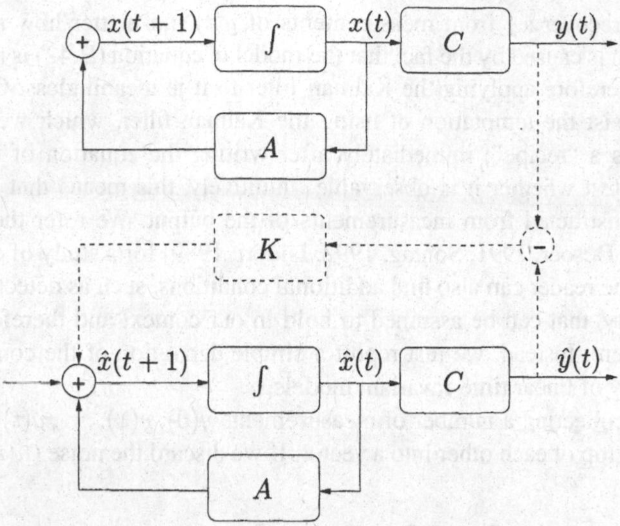


Figure B.2. A naive construction of the Kalman filter. In the absence of noise ($n(t) = 0$), a copy of the original process is created, from which the "state" $\hat{x}(t)$ can be read off. If there is a discrepancy between the measured output of the process $y(t)$ and the estimated output $\hat{y}(t)$, this is fed back to the filter via a (Kalman) gain $K(t)$.

While this naive discussion should by no means convince the reader that this strategy works, the discussion that follows will, provided that certain conditions are met. In general, $e(t)$ becoming small does not guarantee that $\hat{x}(t)$ is close to $x(t)$. Additional properties of the model have to be satisfied, and in particular its *observability*, which we discuss in Section B.2.3, and in Chapter 12 (Section 12.1.1) for the particular model that relates to reconstructing 3-D structure and motion from sequences of images. Furthermore, we need to decide how to choose $K(t)$. This we explain in the next section.

The nice thing about the derivation that follows is that instead of *postulating* a structure of the filter as a linear gain, as we guessed just now in equation (B.41), this structure is going to come naturally as a consequence of the assumptions on the model that generates the data, which we described in the previous subsection.

### B.2.3 Observability

As we have suggested above, having $\hat{y}(t)$ close to $y(t)$ does not guarantee that $\hat{x}(t)$ is close to $x(t)$. For instance, consider the simple linear model

$$\begin{cases} x(t+1) = \begin{bmatrix} 1 & 0 \\ 1 & 1 \end{bmatrix} x(t), \quad x(0) = x_0, \\ y(t) = [1, 0]x(t). \end{cases} \tag{B.42}$$

If we write the trajectory of the output $y(t) = x_1(t)$, we see that it depends only on the first component of the state $x_1$, but not on $x_2$. Therefore, it is impossible to recover $x_2$ from measurements of $y(t)$, no matter how many. This phenomenon is caused by the fact that the model in equation (B.42) is not *observable*, and therefore applying the Kalman filter to it is meaningless. One should therefore resist the temptation of using the Kalman filter, which we are about to derive, as a "recipe"; immediately after writing the equation of the model, one should test whether it is observable. Intuitively, this means that the state $x$ can be reconstructed from measurements of the output. We refer the reader to [Callier and Desoer, 1991, Sontag, 1992, Isidori, 1989] for a study of observability. There, the reader can also find additional conditions, such as detectability and stabilizability, that can be assumed to hold in our context and therefore are not discussed here. Instead, we just report a simple derivation of the conditions for observability of linear time-invariant models.

Imagine collecting a number of measurements $y(0), y(1), \dots, y(t)$ and stacking them on top of each other into a vector. If we discard the noise ($n(t) = 0$), we get

$$\begin{bmatrix} y(0) \\ y(1) \\ \vdots \\ y(t) \end{bmatrix} = \begin{bmatrix} C \\ CA \\ \vdots \\ CA^t \end{bmatrix} x_0 \doteq \mathcal{O} x_0. \tag{B.43}$$

It is clear that if the matrix $\mathcal{O}$ has full rank, then we can solve the linear system of equations above to obtain $x_0$. Once we have $x_0$, all the other values of $x(t)$ are given by integrating (in the absence of noise) $x(t+1) = Ax(t)$ starting from the initial condition $x(0) = x_0$, that is, $x(t) = A^t x_0$.

Although we do not show it here (see [Sontag, 1992], Theorem 16, page 200), the converse is also true; that is, the initial state $x_0$ can be recovered only if $\mathcal{O}$ has full rank $n$. In general, as $t$ increases, the rank of $\mathcal{O}$ increases, and in principle, one does not know how many time steps $t$ are necessary in order to be able to observe $x$. It can be shown easily using the Cayley-Hamilton theorem (see [Callier and Desoer, 1991]) that the rank does not grow beyond $t = n$, and therefore if the test fails after $n$ steps, the model is not observable.

Note that this is a structural condition on the model, which can (and *should*) be tested beforehand. It is a necessary condition for the Kalman filter to return sensible estimates. If it is not satisfied, $e(t)$ may converge to a white zero-mean process, and yet the estimated state $\hat{x}(t)$ wanders freely away from $x(t)$, and its covariance grows to infinity. For a more germane discussion of these topics, we refer the reader to [Jazwinski, 1970]. We now turn to the derivation of the filter.

### *B.2.4 Derivation of the Kalman filter*

Suppose we are given a linear finite-dimensional process that has a realization $(A(t), B(t), C(t), D(t))$ as in equation (B.36). For simplicity we omit the time index in the system matrices $(A, B, C, D)$. While we measure the (noisy) output $y(t)$ of such a realization, we do not have access to its state $x(t)$. The Kalman filter is a dynamical model that accepts as input the output of the process realization, and returns an estimate of its state that has the property of having the least error covariance [Kalman, 1960, Bucy, 1965]. In order to derive the expression for the filter, we write the linear Gaussian model as follows:

$$\begin{cases} x(t+1) = Ax(t) + v(t), \quad x(t_0) = x_0, \\ y(t) = Cx(t) + w(t). \end{cases} \tag{B.44}$$

Here $v(t) = Bn(t)$ is a white zero-mean Gaussian noise with covariance $Q$, $w(t) = Dn(t)$ is a white zero-mean noise with covariance $R$, so that we may write

$$\begin{aligned} v(t) &= \sqrt{Q}n(t), \\ w(t) &= \sqrt{R}n(t), \end{aligned}$$

where $n$ is a unit-variance noise. In general, $v$ and $w$ will be correlated, and in particular, we will write

$$S(t) = E[v(t)w^T(t)]. \tag{B.45}$$

We require that the initial condition $x_0$ be uncorrelated from the noise processes:

$$x_0 \perp \{v\} \text{ and } \{w\}, \quad \forall\, t. \tag{B.46}$$

The first step is to modify the above model so that the model error $v$ is uncorrelated from the measurement error $w$.

*Uncorrelating the model from the measurements*

In order to uncorrelate the model error from the measurement error we can just substitute $v$ with the complement of its projection onto the span of $w$. Let us write

$$\tilde{v}(t) = v(t) - \hat{E}[v(t)|\mathcal{H}_t(w)] = v(t) - \hat{E}[v(t)|w(t)], \tag{B.47}$$

where the last equivalence is due to the fact that $w$ is a white noise. We can now use the results from Section B.1 to conclude that

$$\tilde{v}(t) = v(t) - SR^{-1}w(t), \tag{B.48}$$

and similarly for the covariance matrix

$$\tilde{Q} = Q - SR^{-1}S^T. \tag{B.49}$$

Substituting the expression of $v(t)$ into the model (B.44), we get

$$\begin{cases} x(t+1) = Fx(t) + SR^{-1}y(t) + \tilde{v}(t), \\ y(t) = Cx(t) + w(t), \end{cases} \tag{B.50}$$

where $F = A - SR^{-1}C$. The model error $\tilde{v}$ in the above model is uncorrelated from the measurement noise $w$, at the cost of adding an output-injection term $SR^{-1}y(t)$.

Note that if $S = 0$, then $F = A$. In many applications, for instance in the case study in Chapter 12, there is reason to believe a priori that the model error $v$ is independent of the measurement noise $w$, and therefore $S = 0$.

*Prediction step*

Suppose that at some point in time, we are given a current estimate for the state $\hat{x}(t|t) \doteq \hat{E}[x(t)|\boldsymbol{y}^t]$ and a corresponding estimate of the covariance of the model error $P(t|t) = E[\tilde{x}(t)\tilde{x}(t)^T]$, where $\tilde{x} = x - \hat{x}$. At the initial time $t_0$ we can take $\hat{x}(t_0|t_0) = x_0$ with some bona fide covariance matrix $P(0|0)$. Then it is immediate to compute $\hat{x}(t+1|t) \doteq E[x(t+1)|\boldsymbol{y}^t]$

$$\hat{x}(t+1|t) = F\hat{x}(t|t) + SR^{-1}y(t) + \hat{E}[\tilde{v}(t)|\mathcal{H}_t(y)], \tag{B.51}$$

where the last term is zero, since $\tilde{v}(t) \perp x(s),\ \forall x \leq t$, and $\tilde{v}(t) \perp w(s),\ \forall s$, and therefore $\tilde{v}(t) \perp y(s),\ \forall s \leq t$. The estimation error is therefore

$$\tilde{x}(t+1|t) = F\tilde{x}(t|t) \dotplus \tilde{v}(t), \tag{B.52}$$

where the sum is an orthogonal sum, and therefore it is trivial to compute the covariance as

$$P(t+1|t) = FP(t|t)F^T + \tilde{Q}. \tag{B.53}$$

*Update step*

Once a new measurement is acquired, we can update our prediction so as to take into account the new measurement. The update is defined as $\hat{x}(t+1|t+1) \doteq \hat{E}[x(t+1)|\mathcal{H}_{t+1}(y)]$. Now, as we have seen in Section B.1.4, we can decompose the span of the measurements into the orthogonal sum

$$\mathcal{H}_{t+1}(y) = \mathcal{H}_t(y) \dotplus \{e(t+1)\}, \tag{B.54}$$

where $e(t+1) \doteq y(t+1) - \hat{E}[y(t+1)|\mathcal{H}_t(y)]$ is the innovation process. Therefore, we have

$$\hat{x}(t+1|t+1) = \hat{E}[x(t+1)|\mathcal{H}_t(y)] + \hat{E}[x(t+1)|e(t+1)], \tag{B.55}$$

where the last term can be computed using the results from Section B.1:

$$\hat{x}(t+1|t+1) = \hat{x}(t+1|t) + L(t+1)e(t+1), \tag{B.56}$$

where $L(t+1) \doteq \Sigma_{xe}(t+1)\Sigma_e^{-1}(t+1)$ is called the *Kalman gain*. Substituting the expression for the innovation, we have

$$\hat{x}(t+1|t+1) = \hat{x}(t+1|t) + L(t+1)\left(y(t+1) - C\hat{x}(t+1|t)\right), \tag{B.57}$$

from which we see that the update consists in a linear correction weighted by the Kalman gain (compare with equation (B.41)).

*Computation of the gain*

In order to compute the gain $L(t+1) \doteq \Sigma_{xe}(t+1)\Sigma_e^{-1}(t+1)$ we derive an alternative expression for the innovation,

$$e(t+1) = y(t+1) - Cx(t+1) + Cx(t+1) - C\hat{x}(t+1|t) = w(t+1) + C\tilde{x}(t+1|t), \tag{B.58}$$

from which it is immediate to compute

$$\Sigma_{xe}(t+1) = P(t+1|t)C^T. \tag{B.59}$$

Similarly, we can derive the covariance of the innovation $\Lambda(t+1)$,

$$\Lambda(t+1) \doteq \Sigma_e(t+1) = CP(t+1|t)C^T + R, \tag{B.60}$$

and therefore the Kalman gain is

$$L(t+1) = P(t+1|t)C^T\Lambda^{-1}(t+1). \tag{B.61}$$

*Covariance update*

From the update of the estimation error

$$\tilde{x}(t+1|t+1) = \tilde{x}(t+1|t) - L(t+1)e(t+1) \tag{B.62}$$

we can easily compute the update for the covariance. We first observe that $\tilde{x}(t+1|t+1)$ is by definition orthogonal to $\mathcal{H}_{t+1}(y)$, while the correction term $L(t+1)e(t+1)$ is contained in the history of the innovation, which is by construction equal to the history of the process $y$: $\mathcal{H}_{t+1}(y)$. Then it is immediate to see that

$$P(t+1|t+1) = P(t+1|t) - L(t+1)\Lambda(t+1)L^T(t+1). \tag{B.63}$$

The above equation is not convenient for computational purposes, since it does not guarantee that the updated covariance is a symmetric matrix. An alternative form of the above that does guarantee symmetry of the result is

$$P(t+1|t+1) = \Gamma(t+1)P(t+1|t)\Gamma(t+1)^T + L(t+1)RL(t+1)^T, \tag{B.64}$$

where $\Gamma(t+1) \doteq I - L(t+1)C$. Equation (B.64) is a discrete Riccati equation (DRE).

*Predictor equations*

It is possible to combine the two steps above and derive a single model for the one-step predictor. We summarize the result as follows (compare with equation (B.41)):

$$\begin{aligned} \hat{x}(t+1|t) &= A\hat{x}(t|t-1) + K_s(t)\left(y(t) - C\hat{x}(t|t-1)\right), \\ P(t+1|t) &= F\Gamma(t)P(t|t-1)\Gamma(t)^TF^T + FL(t)RL^T(t)F^T + \tilde{Q}, \end{aligned}$$

where we have defined

$$K_s(t) \doteq FL(t) + SR^{-1} = \left(AP(t|t-1)C^T + S\right)\Lambda^{-1}(t),$$

the Kalman predictor gain.

## B.3 The extended Kalman filter

In this section we derive an approximation of the optimal filter for the case of a nonlinear model, known as the extended Kalman filter (EKF) [Jazwinski, 1970]. It is based on a variational model about the best current trajectory. The system is linearized at each step around the current estimate in order to calculate a correcting gain; the update of the previous estimate is then performed on the original (nonlinear) equations.

The model we address is in the generic form discussed in Chapter 12: we are interested in building an estimator for a process $\{x\}$ that is described by a stochastic difference equation of the form

$$x(t+1) = f(x(t)) + v(t); \quad x(t_0) = x_0,$$

where $v(t) \sim \mathcal{N}(0, Q_v)$ is a white zero-mean Gaussian noise with covariance $Q_v$. Suppose there is a measurable quantity $y(t)$ that is linked to $x$ by the measurement equation

$$y(t) = h(x(t)) + w(t); \quad w(t) \sim \mathcal{N}(0, R_w). \tag{B.65}$$

We will assume throughout that $f, h \in C^r$; $r \geq 1$; the covariance matrix $R_w$ is derived from knowledge of the measurement device. The model we consider is hence of the form

$$\begin{cases} x(t+1) = f(x(t)) + v(t), & x(t_0) = x_0, \\ y(t) = h(x(t)) + w(t). \end{cases} \tag{B.66}$$

*Construction of the variational model about the reference trajectory*

Consider at each time sample $t$ a reference trajectory $\bar{x}(t)$ that solves the difference equation

$$\bar{x}(t+1) = f(\bar{x}(t))$$

and the Jacobian matrix

$$F(\bar{x}(t)) \doteq F(t) = \left(\frac{\partial f}{\partial x}\right)\Big|_{\bar{x}(t)}.$$

The linearization of the measurement equation about the point $\bar{x}(t)$ is

$$h(x(t)) = h(\bar{x}(t)) + C(\bar{x})(x(t) - \bar{x}(t)) + O(\epsilon^2),$$

where

$$\begin{aligned} C(\bar{x}) &\doteq \left(\frac{\partial h}{\partial x}\right)\Big|_{\bar{x}(t)}, \\ \epsilon^2 &\doteq \|x - \bar{x}\|^2, \end{aligned}$$

and the limit implicit in $O$ (which indicates the order of the approximation, not to be confused with the observability matrix) is intended in the mean-square sense.

Setting $\delta x(t) \doteq x(t) - \bar{x}(t)$, we have, up to second-order terms,

$$\delta y(t) \doteq y(t) - h(\bar{x}(t)) = C(\bar{x})\delta x(t) + w(t).$$

*Prediction step*

Suppose at some time $t$ we have available the best estimate $\hat{x}(t|t)$; we may write the variational model about the trajectory $\bar{x}(t)$ defined such that

$$\bar{x}(t+1) = f(\bar{x}(t))\ ;\ \bar{x}(t) = \hat{x}(t|t).$$

For small displacements we may write

$$\delta x(t+1) = F(\bar{x}(t))\delta x(t) + \tilde{v}(t), \tag{B.67}$$

where the noise term $\tilde{v}(t)$ may include a linearization error component.

Note that with such a choice we have $\delta\hat{x}(t|t) = 0$ and $\delta\hat{x}(t+1|t) = F(\bar{x}(t))\delta\hat{x}(t|t) = 0$, from which we can conclude

$$\hat{x}(t+1|t) = \bar{x}(t+1) = f(\bar{x}(t)) = f(\hat{x}(t|t)). \tag{B.68}$$

The covariance of the prediction error $\delta\hat{x}(t+1|t)$ is

$$P(t+1|t) = F(t)P(t|t)F^T(t) + \tilde{Q}, \tag{B.69}$$

where $\tilde{Q}$ is the covariance of $\tilde{v}$. The last two equations represent the prediction step for the estimator and are equal, as expected, to the prediction of the explicit EKF [Jazwinski, 1970].

*Update step*

At time $t+1$ a new measurement becomes available, $y(t+1)$, which is used to update the prediction $\hat{x}(t+1|t)$ and its error covariance $P(t+1|t)$. Exploiting the linearization of the measurement equation about $\bar{x}(t+1) = \hat{x}(t+1|t)$, we obtain, using the shorthand $\hat{x} \doteq \hat{x}(t+1|t)$,

$$\delta y(t+1) \doteq y(t+1) - h(\hat{x}) = C(\hat{x})\delta x(t+1) + n(t+1), \tag{B.70}$$

where the noise term $n(t+1)$ includes both the original measurement noise as well as linearization errors. This, together with equation (B.67), defines a linear model, for which we can finally write the update equation based on the traditional linear Kalman filter as we have done in the previous section:

$$\delta\hat{x}(t+1|t+1) = \delta\hat{x}(t+1|t) + L(t+1)\left(y(t+1) - C(\hat{x})\delta\hat{x}(t+1|t)\right), \tag{B.71}$$

where

$$\begin{aligned}
L(t+1) &= P(t+1|t)C(\hat{x})^T\Lambda^{-1}(t+1),\\
\Lambda(t+1) &= C(\hat{x})P(t|t)C(\hat{x})^T + R_n(t+1),\\
P(t+1|t+1) &= \Gamma(t+1)P(t+1|t)\Gamma^T(t+1) + L(t+1)R_n(t+1)L(t+1)^T,\\
\Gamma(t+1) &= (I - L(t+1)C(\hat{x})).
\end{aligned}$$

Since $\delta\hat{x}(t+1|t) = 0$ and $\delta\hat{x}(t+1|t+1) = \hat{x}(t+1|t+1) - \hat{x}(t+1|t)$, we may write the update equation for the original model:

$$\hat{x}(t+1|t+1) = \hat{x}(t+1|t) + L(t+1)(y(t+1) - h(\hat{x}(t+1|t))). \quad \text{(B.72)}$$

The noise $n$ defined in (B.70) has a covariance that can be computed empirically using a tuning procedure. It is common to approximate $R_n$ with the covariance of the original noise $R_w$, with the diagonal terms inflated to compensate for linearization and other modeling errors: the update of the covariance $P(t+1|t+1)$ is computed from the standard DRE of the linear Kalman filter (B.64), derived in the previous section.

# Appendix C
# Basic Facts from Nonlinear Optimization

*Since the building of all the universe is perfect and is created by the wisdom creator, nothing arises in the universe in which one cannot see the sense of some maximum or minimum.*

– L. Euler

In Appendix B, we discussed how to obtain an optimal *estimate* of a random variable (or process) from its noisy observations. Here we discuss another problem where the notion of optimality also occurs: how to minimize (or maximize) a given deterministic function $f(x)$ where the variable $x$ belongs to some domain $D$, often an open set in $\mathbb{R}^n$. In practice, the function $f(\cdot)$ is often referred to as the *objective function*, and it is usually interpreted as the cost, error, reward, or utility that needs to be optimized. For instance, to compare different 3-D reconstructions from a given set of 2-D images, an objective function $f(\cdot)$ can be contrived to evaluate the "goodness" of each reconstruction $x$. Such a function can be chosen to be the error between the true 3-D structure and that reconstructed or the error between their reprojections on all the images. The choice of this objective function is often determined by a statistical or geometric model used to quantify the error measure. In the previous appendix we have considered the particular cost function $f(\cdot) = E[\|\cdot\|^2]$, which leads to the least-variance estimator. In this appendix we consider the problem where $f(\cdot)$ can be general.

For example, the objective function is sometimes a likelihood function or a geometric distance. But once that step is done, the remaining issue is how to minimize the resulting objective function $f(\cdot)$ and find the optimal solution $x^*$.

We call this an *unconstrained optimization* problem

$$x^* = \arg\min_x f(x), \quad x \in \mathbb{R}^n. \tag{C.1}$$

In many optimization problems that we encounter in this book the variable $x$ is typically not free in a given domain $D$ but is subject to some further restrictions, called constraints. We often consider two forms of constraints on $x$. First, we sometimes know that the optimal $x^*$, in addition to being in $D$, needs to satisfy certain equations, say $h(x^*) = 0$. In this case, the problem of searching for $x^*$ becomes a *constrained optimization*,

$$x^* = \arg\min_x f(x), \quad \text{subject to} \quad h(x) = 0. \tag{C.2}$$

The *Lagrange multiplier method* is an effective tool to convert a constrained optimization problem to an unconstrained one, which we will review briefly below. Second, if the zero level set of $h(x)$ is a smooth section in $D$, we often call it a *differentiable manifold*, typically denoted by $M \subseteq D$. In this case, the problem becomes an *optimization on a manifold*,

$$x^* = \arg\min_x f(x), \quad x \in M. \tag{C.3}$$

Roughly speaking, a differentiable manifold $M$ is a space that only locally resembles $\mathbb{R}^n$ but globally exhibits a nonlinear structure. For instance, a sphere $\mathbb{S}^2 \subset \mathbb{R}^3$ is a 2-D manifold, and the rotation group $SO(3)$ is a 3-D manifold. But neither is an open set of $\mathbb{R}^2$ or $\mathbb{R}^3$. If an analytical description of the manifold $M$ is available, we are able to directly conduct the search for $x^*$ inside this manifold without resorting to the outer domain $D$. For example, the explicit parameterizations studied for the rotation group $SO(3)$ in Chapter 2 enable us to reduce any optimization problem on $SO(3)$ to an unconstrained problem.

In the rest of this appendix, we provide a brief summary of standard results and algorithms for solving both the unconstrained optimization problem and the constrained optimization problem. Since each problem deserves a book of its own, many details about the results and algorithms to be summarized here will be left out without detailed explanation. For interested readers, nevertheless, we recommend [Bertsekas, 1999].

## C.1 Unconstrained optimization: gradient-based methods

In this subsection, we consider the *unconstrained optimization* problem

$$x^* = \arg\min_x f(x), \quad x \in \mathbb{R}^n, \tag{C.4}$$

for a function $f(\cdot) : \mathbb{R}^n \to \mathbb{R}$ that is at least twice continuously differentiable in $\mathbb{R}^n$ (or, formally, $f(\cdot) \in C^2(\mathbb{R}^n)$).

Since $f(\cdot)$ is twice differentiable, for convenience, we often define its *gradient* at $x = [x_1, x_2, \ldots, x_n]^T \in \mathbb{R}^n$, denoted by $\nabla f(x)$, to be the vector[1]

$$\nabla f(x) \doteq \left[\frac{\partial f(x)}{\partial x_1}, \frac{\partial f(x)}{\partial x_2}, \ldots, \frac{\partial f(x)}{\partial x_n}\right]^T \in \mathbb{R}^n, \tag{C.5}$$

and define its *Hessian matrix* at $x$, denoted by $\nabla^2 f(x)$, to be

$$\nabla^2 f(x) \doteq \begin{bmatrix} \frac{\partial^2 f(x)}{\partial x_1 \partial x_1} & \frac{\partial^2 f(x)}{\partial x_1 \partial x_2} & \cdots & \frac{\partial^2 f(x)}{\partial x_1 \partial x_n} \\ \frac{\partial^2 f(x)}{\partial x_2 \partial x_1} & \frac{\partial^2 f(x)}{\partial x_2 \partial x_2} & \cdots & \frac{\partial^2 f(x)}{\partial x_2 \partial x_n} \\ \vdots & \vdots & \ddots & \vdots \\ \frac{\partial^2 f(x)}{\partial x_n \partial x_1} & \frac{\partial^2 f(x)}{\partial x_n \partial x_2} & \cdots & \frac{\partial^2 f(x)}{\partial x_n \partial x_n} \end{bmatrix} \in \mathbb{R}^{n \times n}. \tag{C.6}$$

Note that the gradient vector always points opposite to the direction of steepest descent, and the Hessian matrix is always a symmetric matrix.

### *C.1.1 Optimality conditions*

The following results follow from [Bertsekas, 1999] and are given here without proof.

**Proposition C.1 (Necessary optimality conditions).** *If $x^*$ is an unconstrained local minimum of $f(\cdot)$, then*

$$\nabla f(x^*) = 0, \quad \nabla^2 f(x^*) \geq 0. \tag{C.7}$$

**Proposition C.2 (Second-order sufficent optimality conditions).** *If a vector $x^* \in \mathbb{R}^n$ satisfies the conditions*

$$\nabla f(x^*) = 0, \quad \nabla^2 f(x^*) > 0, \tag{C.8}$$

*then $x^*$ is a strict unconstrained local minimum of $f(\cdot)$.*

So generally speaking, at a (local) minimum $x^*$, the gradient vector $\nabla f(x^*)$ must vanish, and the Hessian matrix $\nabla^2 f(x^*)$ is typically positive definite.

Of course, a local minimum is not necessarily the global minimum that we were looking for in the first place. Unfortunately, the conditions given above are only local criteria, and hence they are not able to distinguish local and global minima. In the special case that the function $f(\cdot)$ is convex, then there is only one minimum, and any (local) minimum therefore must be the global minimum.

[1] Conceptually, the gradient of a function is a covector and hence should be represented by a row vector. However, in this book, to be consistent with our simplifying claim that every vector is a column vector, we define the gradient to be its transposed version.

### *C.1.2 Algorithms*

The general idea of minimizing a nonlinear function $f(\cdot)$ is very simple (unless there is some special knowledge about $f(\cdot)$ that we can use to speed up the search). We start at any initial guess $x = x^0$, and successively update $x$ to $x^1, x^2, \ldots$, such that the value $f(x)$ decreases at each iteration; that is, $f(x^{i+1}) \leq f(x^i)$. Of course, the safest way to ensure a decrease in the value is to follow the "direction of descent," which in our case would be the opposite direction to the gradient vector. This idea gives rise to the *steepest descent method* for searching for the minimum. At each iteration,

$$x^{i+1} = x^i - \alpha^i \nabla f(x^i), \tag{C.9}$$

for some scalar $\alpha^i$, called the *step size*. There exist many different choices for the step size $\alpha^i$, and the simplest one is of course to set it to be a small constant. We will discuss issues related to the choice of the step size again later.

Although the vector $-\nabla f(x^i)$ points to the steepest descent direction locally around $x^i$, it is not necessarily the best choice for searching for the minimum at a larger scale. A modification to the above gradient method is in the form

$$x^{i+1} = x^i - \alpha^i D^i \nabla f(x^i), \tag{C.10}$$

where $D^i \in \mathbb{R}^{n \times n}$ is a positive definite symmetric matrix to be determined in each particular algorithm. The steepest descent method becomes a special case that $D^i \equiv I$. In general, $D^i$ can be viewed as a weight matrix that adjusts the descent direction according to more sophisticated local information about $f(\cdot)$ than the gradient alone. A simple choice for $D^i$ would be a diagonal matrix that scales the descent speed differently in each axial direction. Newton's method below is another classical example of an improved choice for $D^i$.

*Newton's method*

The necessary and sufficient optimality conditions suggest that around a (local) minimum, the function $f(\cdot)$ approximately resembles a quadratic function

$$f(x) \approx f(x^i) + \nabla f(x^i)^T(x - x^i) + \frac{1}{2}(x - x^i)^T \nabla^2 f(x^i)(x - x^i). \tag{C.11}$$

This indicates that when $x^i$ is close to the minimum $x^*$ or the function $f(\cdot)$ is indeed a quadratic function, the minimum $x^*$ is best approximated by the $x$ that makes the derivative of the right-hand side of the above equation vanish (according to Proposition C.1), i.e.

$$\nabla f(x^i) + \nabla^2 f(x^i)(x - x^i) = 0. \tag{C.12}$$

This gives Newton's method

$$x^{i+1} = x^i - \left(\nabla^2 f(x^i)\right)^{-1} \nabla f(x^i). \tag{C.13}$$

A more general iteration uses

$$x^{i+1} = x^i - \alpha^i \left(\nabla^2 f(x^i)\right)^{-1} \nabla f(x^i), \tag{C.14}$$

where a step size $\alpha^i$ is used to control the speed of progress. Note that the Newton's method is just a special case of the general iteration method (C.10) with $D^i = \left(\nabla^2 f(x^i)\right)^{-1}$.

*Gauss-Newton method and Levenberg-Marquardt method*

Computation of the Hessian matrix $\nabla^2 f(x)$ is often expensive and sometimes not possible (e.g., the function $f(\cdot)$ is not twice differentiable). In such cases, alternatives to $D^i$ are often adopted that to some extent approximate the matrix $(\nabla^2 f(x))^{-1}$ and use only the information encoded in the first derivative of $f(x)$. The *Gauss-Newton method* is one such method, which applies, however, only to the problem of minimizing the sum of squares of real-valued functions, say $u(x) = [u_1(x), u_2(x), \ldots, u_m(x)]^T$. For a vector-valued function $u(\cdot)$ we define its *Jacobian matrix* to be

$$\nabla u(x) = \begin{bmatrix} \frac{\partial u_1(x)}{\partial x_1} & \frac{\partial u_2(x)}{\partial x_1} & \cdots & \frac{\partial u_m(x)}{\partial x_1} \\ \frac{\partial u_1(x)}{\partial x_2} & \frac{\partial u_2(x)}{\partial x_2} & \cdots & \frac{\partial u_n(x)}{\partial x_2} \\ \vdots & \vdots & \ddots & \vdots \\ \frac{\partial u_1(x)}{\partial x_n} & \frac{\partial u_2(x)}{\partial x_n} & \cdots & \frac{\partial u_m(x)}{\partial x_n} \end{bmatrix} \in \mathbb{R}^{n \times m}. \tag{C.15}$$

Typically, $m \gg n$, and the Jacobian matrix is always of full rank $n$.

Then, for the objective function $f(\cdot)$ now in the form

$$f(x) = \frac{1}{2}\|u(x)\|^2 = \frac{1}{2}\sum_{i=1}^{m} u_i(x)^2, \tag{C.16}$$

instead of choosing $D^i = \nabla^2 f(x^i)$, the Gauss-Newton method chooses

$$D^i = \left(\nabla u(x^i)\nabla u(x^i)^T\right)^{-1}. \tag{C.17}$$

Since $\nabla f(x) = \nabla u(x) u(x)$, the Gauss-Newton iteration takes the form

$$x^{i+1} = x^i - \alpha^i \left(\nabla u(x^i)\nabla u(x^i)^T\right)^{-1} \nabla u(x^i) u(x^i). \tag{C.18}$$

The Gauss-Newton method is an approximation to Newton's method, especially when the value $\|u(x)\|^2$ is small. Of course, here we use the standard 2-norm to measure $u(x)$. The method needs only to be slightly changed when a different quadratic norm, say $\|u(x)\|_Q = u(x)^T Q u(x)$ for some positive definite symmetric matrix $Q \in \mathbb{R}^{m \times m}$, is used. We leave the details to the reader.

The *Levenberg-Marquardt method* is a slight modification to the Gauss-Newton method that has been widely used in the computer vision literature. Its only difference from the Gauss-Newton method is to set $\alpha^i = 1$ and use instead

$$D^i = \left(\lambda^i I + \nabla u(x^i)\nabla u(x^i)^T\right)^{-1} \in \mathbb{R}^{n \times n}, \tag{C.19}$$

where $\lambda^i > 0$ is a scalar determined by the following rules: it is initially assigned to some small value, and then:

1. If the current value of $\lambda$ results in a decrease in the error, then the iteration is accepted and $\lambda$ is divided by 10 as the initial value for the next iteration.

2. If $\lambda$ results in an increase in the error, then it is multiplied by 10, and the iteration is tried again until a $\lambda$ is found that results in a decrease in the error.

Due to the choice of $D^i$ in (C.19), we see that the Levenberg-Marquardt method still works even if sometimes the Jacobian matrix $\nabla u(x)$ is not of full rank, which occurs often in practice. Moreover, the algorithm tends to adapt its step size (through controlling the value of $\lambda$) based on the history of values of the objective function $f(x)$.

*Choice of the step size*

In the pure Newton method and the Levenberg-Marquardt method, we do not have to specify the step size $\alpha^i$. In other methods, however, one needs to know how to choose $\alpha^i$ properly. The simplest case is to choose the step size $\alpha^i$ to be a constant, but that does not always result in a decrease in the value of $f(x)$ at each iteration. So $\alpha^i$ is often chosen to be the value $\alpha^*$ that is given by solving a line minimization problem:

$$\alpha^* = \arg\min_{\alpha \geq 0} f(x^i - \alpha D^i \nabla f(x^i)).$$

This is called the *minimization rule.*

But finding the optimal $\alpha^*$ at each iteration is computationally costly. A popular choice of $\alpha^i$ without solving such a minimization problem but still ensuring convergence is the *Armijo rule*: for prefixed scalars $s, \beta$, and $\sigma$, with $0 < \beta, \sigma < 1$, we set $\alpha^i = \beta^{k_i} s$, where $k_i$ is the first nonnegative integer $k$ for which

$$f(x^i) - f(x^i - \beta^k s D^i \nabla f(x^i)) \geq -\sigma \beta^k s \nabla f(x^i)^T D^i \nabla f(x^i). \tag{C.20}$$

A typical choice for the scalars is $\sigma \in [10^{-5}, 10^{-1}]$, $\beta \in [0.1, 0.5]$, and $s = 1$.

## C.2 Constrained optimization: Lagrange multiplier method

In this section we consider the optimization problem with equality constraints on the variable $x \in \mathbb{R}^n$:

$$x^* = \arg\min f(x), \quad \text{subject to} \quad h(x) = 0, \tag{C.21}$$

where $h = [h_1, h_2, \ldots, h_m]^T$ is a smooth (multidimensional) function (or map) from $\mathbb{R}^n$ to $\mathbb{R}^m$. For each constraint $h_i(x) = 0$ to be independently effective at the minimum $x^*$, we often assume that their gradients

$$\nabla h_1(x^*), \; \nabla h_2(x^*), \; \ldots, \; \nabla h_m(x^*) \quad \in \mathbb{R}^n \tag{C.22}$$

are linearly independent. If so, the constraints are called *regular*.

## C.2.1 Optimality conditions

For simplicity, we always assume that the functions $f(\cdot)$ and $h(\cdot)$ are at least twice continuously differentiable. Then the main theorem of Lagrange recites:

**Proposition C.3 (Lagrange multiplier theorem; necessary conditions).** *Let $x^*$ be a local minimum of a function $f(\cdot)$ subject to regular constraints $h(x) = 0$. Then there exists a unique vector $\lambda^* = [\lambda_1^*, \lambda_2^*, \ldots, \lambda_m^*]^T \in \mathbb{R}^m$, called Lagrange multipliers, such that*

$$\nabla f(x^*) + \sum_{i=1}^{m} \lambda_i^* \nabla h_i(x^*) = 0. \tag{C.23}$$

*Furthermore, we have*

$$v^T \left( \nabla^2 f(x^*) + \sum_{i=1}^{m} \lambda_i^* \nabla^2 h_i(x^*) \right) v \geq 0 \tag{C.24}$$

*for all vectors $v \in \mathbb{R}^n$ that satisfy $\nabla h_i(x^*)^T v = 0, i = 1, 2, \ldots, m$.*

**Proposition C.4 (Lagrange multiplier theorem; sufficient conditions).** *Assume that $x^* \in \mathbb{R}^n$ and $\lambda^* = [\lambda_1^*, \lambda_2^*, \ldots, \lambda_m^*]^T \in \mathbb{R}^m$ satisfy*

$$\nabla f(x^*) + \sum_{i=1}^{m} \lambda_i^* \nabla h_i(x^*) = 0, \quad h_i(x^*) = 0,\ i = 1, 2, \ldots, m, \tag{C.25}$$

*and furthermore, we have*

$$v^T \left( \nabla^2 f(x^*) + \sum_{i=1}^{m} \lambda_i^* \nabla^2 h_i(x^*) \right) v > 0, \tag{C.26}$$

*for all vectors $v \in \mathbb{R}^n$ that satisfy $\nabla h_i(x^*)^T v = 0, i = 1, 2, \ldots, m$. Then $x^*$ is a strict local minimum of $f(\cdot)$ subject to $h(x) = 0$.*

## C.2.2 Algorithms

There are typically two approaches to solving a constrained optimization problem:

1. Try to use necessary conditions given in Proposition C.3 to solve the minimum (to some extent).

2. Convert or approximate the constrained optimization problem by an unconstrained one.

*The Lagrangian function*

If we define for convenience the *Lagrangian function* $L : \mathbb{R}^{n+m} \to \mathbb{R}$ as

$$L(x, \lambda) \doteq f(x) + \lambda^T h(x), \tag{C.27}$$

then the necessary conditions in Proposition C.3 can be written as

$$\frac{\partial L(x^*, \lambda^*)}{\partial x} = 0, \qquad \frac{\partial L(x^*, \lambda^*)}{\partial \lambda} = 0, \tag{C.28}$$

$$v^T \frac{\partial^2 L(x^*, \lambda^*)}{\partial x^2} v \geq 0, \quad \forall v : v^T \nabla h(x^*) = 0. \tag{C.29}$$

The conditions (C.28) give a system of $n + m$ equations with $n + m$ unknowns: the entries of $x^*$ and $\lambda^*$. If the constraint $h(x) = 0$ is regular, then in principle, this system of equations is independent, and we should be able to solve for $x^*$ and $\lambda^*$ from them. The solutions will contain all the (local) minima, but it is possible that some of them need not be minima at all. Nevertheless, whether we are able to solve these equations or not, they usually provide rich information about the minima of the constrained optimization.

*The augmented Lagrangian function*

If we are not able to solve for the minimum from the equations given by the necessary conditions, we must resort to a brute-force search scheme. The basic idea is try to convert the original constrained optimization to an unconstrained one by introducing an extra *penalty* terms to the objective function. A typical choice is the *augmented Lagrangian function* $L_c : \mathbb{R}^{n+m} \to \mathbb{R}$ defined as

$$L_c(x, \lambda) \doteq f(x) + \lambda^T h(x) + \frac{c}{2}\|h(x)\|^2, \tag{C.30}$$

where $c > 0$ is a positive penalty parameter. It is reasonable to expect that for very large $c$, the location $x^*$ of the global minimum of the unconstrained minimization

$$(x^*, \lambda^*) = \arg\min L_c(x, \lambda) \tag{C.31}$$

should be very close to the global minimum of the original constrained minimization.

**Proposition C.5.** *For $k = 0, 1, \ldots$, let $x^k$ be a global minimum of the unconstrained optimization problem*

$$\min_x L_{c^k}(x, \lambda^k), \tag{C.32}$$

*where $\{\lambda^k\}$ is bounded, $0 < c^k < c^{k+1}$ for all $k$, and $c^k \to \infty$. Then the limit of the sequence $\{x^k\}$ is a global minimum of the original constrained optimization problem.*

# References


[Abel, 1828] Abel, N. H. (1828). Sur la resolution algébrique des équations.

[Adelson and Bergen, 1991] Adelson, E. and Bergen, J. (1991). *Computational models of visual processing. M. Landy and J. Movshon eds.*, chapter "The plenoptic function and the elements of early vision". MIT press. Also appeared as MIT-MediaLab-TR148. September 1990.

[Adiv, 1985] Adiv, G. (1985). Determine three-dimensional motion and structure from optical flow generated by several moving objects. *IEEE Transactions on Pattern Analysis & Machine Intelligence*, 7(4):384–401.

[Aloimonos, 1990] Aloimonos, J. (1990). Perspective approximations. *Image and Vision Computing*, 8(3):179–192.

[Anandan et al., 1994] Anandan, P., Hanna, K., and Kumar, R. (1994). Shape recovery from multiple views: A parallax based approach. In *Proceedings of Int. Conference on Pattern Recognition*, pages A: 685–688.

[Armstrong et al., 1996] Armstrong, M., Zisserman, Z., and Hartley, R. (1996). Self-calibration from image triplets. In *Proceedings of European Conference on Computer Vision*, pages 3–16.

[Åström et al., 1999] Åström, K., Cipolla, R., and Giblin, P. (1999). Generalized epipolar constraints. *Int. Journal of Computer Vision*, 33(1):51–72.

[Avidan and Shashua, 1998] Avidan, S. and Shashua, A. (1998). Novel view synthesis by cascading trilinear tensors. *IEEE Transactions on Visualization and Computer Graphics (TVCG)*, 4(4):293–306.

[Avidan and Shashua, 1999] Avidan, S. and Shashua, A. (1999). Trajectory triangulation of lines: Reconstruction of a 3D point moving along a line from a monocular image sequence. In *Proceedings of Int. Conference on Computer Vision & Pattern Recognition*, pages 2062–2066.

[Avidan and Shashua, 2000] Avidan, S. and Shashua, A. (2000). Trajectory triangulation: 3D reconstruction of moving points from a monocular image sequence. *IEEE Transactions on Pattern Analysis & Machine Intelligence*, 22(4):348–357.

[Ball, 1900] Ball, R. S. (1900). *A Treatise on the Theory of Screws*. Cambridge University Press.

[Bartlett, 1956] Bartlett, M. (1956). *An Introduction to Stochastic Processes*. Cambridge University Press.

[Basri, 1996] Basri, R. (1996). Paraperspective ≡ Affine. *Int. Journal of Computer Vision*, 19(2):169–179.

[Beardsley et al., 1997] Beardsley, P., Zisserman, A., and Murray, D. (1997). Sequential update of projective and affine structure from motion. *Int. Journal of Computer Vision*, 23(3):235–259.

[Bellman, 1957] Bellman, R. (1957). *Dynamic Programming*. Princeton University Press.

[Bergen et al., 1992] Bergen, J. R., Anandan, P., Hanna, K., and Hingorani, R. (1992). Hierarchical model-based motion estimation. In *Proceedings of European Conference on Computer Vision*, volume 1, pages 237–252.

[Berthilsson et al., 1999] Berthilsson, R., Åström, K., and Heyden, A. (1999). Reconstruction of curves in $\mathbb{R}^3$, using factorization and bundle adjustment. In *Proceedings of IEEE International Conference on Computer Vision*, pages 674–679.

[Bertsekas, 1999] Bertsekas, D. P. (1999). *Nonlinear Programming*. Athena Scientific, second edition.

[Bieberbach, 1910] Bieberbach, L. (1910). Über die Bewegungsgruppen des n-dimensionalen Euklidischen Raumes mit einem endlichen Fundamentalbereich. *Gött. Nachr.*, pages 75–84.

[Black and Anandan, 1993] Black, M. J. and Anandan, P. (1993). A framework for the robust estimation of optical flow. In *Proceedings of European Conference on Computer Vision*, pages 231–236.

[Blake and Isard, 1998] Blake, A. and Isard, M. (1998). *Active Contours*. Springer Verlag.

[Born and Wolf, 1999] Born, M. and Wolf, E. (1999). *Electromagnetic Theory of Propagation, Interference and Diffraction of Light*. Cambridge University Press, seventh edition.

[Boufama and Mohr, 1995] Boufama, B. and Mohr, R. (1995). Epipole and fundamental matrix estimation using the virtual parallax property. In *Proceedings of IEEE International Conference on Computer Vision*, pages 1030–1036, Boston, MA.

[Bougnoux, 1998] Bougnoux, S. (1998). From projective to Euclidean space under any practical situation, a criticism of self-calibration. In *Proceedings of Int. Conference on Computer Vision & Pattern Recognition*, pages 790–796.

[Brank et al., 1993] Brank, P., Mohr, R., and Bobet, P. (1993). Distorsions optiques: correction dans un modèle projectif. *Technical Report 1933, LIFIA-INRIA, Rhône-Alpes*.

[Bregler and Malik, 1998] Bregler, C. and Malik, J. (1998). Tracking people with twists and exponential maps. In *Proceedings of Int. Conference on Computer Vision & Pattern Recognition*, pages 141–151.

[Brockett, 1984] Brockett, R. W. (1984). Robotic manipulator and the product of exponentials formula. In *Proceedings of Mathematical Theory of Networks and Systems*, pages 120–129. Springer-Verlag.

[Brodsky et al., 1998] Brodsky, T., Fermüller, C., and Aloimonos, Y. (1998). Self-calibration from image derivatives. In *Proceedings of IEEE International Conference on Computer Vision*, pages 83–89, Bombay, India.

[Broida et al., 1990] Broida, T., Chandrashekhar, S., and Chellappa, R. (1990). Recursive 3-D motion estimation from a monocular image sequence. *IEEE Transactions on Aerospace and Electronic Systems*, 26(4):639–656.

[Broida and Chellappa, 1986a] Broida, T. and Chellappa, R. (1986a). Estimation of object motion parameters from noisy images. *IEEE Transactions on Pattern Analysis & Machine Intelligence*, 8(1):90–99.

[Broida and Chellappa, 1986b] Broida, T. and Chellappa, R. (1986b). Kinematics of a rigid object from a sequence of noisy images: A batch approach. In *Proceedings of Int. Conference on Computer Vision & Pattern Recognition*, pages 176–182.

[Brooks et al., 1997] Brooks, M. J., Chojnacki, W., and Baumela, L. (1997). Determining the ego-motion of an uncalibrated camera from instantaneous optical flow. *Journal of the Optical Society of America*, 14(10):2670–2677.

[Brooks et al., 1996] Brooks, M. J., de Agapito, L., Huynh, D., and Baumela, L. (1996). Direct methods for self-calibration of a moving stereo head. In *Proceedings of European Conference on Computer Vision*, pages II:415–426, Cambridge, UK.

[Bruss and Horn, 1983] Bruss, A. R. and Horn, B. K. (1983). Passive navigation. *Computer Graphics and Image Processing*, 21:3–20.

[Bucy, 1965] Bucy, R. (1965). Nonlinear filtering theory. *IEEE Transactions on Automatic Control*, 10.

[Burt and Adelson, 1983] Burt, P. and Adelson, E. H. (1983). The Laplacian pyramid as a compact image code. *IEEE Transactions on Communication*, 31:532–540.

[Callier and Desoer, 1991] Callier, F. M. and Desoer, C. A. (1991). *Linear System Theory*. Springer Texts in Electrical Engineering. Springer-Verlag.

[Canny, 1986] Canny, J. F. (1986). A computational approach to edge detection. *IEEE Transactions on Pattern Analysis & Machine Intelligence*, 8(6):679–698.

[Canterakis, 2000] Canterakis, N. (2000). A minimal set of constraints for the trifocal tensor. In *Proceedings of European Conference on Computer Vision*, pages I: 84–99.

[Caprile and Torre, 1990] Caprile, B. and Torre, V. (1990). Using vanishing points for camera calibration. *Int. Journal of Computer Vision*, 4(2):127–140.

[Carlsson, 1994] Carlsson, S. (1994). *Applications of Invariance in Computer Vision*, chapter "Multiple image invariance using the double algebra", pages 145–164. Springer Verlag.

[Carlsson, 1998] Carlsson, S. (1998). Symmetry in perspective. In *Proceedings of European Conference on Computer Vision*, pages 249–263.

[Carlsson and Weinshall, 1998] Carlsson, S. and Weinshall, D. (1998). Dual computation of projective shape and camera positions from multiple images. *Int. Journal of Computer Vision*, 27(3):227–241.

[Casadei and Mitter, 1998] Casadei, S. and Mitter, S. (1998). A perceptual organization approach to contour estimation via composition, compression and pruning of contour hypotheses. *LIDS Technical report LIDS-P-2415,MIT*.

[Chai and Ma, 1998] Chai, J. and Ma, S. (1998). Robust epipolar geometry estimation using generic algorithm. *Pattern Recognition Letters*, 19(9):829–838.

[Chan and Vese, 1999] Chan, T. and Vese, L. (1999). An active contours model without edges. In *Proceedings of Int. Conf. Scale-Space Theories in Computer Vision*, pages 141–151.

[Chen and Medioni, 1999] Chen, Q. and Medioni, G. (1999). Efficient iterative solutions to m-view projective reconstruction problem. In *Proceedings of Int. Conference on Computer Vision & Pattern Recognition*, pages II:55–61, Fort Collins, Colorado.

[Chiuso et al., 2000] Chiuso, A., Brockett, R., and Soatto, S. (2000). Optimal structure from motion: local ambiguities and global estimates. *Int. Journal of Computer Vision*, 39(3):195–228.

[Chiuso et al., 2002] Chiuso, A., Favaro, P., Jin, H., and Soatto, S. (2002). Motion and structure causally integrated over time. *IEEE Transactions on Pattern Analysis & Machine Intelligence*, 24 (4):523–535.

[Christy and Horaud, 1996] Christy, S. and Horaud, R. (1996). Euclidean shape and motion from multiple perspective views by affine iterations. *IEEE Transactions on Pattern Analysis & Machine Intelligence*, 18(11):1098–1104.

[Cipolla et al., 1995] Cipolla, R., Åström, K., and Giblin, P. J. (1995). Motion from the frontier of curved surfaces. In *Proceedings of IEEE International Conference on Computer Vision*, pages 269–275.

[Cipolla et al., 1999] Cipolla, R., Robertson, D., and Boye, E. (1999). Photobuilder – 3D models of architectural scenes from uncalibrated images. In *Proceedings of IEEE International Conference on Multimedia Computing and Systems, Firenze*.

[Cohen, 1991] Cohen, L. D. (1991). On active contour models and balloons. *Comp. Vision, Graphics, and Image Processing: Image Understanding*, 53(2):211–218.

[Collins and Weiss, 1990] Collins, R. and Weiss, R. (1990). Vanishing point calculation as a statistical inference on the unit sphere. In *Proceedings of IEEE International Conference on Computer Vision*, pages 400–403.

[Costeira and Kanade, 1995] Costeira, J. and Kanade, T. (1995). A multi-body factorization method for motion analysis. In *Proceedings of IEEE International Conference on Computer Vision*, pages 1071–1076.

[Criminisi, 2000] Criminisi, A. (2000). *Accurate Visual Metrology from Single and Multiple Uncalibrated Images*. Springer Verlag.

[Criminisi et al., 1999] Criminisi, A., Reid, I., and Zisserman, A. (1999). Single view metrology. In *Proceedings of IEEE International Conference on Computer Vision*, pages 434–441.

[Csurka et al., 1998] Csurka, G., Demirdjian, D., Ruf, A., and Horaud, R. (1998). Closed-form solutions for the Euclidean calibration of a stereo rig. In *Proceedings of European Conference on Computer Vision*, pages 426–442.

[Daniilidis and Nagel, 1990] Daniilidis, K. and Nagel, H.-H. (1990). Analytical results on error sensitivity of motion estimation from two views. *Image and Vision Computing*, 8:297–303.

[Daniilidis and Spetsakis, 1997] Daniilidis, K. and Spetsakis, M. (1997). *Visual Navigation*, chapter "Understanding Noise Sensitivity in Structure from Motion", pages 61–68. Lawrence Erlbaum Associates.

[de Agapito et al., 1999] de Agapito, L., Hartley, R., and Hayman, E. (1999). Linear calibration of rotating and zooming cameras. In *Proceedings of Int. Conference on Computer Vision & Pattern Recognition*.

[Debevec et al., 1996] Debevec, P., Taylor, C., and Malik, J. (1996). Modeling and rendering architecture from photographs: A hybrid geometry and image-based approach. In *Proceedings of SIGGRAPH'96*.

[Demazure, 1988] Demazure, M. (1988). Sur deux problèmes de reconstruction. *Technical Report, No. 882, INRIA, Rocquencourt, France.*

[Demirdjian and Horaud, 1992] Demirdjian, D. and Horaud, R. (1992). A projective framework for scene segmentation in the presence of moving objects. In *Proceedings of Int. Conference on Computer Vision & Pattern Recognition*, pages I: 2–8.

[Devernay and Faugeras, 1995] Devernay, F. and Faugeras, O. (1995). Automated calibration and removal of distortion from scenes of structured environments. In *SPIE*, volume 2567.

[Devernay and Faugeras, 1996] Devernay, F. and Faugeras, O. (1996). From projective to Euclidean reconstruction. In *Proceedings of Int. Conference on Computer Vision & Pattern Recognition*, pages 264–269.

[Dickmanns, 1992] Dickmanns, E. D. (1992). A general dynamic vision architecture for UGV and UAV. *Journal of Applied Intelligence*, 2:251–270.

[Dickmanns, 2002] Dickmanns, E. D. (2002). Vision for ground vehicles: history and prospects. *Int. Journal of Vehicle Autonomous Systems*, 1(1):1–44.

[Dickmanns and Christians, 1991] Dickmanns, E. D. and Christians, T. (1991). Relative 3-D-state estimation for autonomous visual guidance of road vehicles. *Robotics and Autonomous Systems*, 7:113–123.

[Dickmanns and Graefe, 1988a] Dickmanns, E. D. and Graefe, V. (1988a). Applications of dynamic monocular machine vision. *Machine Vision and Applications*, 1(4):241–261.

[Dickmanns and Graefe, 1988b] Dickmanns, E. D. and Graefe, V. (1988b). Dynamic monocular machine vision. *Machine Vision and Applications*, 1(4):223–240.

[Dickmanns and Mysliwetz, 1992] Dickmanns, E. D. and Mysliwetz, B. D. (1992). Recursive 3-D road and relative ego-state estimation. *IEEE Transactions on Pattern Analysis & Machine Intelligence*, 14(2):199–213.

[Enciso and Vieville, 1997] Enciso, R. and Vieville, T. (1997). Self-calibration from four views with possibly varying intrinsic parameters. *Journal of Image and Vision Computing*, 15(4):293–305.

[Espiau et al., 1992] Espiau, B., Chaumette, F., and Rives, P. (1992). A new approach to visual servoing in robotics. *IEEE Transactions on Robotics and Automation*, 8(3):313–326.

[Farid and Simoncelli, 1997] Farid, H. and Simoncelli, E. P. (1997). Optimally rotation-equivariant directional derivative kernels. In *Proceedings of Computer Analysis of Images and Patterns*, Kiel, Germany.

[Faugeras, 1992] Faugeras, O. (1992). What can be seen in three dimensions with an uncalibrated stereo rig? In *Proceedings of European Conference on Computer Vision*, pages 563–578. Springer-Verlag.

[Faugeras, 1993] Faugeras, O. (1993). *Three-Dimensional Computer Vision*. The MIT Press.

[Faugeras, 1995] Faugeras, O. (1995). Stratification of three-dimensional vision: projective, affine, and metric representations. *Journal of the Optical Society of America*, 12(3):465–484.

[Faugeras and Laveau, 1994] Faugeras, O. and Laveau, S. (1994). Representing three-dimensional data as a collection of images and fundamental matrices for image synthesis. In *Proceedings of Int. Conference on Pattern Recognition*, pages 689–691, Jerusalem, Israel.

[Faugeras and Luong, 2001] Faugeras, O. and Luong, Q.-T. (2001). *Geometry of Multiple Images*. The MIT Press.

[Faugeras et al., 1992] Faugeras, O., Luong, Q.-T., and Maybank, S. (1992). Camera self-calibration: theory and experiments. In *Proceedings of European Conference on Computer Vision*, pages 321–334. Springer-Verlag.

[Faugeras and Lustman, 1988] Faugeras, O. and Lustman, F. (1988). Motion and structure from motion in a piecewise planar environment. *the International Journal of Pattern Recognition in Artificial Intelligence*, 2(3):485–508.

[Faugeras et al., 1987] Faugeras, O., Lustman, F., and Toscani, G. (1987). Motion and structure from motion from point and line matches. In *Proceedings of IEEE International Conference on Computer Vision*, pages 25–34, London, England. IEEE Comput. Soc. Press.

[Faugeras and Mourrain, 1995] Faugeras, O. and Mourrain, B. (1995). On the geometry and algebra of the point and line correspondences between $n$ images. In *Proceedings of IEEE International Conference on Computer Vision*, pages 951–956, Cambridge, MA, USA. IEEE Comput. Soc. Press.

[Faugeras and Papadopoulos, 1995] Faugeras, O. and Papadopoulos, T. (1995). Grassmann-Cayley algebra for modeling systems of cameras and the algebraic equations of the manifold of trifocal tensors. In *Proceedings of the IEEE workshop of representation of visual scenes*.

[Faugeras and Papadopoulos, 1998] Faugeras, O. and Papadopoulos, T. (1998). A nonlinear method for estimating the projective geometry of three views. In *Proceedings of IEEE International Conference on Computer Vision*, pages 477–484.

[Favaro et al., 2003] Favaro, P., Jin, H., and Soatto, S. (2003). A semi-direct approach to structure from motion. *The Visual Computer*, 19:1–18.

[Fedorov, 1885] Fedorov, E. S. (1885). The elements of the study of figures. *Zapiski Imperatorskogo S. Peterburgskogo Mineralogicheskogo Obshchestva [Proc. S. Peterb. Mineral. Soc.]*, 2(21):1–289.

[Fedorov, 1971] Fedorov, E. S. (1971). *Symmetry of Crystals*. Translated from the 1949 Russian Edition. American Crystallographic Association.

[Felleman and van Essen, 1991] Felleman, D. J. and van Essen, D. C. (1991). Distributed hierarchical processing in the primate cerebral cortex. *Cerebral Cortex*, 1:1–47.

[Fermüller and Aloimonos, 1995] Fermüller, C. and Aloimonos, Y. (1995). Qualitative egomotion. *Int. Journal of Computer Vision*, 15:7–29.

[Fermüller et al., 1997] Fermüller, C., Cheong, L.-F., and Aloimonos, Y. (1997). Visual space distortion. *Biological Cybernetics*, 77:323–337.

[Fischler and Bolles, 1981] Fischler, M. A. and Bolles, R. C. (1981). Random sample consensus: a paradigm for model fitting with application to image analysis and automated cartography. *Comm. of ACM*, 24(6):381–395.

[Fitzgibbon, 2001] Fitzgibbon, A. (2001). Simultaneous linear estimation of multiple view geometry and lens distortion. In *Proceedings of the Int. Conference on Computer Vision and Pattern Recognition, also UK Patent Application 0124608.1.*

[Forsyth and Ponce, 2002] Forsyth, D. and Ponce, J. (2002). *Computer Vision: A Modern Approach*. Prentice Hall.

[François et al., 2002] François, A., Medioni, G., and Waupotitsch, R. (2002). Reconstructing mirror symmetric scenes from a single view using 2-view stereo geometry. In *Proceedings of Int. Conference on Pattern Recognition*.

[Fusiello et al., 1997] Fusiello, A., Trucco, E., , and Verri, A. (1997). Rectification with unconstrained stereo geometry. In *Proceedings of the British Machine Vision Conference*, pages 400–409.

[Galois, 1931] Galois, E. (1931). Mémoire sur les conditions de résolubilité des équations par redicaux. *Ouevres Mathématiques*, pages 33–50.

[Gärding, 1992] Gärding, J. (1992). Shape from texture for smooth curved surfaces in perspective projection. *Journal of Mathematical Imaging and Vision*, 2(4):327–350.

[Gärding, 1993] Gärding, J. (1993). Shape from texture and contour by weak isotropy. *Journal of Artificial Intelligence*, 64(2):243–297.

[Geyer and Daniilidis, 2001] Geyer, C. and Daniilidis, K. (2001). Catadioptric projective geometry. *Int. Journal of Computer Vision*, 43:223–243.

[Gibson, 1950] Gibson, J. (1950). *The Perception of the Visual World*. Houghton Mifflin.

[Golub and Loan, 1989] Golub, G. and Loan, C. V. (1989). *Matrix computations*. Johns Hopkins University Press, second edition.

[Gonzalez and Woods, 1992] Gonzalez, R. and Woods, R. (1992). *Digital Image Processing*. Addison-Wesley.

[Goodman, 2003] Goodman, F. (2003). *Algebra: Abstract and Concrete, Stressing Symmetry*. Prentice Hall, second edition.

[Gortler et al., 1996] Gortler, S. J., Grzeszczuk, R., Szeliski, R., and Choen, M. F. (1996). The lumigraph. In *Proceedings of SIGGRAPH '96*, pages 43–54.

[Gregor et al., 2000] Gregor, R., Lutzeler, M., Pellkofer, M., Siedersberger, K.-H., and Dickmanns, E. D. (2000). EMS-vision: A perceptual systems for autonomous vehicles. In *Proc. of Internat. Symp. on Intelligent Vehicles*, pages 52–57, Dearborn, MI, USA.

[Grünbaum and Shephard, 1987] Grünbaum, B. and Shephard, G. C. (1987). *Tilings and Patterns*. W. H. Freeman and Company.

[Han and Kanade, 2000] Han, M. and Kanade, T. (2000). Reconstruction of a scene with multiple linearly moving objects. In *Proceedings of Int. Conference on Computer Vision & Pattern Recognition*, volume 2, pages 542–549.

[Harris and Stephens, 1988] Harris, C. and Stephens, M. (1988). A combined corner and edge detector. In *Proceedings of the Alvey Conference*, pages 189–192.

[Harris, 1992] Harris, J. (1992). *Algebraic Geometry: A First Course*. Springer-Verlag.

[Hartley, 1994a] Hartley, R. (1994a). Lines and points in three views – a unified approach. In *Proceedings of 1994 Image Understanding Workshop*, pages 1006–1016, Monterey, CA, USA. Omnipress.

[Hartley, 1994b] Hartley, R. (1994b). Self-calibration from multiple views with a rotating camera. In *Proceedings of European Conference on Computer Vision*, pages 471–478.

[Hartley, 1995] Hartley, R. (1995). A linear method for reconstruction from lines and points. In *Proceedings of IEEE International Conference on Computer Vision*, pages 882–887.

[Hartley, 1997] Hartley, R. (1997). In defence of the eight-point algorithm. *IEEE Transactions on Pattern Analysis & Machine Intelligence*, 19(6):580–593.

[Hartley, 1998a] Hartley, R. (1998a). Chirality. *Int. Journal of Computer Vision*, 26(1):41–61.

[Hartley, 1998b] Hartley, R. (1998b). Minimizing algebraic error in geometric estimation problems. In *Proceedings of IEEE International Conference on Computer Vision*, pages 469–476.

[Hartley et al., 1992] Hartley, R., Gupta, R., and Chang, T. (1992). Stereo from uncalibrated cameras. In *Proceedings of Int. Conference on Computer Vision & Pattern Recognition*, pages 761–764, Urbana-Champaign, IL, USA. IEEE Comput. Soc. Press.

[Hartley and Kahl, 2003] Hartley, R. and Kahl, F. (2003). A critical configuration for reconstruction from rectilinear motion. In *Proceedings of Int. Conference on Computer Vision & Pattern Recognition*.

[Hartley and Sturm, 1997] Hartley, R. and Sturm, P. (1997). Triangulation. *Computer Vision and Image Understanding*, 68(2):146–157.

[Hartley and Zisserman, 2000] Hartley, R. and Zisserman, A. (2000). *Multiple View Geometry in Computer Vision*. Cambridge Univ. Press.

[Heeger and Jepson, 1992] Heeger, D. J. and Jepson, A. D. (1992). Subspace methods for recovering rigid motion I: Algorithm and implementation. *Int. Journal of Computer Vision*, 7(2):95–117.

[Heyden, 1995] Heyden, A. (1995). *Geometry and Algebra of Multiple Projective Transformations*. Ph.D. thesis, Lund University.

[Heyden, 1998] Heyden, A. (1998). Reduced multilinear constraints – theory and experiments. *Int. Journal of Computer Vision*, 30(2):5–26.

[Heyden and Åström, 1996] Heyden, A. and Åström, K. (1996). Euclidean reconstruction from constraint intrinsic parameters. In *Proceedings of Int. Conference on Pattern Recognition*, pages 339–343.

[Heyden and Åström, 1997] Heyden, A. and Åström, K. (1997). Algebraic properties of multilinear constraints. *Mathematical Methods in Applied Sciences*, 20(13):1135–1162.

[Heyden and Åström, 1997] Heyden, A. and Åström, K. (1997). Euclidean reconstruction from image sequences with varying and unknown focal length and principal point. In *Proceedings of Int. Conference on Computer Vision & Pattern Recognition*.

[Heyden and Sparr, 1999] Heyden, A. and Sparr, G. (1999). Reconstruction from calibrated cameras – a new proof of the Kruppa Demazure theorem. *Journal of Mathematical Imaging and Vision*, pages 1–20.

[Hockney, 2001] Hockney, D. (2001). *Secret knowledge: rediscovering the lost techniques of the old masters*. Viking Press.

[Hofman et al., 2000] Hofman, U., Rieder, A., and Dickmanns, E. D. (2000). EMS-vision: An application to intelligent cruise control for high speed roads. In *Proc. of Internat. Symp. on Intelligent Vehicles*, pages 468–473, Dearborn, MI, USA.

[Hong et al., 2002] Hong, W., Yang, A. Y., and Ma, Y. (2002). On group symmetry in multiple view geometry: Structure, pose and calibration from single images. *Technical Report, UILU-02-2208 DL-206, September 12.*

[Hoppe, 1996] Hoppe, H. (1996). Progressive meshes. *Computer Graphics*, 30(Annual Conference Series):99–108.

[Horaud and Csurka, 1998] Horaud, R. and Csurka, G. (1998). Self-calibration and Euclidean reconstruction using motions of a stereo rig. In *Proceedings of IEEE International Conference on Computer Vision*, pages 96–103.

[Horn, 1986] Horn, B. (1986). *Robot Vision*. MIT Press.

[Horn, 1987] Horn, B. (1987). Closed-form solution of absolute orientation using unit quaternions. *Journal of the Optical Society of America*, 4(4):629–642.

[Huang et al., 2002] Huang, K., Ma, Y., and Fossum, R. (2002). Generalized rank conditions in multiple view geometry and its applications to dynamical scenes. In *Proceedings of European Conference on Computer Vision*, pages 201–216, Copenhagen, Denmark.

[Huang et al., 2003] Huang, K., Yang, Y., Hong, W., and Ma, Y. (April 14, 2003). Symmetry-based 3-D reconstruction from perspective images (Part II): Matching. *UIUC CSL Technical Report, UILU-ENG-03-2204.*

[Huang and Faugeras, 1989] Huang, T. and Faugeras, O. (1989). Some properties of the E matrix in two-view motion estimation. *IEEE Transactions on Pattern Analysis & Machine Intelligence*, 11(12):1310–1312.

[Hutchinson et al., 1996] Hutchinson, S., Hager, G. D., and Corke, P. I. (1996). A tutorial on visual servo control. *IEEE Transactions on Robotics and Automation*, pages 651–670.

[Huynh, 1999] Huynh, D. (1999). Affine reconstruction from monocular vision in the presence of a symmetry plane. In *Proceedings of IEEE International Conference on Computer Vision*, pages 476–482.

[Irani, 1999] Irani, M. (1999). Multi-frame optical flow estimation using subspace constraints. In *Proceedings of IEEE International Conference on Computer Vision*, pages I: 626–633.

[Isidori, 1989] Isidori, A. (1989). *Nonlinear Control Systems*. Communications and Control Engineering Series. Springer-Verlag, second edition.

[Jazwinski, 1970] Jazwinski, A. H. (1970). *Stochastic Processes and Filtering Theory*. NY: Academic Press.

[Jelinek and Taylor, 1999] Jelinek, D. and Taylor, C. (1999). Reconstructing linearly parameterized models from single views with a camera of unknown focal length. In *Proceedings of Int. Conference on Computer Vision & Pattern Recognition*, pages II: 346–352.

[Jepson and Heeger, 1993] Jepson, A. D. and Heeger, D. J. (1993). Linear subspace methods for recovering translation direction. *Spatial Vision in Humans and Robots, Cambridge Univ. Press*, pages 39–62.

[Jin et al., 2003] Jin, H., Soatto, S., and Yezzi, A. J. (2003). Multi-view stereo beyond Lambert. In *Proceedings of Int. Conference on Computer Vision & Pattern Recognition*.

[Kahl, 1999] Kahl, F. (1999). Critical motions and ambiguous Euclidean reconstructions in auto-calibration. In *Proceedings of IEEE International Conference on Computer Vision*, pages 469–475.

[Kahl and Heyden, 1999] Kahl, F. and Heyden, A. (1999). Affine structure and motion from points, lines and conics. *Int. Journal of Computer Vision*, 33(3):163–180.

[Kalman, 1960] Kalman, R. (1960). A new approach to linear filtering and prediction problems. *Trans. of the ASME-Journal of Basic Engineering.*, 35–45.

[Kanade, 1981] Kanade, T. (1981). Recovery of the three-dimensional shape of an object from a single view. *Journal of Artificial Intelligence*, 33(3):1–18.

[Kanatani, 1985] Kanatani, K. (1985). Detecting the motion of a planar surface by line & surface integrals. In *Computer Vision, Graphics, and Image Processing*, volume 29, pages 13–22.

[Kanatani, 1993a] Kanatani, K. (1993a). 3D interpretation of optical flow by renormalization. *Int. Journal of Computer Vision*, 11(3):267–282.

[Kanatani, 1993b] Kanatani, K. (1993b). *Geometric Computation for Machine Vision*. Oxford Science Publications.

[Kanatani, 2001] Kanatani, K. (2001). Motion segmentation by subspace separation and model selection. In *Proceedings of IEEE International Conference on Computer Vision*, volume 2, pages 586–591.

[Kass et al., 1987] Kass, M., Witkin, A., and Terzopoulos, D. (1987). Snakes: active contour models. *Int. Journal of Computer Vision*, 1:321–331.

[Kichenassamy et al., 1996] Kichenassamy, S., Kumar, A., Olver, P., Tannenbaum, A., and Yezzi, A. (1996). Conformal curvature flows: From phase transitions to active vision. *Arch. Rat. Mech. Anal.*, 134:275–301.

[Konderink and van Doorn, 1991] Konderink, J. J. and van Doorn, A. J. (1991). Affine structure from motion. *Journal of Optical Society of America*, 8(2):337–385.

[Konderink and van Doorn, 1997] Konderink, J. J. and van Doorn, A. J. (1997). The generic bilinear calibration-estimation problem. *Int. Journal of Computer Vision*, 23(3):217–234.

[Kosecká et al., 1997] Kosecká, J., Blasi, R., Taylor, C. J., and Malik, J. (1997). Vision-based lateral control of vehicles. In *Proceedings of Intelligent Transportation Systems Conference*, Boston.

[Kosecká and Ma, 2002] Kosecká, J. and Ma, Y. (2002). Introduction to multiview rank conditions and their applications: A review. In *Proceedings of 2002 Tyrrhenian Workshop on Digital Communications, Advanced Methods for Multimedia Signal Processing*, pages 161–169, Capri, Italy.

[Kosecká and Zhang, 2002] Kosecká, J. and Zhang, W. (2002). Video compass. In *Proceedings of European Conference on Computer Vision*, pages 476–491, Copenhagen, Denmark.

[Kruppa, 1913] Kruppa, E. (1913). Zur Ermittlung eines Objektes aus zwei Perspektiven mit innerer Orientierung. *Sitz.-Ber. Akad. Wiss., Math. Naturw., Kl. Abt. IIa, 122:1939-1948.*

[Kutulakos and Seitz, 1999] Kutulakos, K. and Seitz, S. (1999). A theory of shape by space carving. In *Proceedings of IEEE International Conference on Computer Vision*, pages 307–314.

[Lavest et al., 1993] Lavest, J., Rives, G., and Dhome, M. (1993). 3D reconstruction by zooming. *IEEE Transactions on Robotics and Automation*, pages 196–207.

[Levoy and Hanrahan, 1996] Levoy, M. and Hanrahan, P. (1996). Light field rendering. In *Proceedings of SIGGRAPH '96*, pages 161–170.

[Lhuillier and Quan, 2002] Lhuillier, M. and Quan, L. (2002). Quasi-dense reconstruction from image sequence. In *Proceedings of European Conference on Computer Vision*, volume 2, pages 125–139.

[Li and Brooks, 1999] Li, Y. and Brooks, M. (1999). An efficient recursive factorization method for determining structure from motion. In *Proceedings of IEEE International Conference on Computer Vision*, pages I:138–143, Fort Collins, Colorado.

[Liebowitz and Zisserman, 1998] Liebowitz, D. and Zisserman, A. (1998). Detecting rotational symmetries using normalized convolution. In *Proceedings of Int. Conference on Computer Vision & Pattern Recognition*, pages 482–488.

[Liebowitz and Zisserman, 1999] Liebowitz, D. and Zisserman, A. (1999). Combining scene and autocalibration constraints. In *Proceedings of IEEE International Conference on Computer Vision*, pages I: 293–330.

[Liu and Huang, 1986] Liu, Y. and Huang, T. (1986). Estimation of rigid body motion using straight line correspondences. In *Proceedings of IEEE workshop on motion: Representation and Analysis*, Kiawah Island, SC.

[Liu et al., 1990] Liu, Y., Huang, T., and Faugeras, O. (1990). Determination of camera location from 2-D to 3-D line and point correspondences. *IEEE Transactions on Pattern Analysis & Machine Intelligence*, pages 28–37.

[Longuet-Higgins, 1981] Longuet-Higgins, H. C. (1981). A computer algorithm for reconstructing a scene from two projections. *Nature*, 293:133–135.

[Longuet-Higgins, 1986] Longuet-Higgins, H. C. (1986). The reconstruction of a plane surface from two perspective projections. In *Proceedings of Royal Society of London*, volume 227 of *B*, pages 399–410.

[Longuet-Higgins, 1988] Longuet-Higgins, H. C. (1988). Multiple interpretation of a pair of images of a surface. In *Proceedings of Royal Society of London*, volume 418 of *A*, pages 1–15.

[Lucas and Kanade, 1981] Lucas, B. and Kanade, T. (1981). An iterative image registration technique with an application to stereo vision. In *Proceedings of the Seventh International Joint Conference on Artificial Intelligence*, pages 674–679.

[Luong and Faugeras, 1996] Luong, Q.-T. and Faugeras, O. (1996). The fundamental matrix: theory, algorithms, and stability analysis. *Int. Journal of Computer Vision*, 17(1):43–75.

[Luong and Faugeras, 1997] Luong, Q.-T. and Faugeras, O. (1997). Self-calibration of a moving camera from point correspondences and fundamental matrices. *Int. Journal of Computer Vision*, 22(3):261–289.

[Luong and Vieville, 1994] Luong, Q.-T. and Vieville, T. (1994). Canonical representations for the geometries of multiple projective views. In *Proceedings of European Conference on Computer Vision*, pages 589–599.

[Lutton et al., 1994] Lutton, E., Maitre, H., and Lopez-Krahe, J. (1994). Contributions to the determination of vanishing points using Hough transformation. *IEEE Transactions on Pattern Analysis & Machine Intelligence*, 16(4):430–438.

[Ma, 2003] Ma, Y. (2003). A differential geometric approach to multiple view geometry in spaces of constant curvature. *Int. Journal of Computer Vision.*

[Ma et al., 2002] Ma, Y., Huang, K., and Kosecká, J. (2002). Rank deficiency condition of the multiple view matrix for mixed point and line features. In *Proceedings of Asian Conference on Computer Vision*, Melbourne, Australia.

[Ma et al., 2001a] Ma, Y., Huang, K., Vidal, R., Kosecká, J., and Sastry, S. (2001a). Rank conditions of multiple view matrix in multiple view geometry. *UIUC, CSL Technical Report, UILU-ENG 01-2214 (DC-220).*

[Ma et al., 2003] Ma, Y., Huang, K., Vidal, R., Kosecká, J., and Sastry, S. (2003). Rank conditions on the multiple view matrix. *International Journal of Computer Vision, to appear.*

[Ma et al., 2000a] Ma, Y., Kosecká, J., and Sastry, S. (2000a). Linear differential algorithm for motion recovery: A geometric approach. *Int. Journal of Computer Vision*, 36(1):71–89.

[Ma et al., 2001b] Ma, Y., Kosecká, J., and Sastry, S. (2001b). Optimization criteria and geometric algorithms for motion and structure estimation. *Int. Journal of Computer Vision*, 44(3):219–249.

[Ma et al., 1999] Ma, Y., Soatto, S., Kosecká, J., and Sastry, S. (1999). Euclidean reconstruction and reprojection up to subgroups. In *Proceedings of IEEE International Conference on Computer Vision*, pages 773–780, Corfu, Greece.

[Ma et al., 2000b] Ma, Y., Vidal, R., Kosecká, J., and Sastry, S. (2000b). Kruppa's equations revisited: its degeneracy, renormalization and relations to cherality. In *Proceedings of European Conference on Computer Vision*, Dublin, Ireland.

[Magda et al., 2001] Magda, S., Zickler, T., Kriegman, D. J., and Belhumeur, P. B. (2001). Beyond Lambert: reconstructing surfaces with arbitrary BRDFs. In *Proceedings of IEEE International Conference on Computer Vision*, pages 391–398.

[Malik and Rosenholtz, 1997] Malik, J. and Rosenholtz, R. (1997). Computing local surface orientation and shape from texture for curved surfaces. *Int. Journal of Computer Vision*, 23:149–168.

[Malik et al., 1998] Malik, J., Taylor, C. J., Mclauchlan, P., and Kosecká, J. (1998). Development of binocolar stereopsis for vehicle lateral control. *PATH MOU-257 Final Report.*

[Marr, 1982] Marr, D. (1982). *Vision: a computational investigation into the human representation and processing of visual information.* W.H. Freeman and Company.

[Matthies et al., 1989] Matthies, L., Szelisky, R., and Kanade, T. (1989). Kalman filter-based algorithms for estimating depth from image sequences. *Int. Journal of Computer Vision*, pages 2989–2994.

[Maybank, 1990] Maybank, S. (1990). The projective geometry of ambiguous surfaces. *Philosophical Transactions of the Royal Society.*

[Maybank, 1993] Maybank, S. (1993). *Theory of Reconstruction from Image Motion.* Springer Series in Information Sciences. Springer-Verlag.

[Maybank and Faugeras, 1992] Maybank, S. and Faugeras, O. (1992). A theory of self-calibration of a moving camera. *Int. Journal of Computer Vision*, 8(2):123–151.

[McMillan and Bishop, 1995] McMillan, L. and Bishop, G. (1995). Plenoptic modelling: An image-based rendering system. In *Proceedings of SIGGRAPH'95.*

[Medioni et al., 2000] Medioni, G., Chi-Keung, and Lee, T. M. (2000). *A Computational Framework for Segmentation and Grouping*. Elsevier.

[Meer and Georgescu, www] Meer, P. and Georgescu, B. (www). Edge detection with embedded confidence. *URL: http://www.caip.rutgers.edu/riul/research/robust.html*.

[Menet et al., 1990] Menet, S., Saint-Marc, P., and Medioni, G. (1990). Active contour models: overview, implementation and applications. In *IEEE Inter. Conf. on Systems, Man and Cybernetics*.

[Mitsumoto et al., 1992] Mitsumoto, H., Tamura, S., Okazaki, K., and Fukui, Y. (1992). 3-D reconstruction using mirror images based on a plane symmetry recovering method. *IEEE Transactions on Pattern Analysis & Machine Intelligence*, 14(9):941–946.

[Mohr et al., 1993] Mohr, R., Veillon, F., and Quan, L. (1993). Relative 3D reconstruction using multiple uncalibrated images. In *Proceedings of Int. Conference on Computer Vision & Pattern Recognition*, pages 543–548.

[Morris and Kanade, 1998] Morris, D. and Kanade, T. (1998). A unified factorization algorithm for points, line segements and planes with uncertainty models. In *Proceedings of Int. Conference on Computer Vision & Pattern Recognition*, pages 696–702.

[Mukherjee et al., 1995] Mukherjee, D. P., Zisserman, A., and Brady, J. M. (1995). Shape from symmetry – detecting and exploiting symmetry in affine images. *Phil. Trans. Royal Soc. London*, 351:77–106.

[Mundy and Zisserman, 1992] Mundy, J. L. and Zisserman, A. (1992). *Geometric Invariance in Computer Vision*. MIT Press.

[Murray et al., 1993] Murray, R. M., Li, Z., and Sastry, S. S. (1993). *A Mathematical Introduction to Robotic Manipulation*. CRC press Inc.

[Nagel, 1987] Nagel, H. H. (1987). On the estimation of optical flow: relations between different approaches and some new results. *Artificial Intelligence*, 33:299–324

[Nistér, 2003] Nistér, D. (2003). An efficient solution to the five-point relative pose problem. In *Proceedings of Int. Conference on Computer Vision & Pattern Recognition*, Madison, Wisconsin.

[Ohta et al., 1981] Ohta, Y.-I., Maenohu, K., and Sakai, T. (1981). Obtaining surface orientation from texels under perspective projection. In *Proceedings of the seventh International Joint Çonference on Aritificial Intelligence*, pages 746–751, Vancouver, Canada.

[Oliensis, 1999] Oliensis, J. (1999). A multi-frame structure-from-motion algorithm under perspective projection. *Int. Journal of Computer Vision*, 34:163–192.

[Oliensis, 2001] Oliensis, J. (2001). Exact two-image structure from motion. *NEC Technical Report*.

[Oppenheim et al., 1999] Oppenheim, A. V., Schafer, R. V., and Buck, J. R. (1999). *Discrete-time Digital Signal Processing*. Prentice Hall, second edition.

[Osher and Sethian, 1988] Osher, S. and Sethian, J. (1988). Fronts propagating with curvature-dependent speed: Algorithms based on Hamilton-Jacobi equations. *Journal of Computational Physics*, 79:12–49.

[Papadopoulo and Faugeras, 1998] Papadopoulo, T. and Faugeras, O. (1998). A new characterization of the trifocal tensor. In *Proceedings of European Conference on Computer Vision*.

[Papadopoulo and Faugeras, 1998] Papadopoulo, T. and Faugeras, O. (1998). A new characterization of the trifocal tensor. In *Proceedings of European Conference on Computer Vision*.

[Parent and Zucker, 1989] Parent, P. and Zucker, S. W. (1989). Trace inference, curvature consistency and curve detection. *IEEE Transactions on Pattern Analysis & Machine Intelligence*, 11(8):823–839.

[Philip, 1996] Philip, J. (1996). A non-iterative algorithm for determing all essential matrices corresponding to five point pairs. *Photogrammetric Record*, 15(88):589–599.

[Poelman and Kanade, 1997] Poelman, C. J. and Kanade, T. (1997). A paraperspective factorization method for shape and motion recovery. *IEEE Transactions on Pattern Analysis & Machine Intelligence*, 19(3):206–218.

[Pollefeys, 2000] Pollefeys, M. (2000). 3D model from images. *ECCV tutorial lecture notes, Dublin, Ireland, 2000.*

[Pollefeys and Gool, 1999] Pollefeys, M. and Gool, L. V. (1999). Stratified self-calibration with the modulus constraint. *IEEE Transactions on Pattern Analysis & Machine Intelligence*, 21(8):707–724.

[Pollefeys et al., 1996] Pollefeys, M., Gool, L. V., and Proesmans, M. (1996). Euclidean 3D reconstruction from image sequences with variable focal lengths. In *Proceedings of European Conference on Computer Vision*, pages 31–42.

[Pollefeys et al., 1998] Pollefeys, M., Koch, R., and Gool, L. V. (1998). Self-calibration and metric reconstruction in spite of varying and unknown internal camera parameters. In *Proceedings of IEEE International Conference on Computer Vision*, pages 90–95.

[Ponce and Genc, 1998] Ponce, J. and Genc, Y. (1998). Epipolar geometry and linear subspace methods: a new approach to weak calibration. *Int. Journal of Computer Vision*, 28(3):223–243.

[Ponce et al., 1994] Ponce, J., Marimont, D., and Cass, T. (1994). Analytical methods for uncalibrated stereo and motion reconstruction. In *Proceedings of European Conference on Computer Vision*, pages 463–470.

[Quan, 1993] Quan, L. (1993). Affine stereo calibration for relative affine shape reconstruction. In *Proceedings of British Machine Vision Conference*, pages 659–668.

[Quan, 1994] Quan, L. (1994). Invariants of 6 points from 3 uncalibrated images. In *Proceedings of European Conference on Computer Vision*, pages 459–469.

[Quan, 1995] Quan, L. (1995). Invariants of six points and projective reconstruction from three uncalibrated images. *IEEE Transactions on Pattern Analysis & Machine Intelligence*, 17(1):34–46.

[Quan, 1996] Quan, L. (1996). Self-calibration of an affine camera from multiple views. *Int. Journal of Computer Vision*, 19(1):93–105.

[Quan and Kanade, 1996] Quan, L. and Kanade, T. (1996). A factorization method for affine structure from line correspondences. In *Proceedings of Int. Conference on Computer Vision & Pattern Recognition*, pages 803–808.

[Quan and Kanade, 1997] Quan, L. and Kanade, T. (1997). Affine structure from motion from line correspondences with uncalibrated affine cameras. *IEEE Transactions on Pattern Analysis & Machine Intelligence*, 19(8):834–845.

[Quan and Mohr, 1989] Quan, L. and Mohr, R. (1989). Determining perspective structures using hierarchical Hough transform. *Pattern Recognition Letter*, 9(4):279–286.

[Quan and Mohr, 1991] Quan, L. and Mohr, R. (1991). Towards structure from motion for linear features through reference points. In *Proceedings of IEEE Workshop on Visual Motion*, pages 249–254, Los Alamitos, California, USA.

[Rahimi et al., 2001] Rahimi, A., Morency, L. P., and Darrell, T. (2001). Reducing drift in parametric motion tracking. In *Proceedings of IEEE International Conference on Computer Vision*, pages I: 315–322.

[Rothwell et al., 1997] Rothwell, C. A., Faugeraus, O., and Csurka, G. (1997). A comparison of projective reconstruction methods for pairs of views. *Computer Vision and Image Understanding*, 68(1):37–58.

[Rothwell et al., 1993] Rothwell, C. A., Forsyth, D. A., Zisserman, A., and Mundy, J. L. (1993). Extracting projective structure from single perspective views of 3D point sets. In *Proceedings of IEEE International Conference on Computer Vision*, pages 573–582.

[Rousso and Shilat, 1998] Rousso, B. and Shilat, E. (1998). Varying focal length self-calibration and pose estimation. In *Proceedings of Int. Conference on Computer Vision & Pattern Recognition*, pages 469–474.

[Ruf et al., 1998] Ruf, A., Csurka, G., and Horaud, R. (1998). Projective translation and affine stereo calibration. In *Proceedings of Int. Conference on Computer Vision & Pattern Recognition*, pages 475–481.

[Ruf and Horaud, 1999a] Ruf, A. and Horaud, R. (1999a). Projective rotations applied to a pan-tilt stereo head. In *Proceedings of Int. Conference on Computer Vision & Pattern Recognition*, pages I:144–150.

[Ruf and Horaud, 1999b] Ruf, A. and Horaud, R. (1999b). Rigid and articulated motion seen with an uncalibrated stereo rig. In *Proceedings of IEEE International Conference on Computer Vision*, pages 789–796.

[Sampson, 1982] Sampson, P. D. (1982). Fitting conic section to "very scattered" data: An iterative refinement of the bookstein algorithm. *Computer Vision, Graphics and Image Processing*, 18:97–108.

[Samson et al., 1991] Samson, C., Borgne, M. L., and Espiau, B. (1991). *Robot Control: The Task Function Approach*. Oxford Engineering Science Series. Clarendon Press.

[Sapiro, 2001] Sapiro, G. (2001). *Geometric Partial Differential Equations and Image Processing*. Cambridge University Press.

[Sawhney and Kumar, 1999] Sawhney, H. S. and Kumar, R. (1999). True multi-image alignment and its application to mosaicing and lens distortion correction. *IEEE Transactions on Pattern Analysis and Machine Intelligence*, 21(3):235–243.

[Schaffalitzky and Zisserman, 2001] Schaffalitzky, F. and Zisserman, A. (2001). Viewpoint invariant texture matching and wide baseline stereo. In *Proceedings of the 8th International Conference on Computer Vision, Vancouver, Canada*, pages 636–643.

[Schmidt and Zisserman, 2000] Schmidt, C. and Zisserman, A. (2000). The geometry and matching of lines and curves over multiple views. *Int. Journal of Computer Vision*, 40(3):199–234.

[Seo and Hong, 1999] Seo, Y. and Hong, K. (1999). About the self-calibration of a rotating and zooming camera: Theory and practice. In *Proceedings of IEEE International Conference on Computer Vision*, pages 183–188.

[Shakernia et al., 1999] Shakernia, O., Ma, Y., Koo, J., and Sastry, S. (1999). Landing an unmanned aerial vehicle: Vision based motion estimation and nonlinear control. *Asian Journal of Control*, 1(3):128–145.

[Shakernia et al., 2002] Shakernia, O., Vidal, R., Sharp, C., Ma, Y., and Sastry, S. (2002). Multiple view motion estimation and control for landing an unmanned aerial vehicle. In *Proceedings of International Conference on Robotics and Automation.*

[Sharp et al., 2001] Sharp, C., Shakernia, O., and Sastry, S. (2001). A vision system for landing an unmanned aerial vehicle. In *Proceedings of International Conference on Robotics and Automation.*

[Shashua, 1994] Shashua, A. (1994). Trilinearity in visual recognition by alignment. In *Proceedings of European Conference on Computer Vision*, pages 479–484. Springer-Verlag.

[Shashua and Levin, 2001] Shashua, A. and Levin, A. (2001). Multi-frame infinitesimal motion model for the reconstruction of (dynamic) scenes with multiple linearly moving objects. In *Proceedings of IEEE International Conference on Computer Vision*, volume 2, pages 592–599.

[Shashua and Wolf, 2000] Shashua, A. and Wolf, L. (2000). On the structure and properties of the quadrifocal tensor. In *Proceedings of European Conference on Computer Vision*, pages 711–724. Springer-Verlag.

[Shi and Tomasi, 1994] Shi, J. and Tomasi, C. (1994). Good features to track. In *IEEE Conference on Computer Vision and Pattern Recognition*, pages 593–600.

[Shimoshoni et al., 2000] Shimoshoni, I., Moses, Y., and Lindenbaum, M. (2000). Shape reconstruction of 3D bilaterally symmetric surfaces. *Int. Journal of Computer Vision*, 39:97–112.

[Shizawa and Mase, 1991] Shizawa, M. and Mase, K. (1991). A unified computational theory for motion transparency and motion boundaries based on eigenengergy analysis. In *Proceedings of Int. Conference on Computer Vision & Pattern Recognition*, pages 289–295.

[Shufelt, 1999] Shufelt, J. (1999). Performance evaluation and analysis of vanishing point detection. *IEEE Transactions on Pattern Analysis & Machine Intelligence*, 21(3):282–288.

[Sidenbladh et al., 2000] Sidenbladh, H., Black, M., and Fleet, D. (2000). Stochastic tracking of 3D human figures using 2D image motion. In *Proceedings of European Conference on Computer Vision*, pages II: 307–323.

[Sillion, 1994] Sillion, F. (1994). *Radiosity and Global Illumination*. Morgan Kaufmann Publishers.

[Simoncelli and Freeman, 1995] Simoncelli, E. P. and Freeman, W. T. (1995). The steerable pyramid: A flexible architecture for multi-scale derivative computation. In *Proceedings of the Second IEEE International Conference on Image Processing*, volume 3, pages 444–447. IEEE Signal Processing Society.

[Sinclair et al., 1997] Sinclair, D., Palltta, L., and Pinz, A. (1997). Euclidean structure recovery through articulated motion. In *Proceedings of the 10th Scandinavian Conference on Image Analysis*, Finland.

[Sinclair and Zesar, 1996] Sinclair, D. and Zesar, K. (1996). Further constraints on visual articulated motion. In *Proceedings of Int. Conference on Computer Vision & Pattern Recognition*, pages 94–99, San Francisco.

[Smale, 1997] Smale, S. (1997). Complexity theory and numerical analysis. *Acta Numerica*, 6:523–551.

[Soatto and Brockett, 1998] Soatto, S. and Brockett, R. (1998). Optimal structure from motion: Local ambiguities and global estimates. In *Proceedings of Int. Conference on Computer Vision & Pattern Recognition*, pages 282–288.

[Soatto et al., 1996] Soatto, S., Frezza, R., and Perona, P. (1996). Motion estimation via dynamic vision. *IEEE Transactions on Automatic Control*, 41(3):393–413.

[Sontag, 1992] Sontag, E. (1992). *Mathematical Control Theory*. Springer Verlag.

[Sparr, 1992] Sparr, G. (1992). Depth computations from polyhedral images. In *Proceedings of European Conference on Computer Vision*, pages 378–386.

[Spetsakis, 1994] Spetsakis, M. (1994). Models of statistical visual motion estimation. *CVIPG: Image Understanding*, 60(3):300–312.

[Spetsakis and Aloimonos, 1987] Spetsakis, M. and Aloimonos, J. (1987). Closed form solution to the structure from motion problem using line correspondences. *Technical Report, CAR-TR-274, CS-TR-1798, DAAB07-86-K-F073 (also appeared in Proceedings of AAAI 1987)*.

[Spetsakis and Aloimonos, 1988] Spetsakis, M. and Aloimonos, J. (1988). A multiframe approach to visual motion perception. *Technical Report, CAR-TR-407, CS-TR-2147, DAAB07-86-K-F073 (also appeared in the International Journal of Computer Vision 6, 245-255, 1991)*.

[Spetsakis and Aloimonos, 1990a] Spetsakis, M. and Aloimonos, Y. (1990a). Structure from motion using line correspondences. *Int. Journal of Computer Vision*, 4(3):171–184.

[Spetsakis and Aloimonos, 1990b] Spetsakis, M. and Aloimonos, Y. (1990b). A unified theory of structure from motion. In *Proceedings of DARPA IU Workshop*, pages 271–283.

[Stark and Woods, 2002] Stark, H. and Woods, J. W. (2002). *Probability and Random Processes with Applications to Signal Processing*. Prentice Hall, third edition.

[Stein, 1997] Stein, G. (1997). Lens distortion calibration using point correspondences. In *Proceedings of Int. Conference on Computer Vision & Pattern Recognition*, pages 602–608. IEEE Comput. Soc. Press.

[Strang, 1988] Strang, G. (1988). *Linear Algebra and its Applications*. Saunders, third edition.

[Stroebel, 1999] Stroebel, L. (1999). *View Camera Techniques*. Focal Press, seventh edition.

[Sturm, 1997] Sturm, P. (1997). Critical motion sequences for monocular self-calibration and uncalibrated Euclidean reconstruction. In *Proceedings of Int. Conference on Computer Vision & Pattern Recognition*, pages 1100–1105. IEEE Comput. Soc. Press.

[Sturm, 1999] Sturm, P. (1999). Critical motion sequences for the self-calibration of cameras and stereo systems with variable focal length. In *Proceedings of British Machine Vision Conference*, pages 63–72, Nottingham, England.

[Sturm and Triggs, 1996] Sturm, P. and Triggs, B. (1996). A factorizaton based algorithm for multi-image projective structure and motion. In *Proceedings of European Conference on Computer Vision*, pages 709–720. IEEE Comput. Soc. Press.

[Subbarao and Waxman, 1985] Subbarao, M. and Waxman, A. M. (1985). On the uniqueness of image flow solutions for planar surfaces in motion. In *Proceedings of the third IEEE workshop on computer vision: representation and control*, pages 129–140.

[Svedberg and Carlsson, 1999] Svedberg, D. and Carlsson, S. (1999). Calibration, pose and novel views from single images of constrained scenes. In *Proceedings of the 11th Scandinavian Conference on Image Analysis*, pages 111–117.

[Szeliski and Shum, 1997] Szeliski, R. and Shum, H.-Y. (1997). Creating full view panoramic image mosaics and environment maps. In *Proceedings of SIGGRAPH'97*, volume 31, pages 251–258.

[Tang et al., 1999] Tang, C., Medioni, G., and Lee, M. (1999). Epipolar geometry estimation by tensor voting. In *Proceedings of IEEE International Conference on Computer Vision*, pages 502–509, Kerkyra, Greece. IEEE Comput. Soc. Press.

[Taylor and Kriegman, 1995] Taylor, C. J. and Kriegman, D. J. (1995). Structure and motion from line segments in multiple images. *IEEE Transactions on Pattern Analysis & Machine Intelligence*, 17(11):1021–1032.

[Tell and Carlsson, 2002] Tell, D. and Carlsson, S. (2002). Combining topology and appearance for wide baseline matching. In *Proceedings of European Conference on Computer Vision*, pages 68–81, Copenhagen, Denmark. Springer Verlag.

[Thompson, 1959] Thompson, E. (1959). A rational algebraic formulation of the problem of relative orientation. *Photomgrammetric Record*, 3(14):152–159.

[Tian et al., 1996] Tian, T. Y., Tomasi, C., and Heeger, D. J. (1996). Comparison of approaches to egomotion computation. In *Proceedings of Int. Conference on Computer Vision & Pattern Recognition*, pages 315–320, Los Alamitos, CA, USA. IEEE Comput. Soc. Press.

[Tomasi and Kanade, 1992] Tomasi, C. and Kanade, T. (1992). Shape and motion from image streams under orthography. *Int. Journal of Computer Vision*, 9(2):137–154.

[Torr, 1998] Torr, P. H. S. (1998). Geometric motion segmentation and model selection. *Phil. Trans. Royal Society of London*, 356(1740):1321–1340.

[Torr et al., 1999] Torr, P. H. S., Fitzgibbon, A., and Zisserman, A. (1999). The problem of degenerarcy in structure and motion recovery from uncalibrated image sequences. *Int. Journal of Computer Vision*, 32(1):27–44.

[Torr and Murray, 1997] Torr, P. H. S. and Murray, D. W. (1997). The development and comparison of robust methods for estimating the fundamental matrix. *Int. Journal of Computer Vision*, 24(3):271–300.

[Torr and Zisserman, 1997] Torr, P. H. S. and Zisserman, A. (1997). Robust parameterization and computation of the trifocal tensor. *Image and Vision Computing*, 15:591–605.

[Torr and Zisserman, 2000] Torr, P. H. S. and Zisserman, A. (2000). MLESAC: a new robust estimator with application to estimating image geometry. *Comp. Vision and Image Understanding*, 78(1):138–156.

[Torresani et al., 2001] Torresani, L., Yang, D., Alexander, E., and Bregler, C. (2001). Tracking and modeling non-rigid objects with rank constraints. In *Proceedings of Int. Conference on Computer Vision & Pattern Recognition.*

[Triggs, 1995] Triggs, B. (1995). Matching constraints and the joint image. In *Proceedings of IEEE International Conference on Computer Vision*, pages 338–343, Cambridge, MA, USA. IEEE Comput. Soc. Press.

[Triggs, 1996] Triggs, B. (1996). Factorization methods for projective structure and motion. In *Proceedings of Int. Conference on Computer Vision & Pattern Recognition*, pages 845–851, San Francisco, CA, USA. IEEE Comput. Soc. Press.

[Triggs, 1997] Triggs, B. (1997). Autocalibration and the absolute quadric. In *Proceedings of Int. Conference on Computer Vision & Pattern Recognition*, pages 609–614.

[Triggs, 1998] Triggs, B. (1998). Autocalibration from planar scenes. In *Proceedings of European Conference on Computer Vision*, pages I: 89–105.

[Triggs and Fitzgibbon, 2000] Triggs, P. McLauchlan, R. H. and Fitzgibbon, A. (2000). Bundle adjustement – a modern synthesis. In B. Triggs, A. Z. and Szeliski, R., editors, *Vision Algorithms: Theory and Practice, LNCS vol. 1883*, pages 338–343. Springer.

[Tsai, 1986a] Tsai, R. Y. (1986a). An efficient and accurate camera calibration technique for 3D machine vision. In *Proceedings of Int. Conference on Computer Vision & Pattern Recognition*, IEEE Publ.86CH2290-5, pages 364–374. IEEE.

[Tsai, 1986b] Tsai, R. Y. (1986b). Multiframe image point matching and 3D surface reconstruction. *IEEE Transactions on Pattern Analysis & Machine Intelligence*, 5:159–174.

[Tsai, 1987] Tsai, R. Y. (1987). A versatile camera calibration technique for high-accuracy 3D machine vision metrology using off-the-shelf TV cameras and lenses. *IEEE Transactions on Robotics and Automation*, 3(4):323–344.

[Tsai, 1989] Tsai, R. Y. (1989). Synopsis of recent progress on camera calibration for 3D machine vision. *The Robotics Review*, pages 147–159.

[Tsai and Huang, 1981] Tsai, R. Y. and Huang, T. S. (1981). Estimating 3D motion parameters from a rigid planar patch. *IEEE Transactions on Acoustics, Speech, Signal Processing*, 29(6):1147–1152.

[Tsai and Huang, 1984] Tsai, R. Y. and Huang, T. S. (1984). Uniqueness and estimation of three-dimensional motion parameters of rigid objects with curved surfaces. *IEEE Transactions on Pattern Analysis & Machine Intelligence*, 6(1):13–27.

[Ueshiba and Tomita, 1998] Ueshiba, T. and Tomita, F. (1998). A factorization method for projective and Euclidean reconstruction from multiple perspective views via iterative depth estimation. In *Proceedings of European Conference on Computer Vision*, pages I: 296–310.

[van Trees, 1992] van Trees, H. (1992). *Detection and Estimation Theory*. Krieger.

[Verri and Poggio, 1989] Verri, A. and Poggio, T. (1989). Motion field and optical flow: Qualitative properties. *IEEE Transactions on Pattern Analysis and Machine Intelligence*, 11(5):490–498.

[Vidal et al., 2002a] Vidal, R., Ma, Y., Hsu, S., and Sastry, S. (2002a). Optimal motion estimation from multiview normalized epipolar constraint. In *Proceedings of IEEE International Conference on Computer Vision*, volume I, pages 34–41.

[Vidal et al., 2003] Vidal, R., Ma, Y., and Sastry, S. (2003). Generalized principle component analysis. In *Proceedings of Int. Conference on Computer Vision & Pattern Recognition*.

[Vidal et al., 2002b] Vidal, R., Ma, Y., Soatto, S., and Sastry, S. (May, 2002b). Two-view segmentation of dynamic scenes from the multibody fundamental matrix. Technical report, UCB/ERL M02/11, UC Berkeley.

[Vidal and Sastry, 2003] Vidal, R. and Sastry, S. (2003). Optimal segmentation of dynamic scenes. In *Proceedings of Int. Conference on Computer Vision & Pattern Recognition*.

[Vidal et al., 2002c] Vidal, R., Soatto, S., Ma, Y., and Sastry, S. (2002c). Segmentation of dynamic scenes from the multibody fundamental matrix. In *Proceedings of ECCV workshop on Vision and Modeling of Dynamic Scenes.*

[Vieville et al., 1996] Vieville, T., Faugeras, O., and Luong, Q.-T. (1996). Motion of points and lines in the uncalibrated case. *Int. Journal of Computer Vision*, 17(1):7–42.

[Vieville and Faugeras, 1995] Vieville, T. and Faugeras, O. D. (1995). Motion analysis with a camera with unknown, and possibly varying intrinsic parameters. In *Proceedings of IEEE International Conference on Computer Vision*, pages 750–756.

[VRML Consortium, 1997] VRML Consortium, T. (1997). *The virtual reality modeling language*. ISO/IEC DIS 14772-1.

[Waxman and Ullman, 1985] Waxman, A. and Ullman, S. (1985). Surface structure and three-dimensional motion from image flow kinematics. *Int. Journal of Robotics Research*, 4(3):72–94.

[Waxman et al., 1987] Waxman, A. M., Kamgar-Parsi, B., and Subbarao, M. (1987). Closed form solutions to image flow equations for 3D structure and motion. *Int. Journal of Computer Vision*, pages 239–258.

[Wei and Ma, 1991] Wei, G. and Ma, S. (1991). A complete two-plane camera calibration method and experimental comparisons. In *Proceedings of Int. Conference on Computer Vision & Pattern Recognition*, pages 439–446.

[Weickert et al., 1998] Weickert, J., Haar, B., and Viergever, R. (1998). Efficient and reliable schemes for nonlinear diffusion filtering. *IEEE Transactions on Image Processing*, 7(3):398–410.

[Weinstein, 1996] Weinstein, A. (1996). Groupoids: Unifying internal and external symmetry. *Notices AMS*, 43:744–752.

[Weng et al., 1993a] Weng, J., Ahuja, N., and Huang, T. (1993a). Optimal motion and structure estimation. *IEEE Transactions on Pattern Analysis & Machine Intelligence*, 9(2):137–154.

[Weng et al., 1992a] Weng, J., Cohen, P., and Rebibo, N. (1992a). Motion and structure estimation from stereo image sequences. *IEEE Transactions on Pattern Analysis & Machine Intelligence*, 8(3):362–382.

[Weng et al., 1992b] Weng, J., Huang, T., and Ahuja, N. (1992b). Motion and structure estimation from line correspondences: Closed-form solution, uniqueness and optimization. *IEEE Transactions on Pattern Analysis & Machine Intelligence*, 14(3):318–336.

[Weng et al., 1993b] Weng, J., Huang, T. S., and Ahuja, N. (1993b). *Motion and Structure from Image Sequences*. Springer Verlag.

[Weyl, 1952] Weyl, H. (1952). *Symmetry*. Princeton Univ. Press.

[Wiener, 1949] Wiener, N. (1949). *Cybernetics, or Control and Communication in Men and Machines*. MIT Press.

[Witkin, 1988] Witkin, A. P. (1988). Recovering surface shape and orientation from texture. *Journal of Artificial Intelligence*, 17:17–45.

[Wolf and Shashua, 2001a] Wolf, L. and Shashua, A. (2001a). On projection matrices $\mathcal{P}^k \rightarrow \mathcal{P}^2, k = 3, \ldots, 6$, and their applications in computer vision. In *Proceedings of IEEE International Conference on Computer Vision*, pages 412–419, Vancouver, Canada.

[Wolf and Shashua, 2001b] Wolf, L. and Shashua, A. (2001b). Two-body segmentation from two perspective views. In *Proceedings of Int. Conference on Computer Vision & Pattern Recognition*, pages I: 263–270.

[Wuescher and Boyer, 1991] Wuescher, D. M. and Boyer, K. L. (1991). Robust contour decomposition using a constant curvature criterion. *IEEE Transactions on Pattern Analysis & Machine Intelligence*, 13(1):41–51.

[Xu and Tsuji, 1996] Xu, G. and Tsuji, S. (1996). Correspondence and segmentation of multiple rigid motions via epipolar geometry. In *Proceedings of Int. Conference on Pattern Recognition*, pages 213–217, Vienna, Austria.

[Yacoob and Davis, 1998] Yacoob, Y. and Davis, L. S. (1998). Learned temporal models of image motion. In *Proceedings of IEEE International Conference on Computer Vision*, pages 446–453.

[Yakimovsky and Cunningham, 1978] Yakimovsky, Y. and Cunningham, R. (1978). A system for extracting three-dimensional measurements from a stereo pair of TV cameras. *Computer Graphics and Image Processing*, 7:323–344.

[Yang et al., 2003] Yang, A. Y., Huang, K., Rao, S., and Ma, Y. (April 14, 2003). Symmetry-based 3-D reconstruction from perspective images (Part I): Detection and segmentation. *UIUC CSL Technical Report, UILU-ENG-03-2204, DC-207.*

[Yezzi and Soatto, 2003] Yezzi, A. and Soatto, S. (2003). Stereoscopic segmentation. *Int. Journal of Computer Vision*, 53(1):31–43.

[Zabrodsky et al., 1995] Zabrodsky, H., Peleg, S., and Avnir, D. (1995). Symmetry as a continuous feature. *IEEE Transactions on Pattern Analysis & Machine Intelligence*, 17(12):1154–1166.

[Zabrodsky and Weinshall, 1997] Zabrodsky, H. and Weinshall, D. (1997). Using bilateral symmetry to improve 3D reconstruction from image sequences. *Comp. Vision and Image Understanding*, 67:48–57.

[Zeller and Faugeras, 1996] Zeller, C. and Faugeras, O. (1996). Camera self-calibration from video sequences: the Kruppa equations revisited. *Research Report 2793, INRIA, France.*

[Zhang, 1995] Zhang, Z. (1995). Estimating motion and structure from correspondences of line segments between two perspective views. In *Proceedings of IEEE International Conference on Computer Vision*, pages 257–262, Bombay, India.

[Zhang, 1996] Zhang, Z. (1996). On the epipolar geometry between two images with lens distortion. In *Proceedings of Int. Conference on Pattern Recognition*, pages 407–411.

[Zhang, 1998a] Zhang, Z. (1998a). Determining the epipolar geometry and its uncertainty: a review. *Int. Journal of Computer Vision*, 27(2):161–195.

[Zhang, 1998b] Zhang, Z. (1998b). A flexible new technique for camera calibration. *Microsoft Technical Report MSR-TR-98-71.*

[Zhang, 1998c] Zhang, Z. (1998c). Understanding the relationship between the optimization criteria in two-view motion analysis. In *Proceedings of IEEE International Conference on Computer Vision*, pages 772–777, Bombay, India.

[Zhang et al., 1995] Zhang, Z., Deriche, R., Faugeras, O., and Luong, Q.-T. (1995). A robust technique for matching two uncalibrated images through the recovery of the unknown epipolar geometry. *Artificial Intelligence*, 78:87–119.

[Zhang et al., 1996] Zhang, Z., Luong, Q.-T., and Faugeras, O. (1996). Motion of an uncalibrated stereo rig: self-calibration and metric reconstruction. *IEEE Transactions on Robotics and Automation*, 12(1):103–113.

[Zhao and Chellappa, 2001] Zhao, W. Y. and Chellappa, R. (2001). Symmetric shape-from-shading using self-ratio image. *Int. Journal of Computer Vision*, 45(1):55–75.

[Zhuang and Haralick, 1984] Zhuang, X. and Haralick, R. M. (1984). Rigid body motion and optical flow image. In *Proceedings of the First International Conference on Artificial Intelligence Applications*, pages 366–375.

[Zhuang et al., 1988] Zhuang, X., Huang, T., and Ahuja, N. (1988). A simplified linear optic flow-motion algorithm. *Computer Vision, Graphics and Image Processing*, 42:334–344.

[Zisserman et al., 1995] Zisserman, A., Beardsley, P. A., and Reid, I. D. (1995). Metric calibration of a stereo rig. In *Proceedings of the Workshop on Visual Scene Representation*, pages 93–100, Boston, MA.

[Zisserman et al., 1998] Zisserman, A., Liebowitz, D., and Armstrong, M. (1998). Resolving ambiguities in auto-calibration. *Philosophical Transactions of the Royal Society of London*, 356(1740):1193–1211.

# Glossary of Notation

Frequently used mathematical symbols are defined and listed according to the following categories:

**0.** Set theory and logic symbols

**1.** Sets and linear spaces

**2.** Transformation groups

**3.** Vector and matrix operations

**4.** Geometric primitives in space

**5.** Geometric primitives in images

**6.** Camera motion

**7.** Computer-vision-related matrices

Throughout the book, **every vector is a column vector unless stated otherwise!**

**0. Set theory and logic symbols**

| | |
|---|---|
| $\cap$ | $S_1 \cap S_2$ is the intersection of two sets |
| $\cup$ | $S_1 \cup S_2$ is the union of two sets |
| $\doteq$ | Definition of a symbol |
| $\exists$ | $\exists s \in S, P(s)$ means there exists an element $s$ of set $S$ such that proposition $P(s)$ is true |
| $\forall$ | $\forall s \in S, P(s)$ means for every element $s$ of set $S$, proposition $P(s)$ is true |
| $\in$ | $s \in S$ means $s$ is an element of set $S$ |
| $\Leftrightarrow$ | $P \Leftrightarrow Q$ means propositions $P$ and $Q$ imply each other |
| $\mid$ | $P \mid Q$ means proposition $P$ holds given the condition $Q$ |
| $\Rightarrow$ | $P \Rightarrow Q$ means proposition $P$ implies proposition $Q$ |
| $\setminus$ | $S_1 \setminus S_2$ is the difference of set $S_1$ minus set $S_2$ |
| $\subset$ | $S_1 \subset S_2$ means $S_1$ is a proper subset of $S_2$ |

| | |
|---|---|
| $\{s\}$ | A set consists of elements like $s$ |
| $\rightarrow$ | $f : D \rightarrow R$ means a map $f$ from domain $D$ to range $R$ |
| $\mapsto$ | $f : x \mapsto y$ means $f$ maps an element $x$ in the domain to an element $y$ in the range |
| $\circ$ | $f \circ g$ means composition of map $f$ with map $g$ |

**1. Sets and linear spaces**

| | |
|---|---|
| $\mathbb{C}$ | The set of all complex numbers |
| $\mathbb{C}^n$ | The $n$-dimensional complex linear space |
| $\mathbb{E}^3$ | Three-dimensional Euclidean space, page 16 |
| $\mathbb{H}$ | The set of all quaternions, see equation (2.34), page 40 |
| $\mathbb{P}^n = \mathbb{RP}^n$ | The $n$-dimensional real projective space |
| $\mathbb{R}$ | The set of all real numbers |
| $\mathbb{R}^n$ | The $n$-dimensional real linear space |
| $\mathbb{R}_+$ | The set of all nonnegative real numbers |
| $\mathbb{Z}$ | The set of all integers |
| $\mathbb{Z}_+$ | The set of all nonnegative integers |

**2. Transformation groups**

| | |
|---|---|
| $A(n) = A(n, \mathbb{R})$ | The real affine group on $\mathbb{R}^n$; an element in $A(n)$ is a pair $(A, b)$ with $A \in GL(n)$ and $b \in \mathbb{R}^n$ and it acts on a point $\boldsymbol{X} \in \mathbb{R}^n$ as $A\boldsymbol{X} + b$, page 448 |
| $GL(n) = GL(n, \mathbb{R})$ | The real general linear group on $\mathbb{R}^n$; it can be identified as the set of $n \times n$ invertible real matrices, page 447 |
| $O(n) = O(n, \mathbb{R})$ | The real orthogonal group on $\mathbb{R}^n$; if $U \in O(n)$, then $U^T U = I$, page 448 |
| $SE(n) = SE(n, \mathbb{R})$ | The real special Euclidean group on $\mathbb{R}^n$; an element in $SE(n)$ is a pair $(R, T)$ with $R \in SO(n)$ and $T \in \mathbb{R}^n$ and it acts on a point $\boldsymbol{X} \in \mathbb{R}^n$ as $R\boldsymbol{X} + T$, page 449 |
| $SL(n) = SL(n, \mathbb{R})$ | The real special linear group on $\mathbb{R}^n$; it can be identified as the set of $n \times n$ real matrices of determinant 1, page 448 |
| $SO(n) = SO(n, \mathbb{R})$ | The real special orthogonal group on $\mathbb{R}^n$; if $R \in SO(n)$, then $R^T R = I$ and $\det(R) = 1$, page 449 |

**3. Vector space operations**

| | |
|---|---|
| $\det(M)$ | The determinant of a square matrix $M$ |
| $\langle u, v \rangle \in \mathbb{R}$ | The inner product of two vectors: $\langle u, v \rangle = u^T v$, page 444 |
| $\ker(M)$ | The kernel (or null space) of a linear map, page 451 |

| | |
|---|---|
| null($M$) | The null space (or kernel) of a matrix $M$, page 451 |
| range($M$) | The range or span of a matrix $M$, page 442 |
| span($M$) | The subspace spanned by the column vectors of a matrix $M$, page 442 |
| trace($M$) | The trace of a square matrix $M$, i.e. the sum of all its diagonal entries, sometimes shorthand as tr($M$) |
| rank($M$) | The rank of a matrix $M$, page 451 |
| $\widehat{u} \in \mathbb{R}^{3\times 3}$ | The $3 \times 3$ skew-symmetric matrix associated with a vector $u \in \mathbb{R}^3$, see equation (2.1), page 18 |
| $M^s \in \mathbb{R}^{mn}$ | Stacked version of a matrix $M \in \mathbb{R}^{m\times n}$ obtained by stacking all columns into one vector, page 445 |
| $M^T \in \mathbb{R}^{n\times m}$ | Transpose of a matrix $M \in \mathbb{R}^{m\times n}$ |
| $M_1 \otimes M_2$ | The Kronecker (tensor) product of two matrices or two vectors, page 445 |
| $S^\perp$ | The orthogonal complement of a subspace $S$; in the context of computer vision, it is often used to denote the coimage, page 452 |
| $S_1 \oplus S_2$ | The direct sum of two linear subspaces $S_1$ and $S_2$ |
| $u \sim v$ | Homogeneous equality: two vectors or matrices $u$ and $v$ are equal up to a scalar factor |
| $u \times v \in \mathbb{R}^3$ | The cross product of two 3-D vectors: $u \times v = \widehat{u}v$ |
| $\\|\cdot\\|$ | The standard 2-norm of a vector $\\|v\\| = \sqrt{v^T v}$, also denoted by $\\|\cdot\\|_2$ |

**4. Geometric primitives in space**

| | |
|---|---|
| $\boldsymbol{X}$ | Coordinates $\boldsymbol{X} = [X, Y, Z]^T \in \mathbb{R}^3$ of a point $p$ in space, in the homogeneous representation $\boldsymbol{X} = [X, Y, Z, 1]^T \in \mathbb{R}^4$, page 16 |
| $\boldsymbol{X}_i^j$ | Coordinates of the $j$th point with respect to the $i$th (camera) coordinate frame, shorthand for $\boldsymbol{X}^j(t_i)$ |
| $\boldsymbol{X}_i$ | Coordinates of a point with respect to the $i$th (camera) coordinate frame, shorthand for $\boldsymbol{X}(t_i)$ |
| $p \in \mathbb{E}^3$ | A generic (abstract) point in space, page 16 |
| $L \subset \mathbb{E}^3$ | A generic (abstract) 1-D line in space |
| $P \subset \mathbb{E}^3$ | A generic (abstract) 2-D plane in space |

**5. Geometric primitives in images**

| | |
|---|---|
| $\boldsymbol{x}$ | $(x, y)$-coordinates of the image of a point in the image plane, in homogeneous form: $\boldsymbol{x} = [x, y, z]^T \in \mathbb{R}^3$ |
| $\boldsymbol{x}^\perp$ | Coimage of an image point $\boldsymbol{x}$, typically represented by $\widehat{\boldsymbol{x}}$ |

| | |
|---|---|
| $\boldsymbol{x}_i$ | Coordinates of the $i$th image of a point, shorthand for $\boldsymbol{x}(t_i)$ |
| $\boldsymbol{x}_i^j$ | Coordinates of the $i$th image of the $j$th point, with respect to the $i$th (camera) coordinate frame, shorthand for $\boldsymbol{x}^j(t_i)$ |
| $\boldsymbol{\ell}$ | Coordinates of the coimage of a line, typically in homogeneous form: $\boldsymbol{\ell} = [a, b, c]^T \in \mathbb{R}^3$; and $\boldsymbol{\ell}^T \boldsymbol{x} = 0$ for any (image) point $\boldsymbol{x}$ on this line |
| $\boldsymbol{\ell}^{\perp}$ | Image of a line, typically represented by $\widehat{\boldsymbol{\ell}}$ |
| $\boldsymbol{\ell}_i$ | Coordinates of the $i$th coimage of a line, shorthand for $\boldsymbol{\ell}(t_i)$ |
| $\boldsymbol{\ell}_i^j$ | Coordinates of the $i$th coimage of the $j$th line, shorthand for $\boldsymbol{\ell}^j(t_i)$ |

**6. Camera motion**

| | |
|---|---|
| $(\omega, v)$ | $\omega \in \mathbb{R}^3$ the angular velocity; $v \in \mathbb{R}^3$ the linear velocity, page 32 |
| $(R, T)$ | A rigid-body motion: $R \in SO(3)$ the rotation; $T \in \mathbb{R}^3$ the translation, page 21 |
| $(R_i, T_i)$ | Relative motion (rotation and translation) from the $i$th camera frame to the (default) first camera frame: $\boldsymbol{X}_i = R_i \boldsymbol{X} + T_i$ |
| $(R_{ij}, T_{ij})$ | Relative motion (rotation and translation) from the $i$th camera frame to the $j$th camera frame: $\boldsymbol{X}_i = R_{ij} \boldsymbol{X}_j + T_{ij}$ |
| $g \in SE(3)$ | A rigid-body motion or equivalently a special Euclidean transformation, page 20 |

**7. Computer-vision-related matrices**

| | |
|---|---|
| $\Pi \in \mathbb{R}^{3\times 4}$ | A general $3 \times 4$ projection matrix from $\mathbb{R}^3$ to $\mathbb{R}^2$, but in the multiple-view case it may represent a collection of such matrices, page 57 |
| $\Pi_0 \in \mathbb{R}^{3\times 4}$ | The standard projection matrix $[I, 0]$ from $\mathbb{R}^3$ to $\mathbb{R}^2$, see equation (3.6), page 53 |
| $E \in \mathbb{R}^{3\times 3}$ | The essential matrix, page 113 |
| $F \in \mathbb{R}^{3\times 3}$ | The fundamental matrix, see equation (6.9), page 177 |
| $H$ | The homography matrix, and it usually represents an element in $GL(3)$ or $GL(4)$ |
| $K \in \mathbb{R}^{3\times 3}$ | The camera calibration matrix, also called the intrinsic parameter matrix, see equation (3.13), page 55 |
| $M$ | The multiple-view matrix, and its dimension depends on its type |

# Index

## Interdisciplinary Applied Mathematics

1. *Gutzwiller:* Chaos in Classical and Quantum Mechanics
2. *Wiggins:* Chaotic Transport in Dynamical Systems
3. *Joseph/Renardy:* Fundamentals of Two-Fluid Dynamics: Part I: Mathematical Theory and Applications
4. *Joseph/Renardy:* Fundamentals of Two-Fluid Dynamics: Part II: Lubricated Transport, Drops and Miscible Liquids
5. *Seydel:* Practical Bifurcation and Stability Analysis: From Equilibrium to Chaos
6. *Hornung:* Homogenization and Porous Media
7. *Simo/Hughes:* Computational Inelasticity
8. *Keener/Sneyd:* Mathematical Physiology
9. *Han/Reddy:* Plasticity: Mathematical Theory and Numerical Analysis
10. *Sastry:* Nonlinear Systems: Analysis, Stability, and Control
11. *McCarthy:* Geometric Design of Linkages
12. *Winfree:* The Geometry of Biological Time (Second Edition)
13. *Bleistein/Cohen/Stockwell:* Mathematics of Multidimensional Seismic Imaging, Migration, and Inversion
14. *Okubo/Levin:* Diffusion and Ecological Problems: Modern Perspectives (Second Edition)
15. *Logan:* Transport Modeling in Hydrogeochemical Systems
16. *Torquato:* Random Heterogeneous Materials: Microstructure and Macroscopic Properties
17. *Murray:* Mathematical Biology I: An Introduction (Third Edition)
18. *Murray:* Mathematical Biology II: Spatial Models and Biomedical Applications (Third Edition)
19. *Kimmel/Axelrod:* Branching Processes in Biology
20. *Fall/Marland/Wagner/Tyson (Editors):* Computational Cell Biology
21. *Schlick:* Molecular Modeling and Simulation: An Interdisciplinary Guide
22. *Sahimi:* Heterogeneous Materials I: Linear Transport and Optical Properties
23. *Sahimi:* Heterogeneous Materials II: Nonlinear and Breakdown Properties and Atomistic Modeling
24. *Bloch:* Nonholonomic Mechanics and Control
25. *Beuter/Glass/Mackey/Titcombe (Editors):* Nonlinear Dynamics in Physiology and Medicine
26. *Ma/Soatto/Košecká/Sastry:* An Invitation to 3-D Vision: From Images to Geometric Models